21ˢᵗ Century Nanoscience – A Handbook

21st Century Nanoscience – A Handbook

Nanophysics Sourcebook (Volume One)

Edited by

Klaus D. Sattler

CRC Press
Taylor & Francis Group
Boca Raton London New York

CRC Press is an imprint of the
Taylor & Francis Group, an **informa** business

CRC Press
Taylor & Francis Group
6000 Broken Sound Parkway NW, Suite 300
Boca Raton, FL 33487-2742

First issued in paperback 2022

© 2020 by Taylor & Francis Group, LLC

CRC Press is an imprint of Taylor & Francis Group, an Informa business

No claim to original U.S. Government works

ISBN-13: 978-0-815-38443-4 (hbk)
ISBN-13: 978-1-03-233731-9 (pbk)
DOI: 10.1201/9780367333003

Library of Congress Cataloging-in-Publication Data

Names: Sattler, Klaus D., editor.
Title: 21st century nanoscience : a handbook / edited by Klaus D. Sattler.
Description: Boca Raton, Florida : CRC Press, [2020] | Includes bibliographical references and index. | Contents: volume 1. Nanophysics sourcebook—volume 2. Design strategies for synthesis and fabrication—volume 3. Advanced analytic methods and instrumentation—volume 5. Exotic nanostructures and quantum systems—volume 6. Nanophotonics, nanoelectronics, and nanoplasmonics—volume 7. Bioinspired systems and methods. | Summary: "This 21st Century Nanoscience Handbook will be the most comprehensive, up-to-date large reference work for the field of nanoscience. Handbook of Nanophysics, by the same editor, published in the fall of 2010, was embraced as the first comprehensive reference to consider both fundamental and applied aspects of nanophysics. This follow-up project has been conceived as a necessary expansion and full update that considers the significant advances made in the field since 2010. It goes well beyond the physics as warranted by recent developments in the field"—Provided by publisher.
Identifiers: LCCN 2019024160 (print) | LCCN 2019024161 (ebook) | ISBN 9780815384434 (v. 1 ; hardback) | ISBN 9780815392330 (v. 2 ; hardback) | ISBN 9780815384731 (v. 3 ; hardback) | ISBN 9780815355281 (v. 4 ; hardback) | ISBN 9780815356264 (v. 5 ; hardback) | ISBN 9780815356417 (v. 6 ; hardback) | ISBN 9780815357032 (v. 7 ; hardback) | ISBN 9780815357070 (v. 8 ; hardback) | ISBN 9780815357087 (v. 9 ; hardback) | ISBN 9780815357094 (v. 10 ; hardback) | ISBN 9780367333003 (v. 1 ; ebook) | ISBN 9780367341558 (v. 2 ; ebook) | ISBN 9780429340420 (v. 3 ; ebook) | ISBN 9780429347290 (v. 4 ; ebook) | ISBN 9780429347313 (v. 5 ; ebook) | ISBN 9780429351617 (v. 6 ; ebook) | ISBN 9780429351525 (v. 7 ; ebook) | ISBN 9780429351587 (v. 8 ; ebook) | ISBN 9780429351594 (v. 9 ; ebook) | ISBN 9780429351631 (v. 10 ; ebook)
Subjects: LCSH: Nanoscience—Handbooks, manuals, etc.
Classification: LCC QC176.8.N35 A22 2020 (print) | LCC QC176.8.N35 (ebook) | DDC 500—dc23
LC record available at https://lccn.loc.gov/2019024160
LC ebook record available at https://lccn.loc.gov/2019024161

**Visit the Taylor & Francis Web site at
http://www.taylorandfrancis.com**

**and the CRC Press Web site at
http://www.crcpress.com**

Contents

Editor

Klaus D. Sattler pursued his undergraduate and master's courses at the University of Karlsruhe in Germany. He received his PhD under the guidance of Professors G. Busch and H.C. Siegmann at the Swiss Federal Institute of Technology (ETH) in Zurich. For three years was a Heisenberg fellow at the University of California, Berkeley, where he initiated the first studies with a scanning tunneling microscope of atomic clusters on surfaces. Dr. Sattler accepted a position as professor of physics at the University of Hawaii, Honolulu, in 1988. In 1994, his group produced the first carbon nanocones. His current work focuses on novel nanomaterials and solar photocatalysis with nanoparticles for the purification of water. He is the editor of the sister references, *Carbon Nanomaterials Sourcebook* (2016) and *Silicon Nanomaterials Sourcebook* (2017), as well as *Fundamentals of Picoscience* (2014). Among his many other accomplishments, Dr. Sattler was awarded the prestigious Walter Schottky Prize from the German Physical Society in 1983. At the University of Hawaii, he teaches courses in general physics, solid state physics, and quantum mechanics.

Contributors

O. Adjaoud
Erich Schmid Institute of Materials
   Science
Austrian Academy of Sciences
Leoben, Austria

Markus Arndt
Faculty of Physics
University of Vienna
Vienna, Austria

Mathieu Bauchy
Physics of AmoRphous and Inorganic
   Solids Laboratory (PARISlab)
Department of Civil and
   Environmental Engineering
University of California
Los Angeles, California

K. H. Bennemann
Institute for Theoretical Physics
Freie Universität Berlin
Berlin, Germany

R. Stephen Berry
Department of Chemistry
University of Chicago
Chicago, Illinois

Bing-Yang Cao
Key Laboratory for Thermal Science
   and Power Engineering of Ministry
   of Education
Department of Engineering Mechanics
Tsinghua University
Beijing, P.R. China

Liang Chen
Ningbo Institute of Materials
   Technology & Engineering
Chinese Academy of Sciences
Ningbo, P.R. China

M. F. Ciappina
ELI-Beamlines Project
Institute of Physics of the ASCR
Prague, Czech Republic

Al-Amin Dhirani
Department of Chemistry and
   Department of Physics
University of Toronto
Toronto, Canada

Andrea Droghetti
Nano-Bio Spectroscopy Group
European Theoretical Spectroscopy
Facility and Materials Physics Center
University of the Basque Country
San Sebastian, Spain

S. V. Gantsevich
Solid State Physics Division
A.F. Ioffe Institute
Saint Petersburg, Russia

Stefan Gerlich
Faculty of Physics
University of Vienna
Vienna, Austria

Jafar Ghazanfarian
Faculty of Engineering
Department of Mechanical
   Engineering
University of Zanjan
Zanjan, Iran

Haoran Guo
Ningbo Institute of Materials
   Technology & Engineering
Chinese Academy of Sciences
Ningbo, P.R. China

V. L. Gurevich
Solid State Physics Division
A.F. Ioffe Institute
Saint Petersburg, Russia

Sakander Hayat
Faculty of Engineering Sciences
Institute of Engineering Sciences and
   Technology
Swabi, Pakistan

Takeo Hoshi
Department of Applied Mathematics
   and Physics
Tottori University
Tottori, Japan

Yu-Chao Hua
Key Laboratory for Thermal Science
   and Power Engineering of Ministry
   of Education
Department of Engineering Mechanics
Tsinghua University
Beijing, P.R. China

Yu-Wei Huang
Department of Physics and Centre for
   Quantum Information Science
National Cheng Kung University
Tainan, Taiwan

Md. Humaun Kabir
Department of Physics
Jessore University of Science and
   Technology
Jessore, Bangladesh

Rajiv Kalia
Collaboratory for Advanced
   Computing and Simulations
Department of Physics and
   Astronomy
Department of Computer Science
and
Department of Chemical Engineering
   and Materials Science
University of Southern California
Los Angeles, California

Youqi Ke
School of Physical Science and
   Technology
Shanghaitech University
Shanghai, P.R. China

Stefan Kuhn
Faculty of Physics
University of Vienna
Vienna, Austria

Avanish Kumar
School of Physical Sciences
Jawaharlal Nehru University
New Delhi, India

Kyung Sup Kwak
Department of Information and
    Communication Engineering
Inha University
Incheon, South Korea

Ivi Valentini Lara
Escola de Ciências
Pontifícia Universidade Católica do
    Rio Grande do Sul (PUCRS)
Porto Alegre, RS Brazil

M. Lewenstein
The Barcelona Institute of Science
    and Technology,
ICFO -Institut de Ciencies
    Fotoniques,
Barcelona, Spain

Vivian Machado de Menezes
Campus Laranjeiras do Sul
Universidade Federal da Fronteira Sul
    (UFFS)
Laranjeiras do Sul, P.R. Brazil

Aiichiro Nakano
Collaboratory for Advanced
    Computing and Simulations
Department of Physics and
    Astronomy
Department of Computer Science
and
Department of Chemical Engineering
    and Materials Science
University of Southern California
Los Angeles, California

Vladimir U. Nazarov
Research Center for Applied Sciences
Academia Sinica
Taipei, Taiwan

Akhilesh Pandey
School of Physical Sciences
Jawaharlal Nehru University
New Delhi, India

Chad Priest
Sandia National Laboratories
Albuquerque, New Mexico

Sanjay Puri
School of Physical Sciences
Jawaharlal Nehru University
New Delhi, India

Juan Sebastian Reparaz
Institut de Cincia de Materials de
Barcelona-CSIC
Barcelona, Spain

Ivan Rungger
National Physical Laboratory
Teddington, United Kingdom

Armin Shayeghi
Faculty of Physics
University of Vienna
Vienna, Austria

Kohei Shimamura
Department of Computational Science
Kobe University
Kobe, Japan

Fuyuki Shimojo
Department of Physics
Kumamoto University
Kumamoto, Japan

Zahra Shomali
School of Physics
Institute for Research in Fundamental
    Sciences (IPM)
Tehran, Iran

Tomohiro Sogabe
Department of Applied Physics
Nagoya University
Nagoya, Japan

D. Şopu
Fachgebiet Materialmodellierung
Institut für Materialwissenschaft
Technische Universität Darmstadt
Darmstadt, Germany

Dao-Sheng Tang
Key Laboratory for Thermal Science
    and Power Engineering of Ministry
    of Education
Department of Engineering Mechanics
Tsinghua University
Beijing, P.R. China

Ziqi Tian
Ningbo Institute of Materials
    Technology & Engineering
Chinese Academy of Sciences
Ningbo, P.R. China

Monique Tie
Department of Chemistry and
    Department of Physics
University of Toronto
Toronto, Canada

Priya Vashishta
Collaboratory for Advanced
    Computing and Simulations
Department of Physics and
    Astronomy
Department of Computer Science
and
Department of Chemical Engineering
    and Materials Science
University of Southern California
Los Angeles, California

Markus R. Wagner
Technische Universitat Berlin
Berlin, Germany

Shiyun Xiong
Functional Nano and Soft Materials
    Laboratory (FUNSOM)
Soochow University
Suzhou Shi, P.R. China

Yusaku Yamamoto
Department of Communication
    Engineering and Informatics
The University of
    Electro-Communications
Tokyo, Japan

Pei-Yun Yang
Physics Division
National Center for Theoretical
    Sciences
National Tsing Hua University
Hsinchu, Taiwan

Wei-Min Zhang
Department of Physics and Centre for
    Quantum Information Science
National Cheng Kung University
Tainan, Taiwan

1

Theoretical Atto-nano Physics

M. F. Ciappina
ELI-Beamlines and Institute of Physics of the ASCR

M. Lewenstein
ICFO - Institut de Ciencies Fotoniques, The Barcelona Institute of Science and Technology and ICREA

TWO emerging areas of research, attosecond and nanoscale physics, have recently started to merge. Attosecond physics deals with phenomena occurring when ultrashort laser pulses, with duration on the femtosecond and sub-femtosecond timescales, interact with atoms, molecules or solids. The laser-induced electron dynamics occurs natively on a timescale down to a few hundred or even tens of attoseconds (1 attosecond = 1 as = 10^{-18} s), which is of the order of the optical field cycle. For comparison, the revolution of an electron on a $1s$ orbital of a hydrogen atom is ~ 152 as. On the other hand, the second topic involves the manipulation and engineering of mesoscopic systems, such as solids, metals and dielectrics, with nanometric precision. Although nano-engineering is a vast and well-established research field on its own, the combination with intense laser physics is relatively recent.

We present a comprehensive theoretical overview of the tools to tackle and understand the physics that takes place when short and intense laser pulses interact with nanosystems, such as metallic and dielectric nanostructures. In particular, we elucidate how the spatially inhomogeneous laser-induced fields at a nanometer scale modify the laser-driven electron dynamics. Consequently, this has important impact on pivotal processes such as above-threshold ionization (ATI) and high-order harmonic generation (HHG). The deep understanding of the coupled dynamics between these spatially inhomogeneous fields and matter configures a promising way to new avenues of research and applications.

Thanks to the maturity that attosecond physics has reached, together with the tremendous advance in material engineering and manipulation techniques, the age of atto-nano physics has begun, but it is still in an incipient stage.

1.1 Introduction

This chapter deals with an embryonic field of atomic, molecular and optical (AMO) physics: atto-nano physics. It is an area that combines the traditional and already very mature attosecond physics with the equally well-developed nanophysics. We start our contribution by just presenting the general motivations and description of this new area, restricting ourselves to very basic and well-known concepts and references.

Attosecond physics has traditionally focused on atomic and small molecular targets [53,105]. For such systems, an atomic or molecular electron, once it is ionized by the electric field of a laser pulse, moves in a region that is small compared to the wavelength of the driving laser. Hence, the spatial dependence of the laser field can be safely neglected. In the presence of such so-called 'spatially homogeneous' laser fields, the time-dependent processes occurring on the attosecond timescale have been extensively investigated [5,41]. This subject has come to age based upon well-established theoretical developments and the understanding of numerous nonlinear phenomena (cf. [6,63,94]), as well as the tremendous advances in experimental laser techniques.

For instance, nowadays, experimental measurements with attosecond precision are routinely performed in several facilities around the world.

Simultaneously, bulk matter samples have been downscaled in size to nanometric dimensions, opening the door to study light-matter interaction in a completely new arena. When a strong laser interacts, for instance, with a metallic structure, it can couple with the plasmon modes inducing the ones corresponding to collective oscillations of free charges (electrons). These free charges, driven by the field, generate 'nanospots', of few nanometers size, of highly enhanced near-fields, which exhibit unique temporal and spatial properties. The near-fields, in turn, induce appreciable changes in the local field strength at a scale of the order of tenths of nanometers, and in this way, they play an important role modifying the field-induced electron dynamics. In other words, in this regime, the spatial scale on which the electron dynamics takes place is of the same order as the field variations. Moreover, the near-fields change on a sub-cycle timescale as the free charges respond almost instantaneously to the driving laser. Consequently, we face an unprecedented scenario: the possibility to study and manipulate strong-field-induced processes by rapidly changing fields, which are not spatially homogeneous. In the following subsections, we present a description of these strong field processes, joint with the theoretical tools, particularly modified to tackle them. The emergent field of attosecond physics at the nanoscale marries very fast attosecond processes (1 as $= 10^{-18}$ s), with very small nanometric spatial scales (1 nm $= 10^{-9}$ m), bringing a unique and sometimes unexpected perspective on important underlying strong-field phenomena.

1.1.1 Strong Field Phenomena Driven by Spatially Homogeneous Fields

A particular way of initiating electronic dynamics in atoms, molecules and, recently, solid materials, is to expose these systems to an intense and coherent electromagnetic radiation. A variety of widely studied and important phenomena, which we simply list and shortly describe here, result from this interaction. In order to first put the relevant laser parameters into context, it is useful to compare them with an atomic reference. In the present framework, laser fields are considered intense when their strength is not much smaller or even comparable to the Coulomb field experienced by an atomic electron. The Coulomb field in an hydrogen atom is approximately 5×10^9 V/cm (≈ 514 V/nm), corresponding to an equivalent intensity of 3.51×10^{16} W/cm² – this last value indeed defines the atomic unit of intensity. With regard to time scales, we note that in the Bohr model of hydrogen atom, the electron takes about 150 as to orbit around the proton, thus defining the characteristic time for electron dynamics inside atoms and molecules [28]. Finally, the relevant laser sources are typically in the near-infrared (NIR) regime, and hence, laser frequencies are much below the ionization atomic or molecular potential. In particular,

an 800 nm source corresponds to a photon energy of 1.55 eV (0.057 au), which is much below the ionization potential of hydrogen, given by 1/2 au (13.6 eV). At the same time, laser intensities are in the $10^{13} - 10^{15}$ W/cm² range: high enough to ionize a noticeable fraction of the sample but low enough to avoid space charge effects and full depletion of the ground state.

While the physics of interactions of atoms and molecules with intense laser pulses is quite complex, much can be learnt using theoretical tools developed over the past decades, starting with the original work by Keldysh in the 1960s [2,33,50,86,92]. According to the Keldysh theory, an electron can be freed from an atomic or molecular core either via tunnel or multiphoton ionization. These two regimes are characterized by the Keldysh parameter:

$$\gamma = \omega_0 \frac{\sqrt{2I_p}}{E_0} = \sqrt{\frac{I_p}{2U_p}}, \qquad (1.1)$$

where I_p is the ionization potential, U_p is the ponderomotive energy, defined as $U_p = E_0^2/4\omega_0^2$, where E_0 is the peak laser electric field and ω_0 the laser-carrier frequency. The adiabatic tunnelling regime is then characterized by $\gamma \ll 1$, whereas the multiphoton ionization regime by $\gamma \gg 1$. In the multiphoton regime, ionization rates scale as laser intensity I^N, where N is the order of the process, i.e. the number of photon necessary to overpass the ionization potential.

Many experiments take place in an intermediate or *crossover* region, defined by $\gamma \sim 1$ [61]. Another way to interpret γ is to note that $\gamma = \tau_T/\tau_L$, where τ_T is the Keldysh time (defined as $\tau_T = \frac{\sqrt{2I_p}}{E_0}$) and τ_L is the laser period. Hence, γ serves as a measure of non-adiabaticity by comparing the response time of the electron wave function to the period of the laser field.

When laser intensities approach $10^{13} - 10^{14}$ W/cm², the usual perturbative scaling observed in the multiphoton regime ($\gamma \gg 1$) does not hold anymore and the emission process becomes dominated by tunnelling ($\gamma < 1$). In this regime, a strong laser field bends the binding potential of the atom creating a penetrable potential barrier. The ionization process is thus governed by electrons tunnelling through this potential barrier and subsequently interacting "classically" with the strong laser field far from the parent ion [27,62,98].

This concept of tunnel ionization underpins many important theoretical advances, which have received crystal-clear experimental confirmation with the development of intense ultrashort lasers and attosecond sources over the past two decades. On a fundamental level, theoretical and experimental progress opened the door to the study of basic atomic and molecular processes on the attosecond timescale. On a practical level, this led to the development of attosecond high-frequency extreme ultraviolet (XUV) and X-ray sources, which promise many important applications, such as fine control of atomic and molecular reactions among others. The very fact that we deal here with sources that produce pulses of attosecond duration is remarkable. Attosecond XUV pulses allow, in principle, to capture all processes underlying structural dynamics and chemical reactions, including electronic motion coupled to nuclear dynamics. They also allow to address the basic unresolved and controversial questions in quantum mechanics, such as, for instance, the duration of the strong-field ionization process or the so-called tunnelling time [61,85].

Among the variety of phenomena that take place when atomic systems are driven by coherent and intense electromagnetic radiation, the most notable examples are HHG, ATI and multiple sequential or nonsequential ionization. All these processes present similarities and differences, which we detail briefly below [6,49,63].

HHG takes place whenever an atom or molecule interacts with an intense laser field of frequency ω_0, producing radiation of higher multiples of the fundamental frequency $K\omega_0$, where in the simplest case of rotationally symmetric target K is an odd integer. HHG spectra present very distinct characteristics: there is a sharp decline in conversion efficiency followed by a plateau in which the harmonic intensity hardly varies with the harmonic order K, and eventually an abrupt cutoff. For an inversion symmetric medium (such as all atoms and some molecules), only odd harmonics of the driving field have been observed because of dipole selection rules and the central symmetric character of the potential formed by the laser pulse and the atomic field. The discovery of this plateau region in HHG has made the generation of coherent XUV radiation using tabletop lasers feasible. The abovementioned features characterize a highly nonlinear process [64]. Furthermore, HHG spectroscopy (i.e. the measurement and interpretation of the HHG emission from a sample) has been widely applied to studying the ultrafast dynamics of molecules interacting with strong laser fields (see, e.g. [72]).

Conceptually, HHG is easily understood using the three-step model [27,59,60,62,127]: (i) tunnel ionization due to the intense and low-frequency laser field; (ii) acceleration of the free electron by the laser electric field and (iii) re-collision with the parent atom or molecular ion. The kinetic energy gained by the electron in its travel, under the presence of the laser oscillatory electric field, is converted into a high-energy photon and can be easily calculated starting from semiclassical assumptions.

HHG has received special attention because it configures the workhorse for the creation of attosecond pulses and, simultaneously, it exemplifies a special challenge from a theoretical point of view due to the complex intertwining between the Coulomb and external laser fields. Additionally, HHG is a promising way to provide a coherent tabletop-sized short-wavelength light sources in the XUV and soft x-ray regions of the spectrum. Nonlinear atom–electron dynamics triggered by focusing intense laser pulses onto noble gases generates broadband high photons whose energy reaches the soft X-ray region. This nonlinear phenomenon requires laser intensities in the range of 10^{14} W/cm^2, routinely available from Ti:sapphire femtosecond laser amplifiers [11].

Another widely studied phenomenon is ATI. In fact, and from an historical viewpoint, it was the first one to be considered as a strong nonperturbative laser–matter interaction process [1,75]. Conceptually, ATI is similar to HHG, except the electron does not recombine with the parent atom in the step (iii), but rather it is accelerated away by the laser field, eventually being registered at the detector. Hence, ATI is a much more likely process than HHG, although the latter has opened a venue for a larger set of applications and technological developments. Nevertheless, ATI is an essential tool for laser pulse characterization, in particular, in the few-cycle pulses regime. Unlike in HHG, where macroscopic effects, such as phase matching, often have to be incorporated to reliably reproduce the experiment, single-atom simulations are generally enough for ATI modeling.

In an ordinary ATI experiment, the energy and/or angular distribution of photoelectrons is measured. The ATI spectrum in energy presents a series of peaks given by the formula $E_p = (m + s)\omega_0 - I_p$, where m is the minimum number of laser photons needed to exceed the atomic-binding energy I_p, and s is commonly called the number of 'above-threshold' photons carried by the electron. This picture changes dramatically when few-cycle pulses are used to drive the media and the ATI energy spectra become much richer structurally speaking [73].

In this case, we can clearly distinguish two different regions, corresponding to the direct and rescattered electrons, respectively. The low-energy region, given by $E_k \lesssim 2U_p$, corresponds to direct electrons or electrons which never come back to the vicinity of the parent atom. On the other hand, the high energy part of the ATI spectrum $2U_p \lesssim E_k \lesssim 10U_p$ is dominated by the rescattered electrons, i.e. the electrons that reach the detector after being rescattered by the remaining ion core [84]. The latter are strongly influenced by the absolute phase of a few-cycle pulse, and as a consequence, they are used routinely for laser pulse characterization [83]. These two energy limits for both the direct and rescattered electrons, i.e. $2U_p$ and $10U_p$ can be easily obtained invoking purely classical arguments [7,73,93].

Most of the ATI and HHG experiments use an interacting media multielectronic atoms and molecules and recently condensed and bulk matter. Nevertheless, one often assumes that only one valence electron is active and, hence, determines all the significant features of the strong-field laser–matter interaction. The first observations of two-electron effects in ionization by strong laser pulses go back to the famous Anne L'Huillier's 'knee' [65]. This paper and later the influential Paul Corkum's work [27] stimulated the discussion about sequential versus nonsequential ionization and about a specific mechanism of the latter (shake-off, rescattering, etc.). In the last 20 years, and more recently as well, there has been a growing interest in electron correlations, both in single- and multi-electron ionization regimes, corresponding to lower and higher intensities, respectively (cf. [111,115,128]).

One prominent example where electron correlation plays an instrumental role is the so-called nonsequential double ionization (NSDI) [128]. It stands in contrast to sequential double (or multiple) ionization, i.e. when the process undergoes a sequence of single ionization events, with no correlation between them. NSDI has attracted considerable interest, since it gives direct experimental access to electron–electron correlation – something that is famously difficult to analyze both analytically and numerically (for a recent review see, e.g. [9]).

1.1.2 Introduction to Atto-nano Physics

The interaction of ultrashort strong laser pulses with extended systems has recently received much attention and led to an advancement in our understanding of the attosecond to few-femtosecond electronic and nuclear dynamics. For instance, the interaction of clusters with strong ultrafast laser fields has long been known to lead to the formation of nanoplasmas in which there is a high degree of charge localization and ultrafast dynamics, with the emission of energetic (multiple keV) electrons and highly charged – up to Xe^{40+} – ions with high energy (MeV scale) [29,30,110,116,125]. Most recently, use of short pulses ($\sim$10 fs) has succeeded in isolating the electron dynamics from the longer timescale ion dynamics (which are essentially frozen) revealing a higher degree of fragmentation anisotropy in both electrons and ions compared to the isotropic distributions found from longer pulses ($\sim$100 fs) [114].

Likewise, interactions of intense lasers with nanoparticles, such as micron-scale liquid droplets, lead to hot plasma formation. An important role is found for enhanced local fields on the surface of these droplets driving this interaction via "field hotspots" [31,38,74,118,119,123].

Furthermore, studies of driving bound and free charges in larger molecules, e.g. collective electron dynamics in fullerenes [66] and in graphene-like structures [132], proton migration in hydrocarbon molecules [58] and charge migration in proteins [8,12] could be included in this group. In turn, laser-driven broadband electron wavepackets have been used for static and dynamic diffraction imaging of molecules [10,91,131], obtaining structural information with sub-nanometer resolution.

Tailored ultrashort and intense fields have also been used to drive electron dynamics and electron or photon emission from (nanostructured) solids (for a recent compilation see e.g. [43]). The progress seen in recent years has been largely driven by advances in experimental and engineering techniques (both in laser technology and in nanofabrication). Among the remarkable achievements in just the latest years are the demonstration of driving electron currents and switching the conductivity of dielectrics with ultrashort pulses [101,103], controlling the light-induced electron emission from nanoparticles [122,140] and nanotips [42,56,90], and the sub-cycle-driven photon emission from solids [37, 71,102,126]. Furthermore, the intrinsic electron propagation and photoemission processes have been investigated on their natural, attosecond timescales [14,67,76,78,104].

A key feature of light-nanostructure interaction is the enhancement (amplification) of the electric near-field by several orders of magnitude, and its local confinement on

a sub-wavelength scale [117]. From a theoretical viewpoint, this field localization presents a unique challenge: we have at our disposal strong fields that change on a comparable spatial scale of the oscillatory electron dynamics that are initiated by those same fields. As will be shown throughout this contribution, this singular property entails profound consequences in the underlying physics of the conventional strong-field phenomena. In particular, it defies one of the main assumptions that modeling of strong-field interactions is based upon: the spatial homogeneity of laser fields in the volume of the electronic dynamics under scrutiny.

Interestingly, an exponential growing attention in strong-field phenomena induced by plasmonic-enhanced fields was triggered by the questionable work of Kim et al. [51]. These authors claimed to have been observed efficient HHG from bow-tie metallic nanostructures. Although the interpretation of the outcomes was incorrect, this chapter definitively stimulated a constant interest in the plasmonic-enhanced HHG and ATI [52,80,81,89,112,113].

Within the conventional assumption, both the laser electric field, $E(\mathbf{r}, t)$, and the corresponding vector potential, $A(\mathbf{r}, t)$, are spatially homogeneous in the region where the electron moves and only their time dependence is considered, i.e. $E(\mathbf{r}, t) = E(t)$ and $A(\mathbf{r}, t) = A(t)$. This is an authentic assumption considering the usual electron excursion (estimated classically using $\alpha = E_0/\omega_0^2$) is bounded roughly by a few nanometers in the NIR, for typical laser intensities, and several tens of nanometers for mid-infrared (MIR) sources (note that $\alpha \propto \lambda_0^2$, where λ_0 is the wavelength of the driving laser and $E_0 = \sqrt{I}$, where I is the laser intensity) [11]. Hence, electron excursion is very small relative to the spatial variation of the field in the absence of local (or nanoplasmonic) field enhancement (see Figure 1.1(a)). On the contrary, the fields generated using surface plasmons are spatially dependent on a nanometric region (cf. Figure 1.1(b)). As a consequence, all the standard theoretical tools in the strong-field ionization toolbox (ranging from purely classical to frequently used semiclassical and complete quantum mechanical approaches) have to be repondered. In this contribution, we will therefore concentrate our efforts on how the most important and basic processes in strong-field physics, such as HHG and ATI, are modified in this new setting of strong-field ultrafast phenomena on a nanoscale. Additionally, we discuss how the conventional theoretical tools have to be modified in order to be suitable for this new scenario. Note that the strong-field phenomena driven by plasmonic fields could be treated theoretically within a particular flavor of a non-dipole approximation but completely neglecting magnetic effects.

1.2 Theoretical Approaches

In the next subsections, we describe the theoretical approaches we have developed to tackle strong-field processes driven by spatially inhomogeneous laser fields. We put particular emphasis on HHG and ATI.

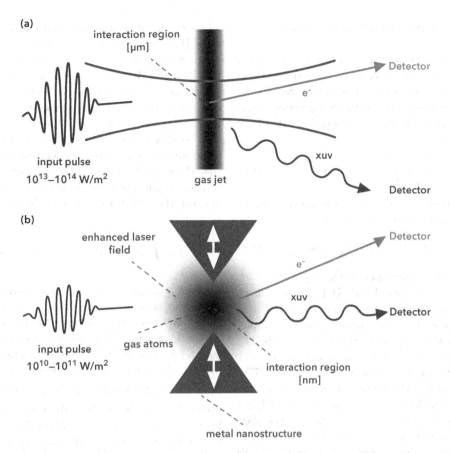

FIGURE 1.1 Sketch of conventional (a) and plasmonic-enhanced (b) strong field processes. Taken with permission from IOP.

1.2.1 Quantum Approaches

The dynamics of a single active atomic electron in a strong laser field takes place along the polarization direction of the field, when linearly polarized laser pulses drive the system. It is then perfectly legitimate to model the HHG and ATI in a 1D spatial dimension by solving the time-dependent Schrödinger equation (1D-TDSE) [18]:

$$i\frac{\partial \Psi(x,t)}{\partial t} = \mathcal{H}(t)\Psi(x,t) \tag{1.2}$$
$$= \left[-\frac{1}{2}\frac{\partial^2}{\partial x^2} + V_a(x) + V_l(x,t)\right]\Psi(x,t),$$

where in order to model an atom in 1D, it is common to use soft core potentials, which are of the form:

$$V_a(x) = -\frac{1}{\sqrt{x^2 + b^2}}, \tag{1.3}$$

where the parameter b allows us to modify the ionization potential I_p of the ground state, fixing it as close as possible to the value of the atom under consideration. We consider the field to be linearly polarized along the x-axis and modify the interaction term $V_l(x,t)$ in order to treat spatially nonhomogeneous fields, while maintaining the dipole character. Consequently we write

$$V_l(x,t) = -E(x,t)\,x \tag{1.4}$$

where $E(x,t)$ is the laser electric field defined thus as

$$E(x,t) = E_0\,f(t)\,(1 + \varepsilon h(x))\,\sin(\omega_0 t + \phi). \tag{1.5}$$

In Eq. (1.5), E_0, ω_0 and ϕ are the laser electric field peak amplitude, the frequency of the laser pulse and the carrier-envelope phase (CEP), respectively. We refer to sin(cos)-like laser pulses where $\phi = 0$ ($\phi = \pi/2$). The pulse envelope is given by $f(t)$, and ε is a small parameter that characterizes the inhomogeneity strength. The function $h(x)$ represents the functional form of the spatial nonhomogeneous field and, in principle, could take any form and be supported by the numerical algorithm (for details see e.g. [17,18]). Most of the approaches use the simplest form for $h(x)$, i.e. the linear term: $h(x) = x$. This choice is motivated by previous investigations [18,46,134], but nothing prevents to use more general functional forms for $h(x)$ [77].

The actual spatial dependence of the enhanced near-field in the surrounding of a metal nanostructure can be obtained by solving the Maxwell equations incorporating both the geometry and material properties of the nanosystem under study and the input laser pulse characteristics (see e.g. [17]). The electric field retrieved numerically is then approximated using a power series $h(x) = \sum_{i=1}^{N} b_i x^i$, where the coefficients b_i are obtained by fitting the real electric field that results from a finite element simulation. Furthermore, in the region relevant for the strong-field physics and electron dynamics and in the range of the parameters we are considering, the electric field can be indeed entirely approximated by its linear dependence.

The 1D-TDSE can be solved numerically by using the Crank–Nicolson scheme in order to obtain the time-propagated electronic wave function $\Psi(x,t)$. Once $\Psi(x,t)$ is found, we can compute the harmonic spectrum $D(\omega)$ by Fourier transforming the dipole acceleration $a(t)$ of the active electron. That is,

$$D(\omega) = \left| \frac{1}{T_p} \frac{1}{\omega^2} \int_{-\infty}^{\infty} dt e^{-i\omega t} a(t) \right|^2, \qquad (1.6)$$

where T_p is the total duration of the laser pulse. $a(t)$ can be obtained by using the commutator relation

$$a(t) = \frac{d^2\langle x \rangle}{dt^2} = -\langle \Psi(t)| \left[\mathcal{H}(t), [\mathcal{H}(t), x] \right] |\Psi(t)\rangle, \qquad (1.7)$$

where $\mathcal{H}(t)$ is the Hamiltonian specified in Eq. (1.2).

One of the main advantages of the 1D-TDSE is that we are able to include any functional form for the spatial variation of the plasmonic field. For instance, we have implemented linear [18] and real (parabolic) plasmonic fields [17], as well as near-fields with exponential decay (evanescent fields) [106] and gaussian-like bounded spatially fields [77].

In order to calculate ATI-related observables, we use the same 1D-TDSE employed for the computation of HHG (Eq. 1.2). In 1D, we are only able to compute the so-called energy-resolved photoelectron spectra $P(E)$, i.e. a quantity proportional to the probability to find electrons with a particular energy E. In order to do so, we use the window function technique developed by Schafer [96,99]. This tool has been widely used, both to calculate angle-resolved and energy-resolved photoelectron spectra [97] and it represents a step forward with respect to the usual projection methods.

An extension of the above-described approach is to solve the TDSE in its full dimensionality and to include in the laser-electron potential the spatial variation of the laser electric field. For only one active electron, we need to deal with three spatial dimensions, and due to the cylindrical symmetry of the problem, we are able to separate the electronic wave function in spherical harmonics, Y_l^m and consider only terms with $m = 0$ (see below).

In particular, the 3D-TDSE in the length gauge can be written as:

$$i\frac{\partial \Psi(\mathbf{r},t)}{\partial t} = H\Psi(\mathbf{r},t)$$
$$= \left[-\frac{\nabla^2}{2} + V_{SAE}(\mathbf{r}) + V_l(\mathbf{r},t) \right] \Psi(\mathbf{r},t), \quad (1.8)$$

where $V_{SAE}(\mathbf{r})$ is the atomic potential in the single active electron (SAE) approximation and $V_l(\mathbf{r},t)$ the laser-electron coupling (see below). The time-dependent electronic wave function $\Psi(\mathbf{r},t)$, can be expanded in terms of spherical harmonics as:

$$\Psi(\mathbf{r},t) = \Psi(r,\theta,\phi,t)$$
$$\approx \sum_{l=0}^{L-1} \sum_{m=-l}^{l} \frac{\Phi_{lm}(r,t)}{r} Y_l^m(\theta,\phi) \qquad (1.9)$$

where the number of partial waves depends on each specific case. Here, in order to assure the numerical convergence, we have used up to $L \approx 250$ in the most extreme case $(I \sim 5 \times 10^{14} \text{ W/cm}^2)$. In addition, due to the fact that the plasmonic field is linearly polarized, the magnetic quantum number is conserved, and consequently, in the following, we can consider only $m = 0$ in Eq. (1.9). This property considerably reduces the complexity of the problem. Here, we consider z as a polarization axis, and we take into account that the spatial variation of the electric field is linear with respect to the position. As a result, the coupling $V_l(\mathbf{r},t)$ between the atomic electron and the electromagnetic radiation reads:

$$V_l(\mathbf{r},t) = \int^{\mathbf{r}} d\mathbf{r}' \cdot \mathbf{E}(\mathbf{r}',t) = E_0 z (1 + \varepsilon z) f(t) \sin(\omega_0 t + \phi), \qquad (1.10)$$

where E_0, ω_0 and ϕ are the laser electric field peak amplitude, the central frequency and the CEP, respectively. As in previous investigations, the parameter ε defines the 'strength' of the inhomogeneity and has units of inverse length (see also [18,46,134]). For modeling short laser pulses in Eq. (1.10), we use a sin-squared envelope $f(t)$ of the form $f(t) = \sin^2\left(\frac{\omega_0 t}{2n_p}\right)$, where n_p is the total number of optical cycles. As a result, the total duration of the laser pulse will be $T_p = n_p \tau_L$ where $\tau_L = 2\pi/\omega_0$ is the laser period. We focus our analysis on a hydrogen atom, i.e. $V_{SAE}(\mathbf{r}) = -1/r$ in Eq. (1.8), and we also assume that before switching on the laser $(t = -\infty)$, the target atom is in its ground state ($1s$), whose analytic form can be found in a standard textbook. Within the SAE approximation, however, our numerical scheme is tunable to treat any complex atom by choosing the adequate effective (Hartree-Fock) potential $V_{SAE}(\mathbf{r})$ and finding the ground state by the means of numerical diagonalization or imaginary time propagation.

Next, we will show how the inhomogeneity modifies the equations which model the laser-electron coupling. Inserting Eq. (1.9) into Eq. (1.8) and considering that

$$\cos\theta Y_l^0 = c_{l-1} Y_{l-1}^0 + c_l Y_{l+1}^0 \qquad (1.11)$$

and

$$\cos^2\theta Y_l^0 = c_{l-2}c_{l-1} Y_{l-1}^0 + (c_{l-1}^2 + c_l^2) Y_l^0 + c_l c_{l+1} Y_{l+2}^0, \qquad (1.12)$$

where

$$c_l = \sqrt{\frac{(l+1)^2}{(2l+1)(2l+3)}}, \qquad (1.13)$$

we obtain a set of coupled differential equations for each of the radial functions $\Phi_l(r,t)$:

$$i\frac{\partial \Phi_l}{\partial t} = \left[-\frac{1}{2}\frac{\partial^2}{\partial r^2} + \frac{l(l+1)}{2r^2} - \frac{1}{2} \right] \Phi_l$$
$$+ \varepsilon r^2 E(t) \left(c_l^2 + c_{l-1}^2 \right) \Phi_l$$
$$+ r E(t) \left(c_{l-1}\Phi_{l-1} + c_l\Phi_{l+1} \right)$$
$$+ \varepsilon r^2 E(t) \left(c_{l-2}c_{l-1}\Phi_{l-2} + c_l c_{l+1}\Phi_{l+2} \right). \quad (1.14)$$

Equation (1.14) is solved using the Crank–Nicolson algorithm considering the additional term, i.e. Eq. (1.12) due to the spatial inhomogeneity. As can be observed, the degree of complexity will increase substantially when a more complex functional form for the spatial inhomogeneous laser electric field is used. For instance, the incorporation of only a linear term couples the angular momenta $l, l \pm 1, l \pm 2$, instead of $l, l \pm 1$, as in the case of conventional (spatial homogeneous) laser fields.

As was already mentioned, typically several hundreds of angular momenta l should to be considered and we could recognize the time evolution of each of them as a 1D problem. We use a Crank–Nicolson method implemented on a splitting of the time-evolution operator that preserves the norm of the wave function for the time propagation, similar to the 1D-TDSE case. The harmonic spectrum $D(\omega)$ is then computed in the same way as in the 1D case but now using the 3D electronic wave functions $\Psi(\mathbf{r}, t)$.

We have also made studies on helium because a majority of experiments in HHG are carried out in noble gases. Nonetheless, other atoms could be easily implemented by choosing the appropriate atomic model potential $V_{SAE}(\mathbf{r})$. After time propagation of the electronic wave function, the HHG spectra can be computed in an analogous way as in the case of the 1D-TDSE. Due to the complexity of the problem, only simulations with nonhomogeneous fields with linear spatial variations along the laser polarization in the 3D-TDSE have been studied. This, however, is enough to confirm that even a small spatial inhomogeneity significantly modifies the HHG spectra (for details see [87]).

For ATI, the utilization of the 3D-TDSE (Eq. 1.8) allows us to calculate not only energy-resolved photoelectron spectra but also angular electron momentum distributions of atoms driven by spatially inhomogeneous fields. As in the 1D case, the nonhomogeneous character of the laser electric field plays an important role on the ATI phenomenon. In addition, our 3D approach is able to model in a reliable way the ATI process both in the tunneling and multiphoton regimes. We show that for the former, the spatial nonhomogeneous field causes significant modifications on the electron momentum distributions and photoelectron spectra, while its effects in the latter appear to be negligible. Indeed, through the tunneling ATI process, one can obtain higher energy electrons as well as a high degree of asymmetry in the momentum space map. In our study, we consider NIR laser fields with intensities in the mid- 10^{14} W/cm^2 range. We use a linear approximation for the plasmonic field, considered valid when the electron excursion is small compared with the inhomogeneity region. Indeed, our 3D simulations confirm that plasmonic fields could drive electrons with energies in the near-keV regime (see e.g. [24]).

Similar to the 1D case, the ATI spectrum is calculated starting from the time-propagated electron wave function, once the laser pulse has ceased. For computing the energy-resolved photoelectron spectra $P(E)$ and two-dimensional electron distributions $\mathcal{H}(k_z, k_r)$, where k_z (k_r) is the electron momentum component parallel (perpendicular) to the polarization direction, we use the window function approach developed in [96,99].

Experimentally speaking, both the direct and rescattered electrons contribute to the energy-resolved photoelectron spectra. It means that for tackling this problem, both physical mechanisms should to be included in any theoretical model. In that sense, the 3D-TDSE, which can be considered as an exact approach to the ATI problem for atoms and molecules in the SAE approximation, appears to be the most suitable tool to predict the $P(E)$ in the whole range of electron energies.

1.2.2 Semiclassical Approach

An independent approach to compute HHG spectra for atoms in intense laser pulses is the Strong-Field Approximation (SFA) or Lewenstein model [62]. The main ingredient of this approach is the evaluation of the time-dependent dipole moment $\mathbf{d}(t)$. Within the SAE approximation, it can be calculated starting from the ionization and recombination transition matrices combined with the classical action of the laser-ionized electron moving in the laser field. The SFA approximation has a direct interpretation in terms of the so-called three-step or simple man's (SM) model [27,62].

Implicitly, the Lewenstein model deals with spatially homogeneous electric and vector potential fields, i.e. fields that do not experience variations in the region where the electron dynamics takes place. In order to consider spatial nonhomogeneous fields, the SFA approach needs to be modified accordingly, i.e. the ionization and recombination transition matrices, joint with the classical action, now should take into account this new feature of the laser electric and vector potential fields (for details see [18,109]).

As for the case of HHG driven by spatially inhomogeneous fields, ATI can also be modeled by using the SFA. In order to do so, it is necessary to modify the SFA ingredients, namely, the classical action and the saddle point equations. The latter are more complex but appear to be solvable for the case of spatially linear inhomogeneous fields (for details see [109]). Within SFA, it is possible to investigate how the individual pairs of quantum orbits contribute to the photoelectron spectra and the two-dimensional electron momentum distributions. We demonstrate that the quantum orbits have a very different behavior in the spatially inhomogeneous field when compared to the homogeneous one. In the case of inhomogeneous fields, the ionization and rescattering times differ between neighboring cycles, despite the field being nearly monochromatic. Indeed, the contributions from one cycle may lead to a lower cutoff, while another may develop a higher cutoff. As was shown both by our quantum mechanical and classical models, our SFA model confirms that the ATI cutoff extends far beyond the semiclassical cutoff, as a function of the inhomogeneity strength. In addition, the angular momentum distributions have very different features compared to the homogeneous case. For the neighboring cycles, the electron momentum distributions do not share the same absolute momentum, and as a consequence, they do not have the same yield.

1.2.3 Classical Framework

Important information, such as the HHG cutoff and the properties of the electron trajectories moving in the oscillatory laser electric field, can be obtained solving the classical one-dimensional Newton–Lorentz equation for an electron moving in a linearly polarized electric field. Specifically, we find the numerical solution of

$$\ddot{x}(t) = -\nabla_x V_l(x,t), \tag{1.15}$$

where $V_l(x,t)$ is defined in Eq. (1.4) with the laser electric field linearly polarized now in the x axis. For fixed values of ionization times t_i, it is possible to obtain the classical trajectories and to numerically calculate the times t_r for which the electron recollides with the parent ion. In addition, once the ionization time t_i is fixed, the full electron trajectory is completely determined (for more details about the classical model see [19]).

The following conditions are commonly set (the resulting model is indeed the classical version of the SM model): i) the electron starts with zero velocity at the origin at time $t = t_i$, i.e. $x(t_i) = 0$ and $\dot{x}(t_i) = 0$; (ii) when the laser electric field reverses its direction, the electron returns to its initial position, i.e. recombines with the parent ion, at a later time, $t = t_r$, i.e. $x(t_r) = 0$. t_i and t_r are known as ionization and recombination times, respectively. The electron kinetic energy at t_r can be obtained from the usual formula $E_k(t_r) = \dot{x}(t_r)^2/2$, and, finding the value of t_r (as a function of t_i) that maximizes this energy, we find that the HHG cutoff is given by $n_c \omega_0 = 3.17 U_p + I_p$, where n_c is the harmonic order at the cutoff, ω_0 is the laser frequency, U_p is the ponderomotive energy and I_p is the ionization potential of the atom or molecule under consideration. It is worth mentioning that the HHG cutoff will be extended when spatially inhomogeneous fields are employed.

From the SM model [27,62], we can describe the physical origin of the ATI process as follows: an atomic electron at a position $x = 0$, is released or *born* at a given ionization time t_i, with zero velocity, i.e. $\dot{x}(t_i) = 0$. This electron now moves only under the influence of the oscillating laser electric field (the residual Coulomb interaction is neglected in this model) and will reach the detector either directly or through a rescattering process. By using the classical equation of motion, it is possible to calculate the maximum energy of the electron for both the direct and rescattered processes.

For the direct ionization, the kinetic energy of an electron released or born at time t_i is

$$E_d = \frac{[\dot{x}(t_i) - \dot{x}(t_f)]^2}{2}, \tag{1.16}$$

where t_f is the end time of the laser pulse. For the rescattering process, in which the electron returns to the core at a

time t_r and reverses its direction, the kinetic energy of the electron yields

$$E_r = \frac{[\dot{x}(t_i) + \dot{x}(t_f) - 2\dot{x}(t_r)]^2}{2}. \tag{1.17}$$

For homogeneous fields, Eqs. (1.16) and (1.17) become $E_d = \frac{[A(t_i) - A(t_f)]^2}{2}$ and $E_r = \frac{[A(t_i) + A(t_f) - 2A(t_r)]^2}{2}$, with $A(t)$ being the laser vector potential $A(t) = -\int^t E(t')dt'$. For the case with $\varepsilon = 0$, it can be shown that the maximum value for E_d is $2U_p$ while for E_r it is $10U_p$ [73]. These two values appear as cutoffs in the energy-resolved photoelectron spectrum.

1.3 HHG Driven by Spatially Inhomogeneous Fields

Field-enhanced HHG, using plasmonics fields generated starting from engineered nanostructures or nanoparticles, requires no extra amplification stages due to the fact that, by exploiting surface plasmon resonances, the input driving electric field can be enhanced by more than 20 dB (corresponding to an increase in the intensity of 2–3 orders of magnitude). As a consequence of this enhancement, the threshold laser intensity for HHG generation in noble gases is largely exceeded and the pulse repetition rate remains unaltered. In addition, the high-harmonic radiation generated from each nanosystem acts as a point-like source, enabling a high collimation or focusing of this coherent radiation by means of (constructive) interference. This fact opens a wide range of possibilities to spatially arrange nanostructures to enhance or shape the spectral and spatial properties of the harmonic radiation in numerous ways [51,81,89].

Due to the nanometric size of the so-called plasmonic 'hot spots', i.e. the spatial region where the electric field reaches its highest intensity, one of the main theoretical assumptions, namely the spatial homogeneity of the driven electric field, should be excluded. As a consequence, both the analytical and numerical approaches to study laser–matter processes in atoms and molecules, in particular HHG, need to be modified to treat adequately this peculiar scenario and allow now for a spatial dependence in both the laser electric and vector potential fields. Several authors have addressed this problem recently [13,15,17–26,32,34–36,40, 45,46,48,68–70,87,106–109,129,130,133,134,136,137,139]. As we will show below, this new characteristic affects considerably the electron dynamics, and this is reflected on the observables, in the case of this subsection, the HHG spectra.

1.3.1 Spatially (Linear) Nonhomogeneous Fields and Electron Confinement

In this subsection, we summarize the study carried out in [18], where it is demonstrated that both the inhomogeneity of the local fields and the constraints in the electron movement play an important role in the HHG process and lead to the generation of even harmonics and a significant increase in the HHG cutoff, more pronounced for longer wavelengths. In order to understand and characterize these new HHG features, we employ two of the approaches mentioned above: the numerical solution of the 1D-TDSE (see panels (a)–(d) in Figure 1.2) and the semiclassical approach known as Strong-Field Approximation (SFA). Both models predict comparable results and describe satisfactorily the new features, but by employing the semiclassical arguments (see panels (e), (f) in Figure 1.2) behind the SFA and time-frequency analysis tools (Figure 1.3), we are able to fully explain the reasons of the cutoff extension.

1.3.2 Spatially (Linear) Nonhomogeneous Fields: The SFA Approach

In this subsection, we summarize the work presented in Reference [108]. In this contribution, we perform a detailed analysis of HHG in atoms within the SFA by considering spatially (linear) inhomogeneous monochromatic laser fields. We investigate how the individual pairs of quantum orbits contribute to the harmonic spectra. To this end, we have modified both the classical action and the saddle points equations by including explicitly the spatial dependence of the laser field. We show that, in the case of a linear inhomogeneous field, the electron tunnels with two different

canonical momenta. One of these momenta leads to a higher cutoff and the other one develops a lower cutoff. Furthermore, we demonstrate that the quantum orbits have a very distinct behavior in comparison to the conventional homogeneous field. A recent study supports our initial findings [138].

We also conclude that in the case of the inhomogeneous fields, both odd and even harmonics are present in the HHG spectra. Within our extended SFA model, we show that the HHG cutoff continues far beyond the standard semiclassical cutoff in spatially homogeneous fields. Our findings are in good agreement both with quantum-mechanical and classical models. Furthermore, our approach confirms the versatility of the SFA approach to tackle now the HHG driven by spatially (linear) inhomogeneous fields.

1.3.3 Real Nonhomogeneous Fields

In this subsection, we present numerical simulations of HHG in an argon model atom produced by the fields generated when a gold bow-tie nanostructure is illuminated by a short laser pulse of long wavelength $\lambda = 1800$ nm (see [17] for more details). The functional form of these fields is extracted from finite element simulations using both the complete geometry of the metal nanostructure and laser pulse characteristics (see Figure 1.4(a)). We use the numerical solution of the TDSE in reduced dimensions to predict the HHG

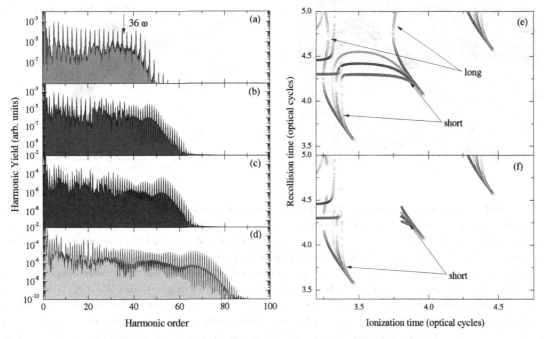

FIGURE 1.2 HHG spectra for a model atom with a ground-state energy, $I_p = -0.67$ a.u. obtained using the 1D-TDSE approach. The laser parameters are $I = 2 \times 10^{14}$ W·cm^{-2} and $\lambda = 800$ nm. We have used a trapezoidal shaped pulse with two optical cycles turn on and turn off, and a plateau with six optical cycles, 10 optical cycles in total, i.e. approximately 27 fs. The arrow indicates the cutoff predicted by the semiclassical model [62]. Panel (a): homogeneous case, (b): $\varepsilon = 0.01$ (100 a.u.), (c): $\varepsilon = 0.02$ (50 a.u) and (d): $\varepsilon = 0.05$ (20 a.u.). The numbers in brackets indicate an estimate of the inhomogeneity region (for more details see e.g [18,46]). The dependence of the semiclassical trajectories on the ionization and recollision times for different values of ε is shown in panel (e) for the non confined case and panel (f) for the confined case. squares: homogeneous case $\varepsilon = 0$; circles: $\varepsilon = 0.01$; triangles: $\varepsilon = 0.02$ and diamonds: $\varepsilon = 0.05$.

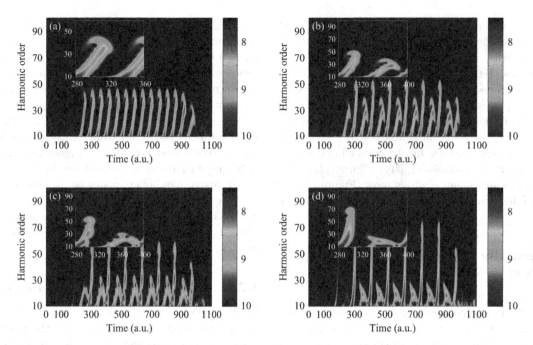

FIGURE 1.3   Panels (a)–(d): Gabor analysis for the corresponding HHG spectra of panels (a)–(d) of Figure 1.2. The zoomed regions in all panels show a time interval during the laser pulse for which the complete electron trajectory, from birth time to recollision time, falls within the pulse plateau. In panels (a)–(d) the scale is logarithmic.

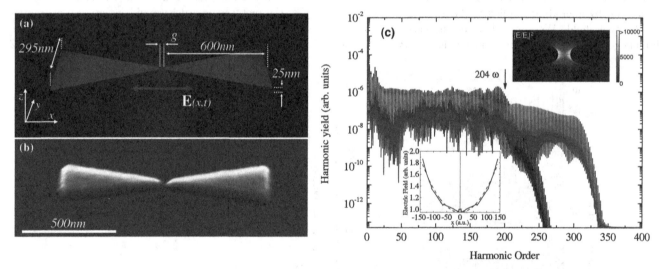

FIGURE 1.4   (a) Schematic representation of the geometry of the considered nanostructure. A gold bow-tie antenna resides on glass substrate (refractive index $n = 1.52$) with superstate medium of air ($n = 1$). The characteristic dimensions of the system and the coordinate system used in the 1D-TDSE simulations are shown. (b) SEM image of a real gold bow-tie antenna. (c) HHG spectra for a model of argon atoms ($I_p = -0.58$ a.u.), driven by a laser pulse with wavelength $\lambda = 1800$ nm and intensity $I = 1.25 \times 10^{14}$ W·cm^{-2} at the center of the gap $x = 0$. We have used a trapezoidal-shaped pulse with three optical cycles turn on and turn off and a plateau with four optical cycles (about 60 fs). The gold bow-tie nanostructure has a gap $g = 15$ nm (283 a.u.). The black line indicates the homogeneous case while the dark gray line indicates the nonhomogeneous case. The arrow indicates the cutoff predicted by the semiclassical model for the homogeneous case [62]. The top left inset shows the functional form of the electric field $E(x, t)$, where the solid lines are the raw data obtained from the finite element simulations and the dashed line is a nonlinear fitting. The top right inset shows the intensity enhancement in the gap region of the gold bow-tie nanostructure.

spectra. A clear extension in the harmonic cutoff position is observed. This characteristic could lead to the production of bright XUV coherent laser sources and open the avenue to the generation of shorter attosecond pulses. It is shown in Figure 1.4(c) that this new feature is a consequence of the combination of a spatial nonhomogeneous electric field, which substantially modifies the electron trajectories and the confinement of the electron dynamics. Furthermore, our numerical results are supported by time-analysis and classical simulations. A more pronounced increase in the harmonic cutoff, in addition to an appreciable growth in conversion efficiency, could be attained by optimizing the nanostructure geometry and its constituent material. These degrees of freedom could pave the way to tailor the harmonic spectra according to specific requirements.

1.3.4 Temporal and Spatial Synthesized Fields

In this subsection, we present a brief summary of the results published in [87]. In short, numerical simulations of HHG in He atoms using a temporal and spatial synthesized laser field are considered using the full 3D-TDSE. This particular field provides a new route for the generation of photons at energies beyond the carbon K-edge using laser pulses at 800 nm, which can be obtained from conventional Ti:Sapphire-based laser sources. The temporal synthesis is performed using two few-cycle laser pulses delayed in time [88]. On the other hand, the spatial synthesis is obtained by using a spatial nonhomogeneous laser field [18,46,134] produced when a laser beam is focused in the vicinity of a metal nanostructure or nanoparticle.

Focusing on the spatial synthesis, the nonhomogeneous spatial distribution of the laser electric field can be obtained experimentally by using the resulting field as produced after the interaction of the laser pulse with nanoplasmonic antennas [18,46,51,134], metallic nanowaveguides [81], metal [120,140] and dielectric nanoparticles [121] or metal nanotips [42,44,55–57,100].

The coupling between the atom and the laser pulse, linearly polarized along the z axis, is modified in order to treat the spatially nonhomogeneous fields, and it can be written as: $V_l(z,t,\tau) = \tilde{E}(z,t,\tau)z$ with $\tilde{E}(z,t,\tau) = E(t,\tau)(1+\varepsilon z)$ and $E(t,\tau) = E_1(t) + E_2(t,\tau)$ the temporal synthesized laser field with τ the time delay between the two pulses (see e.g. [88] for more details). As in the 1D case, the parameter ε defines the strength of the nonhomogeneity, and the dipole approximation is preserved because $\varepsilon \ll 1$.

The linear functional form for the spatial nonhomogeneity described above could be obtained engineering adequately the geometry of plasmonic nanostructures and by adjusting the laser parameters in such a way that the laser-ionized electron feels only a linear spatial variation of the laser electric field when in the continuum (see e.g. [17] and references therein). The harmonic spectrum then obtained in He for $\varepsilon = 0.002$ is presented in Figure 1.5(b). We can observe a considerable cutoff extension up to $12.5U_p$, which is much larger when compared with the double-pulse configuration employed alone (it leads only to a maximum of $4.5U_p$ [88]). This large extension of the cutoff is therefore a signature of the combined effect of the double pulse and the spatial nonhomogeneous character of the laser electric field. For this particular value of the laser peak intensity (1.4×10^{15} W/cm^2), the highest photon energy is greater than 1 keV. Note that the quoted intensity is actually the plasmonic-enhanced intensity,

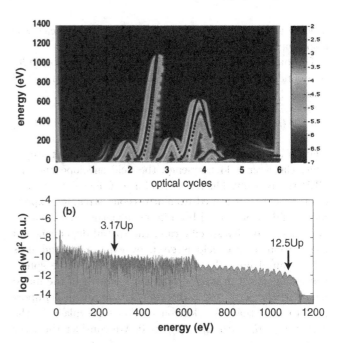

FIGURE 1.5 (a) Time–frequency analysis obtained from the 3D-TDSE harmonic spectrum for a He atom driven by the spatially and temporally synthesized pulse described in the text with $\varepsilon = 0.002$. The plasmonic-enhanced intensity $I = 1.4 \times 10^{15}$ W/cm^2. Superimposed (in dark gray) are the classical rescattering energies and (b) 3D-TDSE harmonic spectrum for the same parameters used in (a).

not the input laser intensity. The latter could be several orders of magnitude smaller, according to the plasmonic enhancement factor (see e.g. [51,81]) and will allow the nanoplasmonic target to survive to the interaction. In order to confirm the underlying physics highlighted by the classical trajectories analysis, we have retrieved the time–frequency distribution of the calculated dipole (from the 3D-TDSE) corresponding to the case of the spectra presented in Figure 1.5(b) using a wavelet analysis. The result is presented in Figure 1.5(a) where we have superimposed the calculated classical recombination energies (in dark gray) to show the excellent agreement between the two theoretical approaches. The consistency of the classical calculations with the full quantum approach is clear and confirms the mechanism of the generation of this $12.5U_p$ cutoff extension. In addition, the HHG spectra exhibit a clean continuum as a result of the trajectory selection on the recombination time, which itself is a consequence of employing a combination of temporally and spatially synthesized laser field.

1.3.5 Plasmonic Near-Fields

This subsection includes an overview of the results reported in [106]. In this contribution, it is shown how the HHG spectra from model Xe atoms are modified by using a plasmonic near-enhanced field generated when a metal nanoparticle is illuminated by a short laser pulse. A setup combining a noble gas as a driven media and metal nanoparticles was also proposed recently in [45,47].

For our near-field, we use the function given by [120] to define the spatial nonhomogeneous laser electric field $E(x,t)$, i.e.

$$E(x,t) = E_0\, f(t)\, \exp(-x/\chi) \sin(\omega_0 t + \phi), \qquad (1.18)$$

where E_0, ω_0, $f(t)$ and ϕ are the electric field peak amplitude, the laser field frequency, the field envelope and the CEP, respectively. The functional form of the resulting laser electric field is extracted from attosecond streaking experiments and incorporated both in our quantum and classical approaches. In this specific case, the spatial dependence of the plasmonic near-field is given by $\exp(-x/\chi)$ and it is a function of both the size and the material of the spherical nanoparticle. $E(x,t)$ is valid for x outside of the metal nanoparticle, i.e. $x \geq R_0$, where R_0 is its radius. It is important to note that the electron motion takes place in the region $x \geq R_0$ with $(x + R_0) \gg 0$. We consider the laser field having a $\sin^2$ envelope: $f(t) = \sin^2\left(\frac{\omega_0 t}{2n_p}\right)$, where n_p is the total number of optical cycles, i.e. the total pulse duration is $\tau_L = 2\pi n_p/\omega_0$. The harmonic yield of the atom is obtained by Fourier transforming the acceleration $a(t)$ of the electronic wavepacket.

Figure 1.6, panels (a)–(c) show the harmonic spectra for model xenon atoms generated by a laser pulse with $I = 2 \times 10^{13}$ W/cm^2, $\lambda = 720$ nm and a $\tau_L = 13$ fs, i.e. $n_p = 5$ (which corresponds to an intensity envelope of ≈ 4.7 fs FWHM) [120]. In the case of a spatial homogeneous field, there are no harmonics beyond the 9th order are observed. The spatial decay parameter χ accounts for the spatial non-homogeneity induced by the nanoparticle and it varies together with its size and the kind of metal employed. Tunning the value of χ is therefore equivalent to choosing the type of nanoparticle used, which allows to overcome the semiclassically predicted cutoff limit and reach higher harmonic orders. For example, with $\chi = 40$ and $\chi = 50$ harmonics in the mid 20s (panel c) and well above the 9th (a clear cutoff at $n_c \approx 15$ is achieved) (panel b), respectively, are obtained. A modification in the harmonic periodicity, related to the breaking of symmetry imposed by the induced nonhomogeneity, is also clearly noticeable.

Now, by means of the semiclassical SM model [27,62], we will study the harmonic cutoff extension. This new effect may be caused by a combination of several factors (for details see [17,18]). As is well known, the cutoff law is $n_c = (3.17 U_p + I_p)/\omega_0$, where n_c is the harmonic order at the cutoff and U_p the ponderomotive energy. We solve numerically Eq. (1.15) for an electron moving in an electric field with the same parameters used in the 1D-TDSE calculations, i.e.

$$\ddot{x}(t) = -\nabla_x V_l(x,t) = -E(x,t)\left(1 - \frac{x(t)}{\chi}\right), \qquad (1.19)$$

and consider the SM model initial conditions: the electron starts at position zero at $t = t_i$ (the ionization time) with zero velocity, i.e. $x(t_i) = 0$ and $\dot{x}(t_i) = 0$. When the electric field reverses, the electron returns to its initial position (i.e. the electron *recollides* or recombines with the

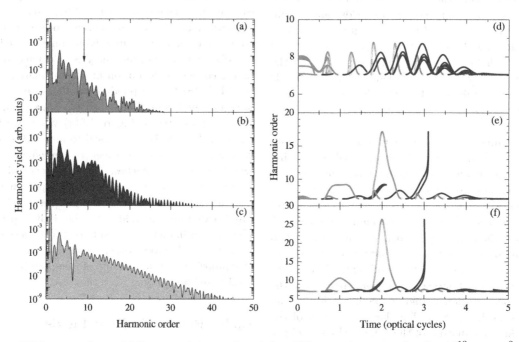

FIGURE 1.6 HHG spectra for model Xe atoms, laser wavelength $\lambda = 720$ nm and intensity $I = 2 \times 10^{13}$ W·cm^{-2}. We use a $\sin^2$ pulse envelope with $n = 5$. Panel (a) represents the homogeneous case, panel (b) $\chi = 50$ and panel (c) $\chi = 40$. The arrow in panel (a) indicates the cutoff predicted by the semiclassical approach [62]. Panels (d)–(f) show the corresponding total energy of the electron (expressed in harmonic order) driven by the laser field calculated from the one-dimensional Newton–Lorentz equation and plotted as a function of the t_i (light gray cycles) or the t_r (dark gray cycles).

parention) at a later time $t = t_r$ (the recombination time), i.e. $x(t_r) = 0$. The electron kinetic energy at the t_r is calculated as usual from: $E_k(t_r) = \frac{\dot{x}(t_r)^2}{2}$ and finding the t_r (as a function of t_i) that maximizes E_k, n_c is also maximized.

Panels (d)–(f) of Figure 1.6 represent the behavior of the harmonic order upon the t_i and t_r, calculated from $n = (E_k(t_{i,r}) + I_p)/\omega_0$ as for the cases (a)–(c) of Figure 1.6, respectively. Panels (e) and (f) show how the nonhomogeneous character of the laser field strongly modifies the electron trajectories towards an extension of the n_c. This is clearly present at $n_c \sim 18\omega_0$ (28 eV) and $n_c \sim 27\omega_0$ (42 eV) for $\chi = 50$ and $\chi = 40$, respectively. These last two cutoff extensions are consistent with the quantum predictions presented in panels (b) and (c) of Figure 1.6.

Classical and quantum approaches predict cutoff extensions that could lead to the production of XUV coherent laser sources and open a direct route to the generation of attosecond pulses. This effect is caused by the induced laser field spatial nonhomogeneity, which modifies substantially the electron trajectories. A more pronounced increment in the harmonic cutoff, in addition to an appreciable growth in the conversion efficiency, could be reached by varying both the radius and the metal material of the spherical nanoparticles. These new degrees of freedom could pave the way to extend the harmonic plateau reaching the XUV regime with modest input laser intensities.

1.4 ATI Driven by Spatially Inhomogeneous Fields

As was mentioned at the outset, ATI represents another fundamental strong field phenomenon. Investigations carried out on ATI, generated by few-cycle driving laser pulses, have attracted much interest due to the sensitivity of the energy and angle-resolved photoelectron spectra to the absolute value of the CEP [73,95]. This feature makes the ATI phenomenon a conceivable tool for laser pulse characterization. In order to characterize the CEP of a few-cycle laser pulse, the so-called backward-forward asymmetry of the ATI spectrum is measured, and from the information collected, the absolute CEP value can be obtained [82,95]. Furthermore, nothing but the high energy region of the photoelectron spectrum appears to be strongly sensitive to the absolute CEP, and consequently, electrons with high kinetic energy are needed in order to describe it [73,82,83].

Nowadays, experiments have demonstrated that ATI photoelectron spectra could be extended further by using plasmon-field enhancement [51,140]. The strong confinement of the plasmonics spots and the distortion of the electric field by the surface plasmons induces a spatial inhomogeneity in the driving laser field, just before the interaction with the corresponding target gas. A related process employing solid-state targets instead of atoms and molecules in gas phase is the so-called above-threshold photoemission (ATP). This laser-driven phenomenon has received

special attention recently due to its novelty and the new physics involved. In ATP, electrons are emitted directly from metallic surfaces or metal nanotips, and they present distinct characteristics, namely higher energies, far beyond the usual cutoff for noble gases and, consequently, the possibility to reach similar electron energies with smaller laser intensities (see e.g. [42,44,54,56,100]). Furthermore, the photoelectrons emitted from these nanosources are distinctly sensitive to the CEP, and consequently, it plays an important role in the angle and energy resolved photoelectron spectra, as in the case of spatial homogeneous fields [3,56,140].

1.4.1 ATI Driven by Spatially Linear Inhomogeneous Fields: The 1D-Case

For our 1D quantum simulations, we employ as a driving field a four-cycle (total duration 10 fs) sin-squared laser pulse with an intensity $I = 3 \times 10^{14}$ W/cm^2 and wavelength $\lambda = 800$ nm. We chose a linear inhomogeneous field and three different values for the parameter that characterizes the inhomogeneity strength, namely $\varepsilon = 0$ (homogeneous case), $\varepsilon = 0.003$ and $\varepsilon = 0.005$. Figure 1.7(a) shows the cases with $\phi = 0$ (a sin-like laser pulse), meanwhile in Figure 1.7(b), $\phi = \pi/2$ (a cos-like laser pulse), respectively. In both the panels, middle gray represents the homogeneous case, i.e. $\varepsilon = 0$, dark gray is for $\varepsilon = 0.003$ and light gray is for $\varepsilon = 0.005$, respectively. For the homogeneous case, the spectra exhibit the usual distinct behavior, namely the $2U_p$ cutoff ($\approx$36 eV for our case) and the $10U_p$ cutoff ($\approx$180 eV), where $U_p = E_0^2/4\omega_0^2$ is the ponderomotive potential. The former cutoff corresponds to those electrons that, once ionized, never return to the atomic core, while the latter one corresponds to the electrons that, once ionized, return to the core and elastically rescatter with it. It is well established using classical arguments that the maximum

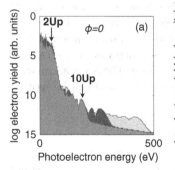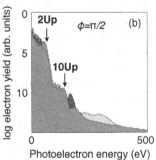

FIGURE 1.7 1D-TDSE energy-resolved photoelectron spectra for a model atom with $I_p = -0.5$ a.u. and for the laser parameters, $I = 3 \times 10^{14}$ W/cm^2, $\lambda = 800$ nm and a sin-squared shaped pulse with a total duration of 4 cycles (10 fs). In middle gray for $\varepsilon = 0$ (homogeneous case), in dark gray for $\varepsilon = 0.003$ and in light gray for $\varepsilon = 0.005$. Panel (a) represents the case for $\phi = 0$ (sin-like pulse) and panel (b) represents the case for $\phi = \pi/2$ (cos-like pulse). The arrows indicate the $2U_p$ and $10U_p$ cutoffs predicted by the classical model [73].

kinetic energies of the *direct* and the *rescattered* electrons are $E_{max}^d = 2U_p$ and $E_{max}^r = 10U_p$, respectively. In a quantum mechanical approach, however, it is possible to find electrons with energies beyond the $10U_p$, although their yield drops several orders of magnitude [73]. The TDSE, which can be considered as an exact approach to the problem, is able to predict the $P(E)$ for the whole range of electron energies. In addition, the most energetic electrons, i.e. those with $E_k \gg 2U_p$, are used to characterize the CEP of few-cycle pulses. As a result, a correct description of the rescattering mechanism is needed.

For the spatial inhomogeneous case, the cutoff positions of both the *direct* and the *rescattered* electrons are extended towards larger energies. For the *rescattered* electrons, this extension is very prominent. In fact, for $\varepsilon = 0.003$ and $\varepsilon = 0.005$, it reaches ≈ 260 eV and ≈ 420 eV, respectively (see Figure 1.7(a)). Furthermore, it appears that the high energy region of $P(E)$, for instance, the region between $200 - 400$ eV for $\varepsilon = 0.005$ (Figure 1.7 in light gray), is strongly sensitive to the CEP. This feature indicates that the high energy region of the photoelectron spectra could resemble a new and better CEP characterization tool. It should be, however, complemented by other well known and established CEP characterization tools, as, for instance, the forward-backward asymmetry (see [73]). Furthermore, the utilization of nonhomogeneous fields would open the avenue for the production of high energy electrons, reaching the keV regime, if a reliable control of the spatial and temporal shape of the laser electric field is attained.

We now concentrate our efforts on explaining the extension of the energy-resolved photoelectron spectra using classical arguments. From the SM model [27,62], we can describe the physical origin of the ATI process as follows: an atomic electron at a position $x = 0$ is released or *born* at a given time, that we call *ionization* time t_i, with zero velocity, i.e. $\dot{x}(t_i) = 0$. This electron now moves only under the influence of the oscillating laser electric field (the residual Coulomb interaction is neglected in this model) and will reach the detector either directly or through a rescattering process. By using the classical equation of motion, it is possible to calculate the maximum energy of the electron for both direct and rescattered processes. The Newton equation of motion for the electron in the laser field can be written as (see Eq. (1.15)):

$$
\begin{aligned}
\ddot{x}(t) &= -\nabla_x V_1(x,t) \\
&= E(x,t) + [\nabla_x E(x,t)]\, x \\
&= E(t)(1 + 2\varepsilon x(t)), \quad\quad (1.20)
\end{aligned}
$$

where we have collected the time-dependent part of the electric field in $E(t)$, i.e. $E(t) = E_0 f(t) \sin(\omega_0 t + \phi)$ and particularized to the case $h(x) = x$. In the limit where $\varepsilon = 0$ in Eq. (1.20), we recover the spatial homogeneous case. Using the classical formalism described in Section 1.2.3, we find the maximum energy for both the direct and rescattered electrons. As can be seen, the electron energy cutoffs now exceed the ones obtained for conventional (spatially homogeneous)

fields (see panels (a) and (b), in middle gray, in Figure 1.7 and the respective arrows).

In Figure 1.8, we present the numerical solutions of Eq. (1.20), which is plotted in terms of the kinetic energy of the direct and rescattered electrons. We employ the same laser parameters as in Figure 1.7. Panels (a)–(c) correspond to the case of $\phi = 0$ (sin-like pulses) and for $\varepsilon = 0$ (homogeneous case), $\varepsilon = 0.003$ and $\varepsilon = 0.005$, respectively. Meanwhile, panels (d)–(f) correspond to the case of $\phi = \pi/2$ (cos-like pulses) and for $\varepsilon = 0$ (homogeneous case), $\varepsilon = 0.003$ and $\varepsilon = 0.005$, respectively. From the panels (b), (c), (e) and (f), we can observe the strong modifications that the nonhomogeneous character of the laser electric field produces in the electron kinetic energy. These are related to the changes in the electron trajectories (for details see e.g. [17,18,134]). In short, the electron trajectories are modified in such a way that now the electron ionizes at an earlier time and recombines later, and in this way, it spends more time in the continuum acquiring energy from the laser electric field. Consequently, higher values of the kinetic energy are achieved. A similar behavior with the photoelectrons was observed recently in ATP using metal nanotips. According to the model presented in [42] the localized fields, modify the electron motion in such a way to allow sub-cycle dynamics. In our studies, however, we consider both direct and rescattered electrons (in [42] only direct electrons are modeled) and the characterization of the dynamics of the photoelectrons is much more complex. Nevertheless, the higher kinetic energy of the rescattered electrons is a clear consequence of the strong modifications of the laser electric field in the region where the electron dynamics takes place, as in the abovementioned case of ATP.

1.4.2 ATI Driven by Spatially Linear Inhomogeneous Fields: The 3D-Case

In the following, we calculate two-dimensional electron momentum distributions for a laser field intensity of $I = 5.0544 \times 10^{14}$ W/cm^2 ($E_0 = 0.12$ a.u). The results are depicted in Figure 1.9 for $\phi = \pi/2$. Here, panels (a)–(d) represent the cases with $\varepsilon = 0$ (homogeneous case), $\varepsilon = 0.002$, $\varepsilon = 0.003$ and $\varepsilon = 0.005$, respectively. By a simple inspection of Figure 1.9, strong modifications produced by the spatial inhomogeneities in both the angular and low-energy structures can be appreciated (see [24] for more details).

However, in the low-intensity regime (i.e. multiphoton regime, $\gamma \gg 1$), the scenario changes radically. In order to study this case, we use a laser electric field with $E_0 = 0.05$ a.u. of peak amplitude ($I = 8.775 \times 10^{13}$ W/cm^2), $\omega_0 = 0.25$ a.u. ($\lambda = 182.5$ nm) and six complete optical cycles. The resulting Keldysh parameter $\gamma = 5$ indicates the predominance of a multiphoton process [4]. In Figure 1.10, we show the corresponding two-dimensional electron distributions. For the homogeneous case, our calculation is identical to the one presented in [4]. We also notice that the two panels present indistinguishable shape and magnitude.

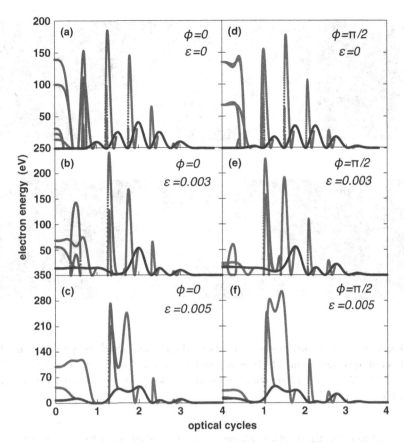

FIGURE 1.8 Numerical solutions of Eq. (1.20) plotted in terms of the direct (dark gray) and rescattered (middle gray) electron kinetic energy. The laser parameters are the same as in Figure 1.7. Panels (a)–(c) correspond to the case of sin-like pulses ($\phi = 0$) and for $\varepsilon = 0$ (homogeneous case), $\varepsilon = 0.003$ and $\varepsilon = 0.005$, respectively. Panels (d)–(f) correspond to the case of cos-like pulses ($\phi = \pi/2$) and for $\varepsilon = 0$ (homogeneous case), $\varepsilon = 0.003$ and $\varepsilon = 0.005$, respectively.

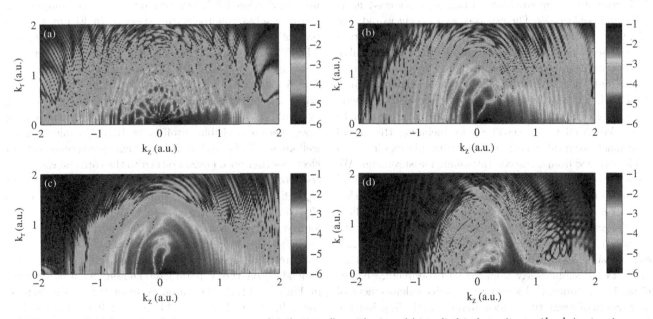

FIGURE 1.9 Two-dimensional electron momentum distributions (logarithmic scale) in cylindrical coordinates (k_z, k_r) using the exact 3D-TDSE calculation for an hydrogen atom. The laser parameters are $I = 5.0544 \times 10^{14}$ W/cm^2 ($E_0 = 0.12$ a.u.) and $\lambda = 800$ nm. We have used a sin-squared shaped pulse with a total duration of four optical cycles (10 fs) with $\phi = \pi/2$. (a) $\varepsilon = 0$ (homogeneous case), (b) $\varepsilon = 0.002$, (c) $\varepsilon = 0.003$ and (d) $\varepsilon = 0.005$.

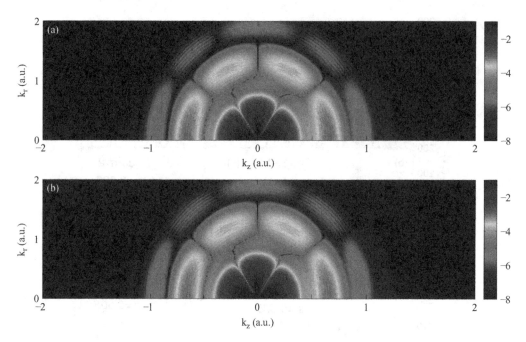

FIGURE 1.10 Two-dimensional electron momentum distributions (logarithmic scale) in cylindrical coordinates (k_z, k_r) using the exact 3D-TDSE calculation for an hydrogen atom. The laser parameters are $E_0 = 0.05$ a.u. ($I = 8.775 \times 10^{13}$ W/cm^2), $\omega_0 = 0.25$ a.u. ($\lambda = 182.5$ nm) and $\phi = \pi/2$. We employ a laser pulse with 6 total cycles. Panel (a) corresponds to the homogeneous case ($\varepsilon = 0$) and panel (b) is for $\varepsilon = 0.005$.

Hence, the differences introduced by the spatial inhomogeneity are practically imperceptible in the multiphoton ionization regime.

1.4.3 Plasmonic Near-Fields

In this section, we put forward the plausibility to perform ATI experiments by combining plasmonic-enhanced near-fields and noble gases. The proposed experiment would take advantage of the plasmonic enhanced near-fields (also known as evanescent fields), which present a strong spatial nonhomogeneous character and the flexibility to use any atom or molecule in gas phase. A similar scheme was previously presented, but now we are interested in generating highly energetic electrons, instead of coherent electromagnetic radiation. We employ the 1D-TDSE by including the actual functional form of metal nanoparticles plasmonic near-fields obtained from attosecond streaking measurements. We have chosen this particular nanostructure since its actual enhanced field is known experimentally, while for the other nanostructures, like bow-ties [51], the actual plasmonic field is unknown. For most of the plasmonic nanostructures, the enhanced field is theoretically calculated using the finite element simulation, which is based on an ideal system that may deviate significantly from actual experimental conditions. For instance, [51] states an intensity enhancement of four orders of magnitude (calculated theoretically), but the maximum harmonic measured was the 17th, which corresponds to an intensity enhancement of only two orders of magnitude (for more details see [17,107]). On the other hand, our numerical tools allow a treatment of a very general

set of spatial nonhomogeneous fields such as those present in the vicinity of metal nanostructures [51], dielectric nanoparticles [140] or metal nanotips [42]. The kinetic energy for both the direct and rescattered electrons can be classically calculated and compared to quantum mechanical predictions (for more details see e.g [25]).

We have employed the same parameters as the ones used in Section 1.3.5, but now our aim is to compute the energy-resolved photoelectron spectra. In Figure 1.11, we present the photoelectron spectra calculated using the 1D-TDSE for Xe atoms and for two different laser intensities, namely $I = 2 \times 10^{13}$ W/cm^2 (Figure 1.11(a)) and $I = 5 \times 10^{13}$ W/cm^2 (Figure 1.11(b)). In Figure 1.11(a), each curve presents different values of χ: homogeneous case ($\chi \to \infty$), $\chi = 40$, $\chi = 35$ and $\chi = 29$. For the homogeneous case, there is a visible cutoff at ≈ 10.5 eV confirming the well-known ATI cutoff at $10U_p$, which corresponds to those electrons that once ionized return to the core and elastically rescatter. Here, U_p is the ponderomotive potential given by $U_p = E_p^2/4\omega_0^2$. On the other hand, for this particular intensity, the cutoff at $2U_p$ (≈ 2.1 eV) developed by the direct ionized electrons is not visible in the spectrum.

For the spatial nonhomogeneous cases, the cutoff of the rescattered electron is far beyond the classical limit $10U_p$, depending on the χ parameter chosen. As it is depicted in Figure 1.11(a), the cutoff is extended as we decrease the value of χ. For $\chi = 40$, the cutoff is at around 14 eV, while for $\chi = 29$, it is around 30 eV. The low energy region of the photoelectron spectra is sensitive to the atomic potential of the target, and one would need to employ the TDSE in full dimensionality in order to model this region

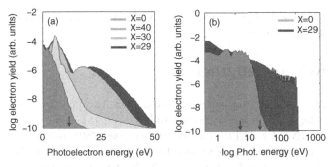

FIGURE 1.11 Energy-resolved photoelectron spectra for Xe atoms driven by an electric enhanced near-field. In panel (a), the laser intensity after interacting with the metal nanoparticles is $I = 2 \times 10^{13}$ W/cm². We employ $\phi = \pi/2$ (cos-like pulses) and the laser wavelength and number of cycles remain unchanged with respect to the input pulse, i.e. $\lambda = 720$ nm and $n_p = 5$ (13 fs in total). Panel (b) shows the output laser intensity of $I = 5 \times 10^{13}$ W/cm² (everything else is the same as in panel (a)). The arrow indicates two conventional classical limits: $2U_p$ at 5.24 eV and $10U_p$ at 26.2 eV, respectively.

adequately. In this chapter, we are interested in the high-energy region of the photoelectron spectra, which is very convenient because it is not greatly affected by the considered atom. Thus, by employing the 1D-TDSE, the conclusions that can be taken from these highly energetic electrons are very reliable.

Figure 1.11(b) shows the photoelectron spectra for the homogeneous case and for $\chi = 29$ using a larger laser field intensity of $I = 5 \times 10^{13}$ W/cm², while keeping all other laser parameters fixed. From this plot, we observe that the nonhomogeneous character of the laser-enhanced electric field introduces a highly nonlinear behavior. For this intensity with $\chi = 29$, it is possible to obtain very energetic electrons reaching values of several hundreds of eV. This is a good indication that the nonlinear behavior of the combined system of metallic nanoparticles and noble gas atoms could pave the way to generate keV electrons with tabletop laser sources. All the above quantum mechanical predictions can be directly confirmed by using classical simulations in the same way as for the HHG case.

Here we propose generation of high-energy photoelectrons using near-enhanced fields by combining metallic nanoparticles and noble gas atoms. Near-enhanced fields present a strong spatial dependence at a nanometer scale, and this behavior introduces substantial changes in the laser-matter processes. We have modified the 1D-TDSE to model the ATI phenomenon in noble gases driven by the enhanced near-fields of such nanostructure. We predict a substantial extension in the cutoff position of the energy-resolved photoelectron spectra, far beyond the conventional $10U_p$ classical limit. These new features are well reproduced by classical simulations. In a combination of metal nanoparticles and gas atoms, each metal nanoparticle configures a laser nanosource with particular characteristics that allow not only the amplification of the input laser field but also

the modification of the laser-matter phenomena due to the strong spatial dependence of the generated coherent electromagnetic radiation.

1.4.4 Emergence of a Higher Energy Structure (HES) in ATI Driven by Spatially Inhomogeneous Laser Fields

Our final example deals with a recent study about the appearance of a higher energy structure (HES) in the energy-resolved ATI photoelectron spectra when the active media is driven by a spatially inhomogeneous laser field [79]. As was discussed throughout this contribution, the theoretical approaches had not considered any spatial dependence in the field (forces) experienced by the laser-ionized electron. On the other hand, the small spatial inhomogeneity introduced by the long-range Coulomb potential has been recently linked to a number of important features in the photoelectron spectrum, such as Coulomb asymmetry, Coulomb focusing, and a distinct set of low-energy structures in the angle-resolved photoelectron spectra. We demonstrated that using an MIR laser source, with a time-varying spatial dependence in the laser electric field, such as that produced in the vicinity of a nanostructure, creates a prominent higher energy peak. This HES originates from direct electrons ionized near the peak of a single half-cycle of the laser pulse. This feature is indeed confirmed both using quantum mechanical, TDSE-based, and classical, classical trajectory Monte Carlo (CTMC), approaches. Interestingly, the HES is well separated from all other ionization events, with its location and energy width strongly dependent on the properties of the spatial inhomogeneous field. As a consequence, the HES can be employed as a sensitive tool for near-field characterization in a regime where the electron's quiver amplitude is on the order to the field decay length. Additionally, the large accumulation of electrons with tuneable energy suggests a promising method for creating a localized source of electron pulses of sub-femtosecond duration using tabletop laser technology.

1.5 Conclusions, Outlook and Perspectives

In this contribution, we have extensively reviewed the theoretical tools to tackle strong field phenomena driven by plasmonic-enhanced fields and discussed a set of relevant results.

Nowadays, for the first time in the history of AMO physics, we have at our disposal laser sources, which, combined with nanostructures, generate fields that exhibit spatial variation at a nanometric scale. This is the native scale of the electron dynamics in atoms, molecules and bulk matter. Consequently, markedly and profound changes occur in systems interacting with such spatially inhomogeneous fields. Using well-known numerical techniques, based on solutions of Maxwell equations, one is able to model both the time and the spatial properties of these laser-induced plasmonic fields. This in the first important step for the subsequent theoretical modeling of the strong-field physical processes driven by them.

Theoretically speaking, in the recent years, there has been a thoughtful and continuous activity in atto-nano physics. Indeed, all of the theoretical tools developed to tackle strong field processes driven by spatially homogeneous fields have beed generalized and adapted to this new arena. Several open problems, however, still remain. For instance, the behavior of complex systems, e.g. multielectronic atoms and molecules, under the influence of spatial inhomogeneous fields is an unexplored area – only few attempts to tackle this problem has been recently reported [16,135,136]. In addition, and just to name another example, it was recently demonstrated that Rydberg atoms could be a plausible alternative as a driven media [124].

Diverse paths could be explored in the future. The manipulation and control of the plasmonic-enhanced fields appears as one of them. From an experimental perspective, this presents a tremendous challenge, considering the nanometric dimensions of the systems, although several experiments are planned in this direction, for instance, combining metal nanotips and molecules in a gas phase. The possibility to tailor the electron trajectories at their natural scale is another path to be considered. By employing quantum control tools, it would be possible, in principle theoretically, to drive the electron following a certain desired 'target',e.g. a one which results with the largest possible velocity, now with a time- and spatial-dependent driving field. The spatial shape of this field could be, subsequently, obtained by engineering a nanostructure.

The quest for HHG from plasmonic nanostructures, joint with an explosive amount of theoretical work, begun with the controversial report of a Korean group on HHG from bow-tie metal nanostructures [51]. Let us mention at the end a recent results of the same group, which clearly seems to be well justified and, as such, opens new perspectives and ways toward efficient HHG in nanostructures. In this recent article, the authors demonstrate plasmonic HHG experimentally by devising a metal-sapphire nanostructure that provides a solid tip as the HHG emitter instead of gaseous atoms. The fabricated solid tips are made of monocrystalline sapphire surrounded by a gold thin-film layer and intended to produce coherent XUV harmonics by the inter- and intraband oscillations of electrons driven by the incident laser. The metal-sapphire nanostructure enhances the incident laser field by means of surface plasmon polaritons (SPPs), triggering HHG directly from moderate femtosecond pulses of 0.1 TW cm^{-2} intensities. Measured XUV spectra show odd-order harmonics up to 60 nm wavelengths without the plasma atomic lines typically seen when using gaseous atoms as the HHG emitter. This experimental outcome confirms that the plasmonic HHG approach is a promising way to realize coherent XUV sources for nanoscale near-field applications in spectroscopy, microscopy, lithography and

attosecond physics [39]. The era of the atto-nano physics has just started.

Acknowledgments

This work was supported by the project Advanced research using high intensity laser produced photons and particles (No. CZ.02.1.01/0.0/0.0/16_019/0000789) from European Regional Development Fund (ADONIS). The results of the project LQ1606 were obtained with the financial support of the Ministry of Education, Youth and Sports as part of targeted support from the National Program of Sustainability II. We acknowledge the Spanish Ministry MINECO (National Plan 15 Grant: FISICATEAMO No. FIS2016-79508-P, SEVERO OCHOA No. SEV-2015-0522, FPI), European Social Fund, Fundació Cellex, Generalitat de Catalunya (AGAUR Grant No. 2017 SGR 1341 and CERCA/Program), ERC AdG OSYRIS and NOQIA, EU FETPRO QUIC, and the National Science Centre, Poland-Symfonia Grant No. 2016/20/W/ST4/00314.

Bibliography

1. P. Agostini, F. Fabre, G. Mainfray, G. Petite, and N. K. Rahman. Free-free transitions following six-photon ionization of xenon atoms. *Phys. Rev. Lett.*, 42:1127, 1979.

2. M. V. Ammosov, N. B. Delone, and V. P. Krainov. Tunnel ionization of complex atoms and of atomic ions in an alternating electromagnetic field. *Sov. Phys. JETP*, 64:1191, 1986.

3. A. Apolonski, P. Dombi, G. G. Paulus, M. Kakehata, R. Holzwarth, Th. Udem, Ch. Lemell, K. Torizuka, J. Burgdörfer, T. W. Hänsch, and F. Krausz. Observation of light-phase-sensitive photoemission from a metal. *Phys. Rev. Lett.*, 92:073902, 2004.

4. D. G. Arbó, J. E. Miraglia, M. S. Gravielle, K. Schiessl, E. Persson, and J. Burgdörfer. Coulomb-volkov approximation for near-threshold ionization by short laser pulses. *Phys. Rev. A*, 77:013401, 2008.

5. A. Baltuska, Th. Udem, M. Uiberacker, M. Hentschel, E. Goulielmakis, Ch. Gohle, R. Holzwarth, V. S. Yakovlev, A. Scrinzi, T. W. Hansch, and F. Krausz. Attosecond control of electronic processes by intense light fields. *Nature*, 421:611–615, 2003.

6. D. Batani, C. J. Joachain, S. Martellucci, and A. N. Chester. *Atoms, Solids, and Plasmas in Super-Intense Laser Fields*. Kluwer Academic/Plenum, New York, 2001.

7. W. Becker, F. Grasbon, R. Kopold, D. B. Milošević, G. G. Paulus, and H. Walther. Above-threshold ionization: From classical features to quantum effects. In B. Bederson and H. Walter, editors, *Advances In Atomic, Molecular, and Optical Physics*, p. 35. Academic Press, San Diego, 2002.

8. L. Belshaw, F. Calegari, M. J. Duffy, A. Trabattoni, L. Poletto, M. Nisoli, and J. B. Greenwood. Observation of ultrafast charge migration in an amino acid. *J. Phys. Chem. Lett.*, 3:3751–3754, 2012.

9. B. Bergues, M. Kübel, N. G. Kling, C. Burger, and M. F. Kling. Single-cycle non-sequential double ionization. *IEEE J. Sel. Top. Quant. Elect.*, 21:8701009, 2015.

10. C. I. Blaga, J. Xu, A. D. DiChiara, E. Sistrunk, K. Zhang, P. Agostini, T. A. Miller, L. F. DiMauro, and C. D. Lin. Imaging ultrafast molecular dynamics with laser-induced electron diffraction. *Nature*, 483:194–197, 2012.

11. T. Brabec and F. Krausz. Intense few-cycle laser fields: Frontiers of nonlinear optics. *Rev. Mod. Phys.*, 72:545–591, 2000.

12. F. Calegari, D. Ayuso, A. Trabattoni, L. Belshaw, S. De Camillis, S. Anumula, F. Frassetto, L. Poletto, A. Palacios, P. Decleva, J. B. Greenwood, F. Martín, and M. Nisoli. Ultrafast electron dynamics in phenylalanine initiated by attosecond pulses. *Science*, 336:346, 2014.

13. X. Cao, S. Jiang, C. Yu, Y. Wang, L. Bai, and R. Lu. Generation of isolated sub-10-attosecond pulses in spatially inhomogenous two-color fields. *Opt. Exp.*, 22:26153–26161, 2014.

14. A. L. Cavalieri, N. Müller, Th. Uphues, V. S. Yakovlev, A. Baltuška, B. Horvath, B. Schmidt, L. Blümel, R. Holzwarth, S. Hendel, M. Drescher, U. Kleineberg, P. M. Echenique, R. Kienberger, F. Krausz, and U Heinzmann. Attosecond spectroscopy in condensed matter. *Nature*, 449:1029–1032, 2007.

15. A. Chacón, M. F. Ciappina, and M. Lewenstein. Numerical studies of light-matter interaction driven by plasmonic fields: The velocity gauge. *Phys. Rev. A*, 92:063834, 2015.

16. A. Chacón, M. F. Ciappina, and M. Lewenstein. Signatures of double-electron recombination in high-order harmonic generation driven by spatial inhomogeneous fields. *Phys. Rev. A*, 94:043407, 2016.

17. M. F. Ciappina, S. S. Aćimović, T. Shaaran, J. Biegert, R. Quidant, and M. Lewenstein. Enhancement of high harmonic generation by confining electron motion in plasmonic nanostrutures. *Opt. Exp.*, 20:26261–26274, 2012.

18. M. F. Ciappina, J. Biegert, R. Quidant, and M. Lewenstein. High-order-harmonic generation from inhomogeneous fields. *Phys. Rev. A*, 85:033828, 2012.

19. M. F. Ciappina, J. A. Pérez-Hernández, and M. Lewenstein. Classstrong: Classical simulations of strong field processes. *Comp. Phys. Comm.*, 185:398–406, 2014.

20. M. F. Ciappina, J. A. Pérez-Hernández, L. Roso, A. Zaïr, and M. Lewenstein. High-order harmonic generation driven by plasmonic fields: a new route

towards the generation of uv and xuv photons? *J. Phys. Conf. Ser.*, 601:012001, 2015.

21. M. F. Ciappina, J. A. Pérez-Hernández, T. Shaaran, J. Biegert, R. Quidant, and M. Lewenstein. Above-threshold ionization by few-cycle spatially inhomogeneous fields. *Phys. Rev. A*, 86:023413, 2012.

22. M. F. Ciappina, J. A. Pérez-Hernández, T. Shaaran, and M. Lewenstein. Coherent xuv generation driven by sharp metal tips photoemission. *Eur. Phys. J. D*, 68:172, 2014.

23. M. F. Ciappina, J. A. Pérez-Hernández, T. Shaaran, M. Lewenstein, M. Krüger, and P. Hommelhoff. High-order harmonic generation driven by metal nanotip photoemission: theory and simulations. *Phys. Rev. A*, 89:013409, 2014.

24. M. F. Ciappina, J. A. Pérez-Hernández, T. Shaaran, L. Roso, and M. Lewenstein. Electron-momentum distributions and photoelectron spectra of atoms driven by an intense spatially inhomogeneous field. *Phys. Rev. A*, 87:063833, 2013.

25. M. F. Ciappina, T. Shaaran, R. Guichard, J. A. Pérez-Hernández, L. Roso, M. Arnold, T. Siegel, A. Zaïr, and M. Lewenstein. High energy photoelectron emission from gases using plasmonic enhanced near-fields. *Las. Phys. Lett.*, 10:105302, 2013.

26. M. F. Ciappina, T. Shaaran, and M. Lewenstein. High order harmonic generation in noble gases using plasmonic field enhancement. *Ann. Phys. (Berlin)*, 525:97–106, 2013.

27. P. B. Corkum. Plasma perspective on strong field multiphoton ionization. *Phys. Rev. Lett.*, 71:1994, 1993.

28. P. B. Corkum and F. Krausz. Attosecond science. *Nat. Phys.*, 3:381–387, 2007.

29. T. Ditmire, R. A. Smith, J. W. G. Tisch, and M. H. R. Hutchinson. High intensity laser absorption by gases of atomic clusters. *Phys. Rev. Lett.*, 78:3121, 1997.

30. T. Ditmire, J. W. G. Tisch, E. Springate, M. B. Mason, N. Hay, R. A. Smith, J. Marangos, and M. H. R. Hutchinson. High-energy ions produced in explosions of superheated atomic clusters. *Nature (London)*, 386:54, 1997.

31. T. D. Donnelly, M. Rust, I. Weiner, M. Allen, R. A. Smith, C. A. Steinke, S. Wilks, J. Zweiback, T. E. Cowan, and T. Ditmire. Hard x-ray and hot electron production from intense laser irradiation of wavelength-scale particles. *J. Phys. B*, 34:L313, 2001.

32. H. Ebadi. Interferences induced by spatially nonhomogeneous fields in high-harmonic generation. *Phys. Rev. A*, 89:053413, 2014.

33. F. H. M. Faisal. *Theory of Multiphoton Processes*. Springer, New York, 1987.

34. L. Feng and H. Liu. Attosecond extreme ultraviolet generation in cluster by using spatially

inhomogeneous field. *Phys. Plasmas*, 22:013107, 2015.

35. L. Feng, M. Yuan, and T. Chu. Attosecond x-ray source generation from two-color polarized gating plasmonic field enhancement. *Phys. Plasmas*, 20:122307, 2013.

36. B. Fetić, K. Kalajdžić, and D. B. Milošević. High-order harmonic generation by a spatially inhomogeneous field. *Ann. Phys. (Berlin)*, 525:107–117, 2012.

37. S. Ghimire, A. D. DiChiara, E. Sistrunk, P. Agostini, L. F. DiMauro, and D. A. Reis. Observation of high-order harmonic generation in a bulk crystal. *Nat. Phys.*, 7:138–141, 2011.

38. E. T. Gumbrell, A. J. Comley, M. H. R. Hutchinson, and R. A. Smith. Intense laser interactions with sprays of submicron droplets. *Phys. Plasmas*, 8:1329, 2001.

39. S. Han, H. Kim, Y. W. Kim, Y.-J. Kim, S. Kim, I.-Y. Park, and S.-W. Kim. High harmonic generation by strongly enhanced femtosecond pulses in metal-sapphire nanostructure waveguide. *Nat. Commun.*, 7:13105, 2016.

40. L. He, Z. Wang, Y. Li, Q. Zhang, P. Lan, and P. Lu. Wavelength dependence of high-order-harmonic yield in inhomogeneous fields. *Phys. Rev. A*, 88:053404, 2013.

41. M. Hentschel, R. Kienberger, C. Spielmann, G. A. Reider, N. Milosevic, T. Brabec, P. B. Corkum, U. Heinzmann, M. Drescher, and F. Krausz. Attosecond metrology. *Nature*, 414:509–513, 2001.

42. G. Herink, D. R. Solli, M. Gulde, and C. Ropers. Field-driven photoemission from nanostructures quenches the quiver motion. *Nature*, 483:190–193, 2012.

43. P. Hommelhoff and M. F. Kling. *Attosecond Nanophysics: From Basic Science to Applications*. Wiley-VCH, Berlin, 2015.

44. P. Hommelhoff, Y. Sortais, A. Aghajani-Talesh, and M. A. Kasevich. Field emission tip as a nanometer source of free electron femtosecond pulses. *Phys. Rev. Lett.*, 96:077401, 2006.

45. A. Husakou and J. Herrmann. Quasi-phase-matched high-harmonic generation in composites of metal nanoparticles and a noble gas. *Phys. Rev. A*, 90:023831, 2014.

46. A. Husakou, S.-J. Im, and J. Herrmann. Theory of plasmon-enhanced high-order harmonic generation in the vicinity of metal nanostructures in noble gases. *Phys. Rev. A*, 83:043839, 2011.

47. A. Husakou, S.-J. Im, K.H. Kim, and J. Herrmann. High harmonic generation assisted by metal nanostructures and nanoparticles. In S. Sakabe, C. Lienau, and R. Grunwald, editors, *Progress in Nonlinear Nano-Optics*, pp. 251–268. Springer International Publishing, Cham, 2015.

48. A. Husakou, F. Kelkensberg, J. Herrmann, and M. J. J. Vrakking. Polarization gating and circularly-polarized high harmonic generation using plasmonic enhancement in metal nanostructures. *Opt. Exp.*, 19:25346–25354, 2011.

49. C. J. Joachain, N. J. Kylstra, and R. M. Potvliege. *Atoms in Intense Laser Fields*. Cambridge University Press, Cambridge, England, 2012.

50. L. V. Keldysh. Ionization in the field of a strong electromagnetic wave. *J. Expt. Theo. Phys.*, 20:1307, 1965.

51. S. Kim, J. Jin, Y.-J. Kim, I.-Y. Park, Y. Kim, and S.-W. Kim. High-harmonic generation by resonant plasmon field enhancement. *Nature*, 453:757–760, 2008.

52. S. Kim, J. Jin, Y.-J. Kim, I.-Y. Park, Y. Kim, and S.-W. Kim. Reply (nature). *Nature*, 485:E1–E3, 2012.

53. F. Krausz and M. Ivanov. Attosecond physics. *Rev. Mod. Phys.*, 81:163–234, 2009.

54. M. Krüger, M. Förster, and P. Hommelhoff. Self-probing of metal nanotips by rescattered electrons reveals the nano-optical near-field. *J. Phys. B*, 47:124022, 2014.

55. M. Krüger, M. Schenk, M. Förster, and P. Hommelhoff. Attosecond physics in photoemission from a metal nanotip. *J. Phys. B*, 45:074006, 2012.

56. M. Krüger, M. Schenk, and P. Hommelhoff. Attosecond control of electrons emitted from a nanoscale metal tip. *Nature*, 475:78–81, 2011.

57. M. Krüger, M. Schenk, P. Hommelhoff, G. Wachter, C. Lemell, and J. Burgdörfer. Interaction of ultrashort laser pulses with metal nanotips: a model system for strong-field phenomena. *New J. Phys.*, 14:085019, 2012.

58. M. Kübel, R. Siemering, C. Burger, N. G. Kling, H. Li, A.S. Alnaser, B. Bergues, S. Zherebtsov, A. M. Azzeer, I. Ben-Itzhak, R. Moshammer, R. de Vivie-Riedle, and M. F. Kling. Steering proton migration in hydrocarbons using intense few-cycle laser fields. *Phys. Rev. Lett.*, 116:193001, 2016.

59. M. Yu. Kuchiev. Atomic antenna. *Sov. Phys. JETP*, 45:404–406, 1987.

60. K. C. Kulander, K. J. Schafer, and J. L. Krause. Dynamics of short-pulse excitation, ionization and harmonic conversion. In B. Piraux, A. L'Huillier, and K. Rzazewski, editors, *Super-Intense Laser-Atom Physics*, pp. 95–110. Plenum, New York, 1993.

61. A. S. Landsman and U. Keller. Attosecond science and the tunnelling time problem. *Phys. Rep.*, 547:1–24, 2015.

62. M. Lewenstein, P. Balcou, M. Y. Ivanov, A. L'Huillier, and P. B. Corkum. Theory of high-harmonic generation by low-frequency laser fields. *Phys. Rev. A*, 49:2117, 1994.

63. M. Lewenstein and A. L'Huillier. Principles of single atom physics: High-order harmonic generation, above-threshold ionization and non-sequential ionization. In T. Brabec, editor, *Strong Field Laser Physics*, pp. 147–183. Springer, New York, 2009.

64. A. L'Huiller, M. Lewenstein, P. Salières, Ph. Balcou, M. Yu. Ivanov, J. Larsson, and C. G. Wahlström. High-order harmonic-generation cutoff. *Phys. Rev. A*, 48:R3433, 1993.

65. A. L'Huiller, L. A. Lompre, G. Mainfray, and C. Manus. Multiply charged ions induced by multi-photon absorption in rate gases at 0.53 μm. *Phys. Rev. A*, 27:2503, 1983.

66. H. Li, B. Mignolet, G. Wachter, S. Skruszewicz, S. Zherebtsov, F. Süssmann, A. Kessel, S. A. Trushin, N. G. Kling, M. Kübel, B. Ahn, D. Kim, I. Ben-Itzhak, C. L. Cocke, T. Fennel, J. Tiggesbumker, K.-H. Meiwes-Broer, C. Lemell, J. Burgdörfer, R. D. Levine, F. Remacle, and M. F. Kling. Coherent electronic wave packet motion in C_{60} controlled by the waveform and polarization of few-cycle laser fields. *Phys. Rev. Lett.*, 114:123004, 2015.

67. R. Locher, L. Castiglioni, M. Lucchini, M. Greif, L. Gallmann, J. Osterwalder, M. Hengsberger, and U. Keller. Energy-dependent photoemission delays from noble metal surfaces by attosecond interferometry. *Optica*, 2:405, 2015.

68. J. Luo, Y. Li, Z. Wang, L. He, Q. Zhang, and P. Lu. Efficient supercontinuum generation by UV-assisted midinfrared plasmonic fields. *Phys. Rev. A*, 89:023405, 2013.

69. J. Luo, Y. Li, Z. Wang, Q. Zhang, P. Lan, and P. Lu. Wavelength dependence of high-order-harmonic yield in inhomogeneous fields. *J. Opt. Soc. Am. B*, 30:2469–2475, 2013.

70. J. Luo, Y. Li, Z. Wang, Q. Zhang, and P. Lu. Ultrashort isolated attosecond emission in mid-infrared inhomogeneous fields without cep stabilization. *J. Phys. B*, 46:145602, 2013.

71. T. T. Luu, M. Garg, S. Yu. Kruchinin, A. Moulet, M. Th. Hassan, and E. Goulielmakis. Extreme ultraviolet high-harmonic spectroscopy of solids. *Nature*, 521:498–502, 2015.

72. J. P. Marangos. Development of high harmonic generation spectroscopy of organic molecules and biomolecules. *J. Phys. B*, 49:132001, 2016.

73. D. B. Milošević, G. G. Paulus, D. Bauer, and W. Becker. Above-threshold ionization by few-cycle pulses. *J. Phys. B*, 39:R203–R262, 2006.

74. L. C. Mountford, R. A. Smith, and M. H. R. Hutchinson. Characterization of a sub-micron liquid spray for laser-plasma x-ray generation. *Rev. Sci. Instrum.*, 69:3780, 1998.

75. H. G. Muller, H. B. van Linden van den Heuvell, and M. J. van der Wiel. Experiments on "above-threshold ionization" of atomic hydrogen. *Phys. Rev. A*, 34:236, 1986.

76. S. Neppl, R. Ernstorfer, E. M. Bothschafter, A. L. Cavalieri, D. Menzel, J. V. Barth, F. Krausz, R. Kienberger, and P. Feulner. Attosecond time-resolved photoemission from core and valence states of magnesium. *Phys. Rev. Lett.*, 109:087401, 2012.

77. E. Neyra, F. Videla, M. F. Ciappina, J. A. Perez-Hernandez, L. Roso, M. Lewenstein, and G. A. Torchia. High-order harmonic generation driven by inhomogeneous plasmonics fields spatially bounded: influence on the cut-off law. *J. Opt.*, 20:034002, 2018.

78. W. A. Okell, T. Witting, D. Fabris, C. A. Arrell, J. Hengster, S. Ibrahimkutty, A. Seiler, M. Barthelmess, S. Stankov, D. Y. Lei, Y. Sonnefraud, M. Rahmani, T. Uphues, S. A. Maier, J. P. Marangos, and J. W. G. Tisch. Temporal broadening of attosecond photoelectron wavepackets from solid surfaces. *Optica*, 2:383–387, 2015.

79. L. Ortmann, J. A. Pérez-Hernández, M. F. Ciappina, J. Schötz, A. Chacón, G. Zeraouli, M. F. Kling, L. Rosoa, M. Lewenstein, and A. S. Landsman. Emergence of a higher energy structure in strong field ionization with inhomogeneous electric fields. *Phys. Rev. Lett.*, 119:053204, 2017.

80. D. J. Park, B. Piglosiewicz, S. Schmidt, H. Kollmann, M. Mascheck, P. Gro, and C. Lienau. Characterizing the optical near-field in the vicinity of a sharp metallic nanoprobe by angle-resolved electron kinetic energy spectroscopy. *Ann. Phys. (Berlin)*, 525:135–142, 2013.

81. I.-Y. Park, S. Kim, J. Choi, D.-H. Lee, Y. J. Kim, M. F. Kling, M. I. Stockman, and S.-W. Kim. Plasmonic generation of ultrashort extreme-ultraviolet light pulses. *Nat. Phot.*, 5:677, 2011.

82. G. G. Paulus, F. Grasbon, H. Walther, P. Villoresi, M. Nisoli, S. Stagira, E. Priori, and S. De Silvestri. Absolute-phase phenomena in photoionization with few-cycle laser pulses. *Nature*, 414:182–184, 2001.

83. G. G. Paulus, F. Lindner, H. Walther, A. Baltuška, E. Goulielmakis, M. Lezius, and F. Krausz. Measurement of the phase of few-cycle laser pulses. *Phys. Rev. Lett.*, 91:253004, 2003.

84. G. G. Paulus, W. Nicklich, X. Huale, P. Lambropoulus, and H. Walter. Plateau in above threshold ionization spectra. *Phys. Rev. Lett.*, 72:2851, 1994.

85. R. Pazourek, S. Nagele, and J. Burgdörfer. Attosecond chronoscopy of photoemission. *Rev. Mod. Phys.*, 87:765, 2015.

86. A. M. Perelomov, V. S. Popov, and M. V. Terentev. Ionization of atoms in an alternating electric field. *Sov. Phys. JETP*, 23:924, 1966.

87. J. A. Pérez-Hernández, M. F. Ciappina, M. Lewenstein, L. Roso, and A. Zaïr. Beyond carbon K-edge harmonic emission using a spatial and temporal synthesized laser field. *Phys. Rev. Lett.*, 110:053001, 2013.

88. J. A. Pérez-Hernández, D. J. Hoffmann, A. Zaïr, L. E. Chipperfield, L. Plaja, C. Ruiz, J. P. Marangos, and L. Roso. Extension of the cut-off in high-harmonic

generation using two delayed pulses of the same colour. *J. Phys. B*, 42:134004, 2009.

89. N. Pfullmann, C. Waltermann, M. Noack, S. Rausch, T. Nagy, C. Reinhardt, M. Kovačev, V. Knittel, R. Bratschitsch, D. Akemeier, A. Hütten, A. Leitenstorfer, and U. Morgner. Bow-tie nano-antenna assisted generation of extreme ultraviolet radiation. *New J. Phys.*, 15:093027, 2013.

90. B. Piglosiewicz, S. Schmidt, D. J. Park, J. Vogelsang, P. Gross, C. Manzoni, P. Farinello, G. Cerullo, and C. Lienau. Carrier-envelope phase effects on the strong-field photoemission of electrons from metallic nanostructures. *Nat. Phot.*, 8:37–42, 2014.

91. M. G. Pullen, B. Wolter, A.-T. Le, M. Baudisch, M. Hemmer, A. Senftleben, C. D. Schröter, J. Ullrich, R. Moshammer, C. D. Lin, and J. Biegert. Imaging an aligned polyatomic molecule with laser-induced electron diffraction. *Nat. Commun.*, 6:7262, 2015.

92. H. R. Reiss. Effect of an intense electromagnetic field on a weakly bound system. *Phys. Rev. A*, 22:1786, 1980.

93. P. Salières, B. Carré, L. Le Déroff, F. Grasbon, G. G. Paulus, H. Walther, R. Kopold, W. Becker, D. B. Milosevic, A. Sanpera, and M. Lewenstein. Feynman's path-integral approach for intense-laser-atom interactions. *Science*, 292:902–905, 2001.

94. P. Salières, A. L'Huillier, P. Antoine, and M. Lewenstein. Study of the spatial and temporal coherence of high-order harmonics. In B. Bederson and H. Walter, editors, *Advances in Atomic, Molecular and Optical Physics. Vol. 41*, pp. 83–142. Academic Press, San Diego, 1999.

95. A. M. Sayler, T. Rathje, W. Müller, K. Rühle, R. Kienberger, and G. G. Paulus. Precise, real-time, every-single-shot, carrier-envelope phase measurement of ultrashort laser pulses. *Opt. Lett.*, 36:1–3, 2011.

96. K. J. Schafer. The energy analysis of time-dependent numerical wave functions. *Comput. Phys. Commun.*, 63:427–434, 1991.

97. K. J. Schafer. Numerical methods in strong field physics. In T. Brabec, editor, *Strong Field Laser Physics*, pp. 111–145. Springer, New York, 2009.

98. K. J. Schafer, B. Yang, L. F. DiMauro, and K. C. Kulander. Above threshold ionization beyond the high harmonic cutoff. *Phys. Rev. Lett.*, 70:1599, 1993.

99. K. J. Schafer and K. C. Kulander. Energy analysis of time-dependent wave functions: Application to above-threshold ionization. *Phys. Rev. A*, 42:5794(R), 1990.

100. M. Schenk, M. Krüger, and P. Hommelhoff. Strong-field above-threshold photoemission from sharp metal tips. *Phys. Rev. Lett.*, 105:257601, 2010.

101. A. Schiffrin, T. Paasch-Colberg, N. Karpowicz, V. Apalkov, D. Gerster, S. Mühlbrandt, M. Korbman, J. Reichert, M. Schultze, S. Holzner, J. V. Barth, R. Kienberger, R. Ernstorfer, V. S. Yakovlev, M. I. Stockman, and F. Krausz. Optical-field-induced current in dielectrics. *Nature*, 493:70–74, 2013.

102. O. Schubert, M. Hohenleutner, F. Langer, B. Urbanek, C. Lange, U. Huttner, D. Golde, T. Meier, M. Kira, S. W. Koch, and R. R. Huber. Sub-cycle control of terahertz high-harmonic generation by dynamical bloch oscillations. *Nat. Phot.*, 8:119–123, 2014.

103. M. Schultze, E. M. Bothschafter, A. Sommer, S. Holzner, W. Schweinberger, M. Fiess, M. Hofstetter, R. Kienberger, V. Apalkov, V. S. Yakovlev, M. I. Stockman, and F. Krausz. Controlling dielectrics with the electric field of light. *Nature*, 493:75–78, 2013.

104. M. Schultze, M. Fiess, N. Karpowicz, J. Gagnon, M. Korbman, M. Hofstetter, S. Neppl, A. L. Cavalieri, Y. Komninos, Th. Mercouris, C. A. Nicolaides, R. Pazourek, S. Nagele, J. Feist, J. Burgdörfer, A. M. Azzeer, R. Ernstorfer, R. Kienberger, U. Kleineberg, E. Goulielmakis, F. Krausz, and V. S. Yakovlev. Delay in photoemission. *Science*, 328:1658–1662, 2010.

105. A. Scrinzi, M. Y. Ivanov, R. Kienberger, and D. M. Villeneuve. Attosecond physics. *J. Phys. B*, 39:R1–R37, 2006.

106. T. Shaaran, M. F. Ciappina, R. Guichard, J. A. Pérez-Hernández, L. Roso, M. Arnold, T. Siegel, A. Zaïr, and M. Lewenstein. High-order-harmonic generation by enhanced plasmonic near-fields in metal nanoparticles. *Phys. Rev. A*, 87:041402(R), 2013.

107. T. Shaaran, M. F. Ciappina, and M. Lewenstein. Estimating the plasmonic field enhancement using high-order harmonic generation: The role of the field inhomogeneity. *J. Mod. Opt.*, 86:1634–1639, 2012.

108. T. Shaaran, M. F. Ciappina, and M. Lewenstein. Quantum-orbit analysis of high-order-harmonic generation by resonant plasmon field enhancement. *Phys. Rev. A*, 86:023408, 2012.

109. T. Shaaran, M. F. Ciappina, and M. Lewenstein. Quantum-orbit analysis of above-threshold ionization driven by an intense spatially inhomogeneous field. *Phys. Rev. A*, 87:053415, 2013.

110. Y. L. Shao, T. Ditmire, J. W. G. Tisch, E. Springate, J. P. Marangos, and M. H. R. Hutchinson. Multi-kev electron generation in the interaction of intense laser pulses with xe clusters. *Phys. Rev. Lett.*, 77:3343, 1996.

111. A. D. Shiner, B. E. Schmidt, C. Trallero-Herrero, H. J. Wörner, S. Patchkovskii, P. B. Corkum, J.-C. Kieffer, F. Légaré, and D. M. Villeneuve. Probing collective multi-electron dynamics in xenon with high-harmonic spectroscopy. *Nat. Phys.*, 7:464–467, 2011.

112. M. Sivis, M. Duwe, B. Abel, and C. Ropers. Nanostructure-enhanced atomic line emission. *Nature*, 485:E1–E3, 2012.

113. M. Sivis, M. Duwe, B. Abel, and C. Ropers. Extreme-ultraviolet light generation in plasmonic nanostructures. *Nat. Phys.*, 9:304–309, 2013.

114. E. Skopalová, Y. C. El-Taha, A. Zaïr, M. Hohenberger, E. Springate, J. W. G. Tisch, R. A. Smith, and J. P. Marangos. Pulse-length dependence of the anisotropy of laser-driven cluster explosions: Transition to the impulsive regime for pulses approaching the few-cycle limit. *Phys. Rev. Lett.*, 104:203401, 2010.

115. O. Smirnova, Y. Mairesse, S. Patchkovskii, N. Dudovich, D. Villeneuve, P. B. Corkum, and M. Yu. Ivanov. High harmonic interferometry of multi-electron dynamics in molecules. *Nature*, 460:972–977, 2009.

116. R. A. Smith, T. Ditmire, and J. W. G. Tisch. Characterization of a cryogenically cooled high-pressure gas jet for laser/cluster interaction experiments. *Rev. Sci. Instrum.*, 69:3798, 1998.

117. M. I. Stockman. Nanoplasmonics: past, present, and glimpse into future. *Opt. Exp.*, 19:22029–22106, 2011.

118. H. A. Sumeruk, S. Kneip, D. R. Symes, I. V. Churina, A. V. Belolipetski, T. D. Donnelly, and T. Ditmire. Control of strong-laser-field coupling to electrons in solid targets with wavelength-scale spheres. *Phys. Rev. Lett.*, 98:045001, 2007.

119. H. A. Sumeruk, S. Kneip, D. R. Symes, I. V. Churina, A. V. Belolipetski, G. Dyer, J. Landry, G. Bansal, A. Bernstein, T. D. Donnelly, A. Karmakar, A. Pukhov, and T. Ditmire. Hot electron and x-ray production from intense laser irradiation of wavelength-scale polystyrene spheres. *Phys. Plasmas*, 14:062704, 2007.

120. F. Süssmann and M. F. Kling. Attosecond measurement of petahertz plasmonic near-fields. *Proc. SPIE*, 8096:80961C, 2011.

121. F. Süssmann and M. F. Kling. Attosecond nanoplasmonic streaking of localized fields near metal nanospheres. *Phys. Rev. B*, 84:121406(R), 2011.

122. F. Süssmann, L. Seiffert, S. Zherebtsov, V. Mondes, J. Stierle, M. Arbeiter, J. Plenge, P. Rupp, C. Peltz, A. Kessel, S. A. Trushin, B. Ahn, D. Kim, C. Graf, E. Rühl, M. F. Kling, and T. Fennel. Field propagation-induced directionality of carrier-envelope phase-controlled photoemission from nanospheres. *Nat. Commun.*, 6:7944, 2015.

123. D. R. Symes, A. J. Comley, and R. A. Smith. Fast-ion production from short-pulse irradiation of ethanol microdroplets. *Phys. Rev. Lett.*, 93:145004, 2004.

124. Y. Tikman, I. Yavuz, M. F. Ciappina, A. Chacón, Z. Altun, and M. Lewenstein. High-order-harmonic generation from rydberg atoms driven by plasmon-enhanced laser fields. *Phys. Rev. A*, 93:023410, 2016.

125. J. W. G. Tisch, T. Ditmire, D. J. Fraser, N. Hay, M. B. Mason, E. Springate, J. P. Marangos, and M. H. R. Hutchinson. Investigation of high-harmonic generation from xenon atom clusters. *J. Phys. B*, 30:L709, 1997.

126. G. Vampa, T. J. Hammond, N. Thiré, B. E. Schmidt, F. Légaré, C. R. McDonald, T. Brabec, and P. B. Corkum. Linking high harmonics from gases and solids. *Nature*, 522:462–464, 2015.

127. H. B. van Linden van den Heuvell and H. G. Muller. Limiting cases of excess-photon ionization. In S. J. Smith and P. L. Knight, editors, *Multiphoton Processes*, p. 25. Cambridge University Press, Cambridge, England, 1988.

128. B. Walker, B. Sheehy, L. F. DiMauro, P. Agostini, K. J. Schafer, and K. C. Kulander. Precision measurement of strong field double ionization of helium. *Phys. Rev. Lett.*, 73:1227, 1994.

129. Z. Wang, L. He, J. Luo, P. Lan, and P. Lu. High-order harmonic generation from rydberg atoms in inhomogeneous fields. *Opt. Exp.*, 22:25909–25922, 2014.

130. Z. Wang, P. Lan, J. Luo, L. He, Q. Zhang, and P. Lu. Control of electron dynamics with a multicycle two-color spatially inhomogeneous field for efficient single-attosecond-pulse generation. *Phys. Rev. A*, 88:063838, 2013.

131. J. Xu, C. I. Blaga, K. Zhang, Y. H. Lai, C. D. Lin, T. A. Miller, P. Agostini, and L. F. DiMauro. Diffraction using laser-driven broadband electron wave packets. *Nat. Commun.*, 5:4635, 2014.

132. V. Yakovlev, M. I. Stockman, F. Krausz, and P. Baum. Atomic-scale diffractive imaging of sub-cycle electron dynamics in condensed matter. *Sci. Rep.*, 5:14581, 2015.

133. I. Yavuz. Gas population effects in harmonic emission by plasmonic fields. *Phys. Rev. A*, 87:053815, 2013.

134. I. Yavuz, E. A. Bleda, Z. Altun, and T. Topcu. Generation of a broadband XUV continuum in high-order-harmonic generation by spatially inhomogeneous fields. *Phys. Rev. A*, 85:013416, 2012.

135. I. Yavuz, M. F. Ciappina, A. Chacón, Z. Altun, M. F. Kling, and M. Lewenstein. Controlling electron localization in H_2^+ by intense plasmon-enhanced laser fields. *Phys. Rev. A*, 93:033404, 2016.

136. I. Yavuz, Y. Tikman, and Z. Altun. High-order-harmonic generation from H_2^+ molecular ions near plasmon-enhanced laser fields. *Phys. Rev. A*, 92:023413, 2015.

137. C. Yu, Y. Wang, X. Cao, S. Jiang, and R. Lu. Isolated few-attosecond emission in a multi-cycle asymmetrically nonhomogeneous two-color laser field. *J. Phys. B*, 47:225602, 2015.

138. C. Zagoya, M. Bonner, H. Chomet, E. Slade, and C. Figueira de Morisson Faria. Different time scales

in plasmonically enhanced high-order-harmonic generation. *Phys. Rev. A*, 93:053419, 2016.

139. C. Zang, C. Lui, and Z. Xu. Control of higher spectral components by spatially inhomogeneous fields in quantum wells. *Phys. Rev. A*, 88:035805, 2013.

140. S. Zherebtsov, T. Fennel, J. Plenge, E. Antonsson, I. Znakovskaya, A. Wirth, O. Herrwerth, F. Süssmann, C. Peltz, I. Ahmad, S. A. Trushin, V. Pervak, S. Karsch, M. J. J. Vrakking, B. Langer, C. Graf, M. I. Stockman, F. Krausz, E. Rühl, and M. F. Kling. Controlled near-field enhanced electron acceleration from dielectric nanospheres with intense few-cycle laser fields. *Nat. Phys.*, 7:656–662, 2011.

2

The de Broglie Wave Nature of Molecules, Clusters and Nanoparticles

Stefan Gerlich, Stefan Kuhn,
Armin Shayeghi, and Markus
Arndt
University of Vienna

2.1 A Century of Matter Waves and Nanoscience

For many decades, nanoscience has studied material properties at length scales where quantum effects can dominate mechanical, thermal, optical, magnetic or electronic properties compared to their macroscopic values. Here we focus on the center-of-mass motion, which can also display quantum behavior when particles are sufficiently small and isolated from their environment. The matter–wave nature of complex composite systems can shed light on the foundations of physics while enabling new measurements of internal characteristics of nanoscale systems.

Since its first conception by Louis de Broglie [1] and its formalization by Erwin Schrödinger [2], the quantum wave nature of matter has become a cornerstone of modern physics. Many nonclassical features of quantum mechanics can be demonstrated in matter–wave interference experiments. This is why Richard Feynman once claimed that double-slit diffraction of massive particles 'contains the only mystery of quantum physics' [3]. Throughout the last century, numerous experiments have endeavored to demonstrate the wave nature of matter with a variety of particles, from electrons [4] and neutrons [5] over atoms [6] and cold quantum gases [7] to highly excited macromolecules

and molecular clusters [8–10]. This research was curiosity driven but inevitably led to important applications in nanotechnology, such as electron microscopy, diffraction and holography or neutron scattering with applications in condensed-matter physics and the life sciences.

Atom interferometry started a few decades later, with the development of nanomechanical diffraction masks and intense narrowband laser light. It has gone a long way from first demonstrations of coherent atom beam splitters [11,12] to full-fledged interferometers [13–15] that can nowadays even delocalize every single atom on the half-meter scale [16]. Such large-scale instruments are being used for advanced tests of general relativity. Smaller and yet precise versions are being developed and commercialized as mobile platforms for inertial sensing and navigation [6,17]. The combination of many atoms into an ultracold quantum degenerate ensemble at nanokelvin temperatures, a Bose–Einstein condensate (BEC) [7,18,19], shows atomic coherence over mesoscopic scales. Such novel quantum states of matter find applications in matter–wave interferometry and in condensed-matter quantum simulations [20].

In the following contribution, we focus on the opposite side of the parameter space: Over the last two decades, matter–wave research has made another leap in complexity, from atomic physics to a combination of quantum optics, nanophysics, physical chemistry and biomolecular physics [21]. Instead of using ultracold, very weakly bound atomic many-body systems, we describe the quantum wave nature of covalently or van der Waals bound clusters and macromolecules, even at temperatures up to 1,000 K. While the number of atoms per molecule can be of the order of $N = 1,000$, comparable to a small BEC, the temperature and binding strength can be 12 orders of magnitude higher. Current quantum experiments can handle objects in the size range of 1–5 nm diameters and masses between 100 and 10,000 amu [22] with prospects of preparing matter waves from even more massive nanoparticles in the near future [23–25].

Nanoparticle quantum interference is driven by two central motivations: First, it is still an open scientific question, whether there is any mass, size and complexity limit to the observation of quantum phenomena. We routinely observe them on the nanoscale but not at all in the macroscopic world.

Various explanations have been put forward to answer this question. They include kinematics and the smallness of Planck's constant of action $\hbar$, decoherence as the quantum mechanical coupling between an object and its environment and even hypothetical corrections to quantum theory itself. A vast range of experimental parameters has remained untested until today, and the prospect of being able to probe quantum mechanics with very massive nanoparticles [24–27] is attracting a growing research community to the interface between nanoscience and quantum optics.

Secondly, it turns out that the matter-wave interferometers, which are built to probe the quantum–classical interface, also serve as extremely sensitive force sensors for measuring inertial, rotational and gravitational forces. Applied to complex many-body systems, they can be used to determine the optical, electronic and magnetic properties of isolated molecules and clusters with unprecedented precision.

The section below is dedicated to illustrating the state of the art in de Broglie optics with macromolecules, clusters and nanoparticles and to highlighting the technological challenges that are caused by the minuscule size of the de Broglie wavelength in such experiments, which scales inversely with the particle's mass m and velocity v. Values of $\lambda_{dB} = h/mv \simeq 10^{-14} - 10^{-11}$ m are common. This is more than million times smaller than in typical cold atom experiments, which explains the need for nanoscience and nanotechnology in the preparation of sources, diffraction elements and detectors for macromolecular matter waves.

2.2 Introduction to Matter–Wave Interference

2.2.1 Basic Diffraction Theory

Since the time-independent Schrödinger equation is mathematically equivalent to the Helmholtz equation of classical electrodynamics, many wave phenomena in light optics find a direct counterpart in matter–wave optics [17]. As in classical optics, one can describe diffraction patterns using Kirchhoff–Fresnel theory [28], as illustrated in Figure 2.1: If a diffraction aperture A is illuminated by a plane wave, the field distribution observed on a screen located at a distance L behind A is then determined by the coherent sum over all spherical wavelets emerging from the aperture. With $t(\xi, \eta)$ being the (complex) transmission function in the aperture coordinates ξ, η and assuming the validity of the paraxial approximation $|\mathbf{r}| \simeq L$, the field amplitude $\psi(x, y)$ on the screen becomes

$$\psi(x,y) \simeq -\frac{i}{L\lambda_{dB}} \int_A e^{-i\mathbf{k}\mathbf{r}} t(\xi,\eta) d\xi d\eta \qquad (2.1)$$

By expanding the radius vector $\mathbf{r}$ between (ξ, η) and (x, y) into a Taylor series and retaining all the terms up to the second order, this can be rewritten as

$$\psi(x,y) \propto \frac{1}{\lambda_{dB}L} \int_A \exp\left[-ik\left(\frac{x\xi+y\eta}{L} + \frac{\xi^2+\eta^2}{2L}\right)\right] t(\xi,\eta) d\xi d\eta \qquad (2.2)$$

Constant phase factors do not contribute to the final probability density $|\psi^2|$ and have therefore been omitted here. For distances $L \gg a^2/\lambda_{dB}$, behind a diffracting element of size a, i.e. in the *far field* or *Fraunhofer* limit, all wavelets can be approximated by plane waves. In this case, the linear phase terms of Eq. (2.2) provide a fair description. The amplitude on the screen is then described by the Fourier transform of the aperture transmission function

$$\psi(x,y) \propto \frac{1}{\lambda_{dB}L} \int_A \exp\left(-\frac{ik}{L}(x\xi+y\eta)\right) t(\xi,\eta) d\xi d\eta. \qquad (2.3)$$

For distances $L \ll a^2/\lambda_{dB}$, the curvature of the wavefronts needs to be taken into account and the quadratic phase terms must be included. Second-order wave phenomena in this *near-field* region are referred to as *Fresnel diffraction*. For a periodic structure with period d, Fresnel diffraction leads to the formation of coherent self-images that recur at integer multiples of the distance $T_L = d^2/\lambda_{dB}$, referred

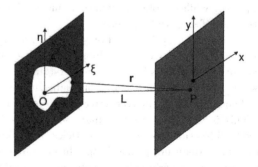

FIGURE 2.1 Diffraction at a plane aperture: Huygens' principle of optics assumes that the final wave function on the screen at position L behind the aperture A of typical dimension a is given by a sum over all possible waves emerging from A.

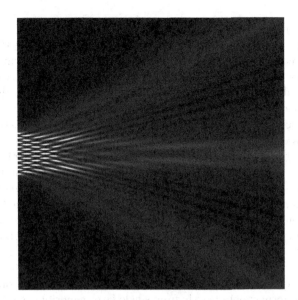

FIGURE 2.2 Simulation of the diffraction behind a coherently illuminated binary transmission grating of aperture width a and period d. Closely behind the grating, self-images of the grating transmission function can be observed at integer multiples of the Talbot length, $L_T = d^2/\lambda_{dB}$. The transition from Fresnel diffraction in the near field to Fraunhofer diffraction in the far field occurs around $L \simeq \pi a^2/\lambda_{dB}$.

to as the *Talbot length*. Figure 2.2 illustrates the transition from these Talbot images to the textbook far-field diffraction pattern, where the angles to the fringe maxima are determined by $\sin \vartheta = n\lambda_{dB}/d$, with $n \in \mathbb{N}$.

2.2.2 Coherence Considerations

The diffraction phenomena described so far rely on sufficient spatial and spectral coherence of the incident particle beam. The transverse coherence must suffice to cover a relevant width of the diffracting element, while the spectral coherence must exceed at least n de Broglie wavelengths to allow the formation of nth-order interference fringes. The van Cittert–Zernike and the Wiener–Khinchin theorem of classical wave optics [28] describe the evolution of transverse coherence behind an incoherently illuminated aperture and the relation of longitudinal coherence with the spectral distribution of the incident radiation, respectively. These principles of light optics also apply to de Broglie optics and can often be simplified by geometrical considerations to practical rules of thumb, which serve well in the lab: The transverse coherence width X_c in a distance L behind a monochromatic but spatially incoherent source of aperture width a can be estimated to be $X_c \simeq 2L\lambda_{dB}/a$. On similar grounds, the longitudinal coherence can be estimated to be $L_c = \lambda_{dB}^2/\Delta\lambda_{dB}$, with λ_{dB} the most probable de Broglie wavelength and $\Delta\lambda_{dB}$ its full width at half maximum. If the various diffraction peaks emerging behind a diffraction mask should be individually resolved, we require in addition the beam collimation angle ϑ_{coll} to be smaller than the diffraction angle $\vartheta_{coll} < \vartheta_{diff}$.

For example, for particles with a de Broglie wavelength of $\lambda_{dB} \simeq 1$ pm diffracted at a $d = 100$ nm grating in 1 m distance behind the source, the aperture must not exceed $a = 10$ μm if we aim at $X_c \simeq 200$ nm, just barely sufficient to illuminate the grating coherently across two neighboring slits. In addition, we would require a divergence angle of $\vartheta_{coll} < 5$ μrad along the grating vector and typically 1 mrad in the orthogonal direction. All present-day macromolecule or nanoparticle experiments rely on beam collimation rather than cooling and therefore cut the signal by a factor of 10^8 compared to the initial molecular beam.

2.3 Diffraction Elements for Large Molecules, Clusters and Nanoparticles

2.3.1 Nanofabricated Masks for Matter–Wave Diffraction

The conceptually most intuitive way to demonstrate the nonlocal wave nature of massive particles is to diffract them at a binary transmission grating, in close analogy to Young's double-slit experiment [29]. Grating diffraction has been demonstrated using electrons [30,31], neutrons [32], atoms [12,33] and small helium clusters [34]. Far-field diffraction of complex molecules was demonstrated for the first time by sending a collimated fullerene C_{60} beam through a nanomechanical grating [8], as sketched in Figure 2.3a. Fullerenes

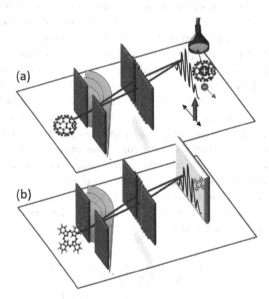

FIGURE 2.3 Far-field diffraction of molecular matter waves with de Broglie wavelengths in the range of $\lambda_{dB} = 5$ pm requires the preparation of nearly plane wavefronts by tight collimation of the molecular beam to typically below 10 μrad, also to ensure transverse coherence across several slits in the nanomechanical masks with a period of $d = 100$ nm. Diffraction angles of several 10 μrad can be resolved (a) by a tightly focused and scanning ionizing laser beam [8,36] or (b) using surface deposition followed by fluorescence microscopy [37].

are especially well suited for that, since they are very stable and can be detected by intense focused laser heating followed by thermionic electron emission. The interference pattern can then be recorded by scanning the position of a tightly focused, intense green laser beam and counting the resulting ions. Interestingly, the quantum fringes persist even for molecular beams emerging from an effusive source as hot as 900 K. At that temperature, many vibrational and rotational modes of the molecules are excited – and yet, high-contrast interference can be observed [35,36]. While this may appear surprising at first glance, since quantum interference always requires indistinguishability of the interfering partial waves, one realizes that this condition is still maintained: Every molecule only interferes with itself, and all intramolecular processes occur in all spatially delocalized branches at the same time, as long as the internal processes do not reveal any 'which-path' information.

Building on that scheme, the wave–particle duality has been recorded and visualized in an even more intuitive way, using phthalocyanine dye molecules, whose high-contrast diffraction pattern was deposited on a quartz slide and then imaged using laser-induced fluorescence (Figure 2.3b). This allows localizing every molecule with 10 nm accuracy, after the molecular wave function must have been delocalized across several hundred nanometers at the nanomachined diffraction grating before [37].

In all these cases, the observed diffraction angles were in good quantitative agreement with the predictions of elementary quantum mechanics. For a rectangular transmission grating, the intensity distribution between the different far-field diffraction orders is well described by a $\mathrm{sinc}(x)$-function, as predicted by Eq. (2.3). With increasing size and complexity of the particles, however, their *internal* structure becomes relevant. Already in experiments with noble gas [38] and alkali [39,40] atoms at 160 nm thick SiN_x gratings with a period of $d = 100$ nm and slit width of $s = 50$ nm, it was noticed that high diffraction orders were populated more intensely than predicted for a matter wave without internal atomic structure. This observation can be explained by taking the attractive Casimir–Polder interaction between the polarizable atom and the dielectric grating bars into account, which adds a phase contribution to the binary transmission mask. The Kirchhoff–Fresnel integral for a one-dimensional slit array is then

$$\psi(x) \propto \frac{1}{\sqrt{\lambda_{\mathrm{dB}} z}} \sum_{n=0}^{N-1} \int_{nd}^{nd+s} \psi(\xi) t(\xi)$$
$$\times \exp\left[-ik\left(\frac{x\xi}{L} + \frac{\xi^2}{2L}\right)\right] d\xi, \qquad (2.4)$$

with the transmission function

$$t(\xi, v_z) = \exp\left(\frac{i}{\hbar v_z} \int_0^b \left(\frac{C_3}{(\xi - nd)^3} - \frac{C_3}{(\xi - nd - s)^3}\right) dz\right). \qquad (2.5)$$

At particle-wall distances up to about 100 nm, the potential is well described by the van der Waals attraction

$V(\xi) = -C_3/(\xi - nd)^3$, referenced to each side of the slit the particle passes through. At larger distances, the effect of the potential is retarded merging into the asymptotic form $V(x) = -C_4/(\xi - nd)^4$. The constants C_3 and C_4 depend on the polarizability of the particle and the dielectric properties of the grating material. The forward velocity v_z enters since the molecule integrates a greater phase shift for a longer interaction time. However, even for gratings machined into 40 nm thin SiN_x, and molecular transit times as short as 200 ps, its attraction to the surface bends the molecular wave function to higher-order fringes, as seen in Figure 2.4a. This effect can be strongly reduced but not entirely eliminated, even in the limit of the thinnest conceivable material mask, i.e. gratings machined into single-layer graphene (Figure 2.4b). This also applies to molecular diffraction at graphene 'nanoharps' – i.e. periodic arrays of graphene nanoscrolls that form spontaneously from single-layer nanoribbons [41]. For polar molecules, such as the functionalized tetraphenlyporphyrin shown in Figure 2.4c is strongly impeded. This is attributed to the almost inevitable presence of local charges that are randomly deposited on the membrane during the manufacturing process [42]. Polar molecules will experience orientation-dependent fringe shifts, which reduce the overall fringe visibility. This is why optical gratings have become an attractive alternative to nanomechanical masks for de Broglie experiments with massive, polarizable and/or even polar particles.

2.3.2　Optical Beam Splitters for Massive Matter Waves

A number of different effects can be utilized to control the particle motion using light. In contrast to neutrons and electrons, atoms offer a rich electronic structure which becomes even more complex for molecules due to their vibrational and rotational level scheme. Even though Kapitza and Dirac [43] considered diffraction of electrons at standing light waves already in 1933, their idea had to wait for 50 years to be realized with atoms [44] and 70 years to be demonstrated with electrons [31] and complex molecules [45].

Beam Splitters for Atoms

A diffraction grating for atoms was first realized with sodium by exploiting the high periodicity of a *near-resonant standing light wave* [44]. The laser field shifts the atomic energy states by $U(x) = \hbar\Omega^2(x)/4\delta$, where the Rabi frequency $\Omega = \left(\Omega_0^2 + \delta^2\right)^{1/2}$ relates to its resonance value $\Omega_0 = d_{eg}E/\hbar$. The detuning $\delta = \omega_L - \omega_{eg}$ measures the difference between the laser frequency ω_L and the frequency ω_{eg} of the dipole transition between the atom's ground and excited state. This establishes a dipole force grating, which shifts the atomic matter–wave phase in proportion to the square of the electric field E in the standing laser wave.

The case of optical phase gratings can be generalized to *far-off-resonant laser fields*. Here it suffices to consider the interaction of the field with the optical polarizability

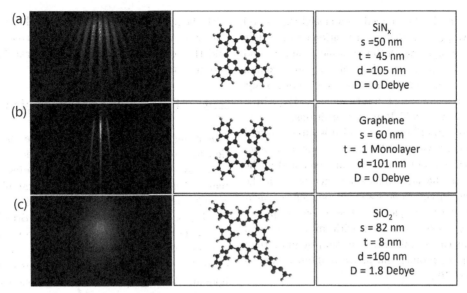

FIGURE 2.4 Far-field diffraction of nonpolar and polar molecules at nanomachined masks of different thickness. (a) When phthalocyanines, which are polarizable but nonpolar, are diffracted at a 40 nm thick SiN_x grating, they experience the effect of the van der Waals potential as a symmetric deflection towards the slit walls. This population of higher diffraction orders does not reveal any which-path information and is therefore fully coherent. One can interpret the results by assuming the slits to be effectively narrowed to about 26 nm [41]. (b) When the thickness of the diffraction is reduced to the extreme limit of a single sheet of graphene, the same PcH2 molecules show again perfect interference contrast, and the Casimir potential is seen to be greatly reduced [41]. (c) The polar molecule 5-(4-methoxycarbonylphenyl)-10,15,20-triphenylporphyrin (MeO-TPP 1.8 Debye) shows no interference fringe when diffracted at an 8 nm thin SiO_2 grating, even when the slit is 82 nm wide. A tentative explanation invokes the presence of local charges in insulating grating material. They will deflect the polar molecules, in different directions for differently oriented molecules [42].

α_{opt} via the dipole potential $V(x,\ z)\ \propto\ \alpha_{opt}E^2(x,\ z)$ $\propto\ \alpha_{opt}P/w_yw_z$. It is determined by the incident power P and the waists w_y, w_z of the Gaussian laser beam transverse to the particle beam. For thin gratings, the Raman–Nath approximation holds when the atomic beam remains parallel to the nodal lines of the grating. The position-dependent phase shift of the matter wave can be calculated by integrating the potential along a straight line. The complex transmission function then reads

$$t_{dip}(x) = \exp\left(\frac{i}{\hbar} \int_{-\infty}^{\infty} V(x, v_z t)dt \right)$$
$$= \exp\left(-i\phi_0 \sin^2\left(\pi\frac{x}{d} \right) \right), \qquad (2.6)$$

with the maximal phase shift at the center of a grating antinode

$$\phi_0 = \sqrt{\frac{8}{\pi} \frac{\alpha_{opt}}{\epsilon_0 \hbar c} \frac{P}{w_y v_z}}. \qquad (2.7)$$

Diffraction at a thin optical grating, also referred to as *Kapitza–Dirac diffraction*, leads to multiple diffraction orders, symmetrically distributed around the central beam.

In a *thick grating*, where the laser waist along the atomic beam exceeds a Talbot length, $(w_z > d^2/\lambda_{dB})$, and in case the kinetic energy of the particle is large enough for the matter wave to cross several nodal lines, quantum interference will select those diffraction orders that fulfill the Bragg condition $n\lambda_{dB} = 2d \sin \vartheta_{Bragg}$. This process is the analogue to x-ray diffraction in material crystals. Bragg diffraction divides the matter wave in just two branches, a deflected

and an unperturbed wavefront. If, on the other hand, the optical potential exceeds the particle's translational energy, a matter wave of sufficient transverse coherence will first be channeled through the optical crystal and then coherently diffracted when it leaves the optical field, in analogy to Kapitza–Dirac diffraction [46].

All grating types described above are known as *wavefront beam splitters*, since they affect different parts of the matter wavefront differently. In contrast to that, *amplitude beam splitters* utilize the coherent dynamics between light and atoms to impart a recoil on a part of the wave amplitude, independent of the particle position. When light is resonantly coupled to two energy levels of an atom, it will drive Rabi oscillations, i.e. coherent population oscillations between its ground state $|g\rangle$ and excited state $|e\rangle$. If the laser intensity and the duration of its interaction with the atom are chosen to be $\pi/2 = \int \Omega(t)dt$, the atom can be transferred into a coherent 50:50 superposition of $|g\rangle$ and $|e\rangle$. The $\pi/2$-beam splitter, thus, transforms the atomic wave function $\psi \propto |g, p\rangle$ into $\psi \propto |g, p\rangle + e^{i\varphi}|e, p + \hbar k\rangle$, where $k = 2\pi/\lambda_L$ is the wave number of the incident light. This state contains maximal entanglement, i.e. inseparable quantum correlation between the internal atomic energy and its momentum. The excited state $|e\rangle$ must have absorbed a photon and received a momentum kick, while the remaining ground state $|g\rangle$ has not. In a sequence of typically four $\pi/2$ atom-laser interaction stages, one can prepare a closed atom interferometer, nowadays often referred to as Ramsey–Bordé interferometer [13].

A similar idea can also be applied to a closed three-level system, where two-photon Raman transitions can couple two hyperfine states $|g_1\rangle$, $|g_2\rangle$ of the same electronic ground state coherently via an intermediate off-resonant excited state $|e\rangle$. Since ground states cannot decay spontaneously, they are well suited for atom interferometry [14]. In a sequence of $\pi/2 - \pi - \pi/2$ Raman transitions, one can create a closed atom interferometer, which is often referred to as either Raman or Kasevich–Chu interferometer. In practice, all these variants profit from being implemented in the time domain, where the laser pulses interact with all the atoms in the cloud in the same way, independent of their velocity. In recent years, sequences of such beam splitter pulses [47,48] have been combined with moving optical lattices [49] to split the atomic wave function by more than 100 photon recoils in momentum space and up to half a meter in real space [16].

Beam Splitters for Clusters and Molecules

Many ideas of atomic beam splitting can be extended to large and complex composite systems. However, their internal energy levels are usually broadened and cannot be individually addressed in coherent Rabi cycles. Except for diatomic and selected small molecules [50–52], the use of amplitude beam splitters is thus precluded, while *wavefront beam splitting* in a periodic potential is universal as it can be realized with almost every kind of particle.

Off-resonant optical phase gratings act on complex molecules and nanoparticles the same way they act on atoms – via the conservative dipole potential. However, while the optical polarizability of atoms may grow by a factor of up to 10^5 close to an electronic resonance, this factor remains typically small and rather constant across the entire visible spectrum of macromolecules such that high laser powers are required. A power of $P = 10$ W focused into a waist of $\omega_y = 1$ mm is required to achieve a matter–wave phase shift of $\Delta\varphi \simeq \pi$ for a molecule with a polarizability of $\alpha_{\mathrm{opt}} = 100$ Å$^3 \times 4\pi\varepsilon_0$ and a velocity of 100–200 m/s.

Optical photo-depletion gratings, on the other hand, couple to the imaginary part of the particle's polarizability. Depending on the laser wavelength and the electronic structure of the particle, the absorption of even a single photon can ionize [53], fragment [54] or internally modify the particle such that it is effectively removed from the molecular beam. If the photo-depletion is already affected by a single photon absorbed from a standing light wave, the transmission function is

$$t(x) = 1 - \exp\left(-\frac{n_0}{2}\sin^2\left(\frac{\pi x}{d}\right)\right). \qquad (2.8)$$

Here, n_0 is defined as the maximum mean number of absorbed photons in the center of an antinode

$$n_0 = \frac{8}{\sqrt{2\pi}}\frac{\sigma_{\mathrm{abs}}\lambda_{\mathrm{L}}}{hc}\frac{P}{w_y v_y}. \qquad (2.9)$$

Here, σ_{abs} denotes the absorption cross section of the particle at λ_{L}. At the same time, the laser field also interacts

with the particles via the electric dipole force. In the limit of thin gratings, the two effects can be treated independently and the total transmission function can be written as the product $t(x) = t_{\mathrm{dip}}(x)t_{\mathrm{abs}}(x)$.

2.4   Talbot Interferometry

Observing high-mass interference is a challenge, since the de Broglie wavelength of massive particles can be extremely small. A spherical particle with a radius of 3.4 nm and a material density of 3.4 g/cm^3, for example, has a mass of $m = 3.4 \times 10^5$ amu. When this particle travels at a velocity of 34 m/s, it is associated with a de Broglie wavelength of $\lambda_{\mathrm{dB}} = 34$ fm. This is 10^5 smaller than the particle size.

Near-field optics have been around for almost two centuries, and it has often been exploited for lensless imaging: When a plane wave traverses a periodic grating, it generates coherent self-images of that structure at integer multiples of the *Talbot distance* $L_{\mathrm{T}} = d^2/\lambda$. That length, named after Henry Fox Talbot [55], depends on the grating period d and the wavelength of the incident radiation. The idea of using near-field interference to demonstrate the quantum nature of matter dates back to John Clauser [56]. He was also the first to demonstrate it with potassium atoms [57]. Meanwhile, it has been shown for electrons [58], metastable atoms [59], BECs [60] and macromolecules in four different interferometer configurations, to be discussed below.

Near-field interference is appealing because it scales favorably with mass. For any given interferometer length L, the grating period required to observe the first Talbot image is $d \propto m^{1/2}$. In high-mass interferometry, i.e. for small de Broglie wavelength λ_{dB}, the square root scaling is experimentally favorable in comparison to the linear far-field dependence $d \propto m$.

2.4.1   Interferometers for Incoherent Particle Beams

The original Talbot effect relies on plane-wave illumination and is subject to the same spatial coherence requirements that renders far-field experiments so demanding for massive molecules. Ernst Lau, however, extended lens-less self-imaging by adding a slit array G_1, which prepares sufficient transverse coherence at the position of the grating G_2, even from initially incoherent sources [61], by confinement and diffraction in each individual slit of G_1 (see Section 2.2).

The first grating G_1 can be regarded as an array of slit sources, each emitting wavelets that evolve to illuminate two or more slits of the second grating G_2 with a well-defined phase relation. Diffraction at G_2 then leads to the formation of an interference pattern around integer multiples of the Talbot distance L_{T}. Each slit in G_1 is thus the source for a fringe pattern. If the distances are chosen right, the individual interference patterns add up even though there is no well-defined phase between the wavelets emitted by any two slits of G_1. Such a *Talbot–Lau* configuration is attractive for all kinds of radiation that lack suitable refractive optical elements, such as x-rays [62] or high-mass nanoparticles.

Throughout recent years, a variety of different molecule interferometers have been successfully demonstrated. They are sketched in Figure 2.5. Their universal character is illustrated by the fact that they enabled quantum experiments with a great variety of large molecules and functionalized nanoparticles, summarized in the gallery of Figure 2.6.

2.4.2 Theoretical Models of Near-Field Interferometry

In the following, we will explore how to describe the quantum-state evolution in a symmetric TLI of length L_T, as illustrated in Figure 2.5b. The underlying concept has been derived and summarized with many extensions in Ref. [72]: The density matrix of a quantum state $\hat{\rho}$ can be represented

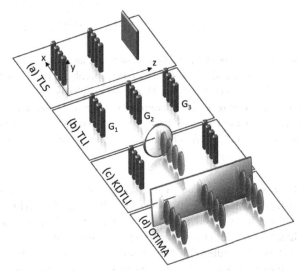

FIGURE 2.5 Talbot–Lau configurations: (a) In a *Talbot–Lau interferometer (TLI) with surface detection* (TLS), the molecular interferogram is deposited in ultra-high vacuum on an atomically flat and clean surface, to be subsequently imaged using scanning probe microscopy [63]. In the experimental realization, the period in the grating and the interference pattern was $d = 257.4$ nm. (b) A TLI was realized with three gold gratings with 480 nm slits etched into a 500 nm thin membrane with a period of $d = 990$ nm [64]. (c) The *Kapitza–Dirac–Talbot–Lau interferometer* 'KDTLI' is built on the insight that van der Waals interactions between macromolecules and mechanical gratings render the instrument too velocity dependent to be practical. The second grating G2 was therefore realized as a standing light-wave phase grating [65] with a period of 266 nm. (d) The *Optical Time-Domain Matter–Wave Interferometer* 'OTIMA' takes this idea one step further by replacing all nanomechanical masks by optical photo-depletion gratings. It was realized using vacuum ultraviolet laser light at $\lambda_L = 157.6$ nm. The controlled interaction time with pulsed gratings eliminates the velocity dependence of various phase shifts, for instance those introduced by gravity [66].

by a Wigner function $W(x, p)$, whose evolution can be conceptually divided into a sequence of grating interactions and free space propagation.

$$W(x,p) = \frac{1}{2\pi\hbar} \int ds\, e^{ips/\hbar} \left\langle x - \frac{s}{2} \,|\hat{\rho}|\, x + \frac{s}{2} \right\rangle. \quad (2.10)$$

When the quantum state hits the first grating with transmission function $t_1(x)$, $W(x, p)$ undergoes the transformation

$$W_1(x,p) = \int dp_0 K_1(x, p - p_0)\, W_0(x, p_0), \quad (2.11)$$

where the kernel K_1 is defined as

$$K_1(x,p) = \frac{1}{2\pi\hbar} \int ds\, e^{ips/\hbar} t_1\left(x - \frac{s}{2}\right) t_1^*\left(x + \frac{s}{2}\right). \quad (2.12)$$

It is part of the beauty of a phase space treatment that the free evolution of the Wigner function transforms as expected for a classical distribution, since it undergoes a shearing transformation during the travel time between the gratings $T = L/v_z$:

$$W_1'(x,p) = W_1\left(x - \frac{pT}{m}, p\right) \quad (2.13)$$

After another grating convolution as in Eq. (2.11), referring to the transmission function of G_2, and another shearing transformation to describe the propagation towards G_3, the spatial probability density that could be captured on a screen is then the marginal of the Wigner function $W_2'(x, p)$, i.e. the integral over all

$$P(x) = \int dp W_2'(x,p) = \sum_{m=-\infty}^{\infty} B_m^*(0) B_{2m}\left(m\frac{T}{T_T}\right)$$
$$\times \exp\left(2\pi i m \frac{x}{d}\right). \quad (2.14)$$

Here, the travel time between the gratings is scaled to the *Talbot time* $T_T = L_T/v = d^2 m/h$ and the *Talbot–Lau coefficients* B_m can be expressed in terms of the Fourier components b_j of the grating transmission function $t(x)$

$$B_m(x) = \sum_{j=-\infty}^{\infty} b_j b_{j-m}^* \exp\left[i\pi(m-2j)x\right], \quad (2.15)$$

with

$$t(x) = \sum_{j=-\infty}^{\infty} b_j \exp\left(2\pi i j \frac{x}{d}\right). \quad (2.16)$$

These equations hold for matter–wave interferometry both in position space and in the time domain, with the replacements $T = L/v$ and $T_T = L_T/v$, respectively. If the interferometer is exposed to a constant external acceleration, such a gravity $\mathbf{a} = \mathbf{g}$ or the Coriolis acceleration $\mathbf{a} = 2\,\mathbf{v} \times \Omega_E$ on Earth rotating at angular velocity Ω_E, the result of Eq. 2.14 still holds, when we replace $x \to x - aT^2$.

In a Talbot–Lau setting, the interference fringe contrast varies with the particle's de Broglie wavelength. The periodicity and phase of the fringe pattern, however, are solely

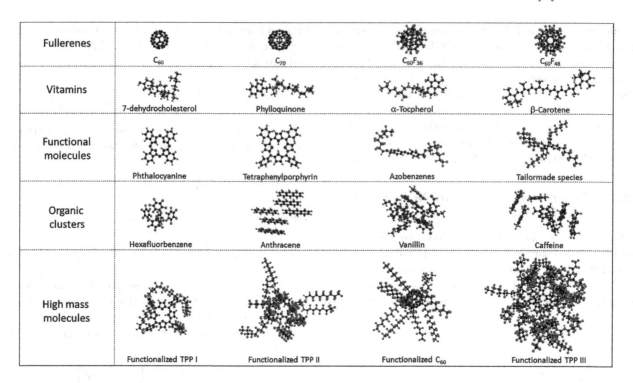

FIGURE 2.6 Quantum interference of hot complex molecules was first seen for fullerenes C_{60} and C_{70} diffracted at nanomechanical [8] and optical [45] gratings. A three-grating TLI was additionally also realized for $C_{60}F_{48}$ and the biodye tetraphenylporphyrin (TPP) [67]. Experiments with functional molecules and high-mass molecules were performed in the KDTLI [65,68]. Matter–wave-enhanced metrology was tested on these functional molecules [69] as well as on vitamins [70]. The de Broglie wave nature of organic and biomolecular clusters was observed with OTIMA interferometry [54,66,71]. The mass record in matter–wave delocalization is currently held with the perfluoroalkyl-functionalized TPP, which showed high-contrast interference in KDTLI [22]. This mass range has recently also been approached by OTIMA interferometry [71].

defined by the grating period and separation. Interference fringes associated with different molecular velocity classes have different contrast but always add up constructively – at least in a broad velocity band. Because of that, Talbot–Lau interferometry can also accept beams of low spectral coherence.

2.4.3 Talbot–Lau Interferometer with Surface Detection, TLS

A direct way to visualize the resulting molecular density pattern $P(x)$ is to capture it on a surface behind G_2 and to image it using scanning tunneling microscopy (STM) [63]. This was realized in a TLI with two nanomechanical masks, as illustrated in Figure 2.7. The gratings with $d = 257$ nm period were etched into a 160 nm thick silicon–nitride waver and placed at a relative distance of $L = 12.5$ mm. A thermal fullerene C_{60} beam was sent across the interferometer and collected on an atomically flat silicon surface. The surface was prepared in ultra-high vacuum by annealing the silicon close to its melting point, where the Si (111) surface is reconstructed in a 7×7 structure. This cleans the top layer, and the resulting dangling bonds immobilize the fullerenes within nanometers of their impact.

The molecular velocity was filtered by a rotating helical turbine which can select the speed down to a few percent,

with special care taken to minimize vibrations of the rotating spindle. Figure 2.7 shows a collection of individual molecules, here marked as a white dot after identification by height discrimination in the STM image. The signal-to-noise level is limited by the small number (ca. 1,100) of molecules in this area. Vertical integration along the grating lines, however, shows a clear fringe pattern with good visibility.

For many years, the quantum optics community has been searching for ways to develop matter–wave lithography as an alternative to more established nanofabrication techniques. The present example shows that stable and nanoscale molecular patterns can be generated but also that the achievable flux, contrast and control lags behind industrial photolithography standards. Using photo-depletion gratings, one could reduce the grating period further and raise the contrast to 100% in future experiments. The Talbot–Lau concept also allows fabricating more complex patterns than just lines, provided they are also encoded in the masks [73].

2.4.4 Talbot–Lau Interferometer with Spatially Resolving Third Grating, TLI

An alternative way of visualizing the molecular quantum fringes is to scan a third grating G_3 across the molecular

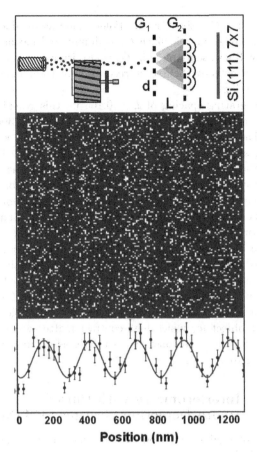

FIGURE 2.7 Deposition of C_{60} molecules on a reconstructed Si (111) 7×7 silicon surface at the end of a TLI with two mechanical gratings of $d = 257.4$ nm period, separated by $L = 12.5$ mm. A velocity selection of $\Delta v/v = 0.05$ on a mean velocity of 115 m/s was achieved using a rotating helix. The interference pattern illustrates the wave–particle duality on the single molecule level here with on average about 1,100 molecules per square micrometer deposited and counted. A vertical sum over all molecules reveals the high-contrast fringe visibility of $V \simeq 36\%$. The continuous line is a sine wave fit to the data [63].

beam along the x-direction, at a position and with a period that matches the expected interference pattern. The result will be a periodic modulation in the transmitted molecular signal as a function of the lateral position of x_3

$$S(x_3) = \int P(x) \, |t_3 \, (x - x_3)|^2 \, dx. \qquad (2.17)$$

This way, G_3 provides the spatial resolution to resolve the closely spaced interference stripes. The TLI permits detection of many fringes at the same time, if an integrating detector, such as a mass spectrometer, counts and identifies all particles behind the third grating. This parallel recording boosts the signal by several orders of magnitude over far-field diffraction at a single grating. The first full interferometer for large hot molecules was realized using three identical gold gratings with a period of $d = 991$ nm each [64]. They were positioned at an equal distance of $L = 22$ cm and aligned in their roll, yaw and pitch angles

to about 1 mrad to each other and to gravity. The thermal molecular beam was detected using either thermionic emission ionization, in the case of fullerenes, or electron-impact ionization quadrupole mass spectrometry (EI-QMS) for all other molecules. The TLI allowed studying the wave nature of C_{60} and C_{70}, the fluorinated fullerenes $C_{60}F_{48}$ as well as the TPP ($C_{44}H_{30}N_4$) [67], and it was successfully used to demonstrate collisional [74] and thermal decoherence [75], for the first time.

In contrast to Fraunhofer diffraction which is impossible to explain by classical physics, the spatial modulation behind two gratings might also arise as a classical Moiré pattern, i.e. as a geometric shadow effect, when the gratings were very close to each other. Such a shadow would, however, be independent of the de Broglie wavelength. A quantitative comparison of the quantum prediction (Eq. 2.16) for the velocity dependence of the fringe contrast with the experiment should thus allow a clear distinction between a classical and a quantum model of nature.

The first TLI experiments, displayed in Figure 2.8, showed that the observed C_{70} interference visibility already deviated from the classical prediction but interestingly also from an idealized quantum model that ignored all internal molecular properties (dotted line). This can be understood, based on the discussion in Section 3.1, since the grating acts on the polarizable particles via the Casimir–Polder interaction which adds a velocity-dependent deflection in both the quantum wave model and for classical ballistic trajectories. The observed velocity dependence of the experimental data is, however, in very good agreement with the extended quantum model (thick continuous line) and it rules out a classical picture also amended by these surface effects (dashed line).

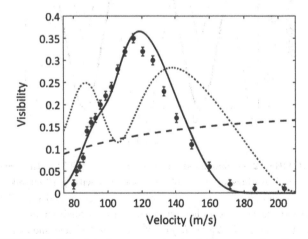

FIGURE 2.8 Interference contrast of C_{70} as a function of velocity. The experimental data (full circles) deviates from both a classical model and is in clear disagreement with the idealized quantum prediction (dotted curve). Only the inclusion of the van der Waals interaction in both the quantum model (continuous line) and the classical shadow model (dashed line) makes it possible to identify the observed fringe contrast as a genuine quantum phenomenon.

2.4.5 Kapitza–Dirac–Talbot–Lau Interferometer, KDTLI

Since the influence of the Casimir–Polder potential grows with molecular polarizability and reduced particle velocity, it is a serious roadblock on the way to quantum experiments with very massive nanoparticles or even polar biomolecules as it renders the Talbot–Lau design again incompatible with beams of low spectral coherence. This is illustrated in the simulation of Figure 2.9, where high fringe visibility can only be observed in narrow wavelength bands, requiring a velocity selection below 1% for most particles beyond 1,000 amu.

Given the state of the art in neutral nanoparticle or molecular beam sources, this is often too demanding. The Casimir–Polder interaction can, however, be eliminated by replacing the central grating G_2 with an off-resonant standing light wave, as described in Section 3.2. The optical grating is formed by retro-reflecting the light of a solid-state laser with up to 18 W power, emitting green light at a wavelength of 532 nm. Due to the electric dipole potential generated in the standing light wave, the particles experience a spatially modulated phase shift. For spatially incoherent sources, it is still imperative that the first grating G_1 prepares the initial transverse coherence and that the third grating G_3 serves as spatially resolving detection mask. This is why they are still implemented as amplitude transmission gratings in SiN_x.

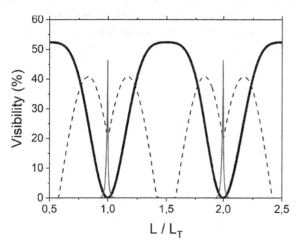

FIGURE 2.9 Simulated fringe visibility in a TLI with three material gratings of $d = 266$ nm, operated with perfluoroalkylated nanospheres ($m = 7,000$ amu, $\alpha = 4\pi\varepsilon_0 \times 200\text{Å}^3$). The velocity dependence is here encoded in the normalized interferometer length, $L/L_T = Lh/mvd^2$. The dashed line shows the expected interference contrast for purely binary transmission gratings. However, the polarizable molecules will be subject to a dispersive phase shift due to their Casimir–Polder interaction with the grating walls. This limits the domain of finite contrast to a narrow range (thin gray curve, peaked at $L/L_T = n \in \mathbb{N}$). When the central grating G_2 is replaced by an optical phase grating, this significantly widens the accessible domain of high-contrast interference (thick solid line) (see also [65]).

At G_1 and G_3, the Casimir–Polder interaction does not play any relevant role since G_1 is illuminated incoherently anyway. Phase scrambling in G_3 is also irrelevant, since the molecule detector behind it records only total molecular transmission.

At a grating spacing of $L = 0.105$ m, this KDTLI can handle de Broglie wavelengths down to $\lambda_{dB} < 300$ fm. This has enabled quantum interference studies with a diverse set of molecular species, also represented in Figure 2.6: The instrument served to demonstrate the wave nature of the fullerenes C_{60} and C_{70}, the functionalized fullerenes $C_{60}F_{36}$ and $C_{60}F_{48}$ [76] and a variety of vitamins including α-tocopherol, phylloquinone, β-carotene and 7-dehydrocholesterol [77].

Even the large functionalized carbon nanospheres [68], depicted in Figure 2.10a, were found to exhibit their quantum wave nature. The perfluoroalkyl-decorated TPP with up to 810 atoms and a mass around 10,123 amu [22] sketched in Figure 2.10b is the most complex and massive object for which high-contrast matter–wave interference has been demonstrated to date, as documented in Figure 2.10c.

2.4.6 Interferometry with Pulsed Photo-Depletion Gratings, OTIMA

Since polarizable and slow nanoparticles may eventually even block the nanomechanical gratings in G_1 and G_3, it is interesting to consider an all-optical matter–wave interferometer [53] with pulsed photo-depletion gratings as illustrated in Figure 2.5d. Pulsed matter–wave beam splitters have become the basis for modern atom interferometry [6,17] because they allow eliminating many dispersive phase shifts. In fixed-length interferometers, the transit time depends on the molecular velocity and the overall interference fringe contrast is reduced when molecules in different velocity classes acquire different phase shifts. In the time-domain interferometry, however, all particles experience the same light pulse at the same time T. Therefore, the phases accumulated between the pulses are the same for all of them, and the contrast remains unaffected by the molecular velocity spread. The only exception for that is the Coriolis shift, which depends explicitly on the particle speed in the rotating reference frame.

In the *optical time-domain matter–wave interferometer*, *OTIMA*, the particles fly past a single stable mirror which retro-reflects three laser pulses to form three standing waves that are separated both in space ($L \simeq 2.5$ cm) and in time T. These gratings act as absorptive masks in G_1 and G_3 to prepare and probe matter–wave coherence, while in G_2, the additional phase contribution is also relevant. For many materials, from organic molecules over semiconducting to metallic clusters, photoionization or photo-fragmentation can efficiently deplete the molecular beam selectively at the antinodes of the standing light waves [53,66]. For that purpose, the light is derived from an F_2-laser emitting in the vacuum ultraviolet wavelength region around 157.6 nm with

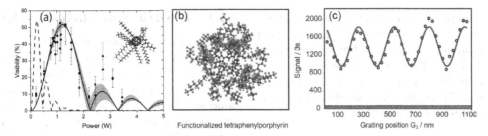

FIGURE 2.10 (a) Fringe visibility of functionalized carbon nanospheres as a function of the diffracting laser power [68]. The black dots mark the experimental data, the error bars correspond to the 1σ-confidence interval of the sinusoidal fits to the interference fringes. The solid line represents the quantum model based on the velocity spread and the estimated polarizability of the molecules. The dashed line depicts the classical expectation for the same parameters. (b) A perfluoroalkyl-functionalized TPP with 10,123 amu. (c) Even for this molecule, composed of 810 atoms covalently bound in a molecule flying at about 500 K, the interference fringe contrast is high and in agreement with quantum mechanics.

a photon energy of 7.9 eV. The delay T between two subsequent gratings is typically adjusted around the Talbot time $T_T = L_T/v = md^2/h$, which is proportional to the particle's mass but independent of the molecular velocity. It amounts to $T_T \simeq 15$ ns per atomic mass unit for a grating period of $d = 78.8$ nm, requiring matter–wave coherence times of 30 ms and velocities around 10 m/s for future experiments with particles around $m \simeq 10^6$ amu [78].

Using a single mirror as a common base for all optical gratings endows the interferometer with exceptional insensitivity to vibrational perturbations, since all tilts and shifts of the entire mirror cancel each other in the total phase shift $\Delta\varphi = (x_1 - 2x_2 + x_3)2\pi/d$. This feat, however, comes at a price: It requires vacuum ultraviolet (VUV) mirrors of exceptional surface quality with less than 10 nm absolute modulation over a total range of 50 mm and a reflectivity higher than 97%. This is the current state of the art in VUV optical technology. The stable single-mirror design requires new ways to visualize successful matter–wave interference, since the quantum fringes cannot be recorded by shifting one grating across the molecular beam. However, matter–wave interference depends on the Talbot time T_T, which is mass dependent. It has proven useful to measure interference by the normalized signal difference $S_N = (S_{\text{res}} - S_{\text{off}})/S_{\text{off}}$, where the *resonant signal* S_{res} is given by the number of molecules transmitted behind G_3, when the two grating pulse delays $\Delta T_{12} = T_2 - T_1 = \Delta T_{23} = T_3 - T_2 \simeq T_T$ are equal and close to the Talbot resonance. Matter–wave interference is very sensitive to asymmetries in the pulse delays and is practically eliminated if $\Delta T_{12} = \Delta T_{23} \pm \tau$. The value of τ is determined by the divergence and tilt of the molecular beam but $\tau \geq 100$ ns is a typical value [71]. The *off-resonant signal* S_{off} recorded in such an asymmetric pulse configuration is taken as a reference in the normalized signal difference above.

At fixed pulse separation time, the quantum resonance is therefore encoded in the mass spectrum of the particles transmitted by all three laser pulses. Figure 2.11a displays the normalized signal difference S_N (or contrast) for anthracene clusters seeded in a room temperature

supersonic neon beam at 930 m/s where the integral over the isotope-resolved mass peak on resonance is normalized to the off-resonant pulse cycle with a pulse delay asymmetry of $\tau = 100$ ns. The curves of Figure 2.11b show the corresponding mass spectra on and off the resonance. This contrast here reaches even above 80%, referring to the sinusoidal visibility.

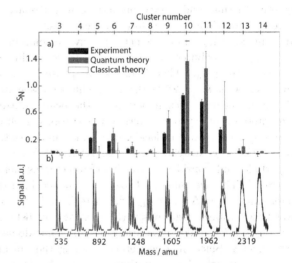

FIGURE 2.11 (a) In OTIMA interferometry, matter–wave interference is encoded as a resonance in the Talbot time, $T_T = d^2m/h$. For any given laser pulse separation time $\Delta T_{12} = \Delta T_{23} \simeq T_T$, one expects constructive interference and enhanced transmission for masses close to $m \simeq hT_T/d^2$ and reduced transmission for others. This is well observed in the interference of anthracene clusters with a mass distribution covering Ac_3–Ac_{12} [66,79]. These clusters were seed in Argon at 620 m/s. (b) Matter–wave interference relies on the timed rephasing of all partial waves and is therefore a resonant phenomenon. It can only be observed around sharply defined times with resonance widths around 100 ns. The difference between the upper red curve and the lower curve is assigned to matter–wave interference. The light grey curve is taken on resonance, with all laser pulse delay times equal to 25.9 μs. The dark gray curve was recorded off-resonance, i.e. with a difference in pulse delays of 100 ns.

The OTIMA concept has proven useful to demonstrate the quantum wave nature of a variety of organic van der Waals clusters, composed of up to a dozen molecules of anthracene [66], vanillin, caffeine or hexafluorobenzene [54], summarized in Figure 2.6. The concept is appealing for high-mass interference [53,71,80] and for molecule metrology in the time domain, as discussed below.

2.5   Quantum-Assisted Molecule Metrology

Matter–wave interferometry imprints nanoscale fringes onto the density distribution of molecular beams which can be read out with nanometer accuracy. This can be exploited for sensitive measurements when the fringe positions are shifted by external perturbations. In atom interferometry, the high potential of coherent metrology has been demonstrated for measuring static polarizabilities [81], fundamental constants [82], the gravitational acceleration or the rotation of the earth [17]. Matter–wave interferometry now opens a new path also to measurements on macromolecules [21] at the interface between quantum optics and (bio)physical chemistry, where information can be retrieved about electronic, magnetic, thermal and optical properties, which relate to electronic, vibrational and conformational states.

In three-grating interferometers, any displacement of G_1–G_3 contributes to the final fringe shift of $\Delta x = x_1 - 2x_2 + x_3$. This leads to a phase shift of $\varphi = k_d \Delta x$, with $k_d = 2\pi/d$ the wave number associated with the momentum kick imparted by the beam splitter. In the presence of constant external accelerations, such as the rotation of the earth or local gravity, one can derive the effective shift by switching into the reference frame of the particle and considering the interferometer as being accelerated. The accumulated fringe shift at the interferometer exit is then $\Delta x_{\mathrm{g,c}} = a_{\mathrm{g,c}} T^2$, where $T = L/v$ is the travel time between two consecutive gratings of distance L or the grating pulse separation time in time-domain interferometry. Since only accelerations along the grating vector are relevant, the gravitational acceleration is $a_{\mathrm{g}} = g \sin \alpha$ with $g = 9.81$ m/s^2 and α the angle between the grating bars and the line of gravity. For the same reason only the x-component of the Coriolis acceleration $a_{\mathrm{c}} = 2(\mathbf{v} \times \Omega)_x$ is relevant for the interference pattern, with Ω_{E} the angular velocity of the rotation of the earth and $\mathbf{v}$ the particle velocity. Modern commercial atom interferometers can determine these accelerations with nine-digit precision for applications in geodesy, inertial navigation and the prospection of natural resources. Advanced research instruments are aiming at 12-digit accuracy to realize modern quantum tests of general relativity [16]. Nanoparticle interferometers have usually lower signal-to-noise ratio than atom interferometers, but they can compare matter of vastly different internal composition and binding energy. This has recently been exploited to demonstrate isotope-selective interferometry with TPP in free fall [71].

2.5.1   Electric and Magnetic Deflectometry

The idea of that technique relates to classical deflectometry of molecular beams, which have been used in physical chemistry [83] to determine electronic and magnetic properties of clusters [84,85] and biomolecules [86,87]. However, Talbot–Lau interferometry adds the advantage of preparing a nanoscale ruler – the interference pattern – which allows measuring forces well below 10^{-25} N. Figure 2.12 shows how electronic properties can be derived by adding a pair of specially designed electrodes to the interferometer, where it is used to create a homogeneous transverse force field $\mathbf{F} \propto (\mathbf{E}\nabla)\mathbf{E}$ on the traversing polarizable particles.

The quantum fringe shift

$$\Delta x \propto \nabla_x E^2 \frac{\chi_{\mathrm{el}}}{m} \frac{1}{v^2} \qquad (2.18)$$

equals the displacement of a classical particle beam. It is proportional to the susceptibility-to-mass ratio χ_{el}/m and inversely proportional to the square of the molecular velocity v. A high-contrast nanoscale fringe pattern improves the spatial resolution over previous classical experiments by orders of magnitude, allowing to run the experiment at lower field strengths and with higher accuracy. This has been used to determine the static *molecular polarizability* [88] from the quadratic interference fringe shift when the external voltage is linearly increased. This has recently been exploited to study different vitamins and pre-vitamins [77], such as α-tocopherol, phylloquinone or β-carotene.

Using the same pair of electrodes, one can study the influence of the molecular *electric dipole moments* [89]: Most molecules are non-isotropic and their rotation axes are randomly oriented. In the presence of an external field gradient, polar molecules will be shifted by $\Delta x \propto -\nabla_x (\mathrm{d} \cdot \mathbf{E})$, i.e. they will be displaced left or right, depending on their initial orientation and the alignment of their rotation axis. For thermal beams with random molecular orientations, the exposure of the particles to a constant electric field gradient induces dephasing. Since the degree of the dephasing depends on temperature, we call it *decoherence metrology*.

At high temperatures, many molecules become floppy. They can change their shape on the nanosecond scale, and the electric dipole moment may fluctuate by a few hundred percent. In that case, the electric susceptibility must be seen as the sum of the static polarizability [90] and a Langevin term that represents the thermal population of the average z-projected dipole moment: $\chi_{\mathrm{el}} = \alpha + \langle d^2 \rangle / 3k_B T$.

By tracking the evolution of the electronic properties, it is possible to derive statements about molecular fragmentation path ways [91], to distinguish structural isomers [69] and to follow conformational dynamics [92]. In a very similar spirit, one can also explore the *magnetic properties of molecules*: their paramagnetism, diamagnetism or Langevin paramagnetism, eventually even the role of nuclear moments. Molecular magnetism is an interesting marker for photo-induced

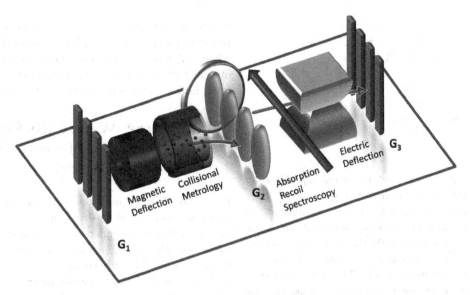

FIGURE 2.12 Talbot–Lau interferometry is extremely sensitive to transverse force fields. Introducing electric and magnetic fields, collision cells or spectroscopy lasers makes it possible to explore a multitude of molecular properties with high precision and during free flight. The electric and magnetic fields are optimized to generate a maximal and most homogeneous $(\mathbf{E}\nabla)\mathbf{E}$ and $(\mathbf{B}\nabla)\mathbf{B}$ field.

electronic or structural changes, since singlet-triplet transitions and aromatic ring openings will change the magnetic response significantly [93]. Even weak diamagnetic effects are predicted to be observable in a strong magnetic field

$$\Delta x \propto \nabla_x B^2 \frac{\chi_{\mathrm{mag}}}{m} \frac{1}{v^2} \qquad (2.19)$$

and using the high spatial resolution of molecular interference patterns. Paramagnetic components, however, will again lead to a reduction of the interference contrast. Finally, photoisomerization and conformational changes may also be tracked by monitoring the change in collisional cross sections via its influence on collisional decoherence [74] (see Figure 2.12).

2.5.2 Optical Absorption Spectroscopy

The high spatial resolution of matter–wave interferometry and the low mass of atoms and molecules makes them particularly useful for monitoring the effect of single-photon recoil. This has already been successfully exploited to derive new precision values for the fine structure constant α through an interferometric measurement of h/m on cesium atoms [94], most recently even with ten-digit precision [82]. A similar idea now bears new fruits when applied to molecular spectroscopy, where one can extract absolute absorption cross sections from the shift or contrast reduction of matter–wave fringes [95]. Optical absorption cross sections can be regarded as fingerprints of the electronic structure of molecules, and vibrational spectra deliver information about the molecular core structure and conformation. Knowing them, therefore, allows benchmarking theoretical models and computational predictions.

The first demonstration of that idea was realized for C_{60} in KDTLI interferometry [96]. By an appropriate

choice of the spectroscopy laser wavelength and beam position, the interference contrast is reduced. The high sensitivity of the method is illustrated by the fact that the absorption cross section can already be determined with percent accuracy when on average as little as 0.2 photons are absorbed per molecule. The method allows measuring *absolute* cross sections without the need to know the particle density, which is important for measurements on very dilute molecular beams. This feat is due to the fact that interference-assisted spectroscopy is self-referencing: Shifted and unshifted interference patterns relate to molecules with or without absorption – both relate to the same molecular beam and both contribute to the same final density pattern. Single-photon recoil spectroscopy is a minimally invasive method to measure cross sections in a nondestructive fashion. It is independent of any 'action', such as subsequent photodissociation, fluorescence or photoionization. This noninvasiveness is expected to enable new studies on weakly bound clusters but also on a wide range of biomolecules whose optical absorption spectra are often hardly accessible in the dilute unsolvated gas phase.

The idea works equally well for interferometry in position space and in the time domain. Combining it with OTIMA interferometry allows using widely tunable high-power lasers, such as pulsed optical parametric oscillators, crossing the molecular beam parallel to the grating vector of the standing light waves, presented in Figure 2.13a as a green running wave [97]. Also, time-dependent processes or studies on molecules with pulsed molecular alignment are facilitated in the time domain. The strongest beam deflection Δx is achieved when the photons are absorbed close to the second grating G_2. The deflection in the n-th Talbot order is then $\Delta x = nd^2/\lambda_{\mathrm{abs}}$. In OTIMA interferometry, a photon with a wavelength of 500 nm would therefore deflect

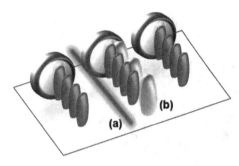

FIGURE 2.13 (a) *Single-photon recoil spectroscopy* conceived for OTIMA interferometry: The spectroscopy laser (straight line) irradiates the molecules near the position of the second grating. The imparted recoil modifies and shifts the interference patterns associated with the absorbing molecule allowing to extract absolute absorption cross sections. The scheme is similar to that demonstrated in the KDTLI setup depicted in Figure 2.12. Nanosecond control over both the grating and the spectroscopy lasers will allow addressing multiphoton excitation, fast internal-state relaxation and the role of molecular alignment. (b) *Optical polarizability spectroscopy:* Information about the UV-VIS polarizability can provide important information about electronic transitions, even without actually transferring the molecule into the excited state. In small molecules, the optical polarizability can be greatly enhanced when the laser frequency is close to that of a molecular resonance. When the spectroscopy laser is shaped into a standing light wave close to G_2, this additional phase grating will scramble the interference pattern. This form or interference dephasing spectroscopy is minimally invasive and can also work on van der Waals clusters [97].

a particle during its first-order Talbot time by 12.3 nm, which is one-sixth of the grating period ($d = 78.5$ nm) and can be resolved.

2.5.3 Optical Polarizability Spectroscopy

The least invasive form of spectroscopy extracts electronic or structural information about the molecule without even absorbing a single photon by measuring the frequency-dependent optical polarizability. Direct spectroscopic observations of the AC-Stark effect still work for small molecules in sufficiently dense samples, and one may also use the dipole force to defocus or deflect a collimated particle beam in proportion to its AC polarizability, as has been demonstrated for C_{60} before [98]. In KDTL interferometry α_{opt} has also been extracted from the dependence of the fringe visibility on the diffraction laser power of the second grating, G_2 [76].

More generally, near-field matter–wave interferometry adds again the advantage of preparing an intrinsic nanoscale ruler, and one may detect the dipole forces via their effect on the matter–wave interference fringe contrast. When a tunable standing light wave acts as a dipole lens array close to G_2, it scrambles the phase of the traversing matter wave and reduces the interference contrast, the more the higher the optical polarizability. All near-field molecule

interferometers can be augmented by this spectroscopic tool, but it is particularly appealing in the OTIMA setup [97], where high-power tunable spectroscopy light can be obtained from a pulsed optical parametric oscillator or dye laser, indicated as red standing light wave in Figure 2.13b.

2.6 Matter Wave Interferometry and the Interface to the Classical World

Even though quantum theory is one of the best tested scientific models and the fundament of modern technologies, we still need to understand its transition into the classical world that surrounds us every day. In the mechanical world view of the 19th century, it was common to think of distinguishable and traceable particles. If we knew their mass m, initial positions x_i and momenta p_i as well as the equations governing their motion in an external potential $V(x, t)$, we could predict their future $x_i(t)$ and $p_i(t)$ for all times with absolute certainty. Quantum theory is radically different from this view as it rather provides a deterministic equation for the probability amplitude of 'wave functions' $|\psi_i\rangle$, which are vectors in a high-dimensional Hilbert space over square integrable complex functions. Uncountable experiments have confirmed Born's rule, stating that $|\psi|^2$ describes the probability density of experimental observations made on an ensemble of identically prepared systems or on a single particle repeatedly prepared in the same way. In nonrelativistic situations, ψ is the solution to the Schrödinger equation

$$i\hbar \frac{\partial}{\partial t}\psi(x,t) = \left[-\frac{\hbar^2}{2m}\frac{\partial^2}{\partial x^2} + V(x,t)\right]\psi(x,t), \qquad (2.20)$$

which is *linear*. If it is solved by ψ_1 and ψ_2, then $\psi = \alpha\psi_1 + \beta\psi_2$ will also be good solution, with $\alpha, \beta \in \mathbb{C}$ and $|\alpha^2| + |\beta^2| = 1$. The formalism allows the wave function to be a sum of two or more solutions which could even be mutually exclusive in a classical mind-set. The measurable consequence is that the probability density $|\alpha\psi_1 + \beta\psi_2|^2$ contains an interference term $\propto \alpha\beta^*\psi_1\psi_2^* \propto e^{i(\varphi_1-\varphi_2)} = e^{i\Delta\varphi}$, where the phase difference $\Delta\varphi = \Delta S/\hbar$ depends on the difference of the normalized action associated with the two solutions [99].

This short reminder of elementary quantum mechanics motivates the ongoing debate on what the wave function ψ 'really' means. If the particle concept of classical physics was wrong, why would we literally see them in the source and the detector, for instance using a microscope, and why would we need to take all internal particle properties into account – mass, polarizability, structure, rotation etc. – also to describe their coherent evolution through a grating? On the other hand, if the wave concept was wrong, how could we understand the availability of nonlocal information that causes the observed interference fringes? How does this relate to the thought experiment of Schrödinger's cat, decoherence and the transition to the classical world?

New experimental tests have recently been proposed for interferometry with $10^6 - 10^8$ amu nanoparticles in free fall, to be performed either in the lab or in a high drop tower [24,25,100]. Interference can also be explored with particles in traps [26] or under microgravity in space [27]. In Section 2.6.1, we briefly review the state of the art, while Section 2.6.2 sketches some of the open challenges in nanoscience underlying these experiments.

2.6.1 Open Challenges on the Road to the Quantum Classical Interface

Schrödinger cat states: High-mass matter–wave studies are related to Schrödinger's famous thought experiment, which illustrates the highly nonclassical nature of quantum states: A cat would be caged in a box together with a radioactive atom that can evolve into a quantum superposition of 'decayed' and 'non-decayed'. If decayed, the emitted gamma quantum would trigger a mechanism that releases a hammer, which shatters a flask, which releases a poison, which kills the cat. If not – the cat stays alive. Since the atom is in a quantum superposition of decayed and not-decayed, the cat would end up in a superposition of being dead and alive. This example illustrates first that it might be possible to prepare quantum superposition states of truly macroscopic, complex, and warm many-body systems, which seem absurd to the classically educated mind. It also shows that such states may be achieved via the entangling interaction between a tiny atom and a macroscopic object. The 10,123 amu massive functionalized porphyrin, shown in Figures 2.6 and 2.10b, can be considered to be the largest Schrödinger cat in matter–wave interferometry today. Here we replace the quality 'dead-and-alive' by 'in the right and in the left slit'. The high internal temperature of $T = 500$ K even exceeds that of macroscopic bodies under ambient conditions. In the experiments [22] The quantum states were clearly distinct for the beam separation of 532 nm compared with a molecule diameter of 5 nm. Since the interference fringe visibility was in good agreement with theory, it is intriguing to explore even higher mass and complexity in such experiments. In the following we discuss different pathways from quantum states to classical observations:

Dynamical phase averaging: Phase averaging is an important experimental constraint for high-mass interferometry. By that, we designate undesired perturbations of the interference pattern that provide, however, neither genuine 'which-path' information nor entanglement with the environment. Uncontrolled phase shifts may be induced by acoustic noise, grating drifts and vibrations, changes in local electric and magnetic fields or tilts of the experiment.

Decoherence: A natural way to explain the transition from quantum phenomena to classical appearances is embedded into quantum theory itself. Decoherence theory recognizes that quantum phenomena become unobservable when a previously isolated particle or 'system' is coupled to a complex many-particle environment. Even though the Schrödinger equation still describes the unitary evolution

of the combined system, one will no longer observe the characteristic quantum phases of the system. They are never lost but diluted into the environment. This can be seen for the simple example of a particle with eigenstates n, which can be prepared in a quantum superposition $\varphi = \sum_n c_n |n\rangle$, with complex phase factors c_n. If this state is coupled to an environment which acts as a 'meter', the combined state can be written as:

$$\left(\sum_n c_n |n\rangle \right) \otimes |\psi_0\rangle \to \sum_n c_n |n\rangle |\psi_n\rangle \qquad (2.21)$$

This state contains entanglement between the particle and the instrument [101,102] since the 'meter' retrieves unambiguous information about the system and transforms its own state $|\psi_0\rangle$ before the interaction into one that is uniquely correlated with an eigenstate $|n\rangle$ after the interaction. This transforms the corresponding density matrix before and after the interaction into

$$\rho_{\text{tot}} = \left(\sum_{m,n} c_m c_n^* |m\rangle \langle n| \right) \otimes |\psi_0\rangle \langle \psi_0|$$
$$\to \sum_{m,n} c_m c_n^* |m\rangle \langle n| \psi_m\rangle \langle \psi_n| , \qquad (2.22)$$

where in all genuine quantum features are encoded in the off-diagonal matrix elements $\rho_{m,n} \neq 0$. If they vanish, the system can also be described by a classical probabilistic mixture of states. Our lack of detailed knowledge about the instrument is formally taken into account by tracing over the instrument's states:

$$\text{Tr}_\psi(\rho) = \sum_k \sum_{m,n} c_m c_n^* \langle \psi_k |m\rangle \langle n|\psi_m\rangle \langle \psi_n |\psi_k\rangle$$
$$= \sum_n |c_n|^2 |n\rangle \langle n| . \qquad (2.23)$$

This represents a particle without any coherences that can be regarded as a classical object. The last step is legitimated by the fact that the meter eigenstates must be orthonormal $\langle \psi_n |\psi_m\rangle = \delta_{nm}$, if the measurement instrument should be capable of distinguishing different particle states unambiguously.

Decoherence theory has been confirmed in dozens of experiments, also in matter–wave interferometry: The effect of spontaneous photon emission was studied with atoms diffracted at a near-resonant standing light wave [103] or exposed to laser excitation inside a Mach–Zehnder interferometer [104,105].

With increasing particle mass, size and complexity, matter waves also become susceptible to decoherence by collisions with residual gas molecules and by emission of thermal radiation [106]. If the wavelength of the emitted photons becomes comparable to the separation of the wave packets in the interferometer, the photons will carry information about the emitter position into the environment, impart a kick to the molecules and reduce the interference contrast.

Collisional [74] and thermal decoherence [75] have been quantitatively observed with fullerenes in Talbot–Lau interferometry, as shown in Figure 2.14: The influence of collisions can be studied by gradually increasing the base pressure in the vacuum chamber until interference vanishes. While high vacuum around 10^{-7} mbar was still sufficient to maintain quantum coherence for C_{70} in this setting, a human being under normal conditions would be localized by about 10^{28} collisions per second. Even ignoring all other effects, this alone would explain the apparent lack of macroscopic delocalization.

Thermal decoherence then became accessible by laser heating C_{70} molecules to microcanonical temperatures beyond 1,500 K, the point at which the blackbody radiation intensity grows and the emitted wavelengths shrink such as to destroy the interference pattern. The radiated thermal power of a human being corresponds to about 10^{22} photons per second, and spreading this much information into the environment clearly suffices to render the quantum nature of

our center or mass state unobservable. Future experiments with nanoparticles will also require low particle temperatures and ultra-high vacuum.

Decoherence is often regarded as the enemy of quantum physics. However, it has been proposed that decoherence can also serve to set new bounds on the existence of ultra-light and strongly coupling dark matter particles [107], where a siderial modulation of decoherence in the absence of any known particles would be taken as a measure for the unknown. The overall sensitivity of such an experiment scales with the square of the interfering mass, and the momentum resolution grows with increasing splitting between the interferometer arms. This is why large-area, high-mass interferometry is an intriguing goal for future research.

Objective wave function collapse: Decoherence is an experimental fact and an important contribution to our understanding why the world appears classical to us. However, within that frame work, the wave function never

(a)

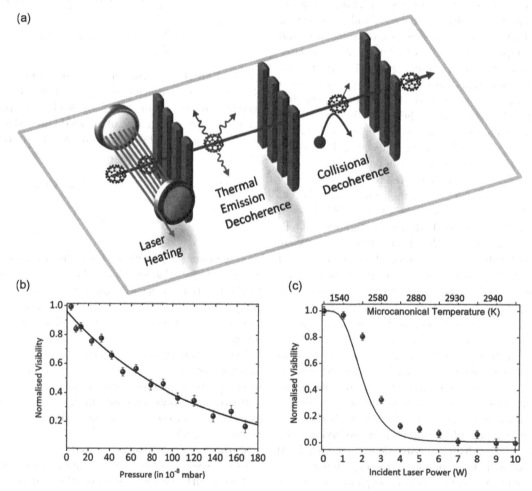

(b)

(c)

FIGURE 2.14 The matter–wave fringes of C_{70} in a TLI that are indicative of their quantum nature. They are observed to vanish – mimicking classical behavior without coherences – when the coupling between the molecule and the environment is gradually increased. The number of collisions with gas molecules can be enhanced by gradually raising the pressure in the vacuum chamber [74]. The probability of thermal photon emission can be increased by heating the fullerenes to high internal temperatures via multiphoton absorption of 532 nm. The delayed emission of visible and near-infrared photons provides a random recoil to the molecule, information for the environment and it entangles the molecular state with that of the photon [75].

assumes the well-defined single value that some researchers still associate with the notion of reality. In an effort to explain the absence of superposition and entanglement in the macro-world, various modifications of the Schrödinger equation have been proposed, which predict an objective, spontaneous and partial localization of the wave function with a rate that grows again quadratically in the particle mass. These extra terms have to be *nonlinear* in order to suppress superpositions and to be *stochastic* to prevent superluminal communication [108].

Continuous Spontaneous Localization theories (CSL) [109–111] assume that an objective wave function collapse occurs with a small rate λ_{CSL} into a packet of small localization width r_C [108]. If this were true, one would expect quantum superpositions to vanish on second time scales for particle masses beyond 10^9 amu. Interestingly, such models may eventually be tested without performing any quantum study at all, since spontaneous wave function collapse entails heating, which violates energy conservation and might be observed in ultra-cold systems even without any interference. In many collapse models, the source of the collapse-inducing fields remains unspecified. Some theories assume it originates from gravity [112–114].

A formal measure of macroscopicity has recently been introduced [80] to assess to what degree an experiment allows discarding even minimal modifications of quantum theory. It is defined as

$$\mu = \log_{10}\left[\left|\frac{1}{\ln V_n}\right|\left(\frac{m}{m_e}\right)^2\frac{t}{1s}\right], \qquad (2.24)$$

with t denoting the observed coherence time, m the mass of the delocalized particle normalized to the electron mass m_e and V_n is the observed interference fringe visibility normalized to the theoretical quantum value. The measure scales again quadratically in mass and linearly in coherence time. It diverges with $V_n \to 1$ because a perfect agreement

between theory and experiment with arbitrary precision can test even the smallest conceivable model differences.

2.6.2 Prospects for High-Mass Matter–Wave Interferometry

The current macroscopicity record of $\mu = 12$ is held by KDTL interferometry with $m = 10^4$ amu [22]. Nanoparticle interference experiments with $m > 10^7$ amu shall explore macroscopicity values beyond $\mu > 20$. They require new source and cooling methods that are capable of preparing a sufficiently coherent particle beam for lab-based interferometers. This is what shall be explored in the following.

Nanoparticle sources in high vacuum: The challenge is to engineer a source that delivers neutral, slow, size- and shape-selected nanoparticles in high vacuum. Laser desorption from a thin pristine silicon wafer offers an interesting first approach as shown in Figure 2.15: Irradiation of the substrate's backside in high vacuum with an energetic (5–20 mJ) nanosecond laser pulse causes surface ablation and substrate deformation by the back action of the desorbed material. Such *laser-induced acoustic desorption* (LIAD) is sufficient to lift off particles from the front side of the substrate and has originally been developed for the mass analysis of organic and biomolecular ions [115,116] up to viruses [117]. At higher pulse energies, the laser-induced thermomechanical stress (LITHMOS) suffices even to break out particles from the substrate's front side itself [23]. This process can even be facilitated by appropriate nanostructuring of the front side. The LITHMOS technique provides a reproducible source of tailored nanoparticles with a well-defined size and shape, from coin-like cylinders to tailored nanorods [118], optimally for masses above 10^6 amu.

Nanoparticle cooling: In recent years, optomechanics has been developed to achieve optical control over the translation and rotation of nanoparticles and microparticles.

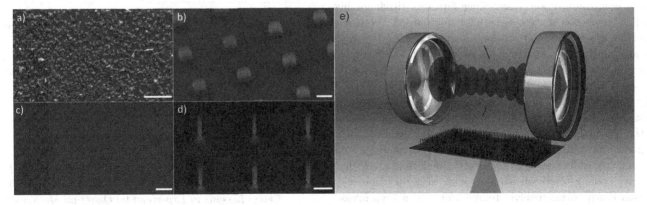

FIGURE 2.15 Preparing particles for launch into ultrahigh vacuum (UHV). (a) Commercial silica particles (diameter 300 nm) spin-coated onto an aluminum foil. (b) Silicon cubes (edge length 300 nm) etched into a silicon on insulator (SOI) wafer. (c) Spherical silicon particles patterned via nanodicing. (d) Silicon nanorods etched into a silicon wafer with a predefined point of rupture. Nanofabrication is carried out at the Tel Aviv University in the group of Fernando Patolsky. (e) Scheme for launching particles through the field of an optical resonator. Scale bars: (a) 5 μm, (b) 400 nm, (c) 2 μm and (d) 400 nm.

This research opens new avenues to single-particle thermodynamics [119], a new search for high-frequency gravity waves [120] or short-range forces [121] as well as new tests of the quantum superposition principle in the limit of high mass [10]. Levitated nanoparticles are also being studied as detectors for tiny accelerations [122] and torques with applications in navigation, pressure sensing [123] and nanoscale magnetometry [124].

Optical cooling of dielectric nanoparticles relies on the electric dipole interaction, as already discussed in the context of optical beam splitters before (Section 2.2). In *active laser feedback cooling,* a dielectric, transparent and highly polarizable particle is harmonically trapped at the center of a focused laser beam, and its scattered light is used to modulate the laser intensity such as to always present a stronger trap potential when the particle is running uphill than when it is moving downhill. This way, single SiO_2 particles between 10^8 and 10^{12} amu were cooled to the millikelvin regime [125–128].

In contrast to that, *cavity cooling* relies on the self-induced optical feedback in a high-finesse infrared cavity [129,130]: When a dielectric nanoparticle is moving along the standing wave of a high-finesse cavity, it changes the optical path length when passing an antinode and decouples again when moving on to a field node. This tunes the cavity in and out of resonance with a red-detuned laser beam that is coupled into the fundamental cavity mode where it creates an optical lattice that attracts the particle to the antinodes of the standing light wave. The motion of the particle thus adjusts the intracavity intensity such as to increase the trapping potential when the particle is leaving the antinode. Cavity cooling has been demonstrated for single atoms [131], BECs [132] and recently also with nanoparticles that were free [23], optically levitated [133] or even electro-dynamically trapped [134].

Matter–wave interferometry and nanoparticle cooling have opposite mass limits. While quantum properties are typically best observed at low mass and long de Broglie wavelength, feedback cooling is easiest for objects with high polarizability and mass. Since direct feedback cooling relies on photon scattering to infer information about the particle's position and motion, it suffers from the $\alpha_{\rm opt}^2$ scaling of Rayleigh scattering. In contrast to that, the dipole coupling of the same particle to a cavity mode, scales only linearly with $\alpha_{\rm opt}$, and its strength can be further boosted by reducing the cavity mode volume. The dipole interaction can be further increased by tailoring the particle geometry, since the polarizability of an oriented nanorod may exceed that of an equivalent sphere of equal volume by up to a factor of five (see Figure 2.15d). Simulations show that cavity cooling of silicon particles with $m \simeq 10^7$ amu may reach down to microkelvin temperatures in a high-finesse microcavity [135].

Elongated nanoparticles also open intriguing possibilities in optomechanical sensing, because the light can exert an optical torque via the dipole force and Rayleigh scattering of circularly polarized photons. This provides a new handle to accelerate, trap and cool nanoparticle rotation and to sense external perturbations by their influence on the rotational state [136]. Fast-rotating levitated nanorods have been shown to change their orientation in synchrony with driven changes in the polarization of the trapping light, and they do this with clock accuracy, i.e. with more than 11-digit precision [123]. On the opposite site of the parameter range, slow rotating nanorods shall facilitate orientation-dependent interferometry [137,138] and enable high-mass tests of quantum physics and wave function collapse in the rotational degrees of freedom [139,140].

The future of nanoparticle quantum interference: Observing de Broglie interference with high-mass nanoparticles requires sufficiently long observation times for the wave function to expand throughout at least one Talbot time. On earth, this time is constrained by the free fall in the gravitational field. The mass limit in a near-field experiment with a height of $H = 1$ m is $m < T_{\rm T}h/d^2 < t_{\rm F}h/d^2 \simeq 10^7$ amu [24,25], if the free fall time $t_{\rm F} = \sqrt{2H/g}$ shall exceed the Talbot time. The total interferometer length further depends on the source preparation, too. The better localized it is, the shorter the setup can be, with the condition that at least two slits of the grating shall be coherently illuminated. Masses up to 10^8 amu are conceivable. Beyond that scale, bouncing or levitating matter waves in optical [140], electric or magnetic traps [26] show a future path. Alternatively, one may install the experiment on a microgravitational platform, such as the international space station or a dedicated satellite [27].

The strictest constraint will be set by collisional decoherence. Since a single scattering event per particle can destroy the superposition, pressures below 10^{-12} mbar are mandatory at 10^7 amu. By choosing a suitable dielectric material, such as silicon, the experiment is predicted not to be harmed by thermal decoherence even at an internal temperature of 700 K [24].

References

1. De Broglie, L. Waves and quanta. *Nature* **112**, 540 (1923).

2. Schrödinger, E. An undulatory theory of the mechanics of atoms and molecules. *Phys. Rev.* **28**, 1049–1070 (1926).

3. Feynman, R. P., Leighton, R. B. & Sands, M. *Lectures on Physics*, vol. III, Addison-Wesley Longman, Amsterdam (1965).

4. Hasselbach, F. Progress in electron- and ion-interferometry. *Rep. Prog. Phys.* **73**, 016101 (2010).

5. Rauch, H. & Werner, S. A. *Neutron Interferometry: Lessons in Experimental Quantum Mechanics, Wave-Particle Duality, and Entanglement*, 2nd edition (Oxford University Press, Oxford, 2015).

6. Cronin, A. D., Schmiedmayer, J. & Pritchard, D. E. Optics and interferometry with atoms and molecules. *Rev. Mod. Phys.* **81**, 1051–1129 (2009).

7. Pethick, C.J. & Smith, H. *Bose-Einstein Condensation in Dilute Gases* (Cambridge University Press, Cambridge 2008).

8. Arndt, M., Nairz, O., Vos-Andreae, J., Keller, C., van der Zouw, G. & Zeilinger, A. Wave-particle duality of C_{60} molecules. *Nature* **401**, 680–682 (1999).

9. Hornberger, K., Gerlich, S., Haslinger, P., Nimmrichter, S. & Arndt, M. Colloquium: Quantum interference of clusters and molecules. English. *Rev. Mod. Phys.* **84**, 157–173 (2012).

10. Arndt, M. & Hornberger, K. Testing the limits of quantum mechanical superpositions. *Nat. Phys.* **10**, 271–277 (2014).

11. Martin, P. J., Gould, P. L., Oldaker, B. G., Miklich, A. H. & Pritchard, D. E. Diffraction of atoms moving through a standing light-wave. *Phys. Rev. A* **36**, 2495–2498 (1987).

12. Keith, D. W., Schattenburg, M. L., Smith, H. I. & Pritchard, D. E. Diffraction of atoms by a transmission grating. *Phys. Rev. Lett.* **61**, 1580–1583 (1988).

13. Bordeé, C. J. Atomic interferometry with internal state labelling. *Phys. Lett. A* **140**, 10–12 (1989).

14. Kasevich, M. & Chu, S. Atomic interferometry using stimulated Raman transitions. *Phys. Rev. Lett.* **67**, 181–184 (1991).

15. Keith, D. W., Ekstrom, C. R., Turchette, Q. A. & Pritchard, D. E. An interferometer for atoms. *Phys. Rev. Lett.* **66**, 2693–2696 (1991).

16. Kovachy, T., Asenbaum, P., Overstreet, C., Donnelly, C. A., Dickerson, S. M., Sugarbaker, A., Hogan, J. M. & Kasevich, M. A. Quantum superposition at the half-metre scale. *Nature* **528**, 530–533 (2015).

17. Tino, G. & Kasevich, M. *Atom Interferometry* (IOS, Varenna, 2014).

18. Ketterle, W. Nobel lecture: When atoms behave as waves: Bose-Einstein condensation and the atom laser. *Rev. Mod. Phys.* **74**, 1131–1151 (2002).

19. Cornell, E. A. & Wieman, C. E. Nobel lecture: Bose-Einstein condensation in a dilute gas, the first 70 years and some recent experiments. *Rev. Mod. Phys.* **74**, 875–893 (2002).

20. Gross, C. & Bloch, I. Quantum simulations with ultra-cold atoms in optical lattices. *Science* **357**, 995–1001 (2017).

21. Arndt, M. De Broglie's meter stick. *Phys. Today* **67**, 33–36 (2014).

22. Eibenberger, S., Gerlich, S., Arndt, M., Mayor, M. & Tüxen, J. Matter–wave interference of particles selected from a molecular library exceeding 10000 amu. *Phys. Chem. Chem. Phys.* **15**, 14696 (2013).

23. Asenbaum, P., Kuhn, S., Nimmrichter, S., Sezer, U. & Arndt, M. Cavity cooling of free silicon nanoparticles in high vacuum. *Nat. Commun.* **4**, 2743 (2013).

24. Bateman, J., Nimmrichter, S., Hornberger, K. & Ulbricht, H. Near-field interferometry of a freefalling nanoparticle from a point-like source. *Nat. Commun.* **5**, 1–16 (2014).

25. Kuhn, S. Cooling and manipulating the ro-translational motion of dielectric particles in high vacuum. PhD thesis, Universität Wien (2017).

26. Romero-Isart, O., Clemente, L., Navau, C., Sanchez, A. & Cirac, J. I. Quantum magnetomechanics with levitating superconducting microspheres. *Phys. Rev. Lett.* **109**, 147205 (2012).

27. Kaltenbaek, R., Aspelmeyer, M., Barker, P. F., Bassi, A., Bateman, J., Christophe, B. & Chwalla, M. Macroscopic Quantum Resonators (MAQRO): 2015 update. *Exp. Astron.* **34**, 123–164 (2016).

28. Born, M. & Wolf, E. *Principles of Optics* (Pergamon Press, Oxford, 1993).

29. Young, T. *Lectures on Natural Philosophy* (J. Johnson, London, 1807).

30. Jönsson, C. Electron diffraction at multiple slits. *Am. J. Phys.* **42**, 4–11 (1974).

31. Freimund, D. L., Aflatooni, K. & Batelaan, H. Observation of the Kapitza-Dirac effect. *Nature* **413**, 142–143 (2001).

32. Zeilinger, A., Gähler, R., Shull, C. G., Treimer, W. & Mampe, W. Single- and double-slit diffraction of neutrons. *Rev. Mod. Phys.* **60**, 1067–1073 (1988).

33. Carnal, O. & Mlynek, J. Young's double-slit experiment with atoms: A simple atom interferometer. *Phys. Rev. Lett.* **66**, 2689–2692 (1991).

34. Schöllkopf, W. & Toennies, J. P. Nondestructive mass selection of small van der Waals clusters. *Science* **266**, 1345–1348 (1994).

35. Nairz, O., Arndt, M. & Zeilinger, A. Experimental challenges in fullerene interferometry. English. *J. Mod. Opt.* **47**, 2811–2821 (2000).

36. Nairz, O., Arndt, M. & Zeilinger, A. Quantum interference experiments with large molecules. English. *Am. J. Phys.* **71**, 319 (2003).

37. Juffmann, T., Milic, A., Müllneritsch, M., Asenbaum, P., Tsukernik, A., Tüxen, J., Mayor, M., Cheshnovsky, O. & Arndt, M. Real-time single-molecule imaging of quantum interference. *Nat. Nanotechnol.* **7**, 297–300 (2012).

38. Brühl, R., Fouquet, P., Grisenti, R. E., Toennies, J. P., Hegerfeldt, G. C., Köhler, T., Stoll, M. & Walter, C. The van der Waals potential between metastable atoms and solid surfaces: Novel diffraction experiments vs. theory. *Europhys. Lett.* **59**, 357–363 (2002).

39. Perreault, J. D. & Cronin, A. D. Observation of atom wave phase shifts induced by van der Waals atom-surface interactions. *Phys. Rev. Lett.* **95**, 133201 (2005).

40. Lonij, V. P. A., Klauss, C. E., Holmgren, W. F. & Cronin, A. D. Atom diffraction reveals the impact of atomic core electrons on atom-surface potentials. *Phys. Rev. Lett.* **105**, 233202 (2010).

41. Brand, C., Sclafani, M., Knobloch, C., Lilach, Y., Juffmann, T., Kotakoski, J., Mangler, C., Winter, A., Turchanin, A., Meyer, J., Cheshnovsky, O. & Arndt, M. An atomically thin matter-wave beamsplitter. *Nat. Nanotechnol.* **10**, 1–5 (2015).

42. Knobloch, C., Stickler, B. A., Brand, C., Sclafani, M., Lilach, Y., Juffmann, T., Cheshnovsky, O., Hornberger, K. & Arndt, M. On the role of the electric dipole moment in the diffraction of biomolecules at nanomechanical gratings. *Fortschr. Phys.* **65**, 1600025 (2017).

43. Kapitza, P. L. & Dirac, P. A. M. The reflection of electrons from standing light waves. *Proc. Camb. Philos. Soc.* **29**, 297–300 (1933).

44. Moskowitz, P. E., Gould, P. L., Atlas, S. R. & Pritchard, D. E. Diffraction of an atomic beam by standing-wave radiation. *Phys. Rev. Lett.* **51**, 370–373 (1983).

45. Nairz, O., Brezger, B., Arndt, M. & Zeilinger, A. Diffraction of complex molecules by structures made of light. *Phys. Rev. Lett.* **87**, 160401 (2001).

46. Oberthaler, M. K., Abfalterer, R., Bernet, S., Keller, C., Schmiedmayer, J. & Zeilinger, A. Dynamical diffraction of atomic matter waves by crystals of light. *Phys. Rev. A* **60**, 456–472 (1999).

47. Chiow, S. W., Kovachy, T., Chien, H. C. & Kasevich, M. A. 102ℏk large area atom interferometers. *Phys. Rev. Lett.* **107**, 1–5 (2011).

48. Müller, H., Chiow, S. W., Long, Q., Herrmann, S. & Chu, S. Atom interferometry with up to 24-photon-momentum-transfer beam splitters. *Phys. Rev. Lett.* **100**, 180405 (2008).

49. Cladé, P., Guellati-Khélifa, S., Nez, F. & Biraben, F. Large momentum beam splitter using bloch oscillations. *Phys. Rev. Lett.* **102**, 240402 (2009).

50. Bordé, C. J., Avrillier, S., Van Lerberghe, A., Salomon, C., Bassi, D. & Scoles, G. Observation of optical Ramsey fringes in the 10 μm spectral region using a supersonic beam of SF$_6$. *Le J. Phys. Colloq.* **42**, C8–15–C8–19 (1981).

51. Bordé, C. J., Courtier, N., du Burck, F., Goncharov, A. N. & Gorlicki, M. Molecular interferometry experiments. *Phys. Lett. A* **188**, 187–197 (1994).

52. Lisdat, C., Frank, M., Knöckel, H., Almazor, M.-L. & Tiemann, E. Realization of a Ramsey-Bordé matter wave interferometer on the K2 molecule. *Eur. Phys. J. D* **12**, 235–240 (2000).

53. Reiger, E., Hackermüller, L., Berninger, M. & Arndt, M. Exploration of gold nanoparticle beams for matter wave interferometry. *Opt. Commun.* **264**, 326–332 (2006).

54. Dörre, N., Rodewald, J., Geyer, P., Von Issendorff, B., Haslinger, P. & Arndt, M. Photofragmentation beam splitters for matter-wave interferometry. *Phys. Rev. Lett.* **113**, 233001 (2014).

55. Talbot, H. LXXVI. Facts relating to optical science. No. IV. *Philos. Mag. Ser. 3* **9**, 401–407 (1836).

56. Clauser, J. Bohmian mechanics and the meaning of the wave function. In: *Experimental Metaphysics* (eds Cohen, R. S., Horne, M. & Stachel, J.), 1–11 (Kluwer Academic, Boston, MA, 1997).

57. Clauser, J. F. & Li, S. Talbot-von Lau atom interferometry with cold slow potassium. *Phys. Rev. A* **49**, R2213 (1994).

58. McMorran, B. J. & Cronin, A. D. An electron Talbot interferometer. *New J. Phys.* **11**, 33021 (2009).

59. Nowak, S., Kurtsiefer, C., Pfau, T. & David, C. High-order Talbot fringes for atomic matter waves. *Opt. Lett.* **22**, 1430–1432 (1997).

60. Deng, L., Hagley, E. W., Denschlag, J., Simsarian, J. E., Edwards, M., Clark, C. W., Helmerson, K., Rolston, S. L. & Phillips, W. D. Temporal, matter-wave-dispersion Talbot effect. *Phys. Rev. Lett.* **83**, 5407–5411 (1999).

61. Lau, E. Beugungserscheinungen an Doppelrastern. *Ann. Phys.* **437**, 417–423 (1948).

62. Pfeiffer, F., Weitkamp, T., Bunk, O. & David, C. Phase retrieval and differential phase-contrast imaging with low-brilliance X-ray sources. *Nat. Phys.* **2**, 258–261 (2006).

63. Juffmann, T., Truppe, S., Geyer, P., Major, A. G., Deachapunya, S., Ulbricht, H. & Arndt, M. Wave and particle in molecular interference lithography. *Phys. Rev. Lett.* **103**, 263601 (2009).

64. Brezger, B., Hackermüller, L., Uttenthaler, S., Petschinka, J., Arndt, M. & Zeilinger, A. Matter wave interferometer for large molecules. *Phys. Rev. Lett.* **88**, 100404 (2002).

65. Gerlich, S., Hackermüller, L., Hornberger, K., Stibor, A., Ulbricht, H., Gring, M., Goldfarb, F., Savas, M., Müri, M., Mayor, M. & Arndt, M. A Kapitza–Dirac–Talbot–Lau interferometer for highly polarizable molecules. *Nat. Phys.* **3**, 711–715 (2007).

66. Haslinger, P., Dörre, N., Geyer, P., Rodewald, J., Nimmrichter, S. & Arndt, M. A universal matter wave interferometer with optical ionization gratings in the time domain. *Nat. Phys.* **9**, 144–148 (2013).

67. Hackermüller, L., Uttenthaler, S., Hornberger, K., Reiger, E., Brezger, B., Zeilinger, A. & Arndt, M. Wave nature of biomolecules and fluorofullerenes. *Phys. Rev. Lett.* **91**, 090408 (2003).

68. Gerlich, S., Eibenberger, S., Tomandl, M., Nimmrichter, S., Hornberger, K., Fagan, P. J., Tüxen, J., Mayor, M. & Arndt, M. Quantum interference of large organic molecules. *Nat. Commun.* **2**, 263 (2011).

69. Tüxen, J., Gerlich, S., Eibenberger, S., Arndt, M. & Mayor, M. Quantum interference distinguishes between constitutional isomers. *Chem. Commun. (Camb).* **46**, 4145–4147 (2010).

70. Mairhofer, L., Eibenberger, S., Cotter, J. P., Romirer, M., Shayeghi, A. & Arndt, M. Quanteninterferenz experimente für die Vermessung von Vitaminen in

der Gasphase. *Angew. Chemie* **129**, 11088–11093 (2017).

71. Rodewald, J., Dörre, N., Grimaldi, A., Geyer, P., Felix, L., Mayor, M., Shayeghi, A. & Arndt, M. Isotope-selective high-order interferometry with large organic molecules in free fall. *New J. Phys.* **20**, 033016 (2018).

72. Nimmrichter, S. *Macroscopic Matter Wave Interferometry,* vol. 279 (Springer, Vienna, 2014).

73. Patorski, K. The self-imaging phenomenon and its applications. *Prog. Opt.* **27**, 1–108 (1989).

74. Hornberger, K., Uttenthaler, S., Brezger, B., Hackermüller, L., Arndt, M. & Zeilinger, A. Collisional decoherence observed in matter wave interferometry. *Phys. Rev. Lett.* **90**, 160401 (2003).

75. Hackermüller, L., Hornberger, K., Brezger, B., Zeilinger, A. & Arndt, M. Decoherence of matter waves by thermal emission of radiation. English. *Nature* **427**, 711–714 (2004).

76. Hornberger, K., Gerlich, S., Ulbricht, H., Hackermüller, L., Nimmrichter, S., Goldt, I. V., Boltalina, O. & Arndt, M. Theory and experimental verification of. *New J. Phys.* **11**, 1–24 (2009).

77. Mairhofer, L., Eibenberger, S., Cotter, J. P., Romirer, M., Shayeghi, A. & Arndt, M. Quantum assisted metrology of neutral vitamins in the gas phase. *Angew. Chemie Int. Ed.* **56**, 10947–10951 (2017).

78. Nimmrichter, S., Haslinger, P., Hornberger, K. & Arndt, M. Concept of an ionizing time-domain matter-wave interferometer. *New J. Phys.* **13**, 75002 (2011).

79. Rodewald, J. Experiments with a pulsed Talbot Lau matter–wave interferometer. PhD thesis, Universität Wien (2017).

80. Nimmrichter, S. & Hornberger, K. Macroscopicity of mechanical quantum superposition states. *Phys. Rev. Lett.* **110**, 160403 (2013).

81. Gregoire, M. D. & Cronin, A. Static polarizability measurements and inertial sensing with nanograting atom interferometry, PhD thesis, University of Arizona (2016).

82. Parker, R. H., Yu, C., Zhong, W., Estey, B. & Müller, H. Measurement of the fine-structure constant as a test of the standard model. *Science* **360**, 191–195 (2018).

83. Bonin, K. & Kresin, V. *Electric-Dipole Polarizabilities of Atoms, Molecules and Clusters* (World Scientific, Singapore, 1997).

84. De Heer, W. A. & Kresin, V. V. in *Handbook of Nanophysics* (ed Sattler, K. D.), 10/1–13 Electric and Magnetic Dipole Moments of Free Nanoclusters (CRC Press, Boca Raton, FL 2011).

85. Heiles, S. & Schäfer, R. *Dielectric Properties of Isolated Clusters*, Springer Briefs in Electrical and Magnetic Properties of Atoms, Molecules, and Clusters, 7–17 (2014).

86. Compagnon, I., Hagemeister, F. C., Antoine, R., Rayane, D., Broyer, M., Dugourd, P., Hudgins, R. R. & Jarrold, M. F. Permanent electric dipole and conformation of unsolvated tryptophan. *J. Am. Chem. Soc.* **123**, 8440–8441 (2001).

87. Antoine, R., Compagnon, I., Rayane, D., Broyer, M., Dugourd, P., Breaux, G., Hagemeister, F. C., Pippen, D., Hudgins, R. R. & Jarrold, M. F. Electric susceptibility of unsolvated glycine-based peptides. *J. Am. Chem. Soc.* **124**, 6737–6741 (2002).

88. Berninger, M., Stefanov, A., Deachapunya, S. & Arndt, M. Polarizability measurements of a molecule via a near-field matter-wave interferometer. *Phys. Rev. A* **76**, 13607 (2007).

89. Eibenberger, S., Gerlich, S., Arndt, M., Tüxen, J. & Mayor, M. Electric moments in molecule interferometry. *New J. Phys.* **13**, 043033 (2011).

90. Compagnon, I., Antoine, R., Rayane, D., Broyer, M. & Dugourd, P. Vibration induced electric dipole in a weakly bound molecular complex. *Phys. Rev. Lett.* **89**, 253001 (2002).

91. Gerlich, S., Gring, M., Ulbricht, H., Hornberger, K., Tüxen, J., Mayor, M. & Arndt, M. Matter wave metrology as a complementary tool for mass spectrometry. *Angew. Chem. Int. Ed. Engl.* **47**, 6195–6198 (2008).

92. Gring, M., Gerlich, S., Eibenberger, S., Nimmrichter, S., Berrada, T., Arndt, M., Ulbricht, H., Hornberger, K., Müri, M., Mayor, M., Böckmann, M. & Doltsinis, N. L. Influence of conformational molecular dynamics on matter wave interferometry. *Phys. Rev. A* **81**, 031604 (2010).

93. Mairhofer, L., Eibenberger, S., Shayeghi, A. & Arndt, M. A quantum ruler for magnetic deflectometry. *Entropy* **20**, 516 (2018).

94. Weiss, D. S., Young, B. C. & Chu, S. Precision measurement of the photon recoil of an atom using atomic interferometry. *Phys. Rev. Lett.* **70**, 2706–2709 (1993).

95. Nimmrichter, S., Hornberger, K., Ulbricht, H. & Arndt, M. Absolute absorption spectroscopy based on molecule interferometry. *Phys. Rev. A* **78**, 063607 (2008).

96. Eibenberger, S., Cheng, X., Cotter, J. P., Arndt, M., Absolute absorption cross sections from photon recoil in a matter-wave interferometer. *Phys. Rev. Lett.* **112**, 250402 (2014).

97. Rodewald, J., Haslinger, P., Dörre, N., Stickler, B. A., Shayeghi, A., Hornberger, K. & Arndt, M. New avenues for matter-wave-enhanced spectroscopy. *Appl. Phys. B Lasers Opt.* **123**, 1–8 (2017).

98. Ballard, A., Bonin, K. & Louderback, J. Absolute measurement of the optical polarizability of C_{60}. *J. Chem. Phys.* **114**, 5732–5735 (2000).

99. Storey, P. & Cohen-Tannoudji, C. The Feynman path integral approach to atomic intermerometry. A tutorial. *J. Phys. II Fr.* **4**, 1999–2027 (1994).

100. Juffmann, T., Nimmrichter, S., Arndt, M., Gleiter, H. & Hornberger, K. New prospects for de Broglie interferometry. *Found. Phys.* **42**, 98–110 (2010).

101. Joos, E. & Zeh, H. D. The emergence of classical properties through interaction with the environment. *Z. Phys. B* **59**, 223–243 (1985).

102. Zurek, W. H. Decoherence and the transition from quantum to classical. *Phys. Today* **44**, 36 (1991).

103. Pfau, T., Spälter, S., Kurtsiefer, C., Ekstrom, C. R. & Mlynek, J. Loss of spatial coherence by a single spontaneous emission. *Phys. Rev. Lett.* **73**, 1223–1226 (1994).

104. Chapman, M. S., Hammond, T. D., Lenef, A., Schmiedmayer, J., Rubenstein, R. A., Smith, E. & Pritchard, D. E. Photon scattering from atoms in an atom interferometer – coherence lost and regained. *Phys. Rev. Lett.* **75**, 3783–3787 (1995).

105. Kokorowski, D. A., Cronin, A. D., Roberts, T. D. & Pritchard, D. E. From single- to multiple-photon decoherence in an atom interferometer. *Phys. Rev. Lett.* **86**, 2191–2195 (2001).

106. Hansen, K. & Campbell, E. Thermal radiation from small particles. *Phys. Rev. E* **58**, 5477–5482 (1998).

107. Riedel, C. J. & Yavin, I. Decoherence as a way to measure extremely soft collisions with dark matter. *Phys. Rev. D* **96**, 023007 (2017).

108. Bassi, A., Lochan, K., Satin, S., Singh, T. P. & Ulbricht, H. Models of wave-function collapse, underlying theories, and experimental tests. *Rev. Mod. Phys.* **85**, 471–527 (2013).

109. Ghirardi, G. C., Rimini, A. & Weber, T. Unified dynamics for microscopic and macroscopic systems. *Phys. Rev. D* **34**, 470–491 (1986).

110. Pearle, P. Combining stochastic dynamical state-vector reduction with spontaneous localization. *Phys. Rev. A* **39**, 2277–2289 (1989).

111. Ghirardi, G. C., Pearle, P. & Rimini, A. Markov processes in Hilbert space and continous spontaneous localization of systems of identical particles. *Phys. Rev. A* **42**, 78 (1990).

112. Karolyhazy, F. Gravitation and quantum mechanics of macroscopic objects. *Nuovo Cim. A* **42**, 390–402 (1966).

113. Diósi, L. Gravitation and quantum mechanical localization of macro objects. *Hung. Acad. Sci.* **3**, 525–532 (1984).

114. Penrose, R. On gravity's role in quantum state reduction. *Gen. Relativ. Gravit.* **28**, 581–600 (1996).

115. Posthumus, M. A., Kistemaker, P. G., Meuzelaar, H. L. C. & de Brauw, M. C. T. N. Laser desorption-mass spectrometry of polar nonvolatile bio-organic molecules. *Anal. Chem.* **50**, 985–991 (1978).

116. Zinovev, A. V., Veryovkin, I. V., Moore, J. F. & Pellin, M. J. Laser-driven acoustic desorption of organic molecules from back-irradiated solid foils. *Anal. Chem.* **79**, 8232–8241 (2007).

117. Peng, W.-P., Yang, Y.-C., Kang, M.-W., Tzeng, Y.-K., Nie, Z., Chang, H.-C., Chang, W. & Chen, C.-H. Laser-induced acoustic desorption mass spectrometry of single bioparticles. *Angew. Chemie* **118**, 1451–1454 (2006).

118. Kuhn, S., Asenbaum, P., Kosloff, A., Sclafani, M., Stickler, B. A., Nimmrichter, S., Hornberger, K., Cheshnovsky, O., Patolsky, F. & Arndt, M. Cavity-assisted manipulation of freely rotating silicon nanorods in high vacuum. *Nano Lett.* **15**, 5604–5608 (2015).

119. Gieseler, J. & Millen, J. Levitated nanoparticles for microscopic thermodynamics—a review. *Entropy* **20**, 326 (2018).

120. Arvanitaki, A. & Geraci, A. A. Detecting high-frequency gravitational waves with optically levitated sensors. *Phys. Rev. Lett.* **110**, 071105 (2013).

121. Geraci, A. A. & Goldman, H. Sensing short range forces with a nanosphere matter-wave interferometer. *Phys. Rev. D* **92** (2015).

122. Ranjit, G., Cunningham, M., Casey, K. & Geraci, A. A. Zeptonewton force sensing with nanospheres in an optical lattice. *Phys. Rev. A* **93**, 053801 (2016).

123. Kuhn, S., Stickler, B. A., Kosloff, A., Patolsky, F., Hornberger, K., Arndt, M. & Millen, J. Optically driven ultra-stable nanomechanical rotor. *Nat. Commun.* **8**, 1670 (2017).

124. Hoang, T. M., Ma, Y., Ahn, J., Bang, J., Robicheaux, F., Yin, Z.-Q. & Li, T. Torsional optomechanics of a levitated nonspherical nanoparticle. *Phys. Rev. Lett.* **117**, 123604 (2016).

125. Li, T., Kheifets, S. & Raizen, M. G. Millikelvin cooling of an optically trapped microsphere in vacuum. *Nat. Phys.* **7**, 527–530 (2011).

126. Gieseler, J., Deutsch, B., Quidant, R. & Novotny, L. Sub-kelvin parametric feedback cooling of a laser-trapped nanoparticle. *Phys. Rev. Lett.* **109**, 103603 (2012).

127. Jain, V., Gieseler, J., Moritz, C., Dellago, C., Quidant, R. & Novotny, L. Direct measurement of photon recoil from a levitated nanoparticle. *Phys. Rev. Lett.* **116**, 243601 (2016).

128. Vovrosh, J., Rashid, M., Hempston, D., Bateman, J., Paternostro, M. & Ulbricht, H. Parametric feedback cooling of levitated optomechanics in a parabolic mirror trap. *J. Opt. Soc. Am. B* **34**, 1421 (2017).

129. Horak, P., Hechenblaikner, G., Gheri, K. M., Stecher, H. & Ritsch, H. Cavity-induced atom cooling in the strong coupling regime. *Phys. Rev. Lett.* **79**, 4974–4977 (1997).

130. Vuletić, V. & Chu, S. Laser cooling of atoms, ions, or molecules by coherent scattering. *Phys. Rev. Lett.* **84**, 3787–3790 (2000).

131. Maunz, P., Puppe, T., Schuster, I., Syassen, N., Pinkse, P. W. H. & Rempe, G. Cavity cooling of a single atom. *Nature* **428**, 50–52 (2004).

132. Schleier-Smith, M. H., Leroux, I. D., Zhang, H., Van Camp, M. A. & Vuletić, V. Optomechanical cavity cooling of an atomic ensemble. *Phys. Rev. Lett.* **107** (2011).

133. Kiesel, N., Blaser, F., Delić, U., Grass, D., Kaltenbaek, R. & Aspelmeyer, M. Cavity cooling of an optically levitated submicron particle. *Proc. Natl. Acad. Sci. U. S. A.* **110**, 14180–14185 (2013).

134. Millen, J., Fonseca, P. Z. G., Mavrogordatos, T., Monteiro, T. S. & Barker, P. F. Cavity cooling a single charged levitated nanosphere. *Phys. Rev. Lett.* **114**, 123602 (2015).

135. Kuhn, S., Wachter, G., Wieser, F.-F., Millen, J., Schneider, M., Schalko, J., Schmid, U., Trupke, M. & Arndt, M. Nanoparticle detection in an open-access silicon microcavity. *Appl. Phys. Lett.* **111**, 253107 (2017).

136. Stickler, B. A., Nimmrichter, S., Martinetz, L., Kuhn, S., Arndt, M. & Hornberger, K. Ro–translational cavity cooling of dielectric rods and disks. *Phys. Rev. A* **94**, 1–6 (2016).

137. Shore, B. W., Dömötör, P., Sadurní, E., Süssmann, G. & Schleich, W. P. Scattering of a particle with internal structure from a single slit. *New J. Phys.* **17**, 013046 (2015).

138. Stickler, B. A. & Hornberger, K. Molecular rotations in matter-wave interferometry. *Phys. Rev. A* **92**, 023619 (2015).

139. Schrinski, B., Stickler, B. A. & Hornberger, K. Collapse-induced orientational localization of rigid rotors. *J. Opt. Soc. Am. B* **34**, C1 (2017).

140. Stickler, B. A., Papendell, B., Kuhn, S., Millen, J., Arndt, M. & Hornberger, K. Orientational quantum revivals of nanoscale rotors. *Phys. Rev. Lett.* **121**, 040401 (2018).

3

Electromagnetic Nanonetworks

Md. Humaun Kabir
Jessore University of Science and Technology

Kyung Sup Kwak
Inha University

3.1 Introduction

Nanonetwork channels are physically separated by up to a few nanometer or/and micrometer and the nodes are assumed to be mobile and quickly deployable. Researchers are studying two types of nanonetworks: electromagnetic and molecular nanonetwork. Generally, nanomachines, act as the most basic functional unit, are able to perform very simple tasks such as computing, data storing, sensing or actuation. A set of interconnected nanomachines, sharing the same medium (e.g., the biological tissue or the fluid flow or other medium) and collaborating for the same task, form a nanonetwork. Possible applications are in healthcare, biomedical field (Freitas, 2005), environmental research (Han, Fu, & Schoch, 2008), military technology (Glenn, 2006) and industrial and consumer goods applications. Nanonetworks expand the number and range of operation envisioned for single nanomachine, since collaborative tasks like coordination, information sharing and fusion can possibly be done by different nanomachines. In molecular nanonetworks, transmission and reception of information are carried out by the motion of molecules, called information molecules. Figure 3.1 shows a typical molecular communication nanonetwork. Based on the type of molecule propagation, molecular communication techniques can be classified into walkaway-based, flow-based or diffusion-based. On the other hand, electromagnetic nanonetworks use electromagnetic waves with the framework of wireless technology. However, wiring a large number of nodes is nearly impractical because of the small size of nanomachine. Besides, it is difficult to integrate the current electromagnetic transceivers into the nanomachines because of their size and complexity. Therefore, the use of carbon structure is a possible way to develop the electronic nano-components. Recent advances in carbon and molecular electronics opened new door to generate electronic nanoscale components such as nanobatteries (Curtright, Bouwman, Wartena, & Swider-Lyons, 2004), nanoscale energy-harvesting systems (Wang, 2008), nano-memories (Bennewitz et al., 2002), logical circuitry in the nanoscale and even nano-antennas (Burke, Li, & Yu, 2006; Burke, Rutherglen, & Yu, 2006, September).

The unique properties observed in nanomaterials would choose specific bandwidths for emission of electromagnetic radiation, the time lag of the emission or the magnitude of the emitted power for a given input energy, among others (Jornet & Akyildiz, 2010a). The intrinsic behavior and characteristics of nanomachines defer from traditional devices working at the macroscale level, and distinguished features at the nanoscale level should be exposed (Akyildiz & Jornet, 2010) as shown in Figure 3.1. In this chapter, we investigate electromagnetic nanocommunications, working in the terahertz band, and foresee the possibility of transmission and reception of electromagnetic radiation from components based on nanomaterials.

This work was supported by the National Research Foundation of Korea (NRF) grant funded by the Korea government (MSIT)-NRF-2017R1A2B2012337.

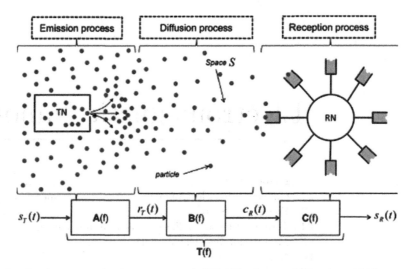

FIGURE 3.1 A typical molecular communication nanonetwork (Akyildiz, Jornet, & Pierobon, 2011).

This chapter is organized as follows. In Section 3.2, we present the challenges of electromagnetic nanocommunication. Section 3.3 describes frequency band used in electromagnetic (EM) nanonetworks. Terahertz band radiation and communication have been discussed in Section 3.4. Modulation and demodulation is presented in Section 3.5, while nanochannel and channel coding are described in Sections 3.6 and 3.7, respectively. Sections 3.8 and 3.9 describe electromagnetic nanoparticles and energy and power, respectively. We place nano-antenna in Section 3.10 and discuss routing protocol for terahertz band communication in Section 3.11. Describing sensing and receiving in Section 3.12 and we finish with the conclusions in Section 3.13.

3.2 Challenges of Electromagnetic Nanocommunication

Since in developing stage, the electromagnetic nanocommunication also has many challenges such as:

- Thermal noise and fading severely affect the communication signals
- New communication techniques must be presented e.g., sub-picosecond or femtosecond long pulses (Jornet, Pujol, & Pareta, 2012), multicarrier modulations and multi-input multi-output (MIMO) nano-antenna
- Regulation and standardization of terahertz band are required. It is imperative to model and develop the future communication standards in terahertz for a very short range such as distances much below one meter
- Information encoding techniques for electromagnetic nanocommunication are needed to be developed that are suitable for the channel characteristics

- New medium access control (MAC) and routing protocols are required to exploit the properties of the terahertz band (Akyildiz & Jornet, 2010; Jornet & Akyildiz, 2011b).

3.3 Frequency Band

Due to the limited computation skills of single nanomachine, communication and signal transmission techniques used in nanonetworks appear to be the most challenging topics. From the communication perspective, the particular characteristics observed in novel nanomaterials will decide the specific bandwidths for the emission of electromagnetic radiation, the time lag of the emission or the magnitude of the emitted power for a given input energy, among others (Jornet & Akyildiz, 2010a). So far, several alternatives have been introduced. Carbon nanotubes (CNTs) and graphene nanoribbons (GNRs) have been proposed for electromagnetic nano-antenna (Abadal et al., 2011). A graphene-based nano-antenna is not only the reduction of a classical antenna but also there are several quantum phenomena that affect the propagation of electromagnetic waves by graphene. The result is that the resonant frequency of these nanostructures can be up to two orders of magnitude below that of their nonocarbon-based counterparts. To determine the frequency band of operation of electromagnetic nanonetworks, it is obligatory to characterize the radiation properties of graphene. Up to date, reports have been found in the literature both from the radiofrequency (Jornet & Akyildiz, 2010b), (Burke et al., 2006), (Zhou, Yang, Xiao, & Li, 2003) and the optical (Da Costa, Kibis, & Portnoi, 2009), (Zhang, Xi, & Lai, 2006), (Kempa et al., 2007) perspectives. The main difference between the two options depends on the interpretation of the radiation in terms of high-frequency resonant waves radiated from nanoscale antennas or low-energy photons radiated from optical nano-emitters. However, even if the difference in origin, both approaches envisage the terahertz band (0.1–10 THz) to become the frequency range of operation for future nano-electromagnetic transceivers (Jornet & Akyildiz, 2010a). An architecture of nanosensor device equipped with nanosensors, nano-actuator, nano-memory, nano-antenna, nano-EM transceiver, nano-processor and nano-power unit has been presented in (Akyildiz & Jornet, 2010). When all these components are integrated, the device can sense, compute or even perform local actuation. Furthermore, the authors predict that nanosensor devices would potentially communicate among them in the terahertz band (i.e., 0.1–10.0 THz).

3.4 Terahertz Band Radiation and Communication

Falling in between infrared radiation and microwave radiation in the electromagnetic spectrum, some properties of terahertz radiation are found to be common with each of these. As in the case of infrared and microwave radiation, terahertz radiation also propagates in a line of sight and is nonionizing. Terahertz radiation can also penetrate a wide variety of nonconducting materials as microwave radiation does. Nonionizing terahertz radiation can pass through clothing, paper, cardboard, wood, masonry, plastic and ceramics. The penetration depth is, in particular, less than that of microwave radiation. Terahertz radiation has limited penetration through fog and clouds and cannot penetrate liquid water or metal. Unlike X-rays, terahertz radiation is not ionizing radiation, and its low photon energies generally do not cause any harm to living tissues and DNA (Deoxyribonucleic acid) (Choudhury, Sonde, & Jha, 2016). It can penetrate some distance through body tissue, so it is of great interest as a replacement for medical X-rays. However, due to its longer wavelength, images made using terahertz have lower resolution than X-rays and need to be enhanced.

The earth's atmosphere is a strong absorber of terahertz radiation making the range of the radiation in air limited to tens of meters. Therefore, terahertz radiation appears to be unsuitable for long-distance communications. However, at distances of the order of 10 m, the band may still allow many useful applications in imaging and construction of high bandwidth wireless networking systems, especially indoor systems (Hanson & De Maagt, 2007; Song & Nagatsuma, 2011). In addition, producing and detecting coherent terahertz radiation remain technically challenging, though inexpensive commercial sources now exist in the 0.3–1.0 THz range (the lower part of the spectrum), including gyrotrons, backward wave oscillators and resonant-tunneling diodes.

3.5 Modulation and Demodulation

One of the fundamental challenges with nanocommunication is the necessity for an appropriate modulation and channel access mechanism. So far, two main alternatives for electromagnetic communication in the nanoscale have drawn interest. First, it was experimentally demonstrated to receive and demodulate an electromagnetic wave by means of a nanoradio, i.e., an electromechanically resonating CNT, which is able to decode an amplitude or frequency-modulated wave (Atakan & Akan, 2010). Second, graphene-based nano-antennas have been analyzed as potential electromagnetic radiators in the terahertz band (Jornet & Akyildiz, 2010b). Time-hopping (TH) combined with M-ary pulse position modulation (PPM) is proposed as a modulation and multiple access scheme for multiuser nanocommunication systems (Singh, Kim, & Jung, 2018). PPM, being the most energy-efficient modulation scheme, with TH would allow nanomachiness to communicate simultaneously with greater efficiency. A modulation and channel-sharing mechanism based on the asynchronous exchange of femtosecond-long pulses transmitted through an on-off keying modulation is introduced for transmitting binary streams among nanomachines of an EM nanonetwork in (Jornet & Akyildiz, 2011c). In (Pujol, Jornet, & Pareta, 2011), a MAC protocol for EM nanonetworks built on the top of the pulse-based communication scheme for the coordination of multiple simultaneous transmissions has been presented. The proposed protocol is designed have the peculiarities of the terahertz band and is constituted by two main stages i.e., (i) the handshaking process and (ii) the transmission process. Finally, authors in (Jornet & Akyildiz, 2012) have developed an energy model for self-powered nanosensor motes, which successfully captures the correlation between the energy-harvesting and the energy-consumption processes.

Electromagnetic nanonetwork is defined as the transmission and reception of electromagnetic radiation from nanoscale components (Rutherglen & Burke, 2009). Moreover, the particular application for which the nanonetworks will be deployed limits the choice on the particular type of nanocommunication. For the time being, several alternatives have been introduced. CNTs and GNRs have been proposed in (Abadal et al., 2011) for electromagnetic nano-antenna. A graphene-based nano-antenna is not only the reduction of a classical antenna but also there are several quantum phenomena that affect the propagation of electromagnetic waves on graphene. The resonant frequency of these nanostructures can be up to two orders of magnitude below that of their noncarbon-based counterparts. However, their radiation efficiency can also be impaired because of this phenomenon. Second, CNTs have also been proposed as the basis of an electromechanical nano-transceiver or nano-radio (Jensen, Weldon, Garcia, & Zettl, 2007), able to modulate and demodulate an electromagnetic wave by means of mechanical resonation.

3.6 Nano-channel

The nanonetwork channels are physically separated by up to a few nano- or/and micrometer. For nanosensor networks, the main challenges of electromagnetic nanocommunications are expressed in terms of terahertz channel modeling, information encoding and protocols. A physical channel model for wireless communication in the terahertz band has been introduced in (Jornet & Akyildiz, 2011b). The authors compute the signal path loss, molecular absorption noise and, ultimately, the channel capacity of EM nanonetworks. Nanomachines working in terahertz band can theoretically have a sufficiently small size that they can be inserted into the human body (Pirmagomedov, Hudoev, Kirichek, Koucheryavy, & Glushakov, 2016). Major drawback of electromagnetic waves in the terahertz band has high losses in the tissues and body fluids of microorganism. During signal propagation in several millimeters, the losses in blood are about 120 dB, in skin – about 90 dB and in fat – about 70 dB (Yang et al., 2015). Therefore, for a distance of a couple of millimeters, the usage of repeaters is required for communication among transceiver nanomachines. The propagation model proposed in (Piro et al., 2015) characterized the performance of terahertz communication in human tissue considering the attenuation of EM waves in human skin tissues. They deduced channel capacity and communication ranges for different physical transmission environment parameters. Path loss and molecular absorption noise temperature were also obtained utilizing optical parameters of human skin tissues and were verified through extensive experimental tests. In addition, SimpleNano, a channel model for wireless nanosensor networks (WNSN) that is capable of transmitting in the terahertz range where a log-distance path loss model together with random attenuation was approximated in media such as the human body (Javed & Naqvi, 2013).

3.7　Channel Coding

Severe path loss and molecular absorption loss occur in electromagnetic communication in terahertz band. Channel coding plays an important role to detect and correct transmission errors. A detailed description of the channel codes has been reported in (Jornet, 2014a). A coding scheme for the purpose of interference mitigation was presented in (Jornet, 2014b). The minimum energy codes for nanonetworks were also introduced in (Chi, Zhu, Jiang, & Tian, 2013). The information capacity is always limited in the existing coding methods, and the network resource is not utilized adequately (Chi et al., 2013; Jornet, 2014b). However, the requirement of high-information data rate has drastically increased in the last three decades (Yao et al., 2015). For instance, wireless data rates have doubled every 18 months and almost approaching the capacity of wired communication systems (Akyildiz, Jornet, & Han, 2014). So it is obligatory to optimize the nanonetworks to improve the network information capacity.

3.8　Electromagnetic Nanoparticles

The fabrication of nanostructure devices received too much interest in the last couple of years. Optical properties of metallic nanoparticles are observed to be very

much appropriate for biomedical applications (Dykman & Khlebtsov, 2012). For instance, depending on the size, shape, geometrical parameters and the surrounding dielectric environment refractive index (i.e., RI), gold nanoparticles have inner electromagnetic properties. Precisely speaking, the strongly enhanced localized surface plasmon resonance (LSPR) of this metal, at optical frequencies, allows them to be good light scatterers and absorbers (Kumar, Boruah & Liang, 2011). Additionally, gold nanoparticles promise to have good biocompatibility, optimal synthesis and conjugation properties (Patra, Bhattacharya, Mukhopadhyay, & Mukherjee, 2010) and are useful tools as contrast agents in cellular and biological imaging (Cho, Glaus, Chen, Welch, & Xia, 2010).

In terms of scattering and absorption cross-section, following assumptions need to be established for explaining the electromagnetic properties of the nanoparticle:

- The particle size should be much smaller than the wavelength in the surrounding medium. In this case, under the limit of electrically small particles, the electromagnetic field is almost unchanged over the particle volume, and then the resonant behavior of the structure can be studied in terms of a quasi-static approximation.

- The considered particle should be considered homogeneous and isotropic. In addition, the surrounding material is also a homogeneous, isotropic and non-absorbing medium.

3.9　Energy and Power

A nanomachine is composed of a power supply, memory, antenna and CPU module, and it behaves like an autonomous node capable of performing simple tasks such as computing, storing, sensing and/or actuating at the nano level (Afsana, Asif-Ur-Rahman, Ahmed, Mahmud, & Kaiser, 2018). A major challenge of nanosensor device is its energy storage capacity (Jornet & Akyildiz, 2012). However, energy harvesting is one of the possible solutions of this specific problem (Cottone, Vocca, & Gammaitoni, 2009; Gammaitoni, Neri, & Vocca, 2009; Wang, 2008; Xu, Hansen, & Wang, 2010a; Xu et al., 2010b). Piezoelectric nanogenerator is experimentally demonstrated in (Xu et al., 2010a). If the energy harvesting and the energy consumption processes are jointly designed, then the lifetime of energy-harvesting networks can considerably be increased (Gorlatova, Wallwater, & Zussman, 2013; Gummeson, Clark, Fu, & Ganesan, 2010). In classical battery-powered devices, the energy decreases until the battery is empty. But self-powered devices have both positive and negative fluctuations (Jornet & Aky-ildiz, 2012). The energy-harvesting process is perceived by means of a piezoelectric nanogenerator, for which a new circuital model is designed that can accurately reproduce existing experimental data. The energy consumption process is due to

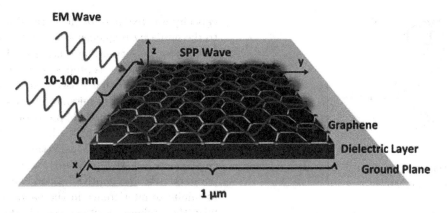

FIGURE 3.2 A graphene-based plasmonic antenna for THz Band communication. (Jornet & Akyildiz, 2013).

the communication among nanosensor motes in the tera-hertz band (Jornet & Akyildiz, 2012). It is judicious to consider that if a nanomachine fully deplenishes its battery and is unable to respond to a communication request, the transmitting nanomachine will attempt to retransmit. This would definitely increase the overall network traffic and multiuser interference, and it ultimately has an impact in the energy of the transmitting nanosensor mote and the neighboring nanomachines (Jornet & Akyildiz, 2012). (Jornet & Akyildiz, 2012) propose an energy model for self-powered nanosensor motes. The model considers both the energy harvesting process by means of a piezoelectric nano-generator and the energy consumption process due to elec-tromagnetic communication in the terahertz band (0.1–10.0 THz) (Jornet & Akyildiz, 2010a, 2011a,b). The model allows to compute the probability distribution of the nanosensor mote energy and to investigate its variations as function of several system and network parameters.

3.10 Nano-antenna

The way in which the nanomachines can communicate depends heavily on the way they are perceived (Akyildiz et al., 2011). Communication options for nanonetworks are very limited due to the size and ability of nanomachines. Graphene is considered as the marvel material of the 21st century. It is a one-atom-thick planar sheet that consists of carbon atoms arranged on a honeycomb crystal lattice. Two derivatives of graphene are CNTs: a folded nanoribbon, and GNR: a thin strip of graphene. Electromagntic nanocommu-nication is impossible in common frequency range, which is from hundreds of megahertz to few gigahertz. Because, in this range, the size of the antenna will be a few centime-ters. If the size of antenna is reduced to a few hundred nanometers, it should operate in extremely high frequency i.e., in terahertz band. This, again, introduces many prob-lems in electromagnetic communication such as high atten-uation. However, nano-antenna, built from graphene, can overcome this limitation. For example, an antenna of few hundred nanometers would lean on the use of very high oper-ating frequencies, which limits the communication range of

nanomachines (Yao et al., 2015). Terahertz band, which spans the electromagnetic spectrum from 0.1THz to 10.0 THz (Llatser et al., 2012), is one of the least explored communication frequency ranges in electromagnetic spec-trum (Jornet et al., 2012). Graphene and its derivatives (Martí et al., 2011), such as CNT (Jornet & Akyildiz, 2010b) and GNRs (Abadal et al., 2011), can be used to radiate at terahertz band. So the terahertz band (0.1–10.0 THz) is recommended as the transmission frequency band of nanonetworks. Figure 3.2 shows how graphene can be used to build novel plasmonic nano-antennas. As depicted in Figure 3.3, by using planar nano-antennas to create ultra-high speed links, the terahertz band can provide effi-cient and scalable means of inter-core communication in wireless on-chip networks. Figure 3.4 refers to a graphene-based nano-patch antenna which analyzes the performance in transmission and reception in terahertz band. The reso-nance frequency of the nano-antenna is calculated as a func-tion of its length and width. The influence of a dielectric substrate with a variable size and the position of the patch with respect to the substrate have also been evaluated. They found that the radiation pattern of a graphene-based nano-patch antenna is very similar to that of an equivalent metallic antenna.

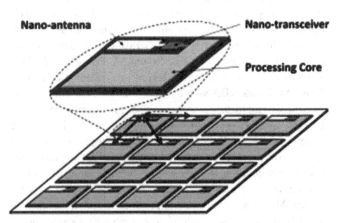

FIGURE 3.3 Wireless on-chip communication. (Abadal et al., 2013).

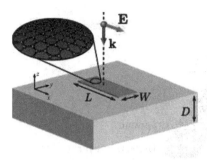

FIGURE 3.4 Schematic diagram of a graphene-based nano-patch antenna. (Llatser et al., 2012).

3.11 Routing Protocol for Terahertz Band Communication

Routing protocol for WNSN is remained in its infancy. The existing communication protocols may not be able to drive the nodes to communicate among themselves at the nanoscale. Therefore, these traditional protocols receive extensive revisions. Based on the transmission medium, different communication protocols have been proposed in course of time, which include acoustic, nanomechanical, molecular and electromagnetic (Akyildiz et al., 2014). A routing framework was introduced based on the peculiarities of the WNSNs, both in terms of terahertz band communication and nanoscale energy harvesting (Pierobon, Jornet, Akkari, Almasri, & Akyildiz, 2014). To reduce the complexity of network operation, a hierarchical cluster-based architecture was considered. This routing framework is based on a hierarchical cluster-based architecture where the WNSN is partitioned into clusters. Within each cluster, a nano-controller, which is a nanodevice with more advanced capabilities than a nanosensor, coordinates the nanosensors and gathers the data they communicate (Akyildiz & Jornet, 2010). An NS-3 module, namely Nano-Sim, modeling WNSNs based on electromagnetic communications in the terahertz band was presented in (Piro, Grieco, Boggia, & Camarda, 2013a). In this preliminary version, the Nano-Sim provides a simple network architecture and a protocol suite for such an emerging technology. In that very year, they have extended the tool by developing a new routing algorithm and a more efficient MAC protocol focusing on a WNSN operating in a health monitoring scenario (Piro, Grieco, Boggia, & Camarda, 2013b). A hierarchical network architecture integrating Body Area Nano-NETwork (BANNET) and an external macroscale healthcare monitoring system was introduced in (Piro, Boggia, & Grieco, 2015). Two different energy-harvesting aware protocol stacks using optimal routing protocol and a greedy routing approach have been formed for dealing with the communication of nanosensors moving uniformly in an environment mimicking human blood vessels. An energy-aware MAC protocol was applied in both strategies to identify the available nanonodes through a handshake mechanism. These two schemes performed well compared with the simple flooding scheme. However, high computational

capacity was required for the optimal scheme. As a solution to the terahertz frequency-selective feature of networking, a channel aware forwarding scheme for EM-based WNSN in the terahertz band was proposed in (Yu, Ng, & Seah, 2015). The authors evaluated classical multihop forwarding and single end-to-end transmission schemes for EM-WNSNs. A Physical Layer Aware MAC protocol for EM nanonetworks in the THz band (PHLAME) has been proposed in (Jornet et al., 2012) where the transmitting and receiving nanomachines were allowed to jointly select the communication parameters in an adaptive fashion. The protocol was found to minimize interference in the nanonetwork and to maximize the probability of successfully decoding the received information. The energy and spectrum-aware MAC protocol was an approach to achieve incessant WNSNs (Wang, Jornet, Malik, Akkari, & Akyildiz, 2013). The objective was to achieve an ample throughput and lifetime optimal channel access by jointly optimizing energy-harvesting and energy-consumption processes in nanosensors. A method was to maximally utilize the harvested energy for nanonodes in perpetual WNSNs that communicate in the terahertz band (Mohrehkesh & Weigle, 2014). They developed an energy model as a Markov decision process considering that energy arrivals follow a stochastic process.

3.12 Sensing and Receiving

Nanomachiness (Akyildiz & Jornet, 2010), equipped with nanosensors, nanoactuator, nano-memory, nano-antenna, nano-EM transceiver, nano-processor and nano-power unit, will be able to exchange information through electromagnetic nanocommunications. In view of communication, the nanomachines will be able to accomplish more complex missions in a cooperative manner. As an example, nanosensors will be able to transmit the sensed information in a multi-hop fashion to a sink or a command centre. For electromagnetic nanonetworks, the exploitation of modulation and channel sharing mechanism based on the asynchronous exchange of femtosecond-long pulses, which are transmitted following an on-off keying modulation spread in time, has been proposed (Jornet & Akyildiz, 2011c). A receiver architecture for electromagnetic nanonetworks has been proposed in (Atakan & Akan, 2008) that makes use of pulse-based modulation. The receiver is designed to be very simple and robust, and it is based on a continuous-time moving average (CTMA) symbol detection scheme. This scheme bases its decision in the received signal power maximum peak after the CTMA, which is implemented with a single low-pass filter. Afterwards, to decode the symbol, this maximum is compared with a previously defined threshold.

3.13 Conclusion

In this chapter, we have addressed recent advances in electromagnetic nanocommunications. Starting from the description of possible applications and challenges, we have presented the state of the art of electromagnetic nanonetworks. We have found that terahertz is the only operating frequency for nanonetwork channel. Terahertz band communication and routing protocols are also addressed. It is found in the literature that an extensive study is required for transmission and routing protocol for terahertz band EM communication. We discuss about graphene-based nano-antenna, the most suited for terahertz communication. Energy harvesting models are also explored.

It can be inferred that nanomaterials are potential candidates for the design of innovative nanomachines. For instance, graphene-based nano-antennas can provide outstanding sensing capabilities, as well as graphene-based transistors are not only smaller but predictably faster too.

References

Abadal, S., Alarcón, E., Cabellos-Aparicio, A., Lemme, M. & Nemirovsky, M., (2013). Graphene-enabled wireless communication for massive multicore architectures. *IEEE Communications Magazine*, 51(11), pp.137–143.

Abadal, S., Jornet, J. M., Llatser, I., Cabellos-Aparicio, A., Alarcón, E., & Akyildiz, I. F. (2011). Wireless nanosensor networks using graphene-based nano-antennas. In *GRAPHENE. Bilbao, Spain*, 1–2.

Afsana, F., Asif-Ur-Rahman, M., Ahmed, M. R., Mahmud, M., & Kaiser, M. S. (2018). An Energy Conserving Routing Scheme for Wireless Body Sensor Nanonetwork Communication. *IEEE Access*, 6, pp. 9186-9200.

Akyildiz, I. F., & Jornet, J. M. (2010). Electromagnetic wireless nanosensor networks. *Nano Communication Networks*, 1(1), 3–19.

Akyildiz, I. F., Jornet, J. M., & Han, C. (2014). Terahertz band: Next frontier for wireless communications. *Physical Communication*, 12, 16–32.

Akyildiz, I. F., Jornet, J. M., & Pierobon, M. (2011). Nanonetworks: A new frontier in communications. *Communications of the ACM*, 54(11), 84–89.

Atakan, B., & Akan, O. B. (2008, November). On molecular multiple-access, broadcast, and relay channels in nanonetworks. In *Proceedings of the 3rd International Conference on Bio-Inspired Models of Network, Information and Computing Sytems* (p. 16). ICST (Institute for Computer Sciences, Social-Informatics and Telecommunications Engineering).

Atakan, B., & Akan, O. B. (2010). Carbon nanotube-based nanoscale ad hoc networks. *IEEE Communications Magazine*, 48(6), 129–135.

Bennewitz, R., Crain, J. N., Kirakosian, A., Lin, J. L., McChesney, J. L., Petrovykh, D. Y., & Himpsel, F. J. (2002). Atomic scale memory at a silicon surface. *Nanotechnology*, 13(4), 499.

Burke, P. J., Li, S., & Yu, Z. (2006). Quantitative theory of nanowire and nanotube antenna performance. *IEEE Transactions on Nanotechnology*, 5(4), 314–334.

Burke, P., Rutherglen, C., & Yu, Z. (2006, September). Carbon nanotube antennas. In *Nanomodeling II* (Vol. 6328, p. 632806). International Society for Optics and Photonics.

Chi, K., Zhu, Y. H., Jiang, X., & Tian, X. (2013). Optimal coding for transmission energy minimization in wireless nanosensor networks. *Nano Communication Networks*, 4(3), 120–130.

Cho, E. C., Glaus, C., Chen, J., Welch, M. J., & Xia, Y. (2010). Inorganic nanoparticle-based contrast agents for molecular imaging. *Trends in Molecular Medicine*, 16(12), 561–573.

Choudhury, B., Sonde, A. R., & Jha, R. M. (2016). Terahertz antenna technology for space applications. In *Terahertz Antenna Technology for Space Applications* (pp. 1–33). Springer, Singapore.

Cottone, F., Vocca, H., & Gammaitoni, L. (2009). Nonlinear energy harvesting. *Physical Review Letters*, 102(8), 080601.

Curtright, A. E., Bouwman, P. J., Wartena, R. C., & Swider-Lyons, K. E. (2004). Power sources for nanotechnology. *International Journal of Nanotechnology*, 1(1-2), 226–239.

Da Costa, M. R., Kibis, O. V., & Portnoi, M. E. (2009). Carbon nanotubes as a basis for terahertz emitters and detectors. *Microelectronics Journal*, 40(4-5), 776–778.

Dykman, L., & Khlebtsov, N. (2012). Gold nanoparticles in biomedical applications: Recent advances and perspectives. *Chemical Society Reviews*, 41(6), 2256–2282.

Freitas, R. A. (2005). Nanotechnology, nanomedicine and nanosurgery. *International Journal of Surgery*, 3(4), 243–246.

Gammaitoni, L., Neri, I., & Vocca, H. (2009). Nonlinear oscillators for vibration energy harvesting. *Applied Physics Letters*, 94(16), 164102.

Glenn, J. C. (2006). Nanotechnology: Future military environmental health considerations. *Technological Forecasting and Social Change*, 73(2), 128–137.

Gorlatova, M., Wallwater, A., & Zussman, G. (2013). Networking low-power energy harvesting devices: Measurements and algorithms. *IEEE Transactions on Mobile Computing*, 12(9), 1853–1865.

Gummeson, J., Clark, S. S., Fu, K., & Ganesan, D. (2010, June). On the limits of effective hybrid micro-energy harvesting on mobile CRFID sensors. In *Proceedings of the 8th International Conference on Mobile Systems, Applications, and Services* (pp. 195–208). ACM.

Han, J., Fu, J., & Schoch, R. B. (2008). Molecular sieving using nanofilters: Past, present and future. *Lab on a Chip*, 8(1), 23–33.

Hanson, G. W., & De Maagt, P. (2007). Guest editorial for the special issue on optical and THz antenna technology. *IEEE Transactions on Antennas and Propagation*, 55(11), 2942–2943.

Javed, I. T., & Naqvi, I. H. (2013, June). Frequency band selection and channel modeling for WNSN applications using simplenano. In *2013 IEEE International*

Conference on Communications (ICC) (pp. 5732–5736). IEEE.

Jensen, K., Weldon, J., Garcia, H., & Zettl, A. (2007). Nanotube radio. *Nano Letters*, 7(11), 3508–3511.

Jornet, J. M. (2014a, May). Low-weight channel codes for error prevention in electromagnetic nanonetworks in the terahertz band. In *Proceedings of ACM The First Annual International Conference on Nanoscale Computing and Communication* (p. 5). ACM.

Jornet, J. M. (2014b). Low-weight error-prevention codes for electromagnetic nanonetworks in the terahertz band. *Nano Communication Networks*, 5(1-2), 35–44.

Jornet, J. M., & Akyildiz, I. F. (2010a, May). Channel capacity of electromagnetic nanonetworks in the terahertz band. In *2010 IEEE International Conference on Communications (ICC)* (pp. 1–6). IEEE.

Jornet, J. M., & Akyildiz, I. F. (2010b, April). Graphene-based nano-antennas for electromagnetic nanocommunications in the terahertz band. In *2010 Proceedings of the Fourth European Conference on Antennas and Propagation (EuCAP)* (pp. 1–5). IEEE.

Jornet, J. M., & Akyildiz, I. F. (2011a, June). Information capacity of pulse-based wireless nanosensor networks. In *2011 8th Annual IEEE Communications Society Conference on Sensor, Mesh and Ad Hoc Communications and Networks (SECON)* (pp. 80–88). IEEE.

Jornet, J. M., & Akyildiz, I. F. (2011b). Channel modeling and capacity analysis for electromagnetic wireless nanonetworks in the terahertz band. *IEEE Transactions on Wireless Communications*, 10(10), 3211–3221.

Jornet, J. M., & Akyildiz, I. F. (2011c, June). Low-weight channel coding for interference mitigation in electromagnetic nanonetworks in the terahertz band. In *2011 IEEE International Conference on Communications (ICC)* (pp. 1–6). IEEE.

Jornet, J. M., & Akyildiz, I. F. (2012). Joint energy harvesting and communication analysis for perpetual wireless nanosensor networks in the terahertz band. *IEEE Transactions on Nanotechnology*, 11(3), 570–580.

Jornet, J. M., Pujol, J. C., & Pareta, J. S. (2012). Phlame: A physical layer aware mac protocol for electromagnetic nanonetworks in the terahertz band. *Nano Communication Networks*, 3(1), 74–81.

Jornet, J.M. & Akyildiz, I.F. (2013) Graphene-based plasmonic nano-antenna for terahertz band communication in nanonetworks. *IEEE Journal on Selected Areas in Communications*, 31(12), pp.685–694.

Kempa, K., Rybczynski, J., Huang, Z., Gregorczyk, K., Vidan, A., Kimball, B., ... & Ren, Z. F. (2007). Carbon nanotubes as optical antennae. *Advanced Materials*, 19(3), 421–426.

Kumar, A., Boruah, B. M., & Liang, X. J. (2011). Gold nanoparticles: Promising nanomaterials for the diagnosis of cancer and HIV/AIDS. *Journal of Nanomaterials*, 2011, 22.

Llatser, I., Kremers, C., Chigrin, D. N., Jornet, J. M., Lemme, M. C., Cabellos-Aparicio, A., & Alarcón, E.

(2012, March). Characterization of graphene-based nano-antennas in the terahertz band. In *2012 6th European Conference on Antennas and Propagation (EUCAP)* (pp. 194–198). IEEE.

Martí, I. L., Kremers, C., Cabellos-Aparicio, A., Jornet, J. M., Alarcón, E., & Chigrin, D. N. (2011, October). Scattering of terahertz radiation on a graphene-based nano-antenna. In *AIP Conference Proceedings* (Vol. 1398, No. 1, pp. 144–146). AIP.

Mohrehkesh, S., & Weigle, M. C. (2014). Optimizing energy consumption in terahertz band nanonetworks. *IEEE Journal on Selected Areas in Communications*, 32(12), 2432–2441.

Patra, C. R., Bhattacharya, R., Mukhopadhyay, D., & Mukherjee, P. (2010). Fabrication of gold nanoparticles for targeted therapy in pancreatic cancer. *Advanced Drug Delivery Reviews*, 62(3), 346–361.

Pierobon, M., Jornet, J. M., Akkari, N., Almasri, S., & Akyildiz, I. F. (2014). A routing framework for energy harvesting wireless nanosensor networks in the Terahertz Band. *Wireless Networks*, 20(5), 1169–1183.

Pirmagomedov, R., Hudoev, I., Kirichek, R., Koucheryavy, A., & Glushakov, R. (2016, October). Analysis of delays in medical applications of nanonetworks. In *2016 8th International Congress on Ultra Modern Telecommunications and Control Systems and Workshops (ICUMT)* (pp. 49–55). IEEE.

Piro, G., Boggia, G., & Grieco, L. A. (2015). On the design of an energy-harvesting protocol stack for Body Area Nano-NETworks. *Nano Communication Networks*, 6(2), 74–84.

Piro, G., Grieco, L. A., Boggia, G., & Camarda, P. (2013a, March). Simulating wireless nano sensor networks in the ns-3 platform. In *2013 27th International Conference on Advanced Information Networking and Applications Workshops (WAINA)* (pp. 67–74). IEEE.

Piro, G., Grieco, L. A., Boggia, G., & Camarda, P. (2013b, March). Nano-Sim: Simulating electromagnetic-based nanonetworks in the network simulator 3. In *Proceedings of the 6th International ICST Conference on Simulation Tools and Techniques* (pp. 203–210). ICST (Institute for Computer Sciences, Social-Informatics and Telecommunications Engineering).

Piro, G., Yang, K., Boggia, G., Chopra, N., Grieco, L. A., & Alomainy, A. (2015). Terahertz communications in human tissues at the nanoscale for healthcare applications. *IEEE Transactions on Nanotechnology*, 14(3), 404–406.

Pujol, J. C., Jornet, J. M., & Pareta, J. S. (2011, April). PHLAME: A physical layer aware MAC protocol for electromagnetic nanonetworks. In *2011 IEEE Conference on Computer Communications Workshops (INFOCOM WKSHPS)* (pp. 431–436). IEEE.

Rutherglen, C., & Burke, P. (2009). Nanoelectromagnetics: Circuit and electromagnetic properties of carbon nanotubes. *Small*, 5(8), 884–906.

Singh, P., Kim, B. W., & Jung, S. Y. (2018). TH-PPM with non-coherent detection for multiple access in electromagnetic wireless nanocommunications. *Nano Communication Networks*, 17, 1–13.

Song, H. J., & Nagatsuma, T. (2011). Present and future of terahertz communications. *IEEE Transactions on Terahertz Science and Technology*, 1(1), 256–263.

Wang, P., Jornet, J. M., Malik, M. A., Akkari, N., & Akyildiz, I. F. (2013). Energy and spectrum-aware MAC protocol for perpetual wireless nanosensor networks in the Terahertz Band. *Ad Hoc Networks*, 11(8), 2541–2555.

Wang, Z. L. (2008). Towards self-powered nanosystems: from nanogenerators to nanopiezotronics. *Advanced Functional Materials*, 18(22), 3553–3567.

Xu, S., Hansen, B. J., & Wang, Z. L. (2010a). Piezoelectric-nanowire-enabled power source for driving wireless microelectronics. *Nature Communications*, 1, 93.

Xu, S., Qin, Y., Xu, C., Wei, Y., Yang, R., & Wang, Z. L. (2010b). Self-powered nanowire devices. *Nature Nanotechnology*, 5(5), 366.

Yang, K., Pellegrini, A., Munoz, M. O., Brizzi, A., Alomainy, A., & Hao, Y. (2015). Numerical analysis and characterization of THz propagation channel for body-centric nano-communications. *IEEE Transactions on Terahertz Science and technology*, 5(3), 419–426.

Yao, X. W., Pan, X. G., Zhao, C., Wang, C. C., Wang, W. L., & Yang, S. H. (2015, September). Pulse position coding for information capacity promotion in electromagnetic nanonetworks. In *Proceedings of the Second Annual International Conference on Nanoscale Computing and Communication* (p. 16). ACM.

Yu, H., Ng, B., & Seah, W. K. (2015, September). Forwarding schemes for EM-based wireless nanosensor networks in the terahertz band. In *Proceedings of the Second Annual International Conference on Nanoscale Computing and Communication* (p. 17). ACM.

Zhang, J., Xi, N., & Lai, K. (2006). Single carbon nanotube infrared detectors. *Science*, 312, 554–556.

Zhou, G., Yang, M., Xiao, X., & Li, Y. (2003). Electronic transport in a quantum wire under external terahertz electromagnetic irradiation. *Physical Review B*, 68(15), 155309.

Nanoscale Energy Transport

Jafar Ghazanfarian
University of Zanjan

Zahra Shomali
Tarbiat Modares University
Institute for Research in Fundamental Sciences
(IPM)

Shiyun Xiong
Soochow University

Simulation of nanoscale heat transport is a vital and important tool for the purpose of optimum design of the future electronic nano-systems. In this chapter, we are to present the definition of various terms and new phenomena in nanoscale thermal transport, and different simulation methods for sub-microscale applications. Concretely, the topics of thermal transport in applications such as heat sinks, hard disks, Joule heating effects, low-dimensional materials, superlattices and microstructures, the metal-oxide semiconductor devices (MOS), the nano-structured carbon materials, carbon nanotubes, graphene and other two-dimensional sheets, solid gases, metal films, and tronco-conical nanowires will be covered.

4.1 Terminology

4.1.1 Phonon

Besides electrons and photons, phonons are the other type of heat carriers in nature. The term "phonon stems" from a Greek word, meaning sound. Long-wavelength or low-frequency phonons give rise to the sound, and shorter-wavelength or higher-frequency phonons (in the range 100 GHz–10 THz) are the origin of the thermal capacity of the solids. Phonons are quanta or units of vibrational energy of the lattice, which can be behaved simultaneously like waves or quasi-particles. Due to the non-rigid relatively strong connections between atoms, the excitation of a group of atoms from their equilibrium state leads to the propagation of vibrational wave pockets through the material.

The quantized behavior of the phonons stems from the existence of a restoring force, which leads to the appearance of a minimum energy state. Based on this fact and due to the free movement of gas molecules, the definition of phonons as heat carriers cannot be extended to gases.

The previously defined phonons are classified into acoustic and optical types based on the distribution of atoms in the lattice and the form of vibration of atoms. Due to the similarities between the propagation of sound in air and the phonons in a lattice consisting of similar atoms, the term "acoustic" is used. Acoustic phonons vibrate along the same direction, while the optical phonons are engaged with the out-of-phase movement of the atoms in the lattices with two or more dissimilar atoms in the primitive cell. The term "optical" comes from the excitation of optical phonons by electromagnetic waves in two-atom dipoles. The speed of propagation of phonons in the material is named as the sound speed or the phonon group velocity, being computed as the mean value of the longitudinal and the transversal sound speeds.

4.1.2 Coherence

Different materials consist of many number of atoms moving randomly in different directions. When an external excitation is received by the material, the collective motion of atoms becomes in harmony with other adjacent phonons. This is the description of coherent (in-phase) phonons in contrary to the incoherent (random-phase) phonons, which appear near interfaces with random distribution of roughness. The wave and the particle natures of phonons are appropriate to describe the motion of coherent and incoherent phonon transports, respectively. A similar definition of coherence can be presented for the motion of electrons, photons, plasmons, and eddies in turbulent flows as well.

4.1.3 Phonon Tunneling

It is found that phonons not only move within the structure of the materials, but also can carry heat through the nanoscale gaps between two materials. This mechanism of heat transport inside the gap between two neighboring lattices is called the phonon tunneling. The phonon tunneling enhances the heat transfer magnitude in the gap, which is also supported by the primary mechanisms including the radiation and the conduction.

4.1.4 Phonon Scattering and Relaxation Times

Scattering means to force a particle, phonon in our case, to deviate from its straight trajectory by one or more paths. Phonons travel inside the material through several mechanisms. These scattering mechanisms are categorized as normal (N) and resistive (R) processes including

- Umklapp (U) phonon–phonon scattering: The name driven from the German origin that means to turn over. This kind of scattering takes place when two or more phonons reach each other. It should be noted that the energy of phonons is always conserved during the U scattering process, but since the phonons can be merged or divided into two or more phonons, the number of phonons is not conserved in this process.
- Phonon-imperfection scattering: The consequence of interaction of phonons with the impurities of the material.
- Boundary–phonon scattering: Becomes important in cryogenic applications and nanoscale structures.
- Phonon–electron scattering: Happens when the material is heavily doped.

The normal processes define a propagation event that transports energy. The resistive scattering is responsible for the extraction of energy and momentum from phonons to an external reservoir.

A relaxation time is defined for each of the abovementioned scattering processes. The relaxation time is the time needed to recover the equilibrium condition after the scattering process occurred. Based on the Matthiessen's rule, the combined scattering rate (the inverse of scattering relaxation time) is the sum of scattering rates of four scattering processes. In addition, it should be noted that the relaxation time is the ratio of phonon's MFP to the group velocity of the heat carriers.

4.1.5 Phonon Engineering

This topic is an active research field especially in the field of nanophononics, which tries to present different approaches of controlling generation, propagation, transportation, scattering, and interaction of phonons with the aim of constructing down-scaled materials with a specific thermal behavior.

4.1.6 Brillouin Zone

Atoms in the lattice excluding microstructures such as impurities have a periodic arrangement. This way we are allowed to reproduce the entire lattice by moving a unit cell or the primitive cell in different directions. The unit cell concept is something similar to the concept of Voronoi cells in mathematics which partitions a plane into distinct regions. This means that in order to investigate the thermal behavior of the whole crystal, we may focus on the unit cell as a representor of the material behavior. The Brillouin zone is equivalent to the momentum space of the unit cell (in the real reciprocal space) obtained by the Fourier transformation and demonstrates the allowed energy regions. The first Brillouin zone consists of points in the real lattice which are located inside the unit cell. Other higher-order Brillouin zones include points in neighboring cells with a larger distance from the main unit cell.

4.1.7 Phonon Photon Interaction

The Brillouin scattering is a process which encounters the inelastic interaction of photons of incoming light with the phonon waves in a host lattice. The photon may lose or gain energy from interaction with phonons in the solid material. The change in the energy of the photon is equal to the energy of the released or absorbed phonon. Another photon scattering process is the Raman scattering. When a photon is scattered from the interface of a lattice, its energy is absorbed by the vibrational behavior of the lattice. For solid materials, the Raman scattering can be employed to detect high-frequency phonons based on the concept of inelastic light scattering by phonons.

4.1.8 Relaxon

Relaxon is a recently proposed concept by Cepellotti and Marzari [1] to reformulate the problem of thermal transport. It is claimed that the phonon description of thermal transport is a limit of a more general relaxon theory in which a new set of vibrational modes called as the relaxons are the superposition of phonon modes. They declared that in cases such as the two-dimensional or the layered materials at room temperature or the three-dimensional crystals at cryogenic temperatures, the phonons cannot be identified as the heat carriers, when most scattering events conserve momentum and do not dissipate heat flux.

4.1.9 Thermal Knudsen Number

The key dimensionless parameter to classify the heat transport in nanoscale systems is the thermal Knudsen number, which is defined as

$$Kn = \frac{\lambda}{L} = \frac{3k}{CvL}, \qquad (4.1)$$

where λ and L are the mean-free-path of heat carriers and the characteristic length of interest, respectively. k is the thermal conductivity, C is the specific heat per unit volume, and v is the average group velocity of phonons. The mean-free-path (MFP) of phonons is the mean traveling distance between two consecutive phonon–phonon scattering processes. The classical Fourier's law is valid at the continuum regime of low Knudsen numbers. By increasing the Knudsen number, the boundary–phonon scattering gradually becomes important, and the no-temperature jump condition fails. At higher Knudsen numbers, for the characteristic length of the geometry in the range 1 mm–1 nm, the continuum assumption breaks down and the particle theory of the Boltzmann transport equation becomes valid. Further, for sub-10-nm structures the molecular dynamics approach is an appropriate choice as a modeling tool. Contemplating even smaller geometries corresponding to higher Kn, the solution of the Schrodinger equation should be used as an ab initio approach.

4.1.10 Bulk/Film Properties

In general, most of the thermal properties of the matter such as the thermal conductivity are thermodynamic properties and can be computed as a function of pressure and temperature. When the size of phononic structure becomes smaller than the phonon MFP, the finite-size effects gradually appear [2]. Hence, the thermal behavior of nanoscale films presents a size-dependent effect. For instance, the film thermal conductivity is lower than its bulk value, and decreases as the thickness of the film shrinks.

4.1.11 Ballistic and Diffusive Regimes

The ballistic and the diffusive regimes of the phonon transport are two limits of heat transfer in phononic structures. The pure diffusive behavior governed by the Fourier's law is valid just for low-Kn geometries and exhibits the heat transfer paradox, the infinite velocity of heat propagation. The pure ballistic limit occurs at high-Kn structures in which the phonons carry energy without internal scattering and altering the direction of motion happens only upon boundary–phonon scattering. The position of phonons in the ballistic regime can be formulated by the product of their speed and time. At the intermediate transitional Knudsen numbers, the phonons experience a combination of ballistic and diffusive transports [3].

4.1.12 Phonon Hydrodynamic Regime

The phonon hydrodynamic regime including the diffuse boundary scattering and the normal scattering is a regime between the ballistic regime including the diffuse boundary scattering without the phonon scattering and the diffusion regime including the umklapp scattering. If the dominant mechanism of phonon scattering is the momentum-conserving normal scattering, a process similar to what happens in flow of a fluid like the Poiseuille flow is observed, sometimes called the phonon Poiseuille flow or the Poiseuille heat conduction. In this regime, the phonons develop a non-zero drift velocity during the heat transport. The phonon hydrodynamic regime takes place in low-temperature applications as well as the low-dimensional materials. In addition, in this regime, a phenomenon similar to the Knudsen minimum which happens in high-Knudsen gas flows in nanochannels is detectable.

4.1.13 Non-Fourier/Anti-Fourier Behaviors

There is a minus sign in the traditional Fourier's law, which implies that heat travels from high-temperature to the low-temperature region. When the reverse of this trend takes place, the heat flows from the cold zone towards the hot zone. An anti-Fourier behavior is reported in Ref. [4] for the case of nanoscale heat transport from a hotspot in a conducting medium. Another report of the anti-Fourier behavior is about the non-equilibrium gas flows through micro/nanochannels [5]. They found that three distinct nanoscale heat transport regimes of complete hot-to-cold, the entire anti-Fourier, and the localized anti-Fourier exist.

Also the Fourier's law states that the relation between the heat flux vector and the temperature gradient is linear. Any other constitutive relation which modifies this relation is a type of non-Fourier models. By increasing the Knudsen number and being far beyond the scope of the thermodynamic equilibrium, the Fourier's law fails to accurately predict the thermal behavior of the material, and the non-Fourier or maybe the anti-Fourier methods should be replaced.

4.1.14 Lévy Flight: Fractal Superdiffusion of Heat

The normal diffusion of heat generates a process which is governed by the Brownian motion of the phonons. The Brownian motion in any stochastic system is the random and irregular movement of the particles. The Lévy dynamics describes a quasi-ballistic motion, which is defined based on the random walk procedure, and generalizes the Brownian motion by means of the fractal dimension parameter (α) in the framework of the fractional calculus [6]. This process is called the superdiffusion of heat based on the Lévy flight concept. The value of α varies between 1 and 2 for pure ballistic and pure Fourier diffusion processes, respectively.

In systems involving small length-scales relative to the phonon mean-free-path, the normal diffusion of heat originating from the Brownian motion gradually vanishes. The generalized governing differential equation to obtain the temperature distribution based on the Lévy flight is

$$\frac{\partial T}{\partial t} = D \frac{\partial^\alpha T}{\partial t^\alpha} \qquad (4.2)$$

where D is the thermal diffusivity, and α in the term of Riesz fractional derivative is the fractal space dimension, which is equal to 2 and 1.6 for the pure Brownian motion and the Lévy flight, respectively. Moreover, the self-affinity nature of the Lévy process leads to the generation of fractal structures in nanoscale thermal transport [7,8].

4.1.15   Thermal Metamaterials

Metamaterials are tunable manipulated materials with artificial structures, which possess some specific thermal characteristics which cannot be seen in common daily life applications. Some of mostly used thermal metamaterials are as follows:

- Thermal cloak: Eliminates the thermal gradient in a specific region and molds the flow of heat around a zone.
- Thermal diode or thermal rectifier: Blocks the flow of heat in one direction. The thermal rectification happens when the heat flows in one direction easier than the opposite direction.
- Thermal rotator: Changes the direction of temperature gradient in a zone.
- Thermal concentrator: Maps the heat transfer from a larger zone to a smaller region.
- Thermal camouflage: Hides the existence of temperature gradient in a region.

Other thermal metamaterials are the thermal shield, the thermal diffuser, and the thermal inverter.

4.1.16   Kapitza Resistance

The Kapitza resistance is defined to measure the thermal interface resistance between a pair of solids or a solid and a liquid. When a phonon tries to pass across an interface, the phonon scattering takes place and the temperature discontinuity appears along the interface as a non-Euclidean surface. The interfacial effects are more dominant in nanoscale geometries since the interface has a great influence on thermal behavior of the bulk material.

4.1.17   Second Sound

Below the lambda point, the superfluidity characteristics appear including zero viscosity, the second velocity of sound, and very high thermal conductivity. The second sound regime originates from fluctuations in the density of phonons. The second sound becomes important in applications such as superfluids, cryogenic liquid helium, heat pulses, and dielectrics [9].

Similar to the wave-like motion of the pressure waves in compressible flows, the second sound as a quantum phenomenon demonstrates the appearance of temperature

waves traveling with a different velocity smaller than the sound speed. So, this phenomenon is called the second sound. From the physical viewpoint, the second sound is observed in the phonon hydrodynamic regime when the dominant phonon–phonon collisions are momentum-conserving. So, the Umklapp phonon–phonon scattering which is a momentum-destroying heat dissipating mechanism is negligible in this regime.

4.1.18   Ab Initio/First Principal Simulation

Simulation methods such as the molecular dynamics or the density functional theory (DFT)-based methods are said to be from first principal or ab initio when the process does not include any extra assumptions rather than the basic laws of physics.

4.2   What is New in Nanoscale?

By decreasing the size of the structure down to the mean-free-path of the heat carriers, the classical theory of Fourier's law loses its accuracy. During the transition from the Fourier to the non-Fourier behaviors in nanoscale heat conduction, some new phenomena occur. In this section, we try to gather together the new aspects of the nanoscale heat transfer, which may be used as a checklist for researchers.

4.2.1   Non-locality in Time

The corrected form of the Fourier's law, known as the dual-phase-lag (DPL) model, reads

$$\vec{q}(t + \tau_q) = -k\nabla T(t + \tau_T), \qquad (4.3)$$

where three characteristic times, including the time in which the energy balance is written (t), the time that the temperature-gradient occurs $(t + \tau_T)$, and the time when the heat flux vector occurs $(t + \tau q)$, are connected. This means that the non-local flow of heat at a specific time is related to the thermal field at two other times [10,11].

4.2.2   Non-locality in Space

A modified convolution-type constitutive law is presented to capture the non-local behavior of the thermal transport in the non-diffusive regime [12].

$$q(x,t) = -\int_0^t dt' \int_{-\infty}^{\infty} dx' k^*(x - x', t - t') \frac{\partial T}{\partial x}(x,t) \quad (4.4)$$

where k^* is the generalized conductivity kernel and is the representative of the spatial and the temporal non-locality of heat flow that acts as an intrinsic material property of the conducting medium. This relation connects the heat flux in a point at the specified time to the temperature-gradient in other times and locations.

Mahan and Claro [13] derived a non-local theory of thermal conductivity for the case in which the heat is carried by the phonons when the phonon mean-free-path is long compared to the distance scale of the variation in temperature gradient. Koh et al. [14] developed an extension of the non-local theory of Claro and Mahan for heat conduction with large thermal gradients and heat fluxes under high-frequency fields. Chen [3] derived a non-local ballistic-diffusive heat equation by decomposing the temperature field to the ballistic and diffusive components. Alvarez et al. [15] expressed the non-local phonon hydrodynamics of the form Guyer–Krumhansl equation as

$$\tau \vec{\dot{q}} + \vec{q} = -k_0 \nabla T + \lambda^2 \nabla^2 \vec{q}. \tag{4.5}$$

where the last term in RHS characterizes the non-local effects.

4.2.3 Boundary Phonon Scattering

In order to capture the phonon–boundary scattering, the following Robin-type boundary condition is presented such that it relates the jumped temperature with the gradient of the temperature on the boundaries.

$$\theta_s - \theta_w = -\alpha Kn \left(\frac{\partial \theta}{\partial n^*} \right)_w \tag{4.6}$$

where θ_s is the dimensionless jumped temperature of the phonons near the boundary, θ_w is the wall temperature, n^* is the normalized wall's outward unit vector, and α is a tunable coefficient. The values of α for 1D [16], 2D [17], 3D [18] nanoscale silicon substrate and the solid argon [19] in the framework of the lagging models have been calculated.

4.2.4 Heat Transfer Paradox: The Wave-Nature of Heat Transport

The parabolic/diffusive nature of the Fourier's law results in the infinite transient heat propagation speed in a continuum media. This fact leads to the generation of unphysical results for the nanoscale and ultra-fast heat transport phenomena, when the continuum mechanics principle fails and the characteristic time is about pico/femto-second. This is called the heat transfer paradox (HTP). So, the use of a modified non-Fourier theory or an atomistic tool is essential for the nanoscale thermal transport modeling in order to add the wave-nature to the results and to eliminate the paradox of heat transfer. Based on the hyperbolic non-Fourier model with the required wave behavior, the infinite speed of sound reduces to the finite value of $\sqrt{\frac{k}{C\tau}}$ where τ is the relaxation time [11].

4.2.5 Rise of Film Properties

When the length-scale of a film is in the order of the magnitude of the phonon mean-free-path, the finite-size effects rise while changing the thermal properties [2]. So, the effective thermal properties of the film can be estimated based on the following formulae:

$$\frac{1}{k_{\text{eff}}} = \frac{1}{k} + \frac{12}{C\nu L}$$
$$\frac{1}{\lambda_{\text{eff}}} = \frac{1}{\lambda} + \frac{1}{L}$$
$$\frac{1}{\nu} = \frac{1}{3} \left(\frac{1}{\nu_L} + \frac{2}{\nu_T} \right) \tag{4.7}$$

where k_{eff}, λ_{eff} are the effective thermal conductivity and the effective mean-free-path, respectively. C is the volumetric heat capacity, ν is the average velocity of sound in the solid, and ν_T and ν_L are the transversal and the longitudinal sound speeds, in the order given.

4.2.6 Phonon-Defect Scattering

When defects are involved, the phonons will be strongly scattered. According to the dimensionality of the defects, they can be classified into four categories: point defects, line defects, planar defects, and volume defects. Usually, defects do not change the group velocity of phonons. However, the phonon relaxation time can be reduced greatly, which is the key mechanism of phonon-defect scattering.

However, the phonon-defect scattering can still be divided into two classes according to the detailed mechanism. If there is bond or/and mass imperfection associate with the defect, the relaxation time is reduced by the phonon-defect scattering. If there is no bond breaking associate with the defect, for example, the screw dislocation, the defect reduces the phonon relaxation time by enhancing the phonon anharmonic effect, i.e., phonon–phonon scattering, due to the defect induced lattice strains. Normally, when there is bond and/or mass imperfections, the above two mechanisms appear simultaneously as it will generate lattice strain around the defect.

4.2.7 Phonon Particle vs. Phonon Wave Effects

Phonons are quantized quasi-particles of lattice vibrations. It possesses the wave-particle duality. When talking about the scattering of phonons, we have actually assumed that the phonon behaves as particles. In this case, it can collide with defects, surfaces, and other phonons. The phonon particle effect is also the basis of many conventional methods in solving the phonon properties, for example, the Boltzmann transport equation.

Except behavior as particles, phonons also show the wave effect, especially for low-frequency phonons whose wavelength is large. Both the phonon particle and the wave effects can be applied to manipulate the phonon transport. Generally speaking, the particle effect can be used to engineer the high-frequency phonon transport benefited from their relatively short wavelength. For the engineering of the low-frequency phonons, the phonon wave effect is more advanced. To apply the wave effect, one needs to design the surface resonances, which can produce resonant modes.

The resonant phonons can interact with the low-frequency propagating modes inside the material due to the anti-cross effect.

4.3 Methodology

There are three common approaches for heat transfer modeling in nanoscale geometries: microscale, mesoscale, and macroscale approaches. Due to the molecular nature of the atomistic methods, the results of such models are accurate for sub-mesoscale simulations. On the other hand, the existence of serious limitations such as high computational costs, involving unknown parameters, and the mathematical and the numerical complexities have made micro- and mesoscale methods too sophisticated for the general-purpose modelings.

So, the traditional continuum models based on the variations of the Fourier's law have been widely replaced the atomistic approaches. Even though, the Fourier-based models are successfully applied to many applications, unfortunately, the validity of the Fourier's law is questionable for the nanoscale transport problems. Consequently, many non-Fourier models have been developed so far, to take into account the anomalies of the Fourier-based models such as the boundary–phonon scattering phenomenon and the well-known heat transfer paradox (HTP) which presumes the infinite speed for heat carriers.

However, the development of the atomistic methods is so significant, since the exact data obtained from such methods helps out the researchers offer rudimentary computationally cheap alternative models. These methods include the direct solution of the Boltzmann transport equation (BTE), the Monte-Carlo technique, the molecular dynamics approach, and the thermal lattice Boltzmann method.

4.3.1 Microscale Modeling: Molecular Level

The molecular dynamics (MD) method has been widely used to predict the thermal transport of nanostructures. The key advantage of the MD technique is that it does not need any assumption on scattering types and involves all orders of anharmonicity. Besides, the interface resistances can be easily obtained with MD. However, MD is classical and does not include any quantum effect. As a result, it cannot be used to evaluate the thermal conductivity at low temperatures.

Thermal Conductivity from Equilibrium Molecular Dynamics Simulation

In equilibrium molecular dynamics simulations, we use the fluctuation-dissipation theorem from linear response theory to provide the connection between the energy dissipation in irreversible processes and the thermal fluctuations in equilibrium [20]. The net flow of heat in a solid, given by the heat current vector **J**, fluctuates around zero at equilibrium.

In the Green–Kubo (GK) method, the thermal conductivity is related to how long it takes these fluctuations to dissipate. In the case of an isotropic material, the conductivity is defined by [20]

$$\kappa = \frac{1}{k_B V T^2} \int_0^\infty \langle J_x(t) J_x(0) \rangle$$
$$= \frac{1}{3 k_B V T^2} \int_0^\infty \langle J(t) \cdot J(0) \rangle \quad (4.8)$$

where V represents the volume of the simulation cell, t is the time, and $J_x(t) J_x(0)$ and $J(t) \cdot J(0)$ are the heat current autocorrelation functions (HCACF) in the x direction and all directions, respectively. In crystals where the fluctuations have long life times (i.e., the mean-free-path of phonons is large), the HCACF decays slowly. The thermal conductivity is related to the integral of the HCACF and is accordingly large. In materials such as amorphous solids, where the mean-free-path of phonons is small, thermal fluctuations are quickly damped, leading to a small integral of the HCACF and a low thermal conductivity [22].

In real computational procedures, instead of integrating up to infinity in Eq. 4.8, the upper limit is a finite but long enough time period that captures the correct statistics. The continuous integral is also replaced by a discrete summation. To remove the arbitrariness on the choice of the upper limit, different methods have been proposed in literature [21–24]. According to the Cattaneo–Vernotte's relation [25,26], Volz et al. derived the time autocorrelation function of the heat flux as [21]

$$\langle J(t) \cdot J(0) \rangle = \langle J(0) \cdot J(0) \rangle \exp(-t/\tau) \quad (4.9)$$

A similar exponential function was used by Li et al. to fit the heat flux autocorrelation function:

$$\frac{\langle J(t) \cdot J(0) \rangle}{3} = g \exp(-t/\tau) \quad (4.10)$$

However, in Li's approach, the single exponential function is not used to fit the whole HCACF curve but only the range $[t_1, t_2]$. This approach is used to determine the tail contribution of HCACF. Instead of using a single exponential function to fit the HCACF in the full-time interval, Che et al. [23] proposed a double exponential function to fit the whole HCACF curve. This approach has also been used by McGaughey et al. [22] for solid argon but with different explanations. The fitting function reads as

$$\frac{\langle J(t) \cdot J(0) \rangle}{3} = A_{sh} \exp(-t/\tau_{sh}) + A_{lg} \exp(-t/\tau_{lg}) \quad (4.11)$$

where the subscripts sh and lg refer to short range and long range, respectively.

Thermal Conductivity from Non-equilibrium Molecular Dynamics Simulation

The non-equilibrium molecular dynamics, also known as the direct method, extracts the thermal conductivity from the

Fourier's law. In this method, one needs to impose a one-dimensional temperature gradient on a simulation cell by allowing thermal power exchange between the heat source and sink and measures the resulting heat flux. The thermal conductivity is then obtained as the ratio of the heat flux and the temperature gradient.

An alternative, but equivalent way consists to induce a heat flux and to measure the resulting temperature gradient. In both cases, the system is first allowed to reach a steady state, after which long simulations are conducted allowing to obtain correct statistical measurements. The non-equilibrium molecular dynamics (NEMD) method is often the method of choice for studies of nanomaterials while for bulk thermal conductivity, particularly high-conductivity materials, the equilibrium method is typically preferred due to less severe size effects [27].

In NEMD simulations, the finite-size effects arise when the length of simulation cell L is not significantly longer than the phonon mean-free-path. This is understood to be a result of the scattering that occurs at the interfaces with the heat source and sink. As a result, the phonon mean-free-path is limited by the system size. To eliminate the size effect, Schelling et al. [28] proposed a method based on the Matthiessen's rule to determine the effective mean-free-path Λ_{eff} when $L \sim \Lambda_{\infty}$, where Λ_{∞} is the mean-free-path for an infinite system. The effective mean-free-path is obtained by the following relation.

$$\frac{1}{\Lambda_{\text{eff}}} = \frac{1}{\Lambda_{\infty}} + \frac{4}{L} \qquad (4.12)$$

Here, the factor of 4 accounts for the fact that as phonons travel along the length of the simulation cell from the source to the sink, its average distance since the last scattering event should be $L/4$. In kinetic theory, the thermal conductivity is given as $\kappa = \frac{1}{3} C_v v \Lambda$, where C_v and v are the specific heat and the phonon group velocity. Combing with Eq. 4.12, the effective thermal conductivity is obtained:

$$\frac{1}{\kappa_{\text{eff}}} = \frac{1}{\kappa_{\infty}} + \frac{12}{C_v v}\frac{1}{L} \qquad (4.13)$$

Eq. 4.13 suggests that a plot of $1/\kappa$ vs. $1/L$ should be linear and that the thermal conductivity of an infinite system can be obtained by extrapolating to $1/L = 0$.

Challenges of Molecular Dynamic Simulations

The molecular dynamics has been widely used to predict various properties in bulk and nanomaterials. It has been regarded as a powerful tool to assist the experimental designs and understand the mechanisms at the atomic scales. Despite its successful applications, it also suffers from several important and fundamental challenges.

The first challenge comes from the simulation size and the computational cost. Nowadays, with the assistance of super-computing, the usual MD simulations contain hundreds to millions of atoms, corresponding to a cubic box with the side length less than 50 nm. However, in some cases, the simulation box could exceed hundreds of nanometers. For example, if one needs to study the grain boundary properties, at least tens of grains should be contained in a simulation box and each grain could have a size of 50 nm in diameter.

The simulation time normally ranges from several picoseconds to hundreds of nanoseconds, which is far from enough in some cases. Examples can be listed from the studies of phase transitions. Most of the phase transitions take place from milliseconds to minutes or even hours in bulk state, which is far beyond the timescales that can be achieved to date. The challenges from the simulation size and time are important but not fundamental and can be solved with the development of future supercomputing.

The most fundamental challenges are that the simulation is classical; that is, the trajectories of the atoms are integrated according to the classical Newtonian mechanics without considering quantum effect. This is usually justified by stating that for most of the elements at room temperature, the atomic mass is sufficiently large and the de Broglie wavelength is considerably smaller than the interatomic distance. This way, the atoms can be treated classically. Nevertheless, due to the classical assumptions, phonons follow the Boltzmann distributions, i.e., $f = A e^{E_i/k_B T}$ instead of the Bose–Einstein distributions, and as a result, the energy equipartition principle applies. This means that at any temperature T, all phonon modes are fully populated and each phonon mode has the energy $k_B T$. This is valid at high temperatures ($T >> T_D$, T_D is the Debye temperature), at which the Bose–Einstein distribution can be simplified to the Boltzmann distribution.

At the temperatures below the Debye temperature, the high-frequency modes will be at least partially unoccupied in contrast to the MD simulations. Consequently, the MD simulations are usually conducted above the Debye temperature (or approximately above $T_D/2$). To overcome this fundamental shortcoming, different quantum corrections in the MD simulations have been tried including a temperature correction [29] and a quantum thermal bath [30].

4.3.2 Mesoscale Modeling: The Boltzmann Equation

Solving the phonon Boltzmann equation is one of the most accurate options to model the heat transfer in nanoscale problems. However, the Boltzmann equation is difficult to be solved in general due to its seven dimensionalities and its complex collision integral. However, the boundary treatment for direct solution of the BTE is irregular especially for the complex curved boundaries. The celebrated phonon Boltzmann transport equation under the single model relaxation time (SMRT) approximation is

$$\frac{\partial f_{\omega}}{\partial t} + \vec{v}_{\omega} \cdot \nabla f_{\omega} = -\frac{f_{\omega} - f_{\omega}^0}{\tau_{\omega}} + g_{e-ph}, \qquad (4.14)$$

where f is the distribution function, f^0 is the equilibrium distribution function of Bose–Einstein, $\vec{v}$ is the heat-carrier group velocity, ω is the angular frequency of heat carriers, and τ is the heat-carrier relaxation time. The term "g_{e-ph}" represents the phonon generation rate due to the electron–phonon scattering [4]. The equilibrium distribution function is given by the Bose–Einstein distribution:

$$f_\omega^0 = \left(e^{\frac{\hbar\omega}{k_b T}} - 1\right)^{-1} \quad (4.15)$$

where $\hbar$ is the Plank's constant divided by 2π, and k_b is the Boltzmann's constant. Integrating the Boltzmann equation over frequency, the energy density of phonons reads

$$e(T) = \int f_\omega \hbar\omega D(\omega) d\omega \quad (4.16)$$

where $D(\omega)$ is the phonon density of state per unit volume. The gray model assumes that all phonons have a same group velocity and relaxation time. So, the subscript ω can be eliminated from the relaxation time and the velocity vector. Therefore, the energy density form of the Boltzmann equation under the gray assumption is

$$\frac{\partial e}{\partial t} + \vec{v} \cdot \nabla e = -\frac{e - e^0}{\tau} + q \quad (4.17)$$

where e^0 and q are the equilibrium phonon energy density and the internal heat generation rate, respectively.

Finite Volume Lattice Boltzmann Method

Equation 4.17 is integrated over finite volumes of the lattice and the velocity of the phonons is discretized along m directions. The finite-volume version of the lattice Boltzmann method is useful to deal with complex geometries and to extend the directional discretization into higher values. Using limited number of directions in velocity discretization leads to generation of unrealistic results. This fact is due to the larger traveling distances for particles on diagonal directions than the distance traveled by particles in the main directions.

For 2D simulations, the face is divided to equal angles; furthermore, equal weighting factors are selected for all directions. For 3D simulations, spatial directions and weighting factors could be obtained using the Legendre equal-weight (PN-EW) quadrature set, which has been proposed by Longoni [31]. The integrated form of the Boltzmann transport equation over finite volumes is

$$\int_{\text{Cell}} \left(\frac{\partial e_\alpha}{\partial t} + \vec{v}_\alpha \cdot \nabla e_\alpha + \frac{e_\alpha - e_\alpha^0}{\tau} - \omega_\alpha q\right) d\forall = 0,$$
$$\alpha = 1, ..., m \quad (4.18)$$

where $e = \Sigma_{\alpha=1}^m e_\alpha$ and ω are the weighting function in each direction. The time derivative term is discretized based on the explicit time advancement scheme. Using the Green's theorem, the final form of the finite-volume lattice Boltzmann equation is

$$e_\alpha^{n+1} = e_\alpha^n - \frac{\delta t}{\forall_{\text{Cell}}} \Sigma_{\text{face}}(\vec{v}_\alpha \cdot \vec{N}_{\text{face}}) e_{\alpha,\text{face}}$$
$$- \frac{\delta t(e_\alpha - e_\alpha^0)}{\tau} + \delta t \omega_\alpha q, \quad (4.19)$$

where n is the time-stet counter, δt is the time-stem size, $\forall_{Cell}$ is the cell volume, "face" denotes the interface between volumes, and $\vec{N}$ is the normal outward vector of the face.

Monte-Carlo Approach

The Monte-Carlo (MC) technique is a method based on the stochastic sampling for solving deterministic problems. The procedure is widely used in nanoscience for the investigation of the electron [32], photon [33], and phonon [34] transports. The MC method is an ideal choice for modelling large dimension problems and complex geometries as it linearly scales with the number of dimensions. Furthermore, the approach is appropriate for modeling phonon transport in combination with the electron transport model.

Phonons are the main heat carriers in many materials such as silicon and two-dimensional systems. The intransitive and important parameters for the phonon Monte-Carlo (PMC) simulation are the phonon dispersion curves, which demonstrate the relation between the frequency ω and the wave vector $\vec{q}$ in the first Brillouin zone (1BZ). These curves reflect the symmetry of the underlying lattice. From the dispersion relation, the properties such as the group velocity and the density of states can be extracted and used in the solution procedure.

In general, phonons are belonged to four different branches of the dispersion relations named as the longitudinal acoustical (LA) and optical (LO), and the transitional acoustical (TA) and optical (TO). Also, low-dimensional materials consist of two more additional bifurcations of Flurexin acoustical (ZA) and optical (ZO). Further, for 2D nanomaterials the contribution from the optical branches is negligible in heat conduction at temperatures up to 600 K and only acoustic branches are taken into account. Besides, the dispersion relation can be considered as the isotropic in which the value of the frequency is direction independent for the materials with the symmetry higher than C_6 [35] or the non-isotropic cases [36]. In case of the isotropic behavior, the dispersion can be modeled with the quadratic fitting to the full dispersion relations via $\omega_b = c_b k^2 + v_b k + \omega_0$.

The MC solution of the BTE assigns three wave vector and three position vector components to each phonon considered as statistical samples. The steps in MC simulation are briefly expressed in the following. At the first stage, the overall number of the phonons per unit volume, N_{actual}, is calculated using the following equation:

$$N_{\text{actual}} = \sum_p \int \left[\exp\left(\frac{\hbar\omega}{k_B T}\right) - 1\right]^{-1} D(\omega_{o,i}) d\omega. \quad (4.20)$$

In this relation, $D(\omega_{o,i})$ is the phonon density of states (PDOS), which generally is computed from the formula $D(\omega_{o,i})d\omega = \frac{1}{2\pi^n}\int_{\omega_q \in \omega + d\omega} d^n q. n$ is the representative of the dimension of the system. For example, the PDOS in two dimension is calculated as $D(\omega_{o,i}) = \frac{1}{2\pi^2}|\frac{d|K|}{d\omega}|$. The integration in Eq. 4.20 is for each phonon polarization over the whole frequency space and is numerically performed by discretizing the frequency space.

The maximum frequency of each dispersion curve, $\omega_{\max}$, is driven, and the frequency interval between 0 and $\omega_{\max}$ is divided into N_{int} intervals. $\omega_{o,i}$ is the central frequency of the bandwidth of the i-th spectral bin. The actual number of phonons calculated from the Eq. 4.20 is mostly a very large number which makes the simulation computationally expensive. Hence, introducing the scaling or a weighting factor, $W = \frac{N_{\text{Actual}}}{N_{\text{prescribed}}}$, the number of simulated phonons is decreased. $N_{\text{prescribed}}$ is the number of phonons (stochastic samples) actually initialized/emitted into the system. In other words, each stochastic sample is the representative of an ensemble of W phonons which should be initialized. As the number of phonons varies for each type of the dispersion relation curve, their W factor is better to be taken differently.

In initializing, first the position and the direction is attributed to each sample. To give an instance, in 3D simulations, the position of each stochastic sample is considered as $\mathbf{r} = R_1 W \hat{x} + R_2 L \hat{y} + R_3 D \hat{z}$ where R_1, R_2 and R_3 satisfying $0 < R_1, R_2, R_3 < 1$ are random numbers. In the second place, the frequency of the phonon ensembles is sampled by calculating the number of phonons in the i-th interval of frequency, ω_i while a normalized frequency cumulative density function, CDF, is constructed as

$$F_i = \frac{\sum_{j=1}^{i} N_j}{\sum_{j=1}^{N_{\text{int}}} N_j}. \qquad (4.21)$$

Since F_i shows the probability of finding a phonon with the frequency less than $\omega_{o,i} + \Delta\omega_i$, the probability of finding a phonon in the i-th frequency interval will be $F_i - F_{i-1}$. Contemplating a random number R_F and by the bisection algorithm, the i-th frequency bin fulfilling the relation $F_{i-1} < R_F < F_i$ the frequency is calculated as $\omega = \omega_{o,i} + (2R_F - 1)\frac{\Delta\omega_i}{2}$. The probability of belonging the frequency to the polarization p of the branch b, in the i-th spectral frequency bin, is given by

$$P_{i,b} = \frac{N_{i,b}}{N_i} + P_{i,p-1}. \qquad (4.22)$$

$N_{i,b}$ is the number of phonons in branch b. Also, N_i is the total number of phonons in the i-th spectral frequency bin corresponding to the all branches. The branch b is prescribed to the polarization number p of 1,2, etc. If $P_{i-1} < R_P < P_i$ with R_P being the randomly chosen number, then the phonon is belonged to the polarization p or branch b. The group velocity of the phonon is then calculated via $v_g = \nabla_k \omega$. The direction is also attributed to the phonon samples using a single random number of R. For instance, in three dimensions, the direction of the phonon is defined as $\hat{s} = \sin\theta\cos\phi\hat{x} + \sin\theta\sin\phi\hat{y} + \cos\theta\hat{z}$ with $\phi = 2\pi R$ and $\cos\theta = 2R - 1$ [37]).

The initialized phonons, then, move from one point to another experiencing impact with other phonons and boundaries. The drift and scattering events are treated separately in the MC technique. The frequency range of the sampled phonon may change while deviating consequently resulting in redistribution of the energy and the temperature in the computational domain. By summing the energy of the all phonons of each cell at the end of each drift, the energy of each spatial frequency interval per unit volume, $\tilde{U}_{\text{cell}}$, is calculated. The pseudo-temperature $\tilde{T}$ is computed via the Newton–Raphson method root-finding method,

$$\tilde{U}_{\text{cell}}.W = \sum_{p} \sum_{i=1}^{N_b} \hbar\omega_{o,i} \left[\exp\left(\frac{\hbar\omega_{o,i}}{k_B \tilde{T}}\right) - 1 \right]^{-1} D(\omega_{o,i})\Delta\omega_i. \qquad (4.23)$$

As the name suggests, the $\tilde{T}$ will be found contemplating the thermodynamic equilibrium situation within each frequency interval. On the other hand, each phonon traveling along the direction $\hat{s}$ may strike the boundaries specularly or diffusively determined by the value of specularity, α. Determining the boundaries at which the phonons hit, the behavior of the phonon ensembles after the collision is investigated. The phonons can be scattered by means of lattice imperfections, the boundaries, or interactions with electrons or other phonons (intrinsic scattering).

Each of these scattering mechanisms results in exchanging energy between different lattice wave vectors. The phonon–phonon scattering is the most important mechanism in the usual temperature range (300–600 K). Three-phonon processes can be the normal (N) scattering with both energy and momentum conserved or the Umklapp (U) process, only preserving the energy but conserving the momentum up to a reciprocal lattice vector.

The normal scatterings do not directly result in thermal resistance and take part in the scattering process via redistributing the phonons. On the other hand, the U scatterings are the origin of the material's thermal resistance. Newly, it is established that while the relaxation-time approximation, a simplified description of the scattering dynamics, is adequately accurate for the phonon–phonon scattering rate of 2D nano-systems with almost high temperatures, it cannot perfectly describe heat transport at low temperatures [38].

Accordingly, at high temperatures, the three-phonon interactions are treated through a scattering rate [39]. The Umklapp phonon–phonon scattering rate is computed from the standard general approximation for dielectric crystals:

$$\tau_{b,U}^{-1}(\omega) = \frac{\hbar\gamma_b^2}{\bar{M}\Theta_b v_{s,b}^2} \omega^2 T e^{-\Theta_b/3T}. \qquad (4.24)$$

In the above equation, $\bar{M}$ is the average atomic mass. The phonon–phonon scattering with the role of four or more phonons gets important at temperatures much higher than the Debye temperature. In the Monte-Carlo simulation, the probability of scattering a phonon with scattering rate of τ_b between time t and $t + \Delta t$ is found via

$$P_{\text{scat}} = 1 - \exp\left(\frac{-\Delta t}{\tau}\right). \qquad (4.25)$$

If the probability of scattering, P_{scat}, is greater than the random number R_{scat}, then the scattering occurs. After the phonon scattering, its frequency, branch, and direction will be re-sampled from the cumulative density function. The time-step Δt is usually chosen to be smaller than the minimum scattering rate of the sampled phonons. As the scattering rates are temperature-dependent, the time-step also relies on temperature. It should be mentioned that in order to keep the rate of formation of the phonons equal to its rate of destruction for a specified state, the distribution function, F_{scat}, has to be altered by the probability of scattering after the scattering

$$F_{\text{scat}}(\tilde{T}) = \frac{\sum_{j=1}^{i} N_j(\tilde{T}) \times P_{\text{scat},j}}{\sum_{j=1}^{N_b} N_j(\tilde{T}) \times P_{\text{scat},j}}. \qquad (4.26)$$

Using the above procedure, the Monte-Carlo simulation is proceeded to investigate the heat transport. The Monte-Carlo has significant advantages over other atomistic simulations. Jorgensen et al. [40] presented that the Monte-Carlo calculations are about two to three times more efficient than the molecular dynamics for the purpose of achieving the equilibrium condition. In other words, the computational cost of the molecular dynamics calculations for attaining the same level of convergence spares is found to be 1.6–3.8 times more than that of the Monte-Carlo simulations.

However, designating the appropriate method depends on the situation conditions. For example, the transport properties such as the viscosity coefficient are only determined via the MD techniques as the MC mostly misses an objective definition of time. Conversely, the MC is easier for the study of interfacial phenomena and simulating the systems with varying particle numbers via the grand canonical Monte-Carlo technique [41]. Moreover, the direct simulation Monte-Carlo (DSMC) method uses the probabilistic MC simulation to solve the Boltzmann equation for finite Knudsen number fluid flows [42]. The DSMC method has been used to solve the problems of flows ranging from evaluation of the space shuttle re-entry aerodynamics, to the simulation of nano-electro-mechanical devices.

Despite the abovementioned advantages, the MC method is computationally expensive and is predisposed to statistical variations. There have been some efforts to overcome these disadvantages like the steady-state approach proposed by Randrianalisoa and Bailis [43] to lessen the computational time and memory for the thermal conductivity calculation of thin films. Besides, Hadjiconstantinou et al. [44] presented the variance-reduced stochastic particle simulation method for solving the relaxation-time model of the BTE resulting in an impressive computational cost.

In brief, the stochastic Monte-Carlo method with different procedures has been used so far to predict the thermal conductivity, to model the phonon transport in porous silicon and other transistor candidate compounds, and to simulate the heat generation process in metal-oxide-semiconductor field-effect transistors (MOSFET) [35,45].

4.3.3 Macroscale Modeling: Non-Fourier Models

Due to complexities and expensive computational costs of atomistic methods, the researchers are trying to develop the semi-classical non-Fourier models in the lieu of the Fourier-based simulations [10]. The well-known non-Fourier methods are classified as follows:

- **Single-phase-lag model or Cattaneo–Vernotte (CV) model**: In order to account for the wave behavior of heat flow which is not captured in the Fourier's law, an additional phase-lag is added to the heat flux vector (τ_q). The resulting equation (Eq. 4.27) is a hyperbolic partial differential equation.

$$\vec{q}(\vec{r}, t + \tau_q) = -k\nabla T(\vec{r}, t). \qquad (4.27)$$

- **Dual-phase-lag model**: As an extension of the CV model, the DPL model adds a second phase-lag parameter to the temperature as well (τ_T). The addition of this new parameter (Eq. 4.28) changes the mathematical nature of the equation to combined parabolic-hyperbolic.

$$\vec{q}(\vec{r}, t + \tau_q) = -k\nabla T(\vec{r}, t + \tau_T). \qquad (4.28)$$

The CV and DPL models overcome the contradiction of infinite velocity of heat carriers in the Fourier's law for nanoscale problems.

- **Thermon gas (thermo-mass concept) model [46]**: This idea uses the classical fluid mechanics equations to describe heat transport by defining the equivalent mass of thermal energy called the thermo-mass based on Einstein's mass-energy relation in relative theory. When the thermal inertia is negligible, the thermo-mass equation reduces to the Fourier's law. Based on the Debye state equation, the thermo-mass pressure can be calculated from

$$P_T = \gamma C T \frac{\rho C T}{c^2}, \qquad (4.29)$$

where P_T and γ are thermo-mass pressure and the Gruneisen constant, and ρ, C, and c are the density, the specific heat of the structure, and the speed of light, respectively. The compressible form of the Navier–Stokes equations are

$$\frac{\partial \rho_T}{\partial t} + \nabla.(\rho_T \vec{u}_T) = 0,$$
$$\rho_T \frac{D\vec{u}_T}{Dt} + \nabla P_T + \vec{f}_T = 0. \qquad (4.30)$$

where ρ_T, $\vec{u}_T = \frac{\vec{q}}{\rho C T}$, and $\vec{f}_T$ are, respectively, the density of thermo-mass, the drift velocity, and the friction force exerted on the thermo-mass per unit volume [47]. $\frac{D}{Dt}$ stands for the material derivative operator.

• **Two-temperature/two-step models**: For the nanoscale applications, the lattice and the electron temperatures are not equal. Free electrons of a thin layer of material can absorb the incoming energy and transfer it to the lattice. This mechanism leads to the creation of a relaxation lag between the excitation and the flow of energy. In rapid procedures, the duration of incoming excitation is shorter than the timescale of the problem. So, the process of energy transport is described in two steps. The parabolic and hyperbolic two-step models originate from the Fourier and the CV models, respectively. The heat conduction in a material can be described regarding two distinct temperatures: the conductive and the thermodynamic temperatures. These temperatures can be equal under certain conditions, but in short-timescale problems, they may be different.

The two-step method heat model is described by

$$C\frac{\partial T_e}{\partial t} = -\nabla.\vec{q_e} - G(T_e - T_l) + S$$
$$C_l\frac{\partial T_l}{\partial T_l} = G(T_e - T_l) \qquad (4.31)$$

where the subscripts l and e stand for the lattice and the electron. G and S are electron-lattice coupling parameter and the heating source term. The hyperbolic version of the two-step model will be obtained if the heat flux vector is substituted by Eq. (4.27).

• **Ballistic-diffusive model**: In the ballistic-diffusive equation (BDE), it is assumed that the internal energy of the heat carriers per unit volume equals the sum of the internal energy of the ballistic ($u_b = CT_b$) and the diffusive ($u_d = CT_d$) components. The constitutive relation for the ballistic and the diffusive internal energy and heat fluxes are

$$\tau\frac{\partial q_d}{\partial t} + q_d = -\frac{k}{C}\nabla u_d$$
$$\tau\frac{\partial u_b}{\partial t} + \nabla \cdot \vec{q_b} = -u_b + \dot{q_h} \qquad (4.32)$$

where $\dot{q_h}$ is the heat generation per unit volume. These constitutive relations should be combined with the energy equation to obtain the governing equation [48].

• **Guyer–Krumhansl (GK) equation**: The GK model is an extension of the CV equation to take into account the second sound and the ballistic propagation in solids. The modified constitutive relation in this case is

$$\tau\frac{\partial \vec{q}}{\partial t} + \vec{q} + k\nabla T - \beta_1\nabla^2\vec{q} - \beta_2\nabla \cdot \nabla\vec{q} = 0 \quad (4.33)$$

where β_1 and β_2 are Guyer–Krumhansl coefficients. This constitutive relation should be

combined with the energy equation to obtain the governing equation [49].

In order to implement a new non-Fourier model, we need a set of exact input data. These data may be obtained using experimental approaches or exact atomistic numerical techniques such as the Monte-Carlo, the MD, and the LBM. The non-Fourier models suffer from the lack of a unified model, which can be applied to various nanomaterials in different multi-dimensional geometries. Although the numerical methods necessary to solve the non-Fourier equations are not sophisticated, the basic mathematical framework should be changed accordingly to include the new nanoscale phenomena.

Further Reading

1. Cepellotti, A. and Marzari, N. (2016). Thermal transport in crystals as a kinetic theory of relaxons. *Physical Review X* 6(4):041013.

2. Shomali, Z., Ghazanfarian, J., and Abbassi, A. (2015). Effect of film properties for non-linear DPL model in a nanoscale MOSFET with high-k material: ZrO2/HfO2/La2O3. *Superlattices and Microstructures* 83:699–718.

3. Chen, G., (2001). Ballistic-diffusive heat-conduction equations. *Physical Review Letters* 86:2297.

4. Samian, R.S., Abbassi, A., and Ghazanfarian, J. (2014). Transient conduction simulation of a nanoscale hotspot using finite volume lattice Boltzmann method. *International Journal of Modern Physics C* 25(4):1350103.

5. Balaj, M., Roohi, E., and Mohammadzadeh, A. (2017). Regulation of anti-Fourier heat transfer for non-equilibrium gas flows through micro/nanochannels. *International Journal of Thermal Sciences* 118: 24–39.

6. Dubkov, A.A., Spagnolo, B., and Uchaikin, V.V. (2008). Lévy flight superdiffusion: An introduction. *International Journal of Bifurcation and Chaos* 18:2649–2672.

7. Vermeersch B, Carrete J, Mingo N, Shakouri A. (2015). Superdiffusive heat conduction in semiconductor alloys. I. Theoretical foundations. *Physical Review B* 91(8):085202.

8. Vermeersch, B., Mohammed, A. M., Pernot, G., Koh, Y. R., Shakouri, A. (2015). Superdiffusive heat conduction in semiconductor alloys. II. Truncated Lévy formalism for experimental analysis. *Physical Review B* 91(8):085203.

9. McNelly, T.F., Rogers, S.J., Channin, D.J., Rollefson, R.J., Goubau, W.M., Schmidt, G.E., Krumhansl, J.A., Pohl, R.O. (1970). Heat pulses in NaF: Onset of second sound, *Physical Review Letters* 24:100.

10. Ghazanfarian, J., Shomali, Z., and Abbassi, A. (2015). Macro- to nanoscale heat and mass transfer: The

lagging behavior. *International Journal of Thermophysics* 36:1416–1467.

11. Tzou, D.Y. (2014). *Macro- to Microscale Heat Transfer: The Lagging Behavior*. John Wiley & Sons, Chicester.

12. Vermeersch, B. and Shakouri, A. (2014). Nonlocality in microscale heat conduction. *arXiv preprint arXiv*:1412.6555.

13. Mahan, G.D. and Claro, F. (1988). Nonlocal theory of thermal conductivity. *Physical Review B* 38: 1963.

14. Koh, Y.K., Cahill, D.G., and Sun, B. (2014). Nonlocal theory for heat transport at high frequencies. *Physical Review B* 90:205412.

15. Alvarez, F.X., Jou, D., and Sellitto, A. (2009). Phonon hydrodynamics and phonon boundary scattering in nanosystems. *Journal of Applied Physics* 105:014317.

16. Ghazanfarian, J. and Abbassi, A. (2009). Effect of boundary phonon scattering on Dual-Phase-Lag model to simulate micro-and nano-scale heat conduction. *International Journal of Heat and Mass Transfer* 52:3706–3711.

17. Ghazanfarian, J. and Shomali, Z. (2012). Investigation of dual-phase-lag heat conduction model in a nanoscale metal-oxide-semiconductor field-effect transistor. *International Journal of Heat and Mass Transfer* 55:6231–6237.

18. Shomali Z., Abbassi A., and Ghazanfarian, J. (2016). Development of non-Fourier thermal attitude for three-dimensional and graphene-based MOS devices. *Applied Thermal Engineering* 104: 616–627.

19. Shomali, Z. and Abbassi, A. (2014). Investigation of highly non-linear dual-phase-lag model in nanoscale solid argon with temperature-dependent properties. *International Journal of Thermal Sciences* 83: 56–67.

20. Kubo, R., Toda, M., and N. Hashitsume, 1985. *Statistical Physics*. Springer, Berlin.

21. Volz, S., Saulnier, J.-B., Lallemand, M., Perrin, B., Depondt, P., and Mareschal, M. (1996). Transient fourier-law deviation by molecular dynamics in solid argon, *Physical Review B* 54:340–347.

22. McGaughey, A. and Kaviany, M. (2004). Thermal conductivity decomposition and analysis using molecular dynamics simulations. Part I. Lennard-Jones Argon. *International Journal of Heat and Mass Transfer* 47:1783–1798.

23. Che, J., Cagin, T., Deng, W., and Goddard, W. A. (2000). Thermal conductivity of diamond and related materials from molecular dynamics simulations. *The Journal of Chemical Physics* 113: 6888–6900.

24. Li, J., Porter, L., and Yip, S. (1998). Atomistic modeling of finite-temperature properties of crystalline β-SiC II. Thermal conductivity and effects of point defects. *The Journal of Chemical Physics* 255:139–152.

25. Vernotte, P. (1958). Les paradoxes de la theorie continue de l'equationde la chaleur.*Comptes Rendus de l' Academie des Sciences* 246: 3154–3155.

26. Cattaneo, C. (1958). Sur une forme de l'equation de la chaleur eliminantle paradoxe d'ine propagation instantanee. *Comptes Rendus de l' Academie des Sciences* 247:431–433.

27. Termentzidis, K. and Merabia, S. (2012). Molecular Dynamics Simulations and Thermal Transport at the Nano-Scale. *InTech*.

28. Schelling, P.K., Phillpot, S.R., and Keblinski, P. (2002). Comparison of atomic-level simulation methods for computing thermal conductivity. *Physical Review B* 65:144306.

29. Wang, C.Z., Chan, C.T., and Ho, K.M. (1990). Tight-binding molecular-dynamics study of phonon anharmonic effects in silicon and diamond. *Physical Review B* 42:11276–11283.

30. Dammak, H., Chalopin, Y., Laroche, M., Hayoun, M., and Greffet, J.-J. (2009). Quantum thermal bath for molecular dynamics simulation. *Physical Review Letters* 103:190601.

31. Samian, R.S., Abbassi, A., and Ghazanfarian, J. (2013). Thermal investigation of common 2D FETs and new generation of 3D FETs using Boltzmann transport equation in nanoscale. *International Journal of Modern Physics C* 24(09):1350064.

32. Jacoboni, C. and Reggiani, L. (1983). The Monte Carlo method for the solution of charge transport in semiconductors with applications to covalent materials. *Reviews of modern Physics* 55(3):645.

33. Wang, L., Jacques, S.L., and Zheng, L. (1995). MCM— Monte Carlo modeling of light transport in multilayered tissues. *Computer Methods and Programs in Biomedicine* 47(2):131–146.

34. Peterson, R.B. (1994). Direct simulation of phonon-mediated heat transfer in a Debye crystal. *Journal of Heat Transfer* 116:815.

35. Shomali, Z., Pedar, B., Ghazanfarian, J., and Abbassi, A. (2017). Monte-Carlo parallel simulation of phonon transport for 3D silicon nano-devices. *International Journal of Thermal Sciences* 114: 139–154.

36. Mei, S., Maurer, L.N., Aksamija, Z., and Knezevic, I. (2014). Full-dispersion Monte Carlo simulation of phonon transport in micron-sized graphene nanoribbons. *Journal of Applied Physics* 116:164307.

37. Shomali, Z., Ghazanfarian, J., and Abbassi, A. (2017). 3-D Atomistic Investigation of silicon MOSFETs. In ICHMT Digital Library Online. Begel House Inc.

38. Cepellotti, A. and Marzari, N. (2017). Boltzmann transport in nanostructures as a friction effect. *Nano Letters* 17(8): 4675.

39. Mazumder, S. and Majumdar, A. (2001). Monte Carlo study of phonon transport in solid thin films including dispersion and polarization. *Journal of Heat Transfer* 123:749.

40. Jorgensen, W.L. and Tirado-Rives, J. (1996). Monte Carlo vs molecular dynamics for conformational sampling. *The Journal of Physical Chemistry* 100(34):14508.

41. Adams, D.J. (1975). Grand canonical ensemble Monte Carlo for a Lennard-Jones fluid. *Molecular Physics* 29(1):307–311.

42. Bird, G.A. (1994). *Molecular Gas Dynamics and the Direct Simulation of Gas Flows*. Clarendon, Oxford.

43. Randrianalisoa, J. and Baillis, D. (2008). Monte Carlo simulation of steady-state microscale phonon heat transport. *Journal of Heat Transfer* 130:072404.

44. Hadjiconstantinou, N.G., Radtke, G.A., and Baker, L.L. (2010). On variance-reduced simulations of the Boltzmann transport equation for small-scale heat transfer applications. *Journal of Heat Transfer* 132:112401.

45. Shomali, Z. and Asgari, R. (2018). Effects of low-dimensional material channels on energy consumption of Nano-devices. *International Communications in Heat and Mass Transfer* 94:77.

46. Guo, Z.-Y. (2006). Motion and transfer of thermal mass-thermal mass and thermon mass. *Journal of Engineering Thermophysics* 4:029.

47. Wang, M. and Guo, Z.-Y. (2010). Understanding of temperature and size dependences of effective thermal conductivity of nanotubes. *Physics Letters A* 374:4312–4315.

48. Yang, R., Chen, G., Laroche, M., and Taur, y. (2005). Simulation of nanoscale multidimensional transient heat conduction problems using ballistic-diffusive equations and phonon Boltzmann equation. *Journal of Heat Transfer* 127:298–306.

49. Guyer, R.A. and Krumhansl, J.A. (1966). Solution of the linearized phonon Boltzmann equation. *Physical Review* 148:766.

5

Coulomb Effects and Exotic Charge Transport in Nanostructured Materials

Monique Tie and Al-Amin
Dhirani
University of Toronto

5.1 Introduction

The effects of strong Coulomb interactions between charge carriers as they delocalize can lead to exotic phases of matter and remarkable behaviours (Basov et al. 2011, Dagotto 2005, Georges et al. 1996, Imada et al. 1998, Wirth and Steglich 2016). For example, in transition metal oxides (such as cuprates which can exhibit high T_c superconductivity), the spatially confined nature of d or f orbitals results in strong electron–electron Coulomb repulsion if multiple electrons occupy the same orbital. These materials also afford means, e.g. via doping or application of high pressure, for tuning the degree to which electron delocalizes. When electron–electron interaction dominates, the materials are insulating and when delocalization, i.e. kinetic energy, dominates, the materials are metallic. In the regime where both interactions and delocalization effects are significant, the materials exhibit an exotic "correlated" phase in which charges move in such a manner so as to accommodate their strong mutual repulsion. It is in this remarkable regime that high T_c superconductivity is observed (Lee et al. 2006). Although the exact mechanism underlying this phenomenon remains an open question, it is widely believed that strong, Coulomb-driven electron–electron interactions combined with electron delocalization play critical roles. Another well-known example of a remarkable phenomenon arising from strong electron–electron interactions is the Kondo effect. This effect arises when an unpaired localized electron interacts with delocalized electrons (Kondo 1964, Gruner and Zawadowski

1974). A delocalized electron cannot pair with the localized electron due to strong Coulomb repulsion, but through virtual processes, the electrons can interact strongly, form a singlet and significantly modify the resistance of the system. An example of a material exhibiting this effect is gold doped with iron atoms (Gruner and Zawadowski 1974, Costi et al. 2009).

5.1.1 Importance of Coulomb Phenomena in Isolated Nanoparticle

Strong Coulomb effects are known to be important in nanostructured materials and have led to interesting optoelectronic and electronic behaviours (Likharev 1999, Talapin et al. 2010, Zabet-Khosousi and Dhirani 2008). To illustrate briefly why this is the case, consider the voltage, V_C, needed to charge a metal nanosphere of radius, R, with a quantum of charge, e. By Coulomb's law, the voltage is given by

$$V_C = \frac{e}{4\pi\epsilon_0 R} \tag{5.1}$$

Taking $R \sim 1$ nm $= 10^{-9}$ m, $4\pi\epsilon_0 \sim 10^{-10}$ N^{-1} m^{-2} C^2 and $e \sim 10^{-19}$ C yields $V_C \sim 1$ V. This energy scale is much larger than that of thermal fluctuations, $k_B T \sim 25$ meV at room temperature, where k_B is Boltzmann's constant, and T is absolute temperature. As a result, the energy required to charge the nanosphere must come from the voltage source. Therefore, connecting the sphere to a voltage source and increasing the electric potential of an initially neutral metal nanoparticle (NP) will result in no electrons hopping on until a critical potential is reached, at which point a single electron will hop on. The suppression of charge transfer onto or off the sphere below this threshold is called Coulomb blockade. The energy required to charge the nanosphere with two electrons is similarly large. Increasing the potential further will result in similar thresholds at which integer multiples of e will occupy the metal sphere (Hanna and Tinkham 1991, Likharev 1999, Ralph et al. 1997). These Coulomb phenomena have been observed in isolated nanostructures and can play an important role in electronic properties of nanostructured materials.

5.1.2 Coulomb Charging Barrier for a NP Embedded in a Film

In addition to being observed in individual NPs, Coulomb effects have also been observed in NP assemblies (Talapin et al. 2010, Zabet-Khosousi and Dhirani 2008). For example, various groups (Brust et al. 1995, Fishelson et al. 2001, Jiang et al. 2012, Musick et al. 2000, Trudeau et al. 2003, Zabet-Khosousi et al. 2006) have found that molecularly linked gold NP films, cross-linked with alkanedithiol of various linker lengths and NP sizes, exhibit Arrhenius T behaviour, and some have observed current suppression in the current–voltage (IV) characteristic reminiscent of Coulomb blockade. Activation energies extracted from IV curves and conductance (g) versus T plots have been found to be 10–50 meV.

Measured activation energies varied with both NP size (Snow and Wohltjen 1998) and linker length (Fishelson et al. 2001, Zabet-Khosousi et al. 2006), and were generally in good agreement with a model considering NPs embedded in an ensemble of other NPs. This picture can be simplified, as shown in Figure 5.1, to that of a metal sphere of radius R_{NP} surrounded by an insulating spacer shell of thickness s_M, which, in turn, is surrounded by a conducting shell. Then, the capacitance is

$$C = 4\pi\epsilon_0\epsilon_r \left(\frac{1}{R_{NP}} - \frac{1}{R_{NP} + s_M} \right)^{-1}$$
$$= 4\pi\epsilon_0\epsilon_r \frac{R_{NP}(R_{NP} + s_M)}{s_M} \qquad (5.2)$$

and the charging energy becomes

$$E_C = \frac{e^2}{8\pi\epsilon_0\epsilon_r} \frac{s_M}{R_{NP}(R_{NP} + s_M)} \qquad (5.3)$$

As NP size decreases, E_C increases, as has been observed by many groups including Brust et al. (1995) and Leibowitz et al. (1999). Additionally, as molecule spacing s_M decreases, so does E_C, which also has been observed by various groups (Fishelson et al. 2001, Zabet-Khosousi et al. 2005) studying charge transport in alkanedithiol-linked gold NP films. Wang et al. (2007) varied both gold NP sizes ($d = 2$–4 nm) and alkanedithiol linker length ($s_M = 0.8$–2 nm) and found good agreement with Eq. 5.3.

5.1.3 Electron Delocalization

Interestingly, a number of groups have observed that the picture of a NP surrounded by an insulating shell breaks down when sufficiently short or conjugated spacer molecules are used. Wessels et al. (2004) studied gold NPs ($d = 4$ nm) cross-linked with organic linkers composed of different cores (benzene or cyclohexane) and different metal-binding functional groups (thiol or dithiocarbamate). NP films cross-linked with molecules with a conjugated core (benzene) were observed to have conductivity one order of magnitude higher than films cross-linked with molecules with a nonconjugated core (cyclohexane). In fact, when using 1,4-phenylene-bis(dithiocarbamate) (PBDT), films exhibited metallic-like behaviour (conductivity decreases with increasing temperature). PBDT is 1.1 nm in length, is highly conjugated and bonds strongly with gold due to the dithiocarbamate. Metallic behaviour in NP films have also been observed when using alkanedithiol linkers with ≤ 4 carbons (i.e. ≤ 0.7 nm) (see Figure 5.2).

There are a number of methods to tune film behaviour from the insulating to metallic regimes. Collier et al. drop cast silver NPs onto a water surface. The NPs were suspended in an organic solvent and capped with an organic ligand. When the organic solvent evaporates, a NP film self-assembles on the water surface with the capping ligands serving as inter-particle molecular spacers. Collier et al. (1997) used this approach with silver NPs to

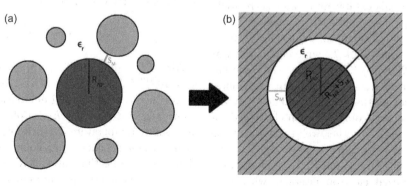

FIGURE 5.1 A model for a metal sphere surrounded by metal spheres in a dielectric medium of dielectric constant ϵ_r (a). The surrounding spheres are approximated as a metal shell (b).

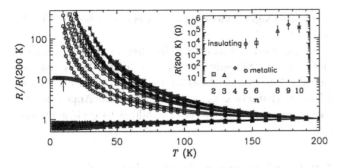

FIGURE 5.2 Normalized resistances of n-alkanedithiol-linked gold NP films of various carbon chain length. Inset: Resistances of films at 200 K of various linker lengths. (Reprinted with permission from Zabet-Khosousi, A. et al. 2006. *Physical Review Letters* 96: 156403. Copyright 2006 American Physical Society.)

form two-dimensional monolayers on a water surface. They further controlled inter-particle distance by compressing the film. They found that at a critical inter-particle distance (<5 angstroms), the films underwent an insulator-to-metal transition and exhibited metallic-like optical behaviour. They used a Mott–Hubbard model (discussed further below) to describe the transition (Mott 1990, Sampaio et al. 2001). In short, the model considers a lattice of atoms with electrons on each lattice site such that the electrons can interact with each other via on-site repulsion (which raises potential energy) and electrons can delocalize via inter-site coupling (which contributes to kinetic energy). Here, silver NPs act as "artificial atoms" occupying the lattice sites. As inter-site coupling increases (i.e. inter-particle spacing decreases), initially localized electrons can become itinerant, resulting in an insulator-to-metal transition.

A transition from insulating to metallic films can also be driven using "layer-by-layer self-assembly". In this method, a functionalized solid substrate is alternately immersed in (i) a solution of NPs which bind to the surface and (ii) a solution of excess bi-functional molecules which replace the capping ligand of the gold NP, leaving one functional end group free. Repeated exposure to solutions of gold NPs and molecular linkers leads to formation of a film containing clusters of molecularly linked gold NPs. Film thickness can be controlled through the number of immersion cycles. Previous low-temperature charge transport studies have shown that, in the limit of thick films, such NP films can be metallic – that is, conductance tends to a nonzero finite value at zero temperature – when sufficiently short or conjugated linkers are used. Examples of such molecular linkers include butanedithiol (Tie et al. 2014), 3-mercaptopropionic acid (Jiang et al. 2012) and PBDT (Wessels et al. 2004). On the other hand, thick films are insulators – that is, conductance tends to zero as temperature goes to zero – when longer or less conjugated linkers such as dodecanedithiol are used (Georges and Kotliar 1992). Focusing on n-alkanedithiols ($C_nS_2H_{2n+2}$) of various lengths as cross-linkers and using gold NPs

($d = 15$ nm), Zabet-Khosousi et al. showed that films were insulating for $n > 5$ and metallic for $n < 5$ (see Figure 5.2). Although such films are not ordered, a number of studies have reported exponential dependence of resistance on n-alkane chain lengths (Dunford et al. 2005, Wessel et al. 2004, Wang et al. 2007, Zabet-Khosousi et al. 2006) and one would expect that varying chain length should change inter-particle coupling in the same spirit as the Mott–Hubbard picture. Furthermore, applying the criterion for critical inter-particle spacing for the Mott–Hubbard metal-insulating transition, Zabet-Khosousi found a value consistent with the observed critical length of $n \sim 5$ for n-alkanedithiol.

An insulator-to-metal transition can also be driven using layer-by-layer self-assembly and percolation. As mentioned, sufficiently short and/or conjugated molecular linkers are observed to allow electrons to delocalize through connected NP pathways in thick films. In the thin film limit, NPs form small isolated molecularly linked clusters on a substrate, and films exhibit thermally activated transport due to Coulomb charging energy barriers associated with charging the clusters with quantized charge carriers during charge transport. As films grow thicker, NP clusters grow and coalesce until at least one sample spanning pathway forms at a so-called "percolation threshold". At this point, electrons remain itinerant even at low temperatures, and the film is just barely metallic. In addition to containing metallic pathways, the film still contains many isolated networks which continue to present Coulomb charging barriers and which contribute to the conductance. As the film becomes thicker still, more isolated networks coalesce to form sample spanning pathways, the film becomes increasingly conducting and metallic pathways eventually dominate.

5.1.4 Strong Electron–Electron Correlations in Nanomaterials

In view of the ability of molecularly linked metal NP films to undergo a percolation-driven insulator-to-metal transition from an insulating phase in which Coulomb barriers inhibit charge transport to a metallic phase in which charges delocalize, the region near the transition is expected to exhibit a phase in which both Coulomb charging and charge delocalization are important, i.e. a phase associated with strongly correlated electron phenomena. It is in such phases that exotic behaviours have been previously observed in transition metal compounds, e.g., suggesting that this regime in nanostructured materials may be a target of opportunity for observing and better understanding such phenomena. This chapter discusses charge transport properties of butanedithiol-linked gold NP films through a percolation-driven metal–insulator transition and, in particular, focuses on the properties near the transition in both the insulating and metallic phases. In these regimes, conductance exhibits signatures of strong Coulomb effects, including electron–electron interactions and a conductance peak (AS resonance) that has previously been observed using transition metal compounds with strongly interacting electrons.

5.2 Observing Signatures of Coulomb Effects in Charge Transport through Nanostructured Materials

5.2.1 Layer-by-Layer Self-Assembled Gold NP Films

Gold NPs can be synthesized using variety of methods. One simple and robust approach is based on the widely used Brust method (Brust et al. 1995). Figure 5.3 shows a UV-Vis spectrum and a transmission electron microscopy (TEM) image of NPs synthesized using this method. The average NP size is 5.3 nm ± 1.1 nm.

Figure 5.4 illustrates self-assembly of NP films for charge transport measurements. Glass slides of desired size (usually ~10 mm × ~10 mm) are cut with a carbide tip scriber from a microscope slide (1 mm thick), cleaned by sonication in methanol and acetone for 10 min each, and then immersed in hot piranha (3:1 H_2SO_4/H_2O_2) for 30 min. The clean slides are then functionalized by immersion into a boiling 40 mM toluene solution of 3-aminopropyldiethoxymethylsilane for 20 min. Gold electrodes of desired pattern are then thermally deposited onto the glass slides using a shadow mask. Indium solder is used

to attach magnet wires onto the electrodes before NP film assembly.

NP films are self-assembled by alternately immersing the slides in Au NP solution for 1–4 h and linker solution (usually 1,4-butanedithiol) for 10–20 min. Figure 5.5 shows UV-V is characterization of Au NP films. Approximately the same amount of material is added to each immersion cycle.

5.2.2 Measuring Conductivity of Nanostructured Materials Excluding Contact Resistance: Four Probe Method

Studies of charge transport through nanostructured materials thus far have primarily focussed on fundamentally two-probe geometries, in which a bias voltage (V) is applied across a device containing nanostructure(s), and the materials' responses (current, I and its derivatives with voltage, $g = dI/dV$, d^2I/dV^2, etc.) combined with contact resistance response are measured. In some cases, a three-terminal or transistor-type geometry has been employed (Tao 2006). These are essentially two-probe devices as well, with a third "gate" electrode capacitively coupling to nanostructures of interest to add or remove charge.

In order to probe both the nanostructured system and contact resistance as well as just the nanostructured system itself without contact resistance, one can use a so-called four-probe method. Figure 5.6a shows the four-probe set-up, and Figure 5.6b shows a simplified resistor network equivalent. R_{F12}, R_{F23} and R_{F34} are film resistances between probes

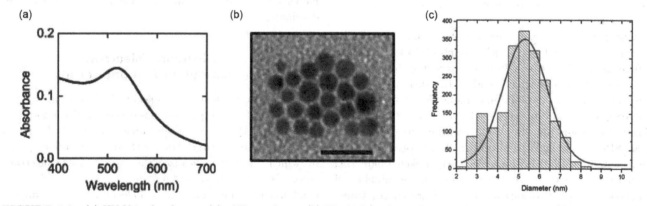

FIGURE 5.3 (a) UV-V is absorbance of Au NP in toluene, (b) TEM of Au NP with a scale bar of 20 nm and (c) histogram of Au NP showing average $d = 5.3$ nm ± 1.1 nm. (Panels (a, b) are reprinted with permission from Tie, M. and Dhirani, A.-A. 2016. *The Journal of Chemical Physics* 145: 104702. Copyright 2016 American Institute of Physics.)

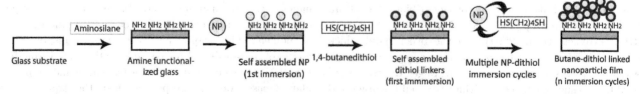

FIGURE 5.4 Cartoon illustrating self-assembly of NP film. (Reprinted with permission from Tie, M. and Dhirani, A.-A. 2015. *Physical Review B* 91: 155131. Copyright 2015 American Physical Society.)

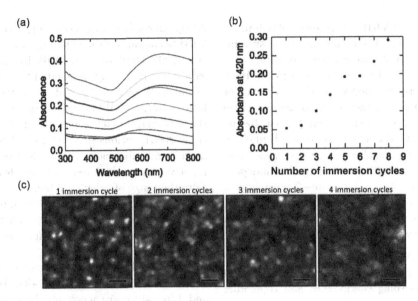

FIGURE 5.5 (a) UV-V is absorbance of Au NP film with immersion cycle, (b) absorbance at 420 nm of Au NP film at immersion cycle 1–8, showing approximately linear growth and (c) SEM of layer-by-layer through immersion cycles. (Panels (a, b) are reprinted with permission from Tie, M. et al. 2014. *Physical Review B* 89: 155117. Copyright 2014 American Physical Society and (c) is reprinted with permission from Tie, M. and Dhirani, A.-A. 2016. *The Journal of Chemical Physics* 145: 104702. Copyright 2016 American Institute of Physics.)

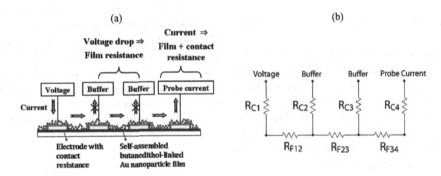

FIGURE 5.6 (a) Four-probe electrode configuration with gold electrodes and a self-assembled butanedithiol-linked gold NP film. (b) A typical four-probe electrode configuration represented by a simplified resistor network. (Reprinted with permission from Joanis, P. et al. 2013. *Langmuir* 29: 1264–1272. Copyright 2013 American Chemical Society.)

1 and 2, 2 and 3, and 3 and 4, respectively. R_{C1}, R_{C2}, R_{C3} and R_{C4} are the contact resistances between film and probes 1, 2, 3 and 4, respectively. A known current, I, flows from probe 1 to 4, through the outer contacts and the film. The total voltage, V_{14}, required to drive this current depends on both the outer contact resistances and the film resistance. By Ohms law, $V_{14} = IR_T$, where R_T is the total resistance of the film plus two contact barriers, $R_T = R_{C1} + R_{F12} + R_{F23} + R_{F34} + R_{C4}$. This would be a standard two-probe measurement which includes contact contributions. Note that a large ratio of contact resistance relative to total resistance leads to a large ratio of voltage drop across the contact resistances relative to the total voltage applied ("voltage divider" effect). In the four-probe method, the voltage difference between probes 2 and 3, V_{23}, is measured using high impedance buffers, which do not allow current to flow across R_{C2} nor R_{C3} but rather force all of the current through the film. As a result, (i)

by Ohms law, the voltage drops across these contact resistances is zero, and V_{23} is also the voltage drop across R_{F23}; and (ii) the current flowing through R_{F23} is I. Notwithstanding the presence of contact resistances, we obtain the film resistance in terms of measured quantities: $R_{F23} = V_{23}/I$. This illustration assumed Ohm's law; that is, I varies linearly with V, and $g = dI/dV$ is a constant. In general, g tends to vary with V. Determining g in this common situation using the four-probe method can be accomplished using lock-in methods as described in reference (Joanis et al. 2013).

5.2.3 Efros–Shklovskii Variable Range Hopping in Insulating NP Films

The interplay between Coulomb barriers and delocalization can strongly influence the temperature dependence of conductance. One example of this is Efros–Shklovskii variable range hopping (ES-VRH). Ignoring Coulomb interactions, Mott argued that VRH can occur in systems just below the percolation-driven insulator-to-metal transition if they possess localized electronic states (Mott 1968, 1990). This is the case for doped semiconductors, for example. According to the model, a charge carrier could absorb thermal energy $k_B T$ and subsequently delocalize by tunnelling to an energetically accessible state. Tunnelling probability between allowed localized states decreases exponentially with distance, r, between states and is independent of temperature; i.e. $P_{\text{tunneling}} \propto \exp(-2\alpha r)$, where α is the tunnelling decay constant. For randomly distributed localized sites, the probability of finding a state within an energy interval ΔW increases with r and T as $P_{\text{absorb}} \propto \exp\left(\frac{-\Delta W(r)}{k_B T}\right)$. The contribution to differential conductance of such a hopping mechanism is proportional to the probability of both absorbing energy ΔW and tunnelling distance r:

$$g \propto \exp\left(\frac{-\Delta W(r)}{k_B T} - 2\alpha r\right). \qquad (5.4)$$

For systems where Coulombic interactions become important, Efros and Shklovskii (ES) (1975) considered

$$\Delta W_{ES} = E_i - E_j + \frac{e^2}{4\pi\epsilon_0\epsilon_r r}, \qquad (5.5)$$

where e is the elementary charge constant, ϵ_0 is the permittivity of vacuum, ϵ_r is the dielectric constant of the insulating material, and E_i and E_j are the energies of the occupied and unoccupied localized states, respectively, separated by distance r. In the limit that $E_i - E_j \ll e^2/(4\pi\epsilon_0\epsilon_r r)$ and solving for the r that maximizes differential conductance $\left(\frac{dg}{dr}(r_{\text{opt}}) = 0\right)$, we find (Efros and Shklovskii 1975, Shklovskii and Efros 1984)

$$g = g_0 \exp\left(-\frac{T_0}{T}\right)^{0.5} \qquad (5.6)$$

where

$$T_0 = \frac{2e^2\alpha}{\pi k_B \epsilon_0 \epsilon_r}, \qquad (5.7)$$

and

$$r_{\text{opt}} = \frac{1}{4\alpha}\left(\frac{T_0}{T}\right)^{1/2} = \frac{e^2}{2\pi\epsilon_0\epsilon_r k_B T_0^{0.5} T^{0.5}} \qquad (5.8)$$

In terms of extracted values,

$$\Delta W_{\text{ES}} = \frac{k_B}{2} T_0^{0.5} T^{0.5} \qquad (5.9)$$

The ES-VRH model has been used to explain g versus T behaviour observed in some bulk-doped semiconductors and semiconductor nanocrystal films (Yu et al. 2004). Comparison of data and the model yield physically reasonable parameters for these systems.

Figure 5.7 plots $\ln(g)$ versus $T^{0.5}$ for four layer-by-layer, self-assembled Au NP films close to but on the insulating side of the insulator-to-metal transition (Dunford et al.

2006). Linear best fit curves on this scale emphasize an ES-VRH behaviour over a wide range of temperatures (Dunford et al. 2005, Dunford et al. 2006). In this study, of 20 samples exhibiting thermally assisted differential conductance, 19 exhibited ES-VRH. T_0 and g_0 coefficients extracted from best fit to ES-VRH behaviour are shown in Table 5.1. Using the ES-VRH model and $\epsilon_r = 2.34$ for 1,4-butanedithiol (Lide 2005), various parameters including optimal hopping energies and distances at a typical temperature ($T = 100$ K) have been calculated and are also listed in Table 5.1. Note the range of calculated hopping and inverse tunnelling decay constant length scales: 60 nm $< r_{\text{opt}} <$ 720 nm and 0.1 μm $< 1/\alpha <$ 15 μm, respectively. These length scales represent tunnelling length scales in the ES-VRH model.

Qualitatively, the observed temperature dependence is consistent with the ES-VRH model. At the same time, r_{opt} and $1/\alpha$ values are much too large to be consistent with simple tunnelling between clusters of linked NPs. Assuming a barrier height of $\varphi \sim 1$ eV for a metal with molecules on its surface, tunnelling distances should be on the order of $\sqrt{\hbar^2/(2m\varphi)} \sim 0.2$ nm, where $\hbar$ is the Planck's constant, and m is the electron mass.

There are indications that the observed large values for r_{opt} and $1/\alpha$ are more appropriately associated with sizes of clusters of linked NPs rather than tunnelling distances. For instance, r_{opt} tends to increase with g_0. This trend reinforces the correlation between r_{opt} and cluster size: as more material is added in constructing the molecularly linked NP films, percolation theory suggests that clusters of linked NPs grow and conductance increases. VRH models, on the other hand,

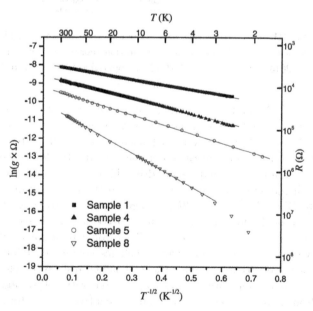

FIGURE 5.7 $\ln(g)$ versus $T^{0.5}$ for four NP films close to but on the insulating side of the insulator-to-metal transition with linear best fit curves and fit parameters given in Table 5.1. (Reprinted with permission from Dunford, J. L. et al. 2006. *Physical Review B* 74: 115417. Copyright 2006 American Physical Society.)

TABLE 5.1 Parameters from "Hopping" Temperature Dependence

Sample #	R (300 K) (kΩ)	$1/g_0$ (kΩ)	T_0 (K)	$1/\alpha$	r_{opt} (100 K) (nm)	ΔW (100 K) (meV)
1	3.36	2.84	8.10	7.05	502	1.23
2	4.47	3.85	3.96	14.43	718	0.86
3	6.25	5.05	9.27	6.17	469	1.31
4	6.25	5.56	16.9	3.37	347	1.77
5	12.4	10.2	25.5	2.24	283	2.18
6	22.4	12.1	75.7	0.754	164	3.75
7	24.5	12.8	82.8	0.690	157	3.92
8	35.7	23.8	84.4	0.677	156	3.96
9	36.3	20.7	105.0	0.546	140	4.41
10	976	520	92.2	0.619	149	4.14
11	4,870	2,240	147.0	0.389	118	5.22
12	6,840	1,630	469.0	0.122	66	9.33
13	44,100	9,930	530	0.108	62	9.92

Measured and fit parameters for 13 samples. Room temperature resistances, $R_{300\text{K}}$, were measured with a multimeter. Intercepts and slopes of linear fits using $\ln(g)$ versus $T^{0.5}$ were used to determine $1/g_0$ and T_0, respectively. The reciprocal of tunnelling decay constants, $1/\alpha$, was determined using Eq. 5.7 with $\epsilon_r = 2.34$ (Lide 2005), and r_{opt} and ΔW were obtained from Eqs. 5.8 to 5.9, respectively. A reference temperature of 100 K was chosen for calculating r_{opt} and ΔW because all samples showed $T^{0.5}$ behaviour in a large range surrounding this temperature.

physically predict an opposite trend: near the percolation-driven insulator-to-metal transition, increasing density of hopping sites should result in shorter hopping distances and higher conductance. We note that similar large r_{opt}'s that are more consistent with grain size than tunnel junction gaps have been observed in evaporated granular metal films (Mandal et al. 1999, Maity et al. 1995). Also, data from these studies point to increasing r_{opt} for samples closer to the insulator-to-metal transition.

Given the likelihood that optimal hopping distances are on the order of sizes of clusters of linked NPs, a question arises as to the interpretation of the observed ES-VRH-like temperature behaviour in molecularly linked NP films and granular metal films in general. In the ES-VRH model, an electron optimizes a probability of thermally overcoming Coulombic interactions between localized states and a probability of tunnelling. The former probability increases as r and T increase; the latter decreases exponentially as r increases and is independent of temperature. An $\exp\left[-(T_0/T)\right]^{0.5}$ dependence is a consequence of the competition between these r and temperature-dependent probabilities. In contrast, in molecularly linked NP and conventionally prepared granular metal films, electrons are "quasilocalized" since they tunnel from cluster to cluster yet can occupy delocalized states on the clusters. These states can exhibit very strong single electron self-charging energy thresholds depending on the sizes of the clusters. Assuming C scales with cluster size, i.e. $C \sim \pi\epsilon_0\epsilon_r L_C$ where L_C is cluster diameter, then

$$E_C = \frac{e^2}{2C} \sim \frac{e^2}{2\pi\epsilon_0\epsilon_r L_C} \qquad (5.10)$$

This form of the charging energy parallels the Coulomb contribution used in the ES-VRH model. The charging energy barrier increases as L_C decreases, which is consistent with observed data. Development of a "quasilocalized hopping" (QLH) model for molecularly linked nanoparticle films (NPs) (and granular metal films in general) – in analogy with the ES-VRH model – would also require a competing mechanism for which the conductance of a cluster drops exponentially with cluster size and is temperature

independent (see Eq. 5.4). One potential mechanism is so-called co-tunnelling in which an electron tunnels onto the cluster while another tunnels off. Another scenario illustrated in Figure 5.8 is that since clusters contain interfaces between NPs, transmission through clusters may decrease exponentially with cluster size (Taylor et al. 2001). If the transmission probability at one interface is Γ (where $\Gamma < 1$) and if there are n' scattering sites per unit length, then the probability, Γ_C, of transmission through cluster is

$$\Gamma_C = \Gamma^{n'L_C} = \exp\left[-n'L_C \ln\left(1/\Gamma\right)\right] \qquad (5.11)$$

which scales exponentially with the size of the cluster. The observed decay constants are very long, as shown in Table 5.1, indicating that backscattering is weak. We note that evidence of elastic scattering in granular materials has been observed via magnetoconductance (Bergmann 1984). In either case, Eqs. 5.10 and 5.11 indicate that there is an optimal choice for cluster size, as illustrated in Figure 5.8.

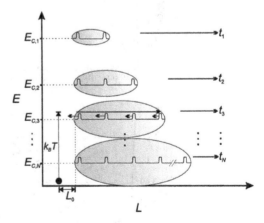

FIGURE 5.8 Quasilocalized hopping (QLH). Illustration of a typical process in the QLH model. An electron can absorb energy $\sim k_B T$ and tunnel a distance L_0 to a neighbouring cluster with charging energy $E_{C,3}$. As the electron traverses a quasi-one-dimensional path along the L_C coordinate, backscattering occurs with transmission Γ at each scattering site. Note that both E_C and total transmission decrease with increased cluster size. (Reprinted with permission from Dunford, J. L. et al. 2005. *Physical Review B* 72: 075441. Copyright 2005 American Physical Society.)

Very small clusters present large charging energy barriers that are difficult to overcome thermally; on the other hand, very large clusters are more difficult to traverse. Following the ES-VRH model outlined above, but with L_C in place of r, the optimized interplay between delocalization and Coulombic barrier results in the $\exp(-T_0/T)^{0.5}$ temperature dependence observed.

5.3 Exotic Behaviour Generated by Strong Electron–Electron Interactions in Nanostructured Materials as Electrons Delocalize

Figure 5.9a shows the evolution of g versus T for a sample through the insulator-to-metal transition. The data were obtained after 12, 18 and 25 alternating immersion cycles in Au NPs and dithiol solutions (S12, S18 and S25, respectively). Since S12's zero-bias g vanishes at low T, it has an energy gap at the Fermi energy and is an insulator. The gap is also evidenced by zero-bias g suppression at low T (Figure 5.9b), which can be attributed to charging barriers (Trudeau et al. 2003) and electron–electron interactions (Teizer et al. 2000). As mentioned, previous studies have shown that NP films with small numbers of immersion cycles can exhibit significant Coulomb barriers and

that charges flow between clusters via a combination of tunnelling and thermally and/or voltage-assisted processes (Beloborodov et al. 2007, Brust et al. 1998, Fishelson et al. 2001, Tran et al. 2005, Trudeau et al. 2003, Wessels et al. 2004).

Figure 5.9a shows that S25 is a metal: g remains nonzero at low T_s, implying the existence of current pathways that have no energy gap at the Fermi level. As an insulating NP film such as S12 is subjected to more immersion cycles, clusters generally grow and charging barriers decrease. Eventually, clusters span the sample, and the film becomes metallic. Figure 5.9c reinforces this picture: g versus bias voltage (V) is nearly Ohmic, with only <1% zero-bias suppression at low T_s.

Figure 5.9a shows that S18 exhibits intermediate behaviour. Nonzero g at the lowest T_s implies that the sample is a metal by definition; however, unlike typical metals, g increases significantly with increasing T_s. This T dependence is "pseudogap-like" in the sense that it can be generated by a density of states (DOS) that is gapless at and increases away from the Fermi level. Quite remarkably, this intermediate sample also exhibits a zero-bias conductance peak that grows rapidly with decreasing T below 4 K (Figure 5.9d). Such a peak is observed in roughly 50% of samples. The data in Figure 5.10c exhibit a remarkably pronounced peak that persists up to almost 20 K. This zero-bias peak is readily observed also in the zero-bias g versus T data (Figure 5.10d).

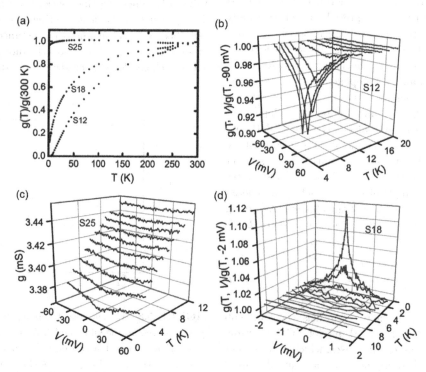

FIGURE 5.9 Four-probe conductance (g) data for an NP film at 12, 18 and 25 immersion cycles (S12, S18 and S25, respectively). (a) Zero-bias g versus T. The data at each immersion cycle are normalized to their respective values at 300 K, 1.1×10^{-5}, 1.9×10^{-4} and 3.5×10^{-3} Ω^{-1} in order of increasing immersion cycles. Corresponding g versus voltage (V) sweeps at various T_s for (b) S12, (c) S25 and (d) S18. The g versus V sweeps at each T are normalized to their respective values at (b) -90 mV and (d) -2 mV. (Reprinted with permission from Tie, M. et al. 2014. *Physical Review B* 89: 155117. Copyright 2014 American Physical Society.)

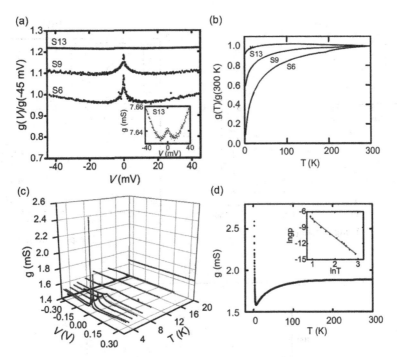

FIGURE 5.10 Four-probe measurements of various samples exhibiting zero-bias conductance peaks. (a) g versus V sweeps normalized at -45 mV and obtained at 2 K using a film at 6, 9 and 13 immersion cycles (S6, S9 and S13, respectively; sweeps offset for clarity). (a, inset) A magnified version of a g versus V sweep for S13. (b) Zero-bias g versus T data for S6, S9 and S13, normalized to their values at 300 K, 5×10^{-4}, 2×10^{-3} and 8×10^{-3} Ω^{-1}, respectively. (c) g versus V sweeps at various T_s for a different film with 11 immersions. (d) The zero-bias g versus T behaviour corresponding to (c) with an inset of log$-$log plot of zero-bias conductance peak (g_p) versus T. The slope yields a power law, $g_p \propto T^{-\nu}$, with exponent $\nu = 3.43$. (Reprinted with permission from Tie, M. et al. 2014. *Physical Review B* 89: 155117. Copyright 2014 American Physical Society.)

In addition to using percolation to tune film conductance, it is also possible to use a combination of percolation and ionic liquid gating. The latter provides a reversible, electronic approach. Application of a gate voltage relative to the NP film causes the ionic liquid to become polarized. The ionic liquid and the NP film (see Figure 5.11) form a capacitor geometry, and due to the ionic nature of the liquid, a strong electric field is concentrated in a nanoscale gap near the NP film. As a result, the gate voltage causes the NP film to charge/discharge and its Fermi level (which depends on charge concentration) to shift significantly.

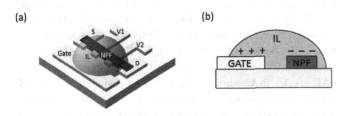

FIGURE 5.11 (a) Schematic of a device with a four-terminal + gate electrode configuration and an ionic liquid droplet. (b) Side view schematic of device showing electrostatic charging upon application of a negative gate voltage. (Reprinted with permission from Tie, M. and Dhirani, A.-A. 2016. *The Journal of Chemical Physics* 145: 104702. Copyright 2016 American Institute of Physics.)

Figure 5.12b shows the influence of ionic liquid gating on a film very close to the metal-insulator transition (MIT). As gate voltage varies and drives the films from the metallic state towards the insulating state, g at 2 K and $V = 0$ V decrease but not quite to zero right away. The bottom curve in fact reveals a zero-bias conductance peak again. A magnified view of the zero-bias region is shown in the bottom right inset for clarity. These data imply that the conductance peaks can be observed near the insulator-to-metal transition using both percolation and gate voltage as independent tuning parameters (Tie et al. 2014).

The origin of the peak remains an open question. To interpret qualitatively the remarkable metallic behaviour exhibited by NP films, one can use the one-band Hubbard Hamiltonian (H) as a guide:

$$H = \sum_{<i,j>,\sigma_s} t_{ij} C_{i,\sigma_s}^{\dagger} C_{j,\sigma_s} + U \sum_i n_{i,\uparrow}^{\dagger} n_{i,\downarrow} \qquad (5.12)$$

where $C_{i,\sigma_s}^{\dagger}$ (C_{j,σ_s}) are the creation (annihilation) operators for electrons at site $i(j)$ and spin σ_s ($\uparrow$ or $\downarrow$), t_{ij} is the transfer energy, U is the on-site Coulomb repulsion energy, $n_{i,\sigma_s} = C_{i,\sigma_s}^{\dagger} C_{i,\sigma_s}$ is the number operator and the first sum is over nearest neighbours and spins. The first sum is the total kinetic energy, and the second is the total potential energy. The Hubbard model is the simplest picture of a lattice with electrons that can transfer from site to site and interact

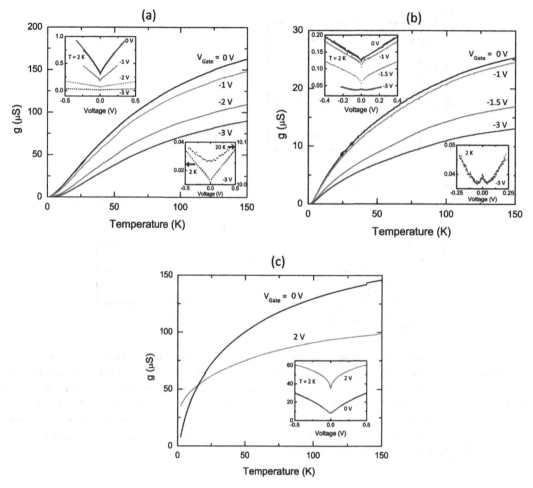

FIGURE 5.12 Data for three different films at different stages near the metal–insulator transition with room temperature conductivities of 25, 33 and 91 Ω^{-1} cm^{-1} corresponding to a film very near MIT, a film near MIT but clearly on the metallic side and a film deeper into the metallic regime, respectively. (Reprinted with permission from Tie, M. and Dhirani, A.-A. 2016. *The Journal of Chemical Physics* 145: 104702. Copyright 2016 American Institute of Physics.)

(Georges and Kotliar 1992, Georges et al. 1996, Jarrell 1992, Kotliar and Vollhardt 2004, Metzner and Vollhardt 1989). Key parameters describing its solution are U and bandwidth (W) determined by the total kinetic energy. In the half-filled Hubbard model, a tendency for on-site repulsion U to yield an insulator competes with a tendency for site-to-site transfer to yield a broadband metal. The model predicts that in the limit that U/W is small, electrons are delocalized, yield a metal and are described by band theory with a band at the Fermi energy with bandwidth W. In the opposite limit that U/W is large, electrons are localized, yield Mott insulators and are described by localized states which yield upper/lower Hubbard bands with bandwidths $\sim W$ located $\pm U/2$ from the Fermi energy. When $U \sim W$, in addition to the upper/lower Hubbard bands, the model predicts the DOS has a third peak or Abrikosov-Suhl (AS) resonance. The peak is located between the upper/lower Hubbard bands at the Fermi energy and appears at T_s below T^*, which characterizes the peak width. The peak describes coherent electrons at T_s below T^*. Above T^*, this coherent phase is thermally destroyed and the peak disappears, leaving behind incoherent transitions from the

lower to the upper Hubbard bands. The peak appears at a critical value of U/W and increases in strength with decreasing U/W. As U/W decreases further, eventually the gap between the upper and lower Hubbard band closes. This peak has been observed in a number of strongly correlated systems ranging from quasi-two-dimensional organic molecular crystals (Merino et al. 2008) to transition metal oxides (Rozenberg et al. 1995, Qazilbash et al. 2007, Mo et al. 2003, Inoue et al. 1995, Fujimori et al. 1992) and chalcogenides (Matsuura et al. 1998). A physical explanation of the peak can be obtained by noting that the Hubbard model, in fact, can be mapped onto the Anderson impurity (AI) model in the limit of infinite dimensions, and the AI model yields the peak in the context of the Kondo effect. The peak reflects an increase in the DOS at the Fermi level arising from interaction between (i) localized, atomic-like states in which charges experience Coulomb repulsion and (ii) delocalized metallic states. Indeed, the AI model itself provides a microscopic perspective of the present system in view of the NPF architecture, and recent studies show that the thiol–gold bond may generate partially filled d-orbitals in the gold (electrons in d-orbitals experience

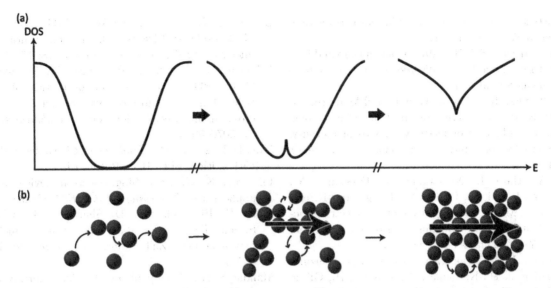

FIGURE 5.13 (a) A schematic of the DOS of a Au NPF and (b) an illustration depicting electron hopping (curved arrows) and metallic transport (straight arrows) through a NPF as NP filling fraction increases and the NPF traverses an insulator-to-metal transition. Due to the disordered nature of NPF, actual DOS may be a mixture of the three scenarios in (a). (Reprinted with permission from Tie, M. and Dhirani, A.-A. 2016. *The Journal of Chemical Physics* 145: 104702. Copyright 2016 American Institute of Physics.)

strong electron–electron Coulomb repulsion as *d*-orbitals are highly localized) (Crespo et al. 2004, Garitaonandia et al. 2008, Trudel 2011). In this picture, Coulomb barriers generated by clusters with partially filled *d*-orbitals interact strongly with a metallic portions of the NP film (see Figure 5.13). In this regime, tunnelling likely plays an important role, and since differential conductance across tunnel junctions contains contributions from DOS, *g* of the film exhibits a peak (the Kondo effect).

5.4 Conclusion

The discussion in this chapter shows that nanostructures (NPs and molecules, for example) can be used to fabricate materials that can exhibit exotic phenomena, including strongly correlated electron motion. Such phenomena lie at the forefront of current materials science. Given inherent control afforded by using nanostructures as building blocks and self-assembly as a means for fabricating materials, the discussion points towards nanostructured materials being a promising approach for generating, controlling, understanding and potentially even harnessing such exotic phenomena.

References

Basov, D. N., Averitt, R. D., van der Marel, D., Dressel, M. and Haule, K. 2011. Electrodynamics of correlated electron materials. *Review of Modern Physics* 83: 471–541.

Beloborodov, I. S., Lopatin, A. V., Vinokur, V. M. and Efetov, K. B. 2007. Granular electronic systems. *Reviews of Modern Physics* 79: 469–518.

Bergmann, B. 1984. Weak localization in thin films: A time-of-flight experiment with conduction electrons. *Physics Reports* 107: 1–58.

Brust, M., Schiffrin, D. J., Bethell, D. and Kiely, C. J. 1995. Novel gold-dithiol nano-networks with non-metallic electronic properties. *Advanced Materials* 7: 795–797.

Brust, M., Bethell, D., Kiely, C. J. and Schiffrin, D. J. 1998. Self-assembled gold nanoparticle thin films with nonmetallic optical and electronic properties. *Langmuir* 14: 5425.

Collier, C. P., Saykally, R. J., Shiang, J. J., Henrichs, S. E. and Heath, J. R. 1997. Reversible tuning of silver quantum dot monolayers through the metal-insulator transition. *Science* 277: 1978–1981.

Costi, T. A., Bergqvist, L., Weichselbaum, A., von Delft, J., Micklitz, T., Rosch, A., Mavropoulos, P., Dederichs, P. H., Mallet, F., Saminadayar, L. and Bäuerle C. 2009. Kondo decoherence: Finding the right spin model for iron impurities in gold and silver. *Physical Review Letters* 102: 056802.

Crespo, P., Litrán, R., Rojas, T. C., Multigner, M., de la Fuente, J. M., Sánchez-López, J. C., García, M. A., Hernando, A., Penadés, S. and Fernández, A. 2004. Permanent magnetism, magnetic anisotropy, and hysteresis of thiol-capped gold nanoparticles. *Physical Review Letters* 93: 087204.

Dagotto, E. 2005. Complexity in strongly correlated electronic systems. *Science* 309: 257–262.

Dunford, J. L., Suganuma, Y., Dhirani A.-A and Statt, B. 2005. Quasilocalized hopping in molecularly linked Au nanoparticle arrays near the metal-insulator transition. *Physical Review B* 72: 075441.

Dunford, J. L., Dhirani, A.-A. and Statt, B. W. 2006. Magnetoconductance of molecularly linked Au nanoparticle arrays near the metal-insulator transition. *Physical Review B* 74: 115417.

Efros, A. L. and Shlovskii, B. I. 1975. Coulomb gap and low-temperature conductivity of disordered systems. *Journal of Physics C* 8: L49–L51.

Fishelson, N., Shkrob, I., Lev, J., Gun, O. and Modestov, A. D. 2001. Studies on charge transport in self assembled gold-dithiol films: Conductivity, photoconductivity, and photoelectrochemical measurements. *Langmuir* 17: 403.

Fujimori, A., Hase, I., Namatame, H., Fujishima, Y., Tokura, Y., Eisaki, H., Uchida, S., Takegahara, K. and de Groot, F. M. F. 1992. Evolution of the spectral function in Mott–Hubbard systems with d^1 configuration. *Physical Review Letters* 69: 1796–1799.

Garitaonandia, J. S., Insausti, M., Goikolea, E., Suzuki, M., Cashion, J. D., Kawamura, N., Ohsawa, H., Gil de Muro, I., Suzuki, K., Plazaola, F., Rojo, T. 2008. Chemically induced permanent magnetism in Au, Ag, and Cu nanoparticles: Localization of the magnetism by element selective techniques. *Nano Letters* 8: 661–667.

Georges, A. and Kotliar, G. 1992. Hubbard model in infinite dimensions. *Physical Review B* 45: 6479.

Georges, A., Kotliar, G., Krauth, W. and Rozenberg, M. J. 1996. Dynamical mean-field theory of strongly correlated fermion systems and the limit of infinite dimensions. *Review of Modern Physics* 68: 13–125.

Gruner, G. and Zawadowski, A. 1974. Magnetic impurities in non-magnetic metals. *Reports on Progress in Physics* 37: 1497–1583.

Hanna, A. E. and Tinkham, M. 1991. Variation of the coulomb staircase in a two-junction system by fractional electron charge. *Physical Review B* 44: 5919–5922.

Imada, M., Fujimori, A. and Tokura, Y. 1998. Metal-insulator transitions. *Reviews of Modern Physics* 70: 1039–1263.

Inoue, I. H., Hase, I., Aiura, Y., Fujimori, A., Haruyama, Y., Maruyama, T. and Nishihara, Y. 1995. Systematic development of the spectral function in the $3d^1$ Mott-Hubbard system $Ca_{1-x}Sr_xVO_3$. *Physical Review Letters* 74: 2539–2542.

Jarrell, M. 1992. Hubbard model in infinite dimensions: A quantum Monte Carlo study. *Physical Review Letters* 69: 168.

Joanis, P., Tie, M. and Dhirani, A.-A. 2013. Influence of low energy barrier contact resistance in charge transport measurements of gold nanoparticle + dithiol-based self-assembled films. *Langmuir* 29: 1264–1272.

Jiang, C.-W., Ni, I., Tzeng, S.-D. and Kuo, W. 2012. Anderson localization in strongly coupled gold-nanoparticle assemblies near the metal–insulator transition. *Applied Physics Letters* 101: 083105.

Kondo, J. 1964. Resistance minimum in dilute magnetic allows. *Progress of Theoretical Physics* 32: 37–49.

Kotliar, G. and Vollhardt, D. 2004. Strongly correlated materials: Insights from dynamical mean-field theory. *Physics Today* 57: 53–59.

Lee, P. A., Nagaosa, N. and Wen, X.-G. 2006. Doping a Mott insulator: Physics of high temperature superconductivity. *Review of Modern Physics* 78: 17–85.

Leibowitz, F. L., Zheng, W., Maye, M. M. and Zhong, C.-J. 1999. Structures and properties of nanoparticle thin films formed via a one-step exchange cross-linking precipitation route. *Analytical Chemistry* 71: 5076–5083.

Lide, D. R. 2005. *CRC Handbook of Chemistry and Physics*, 85th edition. LLC: Boca Raton, FL.

Likharev, K. K. 1999. Single-electron devices and their applications. *Proceedings of the IEEE* 87: 606.

Maity, B., Bhattaeharyya, D., Sharma, S. K., Chaudhuri, S. and Pal, A. K. 1995. Electrical conductivity of nanostructured ZnTe films. *Nanostructured Materials* 5: 717–726.

Mandal, S. K., Gangopadhyay, A., Chaudhuri, S. and Pal, A. K. 1999. Electron transport process in discontinuous silver film. *Vacuum* 52: 485–490.

Matsuura, A. Y., Watanabe, H., Kim, C., Doniach, S., Shen, Z.-X., Thio, T. and Bennett, J. W. 1998. Metal-insulator transition in $NiS_{2-x}Se_x$ and the local impurity self-consistent approximation model. *Physical Review B* 58: 3690–3696.

Merino, J., Dumm, M., Drichko, N., Dressel, M. and McKenzie, R. H. 2008. Quasiparticles at the verge of localization near the Mott metal-insulator transition in a two-dimensional material. *Physical Review Letters* 100: 086404.

Metzner, W. and Vollhardt, D. 1989. Correlated lattice fermions in d = infinity dimensions. *Physical Review Letters* 62: 324.

Mo, S.-K., Denlinger, J. D., Kim, H.-D., Park, J. H., Allen, J. W., Sekiyama, A., Yamasaki, A., Kadono, K., Suga, S., Saitoh, Y., Muro, T., Metacakf, P., Keller, G., Held, K., Eyert, V., Anisimov, V. I. and Vollhardt, D. 2003. Prominent quasiparticle peak in the photoemission spectrum of the metallic phase of V_2O_3. *Physical Review Letters* 90: 186403.

Mott, N. F. 1968. Conduction in glasses containing transition metal ions. *Journal of Non-Crystalline Solids* 1: 1–17.

Mott, N. F. 1990. *Metal-Insulator Transition*, 2nd edition. Taylor and Francis: Philadelphia, PA.

Musick, M. D., Keating, C. D., Lyon, L. A., Botsko, S. L., Pena, D. N., Holliway, W. D., McEvoy, T. M., Richardson, J. N. and Natan, M. J. 2000. Metal films prepared by stepwise assembly. 2. Construction and characterization of colloidal Au and Ag multilayers. *Chemistry of Materials* 12: 2869–2881.

Qazilbash, M. M., Brehm, M., Chae, B.-G., Ho, P.-C., Andreev, G. O., Kim, B.-J., Yun, S. J., Balatsky, A. V., Maple, M. B., Keilmann, F., Kim, H.-T. and Basov, D. N. 2007. Mott transition in VO_2 revealed by infrared spectroscopy and nano-imaging. *Science* 318: 1750–1753.

Ralph, D. C., Black, C. T. and Tinkham, M. 1997. Gate-voltage studies of discrete electronic states in aluminum nanoparticles. *Physical Review Letters* 78: 4087–4090.

Rozenberg, M. J., Kotliar, G. and Kajueter, H. 1995. Optical conductivity in Mott-Hubbard systems. *Physical Review Letters* 75: 105–108.

Sampaio, J. F., Beverly, K. C. and Heath, J. R. 2001. DC transport in self-assembled 2d layers of Ag nanoparticles. *The Journal of Physical Chemistry B* 105: 8797–8800.

Shklovskii, I. and Efros, A. L. 1984. Electronic properties of doped semiconductors. In *Springer Series in Solid-State Sciences,* ed. M. Cardona, P. Fulde and H. J. Qneisser. Springer-Verlag: New York, 202–227.

Snow, A. W. and Wohltjen, H. 1998. Size-induced metal to semiconductor transition in a stabilized gold cluster ensemble. *Chemistry of Materials* 10: 947.

Talapin, D. V., Lee, J.-S., Kovalenko, M. V. and Shevchenko, E. V. 2010. Prospects of colloidal nanocrystals for electronic and optoelectronic applications. *Chemical Reviews* 110: 389–458.

Tao, N. J. 2006. Electron transport in molecular junctions. *Nature Nanotechnology* 1: 173–181.

Taylor, J., Guo, H. and Wang, J. 2001. Ab initio modeling of quantum transport properties of molecular electronic devices. *Physical Review B* 63: 245407.

Teizer, W., Hellman, F. and Dynes, R. C. 2000. Density of states of amorphous GdSi at the metal-insulator transition. *Physical Review Letters* 85: 848–851.

Tie, M. and Dhirani, A.-A. 2015. Conductance of molecularly linked gold NPs across an insulator-to-metal transition: From hopping to strong coulomb electron-electron interactions and correlations. *Physical Review B* 91: 155131.

Tie, M. and Dhirani, A.-A. 2016. Electrolyte-gated charge transport in molecularly linked gold NPs: The transition from a Mott insulator to an exotic metal with strong electron-electron interactions. *The Journal of Chemical Physics* 145: 104702.

Tie, M., Joanis, P., Suganuma, Y. and Dhirani A.-A. 2014. Conductance peaks and exotic metallic behavior in an engineered nanoscale system: Signatures of correlated quasiparticles. *Physical Review B* 89: 155117.

Tran, T. B., Beloborodov, I. S., Lin, X. M., Bigioni, T. P., Vinokur, V. M. and Jaeger, H. M. 2005. Multiple cotunneling in large quantum dot arrays. *Physical Review Letters* 95: 076806.

Trudeau, P.-E., Escorcia, A. and Dhirani, A.-A. 2003. Variable single electron charging energies and percolation effects in molecularly linked nanoparticle films. *Journal of Chemical Physics* 119: 5267.

Trudel, S. 2011. Unexpected magnetism in gold nanostructures: Making gold even more attractive. *Gold Bulletin* 44: 3–13.

Wang, G. R., Wang, L., Rendeng, Q., Wang, J., Luo, J. and Zhong, C.-J. 2007. Correlation between nanostructural parameters and conductivity properties for molecularly-mediated thin film assemblies of gold nanoparticles. *Journal of Materials Chemistry* 17: 457–462.

Wessels, J. M., Nothofer, H.-G., Ford, W. E., Wrochem, F. V., Scholz, F., Vossmeyer, T., Schroedter, A., Weller, H. and Yasuda, A. 2004. Optical and electrical properties of three-dimensional interlinked gold nanoparticle assemblies. *Journal of the American Chemical Society* 126: 3349–3356.

Wirth, S. and Steglich, F. 2016. Exploring heavy fermions from macroscopic to microscopic length scales. *Nature Reviews Materials* 1: 1.

Yu, D., Wang, C., Wehrenberg, B. L. and Guyot-Sionnest, P. 2004. Variable range hopping conduction in semiconductor nanocrystal solids. *Physical Review Letters* 92: 216802.

Zabet-Khosousi, A. and Dhirani A.-A. 2008. Charge transport in nanoparticle assemblies. *Chemical Reviews* 108:4072–4124.

Zabet-Khosousi, A., Suganuma, Y., Lopata, K., Trudeau, P.-E. and Dhirani, A.-A. 2005. Influence of linker molecules on charge transport through self-assembled single-nanoparticle devices. *Physical Review Letters* 94: 096801.

Zabet-Khosousi, A., Trudeau, P.-E., Suganuma, Y. and Dhirani, A.-A. 2006. Metal to insulator transition in films of molecularly linked gold nanoparticles. *Physical Review Letters* 96: 156403.

Spin-Dependent Thermoelectric Currents in Nanostructures (Tunnel Junctions, Thin Films, Small Rings and Quantum Dots)

K.H. Bennemann
Freie Universität Berlin

Spin Currents in Tunnel Junctions for example induced by thermoelectric forces due to temperature and magnetization gradients etc. are analyzed. Using Onsager theory, in particular, for magnetic tunnel junctions, metallic rings and quantum dots yields directly, spin dependently all thermoelectric and thermomagnetic effects like Seebeck and Peltier ones and Josephson like Spin currents driven by the phase gradient of the magnetization. The results can be compared with recent experiments determining the spin-dependent Seebeck effect and other thermoelectric effects. The Onsager theory can be extended towards an electronic theory by expressing the Onsager coefficients by current correlation functions and then calculating these using Lagrange formalism, symmetry and scaling analysis. Note, Onsager theory can also be applied to spin currents in molecules and in magnetic ionic liquids.

6.1 Introduction

Recently, spin-dependent currents in nanostructures and tunnel junctions have been discussed intensively [1–5]. In particular, the spin-dependent thermoelectric and thermomagnetic effects like Seebeck effect and the heat due to spin-dependent currents in ferromagnets (FMs), the spin Peltier effect receives special attention [4,5]. The interdependence of the various currents is most interesting and well described by Onsager theory. Onsager theory $j = LX$ for the currents j driven by the spin-dependent generalized thermodynamical forces X like temperature gradient or magnetization gradient etc. yields the spin-dependent thermoelectric effects [6]. In particular, this holds for

nanostructures like tunnel junctions and metallic rings [2,7] and tunnel currents through molecules and spin currents in magnetic ionic liquids. Note, even if originally one has a homogeneous magnetization, a temperature gradient ΔT will induce a difference ΔM in the magnetization and $\Delta M \propto \Delta T$.

Also, of course, even in a homogeneous FM one gets for itinerant electrons $j_\uparrow \neq j_\downarrow$ for the spin currents due to the spin-dependent density of states (DOS) ($N_\uparrow(\varepsilon) \neq N_\downarrow(\varepsilon)$) etc. Already the Boltzmann equation yields this in a qualitative correct way.

In analogy to the Josephson current in superconductors (SCs) due to the phase difference of the order parameter, one also expects for FMs, magnetic tunnel junctions, with a gradient in the magnetization its magnitude and phase ($M = |M|\exp(i\phi)$) a similar Josephson-like spin current [2,3,6]. Such currents are also expected for metallic rings with inhomogeneous magnetization.

Of course, in inhomogeneous FMs, see Figure 6.1, coupled currents involving spin and charge are expected. This is elegantly described by Onsager theory. Note, the magnetization may result from local magnetic moments (for example, in rare–earth) or from spins of itinerant electrons in transition metals (TMs) or rare–earth.

The Onsager theory may also be applied to describe thermoelectric and thermomagnetic effects in magnetic ionic liquids. Then the spin-dependent pressure effects are expected in case of pressure gradients, possibly interfering with other gradients.

Interesting are, in particular, inhomogeneous systems (nanostructures, tunnel junctions) like ($FM_1|N|FM_2$),

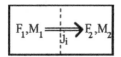

FIGURE 6.1 Illustration of an inhomogeneous FM tunnel junction with temperature gradient $\Delta T = T_1 - T_2$ and magnetization gradient $\Delta M(t) = M_1 - M_2$ and possibly other gradients. Coupled spin dependent currents j_i are expected in accordance with Onsager theory. Note the currents may induce a temperature gradient and thus affect $\Delta M(t)$. In particular one may get Josephson-like spin currents due to a phase difference of the magnetizations $(j_s^J \propto \frac{dM}{dt} \propto \overrightarrow{M_1} \times \overrightarrow{H_{eff}} + \cdots)$ and, for example, from the coupled currents a spin-dependent Seebeck effect $(\Delta(\mu_\uparrow - \mu_\downarrow) \propto \Delta T$ and Peltier effect (heat$\propto$spin current). Clearly the various currents will depend on the magnetic configuration of M_1 and M_2, FM vs. antiferromagnetic (AF) configuration of the two magnets and thus also the electric potential gradient $\Delta\varphi$ depends on $\overrightarrow{M_1}$ and $\overrightarrow{M_2}$, $(\Delta\varphi \propto \Delta T)$.

(FM|SC|FM), etc. for studying spin lifetimes of electron spins injected from a FM into a nonmagnetic metal (N) and for studying spin currents in SCs, analysis of singulet vs. triplet superconductivity (TSC), (FM—SC) interfaces [2,7]. This may be used to test TSC.

As indicated already by the giant Faraday effect in graphene and a few layers of graphene, one expects for graphene structures due to the relatively long spin mean free paths interesting spin-dependent thermoelectric effects (see MacDonald et al. [1], Bennemann and others). For example, for tunnel junctions involving graphene between the two FMs, the Josephson-like spin current driven by a gradient of the phase of the magnetizations on the left and right side of the tunnel junction could be observed.

For tunneling involving TSC and ferromagnetism, the interplay of the order parameters yields novel properties of tunnel junctions. For example, one gets Cooper pair tunneling even for no phase difference between the SCs on both sides of the tunnel junction [7]. Note, regarding the Josephson-like spin current driven by the phase gradient of the magnetization on the left side and right side of the tunnel junction [2,6], see Figure 6.2, this might require relatively long-spin mean free paths. Thus, weak spin–orbit scattering and tunneling, for example, through graphene favors this spin current. Strong spin–orbit scattering is expected to suppress this Josephson spin current.

Spin currents in metallic rings, in particular, persistent ones, are interesting. One expects that the Aharonov–Bohm effect, spin–orbit coupling and interferences of magnetism and superconductivity yield novel behavior [8].

For fluctuating spin currents (in z–direction), one gets according to the Maxwell equations also accompanying electromagnetic fields, see $4\pi j_{s,z} = -4\pi\mu_B \partial_t \langle S_z \rangle = \partial_x E_y - \partial_y E_x$, etc. Generally the connection between spin currents and magnetization dynamics is given by $\partial_t M_i + \partial_\mu j_{i\mu,\sigma} = 0$ [2]. According to Kirchhoff, for example, the emissivity (e) of

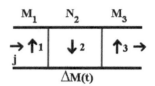

FIGURE 6.2 Nanostructure, tunnel junction composed of FM metals 1 and 2. Due to the magnetization phase difference $\Delta\phi$ between FMs 1 and 3, a spin current proportional to $\sin\Delta\phi$ is expected. Spin relaxation controls this current. Note, if a FM metal 2 is put in between metals 1 and 3 with AF-oriented magnetization causing a GMR, then one may generate an ultrafast oscillating (modulated) current by photon irradiation changing the magnetization in part 2 by $\Delta M(t)$ time dependently. According to Onsager theory, the ΔM–gradients are the driving forces of the spin currents and interference effects may occur. A Josephson-like spin current due to the gradient $\Delta\phi$ may result if N_2 is nonmagnetic and if the spin mean free path is comparable or longer than the thickness of N_2. (treating ϕ as can conjugate variable, N_2: graphene). Magnetization dynamics may cause ultrafast oscillating radiation according to Maxwell equations. As indicated by the GMR or tunnel magnetoresistance (TMR), the tunnel currents depend on the relative orientation of the magnetizations. The Seebeck and Peltier effect will reflect this.

a tunnel junction (or thin film) is related to the magnetization dynamics and magnetic resistance ($\Delta e/e \simeq a(GMR)$, where Δe is the change in the emissivity due to changing the magnetic configuration ($\uparrow/\uparrow$) to ($\uparrow/\downarrow$) of neighboring thin films or tunnel junctions, GMR is the giant magnetoresistance).

In general, nonequilibrium thermodynamics describes the thermoelectric and thermomagnetic effects. The currents j_i, including spin currents, are driven by the spin-dependent thermoelectric forces $X_i = -\frac{\partial\Delta S}{\partial x_i}$, where S is the entropy and $\dot{x}_i = j_i$ (x_i = fluctuations of usual thermodynamical variables, x_i and X_i are conjugate variables). Thus, the currents can be calculated from

$$j_i \propto (1/X_i)d\Delta F/dt, \qquad (6.1)$$

where F is the free energy determined, for example, by an electronic Hamiltonian. Note, Eq. (6.1) is of basic significance, since it relates the currents to the free energy, see F.Bloch, S.de Groot and others [8,9]. Hence, the currents may be calculated from the free energy, and this obviously permits application of scaling theory regarding phase transition behavior.

In case of itinerant electrons, the spin-dependent currents result from the gradients $\Delta\mu_\sigma$ of the spin σ dependent chemical potentials μ_σ. Note, $\mu_\uparrow - \mu_\downarrow \simeq 2\mu_0 H_{eff}$, where H_{eff} is the effective molecular field acting on the itinerant spins [9], $\mu_{\uparrow(\downarrow)} = \mu \mp \mu_0 H_{eff}$.

Above Eq. (6.1) follows from $d\Delta S = \Sigma_i X_i x_i$, $j_i = \dot{x}_i$ and then

$$d\Delta\dot{F} = -\sum_i (Tj_i X_i) + \cdots. \qquad (6.2)$$

This yields, in particular, $j_i\Delta\varphi = -d\dot{F}$ [8]. As discussed later and as quantum mechanically expected, of course, the phase of the driving force $X_i(t)$ plays an important general role, see, for example, Josephson currents in SCs or spin currents in magnets, etc. Clearly, in general, Eq. (6.1) also includes contributions due to time dependencies of phases occurring in the free energy and also applies to SCs.

The following study may be useful to demonstrate how Onsager theory yields the interdependence of the various currents (in nanostructures). Onsager theory is most useful to describe directly all thermoelectric effects etc., spin dependently. Already known results [1,4,5] and new results [2] are presented. This may help to apply studies by F. Bloch [8] and others to spintronics and to new problems. Clearly, tunnel junctions, in particular, yield interesting behavior, by manipulating the magnetic phases.

6.2 Theory

6.2.1 Onsager Theory

As a general framework for deriving the coupled spin-dependent currents in tunnel junctions (and nanostructures in general) driven by thermoelectric forces X_i like magnetization, temperature or chemical potential gradients, one may use Onsager theory, see Kubo, Landau, de Groot et al. [9].

Generally for deriving the spin-dependent thermoelectric and thermomagnetic currents, one may use the Onsager theory. Then the coupled spin-dependent currents j_i are given according to Onsager theory by (expanding $\dot{x}_i = f(x_l) = L_y X_j + \ldots$)

$$j_i(t) = L_{ij}X_j(t) + L_{ijl}X_jX_l + \cdots, \qquad (6.3)$$

with driving forces [9,10] $X_i = -\partial\Delta S/\partial x_i$ and using for the entropy S the expression $d\Delta S = \Sigma_i X_i x_i$. Note, x_i denotes the extensive thermodynamical variables like E, V, e etc. Then from thermodynamics, one gets

$$d\Delta S = \Delta(1/T)dE + \Delta(p/T)dV - \Sigma_\sigma \Delta(\mu_\sigma/T)dN_\sigma$$
$$+ \Delta(H'_{eff}/T)dM_L, \qquad (6.4)$$

where μ_σ is the spin-dependent chemical potential of itinerant electrons, M_L the magnetization of local magnetic moments and H'_{eff} the effective molecular field acting on the local spins, $\mu_\sigma = \mu(\varphi) - \sigma\mu_0 H_{eff}$, $\mu(\varphi) = \mu(0) - e\varphi$ and φ the potential acting on the electron charge, respectively. Note, the term $\Delta\mu_\sigma dN_\sigma$ can also be put into the form $(\Delta\mu dN - \Delta H_{eff})$, where H_{eff} is the molecular field acting on the itinerant electron spins with magnetization M. We may put $X_1 \equiv X_E = \Delta T/T^2$, $X_{2\sigma} = \Delta(\mu_\sigma/T)$, $X_3 = X_M = -\Delta(H'_{eff}/T)$, $X_4 = -\Delta(p/T)$, $X_{5\sigma} = -\Delta(p_\sigma/T)$, the partial pressure of the electrons with spin σ, etc.

Thus, one finds for the coupled currents $j_i = L_{ij}X_j + \cdots$ driven by the forces X_i with $i = 1 = E, i = 2 = e, i = 3 = \uparrow$, etc. the expressions [9]

$$j_E = L_{11}\Delta T/T^2 + \Sigma L_{12}^\sigma\Delta(\mu_\sigma/T) - L_{13}\Delta(H'_{eff}/T)$$
$$- L_{14}^\uparrow\Delta(p_\uparrow/T) + L_{14}^\downarrow\Delta(p_\downarrow/T) + \cdots, \qquad (6.5)$$

$$j_e = L_{21}\Delta T/T^2 + \Sigma L_{22}^\sigma\Delta(\mu_\sigma/T) - L_{23}\Delta(H'_{eff}/T)$$
$$- L_{24}^\uparrow\Delta(p_\uparrow/T) + \cdots, \qquad (6.6)$$

$$j_\uparrow = L_{31}^\uparrow\Delta T/T^2 + L_{32}^\uparrow\Delta(\mu_\uparrow)/T - L_{33}\Delta(H'_{eff}/T)$$
$$- L_{34}^\uparrow\Delta(p_\uparrow/T) + \cdots, \qquad (6.7)$$

and for the local moment magnetization, the current

$$j_{M_L} = L_{41}\Delta T/T^2 + \Sigma L_{42}^\sigma\Delta(\mu_\sigma/T) - L_{43}\Delta(H'_{eff}/T) + \cdots . \qquad (6.8)$$

Note, the replacement $\uparrow\rightarrow\downarrow$ yields $j_\downarrow$. The spin currents $j_\uparrow$ and $j_\downarrow$ may be coupled by spin-flip processes, in particular, spin–orbit interaction. Then a term proportional to $\Delta\mu_\downarrow$ could also contribute to $j_\uparrow$. As usual, symmetries may reduce the number of different Onsager coefficients L_{ij}, for example, $L_{ij}(H_{eff}) = L_{ji}(-H_{eff})$ may hold etc.

The most important and central property of the Onsager equations is the interdependence of the various currents driven by the forces X_i. In particular, the driving force

$$X_{2\sigma} = \Delta(\mu_\sigma/T) \propto -\Delta(H_{eff}/T) + \cdots \propto -\Delta(M/T) + \cdots \qquad (6.9)$$

causes correlated currents due to gradients of the magnetization with respect to phase and magnetization magnitude, respectively. The phase gradient-driven spin currents may be of Josephson type [2,6]. The Onsager equations show that the spin Josephson current is accompanied by a contribution to j_e, j_E, for example, or better ΔM due to a phase gradient also induces a contribution to the other currents, j_e etc. This is immediately obvious from Onsager theory and yields new behavior.

Note, the Onsager equations also apply to SCs and yield different behavior for the single particle currents regarding singulet and triplet SCs, in particular, for j_e and $j_\uparrow$. The current of the Cooper pairs may be added to above Onsager equations. In case of triplet pairing, the spin or angular momentum current of the Cooper pairs is of particular interest.

Also note, the Onsager theory applies to ions (in liquids, gases), in particular, magnetic ones. A special interesting application of Onsager theory may be to a lattice of atoms or molecules and of quantum dots.

The Onsager equations are very useful for directly deriving the thermoelectric and thermomagnetic effects. The Onsager coefficients need to be determined experimentally, by various conductivities, and may be calculated from the free energy using, for example, an electronic theory. Scaling theory may be applied to the coupled currents near phase transitions.

Special situations are easily described by the Onsager equations. For example, decoupling of charge and spin current is described by $j_e = 0$ and $j_\uparrow$ or $j_s = j_\uparrow - j_\downarrow$ not equal to zero.

Taking into account the spatial anisotropy induced by the molecular field $\overrightarrow{H_{eff}}$, by an external magnetic field H, one has j_i^α, $\alpha = x, y, z$, denoting the current of sort i in the direction α. The situation simplifies for $\overrightarrow{H_{eff}} \perp (x, y)$ and isotropic plane (x, y) implying symmetries for coefficients L_{ij} upon transformation $x \rightleftarrows y$, see, for example, de Groot et al. [9]. For spatial anisotropy due to $\overrightarrow{H_{eff}}$, one has $(j^\alpha = (L^\beta) X^\beta + \cdots$, $\alpha, \beta = x, y$, (L^β) as the coefficient matrix.

$$j_i^\alpha = L_{ij} X_j^x + L_{il} X_l^y, \alpha = x, y, \qquad (6.10)$$

$(l > j)$. Symmetry with respect to $x \rightleftarrows y$, $H \to -H$, etc. will reduce the number of different Onsager coefficients as usual. Note, the coupling of responses in x– and y–direction. It is $X_E^\alpha = -1/T^2 \Delta_\alpha T$, etc. . This then, yields spin-dependent galvanomagnetic effects, Hall effect ($\Delta_y \mu$ due to currents j_i^x, etc.), isothermal Nernst effect ($\Delta_y \mu$ due to energy current j_E^x), etc. [9].

Thus, for example, in the presence of the field H (or H_{eff}) perpendicular to the currents in a tunnel junction and taking into account the induced anisotropy, one gets from above Onsager equation

$$j_E^x = \Pi_\uparrow^x j_\uparrow^x + \Pi_\downarrow^x j_\downarrow^x + \cdots, \qquad (6.11)$$

and

$$j_e^x = L_{21}^x \frac{\Delta_x T}{T^2} + L_{21}^y \frac{\Delta_y T}{T^2} + L_{22}^{x\uparrow} \Delta_x (\mu_\uparrow / T)$$
$$+ L22^{x\downarrow} \Delta_x (\mu_\downarrow / T + \cdots \qquad (6.12)$$

and similar equations for j_σ etc. Note, $x \to y$ yields j_E^y, etc.

One gets from these equations as expected that currents induce a spin voltage (Hall effect) etc. For example, the spin current in x–direction induces the spin voltage

$$\Delta_y (\mu_\uparrow - \mu_\downarrow) \propto j_s^x \qquad (6.13)$$

in y–direction.

Some magnetogalvanic effects are discussed later. First spatial anisotropy due to the field $\overrightarrow{H_{eff}}$ is not explicitly taken into account.

Note, one may use Gibbs–Duhem equation, see Kubo et al. [9], to derive per particle

$$s\Delta T - v\Delta p + e\Delta \varphi + \Sigma n_\sigma \Delta \mu_\sigma - m_L \Delta H'_{eff} = 0 \qquad (6.14)$$

and then to write

$$\sum_\sigma n_\sigma \Delta \mu_\sigma = -e\Delta \varphi - s\Delta T + m_L \Delta H'_{eff} + v\Delta p + \cdots \qquad (6.15)$$

Furthermore,

$$H_{eff} = H + H', \qquad (6.16)$$

where H is the external magnetic field and $H' \simeq qM$ the molecular field due to the magnetization acting on itinerant spins and local moments.

As mentioned already, the spin current driven by the phase gradient of the magnetization also results from the continuity equation for the magnetization and from the Landau–Lifshitz (LL) equation [2]. In particular, as discussed already, the gradient of the magnetization phase

yields the Josephson-like spin current between two FMs 1 and 2, in tunnel junctions and at interfaces, see Figure 6.1 and Eqs. (6.2) and (6.3) [11].

Regarding a gradient in the phase of the magnetization, one gets as mentioned already from the gradient of the phase of the magnetization for a tunnel junction or for film multilayers a spin current. Using the continuity equation $\partial_t M_i + \partial_\mu j_{i\mu,\sigma} = 0$ under certain conditions or using the LL equation $dM/dt = a\overrightarrow{M} \times \overrightarrow{H_{eff}} + \cdots$, where $\overrightarrow{H_{eff}}$ refers to the effective molecular field, one may derive a spin current including a Josephson-like spin current j^J of the form

$$j_\sigma = j_\sigma^1(\varphi) + j^J. \qquad (6.17)$$

Here, $j_\sigma^1(\varphi)$ is the spin current due to the electrical potential φ and may result from the spin-dependent DOS. The Josephson-like spin current driven by a phase gradient of the magnetization is given by

$$j^J \propto dM/dt \propto \overrightarrow{M_L} \times \overrightarrow{M_R} + \cdots$$
$$\propto |M_L||M_R| \sin(\phi_L - \phi_R + \cdots), \qquad (6.18)$$

where $(\phi_L - \phi_R)$ is the phase difference of the magnetization on the left and right side of a tunnel junction (or of two films). ϕ and M^z are assumed to be canonical conjugate variables. Note, damping of spin transport may approximately be taken into account, see LL equation or Landau–Lifshitz–Gilbert equation, in the coefficient in front of the term $(\overrightarrow{M_L} \times \overrightarrow{M_R})$ [6].

The Onsager equations for the coupled currents of the itinerant electrons may be rewritten by introducing the spin-dependent Peltier (P) and Seebeck (S) coefficients, respectively,

$$\Pi_\sigma = \frac{j_E}{j_\sigma}, S_\sigma = (1/e)\frac{\Delta \mu_\sigma}{\Delta T}, \qquad (6.19)$$

which expresses that the spin currents carry an energy (heat) current and that the temperature gradient may induce a spin-dependent gradient of the chemical potential. Thus,

$$j_E = j_{E\uparrow} + j_{E\downarrow} = (\frac{j_E}{j_\uparrow})j_\uparrow + (\frac{j_E}{j_\downarrow})j_\downarrow = \Pi_\uparrow j_\uparrow + \Pi_\downarrow j_\downarrow,$$
$$j_\uparrow = [L_{31}^\uparrow / T^2 + L_{32}^\uparrow / T (\frac{\Delta \mu_\uparrow}{\Delta T})]\Delta T + L_{32}^\uparrow \Delta \varphi / T + \cdots,$$
$$j_e = j_\uparrow + j_\downarrow = [\frac{L_{31}^\uparrow + L_{31}^\downarrow}{T^2} + e\frac{L_{32}^\uparrow S_\uparrow + L_{32}^\downarrow S_\downarrow}{T}]\Delta T$$
$$+ L_{32}^\uparrow \Delta \mu_\uparrow / T + L_{32}^\downarrow \Delta \mu_\downarrow / T + \cdots . \qquad (6.20)$$

Here, one may use the expansion $\Delta \frac{\mu_\uparrow}{T} \simeq (1/T)(\Delta \mu_\uparrow(T) + \frac{\Delta \mu_\uparrow}{\Delta T}\Delta T + \cdots)$ [9] and take into account magnetoresistance. Note, the substitution $\uparrow \to \downarrow$ yields the current $j_\downarrow$.

The spin current $j_s = j_\uparrow - j_\downarrow$ is given by

$$j_s = (L_{31}^\uparrow - L_{31}^\downarrow)\frac{\Delta T}{T^2} + (L_{32}^\uparrow - L_{32}^\downarrow)\frac{\Delta(\mu_\uparrow - \mu_\downarrow)}{T}$$
$$+ \cdots = [(L_{31}^\uparrow - L_{31}^\downarrow) + eT(L_{32}^\uparrow S_\uparrow - L_{32}^\downarrow S_\downarrow)]\frac{\Delta T}{T^2}$$
$$+ (L_{32}^\uparrow S_\uparrow - L_{32}^\downarrow)/T\Delta(\mu_\uparrow - \mu_\downarrow)_T + \cdots . \qquad (6.21)$$

Here, we use $\Delta(\mu_\uparrow - \mu_\downarrow) = (\frac{\Delta T}{T})\frac{L_{31}^\uparrow - L_{31}^\downarrow}{L_{33}^\uparrow - L_{33}^\downarrow} = eS_s\Delta T$, where S_s is the Seebeck coefficient resulting from the spin voltage $(\mu_\uparrow - \mu_\downarrow)$.

As usual, the Onsager coefficients are determined by experiments and also can be expressed by transport coefficients [4,9].

In FMs, $\sigma_\uparrow \neq \sigma_\downarrow$ in general and $j_s = j_\uparrow - j_\downarrow \neq 0$ results already from the electric potential difference $\Delta\varphi$ alone, since the electron DOS is spin dependent $(N_\sigma(\varepsilon))$.

Clearly, the gradient of the spin voltage $(\mu_\uparrow - \mu_\downarrow)$ drives a current besides the gradients ΔT and $\Delta\varphi$, etc. (For ionic magnetic liquids pressure effects, the partial pressure p_σ and its gradient should be included in the equations above.)

Note, with the help of Gibbs–Duhem or equivalently by expanding the chemical potential in terms of temperature and pressure etc., one also gets the spin Seebeck effect and [9]

$$\Delta(\mu_\uparrow - \mu_\downarrow) \simeq (\frac{\partial\mu_\uparrow}{\partial n_\uparrow}\Delta n_\uparrow - \frac{\partial\mu_\downarrow}{\partial n_\downarrow}\Delta n_\downarrow) + \frac{\partial\Delta(\mu_\uparrow - \mu_\downarrow)}{\partial\Delta T}\Delta T$$
$$+ \frac{\partial}{\partial p_\uparrow}\mu_\uparrow\Delta p_\uparrow - \frac{\partial}{\partial p_\downarrow}\mu_\downarrow\Delta p_\downarrow + \cdots \quad (6.22)$$

Thus, the gradient of the spin voltage can be expressed by the temperature gradient, partial pressure gradient, etc. The terms with Δn_σ take into account contributions due to concentration gradients of itinerant electrons. The important spin-dependent Seebeck coefficients [9]

$$S_{\uparrow(\downarrow)} = 1/e(\frac{\Delta\mu_{\uparrow(\downarrow)}}{\Delta T}), \quad (6.23)$$

see previous discussion [1,4], express that ΔT induces a spin voltage contribution. Note, then

$$\Delta(\mu_\uparrow - \mu_\downarrow) = eS_s\Delta T + \cdots; \quad S_s = 1/e\frac{\partial}{\partial T}(\mu_\uparrow - \mu_\downarrow). \quad (6.24)$$

This is large in magnetic metals, if the DOS difference $(N_\uparrow(\varepsilon) - N_\downarrow(\varepsilon))$ is large for energies ε around the Fermi energy. Note, one gets directly from the expression for μ_σ that the spin voltage is given by

$$\Delta(\mu_\uparrow - \mu_\downarrow) \propto \Delta H_{eff} + \cdots \quad (6.25)$$

The term $(j_\uparrow + j_\downarrow)$ depends on H_{eff} also via its effect on the DOS.

One gets from above equations for $\Delta T = 0$ and $\Delta\varphi = 0$ that approximately

$$\frac{j_E}{j_\uparrow} = \frac{L_{12}^\uparrow}{L_{32}^\uparrow}(1 + \frac{L_{12}^\downarrow}{L_{12}^\uparrow}\frac{\Delta\mu_\downarrow}{\Delta\mu_\uparrow}) + \cdots = \Pi_\uparrow + \Pi_\downarrow(j_\downarrow/j_\uparrow). \quad (6.26)$$

Above equations allow to exploit the symmetries of the Onsager coefficients L_{ij}. For example, consider dependence on external (magnetic) fields, possibly it is (stationary case) $\Pi_\sigma = TS_\sigma$.

It might be useful to express the Onsager coefficients by the spin-dependent transport coefficients, the electrical conductivity σ_σ, by the thermal conductivity κ_σ, etc. and then to put the Onsager equations into the form [1,2,4,5,9]

$$j_E = \Pi_\uparrow j_\uparrow + \Pi_\downarrow j_\downarrow = -(1/e)[\sigma_\uparrow\Pi_\uparrow\Delta\mu_\uparrow + \sigma_\downarrow\Pi_\downarrow\Delta\mu_\downarrow]$$
$$+ \kappa\Delta T + \cdots,$$
$$j_e = -(1/e)[\sigma_\uparrow\Delta\mu_\uparrow + \sigma_\downarrow\Delta\mu_\downarrow] + [\sigma_\uparrow S_\uparrow + \sigma_\downarrow S_\downarrow]\Delta T + \cdots,$$
$$j_\uparrow = -\sigma_\uparrow[(1/e)\Delta\mu_\uparrow + S_\uparrow\Delta T + \cdots], j_s = j_\uparrow - j_\downarrow. \quad (6.27)$$

Here, $\Pi_\sigma = \frac{j_E}{j_\sigma}$ are the Peltier coefficients, already introduced before. Note, the spin-dependent Seebeck coefficients S_σ and Peltier coefficients Π_σ are important parameters describing the spin-dependent thermoelectric and thermomagnetic effects. The equations above may be related to previous studies using not explicitly Onsager theory, see MacDonald, Maekawa, Uchida, Slichter et al.

As follows from the expression for the heat current, the contribution by spin-up and spin-down itinerant electrons may be different. This will, in general, be the case for the TM, see for example Fe and Ni etc. and in general the case if $N(0)_\uparrow \gg N(0)_\downarrow$ around ε_F. Thus, a spin-polarizing external magnetic field may affect j_E strongly.

Of course, in magnetic tunnel junctions also the usual Seebeck coefficient [9]

$$S = \frac{\Delta\varphi}{\Delta T} \quad (6.28)$$

depends on the magnetic configuration of the tunnel junction, see Figure 6.2 and GMR or TMR. Then,

$$\Delta S = S_{\uparrow\uparrow} - S_{\uparrow\downarrow} \quad (6.29)$$

reflects this and gives the change of the Seebeck effect upon changing the magnetizations on the left and right side of the tunnel junction from $\uparrow\uparrow$ to antiprallel configuration $\uparrow\downarrow$. Using previous equations, one gets

$$S_\sigma = S - \sigma(\mu_0/T)\Delta(H_{eff}/T). \quad (6.30)$$

More and detailed experimental studies are needed to determine Onsager coefficients, to check on previous equations and to determine different Onsager coefficients in external magnetic fields. The Onsager equations again show that, in particular, the Josephson spin current due to ΔM, with respect to its phase gradient, is accompanied by corresponding contributions to j_e, j_E, etc.

Finally, note response theory including nonlinear terms, see $j = LX + L'XX + \ldots$, see second harmonic light (S.H.G.), with $L' = L_{ije}$ implies nonlinear thermoeletric and thermomagnetic effects.

6.2.2 Stationary State $j_e = 0$, $j_s = 0$

It is of interest to analyze for special situations the Onsager equations. For example, it follows from the above equation that for (a) $j_e = 0$:

$$\frac{\Delta\varphi}{\Delta T} = -\frac{[(L_{31}^\uparrow + L_{31}^\downarrow)/T + e(L_{23}^\uparrow S_\uparrow + L_{23}^\downarrow S_\downarrow)]}{L_{32}^\uparrow + L_{32}^\downarrow} + \cdots \quad (6.31)$$

Furthermore, one gets for (b) $j_s = 0$:

$$\Delta(\mu_\uparrow - \mu_\downarrow) = -\frac{L_{31}^\uparrow - L_{31}^\downarrow}{L_{33}^\uparrow - L_{33}^\downarrow}\frac{\Delta T}{T} = eS_s\Delta T = e(S_\uparrow - S_\downarrow)\Delta T$$

$$= -[(L_{31}^\uparrow - (L_{31}^\downarrow) + eT(L_{32}^\uparrow S_\uparrow + L_{32}^\downarrow S_\downarrow)]\frac{\Delta T}{T^2},$$
$$(6.32)$$

note where coefficient $S_s = S_\uparrow - S_\downarrow$ has already been defined before. $(\Delta(\mu_\uparrow - \mu_\downarrow) \propto \Delta H_{eff})$.

The Seebeck coefficient $S = \frac{\Delta\varphi}{\Delta T}$ describes the generation of a voltage by the temperature gradient ΔT, see de Groot et al. [9]. This depends on the magnetic configuration of the tunnel junction as is clear from the TMR. As mentioned already, S is different for parallel ($\uparrow\uparrow$) and antiparallel ($\uparrow\downarrow$) magnetizations on the left and right side of the tunnel junction.

As discussed already, the Seebeck coefficient describes the generation of a spin voltage by a temperature gradient [10]. Note, this is a characteristic result of Onsager theory and which of course can also be derived using an electronic theory. Note again, one gets directly from $\mu_\sigma = \mu(0) - eV - \mu_0\sigma H_{eff}$ that the spin voltage and the Seebeck coefficient is controlled by $H_{eff}(T)$.

Recent studies by others also derived this result, see MacDonald et al. [1,4,5,10]. As discussed, if one expands μ_σ in terms of n_σ, T and pressure etc., one already gets the Seebeck effect. Clearly, S_σ etc. are given by the energy spectrum, DOS $N_\sigma(\varepsilon)$ and electron populations. Note, the form $\Delta(\mu_\uparrow - \mu_\downarrow) = eS_s\Delta T + \cdots$ is already most practial for an electronic calculation.

An external magnetic field $\vec{H}$ affects the spin voltage. Regarding the spatial dependence x of the spin-dependent chemical potential $\mu_\sigma(x,t)$, note for $\frac{dT}{dx} = const.$ the gradient of the spin voltage varies linearly for a (one-dimensional) tunnel junction in x–direction.

As discussed before, the spin Seebeck effect means that a spin current can be induced in a magnetic metal without an electric current ($\Delta\varphi = 0$), since ΔT causes a contribution to the spin voltage $\Delta(\mu_\uparrow - \mu_\downarrow) \neq 0$. The spin current j_s is expressed by L_{ij}^σ and approximately $j_s \propto (N_\uparrow(\varepsilon_F) - N_\downarrow(\varepsilon_F)) + \cdots$ As is clear, this spin current depends on the spin mean free path but might disappear due to spin–flip scattering far less than spin currents injected into metals. For example, in a tunnel junction involving tunneling through graphene (with spin dissipation length $\sim$ nm or more), one might get relatively large spin currents induced by a temperature gradient. This is also the case for the spin currents resulting from the gradient of the phase of the magnetization. Regarding dynamics, the time dependence of the gradient of the magnetization phase is of interest.

To calculate the chemical potential $\mu_\sigma(T, n_\sigma, p_\sigma, \dots)$ and its derivative with respect to temperature etc., using an electronic energy or Hamiltonian is standard. Clearly, the electronic energy spectrum and its population determines $\mu_\sigma(T, p, \dots)$ and its derivatives.

Note, in an FM at nonequilibrium with hot electrons, the chemical potential might change in time t and then $\mu_\sigma(x, t, \dots)$. This is expected to yield interesting dynamical behavior.

6.2.3 Heat Transport due to Spin Currents ($\Delta T = 0$, $\Delta\varphi \neq 0$)

As is evident from the analysis above, the electronic currents carry energy, in particular, also the spin currents. The Peltier coefficients Π_σ describe this. Note,

$$\left(\frac{j_E}{j_e}\right)_{\Delta T = 0} = \Pi, \quad \frac{j_E}{j_\uparrow} = \Pi_\uparrow j_\uparrow + \Pi_\downarrow j_\downarrow, \quad \frac{j_E}{j_s} = \Pi_s \quad (6.33)$$

(a) $\Delta T = 0$: Then it is approximately

$$\frac{j_E}{j_\uparrow} \approx \Pi_\uparrow L_{32}^\uparrow \Delta\mu_\uparrow/T + \Pi_\downarrow L_{32}^\downarrow \Delta\mu_\downarrow/T + \cdots . \quad (6.34)$$

(b) $\Delta\varphi = 0$:

$$j_E \approx -(\mu_0/T)(\Pi_\uparrow L_{32}^\uparrow - \Pi_\downarrow L_{32}^\downarrow)\Delta H_{eff}. \quad (6.35)$$

For a tunnel junction with a magnetic metal A on the left side and a metal B on the right side, see Figure 6.3, one gets in the tunnel medium heat generation

$$j_E^A - j_E^B = (\Pi_{Ai} - \Pi_{Bi})j_i, \quad (6.36)$$

where i refers to an electron current ($i = e$), current for electrons with spin σ ($i = \sigma$) and spin current ($i = s$). The generated heat Π_{AB} is observable in particular if the tunnel medium between A and B is magnetic, for example, FM, or is superconducting. Note, besides heat radiation may also occur.

Also in accordance with magnetoresistance (GMR or TMR), the current discontinuity $\Delta j_E = (j_E^A - j_E^B)$ for electron currents $i = e$ (due to the spin-dependent electron conductivity σ_σ) depends on the direction of the magnetizations of metals A and B.

For $j_\downarrow < j_\uparrow$, it is

$$\frac{j_E}{j_s} = \frac{j_E}{j_\uparrow - j_\downarrow} \approx \frac{j_E}{j_\uparrow}\left(1 + \frac{j_\downarrow}{j_\uparrow}\right). \quad (6.37)$$

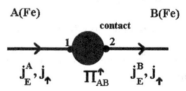

FIGURE 6.3 Peltier effect for two FMs A and B (for example, Fe). Spin-dependent heat Π_σ develops at contact, for example, Cu or a magnetic TM, due to spin current j_σ. Contact 1 may warm up and 2 cool down, and heat $P = \Pi j$ occurs, which should depend on spin polarization, on $(j_\uparrow - j_\downarrow)$. Of course, the Peltier heat will depend on the magnetic configuration of the magnets A and B and is different for FM and AF arrangement of the two magnets. Note, in case of two contacts in sequence and FMs A, B, C, and in series, interesting interferences may occur.

Of interest is to study the Seebeck effect for a tunnel junction consisting of three FMs in series, see Figure 6.2, and to observe the dependence of the spin voltage $\mu_\uparrow - \mu_\downarrow$ on the magnetic configuration of the three FMs, for example, AF vs. FM one.

One may express the heat and spin current within an electronic theory and, thus, obtain an expression for the spin dependence of the Peltier heat suitable for an electronic calculation. It needs to be studied how characteristically the Peltier heat depends on the FM, its magnetization.

In Figure 6.3, the Peltier effect is sketched. As indicated by Figure 6.3, in particular, the Peltier heat of itinerant FMs is expected to depend on the magnetization of the FM and of course on the metals A and B.

All spin-dependent thermoelectric and thermomagnetic effects result from the Onsager currents described above [1,4,9,10]. In the absence of thermal gradients etc., it is

$$j_e \simeq L_{32}^\uparrow \Delta(\mu_\uparrow/T) + L_{32}^\downarrow \Delta(\mu_\downarrow/T) + \cdots, \qquad (6.38)$$

$$j_\uparrow \simeq L_{32}^\uparrow \Delta\mu_\uparrow/T + \cdots \propto L_{32}^\uparrow(\Delta\varphi - \mu_0\Delta H_{eff}) + \cdots \quad (6.39)$$

Furthermore, if the conductivities $\sigma_\uparrow$ and $\sigma \downarrow$ are proportional to each other (the coupling between spin and electric current is described by)

$$j_\sigma \simeq \alpha j_e, (\alpha \leq 1). \qquad (6.40)$$

Of course, spin–orbit coupling affects this relationship.

The stationary case then yields

$$L_{32}^\uparrow \Delta\mu_\uparrow/T + L_{32}^\downarrow \Delta\mu_\downarrow/T \simeq 0 \qquad (6.41)$$

and thus

$$\frac{\Delta\mu_\uparrow}{\Delta\mu_\downarrow} \approx -\frac{L_{32}^\downarrow}{L_{32\uparrow}} = \frac{S^\uparrow}{S^\downarrow} + \cdots, (\frac{\Delta\varphi}{\Delta\mu_\uparrow})_{j_e=0} \simeq -(L_{23}^\uparrow + L_{23}^\downarrow \frac{\Delta\mu_\uparrow}{\Delta\mu_\downarrow}). \qquad (6.42)$$

for the ratio of voltage gradient and gradient of μ_σ of the electrons with spin σ. Note, $\Delta(\mu_\uparrow - \mu_\downarrow) \simeq 2\mu_0\Delta H_{eff}$, $H_{eff} = H + qM$, $\mu_{\uparrow(\downarrow)} \simeq \mu_{\uparrow(\downarrow)}\mu_0 H_{eff}$, $(M = \mu_0[N_\uparrow - N_\downarrow])$, $N = N_\uparrow + (N_\downarrow)$. Here, N_σ refers to the electron DOS at the Fermi energy ε_F. Apparently, the gradient of the chemical potential $(\mu_{+,-})$ or of the effective molecular field H_{eff} causes the electric potential $\Delta\varphi$ for the conducting electrons [1,4,9].

As is clear from Figure 6.1 and from Onsager equations (for $\Delta\varphi = 0$), the temperature gradient ΔT induced by the currents will affect the magnetization $\Delta M = M(T_1) - M(T_2)$ and then change the current driven by ΔM. Approximately, one gets from Onsager equations

$$\left(\frac{\Delta T}{\Delta\varphi}\right)_{j_e=0} \simeq T \frac{L_{22}}{L_{21} - \sum_\sigma L_{23}^\sigma \Delta\mu_\sigma} + \cdots. \qquad (6.43)$$

Regarding magnetic nanostructures, note that the system sketched in Figure 6.2 may yield as mentioned already oscillating currents $j(t)$ and $j_\sigma(t)$ and which are optically induced.

Creating optically, for example, in the FM metal 3 hot electrons then the magnetization M_3 decreases by $\Delta M(t)$ [12]. This changes the magnetoresistance and affects the currents j_e and j_σ. After ultrafast relaxation of the hot electrons, one may repeat the excitation of the electrons. This yields the ultrafast oscillations of the currents. Thus, one may also manipulate the Kirchhoff emission [2].

Of course, one may also apply the Onsager theory to currents in ring structures with gradient forces $X_{i\sigma}$ to obtain interesting thermoelectric and thermomagnetic effects, including Aharonov–Bohm effect etc. [2,8].

6.2.4 Tunnel Junctions Involving SCs

Of interest is also to use SCs as a spin filter, see illustration in Figure 6.4 [10]. As known, a singulet SC may block a spin current and affect the currents driven, for example, by the gradients ΔT, $\Delta M = M_1 - M_2$, etc. Depending on the energy gain due to j_e vs. loss of energy due to (singulet) Cooper pair breaking, one may get that the currents weaken the superconducting state. Note, $\Delta M(t)$ may cause Josephson-like spin current ($j_s \propto \sin\Delta\phi + \cdots$) [13].

If the two FMs are separated by a triplet SC, then the relative orientation of the angular momentum $\overrightarrow{d}$ of the triplet Cooper pairs with respect to the magnetizations $\overrightarrow{M_1}$ and $\overrightarrow{M_2}$ controls the tunnel currents [2,7,13].

Note, $\overrightarrow{d}$ may be oriented via an external magnetic field.

Of particular interest is to study the effect of superconductivity, TSC on the (giant) magnetoresistance in case of two antiferromagnetically (AF) oriented FMs, see Figure 6.4. One expects for parallel configuration of $\overrightarrow{d}$, $\overrightarrow{M_1}$ and $\overrightarrow{M_2}$ the lowest resistance, while the largest one for af configuration of $\overrightarrow{d}$ and magnetizations. Of course Onsager theory can be used to describe the system illustrated in Figure 6.4. Note, $M_2 - M_1$ may act like a magnetization gradient.

Related properties are expected for the tunnel junction shown in Figure 6.5. One may use this to distinguish singlet from TSC. Onsager theory can be used to describe such a system phenomenologically. Josephson currents j_J sensitively characterize such tunnel junctions. The current j_J decreases for increasing thickness d of the FM and

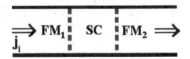

FIGURE 6.4 Illustration of a magnetic tunnel junction consisting of two FM metals $(FM)_1$ and $(FM)_2$ separated by a SC. The electron current j_e as well as $j_\uparrow$ and $j_\downarrow$ and $j_s = \overrightarrow{j_\uparrow} - j_\downarrow$ depend on the relative orientation of the magnetizations $\overrightarrow{M_1}$ and $\overrightarrow{M_2}$ and on the superconducting state, singlet vs. triplet Cooper pairs. Note, $M_2 - M_1 = \Delta M$ may cause Josephson-like spin current, which is particularly affected in case of a triplet SC by the phase of its order parameter. Of course, the spin current is destructively affected by spin-flip scattering.

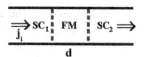

FIGURE 6.5 Tunnel current j between two SCs SC_1 and SC_2 that depends on the relative phase of the order parameter of the two SCs and the phase of the magnetization. Of course, the thickness d of the FM controls the current, in particular, spin-polarized ones. The tunnel current may be manipulated optically via hot electrons in the FM.

for decreasing Cooper pair binding energy (T_c). Also in the spirit of Onsager theory, the difference ($\Delta_2 - \Delta_1$) of the superconducting order parameters acts like a gradient inducing corresponding currents.

In case of TSC, the Josephson current j_J depends in an interesting way on T_c and the angle between the magnetization $\vec{M}$ and direction normal to $\vec{j_J}$. The current should depend on the triplet state and impurity scattering (in particular, spin orbit scattering). Hot electrons in the FM modulate j_J. Generally, the spin polarization of the currents may be manipulated by the gradient $\Delta M(t)$.

In view of the significance of occurrence of TSC in metals, we sketch the situation in Figure 6.6 The current carried by Andreev states is calculated using [7,13]

$$j_J = -(e/\hbar) \sum_i \frac{\partial E_i}{\partial \phi} \tanh(E_i/2kT), \qquad (6.44)$$

where over all Andreev states with energy E_i and mediating the tunneling is summed. Here, ϕ is the phase difference between the Cooper condensates on the left and right side of the tunnel junction. One expects j_J to depend characteristically on the phases of all order parameters, on the

relative orientation of the Cooper pair vectors $\vec{d_L}$, $\vec{d_R}$ and magnetization $\vec{M}$, respectively. The triplet Cooper pairs are described by $\Delta(k) = \sum_l d_l(k)(\sigma_l i\sigma_2)$, $l = 1, 2, 3$ where σ_l are the Pauli spin matrices and d_l are the spin components of the superconducting order parameter, see Bennemann and Ketterson [7,13]. Note, the triplet Cooper pairs have a spin and orbital momentum.

For the transport of angular momentum, obviously the phases of all three order parameters are of importance for tunneling. Even for no phase difference $\phi = \phi_L - \phi_R$ between the triplet Cooper pair condensates on both sides of the tunnel junction, one gets for arbitrary phase of the magnetization of the FM a Josephson current. In the FM, the Andreev states carry the current of the tunneling electrons and the temperature controls its population. Also, of course, the magnitude of the magnetization and electron spin relaxation in the FM matter.

As physically expected, the Josephson current may change sensitively upon rotation of $M_\perp$, change of the direction of $\vec{M}$. Model calculations yield results shown in Figure 6.7(a) [7,13]. This implies that the tunnel junction (TSC—FM—TSC) may act like a switch turning on and off the current j_J. This behavior suggests a sensitive dependence of the current j_J on an external magnetic field.

In Figure 6.7(b), model calculation results, simplifying strongly the influence of the FM metal, are given for the temperature dependence of the Josephson current [7,13]. These should reflect the temperature-controlled occupation of the Andreev states. The change in sign of $j_J(T)$ as a function of T occurs only if Andreev states are nondegenerate and in case of two Andreev states 1 and 2 which derivatives $\frac{\partial E_1}{\partial \phi}$ and $\frac{\partial E_2}{\partial \phi}$ have opposite sign. Also note, the sign change of j_J for increasing temperature may be suppressed by electron scattering in the FM [7,13].

Clearly, in view of the importance studying TSC-improved calculations of the current j_J are needed. The FM tunnel junction metal must be taken into account in a more realistic way.

In case of paramagnetism and $M = \chi H$, where χ denotes the spin susceptibility, the direction of the external magnetic field can be used to manipulate the current.

In view of this rich behavior of (TSC/FM/TSC) tunnel junctions, one also expects interesting behavior for the currents through junctions (FM/TSC/FM). Again the phases of the three order parameters control the currents. One expects

$$j_s^J \simeq A(\Theta) \sin(\Delta\phi + \eta), \qquad (6.45)$$

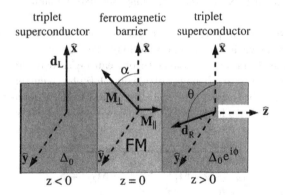

FIGURE 6.6 Illustration of a tunnel junction (TSC/FM/TSC). The phases Θ, ϕ and α of the superconducting order parameter, Cooper pair condensate and of the magnetization $\vec{M}$, respectively, control the tunnel currents. The magnetization may be decomposed into components $M_\perp$ and $M_\parallel$. Due to the spin and angular momentum of the Cooper pairs, one expects that the current through the FM depends sensitively on the relative direction of $\vec{M}$, spin relaxation, spin-flip scattering resulting, for example, from spin–orbit coupling, population of the Andreev states and thickness d of FM.

where Θ refers to the relative phase of the triplet Cooper pairs and $\Delta\phi$ to the phase difference of the magnetization on L side and right side of the junction. Weak spin–orbit scattering and long spin free path for tunneling in the TSC favor the current.

Of course, the Josephson spin current j_s^J can also be manipulated optically by changing the population of the electronic states, by an external magnetic field and by

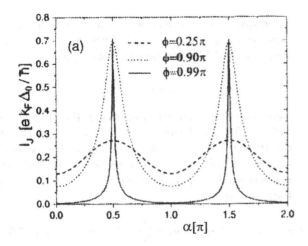

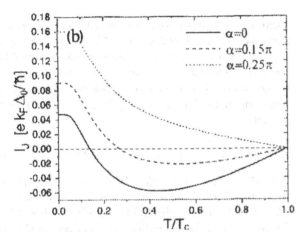

FIGURE 6.7 Results for a tunnel junction (TSC/FM/TSC) sketched in Figure 6.6, see Morr et al. Dependence of the Josephson current on (a) phase α of the magnetization $\vec{M}$ and (b) temperature T for various values of α. Results refer to model calculations simplifying the coupling of the Cooper pairs to the FM. For increasing electron scattering at the FM barrier, j_J does not change sign for increasing temperature any more. The Andreev states carrying the current are determined using Bogoliubov–de Gennes method. T_c is the superconducting transition temperature.

applying a temperature (or pressure) gradient to the tunnel junction.

A spin current $j_s = j_\uparrow - j_\downarrow$ may result due to $N_{\sigma L}(\varepsilon) \neq N_{\sigma R}(\varepsilon)$ for the electronic DOS. Then, $j = j_s + j_s^J + \cdots$.

6.2.5 Galvanomagnetic Effects

Extending as usual the Onsager theory in the presence of an external magnetic field $\vec{H}$ and $\vec{M}$ causing space anisotropy, one obtains the spin-dependent galvanomagnetic effects, the Hall effect, Nernst effect, etc. Then one gets currents

$$j_i^x, j_i^y \qquad (6.46)$$

and in particular currents $j_e^x, j_e^y, j_E^x, j_E^y, j_\sigma^x, j_\sigma^y$, etc. due to forces X_e^α, X_E^α, etc.

For illustration, see Figure 6.8. Note, if M $\perp x, y$–plane, then one also gets for no external magnetic field, $H = 0$, a Hall effect due to gradients ΔT, $\Delta \varphi$, etc.

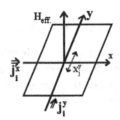

FIGURE 6.8 Currents j_i^α are driven by gradients ΔT^α, $\Delta \varphi^\alpha$, $\Delta \mu_\sigma^\alpha$, ΔM^α, etc. and depend on the relative orientation of the external magnetic field H ($H \perp x, y$ plane) and magnetization $\vec{M}$ which induce spatial anisotropy ($\alpha = x, y$). Indicated is the spin dependent gradient X_j^y induced by currents j_e^x, j_σ^x, etc. The index j may refer to $j = E, \varphi, \sigma, s, \mu_\sigma$, etc. Note, if the magnetization points perpendicular to the x, y–plane, then M may replace H and one gets also for $H = 0$ a Hall potential. For simplicity the magnetization $\vec{M}$ is taken to be in the (x, y)–plane parallel to x or perpendicular to the (x, y)–plane. For example the current j_s^x induces a spin voltage $\Delta_y(\mu_\uparrow - \mu_\downarrow)$ and j_e^x a voltage $\Delta_y \varphi$, j_E^x and gradient $\Delta_x T$ may also cause a spin voltage in y–direction (Nernst–effect). Spin–orbit scattering affects $\nabla_\alpha \mu_\sigma$. In accordance with the Lorentz force (in presence of a molecular field H'), one expects also large Hall effects for 2d-structures (graphene, etc.). Currents should depend sensitively on hot electrons (for example due to light). To simplify the analysis one may assume symmetry with respect to $x \rightleftarrows y$ and for Onsager coefficients $L_{ij}(H_{eff}) = L_{ji}(-H_{eff})$, etc.

To determine the spin-dependent thermomagnetic and galvanomagnetic effects due to the anisotropy resulting from $\vec{H_{eff}}$, one may use previous equations for $j_E^x, j_E^y, j_e^x, j_e^y$, etc. and

$$j_\uparrow^x = L_{31}^\uparrow \frac{\Delta_x T}{T^2} + L_{32}^\uparrow \frac{\Delta_y T}{T^2} + L_{33} \frac{\Delta_x \varphi}{T} + L_{34} \frac{\Delta_y \varphi}{T}$$
$$+ L_{35}^\uparrow \frac{\Delta_x \mu_\uparrow^x}{T} + L_{36}^\uparrow \frac{\Delta_y \mu_\uparrow^y}{T} + \cdots \qquad (6.47)$$

and spin current ($j_s = j_\uparrow - j_\downarrow$)

$$j_s^x = (L_{31}^\uparrow - L_{31}^\downarrow) \frac{\Delta_x T}{T^2} + (L_{32}^\uparrow - L_{32}^\downarrow) \frac{\Delta_y T}{T^2}$$
$$+ (L_{33}^\uparrow - L_{33}^\downarrow) \frac{\Delta_x \varphi}{T} + \cdots$$
$$+ L_{35}^\uparrow \frac{\Delta_x \mu_\uparrow}{T} - L_{35}^\downarrow \frac{\Delta_x \mu_\downarrow}{T} + L_{36}^\uparrow \frac{\Delta_y \mu_\uparrow}{T}$$
$$- L_{36}^\downarrow \frac{\Delta_y \mu_\downarrow}{T} + \cdots \qquad (6.48)$$

It is straightforward to write the other equations for j_e^α, j_E^α, etc. .. It is important to note that the Onsager equations yield coupling of all quantities for certain conditions and in particular coupled currents j_i^x and j_i^y driven by the forces X_l^x and X_l^y, see de Groot [9].

Then rewriting, the Onsager equations as $\Delta_x \mu_\sigma = \cdots$, $\Delta_y \mu_\sigma = \cdots$, $j_E^x = \cdots + L \Delta_x T + \cdots$, etc., thus putting the experimentally controlled quantities on the right side of the equations, one gets directly

(1) the Hall effect, when the current j_s in x–direction generates a spin voltage in y–direction. It is

$$\Delta_y(\mu_\uparrow - \mu_\downarrow) = a(H_{eff})j_s^x + \cdots . \tag{6.49}$$

Thus, the spin current in x–direction induces a spin voltage in y–direction. The analysis uses symmetry with respect to $x \rightleftarrows y$. For $\overrightarrow{H_{eff}} \perp x, y-plane$, using standard analysis one gets

$$a(H_{eff}) \propto H_{eff} + \cdots \tag{6.50}$$

Note, if two magnetic metals are put together parallel to the x–direction in the x, y–plane, then one expects due to TMR (or GMR) -effects a particularly large generation of a spin voltage in y–direction.

Of course, also j_e^x generates $\Delta_y\varphi$, which depends on H_{eff}, $\Delta_y\varphi = \tilde{a}(H_{eff})j_e^x + \cdots$ (usual Hall effect).

(2) Nernst effect: generation of a spin voltage in y–direction due to j_E^x and $\Delta_x T$. It is

$$\Delta_y(\mu_\uparrow - \mu_\downarrow) = b(H_{eff})\Delta_x T + \cdots \tag{6.51}$$

Applying usual symmetry arguments if H_{eff} is perpendicular to the x, y–plane, one gets $b \propto H_{eff}$. Interesting behavior may occur if the external magnetic field H and the molecular field qM are not collinear.

(3) The other effects like Ettinghausen one etc. are obtained similarly from the Onsager equations, see Figure 6.8 for illustration. In the equations above, we neglected for simplicity terms due to the interdependence of the currents j_e^α and j_s^α.

As mentioned, it is of interest to calculate the Onsager coefficients via the corresponding correlation functions [14,15].

Again, of interest are effects due to magnetization gradients. New effects are expected due to a gradient in the phase of the magnetization. One gets for a tunnel junction if ϕ and S^z are conjugate variables, a Josephson-like spin current [6] (j charge = 0).

$$j_s^J = \frac{2E_J}{\mu_B}S^2 \sin\Delta\phi + \ldots, \tag{6.52}$$

with

$$\Delta\phi = 2\mu_B V_S + (\ldots)h. \tag{6.53}$$

h is an external magnetic field. V_S is the spin voltage see Nogueira, Bennemann [6].

In summary, the current j^J across a FM| FM tunnel junction may behave in a similar way as the Josephson current of Cooper pairs. Note, using the general formula $j = -\frac{dF}{d\phi}$, where F is the free energy, one gets

$$j = -(e/\hbar)\sum_i \partial E_i/\partial\phi \tanh(E_i/kT). \tag{6.54}$$

Then using E_i gives the energy difference between opposite directions of the magnetization (molecular field).

Of course, such a Josephson-like spin current is expected on general grounds, if ϕ and S^z are canonical conjugate variables and (approximately) [6]

$$[\phi, S^z] = i. \tag{6.55}$$

Note, this holds for the Heisenberg Hamiltonian as well as for the itinerant magnetism described by the Hubbard Hamiltonian, for example. The commutator relationship suggests in analogy to BCS theory to derive the spin Josephson current from the Hamiltonian

$$H = -E_J S^z \cos(\phi_L - \phi_R) + \frac{\mu_B^2}{2C_s}(S_L^z - S_R^z)^2 + \ldots, \tag{6.56}$$

where again L and R refer to the left-hand and right-hand side of the junction, and C_s denotes the spin capacitance. In general, spin relaxation effects should be taken into account (magnetization dissipation, see the LL equation). Using then the (classical) Hamiltonian equations of motion ($\dot\phi = \partial H/\partial S^z$, $\dot{S}^z = -\partial H/\partial\phi$), one gets

$$\Delta\dot\phi = 2\mu_B V_s, \qquad j_S^J = (2E_J S^2/\mu_B)\sin\Delta\phi. \tag{6.57}$$

Here, $\Delta\phi = \phi_1 - \varphi_2$ and $V_s = (\mu_B/C_s)(S_L^z - S_R^z)$. That $\Delta\phi = 0$ if $M_L\|M_R$ and $\Delta\phi \neq 0$ if M_L is AF aligned relative to M_R can be checked by experiment. It is $J_s^J \sim \sin(\Delta\phi_0 + 4Mt)$. Note, details of the analysis for the ac-like effect are given by Nogueira et al. [6]

Thus, interestingly the spin current in a FM|FM tunnel junction behaves in the same way as the SC Josephson current. Of course, as already mentioned, magnetic relaxation (see the Landau-Lifshitz-Gilbert equation) affects j_S^J. For further details, see again Nogueira and Bennemann [6] and further recent studies by Sudbo and others in *Phys.Rev., Phys.Rev.Lett.,* etc. [6].

Clearly, one expects that junctions involving AF, (AF/F) and (AF/AF) also yield such Josephson currents, since using the order parameter for an AF one also gets that S_q^z and ϕ are conjugate variables. (Treat AF as consisting of two FM sublattices.) Equation 6.56 should hold for both $J > 0$ and $J < 0$, see the Heisenberg Hamiltonian.

Note also, $J_{eff} = J_{eff}(\chi)$ is a functional of the spin susceptibility χ, since the effective exchange coupling between the L and R side of a tunnel junction is mediated by the spin susceptibility of system N_2, see previous figure.

The analysis may be easily extended if an external magnetic field $\vec{B}$ is present. Then from the continuity equation, one gets $j_s \sim \frac{\partial M}{\partial t}$ and $\frac{\partial \vec{M}}{\partial t} = a\overrightarrow{M_L} \times \overrightarrow{M_R} - g\mu_B \overrightarrow{B_L} \times \overrightarrow{S_L} + \ldots$ (and similarly $\frac{\partial \overrightarrow{M_R}}{\partial t} = a\overrightarrow{M_R} \times \overrightarrow{M_L} + \ldots$.) Alternatively, one may use the Hamiltonian–Jacobi equations with the canonical conjugate variables S^z and

$\phi(S^z = \frac{\partial H}{\partial \phi}, \phi = -\frac{\partial H}{\partial S^z})$ and changing the Hamiltonian $H \to H - g\mu_B(\vec{B} \cdot \vec{S_L} + \vec{B} \cdot \vec{S_R})$ to derive the currents j_s and j_s^J.

Note, according to Maxwell's equations, the spin current j_s should induce an electric field E_i given by

$$\partial_x E_y - \partial_y E_x = -4\pi_B \partial_t(S^z) = 4\pi j_s. \qquad (6.58)$$

Here, for simplicity, we assume no voltage and $\partial_t B = 0$ for an external magnetic field B. Of course, in view of the Maxwell equations spin dynamics, the time-dependent tunnel spin currents are accompanied by (polarized) light.

Tunnel junctions with spin and charge current; generally, one gets both a charge current $j_c = -eN_L$ and a spin current $j_s = -\mu_B(S_L^z - S_R^z)$ and these may interfere. This occurs, for example, for SCM|SCM junctions, where SCM refers to nonuniform SCs coexisting with magnetic order (see the Larkin–Ovchinnikov state) and for a (SC/FM/SC) junction with a FM between two SCs. Then one gets after some algebra for the Josephson currents (see Nogueira et al.) [6]

$$j_c^J = (j_1 + j_2 \cos \Delta\varphi) \sin\left(\Delta\phi + \frac{2\pi l}{\phi_0} H_y\right) \qquad (6.59)$$

and

$$j_s^J = j_s \sin \Delta\varphi \cos\left(\Delta\phi + \frac{2\pi l H_y}{\phi_0}\right), \qquad (6.60)$$

Here, ϕ_0 is the elementary flux quantum, H_y an external magnetic field in the y direction, $l = 2\lambda + d$, with λ being the penetration thickness and d the junction thickness. $\Delta\varphi$ and ϕ refer to the phase difference of magnetism and superconductivity, respectively. The magnetic field H_y is perpendicular to the current direction.

The various effects (currents) arising for forces $TX_i^\alpha = -\Delta_\alpha \tilde{\mu}_i$, where $\tilde{\mu}_i = e\varphi + \mu_\sigma$, $\mu_\sigma = \mu(0) - \sigma\mu_0 H_{eff}$, may be summarized by

$$\begin{pmatrix} -\Delta_x \tilde{\mu}_\sigma \\ -\Delta_y \tilde{\mu}_\sigma \\ j_{E\sigma}^x \\ j_{E\sigma}^y \end{pmatrix} = \begin{pmatrix} \sigma_\sigma^{-1} & HR_\sigma & -\epsilon_\sigma & -H\mu_\sigma \\ -HR_\sigma & \sigma_\sigma^{-1} & H\mu_\sigma & -\epsilon_\sigma \\ -T\epsilon_\sigma & -TH\mu_\sigma & -\kappa & -H\kappa L \\ TH\mu_\sigma & -T\epsilon & H\kappa L & -\kappa \end{pmatrix}$$

$$\times \begin{pmatrix} j_\sigma^x \\ j_\sigma^y \\ \Delta_x T \\ \Delta_y T \end{pmatrix}$$

Here, we denote by $L = \frac{\Delta_y T}{H\Delta_x T}$, thermoelectric power $\epsilon_\sigma = -\frac{\Delta_x \tilde{\mu}_\sigma}{\Delta_x T}$, Ettinghausen coefficient $E_\sigma = \frac{\Delta_y T}{H\Delta j_\sigma^x}$, Hall coefficient $R_\sigma = -\frac{\Delta_y \tilde{\mu}_\sigma}{Hj_c^x}$ and Nernst coefficient $\eta_\sigma = -\Delta_y \tilde{\mu}_\sigma / H\Delta_x T$.

6.2.6 Currents in Magnetic Rings

Of special interest is to observe spin electron currents in magnetic rings (see persistent currents [16], and

Aharonov–Bohm effect), in optical lattices or in magnetic quantum dot systems.

For magnetic rings, see Figure 6.9 for illustration, interesting electronic structure may cause special behavior [16–18]. The electron DOS, which for magnetic rings is spin dependent, exhibits oscillations due to the interferences of the most important closed electron orbits (yielding the polygonal paths of the electron current) of the ring, see Stampfli et al. [17] and Figure 6.9. A magnetic field B inside the ring (and directed perpendicular to the ring) causing a flux,

$$\phi = BS, \qquad (6.61)$$

where S is the area enclosed by the electron orbits, yields via the Aharonov–Bohm effect ring currents ($c = 1$) [19]

$$j = -\frac{dF}{d\phi}, j_s = j_\uparrow - j_\downarrow. \qquad (6.62)$$

A spin-polarized current occurs in magnetic rings (also possibly in paramagnetic metals due to the spin polarization by an external field B).

Note, the free energy can be written as $F = F_\uparrow + F_\downarrow$ and the spin-dependent DOS yields F_σ [19].

A spin-polarized current j_s may also be induced by spin–orbit coupling [18]. Note, $H_{so} \sim \vec{\sigma} \bullet \vec{L} \nabla V(r)$ where L is the angular momentum of the electrons in the field $V(r)$. The spin–orbit coupling causes a phase ϕ_{AC} for electrons circling the ring, and thus, $j_\sigma \propto \frac{dF}{dt}$ determines the resultant current (Aharonov–Casher effect). This current may not be very small in case of strong spin–orbit coupling, for example, for topological insulators.

One gets from the theory by Stampfli et al., which is an extension of the Balian–Bloch theory that the electron DOS exhibits interesting structure, $N(\varepsilon, B)$, see Figure 6.10 for typical results. Then, (for usual currents) $j = j_\uparrow + j_\downarrow$, and approximately $j_\sigma = \sigma_\sigma E$ and the conductivity is given by $\sigma_\sigma = e^2 \tau_\sigma(\varepsilon)\overline{vv}N_\sigma(\varepsilon)$, with $\varepsilon \approx \varepsilon_F$. Thus, one expects that

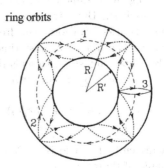

ring orbits

FIGURE 6.9 Most important electron orbits contributing within Balian–Bloch type theory to the electron structure of a ring and the electron DOS. Note, the electron orbits are deformed by the magnetic field inside the ring. This also modifies the Aharonov–Bohm effect. The interferences of the electron paths within the ring cause oscillations in the DOS. The mean free path of the electrons could be spin dependent causing interesting behavior.

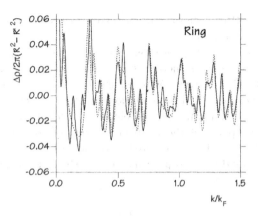

FIGURE 6.10 Electron DOS oscillations for a ring with outer radius R and inner radius R', $R' = 0.3R$, using inside the ring a square well potential $U = -\infty$, see Figure 6.9. The dashed curve refers for comparison with Balian–Bloch type results to quantum mechanical calculations, see Stampfli et al. [17]. The DOS $\Delta\rho$ may be spin split by the molecular field H_{eff}.

the properties of the ring, (spin) currents etc. reflect the structure in the DOS.

The ring current driven by the magnetic flux (Aharonov–Bohm effect) is ($j \propto dF/dt$) approximately given by [17,19]

$$j = -\sum_{t,p,\sigma} \sin(SB)a_{t,p,\sigma}(B), \qquad (6.63)$$

where the coefficients $a_{t,p,\sigma}$ give the contribution to the current of the orbit characterized by t,p, which are the numbers describing how many times the electron orbit circled the center of the ring and how many corners the polygon has, respectively. For illustration, see the Figure 6.9.

The flux area depends on the deformation of the polygonal paths on the magnetic field and $\phi = \pm BS_\pm$, $S_\pm = S_0 \pm \Delta$. Here, $\pm$ refers to clockwise (+) or counterclockwise (−) circling around the center of the ring and Δ is the change of the area due to B [17]. Note, $\frac{dj}{dT}$ analysis or, in general, changing the parameters probing the electronic structure may exhibit the DOS structure.

One may also Fourier transform the free energy. The ring periodicity and invariance against time reversal and $B \to -B$ yields for $B = 0$ inside ring as is the case for the superconducting state $F = \sum_i F_i \cos(2\pi\phi)$ and thus

$$j = \sum_{i\sigma} j_{i\sigma} \sin(2\pi i\phi/(hc/e)) \qquad (6.64)$$

for the current driven by the field B [8]. Note, quantization of the flux in units of (hc/e) [8].

Regarding ring currents, in particular, the persistent currents and Josephson type ones are interesting [2,8,16]. These result if the mean free path of the current-carrying electrons is large enough and comparable to the dimension of the ring and if spin relaxation is weak (weak spin–orbit coupling, …). As discussed, spin–orbit coupling also yields a flux ϕ_{AC} and, thus, persistent currents (Aharonov–Casher

effect). This current is also expected to have structure due to the interference of the main electron orbits in the ring, see Stampfli et al. [17].

Tunnel junctions in rings display interesting behavior, for example, of superconductivity, interplay of superconductivity and magnetism and of singlet and triplet Cooper pairing and $BSC \rightleftarrows BEC$ transition enforced by ring geometry.

Angular momentum conservation may stabilize such persistent currents, in particular, spin-polarized ones ($\sim \vec{j} \times \vec{R}$). From the results above, one may estimate $j \propto \frac{1}{R^n}$ for the currents of a ring with radius R and $n \geq 1$ (the ring current decrease is expected to depend essentially on (l/R), where l is the average length of phase-coherent electron motion). The temperature dependence is given by the expression for the free energy of the electrons [19].

Onsager theory may be used again to study currents driven by the gradients X_i^σ acting (in addition to the Aharonov–Bohm effect) on the itinerant electrons in the ring. For example, the temperature at contacts 1 and 2, see Figure 6.11, may be different and then the resultant gradient ΔT drives a current, similar to gradient $\Delta\mu_{12,\sigma}$, etc. One expects on general grounds that inhomogeneous magnetization present in the ring induces as response spin currents. This is interesting if dissipation length gets comparable to ring length.

If the ring excludes the field B, for example, by becoming superconducting, then as mentioned already, the magnetic flux is quantized [8].

Tunneling through C from a magnetic metal A to a magnetic metal B depends as usual on the configuration of the magnetizations of A and B, see Figure 6.11 for illustration. Thus, via the GMR, the current from 1 to 2 through

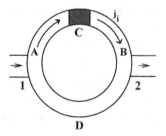

FIGURE 6.11 Illustration of spin currents j_i in a thin ring consisting of two magnetic metals A and B interrupted by a tunnel junction C. The spin dependent DOS $N_\sigma(\varepsilon)$ cause spin polarized currents. In case of contacting the ring at 1 and 2 with external force sources interesting interferences of the currents through A and D may occur. Currents may be driven by applying (spin dependent) gradients $X_i(t)$. Applying an external magnetic field perpendicular to the ring and inside the ring causes Bohm–Aharonov effect. Depending on the electron orbitals carrying the current interesting structure occurs for the DOS and thus for the currents. Of particular interest are persistent (spin) currents arising for relatively large electron mean free path (comparable to the ring dimension) and which are stabilized, for example, by angular momentum conservation.

metals A, C, B may be largely blocked in comparison to the one from 1 to 2 through magnetic metal D ($j_1 = j_A + j_D$), spin dependently. Also note that a spin current entering at 1 may flow (largely) through A, C, B if between 1 and 2 metal D becomes superconducting. In general, via a few parameters, one may manipulate the interference of currents j_A and j_B and thus obtain interesting behavior of the ring currents.

Again, a magnetic current $j_{AB} \propto \frac{dM}{dt} \propto \overrightarrow{M_A} \times \overrightarrow{H_{eff}}(B)$ occurs driven by the magnetic phase gradient [20]. This may yield interesting interferences.

Of interest is also to use a ring of a superconducting metal to study the transition $BSC \to BEC$ due to geometrical restrictions (by changing the width of the ring or narrowing (locally) the ring, etc., causing a corresponding change of the size, radius of the Cooper pairs vs. distance between Cooper pairs, $n_s^{-1/3}$. Note, the coupling strength for Cooper pairing determines the size of the pairs. For Cooper pair size smaller than their distance one expects BEC behavior [21]). It is interesting that this transition may be induced geometrically. Note, this transition BCS⇌BEC is implied by $\Delta \times \Delta\rho \sim h$ and Cooper pair-like ($\hbar V_F/\Delta$) becoming smaller than distance d between Cooper pairs ($d \sim n_s^{-1/2}$, at small superfluid density n_s) [21].

This shows already the many possibilities to study interesting physics using metallic rings.

6.2.7 Currents Involving Quantum Dots

Quantum Dots

The currents between magnetic quantum dots are illustrated in Figures 6.12 and 6.13, see M.Garcia [22]. Spin-dependent currents may result for quantum dots due to the Pauli principle and Coulomb interactions (see Hubbard Hamiltonian) and in particular for magnetic quantum dots and magnetic reservoirs ($\mu_{i\sigma}$).

The spin-dependent electron tunneling, hopping between the magnetic quantum dots, photon assisted, is described by the Hamiltonian

$$H = \sum_{i\sigma} \varepsilon_{i\sigma} c_i^+ c_i + \sum_{ij} T(c_i^+ c_j + h.c.) + H_R, \quad (6.65)$$

where $\varepsilon_{i\sigma} = \varepsilon_{i\sigma}^0 + a(t)\cos(\omega t)$, T gives the electron hopping between quantum dots and H_R describes the coupling to the metallic reservoirs. It is $\varepsilon_{i\sigma}^0(t) = \varepsilon_i^0 + U n_{i\overline{\sigma}}(t) + \cdots$. The time-dependent coupling of the electrons to the photons is given by $a(t)\cos(\omega t)$. The equations of motion for the hopping electrons are given by

$$\dot{c}_{i\sigma} = -i/\hbar [c_{i\sigma}, H]. \quad (6.66)$$

For details of the analysis see Garcia et al. [22]. Results are shown in Figure 6.13. These should be spin dependent for magnetic quantum dots, $j_\sigma \propto \int d\varepsilon N_\sigma(\varepsilon)\dots$. Approximately, the DOS $N_\uparrow(\varepsilon)$ and $N_\downarrow(\varepsilon)$ are split by the molecular field H_{eff}.

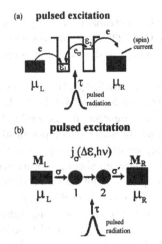

FIGURE 6.12 Illustration of a current between two spin-polarized quantum dots (with electron states described, for example, by $\varepsilon_{i\sigma} = \varepsilon_i^0 + U_i n_i\overline{\sigma} + \cdots$, see Hubbard Hamiltonian) coupled to two magnetic reservoirs (L,R). Optical manipulation of the occupations $n_{i\sigma}$ and of the current and its dynamics is of special interest and offers interesting physical behavior. The photon-assisted tunneling is generally dependent on the energy barrier between the quantum dots, electron spin and light polarization and form of the pulsed radiation field. (a) Illustration of photon-assisted quasi single electron hopping between quantum dots and (b) Sketch of many electron photon-assisted (spin-dependent) tunnel current between two magnetic quantum dots (clusters).

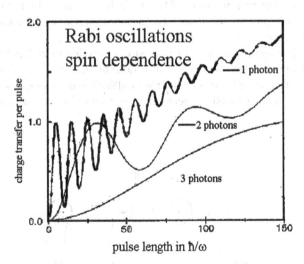

FIGURE 6.13 Spin-dependent tunnel currents between two magnetic quantum dots, note the v. Stckelberg (Rabi) oscillations. The molecular field H_{eff} splits spin-up and spin-down results. Of interest is the dependence of the oscillations on the duration of the photon field, which controls the photon absorption during time of tunneling.

In the spirit of Onsager theory, the field gradients drive the tunnel current. Phases of the magnetizations are also expected to play a role.

In particular, for tunneling involving not many electrons, the currents $j = j_\uparrow + j_\downarrow$ and $j_s = j_\uparrow - j_\downarrow$ may exhibit strong Rabi (v.Stückelberg) oscillations due to back and forth motion of the electrons. Results by M.Garcia et al. are shown in Figure 6.13 [22]. Note, one estimates that approximately the currents $j_\uparrow$ and $j_\downarrow$ are split proportional to the field H_{eff}.

Note, the v. Stückelberg (Rabi) oscillations with frequency approximately given by $\Omega \simeq 2\omega J_N(\frac{a}{\omega})$, with $N\hbar\omega = \sqrt{\Delta\varepsilon^2 + 4\omega^2}$, depend on a general field not periodic in time and also on the shape of the potential of the photon field (and not only on its amplitude). J_N is the Bessel function, and N refers to the number of photons absorbed during hopping between quantum dots in order to fulfill above resonance condition [22]. Note, the oscillations are large for short pulse duration and get damped for long pulses.

Apparently, the interplay of tunnel time and pulse time yields interesting behavior and controls the dynamics. The charge (or spin) transferred between the magnetic quantum dots depends on the number of photons absorbed during tunneling. For long pulse times, this transfer increases when fewer photons are absorbed.

Of course, all thermoelectric and thermomagnetic effects occur also for such a system of quantum dots. For two magnetic quantum dots, see Figure 6.12, the gradients $(T_1 - T_2)$, $(\varepsilon_1 - \varepsilon_2)$ or light-field gradient might drive interesting currents. The fields generated by $j_s(t)$ and charge transfer are given by the Maxwell equations. The magnetic configuration of the quantum dots, parallel magnetizations or antiparallel magnetizations, is likely important for spin or charge transfer (see GMR or TMR). In particular, for intense photon fields, one also expects that the polarization of the photons gets important.

The results can also be related to tunneling in molecules involving a few electronic states determining the tunneling and tunneling between molecules and surface of a (magnetic) solid and tunneling between two molecules or atomic clusters.

anti-dot lattice

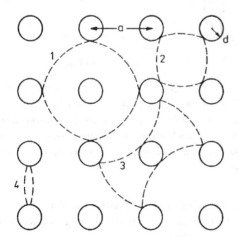

FIGURE 6.14 Polygonal paths 1, 2, 3 etc. in an anti-dot lattice resulting from spin-dependent electron scattering by the repulsive potential of anti-dots (open circles).

Lattice of Quantum Dots

First Anti-Quantum Dot Lattice: Currents in a system of magnetic quantum dots, for example, ensemble of anti-quantum dots arranged as lattice, see Figure 6.14 for illustration, may exhibit interesting behavior. Applying the extension of Balian–Bloch theory by Stampfli et al. [17], one gets for the DOS of the electrons scattered by the anti-quantum dots the expression

$$\Delta N(\varepsilon, B) = \sqrt{2}(a - d)/k_1\pi \sum_{t=1}^{\infty} (\sinh\varphi / \sinh(4t + 1)\varphi)^{1/2}$$
$$\times \cos(BS)\sin\phi_{t,p} \exp{-4tk_2(k_1/B)\varphi_1} + \cdots, \tag{6.67}$$

with ($R_c = k/B$, which is the radius of the corresponding cyclotron radius), $\cosh\varphi = \frac{2a}{d} - 1$, and phase

$$\phi_{t,p} = [k_1(k_1/B)\varphi_1 + B/2(k_1/B)^2(\varphi_1 - \sin\varphi_1) + \delta_{t,4t}] + \pi/2. \tag{6.68}$$

Here, $\varphi_1 = 2\arcsin\frac{a-d}{\sqrt{2}(k_1/B)}$, $S = 2t(a - d)^2$ and phase $\delta_{t,p}$ describing potential scattering at the surface of the quantum dot, $\delta_{t,p} = -\pi$, $for U \to -\infty$, see Stampfli et al. [17].

Roughly, the states of spin-up and spin-down electrons are split by H_{eff}, and thus, the DOS of spin-up and spin-down electrons are shifted by the molecular field H_{eff}. If both B and H_{eff} are perpendicular to the lattice of quantum dots, then the molecular field may act similarly as the external magnetic field B.

The factor $\cos(SB)$ causes periodic oscillations in the electronic DOS, and for small B, it is $R_c \propto 1/B$. For increasing B, one has Landau–level oscillations with periodicity proportional to $(1/B)$, since $S \propto R_c^2$ and flux $\phi_2 \sim SB$ and $\Delta N(\varepsilon, \sigma, B) \propto \cos(SB)$. The DOS changes at T_c, which is the (Curie) ordering temperature of the ensemble of magnetic quantum dots. In strong magnetic fields B, the spin splitting of the electronic levels not only due to H_{eff} but also due to B must be taken into account.

Note, it follows approximately for the electron current using $j \propto dF/d\phi$ that

$$j \sim \sin(BS). \tag{6.69}$$

In Figure 6.15, typical results are shown for the electronic structure of electrons scattered by an ensemble of anti-quantum dots. Such structure is reflected, for example, by the magnetoresistance ρ_{xx} and $\rho_{xx} \propto f[N(\varepsilon, \sigma, B, ..)^2]$.

A potential gradient or thermal gradient etc. may cause a spin-dependent flow, electron current in the anti-quantum dot lattice. Such currents can be described by Onsager theory and may be assisted by a photon field.

Secondly, Lattice of Quantum Dots: In a lattice of quantum dots, see Figure 6.14 for illustration, electron hopping as described by the tight-binding Hamiltonian or tunneling may occur between the quantum dots. Again, this may be assisted by a photon field. Then, a lattice of quantum dots between two metallic reservoirs may act as a switch due

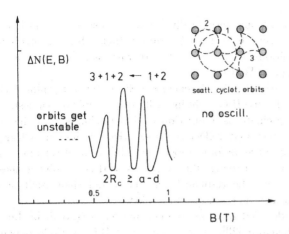

FIGURE 6.15 Schematic illustration of electronic DOS oscillations as a function of magnetic field B of electrons scattered by repulsive (square well) potentials of quantum dots. The oscillations depend on parameters a and d, see Figure 6.14, and molecular field H_{eff} in case of magnetic anti-quantum dots, and magnetic ordering of the anti-quantum dots.

to varying the conductivity of the quantum dot lattice with the help of the photon field. Thus, ultrafast dynamics may occur.

Force gradients $X_{i\sigma}$ induce Spin-dependent currents in a lattice of magnetic quantum dots. Thermoelectric and thermogalvanic effects are expected. In particular, the currents between two (or many) quantum dots having different magnetization, potential, temperature etc. may display interesting behavior.

6.2.8 Magnetooptics

Interesting magnetooptical behavior is exhibited by nanostructures, in particular, magnetic films. In addition to Linear magnetooptics (Faraday effect, Kerr effect), see review by S. Bader et al. in nonlinear optics, K.H. Bennemann [23], S.H.G., in particular, nonlinear magnetooptics (M.S.H.G.) is used intensively in recent years. Note, S.H.G. and M.S.H.G. (note, spin–orbit coupling is the physical origin) are very surface sensitive and reflect the magnetic properties of thin films very well. S.H.G. requires breakdown of inversion symmetry and is thus generated in nanostructures and at the surface and at the interface surface/substrate or at the interface of two films, see Bennemann et al. [24].

Thus, M.S.H.G. is best suited to study multifilm structures, their magnetic configuration and very importantly to study dynamics and nonequilibrium magnetism. Note, in contrast to linear magnetooptics, M.S.H.G. frequently exhibits a much larger Kerr effect and in a more pronounced way magnetism in general.

The symmetry sensitivity of S.H.G. becomes obvious from the expansion of the electric field E-induced polarization $P = \chi^1 E + \chi^2 EE + \cdots$. Here, χ^1 and χ^2 denote the linear and nonlinear susceptibilities. The intensity of S.H.G. light

is given by $I \propto |\chi^2|^2$. S.H.G. depends characteristically on the optical configuration at the surface and on the polarization of the incoming and outgoing light. This is described by the various elements of the susceptibility tensor χ_{ijl}. The optical configuration used for calculating the polarization dependence and dependence on magnetization direction of S.H.G. is illustrated in Figure 6.16.

Figure 6.16 shows the optical configuration and S.H.G. at surfaces (for light $\omega \to 2\omega$). As said, the 2ω–light is characterized by the response function $\chi_{ijl}(\vec{M}, \omega)$. Clearly, since the response depends on $\vec{M}$, note $M_\perp$ and $M_\parallel$ yield different optical response $\chi_{ijl}(\vec{M})$. Hence, the reorientation transition of the magnetization can be studied optically.

For thin films, the nonlinear susceptibility, response function χ^2 may be split into the contribution χ^s from the surface and χ^i from the interface. Then owing to the contribution $\chi^s \chi^i$ to the S.H.G. intensity $I \propto |\chi|^2$, the relative phase of the susceptibilities χ^s and χ^i is important. Furthermore, the magnetic contrast

$$\Delta I(2\omega, M) \propto I(2\omega, M) - I(2\omega, -M) \tag{6.70}$$

will reflect the film magnetism, since the susceptibility has contributions, which are even and odd in M, see Bennemann et al. in Nonlinear Optics, Oxford University Press [23].

Note, high-resolution interference studies are needed to detect Also, for example, lateral magnetic domain structures of films. Also polarized light reflects magnetism and, in particular, the magnetic reorientation transition, see [25].

In magnetooptics (M.S.H.G.), using different combinations of polarization of the incoming and outgoing light, see Figure 6.16, one may analyze

$$\chi_{ijl}(\vec{M}). \tag{6.71}$$

Regarding the dependence on magnetization, for the odd susceptibility in magnetization, one writes

$$\chi^o_{ijl} = \chi_{ijlm} M_m + \cdots \tag{6.72}$$

Here, M_m is the m-th component of the magnetization M. Note, the susceptibility can be split into contributions odd (o) and even (e) in M, respectively, $\chi = \chi^e + \chi^o$. Thus, in particular, the susceptibility as a function of the magnetization, $\chi_{ijl}(M_c)$, changes characteristically for

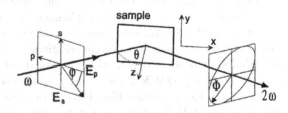

FIGURE 6.16 Optical configuration for incoming ω light and reflected 2ω S.H.G.light. Sketch of polarization of incoming and outgoing light characterized by the angle φ and Φ. z is normal to the surface and the crystal axes x, y are in the surface plane. θ denotes the angle of incidence.

the spatial directions $c = x$, y, or z. For example, for incoming s-polarized light and outgoing p-polarized 2ω light, see Figure 6.16 for illustration, the nonlinear susceptibility $\chi_{ijl}(M_y)$ dominates. Similarly, $\chi_{ijl}(M_z)$ dominates in case of outgoing s–S.H.G. polarization. For general analysis, see Bennemann, in particular Hübner et al. and other chapters, in Nonlinear Optics in Metals, Oxford Univ- Press [23].

Regarding susceptibility χ^e and χ^o note further interesting behavior could result from terms that are higher order in the magnetization M ($\chi^e = \chi_0^e + aM^2 + \cdots$, etc.) and χ_{ijl} also reflects the strength of the spin–orbit interaction. As discussed by Hübner et al., in case of

(a) $\vec{M} \parallel \vec{x}$ (longitudinal configuration), the tensor χ_{ijl} involves χ_{yxx}, χ_{yyy}, χ_{yzz}, χ_{zxx}, χ_{zyy}, χ_{zzz}, $\dots$

(b) $\vec{M} \parallel \vec{z}$ (polar configuration, optical plane x, z), elements χ_{zxx}, χ_{zzz}, χ_{xyz} and χ_{xzx} of susceptibility χ_{ijl} occur.

For longitudinal configuration and polarization combination $s \to p \ \chi_{zyy}$ and for $p \to s \ \chi_{yxx}$, χ_{yzz} are involved. Furthermore, in case of polar configuration and polarization combinations $s \to p$ element χ_{zxx} and for $p \to s$ the element χ_{xyz} occur. This demonstrates clearly that Nolimoke (M.S.H.G.) as well as Moke can observe the magnetic reorientation transition and other interesting magnetic properties of nanostructures, see Bennemann [23].

In Figure 6.16, results by Hübner et al. [23] for the polarization dependence of S.H.G.–light are shown. Note, agreement with experiments is very good. Obviously magnetism is clearly reflected. Similar behavior regarding polarization dependence is expected for the magnetic TMs Cr, Fe and Co. Due to $\langle d \mid z \mid d \rangle \approx \langle d \mid x \mid d \rangle$ for the dipole transition matrix elements, one gets typically same results for s- and p-polarized light. To understand the behavior of Cu, note $\langle s \mid z \mid s \rangle \approx \langle s \mid x \mid s \rangle \approx 0$. Curves (a) and (b) refer to wavelenghts exciting and not exciting the Cu d–electrons, respectively. Generally as physically expected $I_s(p - SH)_{NM} \longrightarrow_\omega I_s(p - SH)_{TM}$. Again for a detailed discussion of the interesting polarization dependence, see magnetooptics and discussion by Hübner et al. [23].

In particular, interesting magnetooptical behavior of S.H.G. and Moke results also from the spin-dependent Q.W.S. occurring in thin films. These states result from the square well potential representing the confinement of the electrons in a thin film. In magnetic films, the resulting electron states are of course spin-split. Characteristic magnetic properties follow. In contrast to band-like states for the electrons in the film, only Q.W.S. show a strong dependence on film thickness. Thus, to study film thickness-dependent (optical) behavior, S.H.G. involving Q.W.S. needs to be studied.

One expects characteristic behavior of the optical response, its magnitude and dependence on light frequency ω and $\vec{M}$. Characteristic oscillations in the M.S.H.G. signal occur, since Q.W.S. involved in resonantly enhanced

S.H.G. occur periodically upon increasing film thickness, see Figure 6.17 for illustration (note, S.H.G. involving transitions $i \to j \to l$ and with one of these states being a Q.W.S.).

Clearly, the Q.W.S. energies shift with varying film thickness. Then S.H.G. light intensity ($I_{2\omega}$) involving these states may oscillate as a function of film thickness, and this, in particular, may reflect the magnetism of the film. Clearly, owing to the periodic appearance of Q.W.S. at certain energies for increasing film thickness, S.H.G. involving these states may be resonantly enhanced and then oscillations occur as a function of film thickness.

A detailed analysis, see calculation of Q.W.S. by Luce, Bennemann [23], shows that the S.H.G. periods depend on the parity of the Q.W.S., the position of the Q.W.S. within the Fermi–see or above, and on the interference of S.H.G. from the surface and interface film/substrate, etc. If this interference is important, then S.H.G. response is sensitive to the parity of Q.W.S., light phase shift at the interface and inversion symmetry of the film. If this interference is not important, then S.H.G. response and oscillations are different, see results for illustration of the interference see [23].

Thus, for an analysis of S.H.G. involving Q.W.S., one may study: (a) $\chi^i \simeq \chi^s$, when interference is important. Different behavior of S.H.G. may result then if 1. Q.W.S. is involved as final state, 2. as intermediate state or 3. both Q.W.S. as intermediate and final state matter. Note, for $\chi^s \chi^i \to (-1)$, for example, due to inversion symmetric films or a phase shift π at the interface, the destructive interference yields no S.H.G. signal. Also, if a final Q.W.S. near the Fermi-energy (and which may set the period of S.H.G. oscillation) has even parity, then in contrast to a Q.W.S. with odd parity, no S.H.G. signal occurs, since for the latter, the product of the three dipole matrix elements is small.

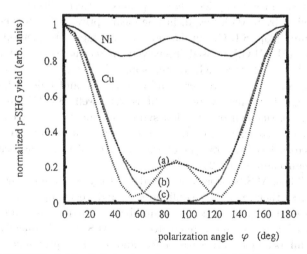

FIGURE 6.17 Polarization dependence of S.H.G. Note the difference between magnetic Ni and nonmagnetic Cu. The angle φ refers to the incoming light polarization. Curves (a)–(c) for Cu result from using different input parameter for the diffraction index $n(\omega)$ and for $k(\omega)$; see calculations by Hübner et al. [23].

(b) $\chi^i \neq \chi^s$ and interference are unimportant. Then different oscillation periods may occur. For example, if a Q.W.S. above the Fermi-energy becomes available at film thickness d_1 for S.H.G., then a first peak in S.H.G. appears and then at film thickness $2d_1$ again at which previous situation is repeated, and so on. Of course, strength of signal depends on wavevector k in the B.Z. (Brillouin zone), the unit cell in reciprocal crystal–lattice space, DOS and frequency ω must fit optical transition. In case of a FM film then the resonantly enhanced S.H.G. transitions are spin dependent and an enhanced magnetic contrast ΔI may occur.

If occupied Q.W.S. below the Fermi–energy are involved, then also oscillations occur, in particular due to DOS, see TMs. If Q.W.S. below and above the Fermi–energy cause oscillations then the period of S.H.G. may result from a superposition, as an example see the behavior of $xCu/Co/Cu(001)$ films.

Characteristic properties of film S.H.G. are listed in Table 6.1. Various properties are listed for the case of no interference ($\chi^i \gg \chi^s$) and strong interference ($\chi^i \approx \chi^s$) of light from surface and interface. The first case is expected for a film system xCu/Fe/Cu(001), for example, since the interface Cu/Fe dominates due to the Q.W.S. of the Cu film and the large DOS of Fe near Fermi energy ε_F (see S.H.G. transitions: $Cu \to_\omega Fe \to_\omega Q.W.S. \to_{2\omega} Cu$). Thus, $\chi^i \gg \chi^s$ and due to the Q.W.S. above ε_F one gets two and more S.H.G. oscillation periods.

The case $\chi^i \approx \chi^s$ is expected for a film xAu/Co(0001)/Au(111), for example, since no Co d–states (see band structure of Co) are available as intermediate states for S.H.G. transitions and the Q.W.S. in Au just below the Fermi–energy controls the S.H.G. contribution, see Figure 6.18 for illustration. The parity of the Q.W.S. causes then oscillations $\Lambda = 2\Lambda_M$ (M refers to Moke), since clearly the 3 dipole (p) matrix elements $\langle d \mid p \mid Q.W.S.\rangle\langle Q.W.S. \mid p \mid d\rangle\langle d \mid p \mid d\rangle$ are much smaller if Q.W.S. has even symmetry (note, p is odd and d is odd). The wave vector k dependence is controlled by the

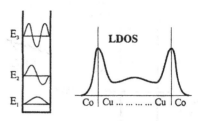

FIGURE 6.18 Illustration of quantum well states (Q.W.S.) due to confinement in a thin film of thickness d. Such states occur, for example, in a Co/10MLCu/Co film system. Note, the DOS of the Q.W.S. is strongly located at the interface. In magnetic films, one gets generally spin-split Q.W.S. The parity of the Q.W.S. changes for increasing energy.

B.Z. structure. (B.Z. denotes unit cell of reciprocal crystal lattice, see books on Solid State Physics, C.Kittel etc.). For a detailed discussion, see Nonlinear Optics in Metals, Oxford University Press [23].

The interesting S.H.G. interference resulting from surface s and interfaces i and reflecting sensitively magnetic properties of the film may be analyzed as follows. The S.H.G. light intensity $I(2\omega)$ is approximately given by

$$I(2\omega) \sim \mid \chi_{ijl} \mid^2, \qquad (6.73)$$

where χ denotes the nonlinear susceptibility. Note, χ may be split as (s:surface, i:interface)

$$\chi_{ijl}(2\omega) = \chi_{ijl}^s + \chi_{ijl}^i. \qquad (6.74)$$

One gets

$$I(2\omega) \sim 2 \mid \chi_{ijl}^s \mid^2 + 2\chi_{ijl}^s \chi_{ijl}^i + \cdots \qquad (6.75)$$

Obviously, the intensity $I(2\omega)$ depends on the resultant phase of the susceptibilities χ^s and χ^i.

Then assuming, for example, that χ^s and χ^i are of nearly equal weight ($\mid \chi^s \mid \sim \mid \chi^i \mid$), it may happen that the 2.term cancels the first one, see, for example, inversion symmetry in films ($\chi_{ijl}^s \to -\chi_{ijl}^i$) or phase shift of the light by π at the interface.

TABLE 6.1 Characteristics of S.H.G. Response from Thin Films

	$\mid \chi^i \mid \gg \mid \chi^\delta \mid$	$\mid \chi^i \mid = \mid \chi^\delta \mid$
k selectivity	(x Cu/Fe/Cu(001) for example)	
	Strong manetic signal due to strong (magnetic) interface contributions	Weak S.H.G. signal, from only few k points and without strong interface contributions
	Sharp S.H.G. peaks due to few contributing k points resulting in strong resonances	Doubled period and additional periods are frequency dependent
	Strong frequency dependence of the S.H.G. oscillation of the Q.W.S. in the $k_\perp$ direction	MOKE period absent; doubled and additional S.H.G. period visible
	MOKE period and larger periods visible; no exact doubling of the MOKE period	
No k selectivity		(x Au/Co(0001)/Au(111) for example)
	Strong magnetic signal, since strong interface contribution Broad S.H.G. peaks, since contributions come from many k points	Smaller magnetic contribution, since interface and (nonmagnetic) surface contributions are of the same magnitude
	Weak frequency dependence of the oscillation period	Broad, smooth peaks, since interference effects do not change magnitude
	MOKE period and larger periods present	S.H.G. oscillation periods rather independent of the frequency, since the S.H.G. signal is caused by the Q.W.S. near E_F
		MOKE period absent, doubled period present

Its Dependence on the Film Thickness Involves Q.W.S. The S.H.G. Oscillations Reflect Magnetic Properties of the Film.

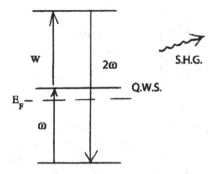

FIGURE 6.19 Illustration of electronic S.H.G. transitions involving a Q.W.S. determining the oscillation period. If the Q.W.S. has even parity then resulting S.H.G. is small. Only Q.W.S. with odd parity cause due to larger dipole matrix elements larger S.H.G. and oscillations. In FM films the Q.W.S. is spin split. Note, parity of Q.W.S. changes for increasing quantum number.

If the interference of light from the surface and interface is Negligible, then different oscillations of the outcoming S.H.G. light as a function of film thickness occur.

The weight of the optical transitions $i \rightarrow j \rightarrow l$ changes as the film thickness increases, see later discussion. Thus, Moke and Nolimoke oscillations occur.

Regarding optical properties, the morphology of the thin film and its magnetic domain structure should play a role in general.

Film multilayers: For magnetic film multilayers, one expects interesting interferences and magnetic optical behavior. Assuming, for example, the structure shown in Figure 6.19, neglecting for simplicity Q.W.S., one gets for S.H.G.

$$I(2\omega) \sim |\chi_s(\vec{M}) + \chi_{int.1} + \chi_{int.2} + \cdots|^2. \quad (6.76)$$

Writing $\chi = \chi^e + \chi^o, \chi^o \sim M$, one has

$$I(2\omega) \sim |\chi_s^e + \chi_s^o(M) + \Sigma_i \chi_{int.i}^e + \cdots|^2 \quad (6.77)$$

for the af structure. In the case of FM aligned films, it is

$$I(2\omega) \sim |\chi_s + \Sigma_i \chi_{int.i}(\vec{M})|^2. \quad (6.78)$$

Note, to detect magnetic domain structure, experiments must achieve high lateral resolution, via interferences, for example, etc. Of course surfaces with a mixture of domains with magnetization $M\perp$ and $M \parallel$ might yield a behavior as observed for the magnetic reorientation transition [24,25].

In summary, the magnetization pattern of a multifilm system is clearly reflected by magnetooptics. The resolution of the optical response can be enhanced using interference effects [23].

6.2.9 Magnetization Dynamics

The time dependence of the magnetization is of great interest. Generally, due to reduced dimension dynamics in nanostructures could be faster than in bulk. For example,

switching of the magnetization in thin films is expected to be faster [26]. Of course, this depends on film thickness, magnetic anisotropy and molecular fields of neighboring magnetic films in a multilayer structure.

To speed up a reversal of the magnetization $(\vec{M} \leftrightarrow -\vec{M})$ within ps times to fs times, one may use thin magnetic films, for example, with many excited electrons, hot electrons, resulting from light irradiation. In heterogeneous magnetic film structures, transfer of angular momentum may be fast so that a fairly fast switching of the topmost film magnetization may occur (Figure 6.20).

The magnetic dynamics is described by (see LL equation)

$$d\vec{M_i}/dt = - \left(\frac{g\mu_B}{\hbar}\right) \vec{M_i} \times \vec{H}_{eff} + \frac{G}{M_s^2} \vec{M_i} \times (\vec{M_i} \times \vec{H}_{eff}). \quad (6.79)$$

Here, i refers to the magnetization of an elementary volume or for heterogeneous structures to a film or a magnetic domain in a nano structured film. $\vec{H}_{eff}$ refers to a molecular field acting on $\vec{M_i}$ and resulting, for example, from neighboring films or magnetic domains. The first term describes precessional motion and the second one relaxation. M_s is the saturation magnetization (Note, using $\vec{M_i} \times \vec{H}_{eff} \rightarrow \partial \vec{M_i}/\partial t$ in the 2. term on the r.h.s. of Eq. (6.79) yields the Gilbert equation with damping parameter G describing spin dissipation)[26].

One gets from Eq. (6.79) (and also from Boltzmann-type theory) that the magnetization in thin films changes during times of the order of (controlled by angular-momentum conservation), see Bennemann et al. [26],

$$\frac{1}{\tau_M} \propto A(T_{el}) \mid V_{\uparrow\downarrow} \mid^2 \overline{NN} + \dots \quad (6.80)$$

Here, in case of excited electrons, T_{el} may refer to the temperature of the hot electrons and $V_{\uparrow\downarrow}$ to the spin-flip scattering potential causing changes in the magnetization (for example, spin-orbit scattering or exchange interactions) and $\overline{N}$ is the average DOS of the electrons.

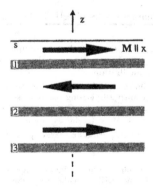

FIGURE 6.20 Multifilm system with AF ordered neighboring films. The shaded areas indicate interface regions important for S.H.G. Note, S.H.G. also results at the surface. Typically, approximately, two atomic layers will contribute. The magnetization is assumed to be parallel to the surface, which normal is in z-direction.

Note, in case of TMs with many hot electrons may yield response times τ_M of the order of 100 fs. Clearly, raising the temperature, $T \to T_{el}$ due to hot electrons, can speed up the magnetic dynamics.

The LL equation is generally important for spin dynamics. One may use the spin-continuity equation $(\partial_t M_i + \partial_\mu j_{i\mu,\sigma} = 0)$ to determine the spin currents $\overrightarrow{j}_\sigma(t)$ (including spin Josephson one) induced by magnetization dynamics and then describe the latter by using the LL equation.

As will be discussed, one gets for magnetic tunnel junctions, of course dependent on the spin mean free path l_s (presumably best if $l_s > l_t$), spin current ($j = -\frac{\partial F}{\partial \phi}$, where F is the free energy and ϕ the phase, see F.Bloch [8]) driven by phase difference $\Delta\Phi$ between the magnetization of neighboring magnets of the form [26,27]

$$\overrightarrow{j_s} \propto \partial \overrightarrow{M}/\partial t \sim \sin(\Delta\Phi). \qquad (6.81)$$

This follows from the LL equation and the spin-continuity equation (see nonequilibrium magnetizm). Note, in accordance with the GMR effect discovered by Grünberg and Fert, the tunnel current depends on the relative orientation of the magnetization on the left and right side of the tunnel junction [25–27].

6.3 Summary

Various experiments can be used to determine the generally spin-dependent currents and Onsager coefficients L^σ. The spin-dependent forces X^σ can generally be manipulated by light creating hot electrons and thus changing the various gradients ΔT, ΔM, etc. For the tunnel system shown, one might expect interesting behavior if, for example, the metal 3 is replaced by Ce which electronic properties, valency changes upon photon induced and controlled population of the s, d and f states. Similarly currents change dramatically if in the figure the material 3 consists of semiconductors like Si, Ge etc. or magnetic semiconductors whose conductivity is strongly affected by hot electrons.

Note, the Onsager coefficients are given by the current correlation function [9,14,15]

$$L_{il}(t) = \langle j_i(t)j_l(0)\rangle, \qquad (6.82)$$

with $j_i(t)$ calculated using response theory, Heisenberg or v.Neumann equation of motion $(\dot\rho \propto [\rho, H] + \cdots)$ or as a functional from the free-energy F [9,14,15], see also Bloch using $jV = dF/dt$ with V being the potential associated with the force $X_{i\sigma}(t)$. Then, (see response theory)

$$j_i(t) = \int_\infty^t dt' L_{il}(t - t') X_l(t'), \qquad (6.83)$$

where the $L_{il}(t)$ are now calculated within an electronic theory. The analysis is simplified if $L_{ij} \propto \delta(t - t')$ (Markov processes, see Kubo [9]).

An electronic theory may be useful in order to apply field theoretical arguments and symmetry considerations,

see, for example, Nogueira [14]. The currents are directly also calculated as derivatives of the electronic free energy [8]. One expects generally the formulae $jV_{th} = dF/dt$, where V_{th} denotes the "potential" associated with the driving force $X_{i\sigma(t)}$ of thermodynamics [9]. Note, for large gradients X_σ nonlinear contributions to the currents may play a role.

In case of magnetic multilayer structures, for example, a FM A on a FM B, one may induce spin currents by shining light on the surface of the thin magnetic film. This creates hot electrons and a temperature gradient and thus induces a spin current etc. Note, according to Maxwell equations, $\Delta\overrightarrow{M(t)}$ will generate electric fields E_i, $i = x, y, z$, which should be reflected in the observed currents.

Onsager theory applies also to currents in (gases) liquids of magnetic ions and in the presence of a polarizing external magnetic field. Separating, for example, such a system by a wall with appropriate holes into two compartments A and B, one may induce charge and spin currents driven by $X_{i\sigma}$, voltage gradient, thermal gradients, magnetic field gradient, etc. One gets approximately for the currents through the wall holes, see, for example, Kubo [9],

$$j_\sigma^\alpha \propto p_\sigma^\alpha/\sqrt{T^\alpha} + \cdots, \alpha = A, B, \qquad (6.84)$$

where p_σ is the partial pressure due to ions with spin σ. Obviously, while for the stationary state $j_i^A = j_i^B$, gradients $X_{i\sigma}$ cause corresponding currents. For example, in case of no pressure difference charge (and spin) may flow from A to B due to a temperature gradient and $T_A < T_B$.

Onsager theory has many applications in thermodynamics. As an example note, in case of magnetostriction thermodynamics yields [9]

$$\Delta M = -\left(\frac{\partial V}{\partial H_{eff}}\right)_{p,T} \Delta p + \cdots \qquad (6.85)$$

Hence, a pressure gradient changes the magnetization and may cause spin current. In particular, a time-dependent pressure $p(t)$ drives a magnetization dynamics.

Onsager theory could also be applied to spin currents in topological insulators and at the interface of semiconducors in the presence of strong magnetic fields. Interplay of spin–orbit coupling and magnetic field and magnetism of substrate of semiconductor should yield interesting results.

Regarding (magnetic) atoms on lattices, including tunnel junctions (structures), spin-dependent gradient forces X_i acting on the atoms may cause currents and novel behavior. Photon assistance of atom or molecule tunneling might be of particular importance.

Coupled currents under the influence of a magnetic field and radiation fields play likely also an important role in interstellar and galactic interactions and could be treated using same theory as above, using Onsager theory for magnetic, ionic gases.

Finally, interesting and novel behavior may occur if (diffusion) currents are accompanied by coupled chemical reactions, see de Groot [9]. Then in magnetic systems spin and magnetization may play a role and cause magnetic effects [28]. It is of interest to use SHG also to study Mie resonance, magnetic vs. non-magnetic clusters.

This discussion demonstrates the many options for inducing spin currents in nanostructures and the powerful general analysis Onsager theory offers.

Acknowledgment

I thank C. Bennemann for help and many useful and critical discussions. This study is dedicated to Prof. J.B. Ketterson (USA) for lifelong help, suggestions and assistance. Similarly, I thank also Prof. V. Bortolani (Italy) for help during many years. Last but not least I thank in particular F. Nogueira, P. Jensen and M.Garcia for ideas and interesting discussions.

Bibliography

1. G.E. Bauer, A.M. MacDonald, S. Maekawa, Spin electronics, Solid State Commun. **150**, 459 (2010).

2. K.H. Bennemann, Magnetic nanostructures, J. Phys. Condens. Matter **22**, 243201 (2010).

3. K.H. Bennemann, Photoinduced phase transitions, J. Phys. Condens. Matter **23**, 073202 (2011).

4. K. Uchida, S. Takahashi et al., Observation of the Spin Seebeck effect, Nature **455**, 778 (2008); F.L. Bakker, A. Slachter, J.P. Adam, B.J. van Wees, Phys. Rev. Lett. (2010); on spin caloritronics : M. Johnson, R.H. Silsbee, Phys. Rev.B **35**, 4959 (1987), M. Johnson, Sol. State Commun. **150**, 543 (2010); L. Gravier, S. Serrano-Giusan, F. Reuse, and J.P. Ansermet, Phys. Rev.B **73**, 024419 (2006).

5. A. Slachter et al., Thermally driven injection from a ferromagnet into a non–magnetic metal, Nat. Phys. **6**, 879 (2010).

6. F. Nogueira and K.H. Bennemann, Europhys. Lett. **67**, 620 (2004). It can be shown that the current j_s^J in the absence of a charge current results from $[\phi, S^z] = i$, (see in analogy to sc. with canonical conjugate variables $[\varphi, N] = i$), thus, $\dot{S}^z = -\partial H/\partial \phi$, $\dot{\phi} = \frac{\partial H}{\partial S^z}$. The current is $j^J = \mu_B \langle \dot{S}_R^z - \dot{S}_L^z \rangle$, where R, L refer to the right and left side of the tunnel junction. Furthermore, it is $H = -E_J S^z \cos \Delta\phi + \frac{\mu_B^2}{2C_s}(S_L^z - S_R^z)^2 + \dots$.

7. B. Kastening, D. Morr, D. Manske, K.H. Bennemann, Phys. Rev. Lett. **96**, 047009 (2006); Phys. Rev. **79**, 144508 (2009).

8. F. Bloch, Phys. Rev. B**2**, 109 (1970).

9. R. Kubo et al., Statistical Mechanics, (North–Holland Publishing Co., Amsterdam, 1965); L.D. Landau and E.M. Lifshitz, Statistical Mechanics (revised by E.M. Lifshitz and L.P. Pitaevski), Pergamon Press 1980; S.R. de Groot, Thermodynamik irreversibler Prozesse, Hochschultaschenbücher, Bd. 18, 1960, and Thermodynamics of Irreversible Processes, North–Holland, 1952.

10. S. Takahashi, H. Imamura, and S. Maekawa, Phys. Rev. Lett. **82**, 3911 (1999).

11. Note, taking into account magnitude and phase of the spin magnetization one gets generally $j_M \propto \Delta|M| + b_1|M|\nabla \exp i\psi$ and similarly for j_σ Josephson like currents. Of course, spin flip scattering plays a role and should be considered.

12. K.H. Bennemann, Ultrafast dynamics in solids, Ann.Phys.(Berlin) **18**, 480 (2009).

13. P.M. Brydon, B. Kastening, D. Morr and D. Manske, Interplay of ferromagnetism and TSC in a Josephson junction, arXiv: 0709.2918v1, 19 Sept 2007; K.H. Bennemann and J. Ketterson, Superconductivity (Springer, 2008), Vol. 2.

14. K.H. Bennemann, F. Nogueira, see Lecture–Notes, FU-Berlin. Of course, expressing the Onsager coefficients L_{ij} by current–current correlation functions and calculating these within Lagrangian theory and electronic Hamiltonian, it is in principle straightforward to get instead of phenomenological Onsager theory results which refer to an electronic Hamiltonian, for example Hubbard hamiltonian. This might be particularly useful for understanding manybody effects, see F. Nogueira, Introduction to the field theory of classical and quantum phase transitions (Lecture Notes, FU–Berlin, September 2010); see also L.D. Landau and E.M. Lifshitz, Kinetics (Vol. 10, Pergamon Press).

15. F. Nogueira and K.H. Bennemann, Current Correlation Functions, FU–Berlin, to be published 2010.

16. A.C. Bleszynski-Jayich, W.E. Shanks, B. Peaudecerf, E. Ginossar, F. von Oppen, L. Glazman, J.G. Harris, Science **326**, 272 (2009); P. Michetti and P. Recher, Bound States and persistent currents in topological insulator rings, arXiv: 1011.5166v1 (2010).

17. B. Tatievski, P. Stampfli, K.H. Bennemann, Ann. der Physik **4**, 202 (1995); Comp. Mat. Sci.**2**, 459 (1994); B. Tatievski, Diploma– thesis, FU–Berlin (1993).

18. E.I. Rashba, Phys. Rev. B **68**, 241315-1 (2003); B.A. Bernevig, Phys. Rev. B **71**, 073201 (2005); A.V. Balatsky, B.L. Altshuler, Phys. Rev. Lett. **70**, 1678 (1993). Spin–orbit coupling causes in a ring a persistent current due to the induced electron phase ϕ_{AC} (Aharonov–Casher effect), see studies in above references. Using Faraday's law $\partial_t \phi_{AC} = -c \oint E_{AC} \bullet dl$ it is $j_{\varphi, \sigma_z} = -(c/2\pi R)\frac{\partial E}{\partial \phi_{AB}}$, $\sigma_z = \pm 1/2$, and electron energies $E_i = E_i(\phi_{AB} + \phi_{AC})$.

19. Assuming particle–hole symmetry it is $j = -\Sigma_i \frac{dE_i}{d\phi} \tanh(\frac{E_i}{2kT})$. Here, i refers to the electron states with energy E_i and for example to

the polygonal orbits of the ring, see Stampfli et al., $E_{t,p} \propto \cos(SB)$ assuming that all orbitals t, p enclose same area S. Then approximately $j = -\int dE N(E) \frac{\partial E}{\partial \phi} \tanh(E/2kT) + \cdots$ and thus at $T = 0$ it is $j = -\Sigma_i \frac{dE_i}{d\phi}$.

20. Note, the Landau–Lifshitz equation, $\frac{d\vec{M}}{dt} = \gamma \vec{M} \times \vec{H_{eff}} + \alpha \vec{M} \times \frac{d\vec{M}}{dt} + \beta \vec{j} \times \partial_t \vec{M}$, includes damping. A time dependent spin current yields in accordance with the Maxwell equations an electric field.

21. A.A. Shanenko, M.D. Critoru, A. Vagov, F.M. Peeters, Phys. Rev. B **78**, 024505 (2008), and more recent publ. by F.M. Peeters et al. This geometrically induced transition has interesting physical consequences.

22. M.E. Garcia, Habilitation thesis, Physik, FU Berlin, 1999.

23. K.H. Bennemann, *Nonlinear Optics in Metals* (Oxford University Press, 1998).

24. P.J. Jensen, K.H. Bennemann, Magnetic Structure of Films, Surface Sience reports **61**, 129 (2006); R. Brinzanik, P.J. Jensen, K.H. Bennemann, Study of nonequilibrium magnetic domain structure during growth of nanostructured ultrathin films, J. of Magnetism and Magnetic Materials **238**, 258 (2002); R. Brinzanik, Ph.D thesis F.U. Berlin (2003), Monte Carlo study of Magnetic nanostructures during Growth; P.J. Jensen, K.H. Bennemann, *Magnetism of 2d-nanostructures*, in book Magnetic Material (Editor A.V. Narlikar, Springer, 2004).

25. P.J. Jensen, D.K. Morr and K.H. Bennemann, Reorientation Transition in Thin ferromagnetic films, Surface Science **307**, 1109 (1994); P.J. Jensen and K.H. Bennemann Temperature Induced Reorientation of Magnetism at Surfaces, Solid State Communic. **100**, 585 (1996).

26. K.H. Bennemann, Ultrafast dynamics in solids, Ann. Physik **18**, 480 (2009).

27. B. Kastening, D.K. Morr, L. Alff and K.H. Bennemann, Charge Transport and Quantum Phase Transition in Tunnel Junctions, Phys. Rev. B **79**, 144508 (2009); B. Kastening, D.K. Morr, D. Manske and K.H. Bennemann, Novel Josephson Effect in Tunnel Junctions, Phys. Rev. Lett. **96**, 47009-1 (2006); F. Nogueira and K.H. Bennemann, Spin Josephson effect in ferromagnetic/ferromagnetic tunnel junctions, Europhysics Lett. **67**.

28. To include chemical reactions occurring in (open) magnetic systems with diffusion currents etc. one may extend Onsager theory taking into account spin and magnetization. For simplicity one may first neglect nonlinear behavior. (However, note frequently this may not be valid.). Then, $j_i = L_{ij} X_j + \cdots$ and chemical reactions are taken into account by the forces $X_j = A_j/T$ with spin dependent chemical affinities $A_j = -\sum_k \nu_{k,j} \mu_k$. Here, the substance k with chemical potential μ_k is also characterized by its spin. The stoichometrical coefficients $\nu_{k,j}$ characterize the chemical reaction which couples to the currents. It is straightforward to work out details of the theory, see de Groot, Landau. Thus one gets spin dependent coupling of diffusion currents and chemical reactions, for example effects due to concentration gradients depending on spin and magnetization, etc.

Joule Heat Generation by Electric Current in Nanostructures

S. V. Gantsevich and V. L. Gurevich

A.F. Ioffe Institute

7.1 Point Contacts and Conducting Channels

The classical and quantum point contacts differ substantially from usual bulk conductors and the experimental and theoretical investigation of their features represents a special problem. For example, one studies a step-like variation of conductance (Ohmic and non-Ohmic), shot noise, thermoelectric properties and a number of other transport effects [1,2].

The resistance of a classical ballistic point contact between two metals was considered at first by Sharvin [3]. The characteristic dimensions of the contact were assumed to be much larger than the de Broglie wavelength. Then Kulik, Shekhter and Omelyanchouk [4] pointed out that the processes leading to electric resistance and heat generation are spatially separated in a classical point contact.

As a further development of the Landauer–Büttiker–Imry [5,6] approach, we consider the heat generation in the case of ballistic Ohmic conduction in semiconductor quantum nanostructures (microcontacts). We assume that the microcontact has a transverse dimension of the order of electron de Broglie wavelength λ. It is joining two reservoirs, each being in independent equilibrium. We show that to measure a quantum conductance, the reservoirs *should be assumed to be classical*. This fact is in accordance with the general quantum theory of measurements.

We wish to emphasize that we do not make any special assumptions concerning the potential distribution along the structure assuming only that it is smooth on the scale of the electron de Broglie wavelength. We consider this a point of some importance as there is a difference between the problem of electrostatic potential distribution in metallic microcontacts and the distribution in semiconductor nanowires. In metals, the potential profile is determined by the spatial distribution of the current and strong screening of the charges (see [7–9]). In semiconductor nanostructures, the electron concentrations are usually rather low and the screening may not be of so much importance. The potential distribution is mainly determined by the arrangement and the potentials of the gate electrodes. Usually there is zero current between the nanostructure and these electrodes. Then the surface of a gate electrode is at a constant potential. By manipulating these potentials, one can in principle create various potential profiles. We consider inelastic electron scattering in the contacts (which, as we have mentioned, are assumed to be classical) and find that for any potential distribution in the nanostructure, the heat generation in both contacts is the same.

We also find that practically all the heat is generated in the classical reservoirs rather than in the quantum contact. This is why we need, along with the equation for the entropy production in the microcontact also, the expressions for the entropy production in classical conductors that will be extensively used throughout the chapter. We give these equations for the various cases of electron and phonon scattering in the first part of the chapter. Although some of them are well known, we are bringing them to the form that can be directly used in the second part.

Now let us discuss the following seemingly paradoxical situation. On the one hand, the rate of Joule heat generation is determined by the relaxation mechanism(s) for the physical system in consideration, or, in other words, by relaxation rates. On the other hand, quite often in nanostructures, there is the so-called collisionless transport where the conductance G is independent of any relaxation rate. This means, in other words, that the resistance and, therefore, the overall heat production does not depend on any mechanism of the electron relaxation. We show by the consideration of heat generation in classical reservoirs how these facts can be reconciled.

It will be convenient to consider an isolated system. One can imagine it as a large capacitor, which is discharged through the conductor of interest. The product RC of the whole system, R and C being the resistance and capacitance, respectively, is much bigger than any relaxation time characterizing the electron or phonon system of the conductor. This means that for all the practical purposes, the conduction process can be looked upon as a stationary one. The total energy of the system, U, is conserved while its total entropy, S, is growing. The rate of heat generation is expressed through $T\partial S/\partial t$ where T is the temperature. So our main purpose will be to calculate the rate of the entropy production for various cases of electric conduction.

7.2 Boltzmann Equations for Electrons and Phonons

Although our main purpose is consideration of nanostructures, to calculate the heat production we need (as has been already mentioned) also the Boltzmann equation for bulk samples since practically all the heat is released in the reservoir regions. Below we briefly discuss its properties bringing them to such a form that can be directly used in the second part of the chapter.

All physical quantities of the electrons and phonons in the quasiclassical approximation can be expressed through their distribution functions $F(\mathbf{p}, \mathbf{r}, t)$ and $N(\mathbf{q}, \mathbf{r}, t)$. The distribution functions give the mean numbers of particles with the given momentum $\mathbf{p}$ or $\mathbf{q}$ in the space point $\mathbf{r}$ at the time t. (In crystal lattice, particles have quasimomentum, but for brevity, the term "momentum" will be used. Also vector variables $\mathbf{p}, \mathbf{q}, \mathbf{r}$ will be frequently written as p, q, r to avoid cumbersome expressions.)

The Boltzmann equation for the particle distribution function can be obtained in the following way. There are two physical reasons for the variations of particle distributions in the coordinate-momentum phase space, namely, the particle movements and their collisions with scatterers. Suppose that the particle momentum $\mathbf{p}$ and its coordinate $\mathbf{r}$ are the functions of time t. The total time derivative dF/dt should be equal to the distribution change due to collisions. Therefore we have,

$$[dF/dt]_{field} = (\partial_t + \dot{\mathbf{r}}\nabla_r + \dot{\mathbf{p}}\nabla_p)F(\mathbf{p}, \mathbf{r}, t)) = [dF/dt]_{col} \tag{7.1}$$

The field term at the left describes the change of occupancy numbers of electrons due to their movement in the phase space according to trajectories $\mathbf{r} = \mathbf{r}(t)$ and $\mathbf{p} = \mathbf{p}(t)$. The term at the right describes the collisions of particles with scatterers after which the particle momentum p becomes p' with some probability $W_{p'p}$ or vice versa. The collision is the swift process of very short duration Δt as compared to the free path of the particle between collisions $\tau \gg \Delta t$. The action of collisions is described by the collision operator, acting on the distribution function $[dF/dt]_{col} \equiv -I_p(F)$. It can be linear or nonlinear. The electron collisions with impurities are described by the linear operator, the electron-phonon collision operator has a linear part proportional to the number of phonons N_q where $\mathbf{q} = \pm(\mathbf{p} - \mathbf{p}')$ and also the (usually small) nonlinear term proportional to $F_p F_{p'}$ without the phonon number N_q. Usually the number of phonon is high $N_q \gg 1$ so one can neglect the nonlinear contributions. The electron–electron scattering is described by the nonlinear collision operator. For simplicity, we consider here at first for electrons only linear (or linearized) collision operator $I_p = I_p^{(imp)} + I_p^{(ph)}$. It has the form

$$I_p F_p = \sum_{p'}(W_{p'p}F_p - W_{pp'}F_{p'}) \tag{7.2}$$

Here $W_{p'p}$ and $W_{pp'}$ are the probabilities of electron transitions $(p \to p')$ and $(p' \to p)$. The collision operator I_p is positive that corresponds to the exponential decay of distribution perturbations in the momentum space:

$$\delta F_p(t) = \exp(-I_p t)\delta F_p(0) \sim \exp(-t/\tau)\delta F_p(0)$$

The field and collision contributions together give the Boltzmann equation for electrons:

$$(\partial_t + \mathbf{v}\nabla_r + \mathbf{f}\nabla_p)F_p + I_p(F) = 0 \tag{7.3}$$

The velocity **v** and the acting force **f** can be found by the equations of motions:

$$\mathbf{v} \equiv \dot{\mathbf{r}} = \nabla_p H(\mathbf{p}, \mathbf{r}), \quad \mathbf{f} \equiv \dot{\mathbf{p}} = -\nabla_r H(\mathbf{p}, \mathbf{r}) \qquad (7.4)$$

where $H(\mathbf{p}, \mathbf{r})$ is the Hamilton function. The velocity and force term in the Boltzmann equation represent the classical Poisson bracket $\{H, F\}$. This bracket has a very useful property. For two space–momentum functions $A(\mathbf{p}, \mathbf{r})$ and $B(\mathbf{p}, \mathbf{r})$, we have,

$$\begin{aligned} \{A, B\} &\equiv (\nabla_p A)(\nabla_r B) - (\nabla_r A)(\nabla_p B) \\ &= \nabla_p (A \nabla_r B) - \nabla_r (A \nabla_p B) \end{aligned} \qquad (7.5)$$

This identity holds since after the differentiation of its right-hand side, two terms with mixed differentiation cancel each other. Note that integration (summation) over p or r with zero or periodic boundaries cancels corresponding gradient terms in (7.5). Such canceling is necessary for the derivation of various continuity equations from Boltzmann equations. The continuity equations connect the densities of physical quantities with their flows.

Now acting in the similar way, we get for phonons:

$$(\partial_t + w\partial_r + J_q)N_q + \Pi_q(N) = 0 \qquad (7.6)$$

Here, $w = (\partial \omega_q / \partial q)$ is the phonon velocity. The phonon frequency ω in general can be a smooth function of the space coordinate r. The linear collision operator $J_q = J_q^{(def)} + J_q^{(e)}$ describes the phonon collisions with electron and lattice defects $J_q = J_q^{(def)} + J_q^{(e)}$. The form of it is similar to I_p:

$$J_q N_q = \sum_{q'}(W_{q'q}N_q - W_{qq'}N_{q'}) \qquad (7.7)$$

The phonon–phonon collisions are described by the nonlinear collision operator $\Pi_q(N)$. It has the form:

$$\begin{aligned} \Pi_q(N) = \sum_{q'q''} \Big\{ &[w_{qq'+q''}[N_q(N_{q'}+1)(N_{q''}+1) \\ &- (N_q+1)N_{q'}N_{q''}] \\ &+ 2w_{q'q+q''}[N_qN_{q'}(N_{q''}+1) \\ &- (N_q+1)(N_{q'}+1)N_{q''}] \Big\} \end{aligned} \qquad (7.8)$$

Here in the first contribution, $w_{qq'+q''} \propto \delta_{qq'+q''}\delta(\omega' + \omega'' - \omega)$ is the decay probability of the selected phonon (q, ω) into the phonons (q', ω') and (q'', ω'') with the participation of already existing phonons $N_{q'}$ and $N_{q''}$. The probability of the reciprocal phonon fusion process is the same. The decay stands for the outcome term in (7.8), while the fusion stands for the income term.

The second double contribution in (7.8) relates to the processes where a selected phonon (q, ω) participates in the decay and fusion processes of phonons (q', ω') and (q'', ω''). In this case, $w_{q'q+q''} \propto \delta_{q'q+q''}\delta(\omega + \omega'' - \omega')$. (Note that terms with $NqN_{q'}N_{q''}$ identically cancel each other in (7.8). They are usually written for more symmetry in the formula).

The collision operators have properties following conservation laws. The linear collision operators I_p and J_q identically vanishe being summed over their indices:

$$\sum_p I_p F_p = \sum_{pp'}[W_{p'p}F_p - W_{pp'}F_{p'}] = \sum_{pp'}[W_{p'p}(F_p - F_p)] \equiv 0 \qquad (7.9)$$

The identity (7.9) means that these types of collisions conserve the number of particles. The phonon–phonon collision operators have no such property due to the processes of phonon merging and phonon splitting. Instead, it conserves the total energy of phonon system and its total momentum:

$$\sum_q \hbar\omega_q \Pi_q(N) \equiv 0, \quad \sum_q \hbar q \Pi_q(N) \equiv 0 \qquad (7.10)$$

Note that phonon momenta are actually (quasi)momenta, so one can add to them an arbitrary vector of reciprocal lattice. The scattering processes with such vectors are called Umklapp processes. They attach the formally independent phonon gas to the lattice atoms. The collisions of electrons with impurities are elastic and conserve the total energy of electrons. The same is true for the phonon collisions with lattice defects. So we have,

$$\sum_p \epsilon_p I_p^{(imp)}(F) = 0, \quad \sum_q \hbar\omega_q J_q^{(def)}(N) = 0 \qquad (7.11)$$

Thus, the distribution of energy between the electron and phonon systems occurs due to electron–phonon collisions. Since the phonon–phonon collisions conserve the total phonon energy, they do not participate in the energy balance. Therefore

$$\sum_p \epsilon_p I_p(F) + \sum_q \hbar\omega_q J_q(N_q) = 0 \qquad (7.12)$$

7.3 The Energy and Charge Conservation and Their Transfer

The total energy of electron system $U^{(e)}$ and the total energy of phonon system is given by

$$U^{(e)} = \sum_p \epsilon_p F_p, \quad U^{(ph)} = \sum_q \hbar\omega_q N_q \qquad (7.13)$$

Using the electron and phonon Boltzmann equations and conservation laws, we obtain a set of phenomenological equations describing the charge and energy transfer in our system. We have the continuity equation:

$$e\frac{\partial n}{\partial t} + \mathrm{div}\mathbf{j} = 0 \qquad (7.14)$$

where $n(\mathbf{r})$ and $\mathbf{j}$ are the electron concentration and current:

$$n(\mathbf{r}) = \sum_p F_p(\mathbf{r}), \quad \mathbf{j} = e\sum_p \mathbf{v}F_p(\mathbf{r}) \qquad (7.15)$$

The physical sense of the continuity equation is simple and transparent. Since the number of electrons is conserved by collisions, the charge density can change only by their movements which creates the corresponding current. The equation (7.15) shows the charge conservation and its transfer in space. The analogous equation takes place for the energy density:

$$\frac{\partial U}{\partial t} + \text{div}\mathbf{Q} = \mathbf{j}\mathbf{E} \tag{7.16}$$

Here $\mathbf{E}$ is the acting electric field. It accelerates electrons producing the energy which enters in the electron–phonon system. This energy later dissolves in the entire system where $U = U^{(e)} + U^{(e)}$ is the total energy density. Though the electron and phonon energies are not conserved separately their sum which is equal to the system total energy remains unchange during electron–phonon interaction. The total energy flow Q is equal to

$$\mathbf{Q} = \sum_p \mathbf{v}\epsilon_p F_p + \sum_q \hbar\omega_q N_q \mathbf{w}_q \tag{7.17}$$

The summation over the phonon branches is implied.

These phenomenological continuity equations are obtained from the Boltzmann equations by its summation over all electron and phonon states. During the summation, one should take into account the conservation laws with the identity (7.5) to get the expressions for the particle and energy flows. For example, using electron Boltzmann equations, we have the following relations:

$$e\sum_p \mathbf{v}\nabla F_p = \nabla(e\sum_p \mathbf{v}F_p) = \text{div}\mathbf{j} \tag{7.18}$$

$$e\sum_p \epsilon_p \mathbf{v}\nabla F_p = \nabla(e\sum_p \epsilon_p \mathbf{v}F_p) = \text{div}\mathbf{Q}^{(\mathbf{e})} \tag{7.19}$$

Analogous transformations is applied to the phonon Boltzmann equation.

7.4 The Entropy and Its Production

The notion of entropy is the basic notion of thermodynamics. Every many-body system can be realized as a set of large number of independent subsystems. All configurations of such subsystems are considered as equivalent provided that they correspond to the same state of a macroscopic body. The entropy S of each degree of freedom under such approach is defined by way of a probability W according to the Boltzmann formula $S = -\ln W$. For a given number of fermion states G and the number of particles N, we have,

$$S = \ln[\frac{G!}{(G-N)!N!}] \simeq G\ln G - (G-N)\ln(G-N) - N\ln N \tag{7.20}$$

Defining the state occupancy number as $F_p = N/G$ and omitting the constant, we get the entropy of the electron state p:

$$s_p^{(e)} = -F_p \ln F_p - (1 - F_p)\ln(1 - F_p) \tag{7.21}$$

For the boson degree of freedom, we have,

$$S = \ln\left[\frac{(G+N)!}{G!N!}\right] \simeq (G+N)\ln(G+N) - G\ln G - N\ln N \tag{7.22}$$

For the phonon state occupancy $N_q = N/G$, we get correspondingly the state entropy $s^{(ph)}$:

$$s_q^{(ph)} = (N_q + 1)\ln(N_q + 1) - N_q \ln N_q \tag{7.23}$$

The entropy has maximum value at equilibrium. Therefore, under given mean energy and mean number of particle, the expressions (7.21) and (7.23) determine the equilibrium distributions for fermions and bosons. For fermions (electrons), we have the Fermi–Dirac distribution function $F_p^{(eq)}$:

$$F_p^{(eq)} = \frac{1}{\exp[(\epsilon_p - \mu)/T] + 1} \tag{7.24}$$

Here, T is the temperature in energy units (the Boltzmann constant $k_B = 1$). The temperature T determines the mean electron energy in the state p and the chemical potential μ their mean numbers. The boson (phonon) state entropy $s^{(ph)}$ is maximal for the Bose–Einstein distribution function for which $\mu = 0$ becomes the Planck function.

$$N_q^{(eq)} = \frac{1}{\exp(\hbar\omega_q/T) + 1} \tag{7.25}$$

The boson Bose–Einstein distribution and the fermion Fermi–Dirac distribution are equilibrium distributions, which determine the equilibrium state of a physical system. The collision operators acting on the equilibrium distribution functions give zero:

$$I_p(F^{(eq)}|N^{(eq)}) = J_q(N^{(eq)}) = \Pi_q(N^{(eq)}) = 0 \tag{7.26}$$

The deviations of equilibrium distributions in the momentum space tend to zero due to collisions. However, the quick collision relaxation determines only the form of distribution in momentum space but not the values of temperature and particle concentrations. Therefore, their local values remain arbitrary so that the total equilibrium is established later by slow diffusion-like processes, which are described by the continuity equations derived above. The energy introduced in the system creates local microscopic nonequilibrium states in momentum space, which quickly relax to the local equilibrium states with some local values of temperature and entropy. Then the slow relaxation processes in coordinate space begin. These processes can be described by phenomenological continuity equations with or without sources.

Now let us consider this general scheme for the case of heat production and its spatial propagation.

The total entropy of the electron and phonon gases can be expressed through the entropy space density as

$$S = \int d^3r[S^{(e)}(\mathbf{r}) + S^{(ph)}(\mathbf{r})] = \int d^3r S(\mathbf{r}) \tag{7.27}$$

where $S^{(e)} + S^{(p)}$ are the sum of the entropy densities of all electron and phonon states:

$$S^{(e)} = \sum_p s_p^{(e)}, \quad S^{(ph)} = \sum_q s_q^{(ph)} \qquad (7.28)$$

The entropy densities $s_p^{(e)}$ and $s_q^{(ph)}$ contain the phonon and electron distribution functions, which generally depend on time and space coordinates. Thus, the total entropy density becomes space and time dependent. Its evolution can be found by using the procedure described above for the particle concentration and energy density. As a result, the following entropy continuity equation takes place

$$\frac{\partial S}{\partial t} + \operatorname{div} \mathbf{J}^{(s)} = \Gamma \qquad (7.29)$$

Here, Γ is the entropy source and $\mathbf{J}^{(s)}$ is the entropy flux, which is the sum of electron and phonon contributions.

$$\mathbf{J}^{(s)} = \sum_p \mathbf{v} s_p^{(e)} + \sum_q \mathbf{w} s_q^{(ph)} \qquad (7.30)$$

The distribution function variations δF_p and δN_q give rise to the entropy density variations:

$$\delta s_p^{(e)} = \ln \frac{1 - F_p}{F_p} \delta F_p, \quad \delta s_q^{(ph)} = \ln \frac{N_q + 1}{N_q} \delta N_q.$$

Thus, the entropy density variations are connected to the distribution function variations by the functional derivatives $(\delta s_p^{(e)}/\delta F_p) = \ln(1 - F_p)/F_p$ and $(\delta s_q^{(ph)}/\delta N_q) = \ln(N_q + 1)/N_q$

The source term $[dS/dt]_{col} \equiv \Gamma$ is given by

$$\Gamma = -\sum_p \ln \frac{1 - F_p}{F_p} I_p(F) - \sum_q \ln \frac{N_q + 1}{N_q} [J_q(N) + \Pi_q(N)] \qquad (7.31)$$

Integrating the continuity equation (7.29) over the entire volume of the system and supposing that the entropy flows at the system boundaries are absent, one obtains the total entropy production within the system:

$$\frac{dS}{dt} = \int d^3 r \Gamma(\mathbf{r}, t) \qquad (7.32)$$

7.5   Heat and Mechanical Energy

Mechanical energy is the maximal amount of work that can be produced if the system finally reaches the state of thermodynamic equilibrium, where all energy is the energy dissipated at all possible freedom degrees in a way that corresponds to the maximal entropy. An isolated macroscopic system generally possesses a certain amount of mechanical energy [10]. The amount of work depends on how the internal state of the system under consideration is changed.

It is well known (see [10]) that

$$\mathcal{E} = U - U_0(S) \qquad (7.33)$$

where $U_0(S)$ is the total energy of the system in thermodynamic equilibrium expressed as a function of its entropy S. Calculating its time derivative, one can write

$$\frac{d\mathcal{E}}{dt} = \frac{dU}{dt} - \frac{dU_0}{dS}\frac{dS}{dt} = \frac{dU}{dt} - T_0\frac{dS}{dt} \qquad (7.34)$$

since $T_0 dS = dU_0$ in equilibrium at constant volume. The temperature T_0 is the temperature of the equilibrium system whose entropy is equal to S. The first term on the right-hand side vanishes because of the conservation of energy of an isolated system. Making use of the expression (7.32), one gets

$$\frac{d\mathcal{E}}{dt} = -T_0\frac{dS}{dt} = -T_0 \int d^3 r \Gamma \qquad (7.35)$$

This means that the mechanical energy of an isolated system, unlike its total energy, is not conserved. The system relaxation processes tend to bring the system into equilibrium, and because of this, the mechanical energy dissipates into heat. The expression (7.34) describes the rate of this processes of energy degradation when mechanical energy transforms into Joule heat.

7.6   Entropy Production Source for Different Scattering Mechanisms

Let us find here the explicit expressions for the entropy production source Γ due to various types of collisions between electrons, phonons and impurities. For the bulk samples, we will either give them without derivation or even just discuss their most important properties. This is an essential part of the chapter as we consider here a nanostructure joining two contacts that are assumed to be classical, or, in other words, are bulk samples. We will see that *it is in the classical contacts that the heat is generated.* To be able to calculate it, it is necessary to give the rate of heat generation in the bulk classical systems, i.e. the systems whose dimensions are much larger than the de Broglie wavelengths of participating particles.

7.7   Collisions of Electrons with Impurities

For simplicity, we will consider a case where a crystal and an impurity have a center of symmetry. Then

$$\Gamma^{(imp)} = \frac{n_i}{2} \sum_{pp'} w_{p'p}(F_p - F_{p'}) \ln \frac{(1 - F_{p'})F_p}{(1 - F_p)F_{p'}} \qquad (7.36)$$

Here, n_i is the impurity concentration, the transition probabilities $w_{pp'} = w_{p'p}$ contain the delta function $\delta(\epsilon_p - \epsilon_{p'})$, which ensures the electron energy conservation since the scattering is supposed to be elastic. One can see that expression (7.36) is nonnegative. It vanishes if F_p is an arbitrary function of electron energy $F_p \equiv F(\epsilon_p)$. Physically this means that collisions of electrons with impurities only redistribute them within a constant energy surface.

7.8   Electron Electron Collisions

Again we will assume that the crystal has a center of symmetry. Then,

$$\Gamma^{(ee)} = \frac{1}{4} \sum_{pp'kk'} W_{kk'}^{pp'} \Phi_{pp'}^{kk'} \ln \frac{F_p F_{p'} (1 - F_k)(1 - F_{k'})}{F_k F_{k'} (1 - F_p)(1 - F_{p'})}$$

$$(7.37)$$

where

$$\Phi_{kk'}^{pp'} = F_p F_{p'} (1 - F_k)(1 - F_{k'}) - F_k F_{k'} (1 - F_p)(1 - F_{p'})$$

$$(7.38)$$

(Note that the terms with $F_p F_{p'} F_k F_{k'}$ vanish in this expression). The quantity $W_{kk'}^{pp'}$ is the probability of an electron-electron collision where (pp') is the state before a collision and (kk') is the state after the collisions. The total energy and momentum of two electrons are conserved:

$$W_{kk'}^{pp'} \propto \delta(\mathbf{k} + \mathbf{k}' - \mathbf{p} - \mathbf{p}') \delta(\epsilon_k + \epsilon_{k'} - \epsilon_p - \epsilon_{p'}).$$

Note that the entropy change $\Gamma^{(ee)}$ vanishes provided that $F_p = F^{(eq)}(\epsilon_p)$ is the equilibrium Fermi–Dirac function with an arbitrary chemical potential and temperature. It also vanishes for the distribution function of the form

$$F_p \equiv F^{(eq)}(\epsilon_p - \mathbf{p}\mathbf{V} - \mu)$$

$$(7.39)$$

where $\mathbf{V}$ is the arbitrary electron-drift velocity. This distribution means that electron gas moves as a whole retaining its equilibrium form of energy distribution. Such situations can take place in semiconductors.

7.9   Electron Phonon Collisions

In electron–phonon systems, there is the permanent exchange of energy and momentum between both subsystems. Therefore, to calculate the entropy production, one should treat simultaneously the electron–phonon and phonon–electron collisions. One obtains [11–13]:

$$\Gamma^{(e-ph)} = \sum_{pp'} W_{p'p}(q) \Phi_{pp'}(q) \ln \frac{F_p (1 - F_{p'})(N_q + 1)}{F_{p'} (1 - F_p)) N_q}$$

$$(7.40)$$

where

$$\Phi_{pp'}(q) = F_p (1 - F_{p'})(N_q + 1) - F_{p'} (1 - F_p) N_q \quad (7.41)$$

Here, $W_{p'p}(q)$ is the probability of an electron–phonon collision with the transition $(p \rightarrow p')$ and the energy conservation $\epsilon_p = \epsilon_{p'} + \hbar \omega_q$. As for the (quasi)momenta of electrons and phonons, they satisfy the conservation law

$$\mathbf{p}' + \hbar \mathbf{q} - \mathbf{p} = \hbar \mathbf{b}$$

$$(7.42)$$

where $\mathbf{b}$ is a vector of the reciprocal lattice. For normal processes, $\mathbf{b} = 0$, and we have

$$\mathbf{p}' + \hbar \mathbf{q} - \mathbf{p} = 0$$

$$(7.43)$$

In the general case of Umklapp processes, expression (7.42) with $\mathbf{b} \neq 0$ is valid.

The quantity $\Gamma^{(e-ph)}$ is nonnegative. It vanishes if F_p is a Fermi function $F^{(eq)}$ of $(\epsilon_p - \mu)/T$ with an arbitrary chemical potential μ and temperature T whereas N_q is a Planck function $N^{(eq)}$ of $\hbar \omega_q / T$ depending on *the same* temperature T. If only normal electron–phonon collisions are of importance it also vanishes for the electron distribution function of the form (7.39) provided that, at the same time, the phonon distribution function is the Planck drift function of $(\omega_\mathbf{q} - \mathbf{q}\mathbf{V})/T$.

7.10   Phonon Phonon Collisions

We will consider three phonon processes due to the cubic anharmonic terms in the interaction Hamiltonian. We have,

$$\Gamma^{(ph)} = \frac{1}{2} \sum_{qq'q''} w_{qq'+q''} \Phi_{qq'q''} \ln \frac{N_q (N_{q'} + 1)(N_{q''} + 1)}{(N_q + 1) N_{q'} N_{q''}}$$

$$(7.44)$$

$$\Phi_{qq'q''} = [N_q (N_{q'} + 1)(N_{q''} + 1) - (N_q + 1) N_{q'} N_{q''}] \quad (7.45)$$

(Note that two terms $N_q N_{q'} N_{q''}$ in $\Phi_{qq'q''}$ identically cancel each other). The probability $w_{qq'+q''}$ in (7.44) is proportional to the product $\delta(\mathbf{q}' + \mathbf{q}'' - \mathbf{q}) \delta(\omega_{q'} + \omega_{q''} - \omega_q)$, which describes the conservation of energy and wave vectors under collisions. Note that phonons in crystals have (quasi)momentum so the wave vector conservation equality generally should be written in this way:

$$\mathbf{q} = \mathbf{q}' + \mathbf{q}'' + \mathbf{b}$$

$$(7.46)$$

where $\mathbf{b}$ is either 0 (normal processes) or a vector of reciprocal lattice (Umklapp processes). If the Umklapp processes are of importance, then the expression (7.44) vanishes provided that N_q is the Planck function with an arbitrary temperature.

7.11   Collisions of Phonons with Lattice Defects

The explicit expression for the source Γ for this case does not differ very much from expression (7.36), and we will not give it here (see [1]). It vanishes for N_q being an arbitrary function of the phonon energy $\hbar \omega_q$.

7.12   Applied Electric Field and Ohm's Law

Let us consider the important case where a weak electric field causes small deviations of the equilibrium distributions. We have,

$$F_p = F^{(eq)}(\epsilon_p) + \Delta F_p, \quad N_q = N^{(eq)}(\omega_q) + \Delta N_q \quad (7.47)$$

where ΔF_p and ΔN_q satisfy the *linearized* Boltzmann equations. The functions $\Delta F_\mathbf{p}$ and $\Delta N_\mathbf{q}$ should be proportional to the external electric field, which we assume to be the source of the deviation of the distribution functions from equilibrium. As the collision terms vanish for the equilibrium parts of the distribution functions, one can express the heat production through the linearized collision operators acting on ΔF_p and ΔN_q. The linear terms in ΔF_p and ΔN_q vanish in the equation for the entropy production. This can be either easily checked directly or established using the following physical considerations. A linear term in ΔF_p or ΔN_q can be of either sign, whereas the rate of heat generation should be nonnegative.

Introducing in (7.36) the functions ΔF_p and $\Delta F_{p'}$, we get [11]

$$\Gamma^{(imp)} = (n_i/2) \sum_{pp'} W_{pp'} \frac{(\Delta F_p - \Delta F_{p'})^2}{F_p(1 - F_{p'})} \quad (7.48)$$

Here, for brevity $F = F^{(eq)}$ and $N = N^{(eq)}$. To give the leading terms of previous expansions for Γ in less cumbersome form, it is convenient to introduce the functions χ_p and ν_q defined as follows:

$$\Delta F_p = \chi_p F_p(1 - F_p), \quad \Delta N_q = \nu_q N_q(N_q + 1) \quad (7.49)$$

We have for the electron–electron, electron–phonon and Phonon–phonon collisions, respectively

$$\Gamma^{(ee)} = \frac{1}{4} \sum_{pp'kk'} W_{pp'}^{kk'} F_k F_{k'}(1 - F_p) \quad (7.50)$$
$$\times (1 - F_{p'})(\chi_p + \chi_{p'} - \chi_k - \chi_{k'})^2$$

$$\Gamma^{(e-ph)} = \sum_{pp'q} W_{pp'}(q) F_{p'}(1 - F_p) N_q(\chi_{p'} - \chi_p + \nu_q)^2 \quad (7.51)$$

$$\Gamma^{(ph)} = \frac{1}{2} \sum_{qq'q''} w_{qq'+q''} N_{q'}(N_{q''} + 1)(\nu_{q'} + \nu_{q''} - \nu_q)^2 \quad (7.52)$$

To get the total rate of heat production, one should integrate these equations over the whole volume of the conductor and multiply the result by temperature T.

We already indicated that expressions (7.48) and (7.50)–(7.52) are quadratic in the deviations of the distribution functions form the equilibrium values. This means that *to calculate the Joule heat, it is sufficient to solve the Boltzmann equation up to the first order in the electric field* **E**. We consider this as an important point as in some papers, to calculate the heat production, the Boltzmann equation is solved up to the second order.

In view of further applications, it is worthwhile to indicate that these equations can be presented in a different form. As an example, we consider the expression (7.51). Then one can write the time variations of ΔF_p and ΔN_q due to collisions using the operator notations $I_p^{e-ph}(\Delta F)$ and $I_q^{ph-e}(\Delta N)$. Since in equilibrium

$$\ln \frac{1 - F_p}{F_p} = \frac{\epsilon_p - \mu}{T}, \quad \ln \frac{1 + N_q}{N_q} = \frac{\hbar \omega_q}{T}, \quad (7.53)$$

one can see that due to energy and electron number conservation laws during electron–phonon collisions, the terms

linear in ΔF_p and ΔN_q vanish. This is as it should be because the linear terms can be of either sign whereas the entropy production should be nonnegative. The quadratic term can be written in the following form (below again $F^{(eq)} = F$ and $N^{(eq)} = N$):

$$\Gamma^{(e-ph)} = \sum_p \frac{\Delta F_p}{F_p(1 - F_p)} I_p^{(e-ph)}(\Delta F_p) \quad (7.54)$$
$$+ \sum_q \frac{\Delta N_q}{N_q(1 + N_q)} J_q^{(ph-e)}(\Delta N_q)$$

The expressions (7.50) and (7.52) can be also transformed to a similar form.

If the relaxation processes within the phonon system are rapid, one can neglect the second term in (7.54) and get

$$\Gamma^{(e-ph)} \simeq \sum_p \frac{\Delta F_p}{F_p(1 - F_p)} I_p^{(e-ph)}(\Delta F) \quad (7.55)$$

Here, $I_p^{(e-ph)}$ is a linearized integral operator acting on ΔF_p. It consists of the outcome and income terms acting on some momentum function x_p in this way:

$$I_p^{(e-ph)} x_p = \frac{x_p}{\tau_p} - \sum_{p'} W_{pp'} x_{p'} \quad (7.56)$$

Here, $1/\tau_p$ is the outcome relaxation time:

$$\frac{1}{\tau_p} \equiv \sum_{p'} W_{p'p} \quad (7.57)$$

For the electron scattering on phonons, the probabilities $W_{pp'}$ and $W_{p'p}$ are not equal (for brevity below $\hbar = 1$):

$$W_{p'p} = \sum_q c_q^2 [\delta(\epsilon_{p'} + \omega_q - \epsilon_p)\delta_{pp'+q}(N_q + 1 + F_{p'})$$
$$+ \delta(\epsilon_p + \omega_q - \epsilon_{p'})\delta_{p'p+q} N_q] \quad (7.58)$$

where c_q is the electron–phonon coupling constant.

$$W_{pp'} = \sum_q c_q^2 [\delta(\epsilon_p + \omega_q - \epsilon_{p'})\delta_{p'p+q}(N_q + 1 + F_p)$$
$$+ \delta(\epsilon_{p'} + \omega_q - \epsilon_p)\delta_{pp'+q} N_q] \quad (7.59)$$

The probabilities W depend on the electron distribution because of original nonlinearity of electron–phonon interaction. It includes the case where two electrons participate in the phonon emission.

In a number of cases, for some physical reasons, the income term of the collision operator can be discarded. For such cases in the relaxation time approximation, we have

$$\Gamma^{e-ph} = \sum_p \frac{(\Delta F_p)^2}{\tau_p F_p(1 - F_p)} \quad (7.60)$$

One can use this expression if ΔF_p has a sharp maximum for some value(s) of **p** and the integral in Eq. (7.60) is dominated by the values of ΔF_p near the maximum (see, for instance, Abrikosov [14], Section 7.3). Such a situation will be considered below.

7.13 Examples

Here, we will briefly discuss several examples of the calculation of heat production in bulk samples in the Ohmic regime. They are a preliminary step for calculation of the heat production by a current through a nanostructure for various types of reservoirs. Although below we will calculate the heat production for one definite type of reservoir, the data given here permit one to generalize the results for other cases.

7.14 Residual Resistance

This is the simplest example of heat generation. It is nevertheless interesting particularly because one can sometimes find in the physical literature statements that only inelastic electron collisions can result in a generation of heat since the collisions with impurities conserve the electron energy.

For weak electric fields, the deviation of electron distribution ΔF_p from equilibrium satisfies the Boltzmann equation with the linear collision operator I_p:

$$I_p \Delta F_p = -e E_\alpha \frac{\partial F}{\partial p_\alpha} \qquad (7.61)$$

Let us show how to solve such equations in a formal way using the inverse collision operator I_p^{-1}. This operator is defined by the relation $I_p^{-1} I_p = I_p I_p^{-1} = 1$. The identity should be fulfilled for any function y_p upon which these operators act. However, for $F_p = F_p^{(eq)}$, we have $I_p F_p = 0$ and the identity does not holds. To avoid the ambiguity any functions proportional to $F_p^{(eq)}$ cannot stand under the signs of I_p^{-1}. For usual algebra, it is convenient that such functions also do not appear under the sign I_p. Therefore, one can unambiguously solve the equation $I_p x_p = y_p$ only in the function space with zero momentum sums. Then its general solution is obtained by the following procedure:

$$I_p x_p = I_p(x_p - F_p \sum_{p'} x_{p'}/N) = y_p,$$

$$x_p = F_p \sum_{p'} x_{p'}/N + I_p^{-1} y_p \qquad (7.62)$$

Here, N is the number of electrons $\sum_p F_p = N$. The right-hand side of equation (7.61) is just the function with the zero sum $\sum_p (\nabla_p F_p) = 0$, as well as the functions ΔF_p. So we have,

$$\Delta F_p = -e\mathbf{E} I_p^{-1} \frac{\partial F}{\partial \mathbf{p}} \qquad (7.63)$$

Now we find the electric current produced by the field E (for simplicity, we consider the unit volume for our system so that the particle number N is equal to the concentration $n = N/V_0$).

$$j_\alpha = -e^2 \sum_p v_\alpha I_p^{-1} \frac{\partial F}{\partial p_\beta} E_\beta = \sigma_{\alpha\beta} E_\beta \qquad (7.64)$$

Here, $\sigma_{\alpha\beta}$ is the conductivity tensor. Using the previous formulae, we get the energy balance in the form

$$T \left[\frac{\partial S}{\partial t} \right]_{col} = \sigma_{\alpha\beta} E_\alpha E_\beta \qquad (7.65)$$

Here $T dS$ is the amount of heat produced by the current as $\mathbf{j E} dt$.

We see that purely elastic collisions have resulted in a heat generation. This has a clear physical meaning. The amount of order in the electron distribution resulting in electric current can bring about mechanical work. For instance, one can let the current flow through a coil and a magnetic rod can be drawn into the coil. In such a way, the electrons transferring the current can execute a work on the rod. As a result of scattering, the amount of order in the electron distribution diminishes and this means that the high- quality mechanical energy converts into the low-quality heat energy with the corresponding increase of entropy.

In this example, inelastic collisions are unnecessary to generate the heat. If the electron contribution to the specific heat is predominant over the lattice (Debye) contribution, the energy will remain within the electron system even when there is some inelastic electron–phonon scattering. For the opposite specific heat ratio, it may eventually go to the lattice. But even in such a case, if the electron-impurity collisions happen more often than the electron–phonon ones, it is they who determine the heat generation. Of course, the inelastic collisions are necessary for the electron system to reach full equilibrium. However, under our assumptions, they should but very little influence the resistance and heat generation.

7.15 Electron–Phonon Scattering

Another possible situation is Joule heat generation under predominantly electron–phonon scattering. We assume that such relaxation processes within the electron–phonon system as phonon–phonon and phonon–defect scattering are so effective that the phonons, with sufficient accuracy, can be considered at equilibrium. Then we have ($F = F^{(eq)}$, $N = N^{(eq)}$):

$$\left[\frac{\partial S}{\partial t} \right]_{coll} = \sum_{pp'q} W_{p'p}(q) F_p (1 - F_{p'}) N_q (\chi_p - \chi_{p'})^2 \quad (7.66)$$

It can be visualized as a result of (quasi)momentum and energy relaxation of electrons due to their collisions with the equilibrium phonons.

This result can also be expressed in a form of (7.65) using the inverse linearized collision operator for electron–phonon scattering [15]. Unlike the scattering by impurities in this case, the electron scattering is inelastic and the produced heat does not remain in the electron system.

7.16 Mutual Electron–Phonon Drag

If normal electron–phonon collisions are predominant, we have $\chi_{p'} - \chi_p = \mathbf{p V}/T$ and $\nu_q = \hbar \mathbf{q V}/T$. Since $\mathbf{p}' - \mathbf{p} = \mathbf{q}$, the entropy production given by expression (7.51) vanishes. This means that all the entropy production is determined by phonon processes although the electrons experience only inelastic collisions with the phonons. In regard to the geometry of heat generation, this case may be opposite to the one considered above. In the residual resistance discussed above, the heat generation repeats the spatial distribution of $\mathbf{E}^2(\mathbf{r})$ (provided that its variation is smooth). To the contrary, the case here may be a nonlocal one as the heat generation is determined by the phonon collisions. It may take place, for instance, *outside* the region of main potential drop.

7.17 Transport in a Quantum Nanostructure

Let us consider a nanostructure (see Figure 7.1) where there is a direction (x-axis) along which the electron motion is infinite and current flows. Along the perpendicular direction, $r_\perp \equiv (y, z)$, the electron motion is quantized. To be definition, we will discuss a 3D case, although one can turn to a 2D case by a simple change of notation. For simplicity, in Figure 7.1, we depicted rectangular boundaries of a nanostructure bridge. Actually there are smooth transitions to the bulk regions with very small curvature variation on a scale of the electron de Broglie wavelength. Moving in such a region electrons change slowly their state parameters without state transitions so that the state occupancies remain the same. Such adiabatic transition regions avoid unnecessary reflections of electron flows back to massive contacts. It means that we will consider the so-called adiabatic transport (see Glazman et al. [16]) where the potential profile varies smoothly along x-axis on the scale of λ (where λ is the electron de Broglie wavelength). We assume that the electron mean free path is much bigger than the characteristic dimensions of the microstructure. Then there is a system of one-dimensional (1D) electron bands (*channels*) describing the electrons' motion in the x-direction both in the microstructure and in the adjoining parts of the contacts. This motion will be considered (quasi)classically. The transverse motion is quantized. In the spirit of an approach by Glazman et al. [16], we assume that the variables x and $r_\perp$ are separable in the adiabatic approximation. This means that for each value of x, the $r_\perp$-dependence of the potential determines the wave function of transverse quantization $\psi_n(x|y, z)$ that depends on x as a parameter. Here n is the quantum number of transverse quantization. The electron spectrum depends on x, and this dependence has the following form

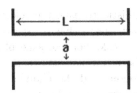

FIGURE 7.1 Nanostructure bridge.

$$\epsilon_n(p|x) = p^2/2m + \epsilon_n(0|x) \qquad (7.67)$$

where m is the electron effective mass and $\epsilon(0, x)$ is the position of the band bottom that depends on x as a parameter. The energy $\epsilon_n(0|x)$ is the solution of the eigenvalue problem for the Hamiltonian

$$-\frac{\hbar^2}{2m}\left(\frac{\partial^2}{\partial y^2} + \frac{\partial^2}{\partial z^2}\right) + e\phi(x, y, z) \qquad (7.68)$$

The electron wave functions can be presented in the following form

$$\psi(\mathbf{r}) = \mathrm{const}\,\frac{1}{\sqrt{p(x)}}\exp\left(i\int p(x')dx'/\hbar\right)\psi_n(x|r_\perp) \qquad (7.69)$$

Here, the wave functions of the transverse quantization ψ_n depend on x as a parameter, n is the corresponding quantum number.

The energy conservation relation for such a system can be written by analogy with expression (7.16) in the following integral form

$$\frac{dU}{dt} = JV \qquad (7.70)$$

where

$$U = \sum_n \int dx \sum_p \epsilon_{np} F_{np} \qquad (7.71)$$

J is the total current across the nanostructure that is given by

$$J = \sum_{np} v_{np} F_{np} \qquad (7.72)$$

Here summation over the spin variable is implied, $v_{np} = \partial\epsilon_{np}/\partial p = p/m$ is the electron group velocity (which does not depend explicitly on n), $F_{np}(x)$ is the electron distribution function that depends on the quantum number n as a parameter, while p (the x-component of the electron (quasi)momentum) and x are classical variables. The function $F_{np}(x)$ satisfies the Boltzmann equation

$$\frac{\partial F_{np}}{\partial t} + v\frac{\partial F_{np}}{\partial x} - \frac{\partial F_{np}}{\partial p}\frac{\partial\epsilon_{np}}{\partial x} + I_{np}(F) = 0 \qquad (7.73)$$

7.18 Calculation of Entropy Production for a Nanostructure

The rate of heat generation for a nanostructure system is given by Eq. (7.35)

$$\frac{dE}{dt} = -T\frac{dS}{dt} \qquad (7.74)$$

For such an electron system interacting with phonons, the entropy production is given by

$$\frac{dS}{dt} = \int dx\left[\frac{\partial S}{\partial t}\right]_{\mathrm{coll}} \qquad (7.75)$$

$$\left[\frac{\partial S}{\partial t}\right]_{coll} = \sum_{np} \ln \frac{F_{np}}{1 - F_{np}} I_{np}(F) + \sum_{q} \ln \frac{N_q}{N_q + 1} J_q(N)$$

$$(7.76)$$

Here the electron collision operator is given by

$$I_{np}(F) = \sum_{n'p'} w_{p'p}^{n'n}(q)[(N_q + 1)F_{np}(1 - F_{n'p'})$$

$$- N_q F_{n'p'}(1 - F_{np})] + w_{pp'}^{nn'}(q)[N_q F_{np}(1 - F_{n'p'})$$

$$- (N_q + 1)F_{n'p'}(1 - F_{np})] \qquad (7.77)$$

where

$$w_{pp'}^{nn'}(q) \propto \delta_{p'p+\hbar q}\delta(\epsilon_{np} + \hbar\omega_q - \epsilon_{n'p'}) \qquad (7.78)$$

$w_{pp'}^{nn'}(q)$ is proportional to the probability of an electron to make a transition from the state specified by the quantum numbers n', p' to the state n, p accompanied by emission of a phonon with wave vector q. In such a general form, this equation is valid for any phonon states. If one considers interaction of the electrons of a nanowire with bulk phonons, then perturbation theory gives (cf. with [17,18])

$$w_{pp'}^{nn'}(q) = \frac{2\pi}{\hbar}|c_{\mathbf{q}}|^2|\langle n| \exp(i\mathbf{q}_\perp \mathbf{r}_\perp)|n'\rangle|^2$$

$$\delta(\epsilon_{np} - \epsilon_{n'p'} + \hbar\omega_q)\delta_{p'p+\hbar q_x} \qquad (7.79)$$

Here, c_q is the matrix element of the interaction of electrons with bulk phonons; $\mathbf{r}_\perp$ denotes y, z; while $\mathbf{q}_\perp$ denotes q_y, q_z. Then, we have

$$J_q^{(ph-e)}(N_q) = \sum_{nn'}\sum_{pp'} w_{pp'}^{nn'}(q)[F_{np}(1 - F_{n'p'})N_q$$

$$- F_{n'p'}(1 - F_{np})(N_q + 1)] \qquad (7.80)$$

Employing (7.76), we finally get,

$$\left[\frac{\partial S}{\partial t}\right]_{coll} = \sum_{nn'}\sum_{pp'q} w_{pp'}^{nn'}(q)\Phi_{pp'}^{nn'}$$

$$\ln \frac{F_{n'p'}(1 - F_{np})(N_q + 1)}{F_{np}(1 - F_{n'p'})N_q}, \qquad (7.81)$$

$$\Phi_{pp'}^{nn'}(q) = F_{n'p'}(1 - F_{np})(N_q + 1) - F_{np}(1 - F_{n'p'})N_q$$

We do not give here the corresponding rate for the interaction of electrons with impurities within the nanowire. The point is that the item of our main interest will be nanowires where the impurity scattering is of no importance. (Where it is of importance, it may be difficult to describe it because the widely used procedure of averaging over the impurity positions which is implied in Eq. (7.36) is not applicable within a nanostructure. However, such a procedure is usually applicable outside the nanostructure, within the reservoir region).

The contribution of the electron–electron interaction to entropy production within a nanowire can be described in principle by a straightforward generalization of Eq. (7.50). However, the possibility of electron–electron collisions is severely restricted for this case (see [19]). Indeed, one can

easily check that because of 1D electron dynamics, the conditions for the energy and (quasi)momentum conservation in the course of electron–electron collisions cannot usually be satisfied. (We can also remark that for sufficiently small values of the relative velocity of two interacting electrons, these collisions cannot be treated by a perturbation theory). This is why we are not giving here the corresponding expression. Again, outside the nanowire, within a contact region one can apply Eq. (7.50) above.

7.19 Heat Generated by Ballistic Current Through Nanostructure

In the spirit of the Landauer–Büttiker-Imry approach [6], we assume the quantum microstructure to be connected with the reservoirs which we call "left" (+) and "right" (−), each of these being in independent equilibrium. We assume that the electrons enter the contacts adiabatically (see [16]). As, however, the width of the classical contacts is much larger than the width of the nanostructure, the number of the channels within the contacts is also much larger. Most of these channels are not current-carrying as the electrons belonging to them are reflected from the wire backward into the corresponding contact. We start with consideration of the current-carrying channels within the nanostructure.

As is shown in [16], unless the bottom of an electron 1D band almost touches the Fermi level (the situation we do not consider in this chapter), the electron transport through a nanostructure is practically reflectionless, which will be assumed here. Let us for example consider the case $p > 0$. Then we are interested in the region $x > 0$ where the relaxation and therefore heat production takes place. The states in the current-carrying channels entering the contact region, i.e. at $x = 0$, will have the distribution function $F^{(eq)}(\epsilon_{pn} - \mu^{(+)})$. For $x < 0$, this distribution function is the same as for all other states at $x < 0$. This means that the current-carrying states with $p > 0$ are at equilibrium with all other states in the region $x < 0$. Indeed, at $x < 0$, all the states with $p > 0$ as well as with $p < 0$ have the distribution function of the same form $F^{(eq)}(\epsilon_{pn} - \mu^{(+)})$. To the contrary, for $x > 0$, the distribution of electrons in the current-carrying channels will be out of equilibrium relative to the rest of electrons [whose distribution is $F^{(eq)}(\epsilon_{pn} - \mu^{(-)})$]. This means that, in this region, the relaxation processes should tend to bring the current-carrying states into equilibrium. It is these processes that we are going to investigate to calculate the overall heat production and its spatial distribution.

We consider ballistic transport of electrons within the nanostructure, which means that the scattering of electrons there is negligibly weak. For the sake of definiteness, let us assume that the electrons relax in the contacts due to collisions with phonons and the phonon relaxation due to anharmonicity and scattering by lattice defects (as well as to the escape of the nonequilibrium phonons from the nanocontact region) is rapid so that Eq. (7.55) is valid, the phonon system being in equilibrium. The electrons to be scattered should penetrate rather deeply into the contact region. Physically, this means that the number of channels in the contact region is large as compared with the number in the nanostructure. As the number of electrons in the contacts is large, they can be treated as a 3D electron gas so that the usual Bloch wave representation can be used instead of the channel representation when this is more convenient. For adiabatic transport, the number of nodes of a function of transverse quantization $\psi_n(\mathbf{r}_\perp|x)$ for a current-carrying channel is invariant in x and is small. Far from the nanostructure, the bottoms of the channels, $\epsilon(0, x)$ are approaching each other. This means that a few electron states that are out of equilibrium have transverse parts of their (quasi)momenta small as compared to the longitudinal part. In other words, their (quasi)momenta have practically the same value $\mathbf{p}$, i.e. only a few electron states of all the states belonging to the contact region are out of equilibrium. This means that the nonequilibrium part of the electron distribution function is a sharp function of the (quasi)momemntum $\mathbf{p}$ and we are dealing with the situation discussed after Eq. (7.60). The income term of the collision operator should be negligibly small as compared to the outcome relaxation time term in (7.60), so one can use the relaxation time approximation [20].

Within this approximation, the nonequilibrium part of the distribution function, ΔF_{np}, satisfies the following equation

$$v_{np}\frac{\partial \Delta F_{np}}{\partial x} + \frac{\Delta F_{np}}{\tau} = 0 \qquad (7.82)$$

Strictly speaking, the boundary condition for this equation should be formulated somewhere near the border of the contact region. As, however, by our assumption, the electrons do not experience collisions within the nanostructure, or, in other words, the electron mean free path is much bigger that the longitudinal dimension of the nanostructure we practically would introduce no mistake by formulating the boundary conditions at $x = 0$

$$\Delta F_{np}(0) = \frac{\partial F_{np}}{\partial \mu}\Delta\mu \qquad (7.83)$$

where $\Delta\mu = \mu^{(+)} - \mu^{(-)} = eV$. We assume here that $eV \ll T$.

The solution of Eq. (7.82) with the boundary condition given by Eq. (7.83) is

$$\Delta F_{np} = \frac{\partial F_{np}}{\partial \mu}\Delta\mu \exp\left(-\frac{x}{v\tau_\mathbf{p}}\right) \qquad (7.84)$$

where $F_{np} = F_{np}^{(eq)}$ is the equilibrium Fermi function. We have neglected the difference between the channel velocity v_{np} and the 3D velocity $v_\mathbf{p}$. Inserting (7.84) into (7.60) we get for spatial distribution of the rate of heat production

$$T\left[\frac{\partial S}{\partial t}\right]_{coll} = 2(eV)^2 \sum_{np} \frac{1}{\tau_p}\frac{\partial F_{np}}{\partial \mu}\exp\left(-\frac{2x}{v\tau_p}\right) \qquad (7.85)$$

Taking into account Eqs. (7.74) and (7.75), we get for the overall heat production within the reservoir

$$\frac{d\mathcal{E}}{dt} = -2(eV)^2 \sum_{np} \int_0^\infty dx \exp\left(-\frac{2x}{v\tau_p}\right)\frac{\partial F_{np}}{\partial \mu}(1/\tau_p) \qquad (7.86)$$

After the summation and integration, we get for the contact region

$$\frac{d\mathcal{E}}{dt} = -\frac{(eV)^2}{2\pi\hbar}\mathcal{N} \qquad (7.87)$$

where $\mathcal{N}$ is the number of active channels, i.e. the channels whose bottoms are below the Fermi level. Calculating the contribution of the region $x < 0$, we find that for the considered case, the full rate of heat generation is the same in both contacts (even though the values of the relaxation times τ_p may be different). This is true irrespective to the actual form of the potential profile. The expression (7.86) gives the spatial distribution of the generated heat. The heat is spread over regions of length equal to the mean free path $v\tau_p$, i.e. well outside the nanostructure.

One can present the total rate of heat generation for the entire system in the form

$$\frac{d\mathcal{E}}{dt} = -GV^2 \qquad (7.88)$$

where for G one has the well-known expression [21]

$$G = \frac{e^2}{\pi\hbar}\mathcal{N} \qquad (7.89)$$

Thus, the calculation of the heat production provides an alternative method to calculate the conductance of a nanostructure. We can also remark that the fact that the relaxation time τ_p was due to the electron–phonon collisions was not crucial for the derivation of expression (7.86). We would have come to the same relation if, for instance, τ_p had been due to another scattering mechanism within the classical contact (for example, the electron–electron collisions).

A crucial point for our derivation was the fact that we have neglected the difference between the channel velocity v_{np} and the 3D velocity $v_\mathbf{p}$. In other words, this means that *we treat the reservoir regions classically*. This fact agrees with the general concept of quantum theory of measurements. To perform a quantum measurement (of a conductance in the present case), one needs a classical measuring device (i.e. classical reservoirs for the present case).

One can obtain the same result using classical Eq. (7.60) for the entropy production. We are going to outline here briefly such a calculation. We have

$$\left[\frac{\partial S}{\partial t}\right]_{coll} = \int_0^\infty dx \int d^2\mathbf{r}_\perp \sum_p \frac{(\Delta F_p)^2}{\tau_p F_p(1 - F_p)} \qquad (7.90)$$

For ΔF, one can use Eq. (7.84) where (as is stated above) one can disregard the dependence on the small transverse components of (quasi)momentum $\mathbf{p}_\perp$ or, in other words, on the channel number n. The integration should be produced over discrete values of $\mathbf{p}_\perp$. As indicated above, the number of terms in such a sum should be equal to the number of open channels, and we again end up with Eq. (7.87) irrespective to the actual scattering mechanism that is responsible for the relaxation time τ. The only dependence of this result on the actual form of the potential profile within the nanostructure is in the number of active channels $\mathcal{N}$.

7.20 Electron–Hole Symmetry of Degenerate Conductors

Two classical reservoirs connected by the bridge with degenerate electrons show equal amount of entropy generation, which means that energy produced by an applied field dissipates equally in both ends of the bridge. These situations surprisingly differ from the energy dissipation in usual electronic devices like vacuum tubes or transistors where one boundary electrode emits charged particles while the another absorbs them after the acceleration by the electric field applied between them. The absorber electrode (anode or collector) receives in such device nearly all energy acquired by carriers and just this electrode requires cooling for stable work. It seems that though the bridge contains only electrons, there are two opposite and equal flows of carriers of different signs. This situation reminds the motion of electrons and holes in a semiconductor with equal number of electrons and holes.

Let us discuss this phenomenon in more detail. At first, one can introduce the quantity $H_{np} = 1 - F_{np}$, which is the distribution function of holes. (For simplicity, consider only first energy level of transverse quantization $n = 1$. For the function H_p, we obtain the collision operator of the same form as for the function F_p:

$$I_p(H) = \sum_{p'} w_{p'p}(q)[(N_q + 1)H_{p'}(1 - H_p) - N_q H_p(1 - H_{p'})]$$
$$+ w_{pp'}[N_q H_{p'}(1 - H_p) - (N_q + 1)H_p(1 - H_{p'})]$$
$$(7.91)$$

Using the same form for electron and hole distribution, we can write an expression for the distribution variation ΔF as well as ΔH.

$$v\frac{\partial \Delta F}{\partial x} + \frac{\Delta F}{\tau} = 0, \quad v\frac{\partial \Delta H}{\partial x} + \frac{\Delta H}{\tau} = 0, \quad (7.92)$$

The first expression in (7.92) relates to one reservoir with the excitation above the Fermi level (electrons), while the second expression relates to the second reservoir with the excitations below this level (holes).

Despite the possible difference in the dissipation mechanisms in both reservoirs, the total amount of absorbed energy is the same since it should be equal to the total gain of energy of both types of carriers that have the same mass and the same velocity provided the numbers of electron and holes are equal. This condition requires the strong inequality $T/\mu \ll 1$ because only narrow energy strip near the Fermi level participates in all transport phenomena, including entropy production. The electron-hole symmetry

holds only if one can neglect the variations of the state density in this strip.

7.21 Conclusion

The methods developed here can also be applicable for other problems. One example is heat release in the contacts in the course of tunneling of electrons.

Above, we considered mostly the Ohmic conduction regime. The simplest example of non-Ohmic conduction is a ballistic resistance for $eV \gg T$ [17,22,23]. Again the heat generation takes place well outside the nanostructure, within the contacts. Its calculation should go along the line developed above.

More complicated are various aspects of the so-called phonon-assisted ballistic resistance, both Ohmic and non-Ohmic [17,18,24]. In the classical regime, nonlinear phenomena in the current-voltage characteristics of point contacts between normal metals were observed and discussed in a pioneering work by Yanson [25]. Here we mean a quantum situation where the conduction electrons experience some scattering within the nanostructure. As indicated in [26], the situation here is quite unlike the usual collisionless transport where the electron–phonon (as well as the electron–electron and electron–impurity) interactions are restricted to the contacts and hence all the heat is released in the contacts only. In the mentioned case, some energy is transferred to phonons and may be released as heat by the phonon system outside both the wire and the contacts.

In the present chapter, we have calculated the Joule heat release in the course of collisionless Ohmic transport of electric current through a quantum nanostructure. We come to the conclusion that to calculate the Joule heat for this case (as well as for any case of Ohmic conduction), it is sufficient to solve the Boltzmann equation up to the first order in the voltage drop across the nanostructure (or in electric field $\mathbf{E}$). We find that the heat is spread over the length of the electron mean free path in the reservoirs. Even if the electron mean free paths in both reservoirs are different, the total rate of heat generation in each reservoir is the same. We indicate that a calculation of the heat generation may provide an alternative method of computation of collisionless conductance.

Finally, as some auxiliary matter let us compare the classical and quantum approaches to the conductivity in a simple qualitative way.

7.22 Elementary Approach to Conducting Channels

The quantum picture of the conductivity based on the concept of quantum channels seemingly has no common with the usual classical picture of classical conductivity where the central point is the free path confines by collisions. There are a number of papers trying to establish a correspondence between both pictures. However, the rigorous approach to this problem meets great mathematical difficulties with no definite results [27,28]. Below we compare both approaches to conductivity using only simple elementary means and try to show that even in diffusion regime of conductivity one can introduce the notion of conducting channels at least at the qualitative level.

In the quantum channel approach, the conductivity is the sum of quantum channel contributions. For simplicity, let us suppose that all channels have the same conductivity with the same transmission coefficients. Therefore, the total sample conductivity can be written as the sum of "conductivity quantum" proportional to the Coulomb velocity $V_C = e^2/\hbar$. This velocity frequently appears in many-body electron system, for example, the ground-state electron energy of a hydrogen atom is equal to $mV_C^2/2$. The velocity V_c is less than light velocity by the factor $\alpha = e^2/\hbar c$. It is convenient for the following estimates to choose the channel transmission coefficient in such a way that each channel contributes $V_C = c\alpha$ to total conductivity. Then from the quantum expression for the conductivity, we have a simple estimate for the inverse resistance:

$$1/R = \sum_j T_j \sigma_j = V_c \beta = c\alpha\beta \qquad (7.93)$$

Here β is the number of conductivity channels with the conductivity $T[2e^2/2\pi\hbar] = e^2/\hbar = V_C$

The quantity β depends on the sample physical characteristics in the following simple way. It is the ratio of two numbers:

$$\beta = N/N_s \qquad (7.94)$$

Here, the first number is the number of carriers N, while the second number N_s may be called the "site number". It is equal to the number of nods on the particle quantum mechanical trajectory ("current channel") or more simply to its path length L_c divided by the de Broglie wavelength $N_s = L_c/\lambda$. The definition of β has a very clear physical meaning, but the real determination of it for arbitrary conductors is rather obscure.

Now let us suppose that we have a sample with the length L and the cross-section S. A d.c. voltage U is applied to the ends of the sample. Let us consider two limiting cases. The first one is the carrier motion without collision or *ballistic regime*, while the second one is the ordinary collision-controlled regime or *diffusion regime*. The word *diffusion* is used to emphasize that there are very great number of collisions, making particle motion similar to diffusion process.

For ballistic motion, we have from the energy conservation law:

$$\frac{mv^2}{2} + eU = \frac{m(v+V)^2}{2} \qquad (7.95)$$

where m is the carrier mass, while v and $v + V$ are its velocity near the ends of the sample. Assuming $v \gg V$, we get for the drift velocity V:

$$V = \frac{eU}{mv} \qquad (7.96)$$

Expressing the current through the drift velocity in the usual way we obtain,

$$J = jS = en_0V = e(N/L)(eU/mv) = (e^2N/mvL)U. \qquad (7.97)$$

Introducing the de Broglie wavelength $\lambda = \hbar/mv$ and the fine structure constant, we can rewrite this formula in such a way:

$$1/R = c(e^2/\hbar c)(\hbar/mv)N/L = c\alpha\beta \qquad (7.98)$$

Here, the coefficient β has the abovementioned meaning and in this case is given by

$$\beta = N/(L/\lambda) = N/N_s \qquad (7.99)$$

Now let us derive the similar expression for the diffusive regime. We have for the inverse resistance of a sample:

$$\frac{1}{R} = \frac{S\sigma}{L} = \frac{e^2N\tau}{mL^2} \qquad (7.100)$$

Here, we use for the conductivity σ the simple Drude formula with a constant relaxation time τ:

$$\sigma = \frac{e^2n_0\tau}{m} \qquad n_0 = N/LS \qquad (7.101)$$

The expression (7.101) can be rewritten as

$$1/R = (e^2/\hbar)(\hbar/mv)(v\tau)N/L^2$$

or (with $l = v\tau$) as

$$1/R = c\alpha[N\lambda l/L^2] \qquad (7.102)$$

Now we should show that the expression in the bracket may be interpreted as the coefficient β in (7.93). Let us suppose that the carrier suffers M collision through its way L. Then,

$$L = \sum_{j=0}^{M} l_j$$

Under the diffusive regime, the path L is a random quantity with zero average and can be measured by its square:

$$L^2 = \sum_{jj'} l_j l_{j'}$$

After averaging, we have

$$L^2 = \sum_{j=0}^{M} l_j^2 = Ml^2 \qquad (7.103)$$

This relation is the central point of the derivation. We have

$$\beta = [N\lambda l/L^2] = N\lambda/Ml = N/(L_c/\lambda) = N/N_s \qquad (7.104)$$

We see that if we take into consideration the actual carrier path length $L_c = Ml$, the inverse resistance formula for the diffusive regime will become the same as for the ballistic regime. Though the carrier trajectory in this case is not a straight line, the ratio $N_s = L_c/\lambda$ has just the same meaning as for the ballistic motion. Thus, we see that from pure physical point of view, there is no principal difference between quantum and classical conductivity. The difference appears only in the methods of description.

References

1. V. L. Gurevich, *Transport in Phonon Systems*. North Holland, New York (1986).

2. Some aspects of the Joule heat generation are briefly discussed in: V. L. Gurevich, *Pis'ma Zh. Eksp. Teoret. Fiz.*, **63**, *1*, 61 (1996) [*Sov. Phys. — JETP Lett.*, **63**, *1*, 70 (1996)].

3. Yu. V. Sharvin, *Zh. Exp. Teor. Fiz.*, **48**, 984 (1965) [*Sov. Phys.—JETP*, **21**, 655 (1965)].

4. I. O. Kulik, R. I. Shekhter and A. N. Omelyanchouk, *Solid State Commun.*, **23**, 301 (1977)

5. R. Landauer, *IBM J. Res. Dev.*, **1**, 233 (1957); **32**(3), 306 (1989).

6. Y. Imry, Physics of mesoscopic systems. In: "*Directions in Condensed Matter Physics*", Ed. by G. Grinstein and G. Mazenko, World Scientific, Singapore (1986) p. 101; M. Büttiker, *Phys. Rev. Lett.*, **57**, 1761 (1986).

7. I. O. Kulik, R. I. Shekhter and A. G. Shkorbatov, *Zh. Eksp. Teoret. Fiz.*, **81**, 2126 (1981) [Sov. Phys. —JETP **54**, (1981)].

8. I. F. Itskovich, I. O. Kulik and R. I. Shekhter, *Solid State Commun.*, **50**, 421 (1984).

9. Y. B. Levinson, *Zh. Eksp. Teoret. Fiz.*, **95**, 2175 (1989) [*Sov. Phys.—JETP*, **68**, 1257 (1989)].

10. L. D. Landau and E. M. Lifshitz, *Statistical Physics*, Part 1, 3rd edition, Pergamon Press, Oxford (1993).

11. J. M. Ziman, *Electrons and Phonons*, Clarendon Press, Oxford (1960).

12. E. M. Lifshitz and L. P. Pitaevskii, *Physical Kinetics*, Pergamon Press, Oxford (1981).

13. A. G. Aronov, Yu. M. Galperin, V. L. Gurevich and V. I. Kozub, *Adv. Phys.*, **30**, 539, 1981. The results for normal metals can be obtained from the equation for superconductors given in this paper by going to the limit $\Delta \to 0$ where Δ is the superconducting gap.

14. A. A. Abrikosov, *Fundamentals of the Theory of Metals*, North-Holland (Elsevier), Amsterdam, 1988.

15. S. V. Gantsevich, V. L. Gurevich and R. Katilius, *Phys. Condens. Matter*, **18**, 165 (1974).

16. L. I. Glazman, G. B. Lesovik, D. E. Khmel'nitskii and R. I. Shekhter, *Pis'ma Zh. Eksp. Teoret. Fiz.*, **48**, 218 (1988) [*Sov. Phys.—JETP Lett.*, **48**, 238 (1988)].

17. V. L. Gurevich, V. B. Pevzner and K. Hess, *Phys. Rev. B* **51**(8), 5219 (1995).

18. V. L. Gurevich, V. B. Pevzner and K. Hess, *J. Phys. Condens. Matter*, **6**, 8363 (1994).

19. J. P. Leburton, *Phys. Rev. B*, **45**, 11022 (1995).

20. One can see that the relaxation time approximation is actually unnecessary for calculation of the overall heat release within the contact region(s). It is, however, essential for calculation of the spatial distribution of the heat.

21. See, for instance, C. W. J. Beenakker and H. van Houten, Quantum transport in semiconductor nanostructures. *Solid State Phys.*, **44** (1991). edited by H. Ehrenreich and D. Turnbull, Academic Press.

22. V. L. Gurevich and V. B. Pevzner, *Phys. Rev. B*, **56**, 13088 (1998).

23. V. L. Gurevich, V. B. Pevzner and G. J. Iafrate, *Phys. Rev. Lett.*, **75**, 1352 (1995).

24. V. B. Pevzner, V. L. Gurevich and E. W. Fenton, *Phys. Rev. B*, **51**, 9465 (1995).

25. I. K. Yanson, *Zh. Eksp. Teoret. Fiz.*, **66** (1974) [*Sov. Phys.—JETP* **39**, 506 (1974)].

26. V. L. Gurevich, V. B. Pevzner and G. J. Iafrate, *J. Phys. Condens. Matter*, **7**, L445 (1995).

27. E. W. Fenton *Phys. Rev. B*, **46**, 3754 (1992).

28. S. G. Agarwal, *Phys. Rev. A*, **45**, 8165 (1992).

Quantum Transport Simulations of Nano-systems: An Introduction to the Green's Function Approach

Andrea Droghetti
University of the Basque Country

Ivan Rungger
National Physical Laboratory

8.1 Introduction

Charge transport properties of materials (in particular metals) at the macro- and microscale are well described according to the Drude model (see, e.g., Chapter 1 in Ref. [2]). The model assumes that electrons in a solid can be treated much like classical particles in a pinball machine instantaneously bouncing off each other, the crystal ions and other degrees of freedom such as the lattice vibrations. In modern days, the Drude model is recovered from the semi-classical Boltzmann equation which treats electrons as particles having an effective mass m^* determined by the material band structure (as well as interactions) and undergoing random quantum scattering (see, e.g., Chapter 17 in Ref. [28]). The most significant result of the Drude model is that it explains the empirical Ohm's law and it relates the conductivity to the mean-free path between scattering

events l. The success of the semi-classical picture relies on the fact that the scattering events associated with l are inelastic, which means that they do not conserve the energy and therefore the quantum mechanical phase of the electron wave-function. In metals at low temperatures, l can be of the order of 100 Å.

Electrons can travel all the way through systems with dimensions smaller than l before their initial phase is destroyed. In this case, the transport is denoted as *coherent*. Coherent transport is nowadays observed in many nano-devices, comprising, for example, two-dimensional semiconductor heterostructures, suspended graphene, nanowires, nanotubes and single molecules. Importantly, not all scattering events are inelastic. In fact, elastic scattering may occur with impurities, defects and interfaces, where the energy is conserved and the transport is coherent. If neither inelastic nor elastic scattering occurs in a device, which is the case when both l and the elastic mean-free path are larger than the device dimension, the transport mechanism is usually referred to as *ballistic*[1].

The description of nano-electronic devices operating in the coherent transport regime requires a full quantum mechanical description that accounts for the wave-like nature of electrons. The development of the theory started already in the 1950s [25,26] and was extended in particular in the 1980s and early 1990s [5,27]. Yet, it has been mostly during the last decade that, thanks to the advancements in computational techniques, the theory has been successfully applied to support and steer experiments in areas such as graphene-based electronics, molecular electronics and spintronics [3,33,36,41,51]. Methods combining transport theory and density functional theory (DFT) [21] are nowadays in the tool-box of any theorist working in those areas and several software for simulating coherent transport in nanodevices from first-principles are freely available.

In this chapter, we review the fundamental theory for coherent transport, and we explain how it can be implemented thanks to the Green's function formalism [48]. Coherent transport implicitly means that the interaction of the electrons either with the other electrons or with other (quasi-)particles (phonons, magnons, etc.) is neglected, because those interactions generally lead to energy non-conserving scattering processes. Importantly, neglecting interaction effects allows us to avoid the use of advanced quantum many-body techniques in the derivation of the main equations. In fact, the chapter requires only a background knowledge of quantum mechanics at the undergraduate level together with basic concepts in solid-state physics.

By guiding the reader through several examples, our goal is to show how transport simulations can be carried out in practice. Moreover, we note that the Green's functions approach as presented here is at the core of state-of-the-art DFT-based transport schemes. Therefore, we dedicate the last part of the chapter to advice the reader on how to use freely distributed software, analyze the results and also understand numerical as well as fundamental limitations.

Finally, we note that a number of text-books dedicated to quantum transport were published during the last few years, for example, Refs. [8,9,11,14,16,32].

8.2   The Linear Combination of Atomic Orbitals Method

We start the chapter by reviewing some basic notions in electronic structure theory. These will be needed in the following sections in order to describe the fundamental concepts and equations in quantum transport. We refer the reader to text-books about electronic structure theory for a more

complete discussion, such as Ref. [50], for a simple, but at the same time detailed, treatment.

8.2.1   General Formalism

In electronic structure theory, the wave-function of an electron $\Psi(\mathbf{r})$ is represented as a linear combination of N orbitals (the *basis set*) $\Phi_\alpha(\mathbf{r}) = \langle \mathbf{r}|\Phi_\alpha\rangle$, where α is a collective variable denoting all relevant quantum numbers. Assuming that the basis set is complete, we have

$$\Psi(\mathbf{r}) = \sum_{\alpha=1}^{N} \langle \mathbf{r}|\Phi_\alpha\rangle\langle\Phi_\alpha|\Psi\rangle = \sum_{\alpha=1}^{N} u_\alpha \Phi_\alpha(\mathbf{r}) \qquad (8.1)$$

with $u_\alpha = \langle\Phi_\alpha|\Psi\rangle$. The orbitals in the basis set may not be orthonormal; therefore, to describe this general case, we define the orbital overlap $S_{\alpha\beta} = \langle\Phi_\alpha|\Phi_\beta\rangle$, which reduces to $S_{\alpha\beta} = \delta_{\alpha\beta}$ for the special case of orthonormal orbitals. The specific choice of the basis set depends on the particular problem of interest. In quantum transport, a widely used choice, that we will also adopt here, is to consider a linear combination of atomic orbitals (LCAO) centered at the positions $\mathbf{r}_A$ and with atomic quantum numbers n, l, m, so that $\alpha = (\mathbf{r}_A, n, l, m)$ and $|\Phi_\alpha\rangle = |\mathbf{r}_A, n, l, m\rangle$, with $\langle\mathbf{r}|\mathbf{r}_A, n, l, m\rangle = R_{nl}(\mathbf{r} - \mathbf{r}_A)Y_{lm}(\theta, \phi)$. Here, $R_{nl}(\mathbf{r} - \mathbf{r}_A)$ is the radial component, which depends on the principal quantum number n and on the orbital quantum number l, and $Y_{lm}(\theta, \phi)$ is the spherical harmonic describing the angular component. Note that in this chapter, we will only consider non-spin-polarized systems. A generalization of all presented equations to the spin-polarized case is straight forward, while we refer to Ref. [44] for a detailed treatment of systems with non-collinear magnetism.

By using the expansion in the basis set, the time-independent Schrödinger equation

$$\hat{H}\,\Psi_n(\mathbf{r}) = E_n\Psi_n(\mathbf{r}) \qquad (8.2)$$

(with $n = 1, ..., N$) reduces to

$$\sum_{\beta=1}^{N} H_{\alpha\beta}\,u_{\beta,n} = E_n \sum_{\beta=1}^{N} S_{\alpha\beta}u_{\beta,n}, \qquad (8.3)$$

where $H_{\alpha\beta} = \langle\Phi_\alpha|\hat{H}|\Phi_\beta\rangle$. Alternatively in a short matrix notation, one can write

$$\mathbf{H}\boldsymbol{\psi}_n = E_n\,\mathbf{S}\,\boldsymbol{\psi}_n. \qquad (8.4)$$

$\boldsymbol{\psi}_n$ is a vector of dimension N and elements $\{u_{\alpha,n}\}$, normalized as $\boldsymbol{\psi}_n{}^\dagger\mathbf{S}\boldsymbol{\psi}_m = \delta_{nm}$. $\mathbf{H}$ and $\mathbf{S}$ are Hermitian matrices of dimension $N \times N$ and elements $(\mathbf{H})_{\alpha\beta} = H_{\alpha\beta}$ and $(\mathbf{S})_{\alpha\beta} = S_{\alpha\beta}$. Note that in this chapter, we indicate both matrices and vectors with bold fonts.

The electron density, $\rho(\mathbf{r})$, is obtained as $\rho(\mathbf{r}) = 2\sum_{n=1}^{N} f_n|\Psi_n(\mathbf{r})|^2$, where the factor 2 takes into account the spin. The occupation probabilities, $f_n = f(E_n)$, are given by the Fermi–Dirac distribution

$$f(E) = \frac{1}{1 + e^{\frac{E - E_F}{k_B \tau}}}, \qquad (8.5)$$

---

[1]Note that coherent and ballistic are sometimes used as synonyms. However here we prefer to distinguish between the two definitions.

which reduces to $\theta(E - E_F)$ at zero temperature. Here, k_B is the Boltzmann constant, τ is the electronic temperature and E_F is the Fermi energy. The electron density can be expressed as $\rho(\mathbf{r}) = \sum_{\alpha,\beta=1}^{N} \rho_{\alpha\beta} \Phi_\beta^*(\mathbf{r}) \Phi_\alpha(\mathbf{r})$, where we have introduced $\rho_{\alpha\beta} = 2 \sum_{n=1}^{N} f_n u_{\alpha,n} u_{\beta,n}^*$, the elements of the density matrix $\boldsymbol{\rho} = 2 \sum_{n=1}^{N} f_n \boldsymbol{\psi}_n \boldsymbol{\psi}_n^\dagger$. The total number of electrons, N_e, is then obtained from $N_e = \int \rho(\mathbf{r})d\mathbf{r} = \sum_{\alpha,\beta=1}^{N} \rho_{\alpha\beta} S_{\beta\alpha} = \mathrm{Tr}[\boldsymbol{\rho S}] = 2 \sum_{n=1}^{N} f_n$. For a given N_e, this relation implicitly determines E_F. We can rewrite the density matrix as

$$\boldsymbol{\rho} = 2 \int_{-\infty}^{\infty} dE \sum_{n=1}^{N} \delta(E - E_n) f(E) \boldsymbol{\psi}_n \boldsymbol{\psi}_n^\dagger, \quad (8.6)$$

and define the density of states (DOS) as the number of states per energy E

$$D(E) = \sum_{n=1}^{N} \delta(E - E_n). \quad (8.7)$$

We then have $N_e = 2 \int_{-\infty}^{\infty} dE f(E) D(E)$. Finally, we introduce the DOS projected over a state of the basis set (PDOS)

$$d_\alpha(E) = \sum_{n=1}^{N} \delta(E - E_n) \sum_{\beta=1}^{N} u_{\alpha,n} u_{\beta,n}^* S_{\beta\alpha} \quad (8.8)$$

so that $D(E) = \sum_{\alpha=1}^{N} d_\alpha(E)$.

8.2.2 Bloch States

For periodic systems, the wave-function is a Bloch function $\Psi_\mathbf{k}(\mathbf{r}) = \langle \mathbf{r} | \Psi_\mathbf{k} \rangle$ where $\mathbf{k}$ is the wave-number inside the Brillouin zone (BZ). $|\Psi_\mathbf{k}\rangle$ is expanded by using the atomic orbital basis set $|\Phi_\alpha\rangle = |\mathbf{R}_i, a\rangle$ as

$$|\Psi_\mathbf{k}\rangle = \sum_{\mathbf{R}_i} \sum_{a=1}^{N_a} e^{i\mathbf{k}\mathbf{R}_i} u_{\mathbf{k}a} |\mathbf{R}_i, a\rangle, \quad (8.9)$$

where $\mathbf{R}_i$ is the vector that identifies the position of the unit cell, and a is a collective index that labels both the atoms inside the unit cell and their quantum numbers and runs from 1 to N_a (see the following examples). The elements of the Hamiltonian and overlap matrices are then denoted as $H_{\alpha\beta} = H_{\mathbf{R}_i,a;\mathbf{R}_j,a'}$ and $S_{\alpha\beta} = S_{\mathbf{R}_i,a;\mathbf{R}_j,a'}$.

Thus, the Schrödinger equation becomes

$$\sum_{a=1}^{N_a} H_{\mathbf{k}a'a} u_{\mathbf{k}a,n} = E_{n\mathbf{k}} \sum_{a=1}^{N_a} S_{\mathbf{k}a'a} u_{\mathbf{k}a,n}, \quad (8.10)$$

where $H_{\mathbf{k}a'a} = \sum_{\mathbf{R}_j} e^{i\mathbf{k}(\mathbf{R}_j - \mathbf{R}_i)} H_{\mathbf{R}_i,a';\mathbf{R}_j,a}$ and $S_{\mathbf{k}a'a} = \sum_{\mathbf{R}_j} e^{i\mathbf{k}(\mathbf{R}_j - \mathbf{R}_i)} S_{\mathbf{R}_i,a';\mathbf{R}_j,a}$. The index n labels the energy bands, and the eigenvalue $E_{n\mathbf{k}}$ plotted against $\mathbf{k}$ defines the band dispersion of the n-th band. The summation over $\mathbf{R}_j$ extends over the whole space, but in practice it is limited to few unit cells around the unit cell at $\mathbf{R}_i$ because when

the distance $|\mathbf{R}_j - \mathbf{R}_i|$ is larger than the spatial extension of the basis set orbitals, the Hamiltonian matrix elements $H_{\mathbf{R}_i,a';\mathbf{R}_j,a}$ become zero. The DOS is given by

$$D(E) = \frac{1}{N_\mathbf{k}} \sum_{n=1}^{N_a} \sum_{\mathbf{k} \in BZ} \delta(E - E_{n,\mathbf{k}}), \quad (8.11)$$

where $N_\mathbf{k}$ is the number of $\mathbf{k}$-points in the BZ. The elements of the density matrix are $\rho_{\mathbf{k}aa'} = 2 \sum_{n=1}^{N_a} u_{\mathbf{k}a,n} u_{\mathbf{k}a',n}^* f_{n,\mathbf{k}}$, or equivalently

$$\rho_{\mathbf{R}_i a;\mathbf{R}_j a'} = \frac{1}{N_\mathbf{k}} \sum_{\mathbf{k} \in BZ} e^{i\mathbf{k}(\mathbf{R}_i - \mathbf{R}_j)} \rho_{\mathbf{k}aa'}. \quad (8.12)$$

We note that the summation over $\mathbf{k}$ is typically carried out in analytical calculations by replacing $\frac{1}{N_\mathbf{k}} \sum_{\mathbf{k} \in BZ} \rightarrow \frac{1}{\Omega_{BZ}} \int_{BZ} d\mathbf{k}$, where Ω_{BZ} is the volume of the BZ.

In the following, we present a few examples of simple electronic systems, whose quantum transport properties will then be studied in Section 8.6.

8.2.3 Infinite Atomic Chain and Square Lattice

We consider an infinitely long atomic chain of hydrogen atoms (Figure 8.1), which we treat as a periodic one-dimensional system. The unit cell contains only one atom with a 1-s orbital. The basis state for a cell located at $\mathbf{R}_l = (l\, d_0, 0, 0)$, with l an integer, is therefore denoted as $|l\rangle \equiv |\mathbf{R}_l, 1\rangle$ (d_0 is the lattice constant, which we set equal to 1 in the following) and we have

$$|\Psi_k\rangle = \sum_l e^{ik\, l} u_{k,l} |l\rangle. \quad (8.13)$$

The on-site matrix elements $H_{l,l} \equiv H_{\mathbf{R}_l,1;\mathbf{R}_l,1}$ are equal to a real number ϵ, and we assume only nearest neighbor hopping t between the atoms in the chain, so that $H_{l,l\pm1} \equiv H_{\mathbf{R}_l,1;\mathbf{R}_{l\pm1},1} = t$.

Eq. (8.10) then reads

$$(\epsilon + e^{ik}t + e^{-ik}t)\, u_{k,l} = E_k\, u_{k,l} \quad (8.14)$$

and the eigenenergy is $E_k = \epsilon + 2t \cos k$ with $-\pi < k < \pi$ that marks the BZ. E_k defines the only energy band of the model. The DOS is computed from Eq. (8.11) and reads

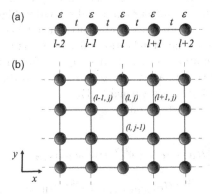

FIGURE 8.1 (a) Infinite chain of hydrogen atoms. (b) Rectangular lattice of hydrogen atoms.

$$D(E) = \frac{1}{\pi} \frac{1}{\sqrt{4t^2 - (E - \epsilon_0)^2}}. \qquad (8.15)$$

This result is obtained after approximating the sum over k with the integral $D(E) = \frac{1}{2\pi} \int_{-\pi}^{\pi} dk\, \delta(E - E_k)$ and subsequently by changing the variable k to E with $(dk/dE)^{-1} = 2t \sin k = \pm \sqrt{4t^2 - (E_k - \epsilon_0)^2}$, where the plus sign applies for $0 < k < \pi$ and the minus sign for $-\pi < k \leq 0$. For a given Fermi energy E_F ($-2t \leq E_F - \epsilon \leq 2t$), we define the Fermi wave-number k_F by solving the equation $E_F = \epsilon_0 + 2t \cos k_F$, which has two solutions between $-\pi$ and π, one positive and one negative with equal modulus. The k-dependent density matrix is $\rho_k = 2\theta(k - |k_F|)$ at zero temperature, and we have

$$\rho_{lj} = \frac{1}{N_k} \sum_k e^{ik(l-j)} \rho_k = \frac{1}{\pi} \int_{-k_F}^{k_F} dk\, e^{ik(l-j)}. \qquad (8.16)$$

$\rho_{ll} = 2k_F/\pi$ is independent of l and represents the atom occupation. In a similar way, we can study an infinite two-dimensional rectangular lattice of hydrogen atoms (Figure 8.1). The coordinates of a unit cell are $\mathbf{R} = (l\, d_0, j\, d_0, 0)$, and we denote the state in the basis set as $|lj\rangle$. The Bloch state is

$$|\Psi_{(k_x,k_y)}\rangle = \sum_{lj} e^{ik_x l} e^{ik_y j} u_{(k_x,k_y),lj} |lj\rangle \qquad (8.17)$$

($d_0 = 1$), and the band dispersion is

$$E_{k_x,k_y} = \epsilon + 2t[\cos k_x + \cos k_y] \qquad (8.18)$$

with $-\pi < k_x, k_y < \pi$.

8.2.4 Model Graphene and Graphene Nano-ribbons

We now examine the honeycomb lattice of graphene. As in the previous examples, we assume only one orbital per atomic site, on-site energy ϵ and nearest neighbor hopping t. The calculation of the band structure for the hexagonal unit cell is presented in many review articles and text-books (see, e.g., p. 70 in Ref. [16]). Here instead, we present the calculation for the rectangular unit cell displayed in Figure 8.2 as this is often a more convenient choice when doing transport calculations as shown in Section 8.6.5. Such rectangular unit cell has four atoms [i.e., a in Eq. (8.9) runs from 1 to 4], and the lattice vectors are $\mathbf{v}_1 = (\sqrt{3}, 0)$ and $\mathbf{v}_2 = (0, 1/2)$ (the lattice spacing d_0 is set equal to 1).

Then, Eq. (8.10) becomes

$$\sum_{a=1}^{4} \Big[H_{\mathbf{R}_i, a'; \mathbf{R}_i, a} + e^{i\mathbf{k}\mathbf{v}_1} H_{\mathbf{R}_i, a'; \mathbf{R}_i + \mathbf{v}_1, a}$$
$$+ e^{-i\mathbf{k}\mathbf{v}_1} H_{\mathbf{R}_i, a'; \mathbf{R}_i - \mathbf{v}_1, a} + e^{i\mathbf{k}\mathbf{v}_2} H_{\mathbf{R}_i, a'; \mathbf{R}_i + \mathbf{v}_2, a}$$
$$+ e^{-i\mathbf{k}\mathbf{v}_2} H_{\mathbf{R}_i, a'; \mathbf{R}_i - \mathbf{v}_2, a} \Big] u_{\mathbf{k}a}^n = E_{n\mathbf{k}} u_{\mathbf{k}a'}^n \qquad (8.19)$$

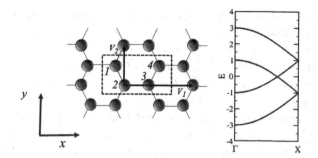

FIGURE 8.2 Rectangular cell for the honeycomb lattice and band structure along the high symmetry line connecting $\Gamma = (0,0)$ and $X = (0,\pi)$ in the BZ calculated by diagonalizing the matrix in Eq. (8.21) with $\epsilon = 0$ and $t = 1$.

or in matrix notation

$$\begin{aligned}
\Bigg[&\begin{pmatrix} \epsilon & t & 0 & 0 \\ t & \epsilon & t & 0 \\ 0 & t & \epsilon_0 & t \\ 0 & t & 0 & \epsilon_0 \end{pmatrix} + \begin{pmatrix} 0 & 0 & 0 & 0 \\ 0 & 0 & 0 & 0 \\ 0 & 0 & 0 & 0 \\ t & 0 & 0 & 0 \end{pmatrix} e^{i\sqrt{3}k_x} \\
&+ \begin{pmatrix} 0 & 0 & 0 & t \\ 0 & 0 & 0 & 0 \\ 0 & 0 & 0 & 0 \\ 0 & 0 & 0 & 0 \end{pmatrix} e^{-i\sqrt{3}k_x} + \begin{pmatrix} 0 & t & 0 & 0 \\ 0 & 0 & 0 & 0 \\ 0 & 0 & 0 & 0 \\ 0 & 0 & t & 0 \end{pmatrix} e^{ik_y} \\
&+ \begin{pmatrix} 0 & 0 & 0 & 0 \\ t & 0 & 0 & 0 \\ 0 & 0 & 0 & 0 \\ 0 & 0 & t & 0 \end{pmatrix} e^{-ik_y} \Bigg] \begin{pmatrix} u_{k_x,k_y,1}^n \\ u_{k_x,k_y,2}^n \\ u_{k_x,k_y,3}^n \\ u_{k_x,k_y,4}^n \end{pmatrix} \\
&= E_{n,k_x,k_y} \begin{pmatrix} u_{k_x,k_y,1}^n \\ u_{k_x,k_y,2}^n \\ u_{k_x,k_y,3}^n \\ u_{k_x,k_y,4}^n \end{pmatrix}
\end{aligned} \qquad (8.20)$$

The eigenenergies for each $\mathbf{k} = (k_x, k_y)$ are then obtained by diagonalizing the matrix

$$\mathbf{H_k} = \begin{pmatrix} \epsilon & t + te^{ik_y} & 0 & te^{-i\sqrt{3}k_x} \\ t + te^{-ik_y} & \epsilon & t & 0 \\ 0 & t & \epsilon_0 & t + te^{-ik_y} \\ te^{i\sqrt{3}k_x} & t + te^{ik_y} & 0 & \epsilon_0 \end{pmatrix}. \qquad (8.21)$$

The resulting four energy bands are plotted in Figure 8.2 along the high symmetry line connecting $\Gamma = (0,0)$ and $X = (0,\pi)$ in the BZ. We note that for this rectangular unit cell, the Dirac-like cone is found at $P = (0, 2\pi/3)$ [and $Q = (0, -2\pi/3)$; not shown].

Let us finally imagine to cut a nano-ribbon out of the infinite model graphene flake (see also Section 2.4 in Ref. [15]). For example, we chose a zig-zag nano-ribbon as shown in Figure 8.3. This is infinite along y, while it is finite along the cross section x, with N_x rectangular unit cells containing four atoms each. Then, Eq. (8.10) reads

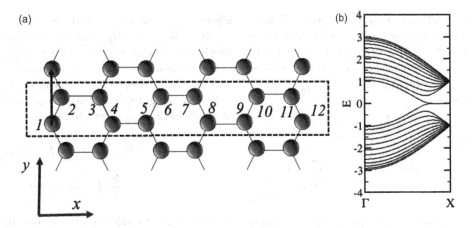

FIGURE 8.3 (a) Unit cell for a model zig-zag graphene nano-ribbon of width $N_x = 3$, that is, 12 atoms. (b) Band structure for a model zig-zag graphene nano-ribbon of width $N_x = 6$.

$$\sum_{a=1}^{4N_x} \Big[H_{\mathbf{R}_i,a';\mathbf{R}_i,a} + e^{i\mathbf{k}\mathbf{v}_2} H_{\mathbf{R}_i,a';\mathbf{R}_i+\mathbf{v}_2,a}$$
$$+ e^{-i\mathbf{k}\mathbf{v}_2} H_{\mathbf{R}_i,a';\mathbf{R}_i-\mathbf{v}_2,a} \Big] u_{\mathbf{k}a}^n = E_{n\mathbf{k}} u_{\mathbf{k}a'}^n \quad (8.22)$$

This equation can be numerically diagonalized for any chosen value of N_x (we call N_x the width of the nano-ribbon). The band structure for a nano-ribbon with $N_x = 6$ is shown in Figure 8.3.

We leave as an exercise for the reader to calculate the band structure of the armchair nano-ribbon (the solution can be found in Section 2.4.1 in Ref. [15]).

8.3 Bond Currents

The first step in order to study transport consists in defining the charge current. Here, we derive an expression for the current from the continuity equation, and, by working in the LCAO framework, we introduce the concept of bond current, which will turn out useful in the following sections.

We start by introducing the symmetrized density matrix $\bar{\boldsymbol{\rho}}$ as

$$\bar{\boldsymbol{\rho}} = \frac{1}{2} \left(\boldsymbol{\rho}\mathbf{S} + \mathbf{S}\boldsymbol{\rho} \right). \quad (8.23)$$

The advantage of using $\bar{\boldsymbol{\rho}}$ instead of $\boldsymbol{\rho}$ is that the total number of electrons, $N_{\text{e,tot}} = \sum_{\alpha=1}^N \bar{\rho}_{\alpha\alpha}$, depends only on the diagonal elements of $\bar{\boldsymbol{\rho}}$. Note that for an orthonormal basis set $S_{\alpha\beta} = \delta_{\alpha\beta}$ and therefore $\bar{\boldsymbol{\rho}} = \boldsymbol{\rho}$. At the same time, for each basis orbital, we can introduce the Mülliken-partitioned local charge as $q_\alpha = e\,\bar{\rho}_{\alpha\alpha}$.

By using the Liouville equation for the density matrix

$$\frac{\partial \boldsymbol{\rho}}{\partial t} = \frac{1}{i\hbar} \left(\mathbf{S}^{-1}\mathbf{H}\boldsymbol{\rho} - \boldsymbol{\rho}\mathbf{H}\mathbf{S}^{-1} \right), \quad (8.24)$$

we derive the continuity equation for the symmetrized density matrix (see Appendix A)

$$\frac{\partial \bar{\boldsymbol{\rho}}}{\partial t} = \frac{e}{i\hbar} \left[\mathbf{H}, \boldsymbol{\rho} \right] + \frac{1}{i\hbar} \left[\mathbf{F}, \mathbf{S} \right]. \quad (8.25)$$

Here

$$\mathbf{F} = \frac{1}{2} \left[\mathbf{S}^{-1}\mathbf{H}\boldsymbol{\rho} + \boldsymbol{\rho}\mathbf{H}\mathbf{S}^{-1} \right] \quad (8.26)$$

is called energy density matrix.

From Eq. (8.25), we have

$$\frac{\partial q_\alpha}{\partial t} = \frac{1}{i\hbar} \sum_{\beta=1}^N \left(H_{\alpha\beta}\rho_{\beta\alpha} - \rho_{\alpha\beta}H_{\beta\alpha} + F_{\alpha\beta}S_{\beta\alpha} - S_{\alpha\beta}F_{\beta\alpha} \right). \quad (8.27)$$

Then, by using the relations

$$\rho_{\alpha\beta}H_{\beta\alpha} = H_{\alpha\beta}^*\rho_{\beta\alpha}^*, \quad (8.28)$$

$$\left(H_{\alpha\beta}\rho_{\beta\alpha} - H_{\alpha\beta}^*\rho_{\beta\alpha}^* \right) = 2i\text{Im}\left[H_{\alpha\beta}\rho_{\beta\alpha} \right], \quad (8.29)$$

$$S_{\alpha\beta}F_{\beta\alpha} = F_{\alpha\beta}^*S_{\beta\alpha}^*, \quad (8.30)$$

$$\left(F_{\alpha\beta}S_{\beta\alpha} - F_{\alpha\beta}^*S_{\beta\alpha}^* \right) = 2i\text{Im}\left[F_{\alpha\beta}S_{\beta\alpha} \right], \quad (8.31)$$

we derive the equation of motion

$$\frac{\partial q_\alpha}{\partial t} = I_\alpha, \quad (8.32)$$

where $I_\alpha = \sum_{\beta=1}^N I_{\alpha\beta}$ is the orbital charge current, and we have introduced the bond current between any two pairs of orbitals α and β

$$I_{\alpha\beta} = \frac{2e}{\hbar}\text{Im}\left[H_{\alpha\beta}\rho_{\beta\alpha} - S_{\alpha\beta}F_{\beta\alpha} \right]. \quad (8.33)$$

We can interpret Eq. (8.32) as the change of local charge due to in- and outflow of electrons from the orbital α. Since in the steady-state $\frac{\partial q_\alpha}{\partial t} = 0$, we have that $I_\alpha = 0$. This means that the in-flow of electrons is equal to the out-flow and therefore the current is conserved.

If we denote with A an arbitrary interface to the current flow, then the total current across this interface, I_A, is obtained by summing up the bond currents between all the orbitals α_L on the left side of this interface and the ones on the right of this interface α_R:

$$I_A = \sum_{\alpha_L, \alpha_R} I_{\alpha_L \alpha_R}. \quad (8.34)$$

As an example to familiarize with the concept of bond current, we consider the case of the infinite atomic chain presented in Section 8.2 (Figure 8.1), and we evaluate the current from the atomic site l to the atomic site $l+1$, that is, $I = I_{l,l+1}$. Since $H_{l,l+1} = t$ and $\rho_{l,l+1} = 2\sum_{-k_F \leq k \leq k_F} e^{ik}$ [cf. Eq. (8.16)], we obtain

$$I_{l,l+1} = \frac{1}{N_k} \sum_{-k_F \leq k \leq k_F} I_{l,l+1}^k = \frac{4e}{\hbar} \frac{1}{N_k} \sum_{-k_F \leq k \leq k_F} \mathrm{Im}\left[te^{ik}\right]$$
$$= \frac{4e}{\hbar} t \frac{1}{N_k} \sum_{-k_F \leq k \leq k_F} \sin k = 2e \frac{1}{N_k} \sum_{-k_F \leq k \leq k_F} v_k,$$
$$(8.35)$$

where in the last equation, we have introduced the group velocity $v_k = \frac{1}{\hbar} \frac{\partial E_k}{\partial k} = 2\frac{t}{\hbar} \sin k$. This results shows that a Bloch state with momentum k carries a net current $I_{l,l+1}^k = 2e\,v_k$. Yet, in order to obtain the whole current, we have to sum over all momenta. By transforming the sum into an integral, we finally find $I_{l,l+1} \propto \int_{-k_F}^{k_F} dk \sin k = 0$ as it has to be for a system in equilibrium. In fact, for any Bloch state with $k > 0$ and with positive group velocity, there is a corresponding Bloch state with $k < 0$ and with negative group velocity.

8.4 Landauer–Büttiker Theory

Let us now consider the atomic chain discussed at the end of previous section and attach it to two leads, also denoted as electrodes, located at the infinitively far left and right extremes (Figure 8.4). This system represents a model two-terminal device. The leads act as charge reservoirs with chemical potentials μ_L and μ_R, while we will refer to the atomic chain as the conductor or conducting region. The chemical potentials of the left and right lead are $\mu_L = E_F + eV_L$ and $\mu_R = E_F + eV_R$, where V_L and V_R are electrostatic potentials such that $\Delta V = V_L - V_R$ defines the bias voltage applied to the device, and E_F is the Fermi energy of the whole system at $\Delta V = 0$. The effect of applying a bias voltage over the leads is therefore that of shifting their chemical potentials. We further assume that the Bloch states of the chain with positive k are in equilibrium with the left lead, while the Bloch states with negative k are in equilibrium with the right lead. Then, the current flowing from the left lead into the conducting region, I_L, is obtained by modifying Eq. (8.35) as

$$I_L = I_L^L + I_L^R = 2e\frac{1}{N_k} \sum_{0 \leq k \leq |k_{F_L}|} v_k + 2e\frac{1}{N_k} \sum_{-|k_{F_R}| \leq k \leq 0} v_k,$$
$$(8.36)$$

where $k_{F_{L(R)}}$ is determined by the relation $\mu_{L(R)} = \epsilon + 2t \cos k_{F_{L(R)}}$. I_L^L (I_L^R) are the contributions to I_L of the Bloch states originating in the left (right) lead. Alternatively, we could calculate the current from the right lead into the conducting region, I_R, which is $I_R = -I_L$.

By changing the momentum summations with energy integrals in Eq. (8.36), we obtain

$$I_L = I_L^L + I_L^R = \frac{2e}{h} \int_{-\infty}^{\infty} dE f_L(E)$$
$$- \frac{2e}{h} \int_{-\infty}^{\infty} dE f_R(E) \qquad (8.37)$$

where the group velocity cancels out because $dk/dE = (\hbar v_k)^{-1}$. $f_{L(R)}(E)$ is the Fermi function with chemical potential $\mu_{L(R)}$

$$f_{L(R)}(E) = \frac{1}{1 + e^{\frac{E - \mu_{L(R)}}{k_B T}}}. \qquad (8.38)$$

Finally, for $\mu_L - \mu_R \to 0^+$ and in the limit of zero temperature, we have

$$I_L = 2\frac{e}{h}(\mu_L - \mu_R). \qquad (8.39)$$

Since, as discussed above, the chemical potential difference is equal to the bias voltage $e\Delta V = (\mu_L - \mu_R)$, the conductance $\mathcal{G} = dI/dV$ is

$$\mathcal{G} = \frac{2e^2}{h} = \mathcal{G}_0, \qquad (8.40)$$

where $\mathcal{G}_0$ is known as *quantum conductance*. Eq. (8.40) is the Landauer formula [25,26], and it means that the conductance through a scattering free device is quantized independently on the nature of the device and of the band dispersion. Furthermore, we stress that the resistance of the device is equal to zero. This means that all energy dissipation happens in the leads, which act as reservoirs, since the charge propagating from one lead eventually loses energy by thermalization when it reaches the other lead and not when it is inside the conductor.

We now generalize our treatment to the case for which the conductor is not completely free of elastic scattering events anymore. The typical device that we study is presented in Fig. 8.5. We now have the energy-dependent transmission and reflection probabilities $T(E)$ and $R(E) = [1 - T(E)]$ across the *scattering region*. The current I_L becomes

$$I_L = I_L^L + I_L^R = \frac{2e}{h} \int_{-\infty}^{\infty} dE\, T(E) f_L(E)$$
$$- \frac{2e}{h} \int_{-\infty}^{\infty} dE\, T(E) f_R(E), \qquad (8.41)$$

FIGURE 8.4 Atomic chain with two leads attached at the two infinitely far left and right extremes. The left (right) lead is in thermal equilibrium with chemical potential $\mu_{L(R)}$.

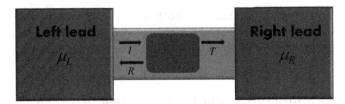

FIGURE 8.5 Schematic description of a nano-device in the Landauer–Büttiker theory. The dark central area represents the scattering potential with the corresponding transmission and reflection probabilities T and $R = 1 - T$.

and the Landauer formula for the conductance in the zero temperature limit is generalized as

$$\mathcal{G} = \frac{2e^2}{h} T(E_F) = \mathcal{G}_0 \, T(E_F), \qquad (8.42)$$

where the transmission must be evaluated at the Fermi energy as $\mu_L - \mu_R \to 0^+$. We remark once more that the key assumption for Eq. (8.42) to be valid is that scattering events are elastic, namely energy conserving, since inelastic scattering would produce decoherence and energy dissipation in the scattering region, which would lead to a modification of this equation [30].

In the case treated so far, where the conductor is an atomic chain, there is only one Bloch state with $k > 0$ for a given energy (and one for $k < 0$). Such Bloch state, which carries a current $2ev_k$, is called *channel*. In general, however, there are several Bloch states available at the same energy. This situation can be understood by looking at the example of the model graphene nano-ribbon in Section 8.2.4. and specifically at the band structure in Figure 8.3, where we recognize several bands with the same energy at different $k_y > 0$ values. Each of these Bloch states carries a current and represents therefore an independent transport channel. For a scattering-free conductor with M channels, the Landauer formula was extended by Büttiker and reads [5]

$$\mathcal{G} = \frac{2e^2}{h} M = \mathcal{G}_0 \, M. \qquad (8.43)$$

This formula states that the zero temperature conductance is obtained by counting the number of channels at the Fermi level and that the contribution of each channel to the conductance is $2e^2/h$ independently from its band dispersion. For instance, in the example of the model graphene nano-ribbon, we see in Figure 8.3 that if we set $E_F = 0$, there is only one channel and $\mathcal{G} = \mathcal{G}_0$, while if we set $E_F = 1$, there are 12 channels, so that $\mathcal{G} = 12\,\mathcal{G}_0$. These remarkably simple results will be re-obtained by using a more involved and general derivation in Section 8.6.5.

Finally, when elastic scattering is present in the conductor with M channels, we have

$$\mathcal{G} = \sum_{ij=1}^{M} \frac{2e^2}{h} T_{ij}(E), \qquad (8.44)$$

where T_{ij} is the probability that the i channel is transmitted through the scattering region in the j channel.

In summary, we have shown here that the Landauer–Büttiker formalism relates via Eq. (8.44) the conductance of a device to the quantum mechanical transmission of the electron wave-functions entering into the device scattering region. This picture relies on four main assumptions: (1) the leads act as charge reservoirs in thermal equilibrium at their own chemical potential; (2) the electrons in the leads as well as those flowing through the scattering region are neither correlated (i.e., the electron–electron interaction is neglected) nor interacting with other degrees of freedom (such as phonons); (3) the difference between μ_L and μ_R is such that $\mu_L - \mu_R \to 0^+$; (4) electrons can enter and leave the scattering region without any scattering.

8.5 Non-equilibrium Green's Functions for Quantum Transport

The key quantity of the Landauer approach to transport is the transmission probability in Eq. (8.44). However, so far we have not discussed how to calculate it. For this purpose, the Green's functions are a very powerful tool. In fact, they allow to obtain in a rather straightforward way the scattering states for a device by knowing the asymptotic states incoming from the leads (see Section 8.5.3). Yet, the use of the Green's functions has also another main advantage: one can go one step beyond the Landauer approach and study systems where a finite bias voltage is applied between the leads; that is, we can release the assumption that $\mu_L - \mu_R \to 0^+$. In fact, the Green's functions encode information about the electronic structure of the device and enable the calculation of the charge density. Eventually, one can therefore introduce a self-consistent procedure, where the charge density is obtained as a function of the applied bias voltage, which in turn will be screened by the charge redistribution.

The approach that we will describe here is usually referred to as non-equilibrium Green's functions (NEGF). For non-interacting systems, as is the case here, the NEGF can be presented without resorting to the complex quantum mechanics for out-of-equilibrium problems. This would not be the case for interacting systems. A more general and rigorous treatment about the NEGF can be found in advanced many-body physics books such as Ref. [48].

To begin this section, we provide the definition of Green's function for a close quantum system at equilibrium, and we introduce its fundamental properties. Then, we will extend the formalism to describe two-terminal devices and will show how to compute the density matrix, the transmission function and the current.

8.5.1 Green's Functions: A Brief Introduction

We consider a closed quantum system in equilibrium described by the Hamiltonian matrix $\mathbf{H}$. We split such

Hamiltonian matrix into two parts as $\mathbf{H} = \mathbf{H}_0 + \mathbf{V}$, where $\mathbf{V}$ can be interpreted as a perturbing potential for a system described by $\mathbf{H}_0$. The unperturbed system has an eigenvector $\boldsymbol{\psi}^0$ $(\mathbf{H}_0\boldsymbol{\psi}^0 = E\mathbf{S}\boldsymbol{\psi}^0)$, while the whole systems satisfy the Schrödinger equation

$$[\mathbf{H}_0 + \mathbf{V}](\boldsymbol{\psi}^0 + \boldsymbol{\psi}^1) = E\mathbf{S}(\boldsymbol{\psi}^0 + \boldsymbol{\psi}^1). \qquad (8.45)$$

$\boldsymbol{\psi}^1$ describes the change of $\boldsymbol{\psi}^0$ caused by $\mathbf{V}$. Thus, we have

$$[E\mathbf{S} - \mathbf{H}^0 - \mathbf{V}]\boldsymbol{\psi}^1 = \mathbf{V}\boldsymbol{\psi}^0, \qquad (8.46)$$

which can be expressed as

$$\boldsymbol{\psi}^1 = \mathbf{G}(E)\mathbf{V}\boldsymbol{\psi}^0. \qquad (8.47)$$

The eigenvector of the whole system then becomes $\boldsymbol{\psi} = \boldsymbol{\psi}^0 + \mathbf{G}(E)\mathbf{V}\boldsymbol{\psi}^0$. Here, we have introduced the Green's function matrix $\mathbf{G}(E)$, defined through the equation

$$[E\mathbf{S} - \mathbf{H}]\mathbf{G}(E) = \mathbf{I}. \qquad (8.48)$$

where $\mathbf{I}$ is the unity matrix, or equivalently as $\mathbf{G}(E) = [E\mathbf{S} - \mathbf{H}]^{-1}$. The Green's function therefore provides a way to describe the response of a system to a perturbation. As we will see in Section 8.5.3, this will be very useful in developing a theoretical approach to quantum transport in nanostructures, where we will calculate the response of the wavefunctions of semi-infinite electrodes when put in contact with a device.

It is important to note that one can obtain two types of solutions for the Green's function, the retarded and the advanced solutions, which describe the response of the system, respectively, forward in time (i.e., causal) and backward in time (i.e., anti-casual). The distinction between the two Green's functions is achieved by replacing $E \rightarrow E + i\delta$ and $E \rightarrow E - i\delta$, where δ is an infinitesimally small positive number $\delta = 0^+$. The retarded and the advanced Green's functions are therefore defined as

$$\mathbf{G}^r(E) = \frac{1}{[(E\mathbf{S} + i\delta) - \mathbf{H}]} \qquad (8.49)$$

and

$$\mathbf{G}^a(E) = \frac{1}{[(E\mathbf{S} - i\delta) - \mathbf{H}]}. \qquad (8.50)$$

The different casual behavior of the advanced and of the retarded Green's functions due to $\pm i\delta$ can be clearly seen by carrying out the Fourier transformation from the energy to the time domain (see, e.g., page 200 in Ref. [9] for a proof). Since $\mathbf{G}^a(E)^\dagger = \mathbf{G}^r(E)$, we will simply denote $\mathbf{G}(E) \equiv \mathbf{G}^r(E)$ and $\mathbf{G}^\dagger(E) \equiv \mathbf{G}^a(E)$ in the following and when referring to the Green's function, we will always mean the retarded Green's function if not stated otherwise.

When the eigenvectors of $\mathbf{H}$ are known, the Green's function can be written in its spectral form as

$$\mathbf{G}(E) = \sum_{n=1}^{N} \frac{1}{E + i\delta - E_n} \boldsymbol{\psi}_n \boldsymbol{\psi}_n^\dagger. \qquad (8.51)$$

We note that if $\mathbf{H}$ and $\mathbf{S}$ are both real, then also the $\boldsymbol{\psi}_n$ can be made real, so that $\mathbf{G}(E)$ is symmetric. We now define the spectral function, $\mathbf{A}(E)$, as

$$\mathbf{A}(E) = i\left[\mathbf{G}(E) - \mathbf{G}^\dagger(E)\right]. \qquad (8.52)$$

By expanding $\mathbf{G}(E)$ in its spectral form, $\mathbf{A}(E)$ can be rewritten as

$$\begin{aligned} \mathbf{A}(E) &= \lim_{\delta \rightarrow 0^+} 2 \sum_{n=1}^{N} \frac{\delta}{(E - E_n)^2 + \delta^2} \boldsymbol{\psi}_n \boldsymbol{\psi}_n^\dagger \\ &= 2\pi \sum_{n=1}^{N} \delta(E - E_n)\, \boldsymbol{\psi}_n \boldsymbol{\psi}_n^\dagger, \qquad (8.53) \end{aligned}$$

which shows that $\boldsymbol{A}(E)$ has a peak at each eigenstate of $\mathbf{H}$, and therefore, $A_\alpha(E) = \frac{1}{2\pi} \sum_{\beta=1}^{N} A_{\alpha\beta}(E) S_{\beta\alpha}$ corresponds to the PDOS in Eq. (8.8). By using the above relation, the density matrix $\boldsymbol{\rho} = 2 \sum_n f_n \boldsymbol{\psi}_n \boldsymbol{\psi}_n^\dagger$ can be written as

$$\boldsymbol{\rho} = \frac{1}{\pi} \int_{-\infty}^{\infty} dE\, f(E)\, \mathbf{A}(E). \qquad (8.54)$$

If one defines the lesser Green's function

$$\mathbf{G}^<(E) = 2if(E)\, \mathbf{A}(E), \qquad (8.55)$$

then $\boldsymbol{\rho}$ can be rewritten as

$$\boldsymbol{\rho} = \frac{1}{2\pi i} \int dE\, \mathbf{G}^<(E). \qquad (8.56)$$

In Green's functions-based numerical calculations, one therefore needs to calculate $\mathbf{G}^<(E)$ and integrate this quantity over energies in order to obtain the electron density.

8.5.2 Device Green's Function and Lead Self-energies

The first step towards a NEGF-based study of transport is to obtain a mathematical representation of the two-terminal device in Figure 8.5. This can be done within the LCAO framework. The scattering region and the left (right) lead are described by the Hamiltonian matrices $\mathbf{H}_S$ and $\mathbf{H}_{L(R)}$, respectively. The coupling between the left (right) leads and the scattering region is accounted for by $\mathbf{H}_{LS(RS)}$, while we require that there is no direct coupling between the two electrodes (Figure 8.6). The Hamiltonian $\mathbf{H}$ of the device has therefore the structure

$$\mathbf{H} = \begin{pmatrix} \mathbf{H}_L & \mathbf{H}_{LS} & 0 \\ \mathbf{H}_{SL} & \mathbf{H}_S & \mathbf{H}_{SR} \\ 0 & \mathbf{H}_{RS} & \mathbf{H}_R \end{pmatrix}. \qquad (8.57)$$

Since $\mathbf{H}$ is Hermitian, we have $\mathbf{H}_{SL} = \mathbf{H}_{LS}^\dagger$ and $\mathbf{H}_{RS} = \mathbf{H}_{SR}^\dagger$. The zero matrices represent the fact that there is no direct interaction between the left and the right lead. $\mathbf{H}_S$ has dimension $N_S \times N_S$, where N_S is the number of basis orbitals included in the scattering region. In contrast, $\mathbf{H}_{L(R)}$ are infinite matrices, since the leads are assumed semi-infinite. The overlap matrix $\mathbf{S}$ has the same block structure as $\mathbf{H}$, and

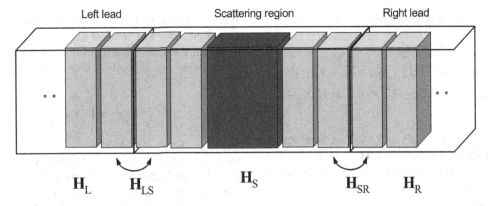

FIGURE 8.6 Schematic representation of the two-terminal setup used in typical quantum transport calculations. A central scattering region is joined to the left and right leads.

therefore, in order to simplify the notation, we define the $\mathbf{K}$ matrix as $\mathbf{K} = \mathbf{H} - (E + i\delta)\mathbf{S}$, which in block form reads

$$\mathbf{K} = \begin{pmatrix} \mathbf{K}_L & \mathbf{K}_{LS} & 0 \\ \mathbf{K}_{SL} & \mathbf{K}_S & \mathbf{K}_{SR} \\ 0 & \mathbf{K}_{RS} & \mathbf{K}_R \end{pmatrix}. \tag{8.58}$$

We note that $K \neq K^\dagger$ for all $\delta \neq 0$, however as we take the limit $\delta \to 0^+$ the matrix becomes Hermitian, so that $\mathbf{K} = \mathbf{K}^\dagger$ for $\delta = 0$.

By using the general result for the inverse of such a block matrix given in Appendix B [Eqs. (8.148) and (8.149)] the Green's function, $\mathbf{G} = [(E + i\delta)\mathbf{S} - \mathbf{H}]^{-1} = -\mathbf{K}^{-1}$, for the whole system can be written as

$$\mathbf{G} = \begin{pmatrix} \mathbf{G}_L + \mathbf{T}_{LS}\mathbf{G}_S\mathbf{T}_{SL} & \mathbf{T}_{LS}\mathbf{G}_S & \mathbf{T}_{LS}\mathbf{G}_S\mathbf{T}_{SR} \\ \mathbf{G}_S\mathbf{T}_{SL} & \mathbf{G}_S & \mathbf{G}_S\mathbf{T}_{SR} \\ \mathbf{T}_{RS}\mathbf{G}_S\mathbf{T}_{SL} & \mathbf{T}_{RS}\mathbf{G}_S & \mathbf{G}_R + \mathbf{T}_{RS}\mathbf{G}_S\mathbf{T}_{SR} \end{pmatrix}. \tag{8.59}$$

Note that here and in the following, we will not indicate explicitly the dependence of the Green's functions on the energy, unless required, in order to keep the notation simple. The matrices

$$\mathbf{G}_L = -\mathbf{K}_L^{-1}, \tag{8.60}$$

$$\mathbf{G}_R = -\mathbf{K}_R^{-1} \tag{8.61}$$

are the Green's functions of the isolated left and right lead, respectively, and

$$\mathbf{T}_{LS} = \mathbf{G}_L \mathbf{K}_{LS}, \tag{8.62}$$

$$\mathbf{T}_{SL} = \mathbf{K}_{SL} \mathbf{G}_L, \tag{8.63}$$

$$\mathbf{T}_{RS} = \mathbf{G}_R \mathbf{K}_{RS}, \tag{8.64}$$

$$\mathbf{T}_{SR} = \mathbf{K}_{SR} \mathbf{G}_R. \tag{8.65}$$

These $\mathbf{T}$-Matrices are generally called "transfer matrices". The Green's function of the scattering region $\mathbf{G}_S$ is given by

$$\mathbf{G}_S = (-\mathbf{K}_S - \mathbf{\Sigma}_L - \mathbf{\Sigma}_R)^{-1}. \tag{8.66}$$

Here, the left lead self-energy $\mathbf{\Sigma}_L$ and the right lead self-energy $\mathbf{\Sigma}_R$ have been introduced. These are defined as

$$\mathbf{\Sigma}_L = \mathbf{K}_{SL} \mathbf{G}_L \mathbf{K}_{LS}, \tag{8.67}$$

$$\mathbf{\Sigma}_R = \mathbf{K}_{SR} \mathbf{G}_R \mathbf{K}_{RS}. \tag{8.68}$$

If we define the effective Hamiltonian of the scattering region as

$$\mathbf{H}_{S,\text{eff}} = \mathbf{H}_S + \mathbf{\Sigma}_L + \mathbf{\Sigma}_R, \tag{8.69}$$

then $\mathbf{G}_S$ can be written as

$$\mathbf{G}_S = [(E + i\delta)\mathbf{S}_S - \mathbf{H}_{S,\text{eff}}]^{-1}. \tag{8.70}$$

The Green's function for the scattering region can therefore be viewed like the one for an isolated system, with the only difference that $\mathbf{H}_S$ has to be replaced by the effective Hamiltonian $\mathbf{H}_{S,\text{eff}}$. This takes into account the effects of the leads onto the scattering region. The self-energies are non-Hermitian matrices and determine how the states in the scattering region are modified by the interaction with the leads. The Hermitian part causes a shift of the scattering region's eigenvalue spectrum, and the antihermitian part leads to a broadening of the energy levels in the scattering region (see Section 8.6.1).

The imaginary part of the left and right self-energies defines the left and right $\mathbf{\Gamma}$-matrices

$$\mathbf{\Gamma}_L = i\left(\mathbf{\Sigma}_L - \mathbf{\Sigma}_L^\dagger\right), \tag{8.71}$$

$$\mathbf{\Gamma}_R = i\left(\mathbf{\Sigma}_R - \mathbf{\Sigma}_R^\dagger\right). \tag{8.72}$$

These are Hermitian, positive semidefinite matrices called "broadening matrices" or "coupling matrices", since they describe the rate of in- and outflow of electrons from/to the leads as we will show in Section 8.5.6.

We will present examples of calculations for the lead self-energies and the $\mathbf{\Gamma}$-matrices of simple systems in Section 8.6. However, computing the lead self-energies is generally a complex task. Different algorithms have been developed during the years. We suggest in particular the one in Ref. [45], which has proved numerically very stable also for large and complex systems.

The spectral function of the scattering region is defined similarly to Eq. (8.52) as

$$\mathbf{A}_S = i(\mathbf{G}_S - \mathbf{G}_S^\dagger). \tag{8.73}$$

$\mathbf{A}_S$ can also be rewritten in terms of the self-energies as

$$\begin{aligned}
\mathbf{A}_S &= i\ \mathbf{G}_S(\mathbf{G}_S^{\dagger-1} - \mathbf{G}_S^{-1})\mathbf{G}_S^{\dagger} \\
&= i\ \mathbf{G}_S[-\mathbf{K}_S^{\dagger} - \mathbf{\Sigma}_L^{\dagger} - \mathbf{\Sigma}_R^{\dagger} + \mathbf{K}_S + \mathbf{\Sigma}_L + \mathbf{\Sigma}_R]\mathbf{G}_S^{\dagger} \\
&= i\ \mathbf{G}_S(\mathbf{\Sigma}_L - \mathbf{\Sigma}_L^{\dagger} + \mathbf{\Sigma}_R - \mathbf{\Sigma}_R^{\dagger})\mathbf{G}_S^{\dagger} + 2\delta\ \mathbf{G}_S\mathbf{S}_S\mathbf{G}_S^{\dagger},
\end{aligned}$$
$$(8.74)$$

where we have used Eq. (8.66) and the fact that for real energies E, and Hermitian matrices $\mathbf{H}_S$ and $\mathbf{S}_S$, we have $\mathbf{K}_S - \mathbf{K}_S^{\dagger} = -2i\delta\mathbf{S}_S$. By using Eqs. (8.71) and (8.72), $\mathbf{A}_S$ can be expressed as

$$\mathbf{A}_S = \mathbf{G}_S(\mathbf{\Gamma}_L + \mathbf{\Gamma}_R)\mathbf{G}_S^{\dagger} + 2\delta\ \mathbf{G}_S\mathbf{S}_S\mathbf{G}_S^{\dagger}, \qquad (8.75)$$

while we define

$$\mathbf{A}_{SS}^R = \mathbf{G}_S\mathbf{\Gamma}_R\mathbf{G}_S^{\dagger} \qquad (8.76)$$

$$\mathbf{A}_{SS}^L = \mathbf{G}_S\mathbf{\Gamma}_L\mathbf{G}_S^{\dagger} \qquad (8.77)$$

as the spectral functions of the scattering region projected over the states incoming from the right and left lead, respectively.

We conclude this subsection with a number of remarks. Since the leads are assumed to be semi-infinite objects, the DOS of the leads is a continuous function of the energy. Therefore, for most systems, the spectrum of the scattering region becomes continuous as well because of the coupling to the leads. This implies that there are no poles in $\mathbf{G}_S$ and the limit $\delta = 0$ can be taken in the matrices $\mathbf{K}_S$, $\mathbf{K}_{LS}$ and $\mathbf{K}_{RS}$. Moreover, the second term in Eq. (8.75) vanishes, since it only contributes when there are poles in $\mathbf{G}_S$. Only if in the scattering region there are localized states (also denoted as "bound states") that are not coupled to the continuous spectrum of the leads, then there will be poles in $\mathbf{G}_S$. In that case, the limit $\delta = 0$ cannot be taken in $\mathbf{K}_S$. A detailed treatment about the bound states is beyond the goals of this introductory chapter, and we forward the interested reader to Ref. [42]. In the remaining part of this chapter, we assume that there are no bound states in the scattering region.

8.5.3 Scattering States

After having developed the Green's function formalism for a two-terminal device, we can now employ it to obtain an expression for the device wave-function as mentioned in Section 8.5.1.

To begin with, we note again that when the leads are decoupled from the scattering region, they have a continuous spectrum. Therefore, for each energy E, there is a finite number $\mathcal{N}_L(E)$ of states $\varphi_{L,n}$ in the left lead satisfying

$$\mathbf{H}_L\ \varphi_{L,n} = E\ \mathbf{S}_L\ \varphi_{L,n} \qquad n \in \{1, \ldots, \mathcal{N}_L(E)\}. \quad (8.78)$$

In the same way, there are a finite number $\mathcal{N}_R(E)$ of states $\varphi_{R,n}$ in the right lead

$$\mathbf{H}_R\ \varphi_{R,n} = E\ \mathbf{S}_R\ \varphi_{R,n} \qquad n \in \{1, \ldots, \mathcal{N}_R(E)\}. \quad (8.79)$$

When connecting the leads to the scattering region, the wave-functions will instead extend over the whole system (scattering region + leads). As matter of notation, we write whole device wave-function $\boldsymbol{\psi}$ as a three components vector

$$\boldsymbol{\psi} = \begin{pmatrix} \boldsymbol{\psi}_L \\ \boldsymbol{\psi}_S \\ \boldsymbol{\psi}_R \end{pmatrix}, \qquad (8.80)$$

where $\boldsymbol{\psi}_L$ ($\boldsymbol{\psi}_R$) is the part of the wave-function extending over the left (right) lead, and $\boldsymbol{\psi}_S$ is the part extending over the scattering region. These components can be computed just as explained in Section 8.5.1 and by considering that the Green's function provides the response of a system to a perturbation, which in this case is represented by the coupling between the scattering region and the leads. In our derivation below, we will loosely follow Ref. [35].

An incoming wave originating from the left lead $\boldsymbol{\psi}_{L,n}$ can be written as

$$\boldsymbol{\psi}_n^L = \begin{pmatrix} \varphi_{L,n} \\ 0 \\ 0 \end{pmatrix} + \boldsymbol{\psi}_n^{\Delta}, \qquad (8.81)$$

where $\boldsymbol{\psi}_n^{\Delta}$ has the dimension of the full infinite system (scattering region+leads). It is then straight forward to show that $\boldsymbol{\psi}_n^{\Delta}$ satisfies the equation

$$(\mathbf{H} - E\mathbf{S})\boldsymbol{\psi}_n^{\Delta} = \begin{pmatrix} 0 \\ -\mathbf{K}_{SL}\ \varphi_{L,n} \\ 0 \end{pmatrix}, \qquad (8.82)$$

where we have used Eq. (8.78) and the definition of $\mathbf{K}$. As already discussed in Section 8.5.1, there are two sets of solutions to this equation. The physical difference between the two solutions is that the first describes waves that flow from the left lead into the scattering region, and the second describes waves that flow from the scattering region into the left lead. If there is no right lead attached to the scattering region, these solutions are identical. Since the goal is to distinguish the solutions originating from a given lead, we use the first set of solutions, so that

$$\boldsymbol{\psi}_n^{\Delta+} = \mathbf{G} \begin{pmatrix} 0 \\ \mathbf{K}_{SL}\ \varphi_{L,n} \\ 0 \end{pmatrix}, \qquad (8.83)$$

and, by using the solution for $\mathbf{G}$ of Eq. (8.59), we obtain the wave-function of the entire system

$$\boldsymbol{\psi}_n^L = \begin{pmatrix} \boldsymbol{\psi}_{L,n}^L \\ \boldsymbol{\psi}_{S,n}^L \\ \boldsymbol{\psi}_{R,n}^L \end{pmatrix} = \begin{pmatrix} \mathbf{I}_L + \mathbf{T}_{LS}\ \mathbf{G}_S\ \mathbf{K}_{SL} \\ \mathbf{G}_S\ \mathbf{K}_{SL} \\ \mathbf{T}_{RS}\ \mathbf{G}_S\ \mathbf{K}_{SL} \end{pmatrix} \varphi_{L,n}, \quad (8.84)$$

where $\mathbf{I}_L$ is a unity matrix with the dimension of the left lead (it is therefore infinite). $\boldsymbol{\psi}_{L,n}^L$, $\boldsymbol{\psi}_{S,n}^L$ and $\boldsymbol{\psi}_{R,n}^L$ are the scattering states. Specifically, $\boldsymbol{\psi}_{L,n}^L$ includes the incoming and the reflected state inside the left lead, and $\boldsymbol{\psi}_{R,n}^L$ is the state transmitted into the right lead.

In an analogous way, we can write an incoming wave originating from the right lead ψ_n^{R} as

$$\psi_n^{\mathrm{R}} = \begin{pmatrix} 0 \\ 0 \\ \varphi_{\mathrm{R},n} \end{pmatrix} + \psi_n^{\Delta}, \qquad (8.85)$$

and we finally obtain

$$\psi_n^{\mathrm{R}} = \begin{pmatrix} \psi_{\mathrm{L},n}^{\mathrm{R}} \\ \psi_{\mathrm{S},n}^{\mathrm{R}} \\ \psi_{\mathrm{R},n}^{\mathrm{R}} \end{pmatrix} = \begin{pmatrix} \mathbf{T}_{\mathrm{LS}} \mathbf{G}_{\mathrm{S}} \, \mathbf{K}_{\mathrm{SR}} \\ \mathbf{G}_{\mathrm{S}} \, \mathbf{K}_{\mathrm{SR}} \\ \mathbf{I}_{\mathrm{R}} + \mathbf{T}_{\mathrm{RS}} \, \mathbf{G}_{\mathrm{S}} \, \mathbf{K}_{\mathrm{SR}} \end{pmatrix} \varphi_{\mathrm{R},n}, \tag{8.86}$$

where $\mathbf{I}_{\mathrm{R}}$ is the unity matrix with the same dimension as the right lead (i.e. infinite).

8.5.4 Density Matrix

As explained at the beginning of Section 8.4, we assume that states originating from the left (right) lead are in local thermal equilibrium with the corresponding lead of chemical potential $\mu_{\mathrm{L(R)}}$ and Fermi function $f_{\mathrm{L(R)}}(E)$ defined in Eq. (8.38). We therefore generalize the definition of the density matrix in Eq. (8.6), and we introduce the full device density matrix

$$\rho^{\mathrm{L(R)}} = 2 \int_{-\infty}^{\infty} dE \sum_{n=1}^{\mathcal{N}_{\mathrm{L(R)}}(E)} \delta(E - E_n) f_{\mathrm{L(R)}}(E) \psi_n^{\mathrm{L(R)}} \psi_n^{\mathrm{L(R)}\dagger}, \tag{8.87}$$

where $\psi_n^{\mathrm{L(R)}}$ are the scattering states in Eqs. (8.84) and (8.86). We then divide $\rho^{\mathrm{L(R)}}$ in blocks

$$\rho_{\mathrm{AB}}^{\mathrm{L(R)}} = 2 \int_{-\infty}^{\infty} dE \sum_{n=1}^{\mathcal{N}_{\mathrm{L(R)}}(E)} \delta(E - E_n) f_{\mathrm{L(R)}}(E) \psi_{\mathrm{A},n}^{\mathrm{L(R)}} \psi_{\mathrm{B},n}^{\mathrm{L(R)}\dagger}, \tag{8.88}$$

where $\mathrm{A}, \mathrm{B} = \mathrm{L}, \mathrm{S}, \mathrm{R}$, while the superscript denotes the origin of the scattering states like in Eqs. (8.84) and (8.86). Here, we consider in particular

$$\rho_{\mathrm{SS}}^{\mathrm{L}} = \int_{-\infty}^{\infty} dE \sum_{n=1}^{\mathcal{N}_{\mathrm{L}}(E)} \delta(E - E_n) f_{\mathrm{L}}(E) \psi_{\mathrm{S},n}^{\mathrm{L}} \psi_{\mathrm{S},n}^{\mathrm{L}\dagger} \tag{8.89}$$

which is the density matrix of the scattering region stemming from the states coming in from the left lead.

Since by using Eq. (8.84), we can write

$$\psi_{\mathrm{S},n}^{\mathrm{L}} \psi_{\mathrm{S},n}^{\mathrm{L}\dagger} = \mathbf{G}_{\mathrm{S}} \mathbf{K}_{\mathrm{SL}} \varphi_{\mathrm{L},n} \varphi_{\mathrm{L},n}^{\dagger} \mathbf{K}_{\mathrm{SL}}^{\dagger} \mathbf{G}_{\mathrm{S}}^{\dagger}, \tag{8.90}$$

we obtain

$$\rho_{\mathrm{SS}}^{\mathrm{L}} = 2 \int_{-\infty}^{\infty} dE \, \mathbf{G}_{\mathrm{S}} \mathbf{K}_{\mathrm{SL}} \left(\sum_{n=1}^{\mathcal{N}_{\mathrm{L}}(E)} \delta(E - E_n) \varphi_{\mathrm{L},n} \varphi_{\mathrm{L},n}^{\dagger} \right)$$
$$\times \mathbf{K}_{\mathrm{SL}}^{\dagger} \mathbf{G}_{\mathrm{S}}^{\dagger} f_{\mathrm{L}}(E). \tag{8.91}$$

We note that the term in parenthesis is proportional to the spectral function for the decoupled left lead $\mathbf{A}_{\mathrm{L}}$ [see Eq. (8.53)]. Thus, we have

$$\rho_{\mathrm{SS}}^{\mathrm{L}} = \frac{1}{\pi} \int_{-\infty}^{\infty} dE \, \mathbf{G}_{\mathrm{S}} \mathbf{K}_{\mathrm{SL}} \mathbf{A}_{\mathrm{L}} \mathbf{K}_{\mathrm{SL}}^{\dagger} \mathbf{G}_{\mathrm{S}}^{\dagger} f_{\mathrm{L}}(E). \tag{8.92}$$

By using Eqs. (8.67) and (8.71), we finally obtain

$$\rho_{\mathrm{SS}}^{\mathrm{L}} = \frac{1}{\pi} \int_{-\infty}^{\infty} dE \, \mathbf{G}_{\mathrm{S}} \mathbf{\Gamma}_{\mathrm{L}} \mathbf{G}_{\mathrm{S}}^{\dagger} f_{\mathrm{L}}(E), \tag{8.93}$$

or equivalently

$$\rho_{\mathrm{SS}}^{\mathrm{L}} = \frac{1}{2\pi} \int_{-\infty}^{\infty} dE \, \mathbf{A}_{\mathrm{SS}}^{\mathrm{L}} \, f_{\mathrm{L}}(E), \tag{8.94}$$

where $\mathbf{A}_{\mathrm{SS}}^{\mathrm{L}}$ was defined in Eq. (8.77). In an analogous way, we can also compute

$$\rho_{\mathrm{SS}}^{\mathrm{R}} = \frac{1}{\pi} \int_{-\infty}^{\infty} dE \, \mathbf{A}_{\mathrm{SS}}^{\mathrm{R}} \, f_{\mathrm{R}}(E). \tag{8.95}$$

The total density matrix of the scattering region is then the sum of the contribution from states originating in the left and right lead

$$\rho_{\mathrm{S}} = \rho_{\mathrm{SS}}^{\mathrm{L}} + \rho_{\mathrm{SS}}^{\mathrm{R}}. \tag{8.96}$$

By introducing the lesser Green's function

$$\mathbf{G}_{\mathrm{S}}^{<} = 2i \mathbf{G}_{\mathrm{S}} \big[f_{\mathrm{L}}(E) \mathbf{\Gamma}_{\mathrm{L}} + f_{\mathrm{R}}(E) \mathbf{\Gamma}_{\mathrm{R}} \big] \mathbf{G}_{\mathrm{S}}^{\dagger}, \tag{8.97}$$

we end up with

$$\rho_{\mathrm{S}} = \frac{1}{2\pi i} \int_{-\infty}^{\infty} dE \, \mathbf{G}_{\mathrm{S}}^{<}(E), \tag{8.98}$$

which is the same definition as in Eq. (8.56). However, we point out that $\mathbf{G}_{\mathrm{S}}^{<}$ in Eq. (8.97) has a more general form than the one defined in Eq. (8.55). The two definitions coincide only in equilibrium when the device occupation function is $f_{\mathrm{L}}(E) = f_{\mathrm{R}}(E)$, while this is not the case anymore when we bring the device out-of-equilibrium by imposing $\mu_{\mathrm{L}} \neq \mu_{\mathrm{R}}$ and $f_{\mathrm{L}}(E) \neq f_{\mathrm{R}}(E)$. Therefore, Eq. (8.97) is a generalization of Eq. (8.55) and its validity out of equilibrium is the reason why we referred to this approach as the *non-equilibrium* Green's functions formalism in Sections 8.5.2, 8.5.3 and 8.5.4.

There are two additional important observations related to the Eqs. (8.97) and (8.98). The first is that for non-interacting systems, as it is the case here, the NEGF returns the density matrix of the scattering region from the equilibrium distribution functions of the leads $f_{\mathrm{L}}(E)$ and $f_{\mathrm{R}}(E)$ with no need for the out-of-equilibrium distribution function of the scattering region. Because of that calculations are possible in practice. This will not be the case anymore for interacting systems (see, e.g., Ref. [12]). Second, the off-diagonal components of the matrices $\mathbf{G}_{\mathrm{S}} \mathbf{\Gamma}_{\mathrm{R}} \mathbf{G}_{\mathrm{S}}^{\dagger}$ and $\mathbf{G}_{\mathrm{S}} \mathbf{\Gamma}_{\mathrm{L}} \mathbf{G}_{\mathrm{S}}^{\dagger}$ are generally complex. Therefore, while in equilibrium ρ_{S} is real because of Eq. (8.75), for the out-of-equilibrium case, that is, $f_{\mathrm{L}}(E) \neq f_{\mathrm{R}}(E)$, ρ_{S} becomes complex. The imaginary part of the density matrix is related to the current (if the Hamiltonian matrix is real), as we have seen previously in Section 8.3 and as we will see in the following.

Finally, we can also compute the other blocks of the density matrix in Eq. (8.88) since they will be needed later in the derivation of the transmission function. For example, by using. Eq. (8.86), we have that

$$\boldsymbol{\rho}_{LS}^R = 2 \int_{-\infty}^{\infty} dE \sum_{n=1}^{\mathcal{N}_R(E)} \delta(E - E_n) f_R(E) \boldsymbol{\psi}_{L,n}^R \boldsymbol{\psi}_{S,n}^{R\,\dagger}$$

$$= \int_{-\infty}^{\infty} dE f_R(E) \mathbf{G}_L \mathbf{K}_{LS} \mathbf{G}_S \mathbf{K}_{SR}$$

$$\times \Big(\sum_{n=1}^{\mathcal{N}_R(E)} \delta(E - E_n) \boldsymbol{\varphi}_{R,n} \boldsymbol{\varphi}_{R,n}^\dagger \Big) \mathbf{K}_{RS} \mathbf{G}_S^\dagger$$

$$= \frac{1}{\pi} \int_{-\infty}^{\infty} dE f_R(E) \mathbf{G}_L \mathbf{K}_{LS} \mathbf{G}_S \mathbf{K}_{SR} \mathbf{A}_R \mathbf{K}_{RS} \mathbf{G}_S^\dagger, \quad (8.99)$$

and therefore

$$\boldsymbol{\rho}_{LS}^R = \frac{1}{\pi} \int_{-\infty}^{\infty} dE f_R(E) \mathbf{G}_L \mathbf{K}_{LS} \mathbf{G}_S \boldsymbol{\Gamma}_R \mathbf{G}_S^\dagger. \quad (8.100)$$

We leave the derivation of the other blocks as an exercise for the reader.

8.5.5 Energy Density Matrix

The device energy density matrix $\mathbf{F}$ in. Eq. (8.26) can be derived in the same way as the density matrix. We first note that the Schrödinger equation for an individual scattering state implies that $\mathbf{S}^{-1}\mathbf{H}\boldsymbol{\psi}_n^{L(R)} = E\boldsymbol{\psi}_n^{L(R)}$ or, equivalently, $\boldsymbol{\psi}_n^{L(R)\,\dagger}\mathbf{H}\mathbf{S}^{-1} = E\boldsymbol{\psi}_n^{L(R)\,\dagger}$, where we have used the fact that $\mathbf{H}$ and $\mathbf{S}$ are Hermitian and that E is real. Then, we have

$$\frac{1}{2}\Big(\mathbf{S}^{-1}\mathbf{H}\boldsymbol{\psi}_n^{L(R)}\boldsymbol{\psi}_n^{L(R)\,\dagger} + \boldsymbol{\psi}_n^{L(R)}\boldsymbol{\psi}_n^{L(R)\,\dagger}\mathbf{H}\mathbf{S}^{-1} \Big)$$

$$= E\boldsymbol{\psi}_n^{L(R)}\boldsymbol{\psi}_n^{L(R)\,\dagger}. \quad (8.101)$$

Inserting this relation and Eq. (8.87) into Eq. (8.26) gives the device energy density matrix:

$$\mathbf{F}^{L(R)} = 2 \int_{-\infty}^{\infty} dE \sum_{n=1}^{\mathcal{N}_{L(R)}(E)} \delta(E - E_n) E f_{L(R)}(E) \boldsymbol{\psi}_n^{L(R)}\boldsymbol{\psi}_n^{L(R)\,\dagger}, \quad (8.102)$$

which can be expressed in a block form similar to Eq. (8.88) for $\boldsymbol{\rho}$, that is

$$\mathbf{F}_{AB}^{L(R)} = 2 \int_{-\infty}^{\infty} dE \sum_{n=1}^{\mathcal{N}_{L(R)}(E)} \delta(E - E_n) E f_{L(R)}(E) \boldsymbol{\psi}_{A,n}^{L(R)}\boldsymbol{\psi}_{B,n}^{L(R)\,\dagger}. \quad (8.103)$$

By proceeding in the very same way as in the previous subsection, we can calculate these blocks, for example [cf. Eqs. (8.93), (8.95) and (8.100)],

$$\mathbf{F}_{SS}^L = \frac{1}{\pi} \int_{-\infty}^{\infty} dE \, E \, \mathbf{G}_S \boldsymbol{\Gamma}_L \mathbf{G}_S^\dagger f_L(E), \quad (8.104)$$

$$\mathbf{F}_{SS}^R = \frac{1}{\pi} \int_{-\infty}^{\infty} dE \, E \, \mathbf{G}_S \boldsymbol{\Gamma}_R \mathbf{G}_S^\dagger f_R(E), \quad (8.105)$$

$$\mathbf{F}_{LS}^R = \frac{1}{\pi} \int_{-\infty}^{\infty} dE \, E \, f_R(E) \mathbf{G}_L \mathbf{K}_{LS} \mathbf{G}_S \boldsymbol{\Gamma}_R \mathbf{G}_S^\dagger. \quad (8.106)$$

Finally, we rewrite $\mathbf{F}$ in terms of the lesser Green's function as

$$\mathbf{F}_S = \frac{1}{2\pi i} \int_{-\infty}^{\infty} dE \, E \, \mathbf{G}_S^<(E). \quad (8.107)$$

8.5.6 Transmission and Current

In order to calculate the current, we employ the charge continuity equation, and we proceed in a similar way to Section 8.3. The electron charge in the scattering region is given by $Q_S = \sum_{\alpha=1}^{N_S} q_\alpha$, where N_S is the number of basis orbitals in the scattering region. Hence, by employing Eq. (8.27), we obtain

$$\frac{\partial Q_S}{\partial t} = \frac{e}{i\hbar} \sum_{\alpha=N_L+1}^{N_L+N_S} \sum_{\beta=1}^{N_L+N_S+N_R} \Big(H_{\alpha\beta}\rho_{\beta\alpha}$$

$$- \rho_{\alpha\beta}H_{\beta\alpha} + F_{\alpha\beta}S_{\beta\alpha} - S_{\alpha\beta}F_{\beta\alpha} \Big), \quad (8.108)$$

where the summation over β runs over the whole (infinite) device space and is not restricted to the scattering region (N_L and N_R are the infinite number of basis orbitals in the left and right lead). The sum of the terms, where both α and β refer to basis orbitals inside the scattering region, is zero so that we obtain

$$\frac{\partial Q_S}{\partial t} = \frac{e}{i\hbar} \sum_{\alpha=N_L+1}^{N_L+N_S} \sum_{\beta=1}^{N_L} \Big(H_{\alpha\beta}\rho_{\beta\alpha} - \rho_{\alpha\beta}H_{\beta\alpha} \Big)$$

$$+ \frac{e}{i\hbar} \sum_{\alpha=N_L+1}^{N_L+N_S} \sum_{\beta=N_L+N_S+1}^{N_R} \Big(H_{\alpha\beta}\rho_{\beta\alpha} - \rho_{\alpha\beta}H_{\beta\alpha} \Big)$$

$$+ \frac{e}{i\hbar} \sum_{\alpha=N_L+1}^{N_L+N_S} \sum_{\beta=1}^{N_L} \Big(F_{\alpha\beta}S_{\beta\alpha} - S_{\alpha\beta}F_{\beta\alpha} \Big)$$

$$+ \frac{e}{i\hbar} \sum_{\alpha=N_L+1}^{N_L+N_S} \sum_{\beta=N_L+N_S+1}^{N_R} \Big(F_{\alpha\beta}S_{\beta\alpha} - S_{\alpha\beta}F_{\beta\alpha} \Big), \quad (8.109)$$

or in matrix notation

$$\frac{\partial Q_S}{\partial t} = \frac{e}{i\hbar} \mathrm{Tr}\Big[\mathbf{H}_{SL}\boldsymbol{\rho}_{LS} - \boldsymbol{\rho}_{SL}\mathbf{H}_{LS} \Big]$$

$$+ \frac{e}{i\hbar} \mathrm{Tr}\Big[\mathbf{H}_{SR}\boldsymbol{\rho}_{RS} - \boldsymbol{\rho}_{SR}\mathbf{H}_{RS} \Big]$$

$$+ \frac{e}{i\hbar} \mathrm{Tr}\Big[\mathbf{F}_{SL}\mathbf{S}_{LS} - \mathbf{S}_{SL}\mathbf{F}_{LS} \Big]$$

$$+ \frac{e}{i\hbar} \mathrm{Tr}\Big[\mathbf{F}_{SR}\mathbf{S}_{RS} - \mathbf{S}_{SR}\mathbf{F}_{RS} \Big]$$

$$= \frac{2e}{\hbar} \mathrm{Im}\Big[\mathrm{Tr}\Big(\mathbf{H}_{SL}\boldsymbol{\rho}_{LS} - \mathbf{S}_{SL}\mathbf{F}_{LS} \Big) \Big]$$

$$+ \frac{2e}{\hbar} \mathrm{Im}\Big[\mathrm{Tr}\Big(\mathbf{H}_{SR}\boldsymbol{\rho}_{RS} - \mathbf{S}_{SR}\mathbf{F}_{RS} \Big) \Big]. \quad (8.110)$$

where the trace runs over the N_S orbitals of the scattering region. By using Eq. (8.34), we can identify the first (second) summand on the right-hand side of the equation as the charge current that flows from the left (right) interface into the scattering region, $I_{L(R)}$, that is,

$$I_L = \frac{2e}{\hbar} \mathrm{Im}\Big[\mathrm{Tr}\Big(\mathbf{H}_{SL}\boldsymbol{\rho}_{LS} - \mathbf{S}_{SL}\mathbf{F}_{LS} \Big) \Big], \quad (8.111)$$

$$I_R = \frac{2e}{\hbar} \text{Im}\left[\text{Tr}\left(\mathbf{H}_{SR}\boldsymbol{\rho}_{RS} - \mathbf{S}_{SR}\mathbf{F}_{RS} \right) \right]. \qquad (8.112)$$

These currents can be further decomposed into the contributions from the states coming from the left and right leads as $I_L = I_L^L + I_L^R$ and $I_R = I_R^L + I_R^R$, with

$$I_L^{L(R)} = \frac{2e}{\hbar} \text{Im}\left[\text{Tr}\left(\mathbf{H}_{SL}\boldsymbol{\rho}_{LS}^{L(R)} - \mathbf{S}_{SL}\mathbf{F}_{LS}^{L(R)} \right) \right], \qquad (8.113)$$

$$I_R^{L(R)} = \frac{2e}{\hbar} \text{Im}\left[\text{Tr}\left(\mathbf{H}_{SR}\boldsymbol{\rho}_{RS}^{L(R)} - \mathbf{S}_{SR}\mathbf{F}_{RS}^{L(R)} \right) \right]. \qquad (8.114)$$

Next, by using Eqs. (8.100) and (8.106), we obtain

$$\text{Tr}\left(\mathbf{H}_{SL}\boldsymbol{\rho}_{LS}^{R} - \mathbf{S}_{SL}\mathbf{F}_{LS}^{R} \right)$$
$$= \frac{1}{\pi} \int_{-\infty}^{\infty} dE f_R(E) \text{Tr}\left(\mathbf{H}_{SL}\mathbf{G}_L\mathbf{K}_{LS}\mathbf{G}_S\boldsymbol{\Gamma}_R\mathbf{G}_S^{\dagger} \right)$$
$$- \frac{1}{\pi} \int_{-\infty}^{\infty} dE f_R(E) \text{Tr}\left(E\mathbf{S}_{SL}\mathbf{G}_L\mathbf{K}_{LS}\mathbf{G}_S\boldsymbol{\Gamma}_R\mathbf{G}_S^{\dagger} \right). \qquad (8.115)$$

Then by noting that $(\mathbf{H}_{SL} - E\mathbf{S}_{SL}) = \mathbf{K}_{SL}$ and by using Eq. (8.67), we obtain

$$\text{Tr}\left(\mathbf{H}_{SL}\boldsymbol{\rho}_{LS}^{R} - \mathbf{S}_{SL}\mathbf{F}_{LS}^{R} \right) = \frac{1}{\pi} \int_{-\infty}^{\infty} dE f_R(E) \text{Tr}\left[\boldsymbol{\Sigma}_L \mathbf{G}_S \boldsymbol{\Gamma}_R \mathbf{G}_S^{\dagger} \right]. \qquad (8.116)$$

In the same way, by recalling the cyclic property of the trace, we have

$$\text{Tr}\left(\boldsymbol{\rho}_{SL}^{R}\mathbf{H}_{LS} - \mathbf{F}_{SL}^{R}\mathbf{S}_{LS} \right) = \frac{1}{\pi} \int_{-\infty}^{\infty} dE f_R(E) \text{Tr}\left[\boldsymbol{\Sigma}_L^{\dagger} \mathbf{G}_S \boldsymbol{\Gamma}_R \mathbf{G}_S^{\dagger} \right]. \qquad (8.117)$$

By inserting these last two equations into Eq. (8.113), we obtain

$$I_L^R = -\frac{2e}{h} \int_{-\infty}^{\infty} dE f_R(E) \, \text{Tr}\left[\boldsymbol{\Gamma}_L \mathbf{G}_S \boldsymbol{\Gamma}_R \mathbf{G}_S^{\dagger} \right]. \qquad (8.118)$$

Similarly, one can show that

$$I_L^L = \frac{2e}{h} \int_{-\infty}^{\infty} dE f_L(E) \, \text{Tr}\left[\boldsymbol{\Gamma}_L \mathbf{G}_S \boldsymbol{\Gamma}_R \mathbf{G}_S^{\dagger} \right]. \qquad (8.119)$$

Since we consider the steady-state situation $\partial Q_S / \partial t = 0$, we have

$$\frac{\partial Q_S}{\partial t} = I_L + I_R = 0 \qquad (8.120)$$

and we can define the current that flows through the scattering region as

$$I = I_L = -I_R. \qquad (8.121)$$

Finally, by adding Eqs. (8.118) and (8.119), we obtain the expression for the current

$$I = \frac{2e}{h} \int_{-\infty}^{\infty} dE \, T(E) \left(f_L(E) - f_R(E) \right), \qquad (8.122)$$

where we have introduced the transmission coefficient

$$T(E) = \text{Tr}\left[\boldsymbol{\Gamma}_L \mathbf{G}_S \boldsymbol{\Gamma}_R \mathbf{G}_S^{\dagger} \right]. \qquad (8.123)$$

Equation (8.122) is a generalization of Eq. (8.41) presented in the Landauer–Büttiker theory.

8.5.7 Introducing the Bias

By deriving Eq. (8.122), we have completed the presentation of the Landauer–Büttiker theory in terms of the NEGF. However, we remind that, by using the NEGF, we do not require anymore that $\mu_L - \mu_R \to 0^+$ as we can calculate the charge density of the scattering region at finite bias by using Eq. (8.98). The Green's functions and the transmission therefore turn into bias-dependent quantities, that is, $\mathbf{G}_S(E, \Delta V)$, $\mathbf{G}_S^{<}(E, \Delta V)$ and $T(E, \Delta V)$ with $e\Delta V = (\mu_L - \mu_R)$. The effect of applying the bias over the leads is that of shifting their relative chemical potentials as discussed at the beginning of Section 8.4. The self-energies therefore become $\boldsymbol{\Sigma}_{L(R)}(E, \Delta V) = \boldsymbol{\Sigma}_{L(R)} \left(E - (\mu_{L(R)} - E_F), \Delta V = 0 \right)$, where E_F is the Fermi energy of the whole system at $\Delta V = 0$.

8.5.8 Transverse Periodic Boundary Conditions

If periodic boundary conditions are used in the plane perpendicular to the transport direction (here set to be z), then $\mathbf{H}_S$, $\mathbf{S}_S$, $\mathbf{G}_S$, $\mathbf{G}_S^{<}$, $\boldsymbol{\Sigma}_{L(R)}$ and $\boldsymbol{\Gamma}_{L(R)}$ depend on $\mathbf{k} = (k_x, k_y)$, so that we have $\mathbf{H}_{\mathbf{k},S}$, $\mathbf{S}_{\mathbf{k},S}$, $\mathbf{G}_{\mathbf{k},S}$, $\mathbf{G}_{\mathbf{k},S}^{<}$, $\boldsymbol{\Sigma}_{\mathbf{k},L(R)}$, $\boldsymbol{\Gamma}_{\mathbf{k},L(R)}$. The equations outlined in the previous sections remain valid for each $\mathbf{k}$-point, and the total density matrix and transmission are given by an integral over the BZ perpendicular to the transport direction as

$$\boldsymbol{\rho}_S(E) = \frac{1}{\Omega_{BZ}} \int_{\Omega_{BZ}} \boldsymbol{\rho}_{\mathbf{k},S}(E) d\mathbf{k}, \qquad (8.124)$$

and

$$T(E) = \frac{1}{\Omega_{BZ}} \int_{\Omega_{BZ}} T_{\mathbf{k}}(E) d\mathbf{k}. \qquad (8.125)$$

8.6   Applications to Model Systems

Here, we present a number of applications of the NEGF method to simple model systems. These examples will help the reader to familiarize with the method and to learn some basics on how it is implemented in practice and to understand fundamental aspects of coherent transport in nanostructures.

8.6.1   Single Orbital Quantum Dot

We consider the system represented in Figure 8.7 (a), where the scattering region comprises only one atom ("quantum dot") with one atomic orbital coupled to two semi-infinite atomic chains forming the leads. The dot has on-site energy ϵ_0 and the hopping to the leads is t_0. Like in Section 8.2, the leads are composed of hydrogen atoms with on-site energy ϵ and nearest-neighbor hopping t.

The first step in solving this model system consists in calculating the lead self-energies. Since the dot is coupled only to the atom 1 in the right lead, from Eq. (8.68) we have $\boldsymbol{\Sigma}_R(E) = t_0^2 (\mathbf{g}_R)_{11}(E) = t_0^2 g_{\Sigma}(E)$, where $g_{\Sigma}(E)$ is the so-called surface Green's function. This can be calculated by noting that the site 1 is in turn a quantum dot

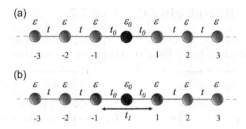

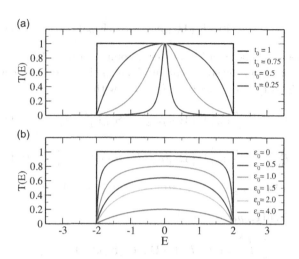

FIGURE 8.7 Model quantum dot attached to two semi-infinite atomic chains. The shaded area represents the electrodes. (a) Only the dot is included into the scattering region. (b) The atoms -1 and 1 are inserted into the scattering region, and they are coupled via a direct hopping term t_1.

coupled to the rest of the right semi-infinite chain via the hopping t. Therefore, by introducing the self-energy $[\Sigma_R]_{11}(E) = t^2(g_R)_{22}(E)$, we have

$$g_\Sigma = \frac{1}{E - \epsilon - t^2(g_R)_{22}}. \tag{8.126}$$

and since $(g)_{R,22} = g_\Sigma$, this equation can be rewritten in a straight-forward way as a quadratic equation

$$t_0^2 g_\Sigma^2 - (E - \epsilon)g_\Sigma^2 + 1 = 0 \tag{8.127}$$

of solution $g_\Sigma = \mathrm{Re}g_\Sigma + i\mathrm{Im}g_\Sigma$ with

$$\mathrm{Re}g_\Sigma = \frac{1}{|t|}[x - \mathrm{sgn}(x)\sqrt{x^2 - 1}\theta(|x| - 1)], \tag{8.128}$$

$$\mathrm{Im}g_\Sigma = -\frac{1}{|t|}\sqrt{1 - x^2}\theta(1 - |x|), \tag{8.129}$$

where $x = (E - \epsilon)/2t$ (the dependence of the surface Green's function on the energy has not been indicated in these equations to keep the notation simple). The self-energy for the quantum dot "0" finally is $\Sigma_R(E) = t_0^2 g_\Sigma(E)$ and is plotted in Figure 8.8. The left self-energy can be obtained in the very same way and is identical to the right self-energy.

Once we have both $\Sigma_R(E)$ and $\Sigma_L(E)$ we are able to compute the Green's function of the dot

$$G_0(E) = \frac{1}{E - \epsilon_0 - \Sigma_L(E) - \Sigma_R(E)}$$

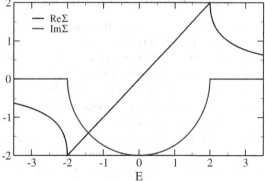

FIGURE 8.8 Real and imaginary part of the lead self-energy for the single orbital quantum dot.

FIGURE 8.9 Transmission function for a single orbital quantum dot computed for different values of t_0 (a) and ϵ_0 (b). E, ϵ_0 and t_0 are in a unit of the chain hopping $t = 1$.

(for $\delta = 0$) and hence its spectral function $A_0(E)$ and its transmission $T(E)$.

The spectral function has the form of a Lorentzian function

$$A_0(E)$$

$$= \frac{\Gamma_L(E) + \Gamma_R(E)}{[E - \epsilon_0 - \mathrm{Re}\Sigma_L(E) - \mathrm{Re}\Sigma_R(E)]^2 - [\Gamma_L(E) + \Gamma_R(E)]^2/4} \tag{8.130}$$

with the (energy-dependent) width $\Gamma_L(E) + \Gamma_R(E)$. From this equation it is therefore evident that the real part of the self-energies causes a shift of the dot energy, while the imaginary part leads to a broadening of the dot level.

$T(E)$ for different values of ϵ_0 and t_0 is displayed in Figure 8.9. When $\epsilon_0 = \epsilon = 0$, T is exactly equal to 1 inside an energy window that extends over the atomic chain bandwidth $2t$. Therefore, for this special case, we have reproduced with the full machinery of the NEGF the special result of the Landauer–Büttiker theory for the atomic chain with a transmission equal to 1 discussed in Section 8.4. Next, when changing t_0, but keeping $\epsilon_0 = 0$, the transmission becomes a sharper peak as the dot state becomes more localized. However, T remains equal to 1 at $E = 0$. In contrast, by misaligning the dot's on-site energy ϵ_0 with respect to that of the leads atoms ϵ, the transmission drops as we create a potential barrier for the electrons to flow.

8.6.2 Wide-Band Limit

If we set $t \to \infty$ and $x \ll t$, we can neglect the real part of the self-energy and the only effect of the leads is the level broadening. Therefore, the self-energy of the right (left) lead becomes $\Sigma_{R(L)}(E) = -i\Gamma_{R(L)}/2$ and is energy-independent over a large energy interval around 0. This limit is usually called the "wide-band limit", and it is invoked in most model studies of quantum dots.

The Green's function of the quantum dot in Figure 8.7 results equal to

$$G_0(E) = (E + i\Gamma)^{-1} \qquad (8.131)$$

for $\epsilon_0 = 0$ and $\Gamma_L = \Gamma_R \equiv \Gamma$. The transmission is

$$T(E) = \frac{\Gamma^2}{E^2 + \Gamma^2}. \qquad (8.132)$$

Next, we define two quantum dots of on-site energies ϵ_{d1} and ϵ_{d2} connected through a hopping integral t_d, and each of them is attached to one lead. The Green's function is then defined via the equation

$$\mathbf{G}^{-1}(E) = \begin{pmatrix} E - \epsilon_{d1} + i\Gamma_L/2 & t_d \\ t_d & E - \epsilon_{d2} + i\Gamma_R/2 \end{pmatrix}. \qquad (8.133)$$

We leave it as an exercise for the reader to invert this matrix by using Eq. (8.151) and to verify that the transmission is

$$T(E) = \frac{\Gamma^2 t_d^2}{(E^2 - \Gamma^2/4 - t_d^2)^2 + (\Gamma E)^2} \qquad (8.134)$$

for $\epsilon_{d1} = \epsilon_{d2} = 0$ and $\Gamma_L = \Gamma_R \equiv \Gamma$. The reader should evaluate the matrices $\mathbf{G}_S \Gamma_R \mathbf{G}_S^\dagger$ and $\mathbf{G}_S \Gamma_L \mathbf{G}_S^\dagger$ noting that the diagonal elements are real and the off-diagonal elements are complex, as stated in Section 8.5.4.

8.6.3 Quantum Interference Effects and Fano Resonance

Here, we modify the quantum dot system by including the atoms 1 and −1 into the scattering region as shown in the bottom panel of Figure 8.7. In this way, we can introduce a direct hopping t_1 between them, and we create two possible paths for the electrons: the first through the dot and the second that implies direct tunneling between the atom 1 and the atom −1. We find in Figure 8.10 that T gets largely suppressed and even becomes zero at some energies. T assumes the so-called Fano shape [31]. This is due to the destructive quantum interference between the

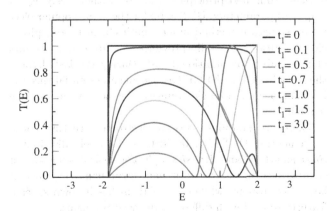

FIGURE 8.10 Transmission function for the system in Figure 8.7 (b) computed for several values of t_1, expressed in the unit of $t = 1$.

two paths. Fano features ascribed to quantum interference effects are often seen in the current versus voltage curves of molecules measured in scanning tunneling microscope or break-junctions experiments, and they represent one of the fascinating manifestations of the coherent quantum nature of electron transport at the nanoscale.

8.6.4 Model Benzene Molecule

Quantum interference can be seen also in transport through molecules, and indeed, many theoretical as well as experimental studies have been published in the last decade on this topic (see, e.g., Refs. [1,6,29] just to mention a few...). Here, we discuss the prototypical example of a model benzene molecule. The scattering region comprises six atoms with only one orbital each of on-site energy $\epsilon = 0$. We assume nearest-neighbor hopping $t_0 = 1$ so that the Hamiltonian matrix $\mathbf{H}_S$ has the only non-zero elements $(\mathbf{H}_S)_{l,l+1} = t_0$, $(\mathbf{H}_S)_{l+1,l} = t_0$ (for $1 \leq l \leq 5$), $(\mathbf{H}_S)_{1,6} = t_0$ and $(\mathbf{H}_S)_{6,1} = t_0$. We then connect the leads in three different relative positions, called *ortho*, *meta* and *para* as shown in Figure 8.11 The only non-zero element of the left lead self-energy is $(\boldsymbol{\Sigma}_L(E))_{11}$ and those of the right lead self-energy are $(\boldsymbol{\Sigma}_R(E))_{44}$, $(\boldsymbol{\Sigma}_R(E))_{33}$ and $(\boldsymbol{\Sigma}_R(E))_{22}$ for the *para-*, *meta-* and *ortho*-benzene, respectively. In order to simplify the problem, we invoke the wide band-limit, and we set the lead self-energies equal to $-i\Gamma/2$ where $\Gamma \gg t_0$.

The transmission function is displayed in Figure 8.11. It shows no interference effect for the *para*-benzene. In contrast, there are three interference features seen as sharp dips at energies equal to 0 and ± 1.5 for the *meta*-benzene and at ± 2.12 and ± 1.5 for the *ortho*-benzene. The difference between the three cases is due to the different paths connecting the left and right lead. A detailed discussion about the system is beyond the goal of this chapter, and we refer the interested readers to the original literature, in particular to Ref. [18].

8.6.5 Graphene and Graphene Nano-ribbons

Here, we investigate coherent transport for the model zig-zag graphene nano-ribbons represented in Section 8.2.4. In order to construct the leads' self-energies, we follow the same reasoning as in Section 8.6.1, but we note that the surface Green's function in Eq. (8.126) becomes a matrix of dimension $4N_x$, defined by [17]

$$[\mathbf{g}_\Sigma^{-1}]_{\mathbf{R}_i,a';\mathbf{R}_i,a} = E - H_{\mathbf{R}_i,a';\mathbf{R}_i,a}$$
$$- \sum_{a'',a'''=1}^{4N_x} H_{\mathbf{R}_i-\mathbf{v}_2,a';\mathbf{R}_i,a''} [\mathbf{g}_\Sigma^{-1}]_{\mathbf{R}_i,a'';\mathbf{R}_i,a'''} H_{\mathbf{R}_i-\mathbf{v}_2,a''';\mathbf{R}_i,a}, \qquad (8.135)$$

where we used the fact that $[\mathbf{g}_\Sigma^{-1}]_{\mathbf{R}_i,a'';\mathbf{R}_i,a'''} = [\mathbf{g}_\Sigma^{-1}]_{\mathbf{R}_i-\mathbf{v}_2,a'';\mathbf{R}_i-\mathbf{v}_2,a'''}$. We can solve Eq. (8.135) self-consistently [17]: starting from a initial guess for $\mathbf{g}_\Sigma$ we use Eq. (8.135) to obtain a new $\mathbf{g}_\Sigma$, which is then fed again

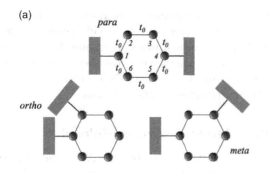
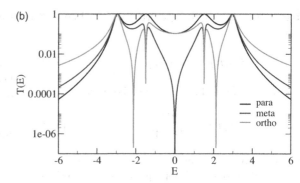
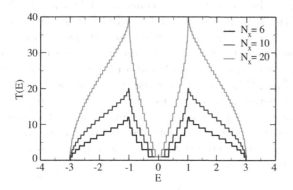
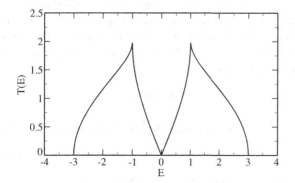

FIGURE 8.11 (a) Schematic representation of the model *para*, *meta* and *ortho* benzene molecular junctions. (b) Transmission function; the energy is in the unit of $t_0 = 1$. Note that $T(E)$ is plotted in logarithmic scale. This is a very useful way to highlight drastic changes in $T(E)$.

into Eq. (8.135) and so on. These operations are repeated several times until convergence is reached.

The transmission function for nano-ribbons of width $N_x = 6$, 10 and 20 is presented in Figure 8.12. We observe that $T(E)$ assumes constant integer values over some energy ranges. In particular, we can focus on the case $N_x = 6$. It is then easy to verify that at each energy, the transmission in Figure 8.12 is exactly equal to the number of bands, that is, channels, that can be found at that energy in the

FIGURE 8.12 Transmission function for model graphene nano-ribbons of width $N_x = 6$, 10 and 20 (cf. Section 8.2.4); the energy is in a unit of the hopping integral $t = 1$.

FIGURE 8.13 Transmission function for model infinite graphene flake (cf. Section 8.2.4); the energy is in unit of the hopping integral $t = 1$.

band structure plot of Figure 8.3. This is exactly what we explained in the Section 8.4 within the framework of the Landauer–Büttiker theory which we have now implemented thanks to the NEGF approach. We leave it as an exercise for the reader to see how $T(E)$ changes when disorder (i.e., elastic scattering) is introduced in the scattering region, for example, by randomizing the on-site energies.

Finally, we note that by increasing the nano-ribbon width, the transmission plateaus become smaller and eventually $T(E)$ merges into the transmission of infinite graphene. For comparison, we present in Figure 8.13 the transmission function of model graphene computed by employing periodic boundary conditions in the direction transverse to the transport in the way described in Section 8.5.8.

8.7 Density Functional Theory Combined with the Non-Equilibrium Green's Functions

One of the great successes of the NEGF formalism as presented here is that it can be readily combined with DFT to obtain predictions of the charge transport in nano-devices from first-principles, that is, without relying on empirical parameters. DFT is by far the most popular electronic structure method among both chemists and physicists because it generally provides accurate enough results at reasonable computational cost. During the last decade, the so-called DFT+NEGF has become an essential theoretical tool for molecular electronics and generally nano-device modeling.

There are nowadays several electronic structure codes implementing DFT+NEGF such as the freely distributed *Smeagol* [41,45] and *Transiesta* [3]. For basic calculations of conductance in molecules or nano-structures, they are all largely equivalent, while each of them has its own specific features to deal with different advanced problems.

This section provides an overview of the method in particular we highlight the numerical and fundamental limitations to be aware of when carrying out DFT+NEGF calculations.

8.7.1 General Idea

DFT is based on the Hohenberg–Kohn theorem [19], which states that the ground state energy of a system of interacting particles is a unique functional of the charge density. In practice, DFT is implemented by mapping the exact functional problem onto a fictitious single-particle Hamiltonian problem, known as the Kohn–Sham (KS) Hamiltonian [22]. This includes a single-particle kinetic energy term, external classical potentials (such as the electron-nuclei interaction), the Hartree term describing the classical Coulomb interaction between electrons, and the exchange-correlation (xc) potential, which accounts for all remaining quantum many-body contributions and which is not known and has to be approximated. The most common xc functionals are based on the local density approximation (LDA) or the semi-local generalized gradient approximation (GGA). A complete description of DFT can be found in a number of dedicated text-books such as Ref. [34] or the more recent Ref. [13].

The energy and the density are computed via a self-consistent procedure. This means that the KS Hamiltonian is diagonalized to obtain the KS eigenvectors and thereafter the density (Figure 8.14), which in turn is used to compute a new Hartree and xc potential to update the KS Hamiltonian. These steps are repeated until the converge between the old and updated density is reached.

In modeling a two-terminal device by DFT, we identify the Hamiltonian $\mathbf{H}$ in Eq. (8.57) as the KS Hamiltonian of the system. The NEGF approach then offers a way for solving the KS problem without calculating the eigenvectors of the KS Hamiltonian (Figure 8.14). If we assume that all the parts of $\mathbf{H}$, except $\mathbf{H}_S$, are independent of the charge density in the scattering region $\boldsymbol{\rho}_S$, we can set up a self-consistent cycle where $\mathbf{G}_S$ is computed from $\mathbf{H}_S$ by using Eq. (8.66), and subsequently, the density matrix $\boldsymbol{\rho}_S$ is obtained through Eq. (8.98). $\boldsymbol{\rho}_S$ is used to update $\mathbf{H}_S$ and to start the calculation over again until convergence is achieved. Finally, the transmission and the current are calculated. These are often called the KS transmission and current to emphasize that they are obtained by using the KS states as scattering states.

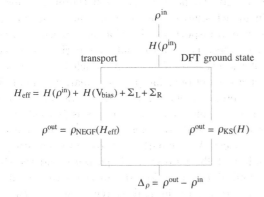

FIGURE 8.14 Schematic diagram illustrating the steps of a standard KS-DFT calculation (right) and of a DFT+NEGF calculation (left).

8.7.2 Some Remarks on Computational Details

Like in standard KS-DFT, the numerical accuracy of DFT+NEGF calculations depends on the employed basis set, on the pseudo-potentials (if used) and on the sampling of the BZ in the plane perpendicular to the transport direction (if transverse periodic boundary conditions are used). Therefore, the quality of each calculation must be assessed with respect to these computational details in the same way as one would do in any standard DFT calculation. Yet we note that there are some numerical issues specific to DFT+NEGF to be aware of. They are related to the way the integral in Eq. (8.98) for the density matrix $\boldsymbol{\rho}_S$ is typically calculated.

In the first place, the key assumption that only the block $\mathbf{H}_S$ in Eq. (8.57) depends on $\boldsymbol{\rho}_S$ is valid only if local and semi-local xc functionals are employed. This has always been the case for standard calculations until now, but such assumption will lead to problems if one wants to extend the method to use hybrid xc functionals, with a fraction of non-local exact exchange. Furthermore, we note that even for local xc functionals, the scattering region has to be chosen large enough, so that changes to the charge density in the center of it do not affect the Hamiltonian at its boundaries. Equivalently, the requirement is that all charge fluctuations in the center of the scattering region are screened once its boundaries are reached. For metallic systems with a short screening-length, this requirement is fulfilled within a few atomic layers, while for semiconducting systems, where the screening-length is very large, the size of the scattering region must be increased accordingly.

Secondly, we point out that Green's functions usually vary rapidly as function of energy, and therefore, the integration of $\mathbf{G}_S^<$ in Eq. (8.98) for $\boldsymbol{\rho}_S$ is problematic. In order to get around this problem, $\mathbf{G}_S^<$ is split into two terms usually called the equilibrium and the non-equilibrium contributions $\mathbf{G}_S^< = \mathbf{G}_{S,\text{eq}}^< + \mathbf{G}_{S,\text{neq}}^<$, so that Eq. (8.98) reads

$$\boldsymbol{\rho}_S = \boldsymbol{\rho}_{S,\text{eq}} + \boldsymbol{\rho}_{S,\text{neq}} = \frac{1}{2\pi i} \int_{-\infty}^{\infty} dE\, \mathbf{G}_{S,\text{eq}}^<(E)$$

$$+ \frac{1}{2\pi i} \int_{-\infty}^{\infty} dE\, \mathbf{G}_{S,\text{neq}}^<(E). \qquad (8.136)$$

After some algebraic manipulations of Eq. (8.97) $\mathbf{G}_{S,\text{eq}}^<$ and $\mathbf{G}_{S,\text{neq}}^<$ are obtained as

$$\mathbf{G}_{S,\text{eq}}^<(E) = -2[\mathbf{G}_S(E) - \mathbf{G}_S^\dagger(E)]$$
$$\times [\zeta f_R(E) + (1-\zeta) f_L(E)] \qquad (8.137)$$

$$\mathbf{G}_{S,\text{neq}}^<(E) = i2\mathbf{G}_S(E)[\zeta\Gamma_L(E) - (1-\zeta)\Gamma_R(E)]\mathbf{G}_S^\dagger(E)$$
$$\times [f_L(E) - f_R(E)], \qquad (8.138)$$

with ζ a real number between 0 and 1. Therefore, $\mathbf{G}_{S,\text{eq}}^<$ is divided into a term proportional to $\mathbf{G}_S$ and one proportional to $\mathbf{G}_S^\dagger$. Each of these terms can be analytically extended to complex energies E, since $\mathbf{G}_S$ ($\mathbf{G}_S^\dagger$) has no poles for $\text{Im}(E) > 0$ [$\text{Im}(E) < 0$]. In evaluating the equilibrium contribution

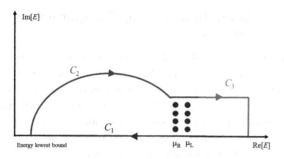

FIGURE 8.15 Path for the contour integral used to evaluate $\rho_{S,eq}$. The dark gray dots represents the poles.

in Eq. (8.136), it is therefore possible to replace the integral over the real energy axis by one over complex energies, where the integrand is much smoother and much less integration points are needed as a consequence. In contrast, the integral over the non-equilibrium $\mathbf{G}_{S,neq}^{<}$ has to be calculated on the real energy axis, since $\mathbf{G}_S$ and $\mathbf{G}_S^{\dagger}$ cannot be separated out. However, the integration range only extends over the bias window, where the Fermi–Dirac distributions of the leads are different.

In Figure 8.15, the integral along C_1, which is the one to evaluate $\rho_{S,eq}$, is given by the contour integral along the closed path minus the integral along C_2 and C_3. To evaluate the contour integral, we use the fact that the only poles inside the contour are the ones of the Fermi–Dirac distributions, which are found where $1 + e^x = 0$, and are therefore at $\text{Re}(E) = \mu_{R(L)}$ and $\text{Im}(E) = (2n + 1)\pi k_B \tau$ ($n \in \mathbb{N}$). The integral over C_3 can be assumed to be zero as the Fermi–Dirac distributions are approximately zero above its lower bound. Therefore, the integral along C_1 is finally given by the sum of the poles minus the integral along C_2, which has to be calculated numerically. As a consequence, the accuracy of $\rho_{S,eq}$ depends on the used number of poles in the Fermi distribution and on the number of energy points along C_2 used for the numerical integration. These are parameters that one usually has to enter to the simulation software to perform a DFT+NEGF calculation. The number of poles specifies how far from the real axis the contour integral will be performed. The further away from the real axis, the smoother the Green's function is. Typically, the number of poles used is of the order of a few tens for τ equal to room temperature, and then, the number of integration point for C_2 is less than 100.

8.7.3 Fundamental Limitations

The Landauer–Büttiker theory strictly applies only to *non-interacting systems,* and it can be extended to *finite-bias problems* by using the NEGF. In contrast, DFT represents a reformulation of the quantum problem for *the ground state of a system of interacting electrons.* The combination of the two theories therefore suffers from a number of conceptual problems that one should keep in consideration. Here, we briefly review them.

As we mentioned above, for the case of systems in equilibrium, DFT+NEGF provides a way to solve the KS

problem for given boundary conditions, set by the self-energies. Strictly speaking observables have to be computed from the density and they represent the observables of the interacting system, while the Kohn–Sham single-particle states, used in DFT+NEGF as scattering states, do not correspond to the true excitation energies. In exact DFT, the only exception is the eigenenergy of the highest occupied state, which is equal to minus the ionization energy [38], but not even this relation is valid for standard local and semi-local functionals. Despite that, for many systems, the KS states provide a good approximate description of the electronic structure (see, for instance, Ref. [23] for a very clear and exhaustive discussion). In fact, the physical nature of the studied system, such as the character of the highest occupied and lowest unoccupied molecular orbitals, is often correctly predicted. One can therefore adopt a practical approach and consider the KS spectrum as good approximation to the true spectrum and with that calculate the transmission. In molecules, for example, most of the shortcomings of the DFT spectrum are related to the size of the fundamental gap and to the alignment between the molecular states and the leads Fermi energy. Different schemes have been proposed in order to correct for these shortcomings, and in general, they are expected to improve the transmission and ultimately the predicted conductance [40]. An example of that is presented below.

The use of DFT+NEGF as an effective single-particle theory to compute transport within the Landauer–Büttiker framework has proven very successful in studies that have focused on qualitative trends, such as the dependence of the conductance of molecules on the conformation, side group functionalization and character of the molecule–electrode bond. Most notably, DFT+NEGF can even achieve quantitatively accurate results if the transport in the studied systems is not affected by strong electronic correlations. Despite that, one may still question whether the KS current is the actual current of the interacting system. By using the time-dependent extension of DFT, it can be shown that this is not exactly the case. Even in linear-response, the interacting system conductance presents an additional contribution over the KS conductance, called dynamical xc correction [20]. So far this has been thoroughly analyzed only in model studies, and it turns out to be important to describe transport in the Coulomb blockade regime. In other cases, it is zero, but the derivative discontinuity of the exact xc functional, which is absent in any approximate functional for practical material simulations, is the key property determining the interacting current. A detailed discussion about these issues is beyond the goal of this chapter, and we refer to Ref. [24] for a comprehensive review. We remark that to date, no study has been able to fully address the importance and the size of the dynamical xc correction and of the derivative discontinuity for transport in real systems. Studies based on model simulations indicate that they are important to describe transport effects driven by strong electron correlation effects.

8.7.4 Example: Au-BDT-Au Molecular Junction

From a historical perspective, the benzene-dithiol (BDT) molecule attached to gold surfaces represents the prototypical molecular junction. Experimental results for the conductance provided a very scattered set of data during early measurements due to the lack of precise control over the atomic details of the molecule–Au contact. Nevertheless, recent experiments are in agreement with an average low-bias value of about 0.01 G_0 [4,53].

The KS transmission at zero bias calculated by using the *Smeagol* DFT+NEGF code [41] is shown in Figure 8.16. The computational details can be found in [43]. When using the LDA, the Fermi energy of the junction E_F lies in the gap between two broad features, which stem from the highest occupied and lowest unoccupied molecular orbital (HOMO and LUMO). Nevertheless, the HOMO is pinned around E_F for LDA. The zero bias conductance, defined as $\mathcal{G} = T(E_F)\mathcal{G}_0$ turns out to be rather large, namely 0.2 $\mathcal{G}_0$. This is due to this pinning of the HOMO and to the broadening induced by the coupling to the gold. The calculated conductance is much larger than the one experimentally measured.

In order to understand the discrepancies between theory and experiments, we observe that one of the main deficiencies of LDA is the self-interaction error [39]. This spurious interaction of an electron with the Hartree and xc potential generated by itself favors charge delocalization and shifts occupied KS states of small molecules to higher energies. In order to deal with this problem, a number of approaches have been proposed. Here, we use the so-called

atomic self-interaction correction (ASIC) [37], which has been shown to systematically improve the position of the HOMO when compared to the ionization potential for molecules in the gas phase. For example, the transmission computed with ASIC in Figure 8.16 shows that the HOMO is drastically shifted towards lower energies and E_F ends up in the middle of a large HOMO–LUMO gap. This results in a reduction of $\mathcal{G}$ by one order of magnitude. The ASIC $\mathcal{G} = 0.02\ \mathcal{G}_0$ is now in much better agreement with the experimental results.

This example clearly illustrates the importance of the having the correct energy level alignment in DFT+NEGF, and we have presented a specific scheme to correct for the LDA failures. We refer the readers to Refs. [46,49,52] for a more detailed discussion about this problem and Au-BDT-Au junctions.

8.8 Conclusions

We presented the Landauer–Büttiker theory for transport, and we explained in detail how to carry out coherent transport calculations by using the NEGF approach. This method is ideally suited to perform device simulations in fields such as graphene and molecular electronics. By going through the various examples, the reader should have acquired all the technical skills needed to set up calculations with tight-binding transport codes that can be extended to deal with real research problems.

In the second part of the chapter, we outlined how to join NEGF with DFT in order to compute transport properties of molecules from first-principles. Most importantly,

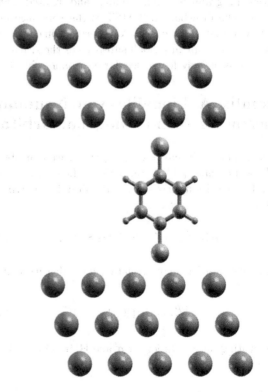

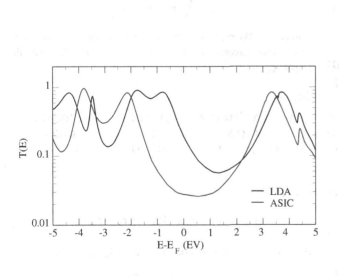

FIGURE 8.16 Transmission function for the Au-BDT-Au junction calculated by using DFT+NEGF with the LDA xc and ASIC.

we presented a compendium of the key issues that should be considered when using commonly available DFT+NEGF software. Despite some conceptual difficulties, DFT+NEGF represents to date the workhorse to support experiments in molecular electronics. In particular, we believe that the real strength of the method is the capability to predict qualitative trends, for example, in the dependence of the conductance of molecules on their conformation or functional groups.

In the near future, most of the research efforts in quantum transport will be dedicated to account for many-body interactions from first-principles. We see two emerging complementary approaches. The first aims at developing a DFT theory for transport, which is based on solid theoretical foundations [47]. The drawback is the difficulty in designing appropriate xc functionals. The second aims at extending techniques, such as many-body perturbation theory (MBPT) (e.g., Ref. [49]) and dynamical mean-field theory (DMFT) (e.g. Refs. [7,10,12]) to transport. When applied to molecular junctions, MBPT is able to account very accurately for fundamental gap of molecules and electronic screening at the molecule–electrode interfaces. DMFT instead captures transport-related phenomena driven by the strong electron–electron interaction. In case of molecular devices, these phenomena include the Kondo effect, Coulomb blockade and inelastic spin excitations, which are not described even at the qualitative level by DFT+NEGF and related schemes. In case of solid state devices, DMFT correctly describes the narrowing of the $3d$ states in transition metals, the insulating nature of Mott and charge-transfer materials and the inelastic scattering by magnetic impurities in graphene nano-ribbons and related nanostructures. The drawback of DMFT is the computational complexity so that, in practice, strong correlation effects can only be accounted for in a small subspace of the scattering region comprising only few molecular or atomic orbitals.

Appendix A Liouville–von Neumann Equation for Non-orthogonal Orbitals

Given a non-orthonormal basis set $\{|\Phi_\alpha\rangle\}$, we can obtain an orthonormal one, $\{|\Phi'_\alpha\rangle\}$, through the Löwdin transformation. The density matrix and the Hamiltonian matrix for the two basis sets are related as

$$\boldsymbol{\rho}' = \mathbf{S}^{1/2}\boldsymbol{\rho}\,\mathbf{S}^{1/2}, \quad \mathbf{H}' = \mathbf{S}^{-1/2}\mathbf{H}\,\mathbf{S}^{-1/2}. \quad (8.139)$$

The Liouville–von Neumann equation for a density matrix $\boldsymbol{\rho}'$ is

$$\frac{\partial \boldsymbol{\rho}'}{\partial t} = \frac{1}{i\hbar}[\mathbf{H}',\boldsymbol{\rho}']. \quad (8.140)$$

By substituting the expression for $\boldsymbol{\rho}'$ and $\mathbf{H}'$ into Eq. (8.140), we obtain

$$\mathbf{S}^{1/2}\frac{\partial \boldsymbol{\rho}}{\partial t}\mathbf{S}^{1/2} = \frac{1}{i\hbar}\left(\mathbf{S}^{-1/2}\mathbf{H}\boldsymbol{\rho}\mathbf{S}^{1/2} - \mathbf{S}^{1/2}\boldsymbol{\rho}\mathbf{H}\mathbf{S}^{-1/2}\right) \quad (8.141)$$

that can be rewritten in the two equivalent ways:

$$\mathbf{S}\frac{\partial \boldsymbol{\rho}}{\partial t} = \frac{1}{i\hbar}\left(\mathbf{H}\boldsymbol{\rho} - \mathbf{S}\boldsymbol{\rho}\mathbf{H}\mathbf{S}^{-1}\right) \quad (8.142)$$

$$\mathbf{S}\frac{\partial \boldsymbol{\rho}}{\partial t} = \frac{1}{i\hbar}\left(\mathbf{S}^{-1}\mathbf{H}\boldsymbol{\rho}\mathbf{S} - \boldsymbol{\rho}\mathbf{H}\right). \quad (8.143)$$

By summing these two equations, we obtain

$$\mathbf{S}\frac{\partial \boldsymbol{\rho}}{\partial t} + \frac{\partial \boldsymbol{\rho}}{\partial t}\mathbf{S} = \frac{1}{i\hbar}\left([\mathbf{H},\boldsymbol{\rho}] + \mathbf{S}^{-1}\mathbf{H}\boldsymbol{\rho}\mathbf{S} - \mathbf{S}\boldsymbol{\rho}\mathbf{H}\mathbf{S}^{-1}\right), \quad (8.144)$$

We now introduce the energy density matrix $\mathbf{F} = \frac{1}{2}\left[\mathbf{S}^{-1}\mathbf{H}\boldsymbol{\rho} + \boldsymbol{\rho}\mathbf{H}\mathbf{S}^{-1}\right]$, so that $\mathbf{S}^{-1}\mathbf{H}\boldsymbol{\rho}\mathbf{S} - \mathbf{S}\boldsymbol{\rho}\mathbf{H}\mathbf{S}^{-1} = 2\mathbf{F}\mathbf{S} - 2\mathbf{S}\mathbf{F} - \boldsymbol{\rho}\mathbf{H} + \mathbf{H}\boldsymbol{\rho}$. With this Eq. 8.144 can be expressed as

$$\frac{1}{2}\left(\mathbf{S}\frac{\partial \boldsymbol{\rho}}{\partial t} + \frac{\partial \boldsymbol{\rho}}{\partial t}\mathbf{S}\right) = \frac{1}{i\hbar}[\mathbf{H},\boldsymbol{\rho}] + \frac{1}{i\hbar}[\mathbf{F},\mathbf{S}]. \quad (8.145)$$

Finally, by using the definition of the symmetrized density matrix in Eq. (8.23), we derive Eq. (8.25), namely

$$\frac{\partial \bar{\boldsymbol{\rho}}}{\partial t} = \frac{1}{i\hbar}[\mathbf{H},\boldsymbol{\rho}] + \frac{1}{i\hbar}[\mathbf{F},\mathbf{S}]. \quad (8.146)$$

Appendix B Special Block Matrix Inversions

We define the matrices $\mathbf{A},\mathbf{B},\mathbf{C},\mathbf{D},\mathbf{E},\mathbf{F},\mathbf{G}$ in such a way that the matrices in each of the sets $\{\mathbf{A},\mathbf{B}\}$, $\{\mathbf{C},\mathbf{D},\mathbf{E}\}$, and $\{\mathbf{F},\mathbf{G}\}$ have the same number of rows, and the number of columns is equal for the matrices in the sets $\{\mathbf{A},\mathbf{C}\}$, $\{\mathbf{B},\mathbf{D},\mathbf{F}\}$ and $\{\mathbf{E},\mathbf{G}\}$. We can then build the block matrix $\mathbf{M}_3$:

$$\mathbf{M}_3 = \begin{pmatrix} \mathbf{A} & \mathbf{B} & \mathbf{0} \\ \mathbf{C} & \mathbf{D} & \mathbf{E} \\ \mathbf{0} & \mathbf{F} & \mathbf{G} \end{pmatrix}, \quad (8.147)$$

where the $\mathbf{0}$s represent zero block matrices of appropriate size. The inverse of such a matrix can be calculated explicitly as

$$\mathbf{M}_3^{-1} =$$
$$\begin{pmatrix} \mathbf{A}^{-1} + \mathbf{A}^{-1}\mathbf{B}\mathbf{L}_3\mathbf{C}\mathbf{A}^{-1} & -\mathbf{A}^{-1}\mathbf{B}\mathbf{L}_3 & \mathbf{A}^{-1}\mathbf{B}\mathbf{L}_3\mathbf{E}\mathbf{G}^{-1} \\ -\mathbf{L}_3\mathbf{C}\mathbf{A}^{-1} & \mathbf{L}_3 & -\mathbf{L}_3\mathbf{E}\mathbf{G}^{-1} \\ \mathbf{G}^{-1}\mathbf{F}\mathbf{L}_3\mathbf{C}\mathbf{A}^{-1} & -\mathbf{G}^{-1}\mathbf{F}\mathbf{L}_3 & \mathbf{G}^{-1} + \mathbf{G}^{-1}\mathbf{F}\mathbf{L}_3\mathbf{E}\mathbf{G}^{-1} \end{pmatrix}, \quad (8.148)$$

where

$$\mathbf{L}_3 = \left(\mathbf{D} - \mathbf{C}\mathbf{A}^{-1}\mathbf{B} - \mathbf{E}\mathbf{G}^{-1}\mathbf{F}\right)^{-1}. \quad (8.149)$$

Similarly for a 2×2 block matrix, $\mathbf{M}_2$ given by

$$\mathbf{M}_2 = \begin{pmatrix} \mathbf{A} & \mathbf{B} \\ \mathbf{C} & \mathbf{D} \end{pmatrix} \quad (8.150)$$

the inverse is

$$\mathbf{M}_2^{-1} = \begin{pmatrix} \mathbf{A}^{-1} + \mathbf{A}^{-1}\mathbf{B}\mathbf{L}_{2,R}\mathbf{C}\mathbf{A}^{-1} & -\mathbf{A}^{-1}\mathbf{B}\mathbf{L}_{2,R} \\ -\mathbf{L}_{2,R}\mathbf{C}\mathbf{A}^{-1} & \mathbf{L}_{2,R} \end{pmatrix}, \quad (8.151)$$

where

$$\mathbf{L}_{2,R} = \left(\mathbf{D} - \mathbf{C}\mathbf{A}^{-1}\mathbf{B}\right)^{-1}. \tag{8.152}$$

The inverse of $\mathbf{M}_2$ can also be written in an equivalent way as

$$\mathbf{M}_2^{-1} = \begin{pmatrix} \mathbf{L}_{2,L} & -\mathbf{L}_{2,L}\mathbf{B}\mathbf{D}^{-1} \\ -\mathbf{D}^{-1}\mathbf{C}\mathbf{L}_{2,L} & \mathbf{D}^{-1} + \mathbf{D}^{-1}\mathbf{C}\mathbf{L}_{2,L}\mathbf{B}\mathbf{D}^{-1} \end{pmatrix}, \tag{8.153}$$

where

$$\mathbf{L}_{2,L} = \left(\mathbf{A} - \mathbf{B}\mathbf{D}^{-1}\mathbf{C}\right)^{-1}. \tag{8.154}$$

Acknowledgments

A. D. was supported from the EU via the Marie Sklodowska-Curie individual fellowship SPINMAN (No. SEP-210189940).

References

1. Arroyo, C., Tarkuc, S., Frisenda, R., Seldenthuis, J. S., Woerde, C., Eelkema, R., and Grozema, F. (2013). Signatures of quantum interference effects on charge transport through a single benzene ring. *Angew. Chem. Int. Ed.*, 52:3152.

2. Ashcroft, N. and Mermin, N. (1976). *Solid State Physics*. Saunders College Publishing, Philadelphia.

3. Brandbyge, M., Mozos, J.-L., Ordejón, P., Taylor, J., and Stokbro, K. (2002). Density-functional method for nonequilibrium electron transport. *Phys. Rev. B*, 65:165401.

4. Bruot, C., Hihath, J., and Tao, N. (2012). Mechanically controlled molecular orbital alignment in single molecule junctions. *Nature Nanotechnol.*, 7:35.

5. Büttiker, M., Imry, Y., Landauer, R., and Pinhas, S. (1985). Generalized many-channel conductance formula with application to small rings. *Phys. Rev. B*, 31:6207.

6. Cardamone, D. M., Stafford, C. A., and Mazumdar, S. (2006). Controlling quantum transport through a single molecule. *Nano Letters*, 6:2422.

7. Chioncel, L., Morari, C., Östlin, A., Appelt, W. H., Droghetti, A., Radonjić, M. M., Rungger, I., Vitos, L., Eckern, U., and Postnikov, A. V. (2015). Transmission through correlated Cu_nCoCu_n heterostructures. *Phys. Rev. B*, 92:054431.

8. Datta, S. (1995). *Electronic Transport in Mesoscopic Systems*. Cambridge University Press, Cambridge, UK.

9. Datta, S. (2005). *Quantum Transport: Atom to Transistor*. Cambridge University Press, Cambridge, UK.

10. David, J. (2015). Towards a full ab initio theory of strong electronic correlations in nanoscale devices. *J. Phys. Conden. Matter*, 27:245606.

11. Di Ventra, M. (2008). *Electrical Transport in Nanoscale Systems*. Cambridge University Press, Cambridge, UK.

12. Droghetti, A. and Rungger, I. (2017). Quantum transport simulation scheme including strong correlations and its application to organic radicals adsorbed on gold. *Phys. Rev. B*, 95:085131.

13. Engel, E. and Dreizler, R. M. (2011). *Density Functional Theory: An Advanced Course*. Springer, Berlin, Heidelberg.

14. Ferry, D., Goodnick, S. M., and Bird, J. (2012). *Transport in Nanostructures - 2nd edition*. Cambridge University Press, Cambridge, UK.

15. Foa Torres, E., Roche, S., and Charlier, J.-C. (2014). *Introduction to Graphene-Based Nanomaterials: From Electronic Structure to Quantum Transport*. Cambridge University Press, Cambridge, UK.

16. Ghosh, A. (2017). *Nanoelectronics: A Molecular View*. World Scientific Publishing Company, Singapore.

17. Golizadeh-Mojarad, R. and Datta, S. (2007). Nonequilibrium green's function based models for dephasing in quantum transport. *Phys. Rev. B*, 75:081301.

18. Hansen, T., Solomon, G., Andrews, D., and Ratner, M. (2009). Interfering pathways in benzene: An analytical treatment. *J. Chem. Phys.*, 131:194704.

19. Hohenberg, P. and Kohn, W. (1964). Inhomogeneous electron gas. *Phys. Rev.*, 136:B864.

20. Koentopp, M., Burke, K., and Evers, F. (2006). Zero-bias molecular electronics: Exchange-correlation corrections to landauer's formula. *Phys. Rev. B*, 73:121403.

21. Kohn, W. (1999). Nobel lecture: Electronic structure of matter-wave functions and density functionals. *Rev. Mod. Phys.*, 71:1253.

22. Kohn, W. and Sham, L. J. (1965). Self-consistent equations including exchange and correlation effects. *Phys. Rev.*, 140:A1133.

23. Kronik, L. and Kümmel, S. (2014). *Gas-Phase Valence-Electron Photoemission Spectroscopy Using Density Functional Theory*, page 137. Springer, Berlin, Heidelberg.

24. Kurth, S. and Stefanucci, G. (2017). Transport through correlated systems with density functional theory. *J. Phys. Condens. Matter*, 29:413002.

25. Landauer, R. (1957). Spatial variation of currents and fields due to localized scatterers in metallic conduction. *IBM J. Res. Dev.*, 1:223.

26. Landauer, R. (1970). Electrical resistance of disordered one-dimensional lattices. *Philos. Mag.*, 21:863.

27. Lang, N. D. (1995). Resistance of atomic wires. *Phys. Rev. B*, 52:5335.

28. Marder, M. (2000). *Condensed Matter Physics*. Wiley Interscience Publication, John Wilet and Sons, INC., New York, USA.

29. Markussen, T., Stadler, R., and Thygesen, K. S. (2010). The relation between structure and quantum interference in single molecule junctions. *Nano Letters*, 10:4260.

30. Meir, Y. and Wingreen, N. (1992). Landauer formula for the current through an interacting electron region. *Phys. Rev. Lett.*, 68:2512.

31. Miroshnichenko, A. E., Flach, S., and Kivshar, Y. S. (2010). Fano resonances in nanoscale structures. *Rev. Mod. Phys.*, 82:2257.

32. Nazarov, Y. and Blanter, Y. M. (2009). *Quantum Transport: Introduction to Nanoscience*. Cambridge University Press, Cambridge, UK.

33. Palacios, J. J., Pérez-Jiménez, A. J., Louis, E., SanFabián, E., and Vergés, J. A. (2002). First-principles approach to electrical transport in atomic-scale nanostructures. *Phys. Rev. B*, 66:035322.

34. Parr, R. and Yang, W. (1989). *Density-Functional Theory of Atoms and Molecules*. Oxford University Press, Oxford, UK.

35. Paulsson, M. (2006). Non equilibrium green's functions for dummies: Introduction to the one particle negf equations. *cond-mat/0210519v2*.

36. Pecchia, A. and di Carlo A. (2004). Atomistic theory of transport in organic and inorganic nanostructures. *Rep. Prog. Phys.*, 67:1497.

37. Pemmaraju, C. D., Archer, T., Sánchez-Portal, D., and Sanvito, S. (2007). Atomic-orbital-based approximate self-interaction correction scheme for molecules and solids. *Phys. Rev. B*, 75:045101.

38. Perdew, J. and Levy, M. (1997). Comment on "significance of the highest occupied kohn-sham eigenvalue". *Phys. Rev. B*, 56:16021.

39. Perdew, J. P. and Zunger, A. (1981). Self-interaction correction to density-functional approximations for many-electron systems. *Phys. Rev. B*, 23:5048.

40. Pertsova, A., Canali, C. M., Pederson, M. R., Rungger, I., and Sanvito, S. (2015). *Adv. At. Mol. Opt. Phys.*, 64:29.

41. Rocha, A., Garcia-Suarez, V., Bailey, S., Lambert, C., Ferrer, J., and Sanvito, S. (2006). Spin and molecular electronics in atomically generated orbital landscapes. *Phys. Rev. B*, 73:085414.

42. Rungger, I. (2008). *Computational methods for electron transport and their application in nanodevices*. Ph.D. Thesis, Trinity College Dublin.

43. Rungger, I., Chen, X., Schwingenschlögl, U., and Sanvito, S. (2010). Finite-bias electronic transport of molecules in a water solution. *Phys. Rev. B*, 81:235407.

44. Rungger, I., Droghetti, A., and Stamanova, M. (2018). *Non-equilibrium Green's Functions Methods for Spin Transport and Dynamics*. Springer, Berlin, Heidelberg.

45. Rungger, I. and Sanvito, S. (2008). Algorithm for the construction of self-energies for electronic transport calculations based on singularity elimination and singular value decomposition. *Phys. Rev. B*, 78:035407.

46. Souza, A., Rungger, I., Pontes, R. B., Rocha, A. R., da Silva, A. J. R., Schwingenschloegl, U., and Sanvito, S. (2014). Stretching of bdt-gold molecular junctions: thiol or thiolate termination? *Nanoscale*, 6:14495.

47. Stefanucci, G. and Kurth, S. (2015). Steady-state density functional theory for finite bias conductances. *Nano Lett.*, 15:8020.

48. Stefanucci, G. and van Leeuwen, R. (2013). *Nonequilibrium Many-Body Theory of Quantum Systems: A Modern Introduction*. Cambridge University Press, Cambridge, UK.

49. Strange, M., Rostgaard, C., Häkkinen, H., and Thygesen, K. S. (2011). Self-consistent GW calculations of electronic transport in thiol- and amine-linked molecular junctions. *Phys. Rev. B*, 83:115108.

50. Sutton, A. P. (1993). *Electronic Structure of Materials*. Clarendon Press, Oxford, UK.

51. Taylor, J., Guo, H., and Wang, J. (2001). Ab initio modeling of quantum transport properties of molecular electronic devices. *Phys. Rev. B*, 63:245407.

52. Toher, C. and Sanvito, S. (2007). Efficient atomic self-interaction correction scheme for nonequilibrium quantum transport. *Phys. Rev. Lett.*, 99:056801.

53. Tsutsui, M., Taniguchi, M., and Kawai, T. (2009). Atomistic mechanics and formation mechanism of metal-molecule-metal junctions. *Nano Lett.*, 9:2433.

9

Transient Quantum Transport in Nanostructures

Pei-Yun Yang
National Tsing Hua University

Yu-Wei Huang and Wei-Min Zhang
National Cheng Kung University

9.1 Introduction to Transient Quantum Transport

Today, there are many practical applications of nanostructures and nanomaterials. For example, the quantum Hall effect now serves as a standard measurement for resistance. Quantum dots are used in many modern application areas including quantum dot lasers in optics, fluorescent tracers in biological and medical settings, and quantum information processing. The theory of nanostructures involves a broad range of physical concepts, from the simple confinement effects to the complex many-body physics, such as the Kondo and fractional quantum Hall effects. Traditional condensed matter and quantum many-body theory all have the role to play in understanding and learning how to control nanostructures as a practically useful device. From the theoretical point of view, electron transport in nanostructures deals mainly with physics of systems consisting of a nanoscale active region (the device system) attached to multi-leads (including source and drain).

The quantum transport theory for these physical systems is based on the following three theoretical approaches. The Landauer–Büttiker approach [7,12,18], because of its simplicity, has often been used to analyze resonant-tunneling diodes [37] and quantum wires [44]. In this approach, electron transport in the device system is simply treated by coherent transport (pure elastic scattering) near thermal equilibrium. However, in order for nanodevices to be functionally operated, it may be subjected to high source-drain voltages and high-frequency bandwidths, in far from equilibrium, highly transient, and highly nonlinear regimes. Thus, a more microscopic theory has been developed for quantum transport in terms of nonequilibrium Green functions [10,17,19,23,38,42,52,53] for the device system. Moreover, the device system exchanges the particles, energy, and information with the leads and is thereby a typical open quantum system. The issues of open quantum systems, such as dissipation, fluctuation and decoherence inevitably arise. The third approach, the master equation approach [16,21, 22,32,41,47,50,57,58], gets the advantage by describing the device system in terms of the reduced density matrix of open systems.

In this chapter, we give a detailed description of the Landauer–Büttiker approach, the nonequilibrium Green function techniques and the master equation approach. The theoretical schemes of these approaches are schematically presented in Figures 9.1(a)–(c). The main differences between these three approaches are the ways of characterizing electron transport flowing through the device system. In the Landauer–Büttiker approach, the device system is depicted as a potential barrier, and all the information of the device system are imbedded in the scattering matrix. The actual structure of the device system is obscure. Comparing to the Landauer–Büttiker approach, the nonequilibrium Green function techniques provide a more microscopic way by describing electrons flowing through the device system

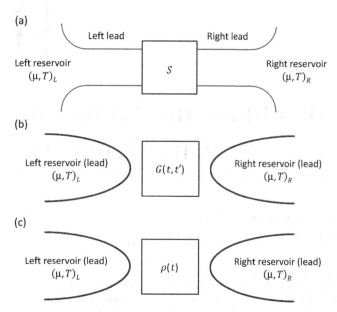

FIGURE 9.1 Theoretical schemes for (a) the Landauer–Büttiker approach, (b) the nonequilibrium Green function techniques and (c) the master equation approach.

with single-particle nonequilibrium Green functions. In the master equation approach, the device system is described by the reduced density matrix, which is the essential quantity for studying quantum coherent and decoherent phenomena. Better than the nonequilibrium Green functions in which the average over the density matrix has been done, quantum coherent dynamics is depicted explicitly by the reduced density matrix.

9.2 Landauer Büttiker Approach

The Landauer–Büttiker formula has been widely utilized to calculate various transport properties in semiconductor nanostructures in the steady-state quantum transport regime. It establishes the fundamental relation between the electron wave functions (scattering amplitudes) of a quantum system and its conducting properties. In the Landauer–Büttiker formula, the transport current is given in terms of transmission coefficients, obtained from the single-particle scattering matrix. This approach was first formulated by Landauer for the single-channel transport [25,26]. Later on, Büttiker et al. extended the formula to multichannel [5] and multiterminal cases [4]. The further development of the Landauer–Büttiker approach is the calculation of the current noise correlations in mesoscopic conductors, the detailed discussion can be found from the review article by Blanter et al. [2].

A typical system considered in the Landauer–Büttiker approach consists of reservoirs (contacts), quantum leads, and a mesoscopic sample (scatterer) (see Figure 9.1(a)). The reservoirs connect to the mesoscopic sample by quantum leads and are always in an equilibrium state in which electrons are always incoherent. However, electron

transport passing through the mesoscopic sample between the reservoirs is phase coherent. Such coherent transport is described by the electron wave function scattered in the mesoscopic sample, which can be characterized by a scattering matrix S. Therefore, the key to describe electron quantum transport in Landauer–Büttiker approach is to determine the scattering matrix S, which relies crucially on the mesoscopic sample structure.

9.2.1 Electron Scattering

We start with the single-channel and two-terminal case. Consider an electron plane wave impinging on a finite potential barrier, respectively, from left ($x < 0$) to right ($x > 0$) and scattered into the reflected and transmitted components. Assume that the energy and momentum are conserved in the scattering process, and the wave function is given, respectively,

$$\psi_L(x) = \begin{cases} e^{ikx} + re^{-ikx} & x < 0, \\ te^{ikx} & x > 0, \end{cases} \tag{9.1}$$

$$\psi_R(x) = \begin{cases} t'e^{-ikx} & x < 0, \\ e^{-ikx} + r'e^{ikx} & x > 0, \end{cases} \tag{9.2}$$

where r (r') and t (t') are respectively the complex reflected and transmitted amplitudes of the wave incoming from the left (right), with $|r|^2$ ($|r'|^2$) and $|t|^2$ ($|t'|^2$) being the reflected and transmitted probabilities. For a general incoming state, $a_L e^{ikx} + a_R e^{-ikx}$, with probability amplitudes $a_{L,R}$, the total wave function should be

$$\Psi(x) = a_L(e^{ikx} + re^{-ikx}) + a_R t'e^{-ikx}$$
$$= a_L e^{ikx} + (a_L r + a_R t')e^{-ikx} \qquad x < 0$$
$$\Psi(x) = a_L te^{ikx} + a_R(e^{-ikx} + r'e^{ikx})$$
$$= a_R e^{-ikx} + (a_L t + a_R r')e^{ikx} \qquad x > 0. \tag{9.3}$$

Introducing probability amplitudes $b_{L,R}$ for the outgoing state, the total wave function simply becomes,

$$\Psi(x) = \begin{cases} a_L e^{ikx} + b_L e^{-ikx} & x < 0, \\ a_R e^{-ikx} + b_R e^{ikx} & x > 0, \end{cases} \tag{9.4}$$

Thus, the incoming and outgoing probability amplitudes are related to each other by the scattering matrix:

$$\begin{pmatrix} b_L \\ b_R \end{pmatrix} = \begin{pmatrix} r & t' \\ t & r' \end{pmatrix} \begin{pmatrix} a_L \\ a_R \end{pmatrix} \equiv S \begin{pmatrix} a_L \\ a_R \end{pmatrix}. \tag{9.5}$$

The coefficients in the scattering matrix (r, t, r', and t') are obtained by solving the Schrödinger equation with the potential that models the mesoscopic sample (scatterer).

The preceding discussion is based on single-channel transport, and it is straightforward to generalize the formalism to multichannel case where there are N_L modes on the left and N_R modes on the right. The incoming and outgoing amplitudes can be written in vectors such that

$$
\boldsymbol{a} = \begin{pmatrix} a_{L1} \\ \vdots \\ a_{LN_L} \\ a_{R1} \\ \vdots \\ a_{RN_R} \end{pmatrix}; \quad \boldsymbol{b} = \begin{pmatrix} b_{L1} \\ \vdots \\ b_{LN_L} \\ b_{R1} \\ \vdots \\ b_{RN_R} \end{pmatrix}. \quad (9.6)
$$

Therefore, the scattering matrix, leading to $\boldsymbol{b} = \boldsymbol{S}\boldsymbol{a}$, is in dimension $(N_L + N_R) \times (N_L + N_R)$ and has the following form,

$$
\boldsymbol{S} = \begin{pmatrix} \boldsymbol{s}_{LL} & \boldsymbol{s}_{LR} \\ \boldsymbol{s}_{RL} & \boldsymbol{s}_{RR} \end{pmatrix} = \begin{pmatrix} \boldsymbol{r} & \boldsymbol{t}' \\ \boldsymbol{t} & \boldsymbol{r}' \end{pmatrix}, \quad (9.7)
$$

where the matrices $\boldsymbol{t}$ $(N_R \times N_L)$ and $\boldsymbol{r}$ $(N_L \times N_L)$ describe, respectively, the transmission and reflection of electrons incoming from the left. The corresponding element $\boldsymbol{t}_{mn}$ and $\boldsymbol{r}_{mn}$ characterize, respectively, the electrons transmitted from the left mode n into the right mode m and the electrons reflected from the left mode n into the left mode m. Similarly, the matrices $\boldsymbol{r}'$ $(N_R \times N_R)$ and $\boldsymbol{t}'$ $(N_L \times N_R)$ represent the reflection and transmission processes for states incoming from the right. The scattering matrix $\boldsymbol{S}$ is unitary due to probability current conservation, i.e.

$$
\boldsymbol{S}^\dagger \boldsymbol{S} = \boldsymbol{S}\boldsymbol{S}^\dagger = \mathbb{1}_{N_L+N_R}. \quad (9.8)
$$

9.2.2 Field Operator of the Lead System

Now, we consider electron transport for multichannel (denoted by English alphabet e.g. m, n...) and multiterminal (denoted by Greek alphabet e.g. α, β...). Lead α is described by the Hamiltonian

$$
H_\alpha = \frac{p_{x_\alpha}^2}{2m^*} + \frac{p_{\perp\alpha}^2}{2m^*} + U(\boldsymbol{r}_{\perp\alpha}), \quad (9.9)
$$

where x_α and $\boldsymbol{r}_{\perp\alpha}$ denote the local coordinates in the longitudinal and transverse directions, respectively, and m^* is the effective mass of the electron in the lead. The motion of electrons in the longitudinal direction is free, but it is quantized in the transverse direction due to the confinement potential $U(\boldsymbol{r}_{\perp\alpha})$. The eigenfunctions of the Hamiltonian H_α can be expressed as

$$
\phi_{\alpha n}^\pm(x_\alpha, \boldsymbol{r}_{\perp\alpha}) = \chi_{\alpha n}(\boldsymbol{r}_{\perp\alpha})e^{\pm i k_{\alpha n} x_\alpha}, \quad (9.10)
$$

where the incoming wave e^{ikx_α} and outgoing wave e^{-ikx_α} characterize the longitudinal motion of elections, and $\chi_{\alpha n}(\boldsymbol{r}_{\perp\alpha})$ satisfies

$$
\left[\frac{p_{\perp\alpha}^2}{2m^*} + U(\boldsymbol{r}_{\perp\alpha}) \right] \chi_{\alpha n}(\boldsymbol{r}_{\perp\alpha}) = \epsilon_{\alpha n} \chi_{\alpha n}(\boldsymbol{r}_{\perp\alpha}), \quad (9.11)
$$

with each transverse mode contributing a transport channel. The dispersion relation is, thus, given by

$$
E_{\alpha n}(k_{\alpha n}) = \frac{\hbar^2 k_{\alpha n}^2}{2m^*} + \epsilon_{\alpha n}. \quad (9.12)
$$

In this case, for an electron from mode m of lead α scattering by the mesoscopic sample, the scattering state of the electron for lead α is

$$
\psi_{\alpha m}(x_\alpha) = \sum_n \left\{ \delta_{mn}\phi_{\alpha n}^+ + \sqrt{\frac{v_{\alpha m}}{v_{\alpha n}}} S_{\alpha\alpha nm}\phi_{\alpha n}^- \right\}, \quad (9.13)
$$

and the scattering state for lead $\beta \neq \alpha$ is

$$
\psi_{\alpha m}(x_\beta) = \sum_n \sqrt{\frac{v_{\alpha m}}{v_{\beta n}}} S_{\beta\alpha nm}\phi_{\beta n}^-, \quad (9.14)
$$

where the scattering matrix with element $S_{\beta\alpha nm}$ represents the amplitude of a state scattered from mode m in lead α to mode n in lead β, and the factor $\sqrt{v_{\alpha m}/v_{\beta n}}$ is introduced to guarantee the probability current conservation, where $v_{\alpha m} = \hbar k_{\alpha m}/m^*$ is the electron velocity.

A general state in the lead system is given by an arbitrary superposition of these scattering states

$$
\Psi(\boldsymbol{r}, t) = \frac{1}{\sqrt{2\pi}} \sum_{\alpha m} \int dk_{\alpha m} \psi_{\alpha m}(\boldsymbol{r}) e^{-i\omega_{\alpha m}(k_{\alpha m})t} a_{\alpha m}(k_{\alpha m}), \quad (9.15)
$$

with $\omega_{\alpha m} = E_{\alpha m}/\hbar$ and $a_{\alpha m}(k_{\alpha m})$ being the probability amplitude of the incoming wave. Now, we extend the formalism from single-particle to many-body physics, the corresponding second quantization field operator is expressed as

$$
\hat{\Psi}(\boldsymbol{r}, t) = \frac{1}{\sqrt{2\pi}} \sum_{\alpha m} \int dk_{\alpha m} \psi_{\alpha m}(\boldsymbol{r}) e^{-i\omega_{\alpha m}(k_{\alpha m})t} \hat{a}_{\alpha m}(k_{\alpha m}), \quad (9.16)
$$

the probability amplitude $a_{\alpha m}$ becomes annihilation operator $\hat{a}_{\alpha m}$, which satisfies the canonical anti-commutation relation,

$$
\{\hat{a}_{\alpha m}(k), \hat{a}_{\beta n}^\dagger(k')\} = \delta_{\alpha\beta}\delta_{nm}\delta(k - k'). \quad (9.17)
$$

By changing integral in Eq. (9.16) from k space to the energy space and defining the incoming operator in the energy space $\hat{a}_{\alpha m}(E) = \hat{a}_{\alpha m}(k)/[\hbar v_{\alpha m}(k)]^{1/2}$, one has

$$
\{\hat{a}_{\alpha m}(E), \hat{a}_{\beta n}^\dagger(E')\} = \delta_{\alpha\beta}\delta_{nm}\delta(E - E'), \quad (9.18)
$$

where $\delta(E - E') = 1/(\hbar v)\delta(k - k')$. The field operator is then given by

$$
\hat{\Psi}(\boldsymbol{r}, t) = \sum_{\alpha m} \int \frac{dE_{\alpha m}}{\sqrt{hv_{\alpha m}(E_{\alpha m})}} \psi_{\alpha m}(r) e^{-i\omega_{\alpha m}t} \hat{a}_{\alpha m}(E_{\alpha m}). \quad (9.19)
$$

With the above field operator, the current flowing from contact α to the mesoscopic sample can be deduced, as shown in the following subsection.

9.2.3 Current Formula in terms of Scattering Matrix

According to the standard quantum mechanics language, the current operator of lead α is given by

$$\hat{I}_\alpha(t) = \int d\boldsymbol{r}_{\perp\alpha}\hat{j}(\boldsymbol{r}_\alpha, t), \qquad (9.20)$$

where the current density operator is expressed as

$$\hat{j}(\boldsymbol{r}, t) = \frac{\hbar}{2m^*i}[\hat{\Psi}^\dagger\nabla\hat{\Psi} - (\nabla\hat{\Psi}^\dagger)\hat{\Psi}]. \qquad (9.21)$$

Substituting the scattering states Eqs. (9.13) and (9.14) into the field operator of Eq. (9.19) gives the following solution,

$$
\begin{aligned}
\hat{\Psi}(\boldsymbol{r}_\alpha, t) &= \sum_m \int \frac{dE_{\alpha m}}{\sqrt{hv_{\alpha m}(E_{\alpha m})}} e^{-i\omega_{\alpha m}t}\hat{a}_{\alpha m}(E_{\alpha m}) \\
&\times \left\{ \sum_n \left[\delta_{mn}\phi_{\alpha n}^+ + \sqrt{\frac{v_{\alpha m}}{v_{\alpha n}}}S_{\alpha\alpha nm}\phi_{\alpha n}^- \right] \right. \\
&\left. + \sum_{\beta\neq\alpha}\sum_n \sqrt{\frac{v_{\alpha m}}{v_{\beta n}}}S_{\beta\alpha nm}\phi_{\beta n}^- \right\} \\
&= \sum_m \int \frac{dE_{\alpha m}}{\sqrt{hv_{\alpha m}(E_{\alpha m})}} e^{-i\omega_{\alpha m}t} \\
&\quad \left[\phi_{\alpha m}^+\hat{a}_{\alpha m}(E_{\alpha m}) + \phi_{\alpha m}^-\hat{b}_{\alpha m}(E_{\alpha m}) \right],
\end{aligned}
\qquad (9.22)
$$

where the contribution of the incoming (the first term) and outgoing (the second term) states are explicitly presented. Using the orthogonal properties of different transverse modes, $\int d\boldsymbol{r}_{\perp\alpha}\chi_{\alpha m}(\boldsymbol{r}_{\perp\alpha})\chi_{\alpha n}(\boldsymbol{r}_{\perp\alpha}) = \delta_{mn}$, the current can be reduced to the following form,

$$
\begin{aligned}
\hat{I}_\alpha(t) = \frac{e}{h}\sum_m \int dEdE' &\left[\hat{a}_{\alpha m}^\dagger(E)\hat{a}_{\alpha m}(E') \right. \\
&\left. - \hat{b}_{\alpha m}^\dagger(E)\hat{b}_{\alpha m}(E') \right]e^{i(E-E')t/\hbar},
\end{aligned}
\qquad (9.23)
$$

where we have used the approximation $v_{\alpha m}(E) = v_{\alpha m}(E')$ which is always valid for a slowly varying function $v(E)$. Using further the scattering relation $\boldsymbol{b} = \boldsymbol{S}\boldsymbol{a}$, we can express the current as,

$$
\begin{aligned}
\hat{I}_\alpha(t) = \frac{e}{h}\sum_{\beta\gamma}\sum_{nk}\int dEdE'\, \hat{a}_{\beta n}^\dagger(E)A_{\beta\gamma}^{nk} \\
\times (\alpha; E, E')\hat{a}_{\gamma k}(E')e^{i(E-E')t/\hbar},
\end{aligned}
\qquad (9.24)
$$

where the matrix A has the following form

$$A_{\beta\gamma}^{nk}(\alpha; E, E') = \delta_{\alpha\beta}\delta_{\alpha\gamma}\delta_{kn} - \sum_m S_{\alpha\beta nm}^\dagger(E)S_{\alpha\gamma mk}(E'). \qquad (9.25)$$

Because contact α is in equilibrium, the average current at lead α is

$$\langle I_\alpha \rangle = \frac{e}{h}\sum_{\beta n}\int dEA_{\beta\beta}^{nn}(\alpha, E, E)f_\beta(E), \qquad (9.26)$$

where $f_\alpha(E) = 1/(e^{(E-\mu_\alpha)/k_BT_\alpha} + 1)$ is the Fermi–Dirac distribution of contact α at the chemical potential μ_α and temperature T_α. Taking the common two-terminal system with the scattering matrix (9.7) as an example, the average current of the left lead becomes

$$\langle I_L \rangle = \frac{e}{h}\int dE\,\mathrm{Tr}[\boldsymbol{t}^\dagger(E)\boldsymbol{t}(E)][f_L(E) - f_R(E)]. \qquad (9.27)$$

This gives the famous Landauer–Büttiker formula [7]. Note that $\boldsymbol{t}(E) = \boldsymbol{t}'(E)$ for a two-terminal system as long as there is no inelastic scattering inside the device.

In the steady-state quantum transport regime, the Landauer–Büttiker approach is a powerful method to calculate various transport properties in semiconductor nanostructures [3,6,39,40]. However, the scattering theory considers the reservoirs connecting to the scatterer (the mesoscopic sample) to be always in equilibrium and electrons in the reservoir are always incoherent. Thus, the Landauer–Büttiker formula becomes invalid to transient quantum transport. The scattering theory method could be extended to deal with time-dependent transport phenomena, through the so-called the Floquet scattering theory [13,34,35], but it is only applicable to the case of the time-dependent quantum transport for systems driven by periodic time-dependent external fields.

9.3 Nonequilibrium Green Function Techniques

Green function techniques are widely used in many-body systems. For equilibrium systems, zero temperature Green functions and Matsubara (finite temperature) Green functions are useful tools for calculating the thermodynamical properties of many-body systems, as well as the linear responses of systems under small time-dependent (or not) perturbations [33]. However, when systems are driven out of equilibrium, nonequilibrium Green functions are utilized [17,53]. The nonequilibrium Green function techniques are initiated by Scwinger [42] and Kadanoff and Baym [23] and popularized by Keldysh [24]. To deal with nonequilibrium phenomena, the contour-ordered Green functions that are defined on complex time contours are introduced such that the equations of motion and perturbation expansions of contour-ordered Green functions are formally identical to that of usual equilibrium Green functions.

In this section, the contour-ordered Green functions defined on two types of contour will be discussed; one is Kadanoff–Baym contour, which takes into account the initial correlations and statistical boundary conditions; the other one is Schwinger–Keldysh contour where the initial correlations is neglected. The real-time nonequilibrium Green functions for both formalisms are deduced from the contour-ordered Green function by analytic continuation. In application, we give a detailed derivation of the steady-state transport current for a mesoscopic system by means of the Keldysh technique. The resulting transport current is formulating in terms of the nonequilibrium Green functions of the device system, which provides a more microscopic picture to the electron transport compared to the Landauer–Büttiker formula. For a more complete description of the nonequilibrium Green function techniques, we refer the readers to [17,23,43].

9.3.1 Contour-Ordered Green Function

The contour-ordered Green function of nonequilibrium many-body theory is defined as

$$
\begin{aligned}
G(\boldsymbol{x},\tau;\boldsymbol{x}',\tau') &= -i \left\langle T_C[\psi(\boldsymbol{x},\tau)\psi^\dagger(\boldsymbol{x}',\tau')]\right\rangle \\
&= -i \operatorname{Tr}\left\{\rho_{tot}(t_0)T_C[\psi(\boldsymbol{x},\tau)\psi^\dagger(\boldsymbol{x}',\tau')]\right\},
\end{aligned}
\tag{9.28}
$$

where $\psi(\boldsymbol{x},\tau)$ and $\psi^\dagger(\boldsymbol{x}',\tau')$ are the fermion field operators in the Heisenberg picture with time variables τ,τ' (denoted by Greek letters) defined on the complex contour C, and T_C is a contour-ordering operator, which orders the operators according to their time labels on the contour:

$$
T_C[\psi(\boldsymbol{x},\tau)\psi^\dagger(\boldsymbol{x}',\tau')] \equiv
\begin{cases}
\psi(\boldsymbol{x},\tau)\psi^\dagger(\boldsymbol{x}',\tau') & \tau >_C \tau', \\
-\psi^\dagger(\boldsymbol{x}',\tau')\psi(\boldsymbol{x},\tau) & \tau <_C \tau'.
\end{cases}
\tag{9.29}
$$

From the above definition, it is straightforward to rewrite the contour-ordered Green function as

$$
\begin{aligned}
G(\boldsymbol{x},\tau;\boldsymbol{x}',\tau') &= \Theta_C(\tau-\tau')G^>(\boldsymbol{x},\tau;\boldsymbol{x}',\tau') \\
&\quad + \Theta_C(\tau'-\tau)G^<(\boldsymbol{x},\tau;\boldsymbol{x}',\tau'),
\end{aligned}
\tag{9.30}
$$

where $\Theta_C(\tau-\tau')$ is the step function defined on the contour, and $G^>$ and $G^<$ are the greater and lesser Green functions, respectively. The configuration of complex contour C is determined by the initial density matrix of the total system $\rho_{tot}(t_0)$. In the following subsections, the Kadanoff–Baym formalism is introduced at first, and then the Keldysh formalism is obtained by neglecting the initial correlations and taking the steady-state limit from the Kadanoff–Baym formalism.

9.3.2 The Kadanoff Baym Contour

The nonequilibrium dynamics considered in the Kadanoff–Baym formalism is formulated as follows. First, we consider a physical system described by a time-independent Hamiltonian,

$$
\mathcal{H} = H_0 + H_i,
\tag{9.31}
$$

where H_0 represents a free Hamiltonian, and H_i is the interaction between the particles. The system is initially assumed at thermal equilibrium, which means the system is in partition-free scheme [11],

$$
\rho_{tot}(t_0) = \frac{\exp(-\beta\mathcal{H})}{\operatorname{Tr}[\exp(-\beta\mathcal{H})]} = \frac{1}{Z}\exp(-\beta\mathcal{H}),
\tag{9.32}
$$

where $\beta = 1/k_B T$, and the particle energies are measured from the chemical potential μ. After $t = t_0$, the system is exposed to external disturbances, e.g. an electric field, a light excitation pulse, or a coupling to contacts at differing (electro) chemical potentials that are described by time-dependent Hamiltonian $H'(t)$. Thus, the total Hamiltonian becomes,

$$
H(t) = \mathcal{H} + H'(t),
\tag{9.33}
$$

where $H'(t < t_0) = 0$. We take the time arguments in contour-ordered Green function (9.28) as real-time variables t and t', the field operator $\psi(\boldsymbol{x},t)$ is then,

$$
\psi(\boldsymbol{x},t) \equiv \psi(1) = \mathcal{S}(t_0,t)\hat{\psi}_{\mathcal{H}}(1)\mathcal{S}(t,t_0),
\tag{9.34}
$$

where we employ the shorthand notation $(1) = (\boldsymbol{x},t)$, and $\mathcal{S}$ is the S-matrix defined in the following form,

$$
\begin{aligned}
\mathcal{S}(t,t') &= e^{i\mathcal{H}(t-t_0)}U(t,t')e^{-i\mathcal{H}(t'-t_0)} \\
&= T\left\{\exp\left[-i\int_{t'}^{t}dt_1 \hat{H}'_{\mathcal{H}}(t_1)\right]\right\},
\end{aligned}
\tag{9.35}
$$

where $U(t,t') = T\exp(-i\int_{t'}^{t}dt_1 H(t_1))$ is the evolution operator. In the above equations, $\hat{\psi}_{\mathcal{H}}(1)$ and $\hat{H}'_{\mathcal{H}}(t)$ are operators in the interaction picture with respect to Hamiltonian $\mathcal{H}$,

$$
\hat{\psi}_{\mathcal{H}}(1) = e^{i\mathcal{H}(t-t_0)}\psi(\boldsymbol{x},t_0)e^{-i\mathcal{H}(t-t_0)},
\tag{9.36}
$$

$$
\hat{H}'_{\mathcal{H}}(t) = e^{i\mathcal{H}(t-t_0)}H'(t)e^{-i\mathcal{H}(t-t_0)}.
\tag{9.37}
$$

The contour-ordered Green function in the Kadanoff–Baym formalism is now written as

$$
\begin{aligned}
iG(1,1') &= \operatorname{Tr}\Big\{\rho_{tot}(t_0)T_C \\
&\quad \times [\mathcal{S}(t_0,t)\hat{\psi}_{\mathcal{H}}(1)\mathcal{S}(t,t_0)\mathcal{S}(t_0,t')\hat{\psi}^\dagger_{\mathcal{H}}(1')\mathcal{S}(t',t_0)]\Big\} \\
&= \operatorname{Tr}\Big\{\rho_{tot}(t_0)T_{C_0}[\mathcal{S}_{C_0}(t_0,t_0)\hat{\psi}_{\mathcal{H}}(1)\hat{\psi}^\dagger_{\mathcal{H}}(1')]\Big\},
\end{aligned}
\tag{9.38}
$$

where $\mathcal{S}_{C_0}(t_0,t_0) = T_{C_0}\{\exp[-i\oint_{C_0}d\tau_1 \hat{H}'_{\mathcal{H}}(\tau_1)]\}$ is the S-matrix defined on the contour C_0 which is called closed-path contour as shown in Figure 9.2.

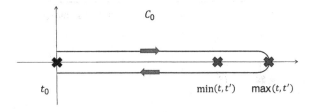

FIGURE 9.2 Closed time path contour C_0

In order to perform the Wick theorem, one needs to further transform the operators $\hat{\psi}_{\mathcal{H}}(1)$, $\hat{\psi}_{\mathcal{H}}^{\dagger}(1')$ in the interaction picture with respect to the free Hamiltonian H_0:

$$\hat{\psi}_{\mathcal{H}}(1) = S(t_0, t)\hat{\psi}(1)S(t, t_0). \tag{9.39}$$

Here, S is the S-matrix for the interaction Hamiltonian H_i,

$$S(t, t') = T\left\{ \exp\left[-i \int_{t'}^{t} dt_1 \hat{H}_i(t) \right] \right\}, \tag{9.40}$$

and $\hat{\psi}(1)$ and $\hat{H}_i(t)$ are given by,

$$\hat{\psi}(1) = e^{iH_0(t-t_0)}\psi(\boldsymbol{x}, t_0)e^{-iH_0(t-t_0)},$$
$$\hat{H}_i(t) = e^{iH_0(t-t_0)}H_i e^{-iH_0(t-t_0)}. \tag{9.41}$$

Furthermore, in terms of S, one can rewrite the factor $\exp(-\beta\mathcal{H})$ in the initial density matrix,

$$\exp(-\beta\mathcal{H}) = \exp(-\beta H_0)S(t_0 - i\beta, t_0). \tag{9.42}$$

Finally, the contour-ordered Green function is reduced to,

$$iG(1, 1') =$$
$$\frac{\text{Tr}\left\{ \rho_0 T_{C_0^*}\left[S_{C_0^*}(t_0 - i\beta, t_0)S_{C_0}(t_0, t_0)\hat{\psi}(1)\hat{\psi}^{\dagger}(1') \right] \right\}}{\text{Tr}\left\{ \rho_0 T_{C_0^*}\left[S_{C_0^*}(t_0 - i\beta, t_0)S_{C_0}(t_0, t_0) \right] \right\}}, \tag{9.43}$$

where $\rho_0 = e^{-\beta H_0}/Z_0$ is the equilibrium density matrix of Hamiltonian H_0, $S_{C_0^*}(t_0 - i\beta, t_0) = T_{C_0^*}\{\exp[-i\int_{C_0^*} d\tau_1 \hat{H}_i(\tau_1)]\}$ is the S-matrix defined on contour C_0^*, and $C_0^* = C_0 \cup [t_0, t_0 - i\beta]$ is the Kadanoff–Baym contour shown in Figure 9.3.

Equation (9.43) is the exact contour-ordered Green function in the Kadanoff–Baym formalism, which is defined in the interaction picture with respect to the free Hamiltonian H_0, so that Wick theorem is always applicable. Thus, the perturbative evaluation of Eq. (9.43) could be put in a form analogous to the usual Feynman diagrammatic technique as in the equilibrium Green function techniques.

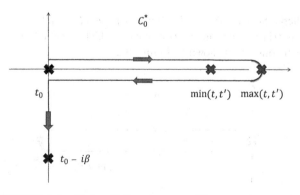

FIGURE 9.3 Kadanoff–Baym contour C_0^*

9.3.3 Analytic Continuation in the Kadanoff Baym Formalism

We have three branches in the Kadanoff–Baym contour which are denoted as $C_i, i = 1, 2, 3$. The branches C_1 and C_2 represent, respectively, the forward and backward time evolution on real time axis, and C_3 is the imaginary time branch $[t_0, t_0 - i\beta]$. First, we consider $\tau, \tau' \in C_1, C_2$, according to Eq. (9.30), the real-time nonequilibrium Green functions can be specified as,

$$G(1, 1') = \begin{cases} G^T(1, 1') = -i\langle T[\psi(1)\psi^{\dagger}(1')]\rangle & \text{if } \tau, \tau' \in C_1, \\ G^<(1, 1') = i\langle \psi^{\dagger}(1')\psi(1)\rangle & \text{if } \tau \in C_1, \tau' \in C_2, \\ G^>(1, 1') = -i\langle \psi(1)\psi^{\dagger}(1')\rangle & \text{if } \tau \in C_2, \tau' \in C_1, \\ G^{\tilde{T}}(1, 1') = -i\langle \tilde{T}[\psi(1)\psi^{\dagger}(1')]\rangle & \text{if } \tau, \tau' \in C_2, \end{cases} \tag{9.44}$$

Here, G^T is the time-ordered Green function with time-ordering operator T on the forward branch, while $G^{\tilde{T}}$ is the anti-time-ordered Green function with anti-time-ordering operator $\tilde{T}$ on the backward branch. Second, if one of the time arguments is on C_3, because the time on branch C_3 is always later than the time on C_1 and C_2, we can only have the following two conditions,

$$G(\tau, \tau') = \begin{cases} G^>(t_0 - i\bar{t}, t') \equiv G^{\lceil}(\bar{t}, t') & \text{if } \tau \in C_3, \tau' \in C_1 \text{ or } C_2, \\ G^<(t, t_0 - i\bar{t}') \equiv G^{\rceil}(t, \bar{t}') & \text{if } \tau \in C_1 \text{ or } C_2, \tau' \in C_3, \end{cases} \tag{9.45}$$

with $\bar{t}$, $\bar{t}'$ being real numbers in the region $[0, \beta]$. Here, the notation $\rceil$ denotes a time argument on the horizontal branch followed by a time argument on the vertical one. Correspondingly, the first argument of $G^{\rceil}$ is real time and the second argument varies in imaginary time. Conversely, the first argument of $G^{\lceil}$ varies in imaginary time and the second argument is real time. For simplicity, we only show the time argument of state variable $(\boldsymbol{x}, \tau)$ hereafter. Finally, if both of τ and τ' are in the branch C_3,

$$G(\tau, \tau') = G(t_0 - i\bar{t}, t_0 - i\bar{t}') \equiv G^M(\bar{t}, \bar{t}'), \tag{9.46}$$

where $-iG^M$ is the Matsubara Green function. If the Green function is a product of other Green functions, we use Langreth theorem [27] to convert the contour integral to a real integral. As an example, a product of two contour-ordered quantities,

$$C(\tau, \tau') = \int_{C_0^*} d\tau_1 A(\tau, \tau_1)B(\tau_1, \tau'). \tag{9.47}$$

Using Eq. (9.30), the above equation can be written as,

$$C(\tau, \tau') = \int_{C_0^*} d\tau_1 [\Theta_C(\tau - \tau_1)\Theta_C(\tau_1 - \tau')A^> B^>$$
$$+ \Theta_C(\tau - \tau_1)\Theta_C(\tau' - \tau_1)A^> B^<$$
$$+ \Theta_C(\tau_1 - \tau)\Theta_C(\tau_1 - \tau')A^< B^>$$
$$+ \Theta_C(\tau_1 - \tau)\Theta_C(\tau' - \tau_1)A^< B^<]. \tag{9.48}$$

Consider $\tau <_C \tau'$, and both of them are real-time variables, we have,

$$
\begin{aligned}
C^<(t,t') &= \int_{t_0}^t dt_1 A^>(t,t_1)B^<(t_1,t') \\
&\quad + \int_{t'}^{t_0-i\beta} d\tau_1 A^<(t,\tau_1)B^>(\tau_1,t') \\
&\quad + \int_t^{t'} dt_1 A^<(t,t_1)B^<(t_1,t') \\
&= \int_{t_0}^t dt_1 [A^>(t,t_1) - A^<(t,t_1)]B^<(t_1,t') \\
&\quad - \int_{t_0}^{t'} dt_1 A^<(t,t_1)[B^>(t_1,t') - B^<(t_1,t')] \\
&\quad + \int_{t_0}^{t_0-i\beta} d\tau_1 A^<(t,\tau_1)B^>(\tau_1,t').
\end{aligned}
\tag{9.49}
$$

By introducing the retarded and advanced Green functions,

$$
\begin{aligned}
G^R(1,1') &= \Theta(t-t')[G^>(1,1') - G^<(1,1')] \\
&= -i\Theta(t-t')\langle\{\psi(1),\psi^\dagger(1')\}\rangle,
\end{aligned}
\tag{9.50a}
$$

$$
\begin{aligned}
G^A(1,1') &= -\Theta(t'-t)[G^>(1,1') - G^<(1,1')] \\
&= i\Theta(t'-t)\langle\{\psi(1),\psi^\dagger(1')\}\rangle,
\end{aligned}
\tag{9.50b}
$$

Eq. (9.49) then becomes,

$$
\begin{aligned}
C^<(t,t') &= \int_{t_0}^\infty dt_1 [A^R(t,t_1)B^<(t_1,t') + A^<(t,t_1)B^A(t_1,t')] \\
&\quad - i\int_0^\beta d\bar{t}_1 A^\rceil(t,\bar{t}_1)B^\lceil(\bar{t}_1,t').
\end{aligned}
\tag{9.51}
$$

One also uses the shorthand notation of matrix multiplication along the real-time axis and the vertical track,

$$
A \cdot B = \int_{t_0}^\infty dt\, A(t)B(t)
$$

$$
A \star B = -i\int_0^\beta d\bar{t}\, A(\bar{t})B(\bar{t}),
\tag{9.52}
$$

so that the equation can be simplified as,

$$
C^< = A^R \cdot B^< + A^< \cdot B^A + A^\rceil \star B^\lceil.
\tag{9.53}
$$

One can also use the same technique to obtain the matrix multiplication form of the other Green functions, e.g. $C^\rceil$, C^M.... We refer the readers to [43].

9.3.4 Schwinger Keldysh Contour

In the Keldysh formalism, many-body physics is only discussed in the steady-state limit, i.e. $t_0 \to -\infty$, where the initial correlation effects are not essential because of memory loss. As a result, the initial correlations are neglected in the Keldysh formalism.

In particular, one neglects the imaginary vertical track $[t_0, t_0 - i\beta]$, and inserts the factors of $\mathcal{S}(t,\infty)\mathcal{S}(\infty,t)$

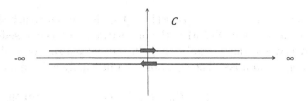

FIGURE 9.4 Schwinger–Keldysh contour

$(\mathcal{S}(t',\infty)\mathcal{S}(\infty,t'))$ and $\mathcal{S}(t,\infty)\mathcal{S}(\infty,t)$ $(\mathcal{S}(t',\infty)\mathcal{S}(\infty,t'))$ with the latest time t (t') and take $t_0 \to -\infty$ in Eq. (9.43). Then, the contour-ordered Green function is given by

$$
\begin{aligned}
iG(1,1') = \mathrm{Tr}\Big\{ \rho_0 T_C \\
\Big[S_C(-\infty,-\infty)\mathcal{S}_C(-\infty,-\infty)\hat{\psi}(1)\hat{\psi}(1') \Big] \Big\},
\end{aligned}
\tag{9.54}
$$

where C is the Schwinger–Keldysh contour, as shown in Figure 9.4. This is the Keldysh formalism.

9.3.5 Analytic Continuation in the Keldysh Formalism

The analytic continuation in Keldysh formalism is simplified correspondingly. The relation between the contour-ordered Green functions and the nonequilibrium Green functions contains only Eq. (9.44). Langreth theorem for the Schwinger–Keldysh contour is the same as for Kadanoff–Baym contour except for the missing imaginary time track. Taking the same example considered in the Kadanoff–Baym formalism, one gets,

$$
C^< = A^R \cdot B^< + A^< \cdot B^A,
\tag{9.55}
$$

with

$$
A \cdot B = \int_{-\infty}^\infty dt\, A(t)B(t).
\tag{9.56}
$$

Now, the lower bound for the integral is extend to $-\infty$. Another useful formula is for a product of three contour-ordered quantities:

$$
D^< = A^R \cdot B^R \cdot C^< + A^R \cdot B^< \cdot C^A + A^< \cdot B^A \cdot C^A.
\tag{9.57}
$$

9.3.6 Dyson Equation

The contour-ordered Green function obeys the following Dyson equation,

$$
G(1,1') = G_0(1,1') + \int d2 \int d3\, G_0(1,2)\Sigma(2,3)G(3,1'),
\tag{9.58a}
$$

$$
G(1,1') = G_0(1,1') + \int d2 \int d3\, G(1,2)\Sigma(2,3)G_0(3,1'),
\tag{9.58b}
$$

where $G_0(1,1') = -i\langle T_C[\hat{\psi}(1)\hat{\psi}^\dagger(1')]\rangle$ is the unperturbed Green function, $\Sigma(2,3)$ is the one particle irreducible self-energy, and the integral sign $\int d2(3)$ denotes a sum over all integral variables. The equations can be simply written as,

$$G = G_0 + G_0 \cdot \Sigma \cdot G, \qquad (9.59a)$$

$$G = G_0 + G \cdot \Sigma \cdot G_0. \qquad (9.59b)$$

The Dyson equation can be regarded as the Schrödinger equation of a particle in the medium subject to the self-energy as the potential. In the Dyson equation, the single-particle Green function is entirely determined by the self-energy which contains all the many-body effects.

9.3.7   Keldysh Green Functions of the Device System

In this and the following subsections, the steady-state transport current in a mesoscopic system is presented with the Keldysh formalism. The steady-state current can be reduced to the Landauer–Büttiker form when the device system is a noninteracting system. Consider a nanostructure consisting of a quantum device coupled with two leads (the source and drain), which can be described by the Fano–Anderson Hamiltonian [14,26,33],

$$H = \sum_{ij} \varepsilon_{ij} a_i^\dagger a_j + \sum_{\alpha k} \epsilon_{\alpha k} c_{\alpha k}^\dagger c_{\alpha k} + \sum_{i\alpha k}[V_{i\alpha k} a_i^\dagger c_{\alpha k} + \text{H.c.}].$$
$$(9.60)$$

Here, $a_i^\dagger$ (a_i) and $c_{\alpha k}^\dagger$ $(c_{\alpha k})$ are creation (annihilation) operators of electrons in the quantum device and lead α, respectively, ε_{ij} and $\epsilon_{\alpha k}$ are the corresponding energy levels, $V_{i\alpha k}$ is the tunneling amplitude between the orbital state i of the device system and the orbital state k of lead α. In the Keldysh approach, the quantum device and the leads are decoupled in the remote past, and the tunneling between them is viewed as a perturbation. Then, the independent Hamiltonian $\mathcal{H}$, time-dependent Hamiltonian $H'(t)$, and the initial density matrix $\rho_{tot}(-\infty)$ of the total system are,

$$\mathcal{H} = H_0 = \sum_{ij} \varepsilon_{ij} a_i^\dagger a_j + \sum_{\alpha k} \epsilon_{\alpha k} c_{\alpha k}^\dagger c_{\alpha k}, \qquad (9.61a)$$

$$H'(t) = \sum_{i\alpha k}[V_{i\alpha k} a_i^\dagger c_{\alpha k} + \text{H.c.}], \qquad (9.61b)$$

$$\rho_{tot}(-\infty) = \rho_0 = \frac{1}{Z}$$
$$\left[e^{-\beta_L(H_L - \mu_L n_L)} \otimes \rho(-\infty) \otimes e^{-\beta_R(H_R - \mu_R n_R)} \right], \qquad (9.61c)$$

where $H_\alpha = \sum_k \epsilon_{\alpha k} c_{\alpha k}^\dagger c_{\alpha k}$ and $n_\alpha = \sum_k c_{\alpha k}^\dagger c_{\alpha k}$ are the Hamiltonian and the total particle number of lead α, respectively. Lead α is initially in thermal equilibrium with inverse temperature β_α and chemical potential μ_α. The initial state of the device system $\rho(-\infty)$ is not important in the Keldysh formalism because of the memory loss. Thus, the total system is in the partitioned scheme [8,9].

In nonequilibrium Green function techniques, the information of the dissipation and fluctuation dynamics of the device system can be extracted from the contour-ordered Green function of the device system $G_{ij}(\tau,\tau')$,

$$G_{ij}(\tau,\tau') = -i\langle T_C[a_i(\tau)a_j^\dagger(\tau')]\rangle$$
$$= \Theta_C(\tau - \tau')G_{ij}^>(\tau,\tau') + \Theta(\tau' - \tau)G_{ij}^<(\tau,\tau'). \qquad (9.62)$$

This Green function obeys the following equations of motion,

$$\sum_l \left[i\frac{\partial}{\partial\tau}\delta_{il} - \varepsilon_{il} \right] G_{lj}(\tau,\tau') = \delta_C(\tau - \tau')\delta_{ij}$$
$$+ \sum_{\alpha k} V_{i\alpha k} G_{\alpha k,j}(\tau,\tau'), \qquad (9.63a)$$

$$\sum_l \left[-i\frac{\partial}{\partial\tau'}\delta_{lj} - \varepsilon_{lj} \right] G_{il}(\tau,\tau') = \delta_C(\tau - \tau')\delta_{ij}$$
$$+ \sum_{\alpha k} G_{i,\alpha k}(\tau,\tau') V_{j\alpha k}^*, \qquad (9.63b)$$

where the mixed contour-ordered Green functions, $G_{\alpha k,j}(\tau,\tau') = -i\langle T_C[c_{\alpha k}(\tau)a_j^\dagger(\tau')]\rangle$ and $G_{i,\alpha k}(\tau,\tau') = -i\langle T_C[a_i(\tau)c_{\alpha k}^\dagger(\tau')]\rangle$. Because the interaction between the device system and the leads is a linear tunneling coupling [17], we have

$$G_{i,\alpha k}(\tau,\tau') = \sum_l \int_C d\tau_1 G_{il}(\tau,\tau_1) V_{l\alpha k} g_{\alpha k}(\tau_1,\tau'), \quad (9.64a)$$

$$G_{\alpha k,j}(\tau,\tau') = \sum_l \int_C d\tau_1 g_{\alpha k}(\tau,\tau_1) V_{l\alpha k}^* G_{lj}(\tau_1,\tau'), \quad (9.64b)$$

where $g_{\alpha k}(\tau,\tau') = -i\langle T_C[\hat{c}_{\alpha k}(\tau)\hat{c}_{\alpha k}^\dagger(\tau')]\rangle$ is the unperturbed contour-ordered Green function of the lead α. Inserting Eq. (9.64) into Eq. (9.63) gives the Dyson equation in a differential form,

$$[i\partial_\tau \mathbb{1}_{N_S} - \varepsilon]G(\tau,\tau') = \mathbb{1}_{N_S}\delta_C(\tau - \tau')$$
$$+ \int_C d\tau_1 \Sigma(\tau,\tau_1)G(\tau_1,\tau'), \qquad (9.65a)$$

$$G(\tau,\tau')[-i\partial_{\tau'}\mathbb{1}_{N_S} - \varepsilon] = \mathbb{1}_{N_S}\delta_C(\tau - \tau')$$
$$+ \int_C d\tau_1 G(\tau,\tau_1)\Sigma(\tau_1,\tau'), \qquad (9.65b)$$

with self-energy,

$$\Sigma_{ij}(\tau,\tau') = \sum_\alpha \Sigma_{\alpha ij}(\tau,\tau') = \sum_{\alpha k} V_{i\alpha k} g_{\alpha k}(\tau,\tau') V_{j\alpha k}^*.$$
$$(9.66)$$

Here, $\mathbb{1}_{N_S}$ is an identity matrix in the dimension of the device system. On the other hand, we have the equation of motion of unperturbed contour-ordered Green function of the device system,

$$[i\partial_\tau \mathbb{1}_{N_S} - \varepsilon]G_0(\tau,\tau') = \mathbb{1}_{N_S}\delta_C(\tau - \tau'), \qquad (9.67a)$$

$$G_0(\tau,\tau')[-i\partial_{\tau'}\mathbb{1}_{N_S} - \varepsilon] = \mathbb{1}_{N_S}\delta_C(\tau - \tau'). \qquad (9.67b)$$

Consequently, we can rewrite the Dyson equation (9.65) in the following form,

$$\boldsymbol{G}_0^{-1} \cdot \boldsymbol{G} = \mathbb{1} + \boldsymbol{\Sigma} \cdot \boldsymbol{G}, \tag{9.68a}$$

$$\boldsymbol{G} \cdot \boldsymbol{G}_0^{-1} = \mathbb{1} + \boldsymbol{G} \cdot \boldsymbol{\Sigma}. \tag{9.68b}$$

Here, the matrix product means a product of all the internal variables (energy level and time). Equation (9.68) produces the integral form of Dyson equation (9.59).

Using the Dyson equation (9.59) and the Langreth theorem, we have

$$\boldsymbol{G}^{R,A} = \boldsymbol{G}_0^{R,A} + \boldsymbol{G}_0^{R,A} \cdot \boldsymbol{\Sigma}^{R,A} \cdot \boldsymbol{G}^{R,A}, \tag{9.69a}$$

$$\boldsymbol{G}^{\gtrless} = \boldsymbol{G}_0^{\gtrless} + \boldsymbol{G}_0^R \cdot \boldsymbol{\Sigma}^R \cdot \boldsymbol{G}^{\gtrless}$$
$$+ \boldsymbol{G}_0^R \cdot \boldsymbol{\Sigma}^{\gtrless} \cdot \boldsymbol{G}^A + \boldsymbol{G}_0^{\gtrless} \cdot \boldsymbol{\Sigma}^A \cdot \boldsymbol{G}^A. \tag{9.69b}$$

One can further iterate Eq. (9.69b) respect to $\boldsymbol{G}^{\gtrless}$ and obtains,

$$\boldsymbol{G}^{\gtrless} = (\mathbb{1} + \boldsymbol{G}_0^R \cdot \boldsymbol{\Sigma}^R) \cdot \boldsymbol{G}_0^{\gtrless} \cdot (\mathbb{1} + \boldsymbol{\Sigma}^A \cdot \boldsymbol{G}^A)$$
$$+ (\boldsymbol{G}_0^R + \boldsymbol{G}_0^R \cdot \boldsymbol{\Sigma}^R \cdot \boldsymbol{G}_0^R) \cdot \boldsymbol{\Sigma}^{\gtrless} \cdot \boldsymbol{G}^A$$
$$+ \boldsymbol{G}_0^R \cdot \boldsymbol{\Sigma}^R \cdot \boldsymbol{G}_0^R \cdot \boldsymbol{\Sigma}^R \cdot \boldsymbol{G}^{\gtrless}. \tag{9.70}$$

After iterate infinite orders, one can get,

$$\boldsymbol{G}^{\gtrless} = (\mathbb{1} + \boldsymbol{G}^R \cdot \boldsymbol{\Sigma}^R) \cdot \boldsymbol{G}_0^{\gtrless} \cdot (\mathbb{1} + \boldsymbol{\Sigma}^A \cdot \boldsymbol{G}^A) + \boldsymbol{G}^R \cdot \boldsymbol{\Sigma}^{\gtrless} \cdot \boldsymbol{G}^A. \tag{9.71}$$

In the Keldysh technique, the first term is neglected because it usually vanishes at steady-state limit. Then,

$$\boldsymbol{G}^{\gtrless} = \boldsymbol{G}^R \cdot \boldsymbol{\Sigma}^{\gtrless} \cdot \boldsymbol{G}^A. \tag{9.72}$$

Equations (9.69a) and (9.72) are the final results of real-time nonequilibrium Green functions in the Keldysh formalism which all the transport properties can be determined by.

9.3.8 Transport Current in the Nonequilibrium Green Function Formalism

The transport current from lead α to the device system is defined as

$$I_\alpha(t) \equiv -e \frac{d}{dt} \langle n_\alpha(t) \rangle = -\frac{ie}{\hbar} \langle [H, n_\alpha] \rangle. \tag{9.73}$$

Calculating the commutation relation $[H, n_\alpha]$ gives,

$$I_\alpha(t) = \frac{2e}{\hbar} \mathrm{Re} \sum_{ik} V_{i\alpha k}^* G_{i,\alpha k}^<(t, t), \tag{9.74}$$

where the mixed lesser Green function, $G_{i,\alpha k}^<(t, t') = i \langle c_{\alpha k}^\dagger(t') a_i(t) \rangle$, can be obtained by applying the Langreth theorem to the mixed contour-ordered Green function (9.64a),

$$G_{i,\alpha k}^<(t, t') = \sum_j \int_{-\infty}^\infty dt_1 \Big[G_{ij}^R(t, t_1) V_{j\alpha k} g_{\alpha k}^<(t_1, t')$$
$$+ G_{ij}^<(t, t_1) V_{j\alpha k} g_{\alpha k}^A(t_1, t') \Big]. \tag{9.75}$$

In the steady-state limit, all the Green functions usually depend only on the differences of time arguments, i.e. $G(t, t') = G(t - t')$ because of time-translation symmetry. Thus, we can express Green function $G_{i,\alpha k}^<(t, t)$ in the frequency domain,

$$G_{i,\alpha k}^<(t, t)$$
$$= \sum_j \int \frac{d\omega}{2\pi} \Big[G_{ij}^R(\omega) V_{j\alpha k} g_{\alpha k}^<(\omega) + G_{ij}^<(\omega) V_{j\alpha k} g_{\alpha k}^A(\omega) \Big], \tag{9.76}$$

Inserting the above equation into Eq. (9.74), the steady-state transport current is reduced to,

$$I_\alpha = \frac{2e}{\hbar} \mathrm{Re} \int \frac{d\omega}{2\pi} \sum_{ijk} V_{i\alpha k}^* V_{j\alpha k}$$
$$\times \Big[G_{ij}^R(\omega) g_{\alpha k}^<(\omega) + G_{ij}^<(\omega) g_{\alpha k}^A(\omega) \Big]$$
$$= \frac{e}{\hbar} \int \frac{d\omega}{2\pi} \sum_{ijk} V_{i\alpha k}^* V_{j\alpha k} \Big\{ \big[G_{ij}^R(\omega) - G_{ij}^A(\omega) \big] g_{\alpha k}^<(\omega) - G_{ij}^<(\omega)$$
$$\times \big[g_{\alpha k}^>(\omega) - g_{\alpha k}^<(\omega) \big] \Big\}$$
$$= \frac{ie}{\hbar} \int \frac{d\omega}{2\pi} \mathrm{Tr} \Big\{ \boldsymbol{\Gamma}_\alpha(\omega) \Big[f_\alpha(\omega)[\boldsymbol{G}^R(\omega) - \boldsymbol{G}^A(\omega)] + \boldsymbol{G}^<(\omega) \Big] \Big\}. \tag{9.77}$$

Here, $\Gamma_{\alpha ij}(\omega) = 2\pi \sum_k V_{i\alpha k} V_{j\alpha k}^* \delta(\omega - \epsilon_{\alpha k})$ is a level-width function, and we have also used the results of the free-particle Green functions,

$$g_{\alpha k}^R(\omega) = \frac{1}{\omega - \epsilon_{\alpha k} + i\delta}, \tag{9.78a}$$

$$g_{\alpha k}^A(\omega) = \frac{1}{\omega - \epsilon_{\alpha k} - i\delta}, \tag{9.78b}$$

$$g_{\alpha k}^<(\omega) = 2\pi i f(\epsilon_{\alpha k}) \delta(\omega - \epsilon_{\alpha k}), \tag{9.78c}$$

$$g_{\alpha k}^>(\omega) = -2\pi i [1 - f(\epsilon_{\alpha k})] \delta(\omega - \epsilon_{\alpha k}). \tag{9.78d}$$

Now, the transport current is fully determined by Green functions of the device system. Besides, for the noninteracting device system, we have,

$$\boldsymbol{G}^R(\omega) - \boldsymbol{G}^A(\omega) = \boldsymbol{G}^>(\omega) - \boldsymbol{G}^<(\omega)$$
$$= \boldsymbol{G}^R(\omega)[\boldsymbol{\Sigma}^>(\omega) - \boldsymbol{\Sigma}^<(\omega)]G^A(\omega)$$
$$= -i\boldsymbol{G}^R(\omega) \sum_\alpha \boldsymbol{\Gamma}_\alpha(\omega) \boldsymbol{G}^A(\omega), \tag{9.79}$$

where the self-energy $\boldsymbol{\Sigma}^{\gtrless}$ has the following form,

$$\Sigma_{ij}^{\gtrless}(\omega) = \sum_{\alpha k} V_{i\alpha k} g_{\alpha k}^{\gtrless}(\omega) V_{j\alpha k}^*. \tag{9.80}$$

Using Eq. (9.79), the steady-state transport current becomes,

$$I_\alpha = \frac{e}{\hbar} \sum_\beta \int \frac{d\omega}{2\pi} T_{\alpha\beta}(\omega)[f_\alpha(\omega) - f_\beta(\omega)], \tag{9.81}$$

and $T_{\alpha\beta} = \text{Tr}\left[\boldsymbol{\Gamma}_\alpha(\omega)\boldsymbol{G}^R(\omega)\boldsymbol{\Gamma}_\beta(\omega)\boldsymbol{G}^A(\omega)\right]$ is the transmission coefficient. This expression of the steady-state transport current reproduces the Landauer–Büttiker formula with the transmission probability derived microscopically. The nonequilibrium Green function techniques based on Keldysh formalism [24,42] has been used extensively to investigate the steady-state quantum transport in mesoscopic systems [17,53].

Wingreen et al. extended Keldysh's nonequilibrium Green function techniques to time-dependent quantum transport under time-dependent external bias and gate voltages [53]. Explicitly, the parameters in Hamiltonian (9.61), controlled by the external bias and gate voltages, become time dependent,

$$\varepsilon_{ij} \to \varepsilon_{ij}(t), \tag{9.82a}$$

$$\epsilon_{\alpha k} \to \epsilon_{\alpha k}(t) = \epsilon_{\alpha k} + \Delta_{\alpha k}(t), \tag{9.82b}$$

$$V_{i\alpha k} \to V_{i\alpha k}(t) \tag{9.82c}$$

Thus, the time-dependent transport current is expressed as,

$$I_\alpha(t) = -\frac{2e}{\hbar}\int_{-\infty}^{t} d\tau \int \frac{d\omega}{2\pi}\,\text{Im}\,\text{Tr}\left\{e^{-i\omega(\tau-t)}\boldsymbol{\Gamma}_\alpha(\omega,\tau,t)\right.$$
$$\left. \times\left[f_\alpha(\omega)\boldsymbol{G}^R(t,\tau) + \boldsymbol{G}^<(t,\tau)\right]\right\}, \tag{9.83}$$

where the level-width function becomes also time dependent,

$$\Gamma_{\alpha ij}(\omega, t_1, t_2) = 2\pi\sum_k V_{i\alpha k}(t_1)$$
$$\exp\left[-i\int_{t_2}^{t_1} ds\,\Delta_{\alpha k}(s)\right]V^*_{j\alpha k}(t_2)\delta(\omega - \epsilon_{\alpha k}). \tag{9.84}$$

In particular, the Green functions in time domain are given by,

$$\boldsymbol{G}^R(t,t') = \boldsymbol{G}_0^R(t,t')$$
$$+ \int dt_1 \int dt_2 \boldsymbol{G}_0^R(t,t_1)\boldsymbol{\Sigma}^R(t_1,t_2)\boldsymbol{G}^R(t_2,t'), \tag{9.85a}$$

$$\boldsymbol{G}^<(t,t') = \int dt_1 \int dt_2 \boldsymbol{G}^R(t,t_1)\boldsymbol{\Sigma}^<(t_1,t_2)\boldsymbol{G}^A(t_2,t'). \tag{9.85b}$$

with self-energy defined as,

$$\Sigma_{ij}^R(t_1,t_2) = \sum_\alpha \Sigma_{\alpha ij}^R(t_1,t_2) = \sum_{\alpha k} V_{i\alpha k}(t_1)g_{\alpha k}^R(t_1,t_2)V^*_{j\alpha k}(t_2)$$
$$= -i\Theta(t_1-t_2)\sum_\alpha \int \frac{d\omega}{2\pi}\Gamma_{\alpha ij}(\omega,t_1,t_2)e^{-i\omega(t_1-t_2)}, \tag{9.86a}$$

$$\Sigma_{ij}^<(t_1,t_2) = \sum_\alpha \Sigma_{\alpha ij}^<(t_1,t_2) = \sum_{\alpha k} V_{i\alpha k}(t_1)g_{\alpha k}^<(t_1,t_2)V^*_{j\alpha k}(t_2)$$
$$= i\sum_\alpha \int \frac{d\omega}{2\pi}f_\alpha(\omega)\Gamma_{\alpha ij}(\omega,t_1,t_2)e^{-i\omega(t_1-t_2)} \tag{9.86b}$$

This gives a general formalism for time-dependent current through the device system valid for nonlinear response, where electron energies can be varied time-dependently by external voltages. However, in the Keldysh formalism, nonequilibrium Green functions are defined with the initial time $t_0 \to -\infty$, where the initial correlations are hardly taken into account. This limits the Keldysh technique to be useful mostly in the nonequilibrium steady-state regime.

9.4 Master Equation Approach

The master equation approach concerns the dynamic properties of the device system in terms of the time evolution of the reduced density matrix $\rho(t) = \text{Tr}_E[\rho_{tot}(t)]$, where Tr_E means tracing over all the environmental (leads) degrees of freedom. The dissipation and fluctuation dynamics of the device system induced by the environment (leads) are fully manifested in the master equation. The transient transport properties can be naturally addressed within the framework of the master equation. Compared to the nonequilibrium Green function technique, the master equation approach manifests the state information of the device system, which is a key element in studying quantum coherent phenomena.

In principle, the master equation for quantum transport can be solved in terms of the real-time diagrammatic expansion approach up to all orders [41]. However, most of the master equations used in nanostructures are obtained by perturbation theory up to the second order of the system-lead couplings, which is mainly applicable in the sequential tunneling regime [31]. A recent development of master equations in quantum transport systems is the hierarchical expansion of the equations of motion for the reduced density matrix [22], which provides a systematical and also quite useful numerical calculation scheme for quantum transport.

A decade ago, we derived an exact master equation for noninteracting nanodevices [21,50], using the Feynman–Vernon influence functional approach [15] in the fermion coherent-state representation [36]. The obtained exact master equation not only describes the quantum state dynamics of the device system but also takes into account all the transient electronic transport properties. The transient transport current is obtained directly from the exact master equation [21], which turns out to be expressed precisely with the nonequilibrium Green functions of the device system [17,53]. This new theory has also been used to study quantum transport (including the transient transport) for various nanostructures recently [20,21,29,30,45–51,54–56]. In the following subsections, we shall give an introduction of this exact master equation approach [21,50] and derive explicitly the transient transport current using the exact master equation.

9.4.1 Fermion Coherent State

In this subsection, we briefly summarize the essential properties of fermion coherent states used in the derivation of the master equation for the reduced density matrix of the device system. For more information of the fermion coherent states, we refer readers to Ref. [36]. Mathematically, a fermion coherent state $|\boldsymbol{\xi}\rangle$ is defined as an eigenstate of annihilation operator a_i associated to eigenvalue ξ_i, that is $a_i|\boldsymbol{\xi}\rangle = \xi_i|\boldsymbol{\xi}\rangle$. Formally,

$$|\boldsymbol{\xi}\rangle \equiv e^{\sum_{i=1}^{N} a_i^\dagger \xi_i}|0\rangle = \prod_{i=1}^{N} e^{a_i^\dagger \xi_i}|0\rangle, \qquad (9.87\text{a})$$

$$\langle\boldsymbol{\xi}| \equiv \langle 0| e^{\sum_{i=1}^{N} \xi_i^* a_i} = \langle 0| \prod_{i=1}^{N} e^{\xi_i^* a_i}. \qquad (9.87\text{b})$$

Here, ξ_i and its conjugation ξ_i^* (for $i = 1\cdots N$) are Grassmann numbers with anti-commuting properties:

$$\{\xi_i, \xi_j\} = \{\xi_i^*, \xi_j^*\} = \{\xi_i, \xi_j^*\} = 0. \qquad (9.88)$$

Particularly, one gets $\xi_i^2 = \xi_i^{*2} = 0$. We denote the anti-commutation between the fermion operators and the Grassmann numbers (e.g. $\{a_i, \xi_j\} = 0$) and the conjugation $(\xi_i a_i^\dagger)^\dagger = a_i \xi_i^*$ in Eq. (9.87).

One can find the differentiation and integration rules in the Grassmann algebra, which are defined to be equivalent. Explicitly,

$$\int d\xi_i^{(*)} 1 = \frac{\partial}{\partial \xi_i^{(*)}} 1 = 0, \quad \int d\xi_i^{(*)} \xi_j^{(*)} = \frac{\partial}{\partial \xi_i^{(*)}} \xi_j^{(*)} = \delta_{ij}. \qquad (9.89)$$

Similar to derivative of ordinary functions but with anti-commuting product rule, the derivative for an analytic function $f(\boldsymbol{\xi})$ in the Grassmann algebra is

$$\left\{ \frac{\partial}{\partial \xi_i}, \xi_j \right\} f(\boldsymbol{\xi}) \equiv \frac{\partial}{\partial \xi_i}(\xi_j f(\boldsymbol{\xi})) + \xi_j \frac{\partial}{\partial \xi_i} f(\boldsymbol{\xi}) = \delta_{ij} f(\xi) \qquad (9.90)$$

From above defined rules, one can easily obtain the following identities:

$$a_i|\boldsymbol{\xi}\rangle = \xi_i|\boldsymbol{\xi}\rangle, a_i^\dagger|\boldsymbol{\xi}\rangle = -\frac{\partial}{\partial \xi_i}|\boldsymbol{\xi}\rangle, \qquad (9.91)$$

$$\langle\boldsymbol{\xi}|a_i^\dagger = \langle\boldsymbol{\xi}|\xi_i^*, \langle\boldsymbol{\xi}|a_i = \frac{\partial}{\partial \xi_i^*}\langle\boldsymbol{\xi}|. \qquad (9.92)$$

The inner product of two coherent states can also be calculated,

$$\langle\boldsymbol{\xi}|\boldsymbol{\xi}'\rangle = e^{\sum_{i=1}^{N} \xi_i^* \xi_i'} = e^{\bar{\boldsymbol{\xi}}\boldsymbol{\xi}'} \qquad (9.93)$$

where $\boldsymbol{\xi}' = (\xi_1', \xi_2', ...)^T$ and $\bar{\boldsymbol{\xi}} = (\xi_1^*, \xi_2^*, ...)$. The fermion coherent states obey the completeness relation,

$$\int d\mu(\boldsymbol{\xi}) |\boldsymbol{\xi}\rangle\langle\boldsymbol{\xi}| = \mathbb{1}_N, \qquad (9.94)$$

where the integration measure is in the form of $d\mu(\boldsymbol{\xi}) = \prod_{i=1}^{N} e^{-\xi_i^* \xi_i} d\xi_i^* d\xi_i$. Also, the trace of an operator A in the Grassmann algebra is in the following form:

$$\text{Tr}[A] = \sum_n \langle n|A|n\rangle = \int d\mu(\boldsymbol{\xi}) \langle -\boldsymbol{\xi}|A|\boldsymbol{\xi}\rangle, \qquad (9.95)$$

where number states $\{|n\rangle\}$ form a complete set in the Fock space. The minus sign reflects the anticyclic property in the Berezin integral, i.e. $\int \prod_i d\xi_i^* d\xi_i \langle\psi|\boldsymbol{\xi}\rangle\langle\boldsymbol{\xi}|\psi'\rangle = \int \prod_i d\xi_i^* d\xi_i \langle -\boldsymbol{\xi}|\psi'\rangle\langle\psi|\boldsymbol{\xi}\rangle$, for $|\psi\rangle$ and $|\psi'\rangle$ in the Fock space. In the fermion coherent-state representation, the Gaussian integral for the Grassmann algebra is a useful tool, in particular, in dealing with time evolution of quantum systems expressed by quadratic Hamiltonians.

$$I(\bar{\boldsymbol{\zeta}}, \boldsymbol{\zeta}') = \int \Big(\prod_{i=1}^{N} d\xi_i^* d\xi_i\Big) e^{-\bar{\boldsymbol{\xi}}M\boldsymbol{\xi}+\bar{\boldsymbol{\xi}}\boldsymbol{\zeta}'+\bar{\boldsymbol{\zeta}}\boldsymbol{\xi}} = (\det M)e^{\bar{\boldsymbol{\zeta}}M^{-1}\boldsymbol{\zeta}'}. \qquad (9.96)$$

With these identities, now we derive the master equation for the reduced density matrix of the device system.

9.4.2 Path Integral Method in Fermion Coherent-State Representation

We consider the same Fano–Anderson Hamiltonian as in Eq. (9.60) for the nonequilibrium Green function techniques, except for the time-dependent coefficients of Hamiltonian in our consideration:

$$H(t) = H_S(t) + H_E(t) + H_I(t),$$

with

$$H_S(t) = \sum_{ij} \varepsilon_{ij}(t) a_i^\dagger a_j,$$

$$H_E(t) = \sum_\alpha H_\alpha(t) = \sum_{\alpha k} \epsilon_{\alpha k}(t) c_{\alpha k}^\dagger c_{\alpha k},$$

$$H_I(t) = \sum_{i\alpha k} \Big[V_{i\alpha k}(t) a_i^\dagger c_{\alpha k} + V_{i\alpha k}^*(t) c_{\alpha k}^\dagger a_i \Big], \qquad (9.97)$$

where energy levels $\varepsilon_{ij}(t)$ and $\epsilon_{\alpha k}(t)$ and the tunneling amplitude $V_{i\alpha k}(t)$ are time-dependent parameters that can be manipulated by external bias and gate voltages in experiments (see Figure 9.5).

The nonequilibrium electron dynamics of the device system is completely determined by the reduced density

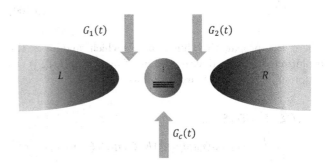

FIGURE 9.5 A schematic plot of a nanoscale quantum device in which the bias voltage is applied to the source and the drain electrode leads labeled L and R, and other gates labeled G_c, G_1, G_2 control the energy levels of the central region as well as the couplings between the central region and the leads.

matrix: $\rho(t) = \text{Tr}_E[\rho_{tot}(t)]$. In the fermion coherent-state representation, the reduced density matrix at an arbitrary later time t is expressed as,

$$\langle \xi_t | \rho(t) | \xi_t' \rangle = \int d\mu(\eta_t) \langle \xi_t, -\eta_t | \rho_{tot}(t) | \xi_t', \eta_t \rangle$$
$$= \int d\mu(\eta_t) \langle \xi_t, -\eta_t | U(t,t_0) \rho_{tot}(t_0) U^\dagger(t,t_0) | \xi_t', \eta_t \rangle,$$

(9.98)

where the total system, which is a closed quantum system, follows unitary evolution, with the evolution operator $U(t,t_0) = T \exp\{-i \int_{t_0}^t H(t')dt'\}$, where T is the time-ordering operator. Here, we denote the fermion coherent state for the total system as $|\xi, \eta\rangle$, with ξ_i and $\eta_{\alpha k}$ being, respectively, the eigenvalues of the annihilation operators for the device system a_i and the leads $c_{\alpha k}$.

As usual, in the master equation approach, the device system is assumed to be uncorrelated with the environment (leads) before the tunneling couplings are turned on [28]:

$$\rho_{tot}(t_0) = \rho(t_0) \otimes \rho_E(t_0),$$
$$\rho_E(t_0) = \frac{1}{Z} e^{-\sum_\alpha \beta_\alpha (H_\alpha(t_0) - \mu_\alpha n_\alpha)}.$$

(9.99)

Here, the system can be in an arbitrary state $\rho(t_0)$, and the leads are initially in thermal equilibrium, where $\beta_\alpha = 1/(k_B T_\alpha)$ is the inverse temperature of lead α at initial time t_0, $n_\alpha = \sum_k c_{\alpha k}^\dagger c_{\alpha k}$ is the total particle number for lead α, and $Z = \text{Tr}_E[e^{-\sum_\alpha \beta_\alpha (H_\alpha(t_0) - \mu_\alpha n_\alpha)}]$ is the partition function. In other words, the system is in the so-called partitioned scheme [8,9] as in the Keldysh framework.

After substitution of $\rho_{tot}(t_0)$ into Eq. (9.98) and insertion of the completeness relations, the reduced density matrix becomes,

$$\langle \xi_t | \rho(t) | \xi_t' \rangle$$
$$= \int d\mu(\xi_0) d\mu(\xi_0') \langle \xi_0 | \rho(t_0) | \xi_0' \rangle \, \mathcal{J}(\bar{\xi}_t, \xi_t', t | \xi_0, \bar{\xi}_0', t_0),$$

(9.100)

where the propagating function $\mathcal{J}$, which manifests the evolution for the device system from $\langle \xi_0 | \rho(t_0) | \xi_0' \rangle$ to $\langle \xi_t | \rho(t) | \xi_t' \rangle$, is explicitly expressed as,

$$\mathcal{J}(\bar{\xi}_t, \xi_t', t | \xi_0, \bar{\xi}_0', t_0)$$
$$= \int d\mu(\eta_t) d\mu(\eta_0) d\mu(\eta_0') K(\bar{\xi}_t, \bar{\eta}_t, t; \xi_0, \eta_0, t_0)$$
$$\times \langle \eta_0 | \rho_E(t_0) | \eta_0' \rangle \bar{K}'(\xi_t', \eta_t', t; \bar{\xi}_0', \bar{\eta}_0', t_0),$$

(9.101)

with the forward and backward transition amplitudes K and $\bar{K}'$ which can be exactly solved using path integral for Hamiltonian (9.97). Explicitly,

$$K(\bar{\xi}_t, \bar{\eta}_t, t; \xi_0, \eta_0, t_0)$$
$$= \langle \xi_t, -\eta_t | U(t,t_0) | \xi_0, \eta_0 \rangle$$
$$= \langle \xi_t, -\eta_t | \left[\prod_{n=0}^{N-1} U(t_{n+1}, t_n) \right] | \xi_0, \eta_0 \rangle$$
$$= \langle \xi_N, \eta_N | \left[\prod_{n=1}^{N-1} U(t_{n+1}, t_n) \int d\mu(\xi_n) d\mu(\eta_n) | \xi_n, \eta_n \rangle \right.$$
$$\left. \langle \xi_n, \eta_n | \right] U(t_1, t_0) | \xi_0, \eta_0 \rangle$$
$$= \int \left[\prod_{n=1}^{N-1} d\mu(\xi_n) d\mu(\eta_n) \right]$$
$$\left[\prod_{n=0}^{N-1} K(\bar{\xi}_{n+1}, \bar{\eta}_{n+1}, t_{n+1}; \xi_n, \eta_n, t_n) \right],$$

(9.102)

where the forward transition amplitude K has been written as path integral with N infinitesimal time intervals $\{t_0, t_1, ..., t_N = t\}$ with,

$$K(\bar{\xi}_{n+1}, \bar{\eta}_{n+1}, t_{n+1}; \xi_n, \eta_n, t_n)$$
$$\approx \langle \xi_{n+1}, \eta_{n+1} | \left[1 - \frac{i}{\hbar} H(t_{n+1/2}) \Delta t \right] | \xi_n \eta_n \rangle$$
$$= \left[1 - \frac{i}{\hbar} H(\bar{\xi}_{n+1}, \bar{\eta}_{n+1}, \xi_n, \eta_n; t_{n+1/2}) \Delta t \right]$$
$$\times \exp(\bar{\xi}_{n+1} \xi_n + \bar{\eta}_{n+1} \eta_n)$$
$$\approx \exp\left(\bar{\xi}_{n+1} \xi_n + \bar{\eta}_{n+1} \eta_n - \frac{i}{\hbar} H \right.$$
$$\left. \times (\bar{\xi}_{n+1}, \bar{\eta}_{n+1}, \xi_n, \eta_n; t_{n+1/2}) \right).$$

(9.103)

Note that since Hamiltonian (9.97) is normal ordered, one can just replace the operators for the device system and the leads with the corresponding eigenvalues in the fermion coherent-state representation. Now, put Eq. (9.103) back into Eq. (9.102) and use continuous approach with fixed boundaries $\{\bar{\xi}_N = \bar{\xi}_t, \bar{\eta}_N = -\bar{\eta}_t, \xi_0, \eta_0\}$, we can get,

$$K(\bar{\xi}_t, \bar{\eta}_t, t; \xi_0, \eta_0, t_0) =$$
$$\int \mathcal{D}[\bar{\xi}\xi; \bar{\eta}\eta] \exp\left\{ \frac{i}{\hbar} \left(S_S[\bar{\xi}, \xi] + S_E[\bar{\eta}, \eta] + S_I[\bar{\xi}, \xi, \bar{\eta}, \eta] \right) \right\},$$

(9.104)

where

$$S_S[\bar{\xi}, \xi] = \frac{\hbar}{2i} \left[\bar{\xi}_t \xi(t) + \bar{\xi}(t_0) \xi_0 \right]$$
$$+ \int_{t_0}^t dt' \left[\frac{\hbar}{2i} \left(\dot{\bar{\xi}}(t') \xi(t') - \bar{\xi}(t') \dot{\xi}(t') \right) - H_S(\bar{\xi}(t'), \xi(t')) \right]$$

$$S_E[\bar{\eta}, \eta] = \frac{\hbar}{2i} \left[-\bar{\eta}_t \eta(t) + \bar{\eta}(t_0) \eta_0 \right]$$
$$+ \int_{t_0}^t dt' \left[\frac{\hbar}{2i} \left(\dot{\bar{\eta}}(t') \eta(t') - \bar{\eta}(t') \dot{\eta}(t') \right) - H_E(\bar{\eta}(t'), \eta(t')) \right]$$

$$S_I[\bar{\xi}, \xi, \bar{\eta}, \eta] = - \int_{t_0}^t dt' H_I(\bar{\xi}(t'), \xi(t'), \bar{\eta}(t'), \eta(t')),$$

(9.105)

are forward actions for the device system, the leads, and the interaction between them.

Similarly, the backward transition amplitude has a similar but conjugate form, Then, the propagating function is given as

$$\mathcal{J}(\bar{\xi}_t, \xi_t', t|\xi_0, \bar{\xi}_0', t_0) = \int \mathcal{D}[\bar{\xi}\xi; \bar{\xi}'\xi'] e^{\frac{i}{\hbar}(S_S[\bar{\xi},\xi] - S_S'[\bar{\xi}'\xi'])}$$
$$\mathcal{F}[\bar{\xi}\xi; \bar{\xi}'\xi']$$
$$= \int \mathcal{D}[\bar{\xi}\xi; \bar{\xi}'\xi'] e^{\frac{i}{\hbar}S_{eff}[\bar{\xi}\xi;\bar{\xi}'\xi']}, \quad (9.106)$$

where $\mathcal{F}[\bar{\xi}\xi; \bar{\xi}'\xi']$ is defined as the influence functional which takes fully into account the back-action effects of the environment (leads) to the device system, it modifies the original action of the device system into an effective one, which dramatically changes the dynamics of the device system. Explicitly,

$$\mathcal{F}[\bar{\xi}\xi; \bar{\xi}'\xi'] = \int d\mu(\boldsymbol{\eta}_f) d\mu(\boldsymbol{\eta}_0) d\mu(\boldsymbol{\eta}_0') \langle \boldsymbol{\eta}_0|\rho_E(t_0)|\boldsymbol{\eta}_0'\rangle$$
$$\times \int \mathcal{D}[\bar{\boldsymbol{\eta}}\boldsymbol{\eta}; \bar{\boldsymbol{\eta}}'\boldsymbol{\eta}'] \exp \frac{i}{\hbar} \Big(S_E[\bar{\boldsymbol{\eta}}, \boldsymbol{\eta}] - S_E'[\bar{\boldsymbol{\eta}}', \boldsymbol{\eta}']$$
$$+ S_I[\bar{\boldsymbol{\xi}}, \boldsymbol{\xi}, \bar{\boldsymbol{\eta}}, \boldsymbol{\eta}] - S_I'[\bar{\boldsymbol{\xi}}', \boldsymbol{\xi}', \bar{\boldsymbol{\eta}}', \boldsymbol{\eta}'] \Big). \quad (9.107)$$

For initially equilibrium source and drain, one can find,

$$\langle \boldsymbol{\eta}_0|\rho_E(t_0)|\boldsymbol{\eta}_0'\rangle = \exp(\bar{\boldsymbol{\eta}}_0 \boldsymbol{\Omega} \boldsymbol{\eta}_0') \det[\mathbb{1}_{N_E} + \boldsymbol{\Omega}]^{-1} \quad (9.108)$$

where $\boldsymbol{\Omega}$ is a diagonal matrix with element $\Omega_{\alpha k, \alpha k} = e^{-\beta_\alpha(\epsilon_{\alpha k} - \mu_\alpha)}$. Because the actions in Eq. (9.107) are in quadratic form, the path integral can be exactly carried out using the stationary-path approach. Then, by using Gaussian integral of Eq. (9.96) to integrate out all the environmental degrees of freedom, the influence functional has the following form,

$$\mathcal{F}[\bar{\xi}\xi; \bar{\xi}'\xi'] = \exp \Bigg\{ -\sum_\alpha \int_{t_0}^t dt' \Bigg[\int_{t_0}^{t'} dt'' [\bar{\xi}(t') \boldsymbol{g}_\alpha(t', t'') \xi(t'')$$
$$+ \bar{\xi}'(t'') \boldsymbol{g}_\alpha(t'', t') \xi'(t')]$$
$$+ \int_{t_0}^t dt'' \Big\{ \bar{\xi}'(t') \boldsymbol{g}_\alpha(t', t'') \xi(t'')$$
$$- [\bar{\xi}(t') + \bar{\xi}'(t')] \widetilde{\boldsymbol{g}}_\alpha(t', t'') [\xi(t'') + \xi'(t'')] \Big\} \Bigg] \Bigg\}.$$
$$(9.109)$$

In the above equation, the time nonlocal integral kernels, $\boldsymbol{g}_\alpha(t', t'')$ and $\widetilde{\boldsymbol{g}}_\alpha(t', t'')$ are defined as

$$g_{\alpha ij}(t', t'') = \frac{1}{\hbar^2} \sum_k V_{i\alpha k}(t') V_{j\alpha k}^*(t'') e^{-\frac{i}{\hbar} \int_{t''}^{t'} d\bar{t} \epsilon_{\alpha k}(\bar{t})},$$
$$(9.110a)$$

$$\widetilde{g}_{\alpha ij}(t', t'') = \frac{1}{\hbar^2} \sum_k V_{i\alpha k}(t') V_{j\alpha k}^*(t'') f_\alpha(\epsilon_{\alpha k}) e^{-\frac{i}{\hbar} \int_{t''}^{t'} d\bar{t} \epsilon_{\alpha k}(\bar{t})},$$
$$(9.110b)$$

where $f_\alpha(\epsilon_{\alpha k}) = 1/(e^{\beta_\alpha(\epsilon_{\alpha k} - \mu_\alpha)} + 1)$ is the Fermi–Dirac distribution function of lead α at initial time t_0.

Inserting Eq. (9.109) into Eq. (9.106), because the effective action is quadratic in terms of $\{\xi_i, \xi_i^*, \xi_i', \xi_i'^*\}$, we use the stationary-path approach again. The resulting propagating function is

$$\mathcal{J}(\bar{\xi}_t, \xi_t', t|\xi_0, \bar{\xi}_0', t_0) = A(t) \exp \Big\{ \frac{1}{2} \Big[\bar{\xi}_t \xi(t) + \bar{\xi}(t_0)\xi_0$$
$$+ \bar{\xi}'(t)\xi_t' + \bar{\xi}_0' \xi'(t_0) \Big] \Big\} \quad (9.111)$$

where $A(t)$ is a time-dependent factor coming from stationary path approach which can be easily determined by the normalization of the reduced density matrix, and $\{\boldsymbol{\xi}(t), \bar{\boldsymbol{\xi}}(t), \bar{\boldsymbol{\xi}}'(t), \boldsymbol{\xi}'(t)\}$ obey the following equations of motion,

$$\dot{\boldsymbol{\xi}}(t') = -\frac{i}{\hbar}\boldsymbol{\varepsilon}(t')\boldsymbol{\xi}(t') - \int_{t_0}^{t'} dt'' \sum_\alpha \boldsymbol{g}_\alpha(t', t'')\boldsymbol{\xi}(t'')$$
$$+ \int_{t_0}^t dt'' \sum_\alpha \widetilde{\boldsymbol{g}}_\alpha(t', t'')[\boldsymbol{\xi}(t'') + \boldsymbol{\xi}'(t'')], \quad (9.112a)$$

$$\dot{\boldsymbol{\xi}}'(t') = -\frac{i}{\hbar}\boldsymbol{\varepsilon}(t')\boldsymbol{\xi}'(t') + \int_{t'}^t dt'' \sum_\alpha \boldsymbol{g}_\alpha(t', t'')\boldsymbol{\xi}'(t'')$$
$$+ \int_{t_0}^t dt'' \sum_\alpha \boldsymbol{g}_\alpha(t', t'')\boldsymbol{\xi}(t'')$$
$$- \int_{t_0}^t dt'' \sum_\alpha \widetilde{\boldsymbol{g}}_\alpha(t', t'')[\boldsymbol{\xi}(t'') + \boldsymbol{\xi}'(t'')], \quad (9.112b)$$

$$\dot{\bar{\boldsymbol{\xi}}}'(t') = \frac{i}{\hbar}\bar{\boldsymbol{\xi}}'(t')\boldsymbol{\varepsilon}(t') - \int_{t_0}^{t'} dt'' \sum_\alpha \bar{\boldsymbol{\xi}}'(t'')\boldsymbol{g}_\alpha(t'', t')$$
$$+ \int_{t_0}^t dt'' \sum_\alpha [\bar{\boldsymbol{\xi}}'(t'') + \bar{\boldsymbol{\xi}}(t'')]\widetilde{\boldsymbol{g}}_\alpha(t'', t'), \quad (9.112c)$$

$$\dot{\bar{\boldsymbol{\xi}}}(t') = \frac{i}{\hbar}\bar{\boldsymbol{\xi}}(t')\boldsymbol{\varepsilon}(t') + \int_{t'}^t dt'' \sum_\alpha \bar{\boldsymbol{\xi}}(t'')\boldsymbol{g}_\alpha(t'', t')$$
$$+ \int_{t_0}^t dt'' \sum_\alpha \bar{\boldsymbol{\xi}}'(t'')\boldsymbol{g}_\alpha(t'', t')$$
$$- \int_{t_0}^t dt'' \sum_\alpha [\bar{\boldsymbol{\xi}}'(t'') + \bar{\boldsymbol{\xi}}(t'')]\widetilde{\boldsymbol{g}}_\alpha(t'', t'). \quad (9.112d)$$

It is worth to note that, because of the boundary condition, $\bar{\boldsymbol{\xi}}(t) = \bar{\boldsymbol{\xi}}_t, \boldsymbol{\xi}(t_0) = \boldsymbol{\xi}_0, \bar{\boldsymbol{\xi}}'(t_0) = \bar{\boldsymbol{\xi}}_0'$, and $\boldsymbol{\xi}'(t) = \boldsymbol{\xi}_t'$, we get $(\xi_i(t'))^* = \xi_i'^*(t')$ and $(\xi_i'(t'))^* = \xi_i^*(t')$. By introducing the following transformation,

$$\boldsymbol{\xi}(t') = \boldsymbol{u}(t', t_0)\boldsymbol{\xi}(t_0) + \boldsymbol{v}(t', t)[\boldsymbol{\xi}(t) + \boldsymbol{\xi}'(t_0)], \quad (9.113a)$$
$$\boldsymbol{\xi}(t') + \boldsymbol{\xi}'(t') = \boldsymbol{u}^\dagger(t, t')[\boldsymbol{\xi}(t) + \boldsymbol{\xi}'(t)], \quad (9.113b)$$

and normalize the reduced density matrix, the propagating function becomes,

$$\mathcal{J}(\bar{\xi}_t, \xi_t', t|\xi_0, \bar{\xi}_0', t_0) = \frac{1}{\det[\boldsymbol{w}(t)]} \exp\{\bar{\xi}_t \boldsymbol{J}_1(t)\xi_0 + \bar{\xi}_t \boldsymbol{J}_2(t)\xi_t'$$
$$+ \bar{\xi}_0' \boldsymbol{J}_3(t)\xi_0 + \bar{\xi}_0' \boldsymbol{J}_1^\dagger(t)\xi_t'\}. \quad (9.114)$$

The time-dependent coefficients are given explicitly as,

$$\boldsymbol{J}_1(t) = \boldsymbol{w}(t)\boldsymbol{u}(t, t_0),$$
$$\boldsymbol{J}_2(t) = \boldsymbol{w}(t) - \mathbb{1}_{N_S},$$
$$\boldsymbol{J}_3(t) = \boldsymbol{u}^\dagger(t, t_0)\boldsymbol{w}(t)\boldsymbol{u}(t, t_0) - \mathbb{1}_{N_S}, \qquad (9.115)$$

with $\boldsymbol{w}(t) = [\mathbb{1}_{N_S} - \boldsymbol{v}(t, t)]^{-1}$. As one can see, the propagating function is determined by the two matrix functions $\boldsymbol{u}(t, t_0)$ and $\boldsymbol{v}(t, t)$, which are $N_S \times N_S$ matrix with N_S being the total number of single-particle energy levels in the device system. They satisfy the following integrodifferential equations,

$$\frac{d}{dt'}\boldsymbol{u}(t', t_0) + i\boldsymbol{\varepsilon}(t')\boldsymbol{u}(t', t_0)$$
$$+ \sum_\alpha \int_{t_0}^{t'} dt'' \boldsymbol{g}_\alpha(t', t'')\boldsymbol{u}(t'', t_0) = 0, \qquad (9.116a)$$

$$\frac{d}{dt'}\boldsymbol{v}(t', t) + i\boldsymbol{\varepsilon}(t')\boldsymbol{v}(t', t) + \sum_\alpha \int_{t_0}^{t'} dt'' \boldsymbol{g}_\alpha(t', t'')\boldsymbol{v}(t'', t)$$
$$= \sum_\alpha \int_{t_0}^{t} dt'' \widetilde{\boldsymbol{g}}_\alpha(t', t'')\boldsymbol{u}^\dagger(t, t''), \qquad (9.116b)$$

subject to the boundary conditions $\boldsymbol{u}(t_0, t_0) = \mathbb{1}_{N_S}$ and $\boldsymbol{v}(t_0, t) = \boldsymbol{0}_{N_S}$ with $t_0 \leq t' \leq t$. Actually, $\boldsymbol{u}(t', t_0)$ and $\boldsymbol{v}(t', t)$ are related to the nonequilibrium Green functions of the device system as we will show in the next subsection. The convolution equation (9.116) shows that the time nonlocal integral kernels $\boldsymbol{g}_\alpha(t', t'')$ and $\widetilde{\boldsymbol{g}}_\alpha(t', t'')$ characterize the memory effects between the device system and the lead α.

Taking the time derivative of the reduced density matrix (9.100) with the solution of the propagating function (9.114), together with the Grassmann algebra of the fermion creation and annihilation operators in the fermion coherent-state representation (see Eq. (9.91)), we arrive at the final form of the exact master equation,

$$\frac{d\rho(t)}{dt} = -i\Big[H'_S(t, t_0), \rho(t)\Big]$$
$$+ \sum_{ij} \Big\{ \gamma_{ij}(t, t_0)\Big[2a_j\rho(t)a_i^\dagger - a_i^\dagger a_j\rho(t) - \rho(t)a_i^\dagger a_j\Big]$$
$$+ \widetilde{\gamma}_{ij}(t, t_0)\Big[a_i^\dagger\rho(t)a_j - a_j\rho(t)a_i^\dagger$$
$$+ a_i^\dagger a_j\rho(t) - \rho(t)a_j a_i^\dagger\Big]\Big\}$$
$$= -i\Big[H_S(t), \rho(t)\Big] + \sum_\alpha \Big[\mathcal{L}_\alpha^+(t, t_0) + \mathcal{L}_\alpha^-(t, t_0)\Big]\rho(t). \qquad (9.117)$$

The first term describes the unitary evolution of electrons in the device system, where the renormalization effect, after integrating out all the lead degrees of freedom, has been fully taken into account. The resultant renormalized Hamiltonian is $H'_S(t) = \sum_{ij} \varepsilon'_{ij}(t, t_0)a_i^\dagger a_j$, with $\varepsilon'_{ij}(t, t_0)$ being the corresponding renormalized energy matrix of the device system, including the energy shift of each level and the lead-induced

couplings between different levels. The remaining terms give the non-unitary dissipation and fluctuation processes induced by back actions of electrons from the leads, and are described by the dissipation and fluctuation coefficients $\boldsymbol{\gamma}(t, t_0)$ and $\widetilde{\boldsymbol{\gamma}}(t, t_0)$, respectively. On the other hand, the current superoperators of lead α, $\mathcal{L}_\alpha^+(t, t_0)$ and $\mathcal{L}_\alpha^-(t, t_0)$, determine the transport current flowing from lead α into the device system:

$$I_\alpha(t) = -e\left\langle \frac{dn_\alpha(t)}{dt} \right\rangle = \frac{e}{\hbar}\mathrm{Tr}\Big[\mathcal{L}_\alpha^+(t, t_0)\rho(t)\Big]$$
$$= -\frac{e}{\hbar}\mathrm{Tr}\Big[\mathcal{L}_\alpha^-(t, t_0)\rho(t)\Big], \qquad (9.118)$$

where $n_\alpha(t) = \sum_k c_{\alpha k}^\dagger(t)c_{\alpha k}(t)$ is the total particle number of lead α.

All the time-dependent coefficients in Eq. (9.117) are found to be

$$\varepsilon'_{ij}(t, t_0) = \frac{i}{2}\Big[\dot{\boldsymbol{u}}(t, t_0)\boldsymbol{u}^{-1}(t, t_0) - \mathrm{H.c.}\Big]_{ij}$$
$$= \varepsilon_{ij}(t) - \frac{i}{2}\sum_\alpha [\boldsymbol{\kappa}_\alpha(t, t_0) - \boldsymbol{\kappa}_\alpha^\dagger(t, t_0)]_{ij}, \qquad (9.119a)$$

$$\gamma_{ij}(t, t_0) = -\frac{1}{2}\Big[\dot{\boldsymbol{u}}(t, t_0)\boldsymbol{u}^{-1}(t, t_0) + \mathrm{H.c.}\Big]_{ij}$$
$$= \frac{1}{2}\sum_\alpha [\boldsymbol{\kappa}_\alpha(t, t_0) + \boldsymbol{\kappa}_\alpha^\dagger(t, t_0)]_{ij}, \qquad (9.119b)$$

$$\widetilde{\gamma}_{ij}(t, t_0) = \frac{d}{dt}v_{ij}(t, t) - [\dot{\boldsymbol{u}}(t, t_0)\boldsymbol{u}^{-1}(t, t_0)\boldsymbol{v}(t, t) + \mathrm{H.c.}]_{ij}$$
$$= -\sum_\alpha [\boldsymbol{\lambda}_\alpha(t, t_0) + \boldsymbol{\lambda}_\alpha^\dagger(t, t_0)]_{ij}, \qquad (9.119c)$$

The current superoperators of lead α, $\mathcal{L}_\alpha^+(t)$ and $\mathcal{L}_\alpha^-(t)$, are also explicitly given by

$$\mathcal{L}_\alpha^+(t, t_0)\rho(t) = -\sum_{ij}\Big\{ \lambda_{\alpha ij}(t, t_0)\Big[a_i^\dagger a_j\rho(t) + a_i^\dagger\rho(t)a_j\Big]$$
$$+ \kappa_{\alpha ij}(t, t_0)a_i^\dagger a_j\rho(t) + \mathrm{H.c.}\Big\}, \qquad (9.120a)$$

$$\mathcal{L}_\alpha^-(t, t_0)\rho(t) = \sum_{ij}\Big\{ \lambda_{\alpha ij}(t, t_0)\Big[a_j\rho(t)a_i^\dagger + \rho(t)a_j a_i^\dagger\Big]$$
$$+ \kappa_{\alpha ij}(t, t_0)a_j\rho(t)a_i^\dagger + \mathrm{H.c.}\Big\}. \qquad (9.120b)$$

The functions $\boldsymbol{\kappa}_\alpha(t, t_0)$ and $\boldsymbol{\lambda}_\alpha(t, t_0)$ in Eqs. (9.119) and (9.120) are solved from Eq. (9.116),

$$\boldsymbol{\kappa}_\alpha(t, t_0) = \int_{t_0}^{t} dt' \boldsymbol{g}_\alpha(t, t')\boldsymbol{u}(t', t_0)[\boldsymbol{u}(t, t_0)]^{-1}, \qquad (9.121a)$$

$$\boldsymbol{\lambda}_\alpha(t, t_0) = \int_{t_0}^{t} dt'[\boldsymbol{g}_\alpha(t, t')\boldsymbol{v}(t', t) - \widetilde{\boldsymbol{g}}_\alpha(t, t')\boldsymbol{u}^\dagger(t, t')]$$
$$- \boldsymbol{\kappa}_\alpha(t, t_0)\boldsymbol{v}(t, t). \qquad (9.121b)$$

As one sees, the master equation (9.117) takes a convolutionless form, so the non-Markovian dynamics are fully encoded in the time-dependent coefficients (9.119). These coefficients are determined by the functions $\boldsymbol{u}(t', t_0)$ and $\boldsymbol{v}(t', t)$ which obey the time-convolution equations of motion(9.116), so that the integral kernels (9.110) manifest the non-Markovian memory effects. The master equation is derived exactly so that the positivity, Hermiticity of the trace of the reduced density matrix are guaranteed. It is also worth mentioning that the master equation (9.117) is valid for various nano-devices coupled to various surroundings through particle tunnelings, even when initial correlations are presented as long as the electron–electron interaction can be ignored (see Ref. [54], where the partition-free scheme is reproduced).

9.4.3 Transient Transport Current Derived from the Master Equation

From Eqs. (9.117–9.118), the transient transport current is given explicitly as follows:

$$
I_\alpha(t) = -\frac{e}{\hbar}\,\mathrm{Tr}[\boldsymbol{\lambda}_\alpha(t,t_0) + \boldsymbol{\kappa}_\alpha(t,t_0)\boldsymbol{\rho}^{(1)}(t,t) + H.c.]
$$
$$
= -\frac{2e}{\hbar}\,\mathrm{Re}\int_{t_0}^{t} dt'\,\mathrm{Tr}
$$
$$
\times\left[\boldsymbol{g}_\alpha(t,t')\boldsymbol{\rho}^{(1)}(t',t) - \widetilde{\boldsymbol{g}}_\alpha(t,t')\boldsymbol{u}^\dagger(t,t')\right]. \quad (9.122)
$$

In Eq. (9.122), the single-particle correlation function of the device system $\boldsymbol{\rho}^{(1)}(t',t)$ is given by

$$
\rho_{ij}^{(1)}(t',t) = \left[\boldsymbol{u}(t',t_0)\boldsymbol{\rho}^{(1)}(t_0)\boldsymbol{u}^\dagger(t,t_0) + \boldsymbol{v}(t',t)\right]_{ij}, \quad (9.123)
$$

where $\rho_{ij}^{(1)}(t_0) = \mathrm{Tr}_S[a_j^\dagger a_i \rho(t_0)]$, is the initial single-particle density matrix. The transient transport current obtained from the master equation actually has exactly the same formula as the one using the nonequilibrium Green function techniques [53], except for the first term of the single-particle correlation function (9.123) that is originated from the initial occupation $\rho_{ij}^{(1)}(t_0)$ in the device system, which was missing in Ref. [53]. We will derive this connection in the next subsection.

9.4.4 Unify the Master Equation Approach and the Nonequilibrium Green Function Techniques

As one sees, both the master equation (9.117) and the transient current (9.122) are completely determined by the propagating matrices $\boldsymbol{u}(t,t_0)$ and $\boldsymbol{v}(t',t)$, which are introduced in Eq. (9.113). The equations (9.113) show that $\boldsymbol{u}(t',t_0)$ describes the electron forward propagation from time t_0 to time t', $\boldsymbol{u}^\dagger(t,t')$ describes the electron backward propagation from time t to time t', and $\boldsymbol{v}(t',t)$ describes the electron propagation mixing the forward and backward paths. These propagating matrices satisfy the integrodifferential equations (9.116). Solving inhomogeneous equation (9.116b) with initial condition $\boldsymbol{v}(t_0,t) = \boldsymbol{0}_{N_S}$, we obtain

$$
\boldsymbol{v}(t',t) = \sum_\alpha \int_{t_0}^{t'} dt_1 \int_{t_0}^{t} dt_2\,\boldsymbol{u}(t',t_1)\widetilde{\boldsymbol{g}}_\alpha(t_1,t_2)\boldsymbol{u}^\dagger(t,t_2),
$$
$$
(9.124)
$$

with $\boldsymbol{u}(t,t')$ satisfying the following equation of motion,

$$
\frac{d}{dt}\boldsymbol{u}(t,t') + i\boldsymbol{\varepsilon}(t)\boldsymbol{u}(t,t') + \sum_\alpha \int_{t'}^{t} dt_1 \boldsymbol{g}_\alpha(t,t_1)\boldsymbol{u}(t_1,t') = 0,
$$
$$
(9.125)
$$

and subject to the boundary condition $\boldsymbol{u}(t',t') = \mathbb{1}_{N_S}$.

It is easy to infer that

$$
u_{ij}(t,t') = \langle\{a_i(t), a_j^\dagger(t')\}\rangle = i[\boldsymbol{G}^R(t,t') - \boldsymbol{G}^A(t,t')]_{ij},
$$
$$
(9.126)
$$

which is the spectral function in the nonequilibrium Green function techniques, with

$$
\boldsymbol{g}_\alpha(t,t') = i[\boldsymbol{\Sigma}_\alpha^R(t,t') - \boldsymbol{\Sigma}_\alpha^A(t,t')]
$$
$$
= \int \frac{d\omega}{2\pi}\boldsymbol{\Gamma}_\alpha(\omega,t,t')e^{-i\omega(t-t')}. \quad (9.127)
$$

As a result, matrix function $\boldsymbol{v}(t',t)$ in (9.124) can be written in terms of the nonequilibrium Green functions,

$$
\boldsymbol{v}(t',t) = -i\int_{t_0}^{t'} dt_1 \int_{t_0}^{t} dt_2\,\boldsymbol{G}^R(t',t_1)\boldsymbol{\Sigma}^<(t_1,t_2)\boldsymbol{G}^A(t_2,t),
$$
$$
(9.128)
$$

where

$$
\widetilde{\boldsymbol{g}}_\alpha(t,t') = -i\boldsymbol{\Sigma}_\alpha^<(t,t')
$$
$$
= \int \frac{d\omega}{2\pi} f_\alpha(\omega)\boldsymbol{\Gamma}_\alpha(\omega,t,t')e^{-i\omega(t-t')}. \quad (9.129)
$$

Comparing Eq. (9.128) with Eq. (9.85b), we find that $\boldsymbol{v}(t',t)$ exactly has the same form as the lesser Green function in the Keldysh formalism. However, when one considers transient electron dynamics, the general solution of the lesser Green function is related to the single-particle correlation function in the master equation approach,

$$
\boldsymbol{G}^<(t',t) = i\boldsymbol{\rho}^{(1)}(t',t) = i\left[\boldsymbol{u}(t',t_0)\boldsymbol{\rho}^{(1)}(t_0)\boldsymbol{u}^\dagger(t,t_0) + \boldsymbol{v}(t',t)\right]
$$
$$
= \boldsymbol{G}^R(t',t_0)\boldsymbol{G}^<(t_0,t_0)\boldsymbol{G}^A(t_0,t)
$$
$$
+ \int_{t_0}^{\infty} dt_1 \int_{t_0}^{\infty} dt_2\,\boldsymbol{G}^R(t',t_1)\boldsymbol{\Sigma}^<(t_1,t_2)\boldsymbol{G}^A(t_2,t').
$$
$$
(9.130)
$$

The first term that depends on the initial occupation of the device system is usually omitted when one discusses the quantum transport in the steady-state limit. According to the above results, we can express the transient transport current in terms of the nonequilibrium Green functions:

$$
I_\alpha(t) = -\frac{2e}{\hbar}\,\mathrm{Re}\int_{t_0}^{t} dt'\,\mathrm{Tr}\left[\boldsymbol{g}_\alpha(t,t')\boldsymbol{\rho}^{(1)}(t',t)\right.
$$
$$
\left. -\widetilde{\boldsymbol{g}}_\alpha(t,t')\boldsymbol{u}^\dagger(t,t')\right]
$$
$$
= -\frac{2e}{\hbar}\,\mathrm{Re}\int_{t_0}^{t} dt'\,\mathrm{Tr}\left\{[\boldsymbol{\Sigma}_\alpha^R(t,t') - \boldsymbol{\Sigma}_\alpha^A(t,t')]\boldsymbol{G}^<(t',t)\right.
$$
$$
\left. -\boldsymbol{\Sigma}_\alpha^<(t,t')[\boldsymbol{G}^R(t',t) - \boldsymbol{G}^A(t',t)]\right\}
$$

$$= -\frac{2e}{\hbar} \, \mathrm{Re} \int_{t_0}^{t} dt' \, \mathrm{Tr} \left[\mathbf{\Sigma}_{\alpha}^{R}(t,t') \mathbf{G}^{<}(t',t) \right.$$

$$\left. + \mathbf{\Sigma}_{\alpha}^{<}(t,t') \mathbf{G}^{A}(t',t) \right]. \qquad (9.131)$$

It is easy to check the consistency between Eqs. (9.131) and (9.83), except that in Eq. (9.83), the first term in Eq. (9.130) is missed. Thus, we have proved that the transient transport current obtained from the master equation has exactly the same formula as the one using the nonequilibrium Green function techniques [53], except for the term that is originated from the initial occupation $\rho_{ij}^{(1)}(t_0)$ in the device system. This also indicates further that the Keldysh's nonequilibrium Green function technique is mostly valid in the steady-state limit.

References

1. Anderson, P. W. 1958. Absence of diffusion in certain random lattices. *Phys. Rev.* 109: 1492–505.

2. Blanter, Ya. M. and Büttiker, M. 2000. Shot noise in mesoscopic conductors. *Phys. Rep.* 336: 1–166.

3. Büttiker, M. 1988. Absence of backscattering in the quantum Hall effect in multiprobe conductors. *Phys. Rev. B* 38: 9375–89.

4. Büttiker, M. 1986. Four-terminal phase-coherent conductance. *Phys. Rev. Lett.* 57: 1761–4.

5. Büttiker, M., Imry, Y., Landauer, R., and Pinhas, S. 1985. Generalized many-channel conductance formula with application to small rings. *Phys. Rev. B* 31: 6207–15.

6. Büttiker, M. 1990. Quantized transmission of a saddle-point constriction. *Phys. Rev. B* 41: 7906–9.

7. Büttiker, M. 1992. Scattering theory of current and intensity noise correlations in conductors and wave guides. *Phys. Rev. B* 46: 12485–507.

8. Caroli, C., Combescot, R., Nozeres, P., and Saint-James, D. 1971. A direct calculation of the tunnelling current. II. Free electron description. *J. Phys. C* 4: 2598–610.

9. Caroli, C., Combescot, R., Nozeres, P., and Saint-James, D. 1971. Direct calculation of the tunneling current. *J. Phys. C* 4: 916–29.

10. Chou, K. C., Su, Z.-B., Hao, B.-L., and Yu, L. 1985. Equilibrium and nonequilibrium formalisms made unified. *Phys. Rep.* 118: 1–131.

11. Cini, M. 1980. Time-dependent approach to electron transport through junctions: General theory and simple applications. *Phys. Rev. B* 22: 5887–99.

12. Datta, S. 1995. *Electronic Transport in Mesoscopic Systems*. Cambridge: Cambridge University Press.

13. Entin-Wohlman, O., Aharony, A., and Levinson, Y. 2002. Adiabatic transport in nanostructures. *Phys. Rev. B* 65: 195411.

14. Fano, U. 1961. Effects of configuration interaction on intensities and phase shifts. *Phys. Rev.* 124: 1866–78.

15. Feynman, R. P. and Vernon, F. L. 1963 The theory of a general quantum system interacting with a linear dissipative system. *Ann. Phys. (N.Y.)* 24: 118–73.

16. Gurvitz, S. A. and Prager, Ya. S. 1996. Microscopic derivation of rate equations for quantum transport. *Phys. Rev. B* 53: 15932–43.

17. Haug, H. and Jauho, A. P. 2008. *Quantum Kinetics in Transport and Optics of Semiconductors*. Springer Series in Solid-State Sciences, Vol. 123. Berlin: Springer.

18. Imry, Y. 2002. *Introduction to Mesoscopic Physics*. 2nd Ed. Oxford: Oxford University Presss.

19. Jauho, A. P., Wingreen, N. S., and Meir, Y. 1994. Time-dependent transport in interacting and noninteracting resonant-tunneling systems. *Phys. Rev. B* 50: 5528–44

20. Jin, J. S., Tu, M. W.-Y., Wang, N. E., and Zhang, W. M. 2013. Precision control of charge coherence in parallel double dot systems through spin-orbit interaction, *J. Chem. Phys.* 139: 064706.

21. Jin, J. S., Tu, M. W.-Y., Zhang, W. M., and Yan, Y. J. 2010. Non-equilibrium quantum theory for nanodevices based on the Feynman–Vernon influence functional. *New J. Phys.* 12: 083013.

22. Jin, J. S., Zheng, X., and Yan, Y. J. 2008. Exact dynamics of dissipative electronic systems and quantum transport: Hierarchical equations of motion approach. *J. Chem. Phys.* 128: 234703.

23. Kadanoff, L. P. and Baym, G. 1962. *Quantum Statistical Mechanics*. New York: Benjamin.

24. Keldysh, L. V. 1964. Diagram technique for nonequilibrium processes. *Zh. Eksp. Teor. Fiz.* 47: 1515–27 [1965. *Sov. Phys. JETP* 20: 1018–26]

25. Landauer, R. 1970. Electrical resistance of disordered one-dimensional lattices. *Philos. Mag.* 21: 863–67.

26. Landauer, R. 1957. Spatial variation of currents and fields due to localized scatterers in metallic conduction. *IBM J. Res. Dev.* 1: 223–31.

27. Langreth, D. C. 1976. *Linear and Nonlinear Electron Transport in Solids*, ed. by J. T. Devreese, E. Van Doren, New York: Plenum.

28. Leggett, A. J., Chakravarty, S., Dorsey, A. T., Fisher, M. P. A., Garg, A., and Zwerger, W. 1987. Dynamics of the dissipative two-state system. *Rev. Mod. Phys.* 59: 1–85.

29. Lin, C. Y. and Zhang, W. M. 2011. Single-electron turnstile pumping with high frequencies. *Appl. Phys. Lett.* 99: 072105.

30. Liu, J. H., Tu, M. W. Y., and Zhang, W. M. 2016. Quantum coherence of the molecular states and their corresponding currents in nanoscale Aharonov-Bohm interferometers. *Phys. Rev. B* 94: 045403.

31. Li, X. Q., Luo, J., Yang, Y. G., Cui, P., and Yan, Y. J. 2005. Quantum master-equation approach to quantum transport through mesoscopic systems. *Phys. Rev. B* 71: 205304.

32. Li, X. Q. 2016. Number-resolved master equation approach to quantum measurement and quantum transport. *Front. Phys.* 11: 110307.

33. Mahan, G. D. 1990. *Many Particle Physics.* 2nd ed. New York: Plenum.

34. Moskalets, M. and Büttiker, M. 2004. Adiabatic quantum pump in the presence of external ac voltages. *Phys. Rev. B* 69: 205316.

35. Moskalets, M. and Büttiker, M. 2007. Time-resolved noise of adiabatic quantum pumps. *Phys. Rev. B* 75: 035315.

36. Negele, J. W. and Orland, H. 1988. *Quantum Many-Particle Systems.* Boulder: Westview Press.

37. Ohnishi, H., Inata, T., Muto, S., Yokoyama, N., and Shibatomi, A. 1986. Selfconsistent analysis of resonant tunneling current. *Appl. Phys. Lett.* 49: 1248–50.

38. Rammer, J. and Smith, H. 1986. Quantum field-theoretical methods in transport theory of metals. *Rev. Mod. Phys.* 58: 323–59.

39. Rothstein, E. A., Entin-Wohlman, O., and Aharony, A. 2009. Noise spectra of a biased quantum dot. *Phys. Rev. B* 79: 075307.

40. Samuelsson, P. and Büttiker, M. 2006. Quantum state tomography with quantum shot noise. *Phys. Rev. B* 73: 041305(R).

41. Schoeller, H. and Schön, G. 1994. Mesoscopic quantum transport: Resonant tunneling in the presence of a strong Coulomb interaction. *Phys. Rev. B* 50: 18436–52.

42. Schwinger, J. 1961. Brownian motion of a quantum oscillator. *J. Math. Phys. (N.Y.)* 2: 407–32.

43. Stefanucci, G. and Almbladh, C.-O. 2004. Time-dependent partition-free approach in resonant tunneling systems. *Phys. Rev. B* 69: 195318.

44. Szafer, A. and Stone, A. D. 1989. Theory of quantum conduction through a constriction. *Phys. Rev. Lett.* 62: 300–3.

45. Tu, M. W.-Y., Aharony, A., Entin-Wohlman, O., Schiller, A., and Zhang, W. M. 2016. Transient probing of the symmetry and the asymmetry of electron interference. *Phys. Rev. B* 93: 125437.

46. Tu, M. W.-Y., Aharony, A., Zhang, W. M., and Entin-Wohlman, O. 2014. Real-time dynamics of spin-dependent transport through a double-quantum-dot Aharonov-Bohm interferometer with spin-orbit interaction. *Phys. Rev. B* 90: 165422.

47. Tu, M. W.-Y., Lee, M.-T., and Zhang, W. M. 2009. Exact master equation and non-markovian decoherence for quantum dot quantum computing. *Quantum Inf. Process.* (Springer) 8: 631–46.

48. Tu, M. W.-Y., Zhang, W. M., Jin, J. S., Entin-Wohlman, O., and Aharony, A. 2012. Transient quantum transport in double-dot Aharonov-Bohm interferometers. *Phys. Rev. B* 86: 115453.

49. Tu, M. W.-Y., Zhang, W. M., and Jin, J. S. 2011. Intrinsic coherence dynamics and phase localization in nanoscale Aharonov–Bohm interferometers. *Phys. Rev. B* 83: 115318

50. Tu, M. W.-Y. and Zhang, W. M. 2008. Non-Markovian decoherence theory for a double-dot charge qubit. *Phys. Rev. B* 78: 235311-1–27.

51. Tu, M. W.-Y., Zhang, W. M., and Nori, F. 2012. Coherent control of double-dot molecules using Aharonov–Bohm magnetic flux. *Phys. Rev. B* 86: 195403.

52. Wang, J. S., Agarwalla, B. K., Li, H., and Thingna, J. 2014. Nonequilibrium Greens function method for quantum thermal transport. *Front. Phys.* 9: 673–97.

53. Wingreen, N. S., Jauho, A. P., and Meir, Y. 1993. Time-dependent transport through a mesoscopic structure. *Phys. Rev. B* 48: 8487–90.

54. Yang, P. Y., Lin, C. Y., and Zhang, W. M. 2015. Master equation approach to transient quantum transport in nanostructures incorporating initial correlations. *Phys. Rev. B* 92: 165403.

55. Yang, P. Y., Lin, C. Y., and Zhang, W. M. 2014. Transient current-current correlations and noise spectra. *Phys. Rev. B* 89: 115411.

56. Yang, P. Y. and Zhang, W. M. 2018. Buildup of Fano resonances in the time domain in a double quantum dot Aharonov–Bohm interferometer. *Phys. Rev. B* 97: 054301.

57. Yang, P. Y. and Zhang, W. M. 2017. Master equation approach to transient quantum transport in nanostructures. *Front. Phys.* 12: 127204.

58. Yan, Y. J., Jin, J. S., Xu, R. X., and Zheng, X. 2016. Dissipation equation of motion approach to open quantum systems. *Front. Phys.* 11: 110306.

10

Thermal Transport in Nanofilms

Yu-Chao Hua, Dao-Sheng Tang,
and Bing-Yang Cao
Tsinghua University

10.1 Introduction

Nanofilm, of which the width ranges from one to several hundred nanometers, is a basic unit of functional devices such as electronic devices (Mishra et al., 2002) and composite nanomaterials (Richardson et al., 2015). In this chapter, we will discuss the thermal transport in nanofilms and mainly focus on thermal transport in nanofilms of dielectric materials where phonon transport dominates.

Due to the very small characteristic size, the requirement that characteristic length is much larger than phonon wavelength and phonon mean free path (MFP) contained in Fourier's law cannot be satisfied anymore in nanofilms. In this case, non-Fourier thermal transport occurs. In detail, when characteristic length of the nanofilm is comparable with phonon MFP and both of them are much larger than phonon wavelength, phonon boundary scatterings and phonon interface scatterings dominate in phonon transport process, and phonon transport is in ballistic-diffusive regime (Majumdar, 1993), which can be well described by phonon Boltzmann transport equation. When the nanofilm is very thin and its characteristic length is comparable with phonon wavelength, the assumption that phonon only shows its particle property should be modified and phonon wave property dominates (Luckyanova et al., 2012). At this condition, phonon coherence has to be considered in thermal transport. Concretely, the effect of phonon phase position in phonon–phonon interactions could not be ignored, and this effect may influence the phonon spectrum, phonon dispersion relation and other related phonon properties.

For ballistic-diffusive thermal transport, many non-Fourier phenomena emerge, such as boundary temperature jump (Hua and Cao, 2014), boundary heat flux slip (Hua and Cao, 2017b), size dependence as well as anisotropy of effective thermal conductivity (Hua and Cao, 2016b, 2017c), and thermal waves (Tang et al., 2016a,b; Tang and Cao, 2017a,b). Moreover, the ways to measure thermal conductivity influence the effective thermal conductivity values (Hua and Cao, 2016a). Understanding these phenomena is very helpful to tune thermal transport in nanostructures. One of the excellent examples is tuning thermal conductivity of nanofilms by interface. Different from that interface always adds the thermal resistance in thermal transport of Fourier's framework, the interface can both increase and decease the thermal conductivity of the nanofilm systems, that is, two-way tuning, when phonon transport is in ballistic-diffusive regime (Hua and Cao, 2018).

Here, five topics relevant to nanofilm thermal transport are discussed, and they are *multiply-constraint nanofilms, nanofilms with internal heat source (IHS), periodic nanoporous films, ultra-fast thermal transport in nanofilms,* and *interface-based two-way tuning of nanofilm thermal conductivity.* We believe that those topics could have broad interests for research community and potential applications in electronic device cooling and thermoelectrics.

10.2 Multiply-Constraint Nanofilms

There have been many studies on the size-dependent thermal conductivity of nanofilms with some simple boundary constraints, such as the in-plane and cross-plane nanofilms (Liu and Asheghi, 2004; Hua and Cao, 2016b; Wolf et al., 2014). However, in practice, a nanofilm can have more than one boundary constraints, for example, the fin structure within FinFETs (Pham et al., 2006). In this sense, the study on the thermal transport process within nanofilms needs to be extended to the multiply-constraint case. In fact, different boundary constraints can impact on the thermal transport in different ways. In the direction parallel to the heat flow, the phonon ballistic transport can cause temperature jumps at the boundaries in contact with the phonon baths. By contrast, for the lateral boundary constraint, the heat flux is reduced near the boundaries due to diffusive phonon-boundary scatterings.

Figure 10.1 illustrates a representative geometry of multiply-constrained nanofilms, which can reduce to other typical structures, that is, the cross-plane and in-plane nanofilms. The left and right ends are in contact with the hot (T_1) and cold (T_2) phonon baths, respectively. A heat flux, q, caused by the temperature difference (TD) is along the x direction, and the length in this direction is denoted by L_x. The lateral boundaries are adiabatic, and the phonons scatter at them with a specular scattering rate, P. In this case, the effective thermal conductivity should depend on both the longitudinal length, L_x, and the lateral length, L_y.

Actually, the thermal conductivity of a multiply-constrained nanostructure can be calculated by solving the phonon Boltzmann transport equation (BTE) (Peraud and Hadjiconstantinou, 2012; Hua and Cao, 2017d). However, it is rather difficult to derive an analytical solution of the phonon BTE when simultaneously considering multiple boundary constraints. Another way to solve this problem is to deal with the boundary constraints separately and then combine their effects on the effective thermal conductivity together (Hua and Cao, 2016b).

The total thermal resistance of the multiply-constraint nanofilms shown in Figure 10.1 can be written as

$$R_t = \frac{L_x}{S\kappa_{\text{eff}}} = R_0 + R_x + R_l, \quad (10.1)$$

in which, S is the cross-sectional area, R_0 is the intrinsic thermal resistance, $R_0 = L_x/S\kappa_0$, R_x is the thermal resistance merely caused by the longitudinal boundary constraint, and R_l is the thermal resistance merely caused by the lateral boundary constraint. The thermal resistances caused by boundary constraints can be derived from the phonon BTE separately and then be combined using Eq. (10.1) to derive the effective thermal conductivity.

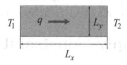

FIGURE 10.1 A representative geometry of multiply-constrained nanofilms.

10.2.1 Longitudinal Boundary Constraint

First consider the longitudinal constraint. In this case, the multiply-constraint nanofilm shown in Figure 10.1 is reduced to the cross-plane nanofilm, with the corresponding one-dimensional BTE given by

$$\mu v_{g\omega} \frac{\partial f_\omega}{\partial x} = \frac{f_{0\omega} - f_\omega}{\tau_\omega} \quad (10.2)$$

in which, f_ω is the phonon distribution function, f_0 is the equilibrium distribution function, v_g is the phonon group velocity, and τ is the relaxation time. Using the differential approximation with boundary temperature jump conditions, the model for the cross-plane effective thermal conductivity can be derived (Majumdar, 1993)

$$\frac{\kappa_{\text{eff\_cr}}}{\kappa_0} = \frac{1}{1 + \frac{4}{3}Kn_x} \quad (10.3)$$

with the Knudsen number, $Kn_x = l_0/L_x$, where l_0 is the phonon MFP. Thus, the total thermal resistance of the cross-plane nanofilms is derived,

$$R_{cr} = \frac{L_x}{S\kappa_{\text{eff\_cr}}} = R_0 \left(1 + \frac{4}{3}Kn_x\right) = R_0 + R_x. \quad (10.4)$$

Therefore, the thermal resistance merely caused by the longitudinal boundary constraint is obtained from Eq. (10.4), that is,

$$R_x = R_{cr} - R_0 = \frac{4}{3}Kn_x R_0. \quad (10.5)$$

10.2.2 Lateral Boundary Constraint

In this case, the total thermal resistance becomes

$$R_{in} = \frac{L_x}{S\kappa_{\text{eff\_in}}} = R_0 + R_l \quad (10.6)$$

The corresponding phonon BTE for the lateral boundary constraints is given by

$$v_{g\omega x} \frac{\partial f_{0\omega}}{\partial T} \frac{\partial T}{\partial x} + v_{g\omega y} \frac{\partial \Delta f_\omega}{\partial y} + v_{g\omega z} \frac{\partial \Delta f_\omega}{\partial z} = -\frac{\Delta f_\omega}{\tau_\omega} \quad (10.7)$$

with $\Delta f_\omega = f_\omega - f_{0\omega}$. The effective thermal conductivity can be derived from Eq. (10.7), that is (Hua and Cao, 2016b),

$$\frac{\kappa_{\text{eff\_in}}}{\kappa_0} = 1 - \frac{3}{2\pi S} \int_S dS \int_0^{2\pi} \int_{\pi/2}^0$$

$$\times \left[\frac{(1-P)\exp\left(-\frac{|\mathbf{r}-\mathbf{r}_B|}{l_0\sqrt{1-\mu^2}}\right)}{1-P\exp\left(-\frac{|\mathbf{r}-\mathbf{r}_B|}{l_0\sqrt{1-\mu^2}}\right)} \right] \mu^2 d\mu d\varphi$$

$$= \frac{1}{G} \quad (10.8)$$

where $|\mathbf{r} - \mathbf{r}_B|$ is the distance from a specific point $\mathbf{r}$ on the cross section to a point $\mathbf{r}_B$ at the lateral boundary.

For clarity, we denote $\kappa_{\text{eff\_in}}/\kappa_0$ as G^{-1}. Particularly for nanofilms, G^{-1} becomes

$$G_{\text{film}}^{-1} = 1 - \frac{3}{2} Kn_y \int_0^1 \left[1 - \exp\left(-\frac{1}{Kn_y \sqrt{1-\mu^2}} \right) \right] \mu^3 d\mu \tag{10.9}$$

with the Knudsen number, $Kn_y = l_0/L_y$. According to Eq. (10.6), the thermal resistance merely caused by the lateral boundary constraint is obtained

$$R_{\text{l}} = R_0 \left[G(l_0) - 1 \right]. \tag{10.10}$$

10.2.3 Multiply-Constraint Effective Thermal Conductivity

Combination of Eqs. (10.1), (10.5), and (10.10) leads to the total thermal resistance of the multiply-constraint nanofilms

$$R_{\text{t}} = \frac{L_x}{S\kappa_{\text{eff}}} = R_0 + \frac{4}{3} Kn_x R_0 + R_0 \left[G(l_0) - 1 \right]$$
$$= R_0 \left(\frac{4}{3} Kn_x + G(l_0) \right), \tag{10.11}$$

which corresponds to the multiply-constraint effective thermal conductivity

$$\frac{\kappa_{\text{eff}}}{\kappa_0} = \frac{1}{\frac{4}{3} Kn_x + G(l_0)}. \tag{10.12}$$

We note that Eq. (10.12) is in the gray approximation, and it is straightforward to extend the model to the non-gray case, that is,

$$\kappa_{\text{eff}} = \frac{1}{3} \sum_j \int_0^{\omega_{mj}} \hbar\omega \frac{\partial f_{\text{BE}}}{\partial T} v_{g\omega j} l_{Bj}(\omega) \text{DOS}_j(\omega) d\omega \tag{10.13}$$

with

$$l_{Bj} = \frac{l_{\text{int}j}}{\frac{4l_{\text{int}j}}{3L_x} + G(l_{\text{int}j})}.$$

The phonon Monte Carlo (MC) technique is used to directly solve the phonon BTE (Hua and Cao, 2017d). In addition, the effective thermal conductivity model proposed by Alvarez and Jou (2007) is given for comparison

$$\frac{\kappa_{\text{eff}}}{\kappa_0} = \frac{L_{\text{eff}}^2}{2\pi^2 l_0^2} \left(\sqrt{1 + \frac{4\pi^2 l_0^2}{L_{\text{eff}}^2}} - 1 \right) \tag{10.14}$$

where $L_{\text{eff}}^{-2} = L_x^{-2} + L_y^{-2}$ is an effective characteristic length. Figure 10.2 compares the thermal conductivities predicted by the present model and the MC simulations. For a given Kn_x, the effective thermal conductivity decreases with increasing Kn_y and vice versa. The present model agrees better with the MC simulations than Alvarez and Jou's model, because the effective characteristic length in Alvarez and Jou's model assumes that the constraints have the same effect regardless of their different influences on the thermal transport process. Although in some cases, Alvarez

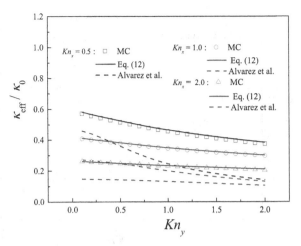

FIGURE 10.2 Effective thermal conductivities predicted by the model, Eq. (10.12), and the MC simulations.

and Jou's model can slightly deviate from the MC simulations, it is also helpful and convenient in the practical applications due to its straightforward physical meaning and simple expression. The agreement between the MC simulations and present model indicates the validity of the present model.

The predictions of the present model are compared with available experimental data for silicon nanofilms at room temperature (Asheghi et al., 1997; Ju and Goodson, 1999; Liu and Asheghi, 2004; Ju, 2005; Hopkins et al., 2011) as shown in Figure 10.3. The experimental data is converted to dimensionless units using a bulk thermal conductivity of 150 W/m K and an MFP for bulk silicon of 210 nm according to references (Chen, 1998). With Kn_x equal to 0, the present model gives the in-plane thermal conductivity of a nanoflim. In this case, according to Figure 10.3a, the predictions by the present model agree well with the experimental data and the MC simulations. By contrast, as Kn_y vanishes, the present model gives the cross-plane thermal conductivity. Figure 10.3b shows that the model also agrees well with the MC simulations, but the experimental data from Hopkins et al. is much lower than the model predictions. A similar result was also reported by McGaughey et al. (2011).

10.3 Nanofilms with IHS

When studying the size dependence of thermal conductivity, it generally assumes that nanofilm is in contact with two heat sinks of different temperatures, resulting in a heat flow through the structure, and thus, the effective thermal conductivity can be calculated using the Fourier's law. Using this TD scheme, the several predictive models of the effective thermal conductivity have been derived from the phonon BTE (Hua and Cao, 2017b; Wolf et al., 2014). Moreover, the TD scheme has been widely adopted in simulations and experiments (Lacroix et al., 2005; Hsiao et al., 2013). In fact, except for the TD scheme, the IHS scheme has another choice (Liu and Asheghi, 2004; Li and Cao, 2011). In the

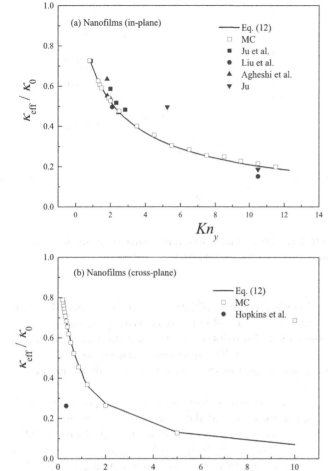

FIGURE 10.3 Comparison of the thermal conductivities predicted by the model, Eq. (10.12), with the experimental data of silicon nanofilms: (a) nanofilms (in-plane) and (b) nanofilms (cross-plane).

IHS scheme, an IHS is introduced within nanofilm, and the resulting temperature rise is calculated or measured; then, the effective thermal conductivity can be obtained by fitting the calculated or measured result with the prediction by the diffusive heat conduction equation. For example, Liu and Asheghi (2004) measured the in-plane thermal conductivity of silicon layers by introducing a steady-state uniform IHS (Joule heating), while the theoretical model obtained from the TD scheme was employed to analyze the experimental data. Although both the TD and IHS schemes have been widely adopted, it is still ambiguous whether the effective thermal conductivity obtained by the TD scheme is the same as that by the IHS scheme even for the identical nanofilm. Li and Cao (2011) studied the effective thermal conductivity of nanotubes with IHS by the nonequilibrium molecular dynamics (MD) simulations and found that the effective thermal conductivity in the IHS scheme was significantly lower than that in the TD scheme.

Phonons emit from the heat sinks at the boundaries in the TD scheme, while in the IHS scheme, phonons originate

within the media. The different phonon-emitting patterns can lead to different boundary-confined effects on phonon transport process. Importantly for the electronic devices where self-heating does exist, the accurate prediction to the effective thermal conductivity of nanofilms with IHS is of essential importance.

10.3.1 TD Scheme

In the TD scheme as shown in Figure 10.4a, the nanofilm is in contact with two heat sinks of temperatures (T_1 and T_0), and the TD is ΔT, causing an x-directional heat flux q. Using the Fourier's law, the effective thermal conductivity is calculated as

$$\kappa_T = \frac{qL_x}{\Delta T}. \tag{10.15}$$

10.3.2 IHS Scheme

The IHS scheme is illustrated in Figure 10.4b. A steady-state uniform IHS $\dot{S}$ is introduced in the nanofilm in contact with two heat sinks of the reference temperature T_0. Therefore, in the diffusive limit, the heat conduction can be regarded as one dimensional, and the temperature profile derived from the Fourier's law is given by

$$T(x) = \frac{\dot{S}}{2\kappa}\left(L_x - x\right)x + T_0 \tag{10.16}$$

The effective thermal conductivity is then extracted from the mean temperature $\overline{\Delta T}$ of the nanofilm

$$\kappa_I = \frac{L_x^2 \dot{S}}{12\overline{\Delta T}} \tag{10.17}$$

with

$$\overline{\Delta T} = \frac{1}{L_x}\int_{L_x} T dx - T_0$$

10.3.3 Phonon BTE-Based Effective Thermal Conductivity Model

The 1D phonon BTE can be applied to characterize the cross-plane phonon transport in nanofilms with IHS

$$\mu v_{g\omega}\frac{\partial f_\omega}{\partial x} = \frac{f_{0\omega} - f_\omega}{\tau_\omega} + \dot{S}_\Omega \tag{10.18}$$

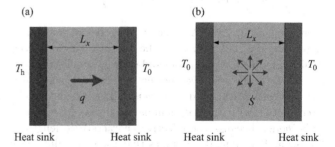

FIGURE 10.4 Schematics of ballistic-diffusive heat conduction in cross-plane nanofilms: (a) TD scheme and (b) IHS scheme.

where $\dot{S}_\Omega$ is the intensity of phonon source. The distirbution function can be divided into two parts

$$f_\omega = f_{\omega\_s} + f_{\omega\_d} \qquad (10.19)$$

where f_{ω_s} is the source-induced part and f_{ω_d} is the diffusive part. The governing equation of f_{ω_s} is given by

$$v_g\mu\frac{\partial f_s}{\partial x} = -\frac{f_s}{\tau} + \dot{S}_\Omega. \qquad (10.20)$$

By contrast, the diffusive part is characterized by

$$\mu v_{g\omega}\frac{\partial f_{\omega\_d}}{\partial x} = \frac{-f_{\omega\_d} + f_{\omega0}}{\tau_\omega}. \qquad (10.21)$$

Solving Eqs. (10.20) and (10.21) with the gray approximation (Hua and Cao, 2017b), an anlytical model for effecctive thermal conductiviy in this case can be derived (Hua and Cao, 2016a),

$$\frac{\kappa_{\text{IHS}}}{\kappa_0} = \frac{1}{1+4Kn_x+\frac{Kn_x^2}{2}\left[1+\exp\left(-\frac{2}{Kn_x}\right)\right]+\frac{Kn_x^3}{2}\left[\exp\left(-\frac{2}{Kn_x}\right)-1\right]}$$
$$\approx \frac{1}{1+4Kn_x}. \qquad (10.22)$$

By contrast, the cross-plane effective thermal conductivity of nanofilms in the TD scheme is given by

$$\frac{\kappa_{\text{DT}}}{\kappa_0} = \frac{1}{1+\frac{4}{3}Kn_x}. \qquad (10.23)$$

Figure 10.5 shows the cross-plane effective thermal conductivity of nanofilms. The effective thermal conductivity decreases with the increasing Kn_x in both the IHS and TD schemes. The effective thermal conductivity in the IHS scheme is significantly lower than that in the TD scheme. Moreover, the results by the present model, Eq. (10.22), well agrees with those by the MC simulations, especially in the regime of small Knudsen number. The models and MC simulations demonstrate that the phonon-emitting patterns can significantly impact on the effective thermal conductivity of nanofilms.

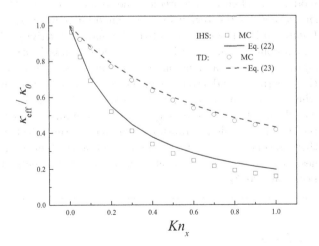

FIGURE 10.5 Cross-plane effective thermal conductivities of nanofilms vs. Kn_x.

10.4 Periodic Nanoporous Films

In this section, we turn to analyze the thermal transport in periodic nanoporous films as shown in Figure 10.6a. It has been found that by etching periodic nanoscale holes in silicon films, the effective thermal conductivity can be greatly reduced without adversely influencing the electrical transport ability, which provides a promising way to develop highly efficient thermoelectric devices (Galli and Donadio, 2010; Lee et al., 2015; Yu et al., 2010). For instance, Yu et al. (2010) found that the in-plane effective thermal conductivity for a 22-nm-thick periodic nanoporous film with pore diameter of 11 nm and period of 34 nm could only be 2 W/(m K), which is about two orders of magnitude smaller than the silicon bulk value (148 W/m K) at room temperature. It has been found that the effective thermal conductivity of nanoporous thin films can significantly reduce as compared to that of macroporous materials and increase with the increasing characteristic lengths, for example, film thickness and pore radius. Moreover, for a 2D periodic nanoporous film, its strong columnar microscopic structure indicates the high anisotropy of the effective thermal conductivity that determines which direction (in-plane or cross-plane) is more efficient for thermoelectric applications. Numerical and experimental results uncovered the variation rules of the effective thermal conductivity of nanoporous silicon structures and promoted the development of relevant theoretical models (Dechaumphai and Chen, 2012; Ravichandran and Minnich, 2014). It should be noted that most of the modeling works at present are based on the phenomenological methodologies and cannot simultaneously take multiple geometrical constraints into account.

10.4.1 Cross-Plane Periodic Nanoporous Thin Films

In the classical Fourier's law-based model for macroporous materials, the effective thermal conductivity, κ_{eff0}, for the given pore shape and heat flux direction, is only a function of porosity, $\kappa_{\text{eff0}} = \kappa_0 F(\varepsilon)$ with $F(\varepsilon)$ describes the porosity dependence of effective thermal conductivity. In the cross-plane case shown in Figure 10.6b, according to the effective medium approach (EMA) (Nan et al., 1997), we have

$$F = 1 - \varepsilon$$
$$\kappa_{\text{eff0\_cr}} = [1 - \varepsilon]\kappa_0, \qquad (10.24)$$

which indicates that the cross-plane effective thermal conductivity linearly decreases with the increasing porosity. By contrast, for nanoporous thin films where the scale of temperature gradient can be comparable to the phonon MFP, ballistic transport also leads to the thickness dependence of effective thermal conductivity. Therefore, the effective thermal conductivity of nanoporous thin films should be expressed as

$$\kappa_{\text{eff\_cr}} = [1 - \varepsilon]\kappa_{\text{m\_cr}}\left(R_{\text{p}}, \varepsilon, L_x\right), \qquad (10.25)$$

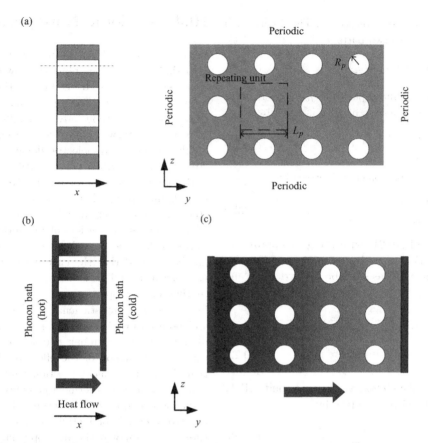

FIGURE 10.6 (a) Two-dimensional periodic silicon nanoporous film: the period is denoted by L_p and the pore radius is R_p. (b) Cross-plane heat conduction: the heat flow is along pore axis (along x-direction). (c) In-plane heat conduction: the heat flow is perpendicular to pore axis (along y-direction). (Reprinted (adapted) with permission from Hua, Y.-C & Cao, B.-Y 2017c. Anisotropic heat conduction in two-dimensional periodic silicon nanoporous films. *Journal of Physical Chemistry* C, 121, 5293–5301. Copyright (2018) American Chemical Society.)

in which the thermal conductivity of host material, $\kappa_{m\_cr}(R_p, \varepsilon, L_x)$, depends on the pore radius, the porosity, and the film thickness simultaneously. An analytical model for $\kappa_{m\_cr}(R_p, \varepsilon, L_x)$ can be derived from the phonon BTE (Hua and Cao, 2017a),

$$\frac{\kappa_{m\_cr}}{\kappa_0} = \frac{1}{G_{NP\_cr} + \frac{4}{3}Kn_x}, \qquad (10.26)$$

in which the Knudsen number Kn_x characterizes the thickness dependence, and the function, G_{NP_cr}, describes the effect of pore boundary scattering on κ_{m_cr}. G_{NP_cr} is given by

$$G_{NP\_cr}^{-1} = 1 - \frac{3}{\pi R_p^2(\varepsilon^{-1} - 1)} \int_{R_p}^{R_p/\sqrt{\varepsilon}} r\,dr \int_0^{2\pi} \int_0^1$$
$$\times \exp\left(-\frac{L_{rp}}{l_0\sqrt{1-\mu^2}}\right) \mu^2\,d\mu\,d\varphi \qquad (10.27)$$

with

Figure 10.7a shows the effective thermal conductivity of nanoporous thin films with the given porosity ($\varepsilon = 0.5$) as a function of film thickness (Kn_x). With the increasing Kn_x, the scale of temperature gradient becomes comparable to the phonon MFP, and the influence of ballistic transport reduces the effective thermal conductivity. The model can slightly overpredict the results by the MC simulations in the region of large Knudsen number Kn_x. Figure 10.7b shows the effective thermal conductivity of nanoporous thin films with the given pore radius ($Kn_R = l_0/R_p = 1.0$) as a function of film thickness. The effective thermal conductivity can be significantly reduced when compared to the results predicted by the EMA model. In the region of larger Knudsen number where the phonon ballistic transport dominates, the effective thermal conductivity is mainly dependent on the film thickness, which has been demonstrated in the experiments of Lee et al. (2015).

$$L_{rp} = \begin{cases} r\left[|\cos\varphi| - \left(\left(\frac{R_p}{r}\right)^2 - \sin^2\varphi\right)^{1/2}\right] & \pi - \varphi_{c1} \leq \varphi \leq \pi + \varphi_{c1} \\ \frac{R_p}{\sqrt{\varepsilon}}\frac{\sin(\varphi-\varphi_2)}{\sin(\varphi)} + \frac{R_p}{\sqrt{\varepsilon}}\left[|\cos\varphi_2| - \left(\varepsilon - \sin^2\varphi_2\right)^{1/2}\right] & 0 \leq \varphi \leq \varphi_{c1} \text{ or } 2\pi - \varphi_{c1} < \varphi \leq 2\pi \end{cases}$$

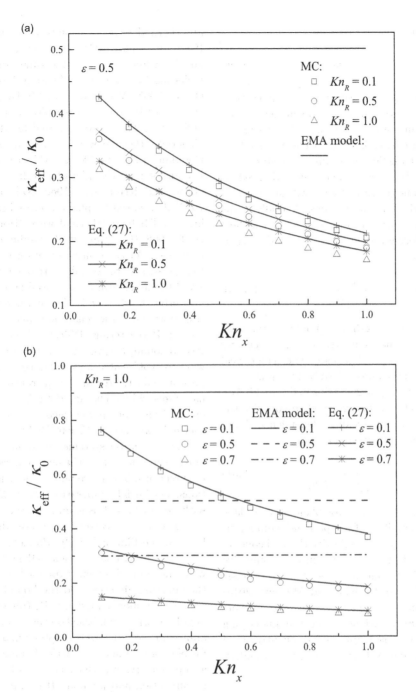

FIGURE 10.7 (a) Cross-plane effective thermal conductivity of nanoporous thin films with different pore radii ($Kn_R = 0.1, 0.5, 1.0$) and a given porosity ($\varepsilon = 0.5$) along the pore axial direction as a function of thickness (Kn_x) and (b) cross-plane effective thermal conductivity of nanoporous thin films with different porosities ($\varepsilon = 0.1, 0.5, 0.7$) and a given pore radius ($Kn_R = 1.0$) along the pore axial direction as a function of thickness (Kn_x).

In fact, the function, G_{NP_cr} can be simplified to a form derived from Matthiessen's rule (Hua and Cao, 2017c), that is,

$$G_{NP\_cr} \to 1 + \frac{l_0}{\alpha_{cr} L_{ch}} \tag{10.28}$$

where L_{ch} the characteristic length that depends on the period length L_p and the pore radius R_p, and α_{cr} is the geometrical factor. In this case, the characteristic length can be set as the neck length, $L_p - 2R_p$ as the neck length, and the geometrical factor can be obtained by fitting with

the above model's predictions or the simulation results. According to Hua and Cao (2017c), the geometrical factor for the cross-plane 2D periodic nanoporous structure is $\alpha_{cr} = 4.65$.

10.4.2 In-Plane Periodic Nanoporous Thin Films

At the macroscale, the in-plane effective thermal conductivity, for which the heat flow is perpendicular to the pore axis as shown in Figure 10.6c, is given by

$$\kappa_{\text{eff0\_in}} = \frac{1 - \varepsilon}{1 + \varepsilon}\kappa_0. \qquad (10.29)$$

We note that, when considering the pore boundary scattering, it is not easy to derive an analytical model directly from the phonon BTE in this case. Although Prasher (2006) proposed an approximate ballistic-diffusive effective medium model by adding the ballistic and diffusive resistances, it has been found that this model significantly underpredicts the effective thermal conductivities by the MC simulations, and the maximum deviation is about 40%. Actually, using Matthiessen's rule, the thermal conductivity of the host material merely considering the influence of boundary scattering can be written as

$$\frac{\kappa_{\text{m\_in}}}{\kappa_0} = \frac{1}{1 + \frac{l_0}{\alpha_{\text{in}} L_{\text{ch}}}}, \qquad (10.30)$$

in which the characteristic length L_{ch} should be the neck length, $L_{\text{p}} - 2R_{\text{p}}$, but the geometrical factor becomes different from that in the cross-plane case. The geometrical factor in the in-plane case can be derived by fitting with the MC simulation results and $\alpha_{\text{in}} = 2.25$ (Hua and Cao, 2017c). Furthermore, Eq. (10.30) can be extended to consider the pore boundary scattering and thickness dependence simultaneously using the methodology stated above

$$\frac{\kappa_{\text{m\_in}}}{\kappa_0} = \frac{1}{1 + \frac{l_0}{\alpha_{\text{in}} L_{\text{ch}}} + \frac{4}{3}\text{Kn}_x}. \qquad (10.31)$$

Figure 10.8a and b compares the cross-plane and in-plane effective thermal conductivities of 2D periodic nanoporous films. The strong anisotropy of the effective thermal conductivity can be identified, that is, the in-plane effective thermal conductivity is significantly less than the cross-plane one. The experiments by Kim and Murphy (2015) demonstrated the same phenomenon, but the authors did not give enough explanations about the underlying mechanism. The strong anisotropy should be attributed to both the material removal and the pore boundary scattering.

10.5   Ultra-Fast Thermal Transport in Nanofilms

For transient heat conduction in nanofilms, where the characteristic length and time are comparable to the phonon MFP and relaxation time, respectively, phonons propagate as thermal waves in ballistic-diffusive regime (Joseph and Preziosi, 1989). Studies on thermal waves are mainly focused on theoretical modeling and simulations due to the difficulty in experimental research. In the past decades, several phenomenological models have been proposed, such as the famous C-V model (Cattaneo, 1948; Vernotte, 1958), dual-phase-lagging model (Tzou and Puri, 1997), ballistic-diffusive heat conduction equation (Chen, 2001),

and thermal wave equation based on the thermomass model (Cao and Guo, 2007; Dong et al., 2011; Zhang et al., 2014). Almost at the meantime, simulation approaches, including MD simulation (Tsai and MacDonald, 1976; Kim et al., 2007; Yao and Cao, 2014), MC simulation (Hua et al., 2013), and lattice Boltzmann method (LBM) (Xu and Wang, 2004), are also used for thermal wave research. Generally, thermal wave is often involved in second sound (Landau, 1941; Peshkov, 1944, 1960; Lane et al., 1947; Prohofsky and Krumhansl, 1964; Ackerman et al., 1966; Guyer and Krumhansl, 1966; Narayanamurti and Dynes, 1972) dominated by phonon normal scatterings (N scatterings) and ballistic thermal wave (Hoevers et al., 2005; Du et al., 2008; Hua et al., 2013) dominated by phonon ballistic transport. For the generation of second sound, it is required that $\tau_{\text{N}} \ll t \ll \tau_{\text{R}}$ to make sure of enough N scatterings and avoid the momentum loss caused by resistive scatterings (R scatterings), where t is the characteristic time of thermal transport, τ_{N} is the relaxation time of the N scattering and τ_{R} the R scattering. Different from the time window for second sound, $t \ll \tau_{\text{R}} \ll \tau_{\text{N}}$ is required for ballistic thermal wave to avoid scatterings and make sure that phonons transport in ballistic regime. In ultrafast thermal transport in nanofilms, where characteristic length and time are in the order of nanometers and picoseconds, respectively, ballistic thermal wave occurs (Chen, 2005).

The simulation systems are shown in Figure 10.9. Heat pulse propagates along the normal direction of the boundary (x-direction). There are no confinements in y- and z- directions. The initial temperature, T_0, is 300 K, and a heat pulse with period $t_{\text{p}} = 2$ ps is input to the left side of the film at $t = 0$. Phonons scatter at the boundary and are reflected back to the film diffusively. Phonons emit in two different ways: the directional emission (DE) (all phonons are emitted with the same direction expressed by the angle θ between the emission direction and the boundary normal direction) and the Lambert emission (LE) (phonons are emitted with angular distribution based on the Lambert cosine law). Gray approximation and Debye approximation are adopted here for phonons (Hua et al., 2013; Hua and Cao, 2014). Phonon group velocity, v_{g}, which is set to be the phonon velocity in simulation and phonon MFP, l, are set to be 5,000 m/s and 56.2 nm, respectively (Chen, 2005). The phonon relaxation time is then calculated to be 11.2 ps. The widely adopted heat flux boundary conditions in experimental and numerical studies (Wang et al., 2011; Zhang et al., 2013) are selected in the simulations.

Temperature profiles in sinusoidal pulse case calculated by the MC simulations with LE and DE ($\theta = 0°$) are shown in Figure 10.10, respectively. The C-V model predicts a nondispersive dissipative thermal wave with dispersion relation

$$\omega_t = \frac{\sqrt{3}}{3} v_{\text{g}} k_t, \qquad (10.32)$$

where ω_t and k_t are the frequency and the wave vector of the thermal wave, respectively, and the velocity of the wave front and the wave peak are both equal to $\sqrt{3}v_{\text{g}}/3$. The C-V

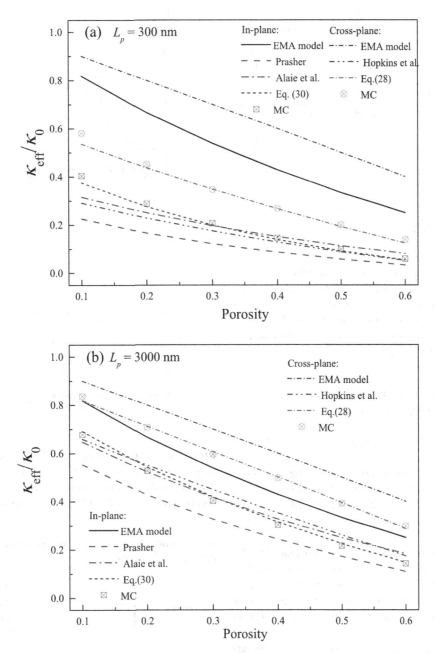

FIGURE 10.8 In-plane and cross-plane effective thermal conductivities of 2D periodic nanoporous silicon films with various periods (L_p = 300, 3,000 nm) as a function of porosity. (Reprinted (adapted) with permission from Hua, Y.-C & Cao, B.-Y 2017c. Anisotropic heat conduction in two-dimensional periodic silicon nanoporous films. *Journal of Physical Chemistry* C, 121, 5293–5301. Copyright (2018) American Chemical Society.)

model predicts a monochromatic dissipative wave with wave velocity and dispersion relation $\omega_t = \sqrt{3}v_g k_t/3$ obtained by making spatial and temporal Fourier transform on it (Tang et al., 2016b). However, the MC simulation with LE predicts a dispersive dissipative thermal wave with dispersion relation

$$\omega_t = v_g \cos(\theta) k_t, 0° < \theta < 90°, \qquad (10.33)$$

and the velocity of the wave front equates to v_g. In the MC simulations, considering the phonon Boltzmann transport

equation without scattering term, we have

$$\frac{\partial f}{\partial t} + v_g \cos(\theta)\frac{\partial f}{\partial x} = 0, \qquad (10.34)$$

By taking spatial and temporal Fourier transform, dispersion relation of Eq. (10.34) is obtained as shown in Eq. (10.33) in which θ distributes according to the Lambert cosine law. Obviously, resulting from the superposition of phonon waves with different direction, ω_t is not the single value function of k_t, which is different from the simple

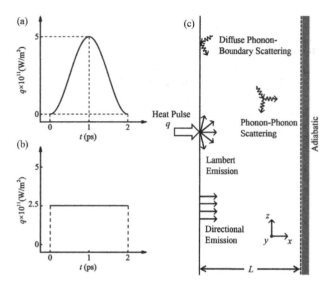

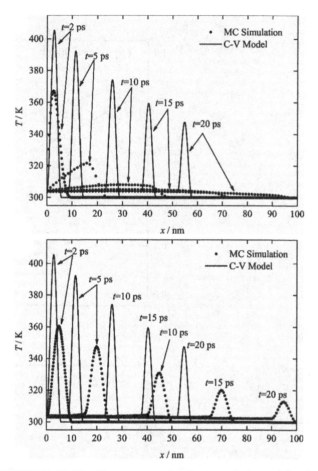

FIGURE 10.9 Schematics of heat flux boundary conditions and simulation system: (a) sinusoidal heat pulse, (b) rectangle heat pulse, and (c) simulation system including the regimes of phonon emission and scattering. (Reproduced from *Journal of Applied Physics* 119, 124301 (2016); doi: 10.1063/1.4944646, with the permission of AIP Publishing.)

dispersion relations. For the MC simulations with DE, $\theta = \theta_0$, the dispersion relation becomes

$$\omega_t = v_g \cos(\theta_0) k_t. \qquad (10.35)$$

Waveform of the heat pulse is not kept during the propagation process and barely influences the shapes of the temperature profiles in MC simulations with LE as shown in Figure 10.10a. By contrast, for the C-V model energy is obviously concentrated near the wave peak leading to higher max temperature in practical applications. In MC simulations, temperature profiles become much different as the phonon emission is changed to be DE ($\theta = 0°$), as illustrated in Figure 10.10b. Heat pulse propagates as a monochromatic dissipative thermal wave with dispersion relation

$$\omega_t = v_g \cos(\theta) k_t, \quad \theta = 0°, \qquad (10.36)$$

and the velocity of the wave front and the wave peak are both equal to v_g.

Energy mean square displacement (MSD) profiles with respect to time can well reflect the phonon transport regimes, which is a very meaningful supplement to phonon dynamic analysis above. The definition of energy MSD is (Zhang and Li, 2005)

$$\sigma^2(t) = \frac{\int (E(x,t) - E_0)(x - x_0)^2 \, dx}{\int (E(x,t) - E_0) \, dx}, \qquad (10.37)$$

where x is the position, x_0 is the position of heat pulse, $E(x,t)$ is the internal energy of the film at time t, and position x and E_0 is the initial internal energy of the thin film. From the MSD–time relations, we can deduce

FIGURE 10.10 Temperature distributions of sinusoidal pulse case calculated by MC simulations and the C-V model. (Reproduced from *Journal of Applied Physics* 119, 124301 (2016); doi: org/10.1063/1.4944646, with the permission of AIP Publishing.)

the regime of phonon transport with the following rule (Zhang and Li, 2005)

$$< \sigma^2(t) > \sim \begin{cases} t, & \text{Diffusive} \\ t^\alpha (1 < \alpha < 2), & \text{Ballistic-diffusive/} \\ & \text{superdiffusive} \\ t^2, & \text{Ballistic} \end{cases}$$
$$(10.38)$$

where $<.>$ means ensemble averaging. MSD-time relations for MC simulation with DE and LE (under excitation of sinusoidal pulse) are shown in Figure 10.11a and b. The indexes in MC simulations with LE, that is, 1.88, 1.68 and 1.58, respectively, are larger than 1 but less than 2 and decrease as time increases, indicating that thermal wave propagates in phonon ballistic-diffusive regime and it evolves from ballistic to diffusive when enough resistive scatterings happen.

Additionally, the superballistic characteristics can emerge in this ultrafast phonon thermal transport process. Prior to discussion, it is necessary to first explain the method employed for calculating the energy MSD during the heat-pulse input process. Since the total energy in the nanofilm increases during the heat input process, Eq. (10.37) cannot

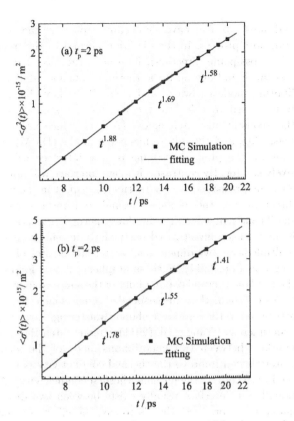

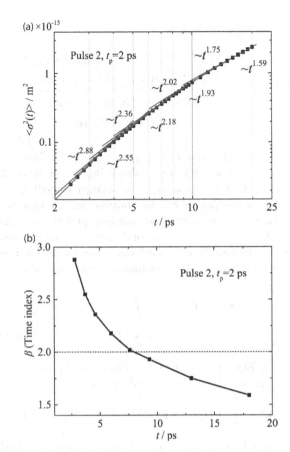

FIGURE 10.11 MSD-time relations in sinusoidal pulse cases by MC simulation with (a) LE and (b) DE.

FIGURE 10.12 Superballistic characteristics obtained after the process of inputting heat pulse 2 into the nanofilm: (a) energy MSD-time relations and (b) time index. (Reproduced from *Applied Physics Letters* 111, 113109 (2017); doi: org/10.1063/1.5003639, with the permission of AIP Publishing.)

reflect the increase of the total energy and the normalization procedure will affect the final results of β. Therefore, we redefine the energy MSD in the input process of heat pulse as

$$\langle \sigma^2(t) \rangle = \int_0^X (E(x,t) - E_0)(x - x_0)^2\, dx, \qquad (10.39)$$

where the normalization procedure has been eliminated.

The energy MSD–time relations calculated after the in put of a sinusoidal heat pulse into the nanofilm, that is, for $t > 2$ ps are shown in Figure 10.12a, with the corresponding values of β shown in Figure 10.12b. Superballistic characteristics are presented with $\beta > 2$ in the early period of heat pulse propagation. Because the time range of the heat pulse propagation process is much greater than that of the heat-pulse input process, we can observe that β changes from a state reflecting superballistic characteristics to that reflecting ballistic-diffusive characteristics, that is, from $\beta < 2$ to $\beta > 2$. The previous investigations related to MSD-time relations have focused on delta-function heat pulse propagation (Zhang and Li, 2005). However, a heat pulse cannot be effectively treated as a delta-function pulse when the heat pulse propagation time is not sufficiently large compared with the heat pulse period t_p.

The effect of a non-negligible value of heat pulse period can also be called the superposition effect of the delta-function heat pulses since the heat pulse with a non-negligible period can be regarded as the superposition of infinite delta-function heat pulses. For clarity, we assume a

process containing the propagation of two identical delta-function heat pulses, where one heat pulse is input into the nanofilm at $t = 0$ and the other is input at $t = t_0$. Then, we consider these two heat pulses collectively and calculate the overall energy MSD-time relations at t ($t > t_0$). For the first heat pulse (with values indicated by subscripts 1), the energy MSD is given as

$$\langle \sigma_1^2(t) \rangle = \frac{\int_0^{X_1} (E_1(x,t) - E_{01})(x - x_{01})^2\, dx}{\int_0^{X_1} (E_1(x,t) - E_{01})\, dx} = 2D_1 t^{\beta_1}. \qquad (10.40)$$

Meanwhile, for the second pulse, this is given as

$$\langle \sigma_2^2(t) \rangle = \frac{\int_0^{X_2} (E_2(x,t) - E_{02})(x - x_{02})^2\, dx}{\int_0^{X_2} (E_2(x,t) - E_{02})\, dx}$$
$$= 2D_2 (t - t_0)^{\beta_2}. \qquad (10.41)$$

D_1 and D_2 are corresponding diffusive coefficients. Then, the energy MSD for both heat pulses collectively is given as

$$\langle \sigma^2(t) \rangle = \frac{\int_0^{\max(X_1, X_2)} (E(x,t) - E_0)(x - x_0)^2\, dx}{\int_0^{\max(X_1, X_2)} (E(x,t) - E_0)\, dx}, \qquad (10.42)$$

which can be written as

$$\langle \sigma^2(t) \rangle = \frac{\int_0^{\max(X_1, X_2)} (E_1(x,t) - E_{01})(x - x_{01})^2 + (E_2(x,t) - E_{02})(x - x_{02})^2 \, dx}{\int_0^{\max(X_1, X_2)} (E_1(x,t) - E_{01}) + (E_2(x,t) - E_{02}) \, dx}.$$

(10.43)

Here, E_{01} and E_{02} are set to $E_{01} = E_{02} = 0$ for simplicity. The location where the heat pulse is input represents the zero point, that is, $x_{01} = x_{02} = 0$. A larger spreading area and greater dissipation will be obtained for the first heat pulse because it propagates for a longer period, such that $X_1 > X_2$ and $\beta_1 < \beta_2$. The total energy of the two heat pulses in the nanofilm are equivalent once both have been completely input into the nanofilm at time t, such that

$$\int_0^{X_1} (E_1(x,t) - E_{01}) \, dx = \int_0^{X_2} (E_2(x,t) - E_{02}) \, dx = \sum E.$$

(10.44)

Based on Eqs. (10.40)–(10.44), we can simplify the energy MSD of these two heat pulses collectively as

$$\langle \sigma^2(t) \rangle = \left(D_1 t^{\beta_1} + D_2 (t - t_0)^{\beta_2} \right) \sim t^{\beta}.$$

(10.45)

Under specific conditions, curve fitting to the energy MSD values given by Eq. (10.45) using a power function of time, t, may lead to $\beta > \beta_2$. As a result, a larger value of β may be obtained owing to the superposition of the delta-function heat pulses. For completely ballistic transport, $\beta_1 = \beta_2 = 2$, such that $\beta > 2$. A dimensionless parameter denoted as the relative superposition time (RST) can be defined as

$$\text{RST} = t_s / t,$$

(10.46)

where t_s is the time of superposition, which is given as t_0 in the case of the propagation of two delta-function heat pulses and t_p in the case of the propagation of a heat pulse with a non-negligible period. For heat pulses with a non-negligible period, RST is always less than 1 in the heat-pulse propagation process and decreases during propagation. For a delta-function heat pulse, RST = 0. The effect of superposition increases with increasing RST and vanishes with RST = 0. Furthermore, to observe the superballistic characteristics in experiments of transient phonon transport, both superposition effect and phonon ballistic-diffusive transport are required, which means that both RST and time Knudsen number ($Kn_t = \tau/t$) should be large enough.

10.6 Interface-Based Two-Way Tuning of Nanofilm Thermal Conductivity

This section will discuss the thermal transport in nanofilms on substrate which widely exists in electronic devices (Moore and Shi, 2014). In this case, interface between nanofilm

and substrate can have significant impact on the thermal transport process. In the direction normal to the interface (the cross-plane direction), interface introduces a thermal resistance and degrades the thermal transport ability. Two limiting models (Swartz and Pohl, 1989) have been long-termly used to calculate the interface resistance, that is, the acoustic mismatch model (AMM), which assumes no scattering, and the diffuse mismatch model (DMM), which assumes the phonons striking on the interface will diffusively scatter. By contrast, although interface will not cause a thermal resistance to the thermal transport in the parallel direction (i.e., the in-plane thermal transport), it can also significantly influence the in-plane thermal transport. This issue has been investigated particularly for single-layer and multiple-layer two-dimensional materials (Seol et al., 2010; Ong and Pop, 2011). Such as graphene, it was found that the in-plane thermal conductivity of the supported graphene is lower than that of suspended one, which has been attributed to the interlayer phonon scattering. However, the experiments of Yang et al. (2011) demonstrated that a vdW interface between two nanoribbons made of the identical materials can improve the thermal conductivity parallel to the interface (i.e., the in-plane thermal conductivity). Additionally, an interface usually exists between two disparate materials in practice, that is to say, the effects of phonon property dissimilarity also impact on the thermal transport. By using MD simulations, Zhang et al. (2015) found that the thermal conductivity of silicene on the substrate could be either enhanced or suppressed by changing the property of the substrate. Indeed, it could still be an open question whether the thermal transport improvement by interface can be achieved even in the case with phonon property dissimilarity at the interface. In fact, the interface-based tuning of the thermal transport should depend on three major factors: (i) interface adhesion energy, (ii) interface roughness, (iii) phonon property dissimilarity. Here, we focus on the interfacial effect on the in-plane thermal transport in the nanofilms.

Figure 10.13 shows a bilayer nanofilm with an interface. A vdW interface is between these two layers. The phonons in layer "a" can scatter on the interface or pass through the interface and enter layer "b". The phonons in layer "b" also undergoes the identical process. The phonon BTEs for the layers labelled by "a" and "b" are given by

$$v_{g\omega y a} \tau_{\omega a} \frac{\partial \Delta f_{\omega a}}{\partial y} + \Delta f_{\omega a} = -v_{g\omega x a} \tau_{\omega a} \frac{\partial f_{0\omega a}}{\partial T} \frac{dT}{dx}$$

(10.47)

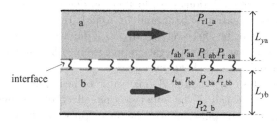

FIGURE 10.13 Schematic for a typical bilayer nanofilm with interface.

and

$$v_{g\omega y b}\tau_{\omega b}\frac{\partial \Delta f_{\omega b}}{\partial y} + \Delta f_{\omega b} = -v_{g\omega x b}\tau_{\omega b}\frac{\partial f_{0\omega b}}{\partial T}\frac{dT}{dx} \quad (10.48)$$

with the interface and boundary conditions

$$\Delta f_{\omega a}\left(L_{ya}, v_{g\omega y a} < 0\right) = P_a \Delta f_{\omega a}\left(L_{ya}, v_{g\omega y a}\right) > 0;$$

$$\Delta f_{\omega a}\left(0, v_{g\omega y a} > 0\right) = P_{ab}\left[r_{aa}\Delta f_{\omega a}\left(0, v_{g\omega y a} < 0\right)\right.$$
$$\left. + t_{ba}\Delta f_{\omega b}(0, v_{g\omega y b} > 0)\right],$$

$$\Delta f_{\omega b}\left(0, v_{g\omega y b} < 0\right) = P_{ab}\left[r_{bb}\Delta f_{\omega b}\left(0, v_{g\omega y b} > 0\right)\right.$$
$$\left. + t_{ab}\Delta f_{\omega a}(0, v_{g\omega y a} < 0)\right];$$

$$\Delta f_{\omega b}\left(-L_{yb}, v_{g\omega y b} > 0\right) = 0. \quad (10.49)$$

In order to describe the phonon transmission and reflection probabilities at the interface, two transmissivities (t_{ab}, t_{ba})

as well as the two corresponding reflectivities (r_{aa}, r_{bb}) were introduced and $t_{ab} = 1 - r_{aa}$, $t_{ba} = 1 - r_{bb}$. According to the phonon transmissivity model by Prasher[19], the influence of the vdW interaction strength (or the interface adhesion energy) can be involved by the phonon transmissivity. The increasing vdW interaction strength can enhance the phonon transmissivity at the interface. Four reflection specularity parameters (P_{r1_a}, P_{r_aa}, P_{r_bb}, P_{r2_b}) are used to describe the phonon reflection mode (specularly or diffusively) at the boundaries and the interface. According to Li and McGaughey (2015), at the interface, the specularity parameter of the transmitted phonons can differ from that of the reflected phonons; thus, two transmission specularity parameters (P_{t_ab}, P_{t_ba}) are introduced to characterize the phonon transmission mode through the interface.

Then, combining Eqs. (10.47)–(10.49), we can have (Hua and Cao, 2018)

$$\kappa_{\text{eff\_a}} = -\frac{1}{(dT/dx)L_{ya}}\int_0^{L_{ya}} q_{xa}(y)dy = \frac{1}{3L_{ya}}\int_0^{\omega_{ma}} v_{g\omega\_a}l_{0\omega a}\omega\hbar\frac{\partial f_{BE}}{\partial T}\text{DOS}(\omega)d\omega$$

$$\int_0^{L_{ya}}\left[\begin{array}{c}1-\\ \frac{3}{4}\int_0^1\left(\begin{array}{c}G_a^+\exp\left(-\frac{y}{l_{0\omega a}\mu}\right)\\ + G_a^-\exp\left(\frac{y}{l_{0\omega a}\mu}\right)\end{array}\right)\left(1-\mu^2\right)d\mu\end{array}\right]dy \quad (10.50)$$

and

$$\kappa_{\text{eff\_b}} = -\frac{1}{(dT/dx)L_{yb}}\int_0^{L_{yb}} q_{xb}(y)dy = \frac{1}{3L_{ya}}\int_0^{\omega_{mb}} v_{g\omega\_b}l_{0\omega b}\omega\hbar\frac{\partial f_{BE}}{\partial T}\text{DOS}(\omega)d\omega$$

$$\int_0^{L_{yb}}\left[\begin{array}{c}1\\ -\frac{3}{4}\int_0^1\left(\begin{array}{c}G_b^+\exp\left(-\frac{y}{l_{0\omega b}\mu}\right)\\ + G_b^-\exp\left(\frac{y}{l_{0\omega b}\mu}\right)\end{array}\right)\left(1-\mu^2\right)d\mu\end{array}\right]dy \quad (10.51)$$

with

$$G_a^+ = \frac{\left\{\begin{array}{l}\left(1-P_{r1\_a}\right)P_{r\_aa}r_{aa}\exp\left(-\frac{L_{ya}}{l_{0\omega a}\mu}\right) + \exp\left(-\frac{L_{yb}}{l_{0\omega b}\mu}\right)\gamma\left(1-P_{r2\_b}\right)P_{t\_ba}t_{ba}\\ + \left(1 - P_{r\_aa}r_{aa} - \gamma P_{t\_ba}t_{ba}\right)\\ - \left(1-P_{r1\_a}\right)P_{r2\_b}\left(P_{r\_aa}P_{r\_bb}r_{aa}r_{bb} - P_{t\_ab}P_{t\_ba}t_{ab}t_{ba}\right)\exp\left(-2\frac{L_{yb}}{l_{0\omega b}\mu} - \frac{L_{ya}}{l_{0\omega a}\mu}\right)\\ + \exp\left(-2\frac{L_{yb}}{l_{0\omega b}\mu}\right)P_{r2\_b}\left(\gamma P_{t\_ba}t_{ba} - P_{r\_bb}r_{bb} + P_{r\_aa}P_{r\_bb}r_{aa}r_{bb} - P_{t\_ab}P_{t\_ba}t_{ab}t_{ba}\right)\end{array}\right\}}{1-\exp\left(-\frac{2L_{ya}}{l_{0\omega a}\mu}\right)P_{r1\_a}P_{r\_aa}r_{aa} - \exp\left(-\frac{2L_{yb}}{l_{0\omega b}\mu}\right)P_{r2\_b}P_{r\_bb}r_{bb}} \\ + P_{r1\_a}P_{r2\_b}\left(P_{r\_aa}P_{r\_bb}r_{aa}r_{bb} - P_{t\_ab}P_{t\_ba}t_{ab}t_{ba}\right)\exp\left(-\frac{2L_{ya}}{l_{0\omega a}\mu} - \frac{2L_{yb}}{l_{0\omega b}\mu}\right)},$$

$$G_a^- = \frac{\exp\left(-\frac{L_{ya}}{l_{0\omega a}\mu}\right)\left\{\begin{array}{l}\left(1-P_{r1\_a}\right) - \exp\left(-2\frac{L_{yb}}{l_{0\omega b}\mu}\right)\left(1-P_{r1\_a}\right)P_{r2\_b}P_{r\_bb}r_{bb}\\ + \exp\left(-\frac{L_{ya}}{l_{0\omega a}\mu} - \frac{L_{yb}}{l_{0\omega b}\mu}\right)\gamma P_{r1\_a}\left(1-P_{r2\_b}\right)P_{t\_ba}t_{ba}\\ + \exp\left(-\frac{L_{ya}}{l_{0\omega a}\mu} - 2\frac{L_{yb}}{l_{0\omega b}\mu}\right)P_{r1\_a}P_{r2\_b}\left[\begin{array}{c}\gamma P_{t\_ba}t_{ba} - P_{r\_bb}r_{bb}+\\ P_{r\_bb}P_{r\_aa}r_{aa}r_{bb} - P_{t\_ab}P_{t\_ba}t_{ab}t_{ba}\end{array}\right]\\ + \exp\left(-\frac{L_{ya}}{l_{0\omega a}\mu}\right)P_{r1\_a}\left(1 - P_{r\_aa}r_{aa} - \gamma P_{t\_ba}t_{ba}\right)\end{array}\right\}}{\left\{\begin{array}{l}1-\exp\left(-\frac{2L_{ya}}{l_{0\omega a}\mu}\right)P_{r\_a}P_{r\_aa}r_{aa} - \exp\left(-\frac{2L_{yb}}{l_{0\omega b}\mu}\right)P_{r2\_b}P_{r\_bb}r_{bb}\\ + P_{r1\_a}P_{r2\_b}\left(P_{r\_aa}P_{r\_bb}r_{aa}r_{bb} - P_{t\_ab}P_{t\_ba}t_{ab}t_{ba}\right)\exp\left(-\frac{2L_{ya}}{l_{0\omega a}\mu} - \frac{2L_{yb}}{l_{0\omega b}\mu}\right)\end{array}\right\}},$$

$$
G_{\mathrm{b}}^{+} = \frac{\exp\left(-\frac{L_{yb}}{l_{0\omega b}\mu}\right)\left\{\begin{array}{l}(1 - P_{\mathrm{r2\_b}}) - \exp\left(-2\frac{L_{ya}}{l_{0\omega a}\mu}\right)P_{\mathrm{r1\_a}}P_{\mathrm{r\_aa}}\left(1 - P_{\mathrm{r2\_b}}\right)r_{\mathrm{aa}}\\[4pt] + \exp\left(-\frac{L_{ya}}{l_{0\omega a}\mu} - \frac{L_{yb}}{l_{0\omega b}\mu}\right)\gamma^{-1}\left(1 - P_{\mathrm{r1\_a}}\right)P_{\mathrm{r2\_b}}P_{\mathrm{t\_ab}}t_{\mathrm{ab}}\\[4pt] + P_{\mathrm{r1\_a}}P_{\mathrm{r2\_b}}\left(\begin{array}{l}\gamma^{-1}P_{\mathrm{t\_ab}}t_{\mathrm{ab}} - P_{\mathrm{r\_aa}}r_{\mathrm{aa}}\\ + P_{\mathrm{r\_aa}}P_{\mathrm{r\_bb}}r_{\mathrm{aa}}r_{\mathrm{bb}} - P_{\mathrm{t\_ab}}P_{\mathrm{t\_ba}}t_{\mathrm{ab}}t_{\mathrm{ba}}\end{array}\right)\exp\left(-2\frac{L_{ya}}{l_{0\omega a}\mu} - \frac{L_{yb}}{l_{0\omega b}\mu}\right)\\[4pt] + \exp\left(-\frac{L_{yb}}{l_{0\omega b}\mu}\right)P_{\mathrm{r2\_b}}\left(1 - P_{\mathrm{r\_bb}}r_{\mathrm{bb}} - \gamma^{-1}P_{\mathrm{t\_ab}}t_{\mathrm{ab}}\right)\end{array}\right\}}{1 - \exp\left(-\frac{2L_{ya}}{l_{0\omega a}\mu}\right)P_{\mathrm{r1\_a}}P_{\mathrm{r\_aa}}r_{\mathrm{aa}} - \exp\left(-\frac{2L_{yb}}{l_{0\omega b}\mu}\right)P_{\mathrm{r2\_b}}P_{\mathrm{r\_bb}}r_{\mathrm{bb}} + P_{\mathrm{r1\_a}}P_{\mathrm{r2\_b}}\left(P_{\mathrm{r\_aa}}P_{\mathrm{r\_bb}}r_{\mathrm{aa}}r_{\mathrm{bb}} - P_{\mathrm{t\_ab}}P_{\mathrm{t\_ba}}t_{\mathrm{ab}}t_{\mathrm{ba}}\right)\exp\left(-\frac{2L_{ya}}{l_{0\omega a}\mu} - \frac{2L_{yb}}{l_{0\omega b}\mu}\right)},
$$

$$
G_{\mathrm{b}}^{-} = \frac{\left\{\begin{array}{l}\exp\left(-\frac{L_{yb}}{l_{0\omega b}\mu}\right)\left(1 - P_{\mathrm{r2\_b}}\right)P_{\mathrm{r\_bb}}r_{\mathrm{bb}}\\[4pt] + \exp\left(-\frac{L_{ya}}{l_{0\omega a}\mu}\right)\gamma^{-1}\left(1 - P_{\mathrm{r1\_a}}\right)P_{\mathrm{t\_ab}}t_{\mathrm{ab}} + \left(1 - P_{\mathrm{r\_bb}}r_{\mathrm{bb}} - \gamma^{-1}P_{\mathrm{t\_ab}}t_{\mathrm{ab}}\right)\\[4pt] - P_{\mathrm{r1\_a}}\left(1 - P_{\mathrm{r2\_b}}\right)\left(P_{\mathrm{r\_aa}}P_{\mathrm{r\_bb}}r_{\mathrm{aa}}r_{\mathrm{bb}} - P_{\mathrm{t\_ab}}P_{\mathrm{t\_ba}}t_{\mathrm{ab}}t_{\mathrm{ba}}\right)\exp\left(-2\frac{L_{ya}}{l_{0\omega a}\mu} - \frac{L_{yb}}{l_{0\omega b}\mu}\right)\\[4pt] + \exp\left(-2\frac{L_{ya}}{l_{0\omega a}\mu}\right)P_{\mathrm{r1\_a}}\left(\gamma^{-1}P_{\mathrm{t\_ab}}t_{\mathrm{ab}} - P_{\mathrm{r\_ab}}r_{\mathrm{aa}} + P_{\mathrm{r\_aa}}P_{\mathrm{r\_bb}}r_{\mathrm{aa}}r_{\mathrm{bb}} - P_{\mathrm{t\_ab}}P_{\mathrm{t\_ba}}t_{\mathrm{ab}}t_{\mathrm{ba}}\right)\end{array}\right\}}{1 - \exp\left(-\frac{2L_{ya}}{l_{0\omega a}\mu}\right)P_{\mathrm{r\_a}}P_{\mathrm{r\_aa}}r_{\mathrm{aa}} - \exp\left(-\frac{2L_{yb}}{l_{0\omega b}\mu}\right)P_{\mathrm{r\_b}}P_{\mathrm{r\_bb}}r_{\mathrm{bb}} + P_{\mathrm{r1\_a}}P_{\mathrm{r2\_b}}\left(P_{\mathrm{r\_aa}}P_{\mathrm{r\_bb}}r_{\mathrm{aa}}r_{\mathrm{bb}} - P_{\mathrm{t\_ab}}P_{\mathrm{t\_ba}}t_{\mathrm{ab}}t_{\mathrm{ba}}\right)\exp\left(-\frac{2L_{ya}}{l_{0\omega a}\mu} - \frac{2L_{yb}}{l_{0\omega b}\mu}\right)}.
$$

In practice, the phonon property dissimilarity can mainly be reflected by the MFP ratio between the materials; for example, for the graphene on the silica substrate (Seol et al., 2010), the MFP ratio between graphene and silica can reach about 100. According to Eqs. (10.50) and (10.51), the MFP ratio, γ, has been involved in the above model. To well clarify the influence of the MFP ratio, it is assumed that layer "b" is made of a virtual material that holds the identical phonon properties as with the material of layer "a" except for its intrinsic MFP. Therefore, we can set the intrinsic MFP of layer "a" fixed and vary that of layer "b" to focus on the influence of the MFP ratio on the in-plane thermal transport within layer "a". It is noted that this model is actually capable to handle the in-plane thermal transport with an interface between two practical materials, but this can make the analysis process become too trivial to clarify the influence of the MFP ratio.

The in-plane thermal conductivities of the nanofilms with the partially specular interface are calculated as the function of the phonon transmissivity. Several researchers highlighted the influence of the phonon transmissivity that can involve the influence of the interface adhesion energy (Prasher, 2009). For instance, Yang et al. (2011) concluded that the emergence of the thermal conductivity improvement by the interfacial effect mainly requires a strong interface adhesion energy, that is, a high phonon transmissivity through the interface. In Figure 10.14a, the Knudsen number is 1, the transmission specularity parameter is 1, the reflection specularity parameters are 0.5, and the MFP ratios are 0.5, 1.0, 1.5, respectively. When the MFP ratio is 1 or 1.5, the thermal conductivity improvement phenomena occurs, and it is enhanced with the increasing phonon transmissivity.

This could be the case that Yang et al. (2011) discussed. By contrast, in the case of the MFP ratio less than 1.0 (such as the multilayer graphene film on a silica substrate (Sadeghi et al., 2013)), the behavior of the thermal conductivity variation becomes much different. In Figure 10.14b, the MFP ratio is set as 0.5, and the reflection specularity parameters are 0.2, 0.5, 0.8, respectively. Both the model and the simulations show that when the reflection specularity parameter is equal to 0.2, the thermal conductivity increases with the increasing phonon transmissivity; however, when the reflection specularity parameter is 0.8, the thermal conductivity can decrease with the increasing phonon transmissivity. Therefore, both the model and the MC simulations indicate that the two-way tuning of the in-plane thermal transport can be achieved using interfacial effect.

10.7   Conclusions

1. For the multiply-constraint nanofilms, different boundary constraints have different impacts on the thermal transport process. Also, it has been proven that boundary constraints can be dealt with separately and then combined to derive the effective thermal conductivity in this case.

2. Except for size and geometry dependence, the effective thermal conductivity of a nanofilm can also depend on its heating conditions. Both the simulations and models demonstrate that the effective thermal conductivity of a cross-plane nanofilm obtained by IHS scheme can be lower than that obtained by TD scheme.

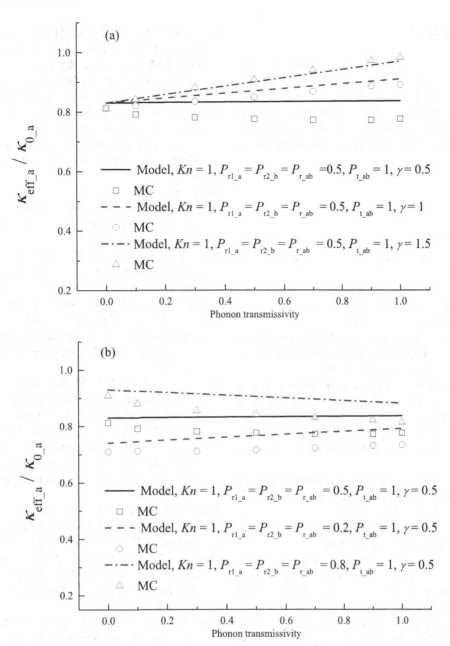

FIGURE 10.14 In-plane thermal conductivity of nanofilms with the partially specular interface as a function of the phonon transmissivity: (a) the transmission specularity parameters are 1; the reflection specularity parameters are 0.5 and the MFP ratios are 0.5, 1.0 and 1.5, respectively and (b) the transmission specularity parameters are 1; the MFP ratios are 0.5, and the reflection specularity parameters are 0.2, 0.5 and 0.8, respectively.

3. The effective thermal conductivity of periodic nanoporous films is simultaneously dependent on the thickness and the characteristic length of pore structure. Additionally, the thermal transport in the 2D periodic nanoporous films holds a significant anisotropy due to the anisotropic pore structure, which can be well characterized using the models derived based on Matthiessen's rule.

4. In nanofilms, ultrashort heat pulse propagates with a finite velocity as ballistic thermal wave, which is further confirmed using energy MSD relations with respect to time. And, we have demonstrated that transient phonon ballistic-diffusive transport can include the superballistic characteristics both during and after the input of a heat pulse into a nanofilm.

5. Interface between nanofilm and substrate significantly impacts on the thermal transport process and, thus, can be used to effectively tune the effective thermal conductivity of nanofilms. It is found that the interface can achieve a two-way (i.e., enhanced or reduced) tuning of the in-plane effective thermal conductivity of a nanofilm on substrate.

Acknowledgments

This work is financially supported by National Natural Science Foundation of China (No. 51825601, 51676108), the Initiative Postdocs Supporting Program of China Postdoctoral Science Foundation (No. BX20180155), Project funded by China Postdoctoral Science Foundation (No. 2018M641348), Science Fund for Creative Research Group (No. 51321002).

References

Ackerman, C. C., Bertman, B., Fairbank, H. A. & Guyer, R. 1966. Second sound in solid helium. *Physical Review Letters*, 16, 789.

Alvarez, F. X. & Jou, D. 2007. Memory and nonlocal effects in heat transport: From diffusive to ballistic regimes. *Applied Physics Letters*, 90, 083109.

Asheghi, M., Leung, Y. K., Wong, S. S. & Goodson, K. E. 1997. Phonon-boundary scattering in thin silicon layers. *Applied Physics Letters*, 71, 1798–1800.

Cao, B.-Y. & Guo, Z. 2007. Equation of motion of a phonon gas and non-Fourier heat conduction. *Journal of Applied Physics*, 102, 053503.

Cattaneo, C. 1948. Sulla conduzione del calore. *Atti del Seminario Fisico Matematicoe dell'Uniersità di Modena*, 3, 83–101.

Chen, G. 1998. Thermal conductivity and ballistic-phonon transport in the cross-plane direction of superlattices. *Physical Review B*, 57, 14958–14973.

Chen, G. 2001. Ballistic-diffusive heat-conduction equations. *Physical Review Letters*, 86, 2297.

Chen, G. 2005. *Nanoscale Energy Transport and Conversion: A Parallel Treatment of Electrons, Molecules, Phonons, and Photons*. New York: Oxford University Press.

Dechaumphai, E. & Chen, R. 2012. Thermal transport in phononic crystals: The role of zone folding effect. *Journal of Applied Physics*, 111, 073508.

Dong, Y., Cao, B.-Y. & Guo, Z. 2011. Generalized heat conduction laws based on thermomass theory and phonon hydrodynamics. *Journal of Applied Physics*, 110, 063504.

Du, X., Skachko, I., Barker, A. & Andrei, E. Y. 2008. Approaching ballistic transport in suspended graphene. *Nature Nanotechnology*, 3, 491.

Galli, G. & Donadio, D. 2010. Thermoelectric materials: Silicon stops heat in its tracks. *Nature Nanotechnology*, 5, 701–702.

Guyer, R. & Krumhansl, J. 1966. Thermal conductivity, second sound, and phonon hydrodynamic phenomena in nonmetallic crystals. *Physical Review*, 148, 778.

Hoevers, H., Ridder, M., Germeau, A., Bruijn, M., de Korte, P. & Wiegerink, R. 2005. Radiative ballistic phonon transport in silicon-nitride membranes at low temperatures. *Applied Physics Letters*, 86, 251903.

Hopkins, P. E., Reinke, C. M., Su, M. F., Olsson, R. H., Shaner, E. A., Leseman, Z. C., Serrano, J. R., Phinney, L. M. & El-Kady, I. 2011. Reduction in the thermal conductivity of single crystalline silicon by phononic crystal patterning. *Nano Letters*, 11, 107–112.

Hsiao, T. K., Chang, H. K., Liou, S. C., Chu, M. W., Lee, S. C. & Chang, C. W. 2013. Observation of room-temperature ballistic thermal conduction persisting over 8.3 microm in SiGe nanowires. *Nature Nanotechnology*, 8, 534–538.

Hua, Y.-C. & Cao, B.-Y., 2014. Phonon ballistic-diffusive heat conduction in silicon nanofilms by Monte Carlo simulations. *International Journal of Heat Mass Transfer*, 78, 755–759.

Hua, Y.-C. & Cao, B.-Y. 2016a. The effective thermal conductivity of ballistic–diffusive heat conduction in nanostructures with internal heat source. *International Journal of Heat and Mass Transfer*, 92, 995–1003.

Hua, Y.-C. & Cao, B.-Y. 2016b. Ballistic-diffusive heat conduction in multiply-constrained nanostructures. *International Journal of Thermal Sciences*, 101, 126–132.

Hua, Y.-C. & Cao, B.-Y. 2017a. Cross-plane heat conduction in nanoporous silicon thin films by phonon Boltzmann transport equation and Monte Carlo simulations. *Applied Thermal Engineering*, 111, 1401–1408.

Hua, Y.-C. & Cao, B.-Y. 2017b. Slip boundary conditions in ballistic–diffusive heat transport in nanostructures. *Nanoscale and Microscale Thermophysical Engineering*, 21, 159–176.

Hua, Y.-C. & Cao, B.-Y. 2017c. Anisotropic heat conduction in two-dimensional periodic silicon nanoporous films. *Journal of Physical Chemistry C*, 121, 19080–19086.

Hua, Y.-C. & Cao, B.-Y. 2017d. An efficient two-step Monte Carlo method for heat conduction in nanostructures. *Journal of Computational Physics*, 342, 253–266.

Hua, Y.-C. & Cao, B.-Y. 2018. Interface-based two-way tuning of the in-plane thermal transport in nanofilms. *Journal of Applied Physics*, 123, 114304.

Hua, Y.-C, Dong, Y. & Cao, B.-Y. 2013. Monte Carlo simulation of phonon ballistic diffusive heat conduction in silicon nanofilm. *Acta Physica Sinica*, 62, 244401.

Joseph, D. D. & Preziosi, L. 1989. Heat waves. *Review of Modern Physics*, 61, 41.

Ju, Y. S. 2005. Phonon heat transport in silicon nanostructures. *Applied Physics Letters*, 87, 153106.

Ju, Y. S. & Goodson, K. E. 1999. Phonon scattering in silicon films with thickness of order 100 nm. *Applied Physics Letters*, 74, 3005–3007.

Kim, K. & Murphy, T. E. 2015. Strong anisotropic thermal conductivity of nanoporous silicon. *Journal of Applied Physics*, 118, 154304.

Kim, T., Osman, M. A., Richards, C. D., Bahr, D. F. & Richard, R. F. 2007. Molecular dynamic simulation of heat pulse propagation in multiwall carbon nanotubes. *Physical Review B*, 76, 155424.

Lacroix, D., Joulain, K. & Lemonnier, D. 2005. Monte Carlo transient phonon transport in silicon and germanium at nanoscales. *Physical Review B,* 72, 064305.

Landau, L. 1941. Theory of the superfluidity of helium II. *Physical Review,* 5, 71.

Lane, C., Fairbank, H. A. & Fairbank, W. M. 1947. Second sound in liquid helium II. *Physical Review,* 71, 600.

Lee, J., Lim, J. & Yang, P. 2015. Ballistic phonon transport in holey silicon. *Nano Letters,* 15, 3273–3279.

Li, D. & McGaughey, A. J. H. 2015. Phonon dynamics at surfaces and interfaces and its implications in energy transport in nanostructured materials—an opinion paper. *Nanoscale and Microscale Thermophysical Engineering,* 19, 166–182.

Li, Y.-W. & Cao, B.-Y. 2011. Thermal conductivity of single-walled carbon nanotube with internal heat source studied by molecular dynamics simulation. *International Journal of Thermophysics,* 34, 2361–2370.

Liu, W. & Asheghi, M. 2004. Phonon–boundary scattering in ultrathin single-crystal silicon layers. *Applied Physics Letters,* 84, 3819–3821.

Luckyanova, M. N., Garg, J., Esfarjani, K., Jandl, A., Bulsara, M. T., Schmidt, A. J., Minnich, A. J., Chen, S., Dresselhaus, M. S. & Ren, Z. 2012. Coherent phonon heat conduction in superlattices. *Science,* 338, 936–939.

Majumdar, A. 1993. Microscale heat conduction in dielectric thin films. *Journal of Heat Transfer,* 115, 7–16.

McGaughey, A. J. H., Landry, E. S., Sellan, D. P. & Amon, C. H. 2011. Size-dependent model for thin film and nanowire thermal conductivity. *Applied Physics Letters,* 99, 131904.

Mishra, U. K., Parikh, P. & Wu, Y. 2002. AlGaN/GaN HEMTs-an overview of device operation and applications. *Proceedings of the IEEE,* 90, 1022–1031.

Moore, A. L. & Shi, L. 2014. Emerging challenges and materials for thermal management of electronics. *Materials Today,* 17, 163–174.

Nan, C.-W., Birringer, R., Clarke, D. R. & Gleiter, H. 1997. Effective thermal conductivity of particulate composites with interfacial thermal resistance. *Journal of Applied Physics,* 81, 6692–6699.

Narayanamurti, V. & Dynes, R. 1972. Observation of second sound in bismuth. *Physical Review Letters,* 28, 1461.

Ong, Z.-Y. & Pop, E. 2011. Effect of substrate modes on thermal transport in supported graphene. *Physical Review B,* 84, 075471.

Peraud, J. P. M. & Hadjiconstantinou, N. G. 2012. An alternative approach to efficient simulation of micro/nanoscale phonon transport. *Applied Physics Letters,* 101, 205331.

Peshkov, V. 1944. The second sound in Helium II. *Journal of Physics (USSR),* 8, 381.

Peshkov, V. 1960. Second sound in helium II. *Journal of Soviet Physics,* 11, 580.

Pham, D., Larson, L. & Ji-Woon, Y. 2006. FINFET device junction formation challenges. *2006 International Workshop on Junction Technology,* Shanghai, China 73–77.

Prasher, R. 2006. Transverse thermal conductivity of porous materials made from aligned nano- and micro-cylindrical pores. *Journal of Applied Physics,* 100, 064302.

Prasher, R. 2009. Acoustic mismatch model for thermal contact resistance of van der Waals contacts. *Applied Physics Letters,* 94, 041905.

Prohofsky, E. & Krumhansl, J. 1964. Second-sound propagation in dielectric solids. *Physical Review,* 133, A1403.

Ravichandran, N. K. & Minnich, A. J. 2014. Coherent and incoherent thermal transport in nanomeshes. *Physical Review B,* 89, 205432.

Richardson, J. J., Bj Rnmalm, M. & Caruso, F. 2015. Technology-driven layer-by-layer assembly of nanofilms. *Science,* 348, aaa2491.

Sadeghi, M. M., Jo, I. & Shi, L. 2013. Phonon-interface scattering in multilayer graphene on an amorphous support. *Proceedings of the National Academy of Sciences of the United States of America,* 110, 16321–16326.

Seol, J. H., Jo, I., Moore, A. L., Lindsay, L. & Aitken, Z. H. 2010. Two-dimensional phonon transport in supported graphene. *Science,* 328, 213–216.

Swartz, E. T. & Pohl, R. O. 1989. Thermal boundary resistance. *Reviews of Modern Physics,* 61, 605–668.

Tang, D.-S. & Cao, B.-Y. 2017a. Ballistic thermal wave propagation along nanowires modeled using phonon Monte Carlo simulations. *Applied Thermal Engineering,* 117, 609–616.

Tang, D.-S. & Cao, B.-Y. 2017b. Superballistic characteristics in transient phonon ballistic-diffusive transport. *Applied Physics Letters,* 111, 113109.

Tang, D.-S., Hua, Y. & Cao, B.-Y. 2016a. Thermal wave propagation through nanofilms in ballistic-diffusive regime by Monte Carlo simulations. *International Journal of Thermal Sciences,* 109, 81–89.

Tang, D.-S., Hua, Y.-C., Nie, B.-D. & Cao, B.-Y. 2016b. Phonon wave propagation in ballistic-diffusive regime. *Journal of Applied Physics,* 119, 793.

Tsai, D. & MacDonald, R. 1976. Molecular-dynamical study of second sound in a solid excited by a strong heat pulse. *Physical Review B,* 14, 4714.

Tzou, D. & Puri, P. 1997. Macro-to microscale heat transfer: The lagging behavior. *Applied Mechanics Reviews,* 50, B82.

Vernotte, P. 1958. Les paradoxes de la theorie continue de l'equation de la chaleur. *Journal of Computer Rendus,* 246, 3154–3155.

Wang, H., Ma, W., Zhang, X., Wang, W. & Guo, Z. 2011. Theoretical and experimental study on the heat transport in metallic nanofilms heated by ultra-short pulsed laser. *International Journal of Heat Mass Transfer,* 54, 967–974.

Wolf, S., Neophytou, N. & Kosina, H. 2014. Thermal conductivity of silicon nanomeshes: Effects of porosity and roughness. *Journal of Applied Physics,* 115, 718–721.

Xu, J. & Wang, X. 2004. Simulation of ballistic and non-Fourier thermal transport in ultra-fast laser heating. *Journal of Physica B: Condensed Matter,* 351, 213–226.

Yang, J., Yang, Y., Waltermire, S. W., Wu, X., Zhang, H., Gutu, T., Jiang, Y., Chen, Y., Zinn, A. A., Prasher, R., Xu, T. T. & Li, D. 2011. Enhanced and switchable nanoscale thermal conduction due to van der Waals interfaces. *Nature Nanotechnology,* 7, 91–95.

Yao, W. & Cao, B.-Y 2014. Thermal wave propagation in graphene studied by molecular dynamics simulations. *Chinese Science Bulletin,* 59, 3495–3503.

Yu, J. K., Mitrovic, S., Tham, D., Varghese, J. & Heath, J. R. 2010. Reduction of thermal conductivity in phononic nanomesh structures. *Nature Nanotechnology,* 5, 718–721.

Zhang, G. & Li, B. 2005. Anomalous vibrational energy diffusion in carbon nanotubes. *The Journal of Chemical Physics,* 123, 014705.

Zhang, M., Cao, B.-Y & Guo, Y. 2013. Numerical studies on dispersion of thermal waves. *International Journal of Heat Mass Transfer,* 67, 1072–1082.

Zhang, M., Cao, B.-Y. & Guo, Y. 2014. Numerical studies on damping of thermal waves. *International Journal of Thermal Sciences,* 84, 9–20.

Zhang, X., Bao, H. & Hu, M. 2015. Bilateral substrate effect on the thermal conductivity of two-dimensional silicon. *Nanoscale,* 7, 6014–6022.

Thermal Transport and Phonon Coherence in Phononic Nanostructures

Juan Sebastian Reparaz
Institut de Ciència de Materials de Barcelona-CSIC

Markus R. Wagner
Technische Universität Berlin

11.1 Introduction

In the last century, a large diversity of research fields emerged targeting the fundamental and applied properties of the collective vibrations of lattice atoms of materials (phonons). Many works were driven by the discovery of inelastic light scattering as theoretically predicted by Brillouin[1] and experimentally shown by Raman[2] in the 1920s. After these discoveries, many works focused on unraveling the detailed structure of the vibrational spectrum of materials. As phonons are the main heat carriers in most nonmetallic materials, the study of the phononic properties of materials is closely linked to the study of the thermal properties and the propagation of heat. Although heat transport has been investigated for centuries, it is only with the advent of nanotechnology that advanced techniques have been developed to study its properties at the nanoscale. On macroscopic length scales and at temperatures over the Debye temperature (i.e., large phonon occupation), heat transport is usually explained using Fourier's law, $Q = -k\nabla T$, where Q is the heat flux, κ is the thermal conductivity, and T is the temperature. However, at the nanoscale or at low temperatures (low-phonon occupation), deviations from this classic behavior occur due to the ballistic component of heat propagation, i.e., when the typical dimensions of the ρ material are comparable to the thermal phonon mean free path Λ_{th}, which defines the average distance that thermal phonons can travel without being scattered.

In recent years, a fundamental problem in our understanding of heat has become apparent, partly driven by the successful fabrication of nanoscale devices for which heat plays a fundamental role. We were forced to address the problem of heat propagation through interfaces, at nanoscale hotspots, and at spatial scales below the thermal phonon mean free path Λ_{th}, i.e., where the heat flux Q is far from an equilibrium situation. In other words, temporal scales prior to the average thermal phonon relaxation time (τ_{th}) have to be considered. This nonequilibrium situation complicates the problem of addressing heat propagation.

Understanding the fundamentals of thermal transport at the nanoscale remains an open challenge for several reasons. On the technological front, the ability to improve the control of heat propagation would result in more efficient thermal designs, e.g., thermal insulation,[3] optical phase-change memory devices,[4] and thermoelectric energy conversion.[5] On the other hand, from a purely scientific perspective, aspects, such as the transition from the ballistic to the diffusive thermal transport regimes or the spectral distribution of the phonon mean free path, are examples where a deeper understanding is necessary. Partially, this lack of understanding probably arises from the technical limitations imposed by the experimental methods available to study thermal transport. For example, the concept of a phonon spectrometer (in analogy with an optical spectrometer) still remains an experimental challenge. Despite the fact that this lack of fundamental knowledge still poses challenges for the progress of experimental and theoretical research, many proposals have recently appeared envisioning applications similar to those already achieved in optoelectronics, e.g., thermal diodes,[6,7] thermal cloaking,[8-11] and phonon waveguiding.[12,13] As promising as many of these applications may be, there is still a rather steep path from the early proof of concepts to their application in commercial devices.

In this chapter, we discuss the latest advances in nanoscale thermal transport with special focus on the last 5–10 years. In the first part, we introduce the reader into the state-of-the-art experimental methodologies. We discuss the main principles and applications of several contact and contactless (optical) techniques describing some of their advantages and drawbacks. We also provide selected examples that demonstrate the successful applications of these techniques.

In the second part, we provide a brief overview of the mechanisms that can modify the thermal conductivity and describe the different phonon scattering processes. We then focus on the influence of dimensionality on the thermal conductivity reduction. In particular, we discuss the cases of suspended thin films, two-dimensional phononic crystals, and one-dimensional superlattice (SL) structures and elucidate the difference between coherent and noncoherent phonon processes.

11.2 Experimental Techniques for Nanoscale Thermal Transport Studies

In the last decade, several novel research methodologies were developed to study thermal transport at the nanoscale. In this section, we aim to give a short and comparative overview of the state of the art of these methodologies highlighting the most suitable experimental approach for samples with different dimensionality (D) at the nanoscale. Thin films are among the most investigated geometries besides bulk samples. The 3-omega method[14] (Section 11.2.1) was developed in the 1990s and sets its place as the most established and accurate technique to study thermal transport in thin films and bulk samples. In addition, contactless techniques received considerable attention due to their experimental versatility and their ability to avoid thermal contact resistances arising from electrical contacts. The most prominent examples are time- and frequency-domain thermoreflectance (TDTR and FDTR)[15,16] which use a metallic transducer to monitor a laser-induced temperature rise by its temperature-dependent optical reflectivity (Section 11.2.2). Raman thermometry[17,18] in its one and two-laser versions (Section 11.2.3) has also proven to be of great value for 2D systems and for suspended thin films (quasi-2D). Lower dimensional systems such as quantum dots (0D) and nanowires (1D) are usually investigated using scanning thermal microscopy (SThM)[19] (Section 11.2.4) and microchip-suspended platforms (Section 11.2.5). SThM achieves the highest spatial resolutions since it is based on atomic force microscopy. Finally, the microchip-suspended platform approach is specifically suitable for nanowires, 2D and quasi-2D systems, although this approach usually requires a considerable amount of fabrication efforts. In the following sections, we give a short overview of the operational principle of each of these methodologies including their advantages and disadvantages for each studied case.

For a full picture of the state of the art, the reader may refer to the original publications as well as a number of reviews discussing each approach in detail.[20–29]

11.2.1 The 3-Omega Method

The 3-omega technique is based on electrically heating a thin planar metallic resistor with length L, using an AC harmonic current at a frequency ω, and subsequently measuring the resulting voltage drop at the first ($V_{1\omega}$) and third ($V_{3\omega}$) harmonics of the excitation current. The key concept behind this approach is the occurrence of a voltage component at the third harmonic ($V_{3\omega}$) of the excitation current ($I_{1\omega}$), which arises from the convolution of the current oscillating at 1ω, with the resistance of the metallic strip oscillating at 2ω, i.e., heat dissipation on the metallic strip is independent on the sign (or direction) of the excitation current. In simple words, upon heating a thin resistor using an AC current, the temperature of the resistor will increase due to Joule effect with a magnitude that depends on the thermal conductivity (k) of the underlying sample. The power dissipated by the harmonic current, $P(t) = V(t)I(t)$, leads to a time-dependent temperature rise of the resistor consisting of a DC part (T_{DC}), which can be related to the root-mean-square (RMS) value of $P(t)$, and an AC part (T_{AC}) which oscillates around T_{DC} at odd harmonics of the fundamental excitation frequency. The amplitude of the AC oscillations $|T_{AC}|$ can be expressed in terms of the third harmonic of the voltage signal and through the temperature coefficient of resistance of the resistor, $\beta = (1/R_0)[dR/dT]$, as

$$|\Delta T_{AC}| = \frac{2V_{3\omega}}{\beta V_{1\omega}} \qquad (11.1)$$

where R and R_0 are the resistance of the samples at T and T_0, respectively. The analytic solution to this problem can be approximated solving the heat equation for the case of an infinitely long and narrow heat source (i.e., the metallic strip in our case) over a semi-infinite substrate. Following the solution by S. Carslaw and D. G. Jaeger,[30] the temperature distribution over the heat source is obtained solving the heat equation as follows:

$$\Delta T = \frac{P_0}{\pi L k} K_0(qr), \qquad (11.2)$$

where K_0 is the zeroth-order modified Bessel function, $1/q = (\alpha/i2\omega)^{1/2}$ is the wavelength of the induced thermal wave (or thermal penetration depth), α is the thermal diffusivity of the specimen, ω is the angular frequency, P_0 is the power dissipated in the resistor, and L is the length of the resistor. In the limit with $|qr| \ll 1$, i.e., large thermal penetration depth, the previous equation can be approximated as

$$\Delta T = \frac{P_0}{\pi L k} \left[A - \frac{1}{2} \log\left(2\omega\right) - \frac{i\pi}{4} \right], \qquad (11.3)$$

where A is a constant. Note that the magnitude of the power component which drives the AC temperature oscillations is

$P_{\mathrm{RMS}} = P_0/2$. Whereas half of the power is responsible for a DC temperature rise, the other half drives the temperature oscillations, which are manifested in $V_{3\omega}$. Finally, introducing the measured temperature rise into the previous equation, we can obtain the thermal conductivity of the sample since all remaining quantities are known.

So far, we have discussed the basics of the 3-omega methodology for the case of semi-infinite substrates. The solution for the case of thin films is rather easy to obtain and was derived by D. G. Cahill[14] already at the beginning of the 1990s. The key idea is to consider that the thin film under investigation behaves like an interface resistance and, thus, contributes to the temperature rise of the heater, which itself is frequency independent. This approximation is expressed by

$$|\Delta T_{\mathrm{AC}}| = |\Delta T_{\mathrm{AC}}|_{\mathrm{substrate}} + |\Delta T_{\mathrm{AC}}|_{\mathrm{TF}} \qquad (11.4)$$

In order to find the explicit expression for this last term, we consider a thin film on a substrate with a film thickness d and a thermal conductivity k_{TF}, which is deposited over a substrate with a thermal conductivity k_{s}. The resistor on top of the thin film is defined by a width $2b$ and a length L. Applying the 1D form of Fourier's law along the cross-plane direction, i.e., parallel to the thickness d of the thin film, we obtain

$$\frac{P}{2bL} = -k_{\mathrm{TF}}\frac{|\Delta T|_{\mathrm{TF}}}{d} \qquad (11.5)$$

The previous expression gives the thermal conductivity of the thin film from the measured temperature offset with respect to the reference measurement (substrate). This equation constitutes a valid approximation if the substrate and thin film fulfill the following criteria:

- $2b \gg d$: the physical meaning of this approximation is to ensure that the 1D form of Fourier's law can be applied.

- $1/q = (\alpha/i2\omega)^{1/2} \gg d$: this means that the thermal penetration depth must be large as compared to the thickness of the thin film, which implies that the thermal properties of the thin film are frequency independent. Note that in Eq. (11.3), only the contribution arising from the substrate is frequency dependent.

- $k_s > 10k_{\mathrm{TF}}$: this relation comes mostly from experience and ensures that heat flows in the cross-plane direction through the thin film with negligible in-plane contribution.

Figure 11.1 shows two examples of data measured using the 3-omega method in a bulk Pyrex 7740 substrate (a) as well as for the case of a 200 nm thick SiO_2 thin film deposited over a Si substrate (b). In both cases, the temperature rise in the metallic transducer is plotted as a function of the excitation frequency. In the bulk case, the real part of the temperature oscillations is fit using a linear function, whereas in the case of the thin film, a reference sample containing only the substrate must be measured. The thermal conductivity is obtained by the temperature offset between the substrate and the thin film plus the substrate.

11.2.2 Time- and Frequency-Domain Thermoreflectance

TDTR and FDTR, respectively, set their place as versatile all-optical techniques for the measurement of thermal conductivity, heat capacity, and thermal boundary resistance. The contactless nature of both techniques simplifies the measurements and the required fabrication steps as compared to other methods such as the 3-omega method, SThM, and the microchip-suspended platforms. TDTR and FDTR are also suitable to study electrically conductive samples, which is usually more complicated using, e.g., the 3-omega technique due to the need of an insulating

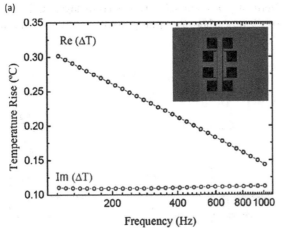
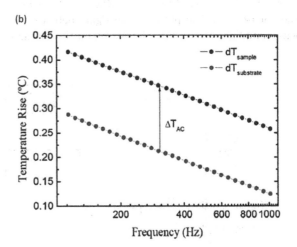

FIGURE 11.1 (a) A typical example of the output obtained from 3-omega measurements in a bulk Pyrex 7740 substrate. The in-phase and out-of-phase temperature oscillations are shown as well as an optical micrograph of the metallic resistor deposited with 1 mm length between the inner pads (inset). (b) Temperature rise of the resistor for thin films with respect to the reference substrate for a 200 nm SiO_2/Si thin film. The thermal conductivity is extracted from the temperature offset averaging through the entire frequency range.

layer that prevents current leakage from the metallic transducer to the sample under investigation. In addition, they provide relatively high spatial resolution near the diffraction limit. In order to present the basic concepts of both the techniques, which share the same physical principles but in different domains (time and frequency), we follow the derivation by Cahill[15] and Schmidt et al.[16] The operational principle of these techniques is based on a two-laser approach similar to a pump-and-probe experiment (Figure 11.2). A comparatively low-power laser, the probe laser, is focused on the surface of the sample (or transducer), whereas a higher power laser, the pump laser, creates a nonequilibrium thermal situation. The reflectivity as a function of time (frequency) is probed, leading to a time (frequency)-dependent temperature rise in the spot region. However, the temperature rise is usually a noisy variable with the noise being introduced by laser power fluctuations (typically <1%).

In the case of FDTR, the pump laser is modulated using an electro-optical modulator (EOM), acousto-optical modulator (AOM), or simply directly modulating the laser intensity intracavity to generate the thermal excitations. Since the pump laser is modulated at a frequency ω_0, the reflectivity of the sample will oscillate with the same frequency. However, due to the finite response time of the sample (or finite thermal diffusivity), the reflectivity oscillations will usually exhibit a different phase than the excitation, i.e., the reflectivity oscillations are retarded with respect to the pump laser. This quantity is named as "phase lag" and can be easily probed by using a secondary laser with a different wavelength (chosen to facilitate spectral filtering of the pump laser). In order to measure this "phase lag", the most convenient approach is to use a lock-in amplifier, which selectively measures the signal at the reference (excitation) frequency, thus, damping the noise arising from other frequencies. Thus, if the intensity of the pump laser is $I_{\text{pump}} \alpha e^{-i\omega_0 t}$, then the system response at the probe wavelength will be $I_{\text{pr}} = Ae^{-i(\omega_0 t + \varphi)} = Z(\omega_0)e^{-i\omega_0 t}$, where φ is the "phase lag". The phase of the signal detected by the lock-in amplifier carries the information on the thermal

properties of the sample. Two solutions can be derived for a multilayered system depending on the choice of the probe laser. The phase lag (φ) can be expressed as

$$\varphi = \tan^{-1} \left\{ \text{Im} \left[Z \left(\omega_0 \right) \right] / \text{Re} [Z (\omega_0)] \right\} \quad (11.6)$$

We then obtain for a continuous wave (cw) laser

$$Z \left(\omega_0 \right) = \beta \frac{A_0}{2\pi} \int_0^\infty x \left[\frac{-D(\omega_0)}{C(\omega_0)} \right] \frac{e^{-x^2(r_{\text{pump}}^2 + r_{\text{probe}}^2)}}{8} dx \quad (11.7)$$

and for a pulsed laser with repetition rate ω_s and delay between pulses τ

$$Z \left(\omega_0 \right) = \beta \frac{A_0}{2\pi} \sum_{j=-\infty}^{\infty} \left\{ \int_0^\infty x \left[\frac{-D(\omega_0 + j\omega_s)}{C(\omega_0 + j\omega_s)} \right] \right. $$
$$\left. \times \frac{e^{-x^2(r_{\text{pump}}^2 + r_{\text{probe}}^2)}}{8} dx \right\} e^{-i\omega_s j\tau} \quad (11.8)$$

where β is proportional to the thermal coefficient of the surface, A_0 is the power of the pump laser, r_{pump} and r_{probe} are the radius of the pump-and-probe lasers, and C and D are coefficients of the transfer matrix, with $C = k_z q \sinh(qd)$ and $D = \cosh(qd)$. For the previous coefficients (C and D), the quantity q is given by $q^2 = (k_r x^2 + \rho C_v i\omega)/k_z$, with k_r and k_z being the thermal conductivities in the radial and cross-plane directions, ρ and C_v are the density and specific heat capacity of the material, d is the thickness of the sample, and $\omega = \omega_0$ or $\omega = \omega_0 + j\omega_s$, depending on the probe laser being cw or pulsed, respectively. The previous approach is very convenient for multilayered structures. In this case, the two previous equations are still valid, and the transfer matrix of the system is obtained as a product of the individual matrices

$$M = M_n \dots M_1 = \begin{pmatrix} A_n & B_n \\ C_n & D_n \end{pmatrix} \dots \begin{pmatrix} A_1 & B_1 \\ C_1 & D_1 \end{pmatrix}$$

Here, M_n accounts for the bottom layer and M_1 for the top layers where the pump laser is incident. The previous

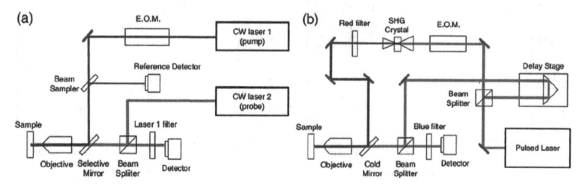

FIGURE 11.2 (a) Schematics of the FDTR setup based on two continuous wave lasers. The pump laser is modulated using an electro-optical modulator (EOM). (b) Schematics of the TDTR setup. A pulsed laser is used as pump-and-probe laser. The pump laser is frequency doubled to facilitate spectral filtering of pump and probe signals, and the probe laser is time delayed by a mechanical delay stage. (Taken from Reference[16] Reprinted from *Rev. Sci. Instrum.* 80, 094901 (2009) with the permission of AIP Publishing.)

approach is valid when the bottom layer is in thermal equilibrium; however, if this is not a valid assumption, the boundary condition must be included in the previous set of equations. We note that, in general, it is valid to assume that the bottom layer (or the substrate) is in thermal equilibrium.

Finally, the solution for the transient case as required by the TDTR approach is simply a generalization of Eq. (11.8) by substituting τ by the temporal variable t. In addition, since usually the lock-in technique is also used to increase the signal-to-noise ratio, the real and imaginary parts of Eq. (11.8) are directly related to the in-phase and out-of-phase components measured by the lock-in amplifier. Figure 11.3 displays two typical examples of the data collected by FDTR (cw and pulsed excitation) and TDTR (pulsed excitation). In order to obtain good resolution, the FDTR measurements must be conducted in a wide frequency range, preferably, scanning up to frequencies over 1 MHz. For the case of TDTR, we display in Figure 11.3b, the time evolution of the real and imaginary part of the measured voltage.

We summarize the main capabilities and advantages of TDTR and FDTR:

- Both techniques are suitable to measure the thermal conductivity, heat capacity, and thermal boundary resistances with rather high accuracy.

- Their contactless fashion makes them suitable to measure electrically conductive samples, thus, avoiding potential errors arising from current leakage as in the case of the 3-omega method.

- To obtain the previous quantities, it is NOT necessary to know the power absorbed by the sample in contrast to, e.g., Raman thermometry (Section 11.2.3) or the 3-omega method (Section 11.2.1).

11.2.3 Raman Thermometry (One- and Two-Laser Version)

The one-laser version of Raman thermometry is a contactless optical technique for which a laser beam is focused onto the surface of a sample and the Raman shift of any Raman active optical mode is monitored as a function of the absorbed power. With increasing absorbed power, the laser-induced heating results in a lattice expansion of the material, which causes an increasing red shift of the Raman mode. Thus, the lattice temperature can be directly determined provided that the spectral position of the selected Raman mode as a function of temperature is available for calibration. Such a calibration can be easily obtained by, e.g., temperature-dependent low-power Raman measurements in a cryostat or heating chamber. The thermal conductivity k of the sample can then be extracted if the absorbed power by the incident laser is known. With the simultaneous knowledge of the local spot temperature, the laser power absorbed by the specimen, and the laser spot size and shape, it is possible to solve the 2D heat equation by using appropriate boundary conditions. This technique can be applied to bulk samples and supported thin films; however, it is for the case of 2D materials and suspended thin films (quasi-2D) where it provides the best results. The main reason is that a quasi-2D geometry considerably simplifies the applied heat flow model as well as the modeling of the laser spot shape, particularly in its depth profile. In a 2D material, as well as in suspended thin films, the heat source given by the laser spot can be considered as circular and homogeneous in the out-of-plane direction.

This approach has proven to be of great value for cases where other methodologies are experimentally challenging, e.g., the case of graphene and other 2D materials. Figure 11.4 displays an example of the successful

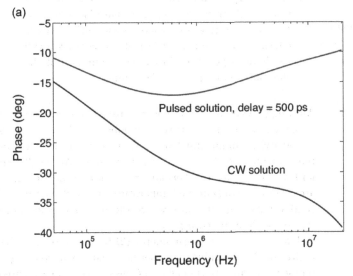
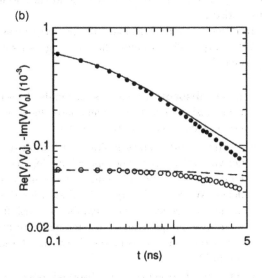

(a)

(b)

FIGURE 11.3 (a) Calculated phase lag response for FDTR in the cases of cw and pulsed probe lasers for a sapphire substrate coated with 100 nm of Al.[16] (Reprinted from *Rev. Sci. Instrum.* 80, 094901 (2009) with the permission of AIP Publishing.) (b) Calculated (solid lines) and experimental data (symbols) of the imaginary and real part of the temperature rise as a function of time for a TiN/MgO(001) epitaxial layer.[15] (Reprinted from *Rev. Sci. Instrum.* 75, 5119 (2004) with the permission of AIP Publishing.)

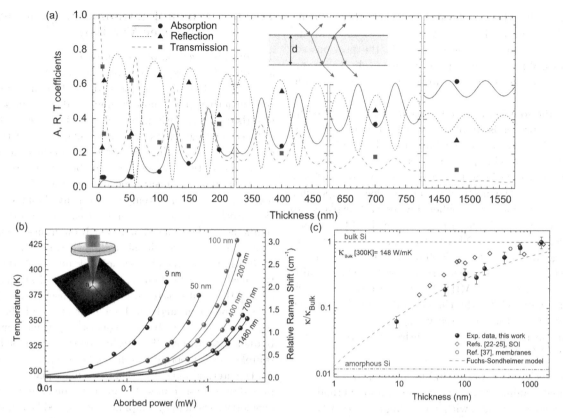

FIGURE 11.4 (a) Measured absorption (circles), reflection (triangles), and transmission (squares) coefficients as a function of thickness of Si thin films and corresponding calculations solving the Maxwell equations (lines). (b) Temperature of the spot as function of the absorbed laser power. (c) Thermal conductivity obtained solving the heat equation.[17] (Reprinted from *APL Mater.* 2, 012113. Copyright (2014) licensed under CC BY 4.0.)

application of this technique. The systems under investigation are suspended Si thin films with thicknesses between 10 and 1,000 nm. The thermal conductivity of the thin films decreases with decreasing thickness due to boundary scattering (geometrical phonon scattering) at the surfaces of the suspended films.

In order to conduct Raman thermometry measurements in 2D and quasi-2D systems, the following issues need to be considered:

- **Laser absorption:** The systems under investigation must absorb the incident laser light, which grants local heating of the samples.
- **Raman mode:** The samples must exhibit Raman activity of at least one optical mode, and this mode must exhibit a strong temperature dependence. We recall that typically first-order Raman modes in semiconductors exhibit a redshift for increasing temperatures with temperature coefficients ranging approximately between 20 and 65 K/cm.[18]
- **Absorbed power:** The absorbed power must be experimentally determined with the highest possible accuracy. A common source of large errors in the determination of k is to rely solely on calculations of the absorbed power based on

material composition and dimensions. However, it must be noted that the absorbed power also depends strongly on the local temperature of the laser spot on the sample.
- **Boundary conditions:** The area of the suspended thin film or 2D material must be large enough to ensure thermal equilibrium at the edges of the sample. Neglect of this condition will result in large errors in the determination of the thermal conductivity of the sample.

Although Raman thermometry, based on a one-laser approach, is an interesting technique mainly due to its simplicity, its main drawback is that only one point of the temperature field is probed to determine k. We recall that for the case of diffusive thermal transport, at least two points are required to obtain the temperature field and the thermal conductivity, e.g., in cases where the boundary conditions cannot be determined a priori.

Two-laser Raman thermometry (2LRT)[18] overcomes this limitation by mapping the thermal field upon creating a temperature distribution using a Gaussian laser spot focused onto a sample. While a heating laser with wavelength λ_1 is used to produce a thermal hotspot, a thermometer laser with wavelength λ_2 measures the spatial distribution of the local temperature through the temperature-dependent redshift of a Raman mode of the sample. The main advantage of this technique as compared to other contactless steady-state methods such as infrared thermometry is its submicron spatial resolution. Techniques based on scanning probes such as SThM or infrared scanning near-field optical microscopy (IR-SNOM) can even surpass its spatial resolution; however, the difficulties in modeling the thermal response of the tips constitute a major drawback. On the other hand, short-pulse optical techniques such as TDTR or transient thermal grating (TTG) only provides indirect access to k through the thermal diffusivity (α).

We next derive the analytical solution for the temperature field in the case of a free-standing isotropic membrane. A temperature distribution is created upon excitation with a point-like laser source as shown in Figure 11.5. The solution is simply given by integrating Fourier's equation: $P_{\text{abs}}/(2\pi r) = -k\nabla T$, where P_{abs} is the power absorbed by the system, $2\pi r d$ is the cross-sectional area of the heat flux, k is the thermal conductivity, and T is the temperature. Integration of the Fourier's equation leads to the following solutions for the temperature field.

$$T(r) = T_0 - \left[\frac{P_{\text{abs}}}{2\pi d k_0}\right] \ln(r/r_0) \rightarrow k = k_0 \qquad (11.9)$$

$$T(r) = T_0 - \left[\frac{r}{r_0}\right]^{-\frac{P_{\text{abs}}}{2\pi d A}} \rightarrow k(T) = \frac{A}{T} \qquad (11.10)$$

where we have considered a temperature independent thermal conductivity (k_0) as well as a temperature-dependent thermal conductivity $k = A/T$, which is typical for most semiconductors at elevated temperatures. It is interesting to note that from a single temperature map, it is possible to obtain the temperature dependence of the thermal conductivity in a wide temperature range.

Figure 11.5 displays the schematics of the experimental arrangement. A heating laser with $\lambda_1 = 405$ nm is focused onto the lower surface of the Si membranes, whereas a probe laser with $\lambda_2 = 488$ nm is scanned over its upper surface to obtain the local temperature. We note that while relatively high powers are required for the heating laser (λ_1), in order to create a spatially dependent thermal field, low powers are important for the probe laser (λ_2) to avoid an additional

thermal perturbation. A ratio of 10:1 between the heating and probe lasers, respectively, is typically used. Both lasers are focused onto the samples using long-distance objectives with numerical aperture of NA ≥ 0.5. Although these objectives provide lower spatial resolution than short distances objectives (which typically have NA ≥ 0.8), they allow to perform the thermal maps in a controlled environment or in variable temperature conditions. Figure 11.5b displays a 2D temperature map of a $250 - $nm-thick suspended Si film. The maximum temperature at the center of the map, i.e., heating and probe lasers focused to the same position in the sample, is $T \approx 800$ K, and a thermal decay with radial symmetry is observed in the temperature projection plane. Although this radial symmetry arises from the isotropic nature of Si at room temperature, it is expected that materials with a directional-dependent thermal conductivity (k_{ij}) will exhibit an asymmetric thermal decay. It is interesting to note that the temperature field does not fully decay to the thermal bath temperature of 294 K as expressed by the measured temperature of $T \approx 400$ K at a distance of 150 μm from the heating laser. The reason for this far-reaching thermal decay is the relatively high thermal conductivity of about 80 W/mK for a Si film thickness of 250 nm.[17,18,31]

11.2.4 Scanning Thermal Microscopy

This approach is based on the scanning probe microscopy (SPM) previously developed for topography imaging with high spatial resolution. In fact, the operational principle of SThM is simply to place a heating/sensing element in a tip whose spatial position with respect to the sample can be controlled using, e.g., piezoelectric scanners. The main advantage of this technique as compared to optical approaches, such as TDTR, FDTR, and Raman thermometry, is the superior lateral spatial resolution that can be achieved, which is only limited by the size of the tips and the measurement conditions (vacuum or ambient). Intense research work has been conducted since the invention of SThM in 1986.[19] This technique has been actively developed and successfully applied to various areas such as microelectronics, optoelectronics, polymer science, and carbon allotropes (CNTS, graphene, etc.). A large number of approaches have been developed to probe temperature and to create local hotspots using SPM tips. The most common approaches are as follows:

- **Thermovoltage-based methods** make use of the thermoelectric voltage generated at the junction between two electrodes. The electrodes can be chosen as the tip-sample interface or built-in sensors within the tip such as a thermocouple and/or Schottky diodes.

- **Thermoresistive probes methods** are typically based on metallic or Si-doped tips, where the temperature coefficient of resistance is the mechanism used to sense the temperature: $R(T) = R_0(1 + \beta\Delta T)$.

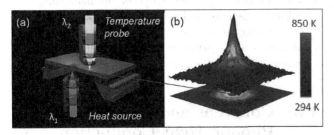

FIGURE 11.5 (a) Schematics of the experimental setup, 2LRT. (b) Representative thermal map of a 250-nm-thick Si thin films. A projection of the thermal field is also shown in a lower plane. The arrow indicates the heating spot in both panels. (Adapted based on Reparaz, J. S. et al. *Rev. Sci. Instrum.* 85, 034901 (2014).)

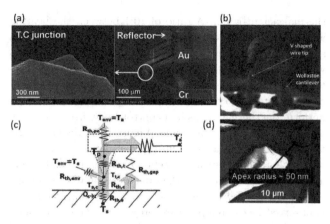

FIGURE 11.6 (a) Scanning electron microscopy (SEM) images of an Au-Cr thermocouple SThM probe and its magnification.[32] (Reprinted from *Appl. Phys. Lett.* 93, 203115 (2008) with the permission of AIP Publishing.) SEM images of a Wollaston wire probe (b) and a palladium probe (d). (c) Schematics of the thermal model to account for the most important heat losses.[33] (Reprinted with permission from *Phys. Stat. Sol. A* 212, 477. Copyright (2015) with the permission of John Wiley & Sons, Inc.)

Figure 11.6 shows selected SEM images of probes used for SThM. Although SThM is a promising technique in terms of lateral spatial resolution, its main drawback is the data-reduction[1] procedure. In order to obtain the local temperature and/or the thermal conductivity, a number of assumptions must be made in order to model the conditions at which the experiment is conducted. For example, the influence of air and water (water meniscus) or the detailed geometrical shape of the tip can have substantial influence on the data reduction. We recall that usually a voltage is measured at the tip apex, which must be converted to a local temperature or thermal conductivity. Heat transport due to convection through the surrounding air, heat losses through the tip itself, the influence of the contact resistance between the tip and the sample, heat radiation losses, etc., are some of the boundary conditions that must be considered and modeled in order to obtain reliable quantitative thermal data of the specimen. The complexity of such a modeling is schematically illustrated in the bottom left of Figure 11.6. These and related problems of tip-based thermal measurement techniques are currently stimulating intense research effects of several groups around the world.

11.2.5 Microchip-Suspended Platforms

Suspended platforms constitute a unique alternative to measure the thermal conductivity of 1D and 2D materials. Typically, these platforms consist of a metallic heater/thermometer deposited over a SiN_x membrane.

FIGURE 11.7 (a) Scanning electron microscope image of a suspended platform used to measure a single NW.[34] (Reprinted with permission from *ACS Nano* 2011, 5, 3954. Copyright (2011) American Chemical Society.) (b) SEM image of a suspended platform to measure a graphene flake.[35] (Reprinted with permission from *Nano Lett.* 2011, 11, 113. Copyright (2011) American Chemical Society.)

A gap is intentionally etched in the middle of the suspended membrane, where the sample under investigation is placed. Figure 11.7 displays an example of a microchip thermal sensing platform. The methodology resembles the steady-state approach based on Fourier's law, i.e., a known heat flux is induced through the nanostructure, and the temperature gradient is obtained using the metallic thermometers deposited in each side of the membrane. The resolution of the suspended platform approach in thermal conductance is of the order of 1 nW/K at room temperature. However, recent improvements based on a Wheatstone bridge circuit have demonstrated a significant reduction of the noise to values of 10 pW/K. Suspended structures have been used to measure the thermal conductance of a variety of nanostructures such as single- and multi-walled carbon nanotubes; graphene as well as Si, SiC, Si/SiGe, Bi, Bi_2Te_3, InAs, PbS, PbSe, PbTe, and ZnO nanowires. This approach, however, is prone to large experimental errors if the thermal leakages within the suspended platform are not carefully taken into account. We note that usually a rather small heat flux can be induced through the nanostructures, and thus, heat leakages in any form are relevant.

In summary, microchip platforms present the following characteristics:

- Their design is excellent to study 1D and 2D structures. In particular, part of the generated heat flows strictly through the nanostructure avoiding the influence of substrates or interface resistances.

- The fabrication process is one of the main difficulties. Expensive equipment and facilities such as clean rooms are necessary to fabricate the sensing platforms. In addition, in some cases, placing the nanostructure in the gap may be challenging.

11.3 Coherent and Noncoherent Phonon Heat Conduction

Following the review of various experimental techniques for the investigation of nanoscale thermal properties, this section discusses the effects of geometry and artificial periodicity on the thermal properties of materials with

---

[1]Data reduction is defined as the process of transforming measurable variables into physical variables, e.g., an increase in resistance can be interpreted as a temperature rise given a suitable model.

reduced dimensionality. The text focusses on nonmetallic nanostructures where phonons constitute the main particles responsible for thermal transport and heat propagation. The first subsection briefly summarizes the main effects that influence the thermal conductivity in semiconductor nanostructures including both particle-like and wave-like phonon effects.[36] Subsequently, we elucidate thermal transport properties in semiconductors with reduced dimensionality (thin films, suspended membranes) and second-order periodicity (two-dimensional phononic crystals, SLs). We thereby focus on recent progress in the understanding and control of phonon-mediated heat propagation within the last 5–10 years. For a more general discussion, the reader may consult a variety of comprehensive reviews on nanoscale thermal transport.[22,23,25,26,28,37−41] For an in-depth discussion of the various aspects of phononic crystals with focus on the propagation of acoustic wave, we refer the reader to Reference[42].

11.3.1 Mechanisms for Thermal Conductivity Modification at the Nanoscale

The continuous reduction of dimensions during the last decades has opened the window to many novel physico-chemical properties of materials at the micro- and nanoscale. With decreasing dimensions, the presence of heat in the form of atomic vibrations can drastically modify material properties and, in some cases, even structurally change nanomaterials as, e.g., in the case of phase-change materials.[4] This fact constitutes a major challenge for the design of nanoscale devices due to the complexity of controlling heat flow at the nanoscale. One of the main reasons lies within the complex frequency spectrum of phonons that generally consists of many different possible vibrations (or branches) depending on the structural symmetry of the underlying atomic lattice. Furthermore, phonons experience anharmonic decay, which complicates the control of their individual propagation as they convert to lower frequency phonons through *Normal* and *Umklapp* scattering within short timescales. Thus, two key questions that summarize the importance of this topic are as follows:

- Which are the leading mechanisms that influence heat transport at the nanoscale?
- To what extent can we control these mechanisms in order to influence thermal transport at the nanoscale?

Nanoscale heat transport is rather complicated to fully address due to the large number of different processes involved. However, we can divide phonon-related thermal processes into two main categories with fundamentally different physical origin. These are noncoherent particle-like and coherent wave-like processes that arise from the particle/wave duality of phonons. It should be noted that most strategies to modify the thermal conductivity by nanostructuring usually lead to a reduction of k as compared

to the bulk value of a given material. Increasing the thermal conductivity well beyond a given bulk value is not easy to achieve, and it is a somehow unknown field. The discovery of a thermal superconductor in analogy to its electrical counterpart remains an open scientific question, which has received little attention, although no fundamental law forbids its existence. Perhaps the key strategy to substantially increase the thermal conductivity is based on obtaining "purer materials", e.g., reducing the number of isotopes present naturally reduces isotope mass scattering and would result in larger values of the thermal conductivity. On the other hand, the reduction of the thermal conductivity by incoherent scattering processes can be summarized as follows:

- **Boundary scattering** is the scattering of phonons at interfaces. These interfaces can be the surface of the sample, interfaces in multilayer materials, or grain boundaries in polycrystalline materials.
- **Impurity scattering** due to the introduction of a new chemical element in the form of impurities (low concentrations).
- **Alloy scattering** due to the introduction of a new chemical element at higher concentrations that forms an alloy. The distribution of the alloying element within the lattice determines the disorder degree and affects the thermal conductivity.
- **Isotope scattering** arises from the mass difference between different isotopes, which naturally leads to broader phonon modes, i.e., shorter lifetimes and, thus, lower thermal conductivity.

In contrast to these incoherent processes, coherent scattering arises from the wave nature of phonons, which results in the modification of the phonon dispersion relation and influences the coupling between different phonon branches, phonon group velocities, and phonon lifetimes. Two typical examples of this case are phononic crystals (Section 11.3.3), and SLs (Section 11.3.4) were the phonon dispersion can be tailored by appropriate choice of materials, lattice symmetry, and dimensions. It should be noted that while incoherent scattering can rather easily be introduced and tuned using the previously described mechanisms, coherent heat manipulation is still a developing field that requires further fundamental and applied research to realize practical applications.

11.3.2 Thin Films and Membranes

Thin films are among the most studied nanostructures and are the base platform to fabricate phononic crystals. We have chosen silicon-supported and suspended thin films as examples to show how the reduction of dimensionality (3D → quasi-2D) leads to incoherent boundary scattering and, thus, a substantial reduction of the thermal conductivity. One possible application of nanostructured crystalline

Si thin films is their use in thermoelectric devices.[43,44] In order to achieve this goal, a rather large reduction (at least two orders of magnitude) of the thermal conductivity is desirable due to its large value of about 148 Wm^{-1} K^{-1} for bulk Si at room temperature.[45] Pioneering studies of supported Si thin films have revealed clear size effects of the thermal conductivity, i.e., the in-plane and cross-plane components of k were found to decrease as function of film thickness.[46–48] These results were later confirmed by extensive studies based on contactless methodologies such as Raman thermometry in its one-laser[17,49] and two-laser version,[18,50] TDTR and FDTR,[16,51] and transient thermal grating (TTG).[52,53]

The origin of the thermal conductivity reduction in supported and suspended Si thin films is simply scattering of phonons at the surface of the films. In other words, the thickness of the films sets an upper limit for the mean free path of thermal phonons Λ_{th}. Thus, if phonons with mean free path larger than the thickness of the samples are present, these will mostly scatter diffusively at the surface and, therefore, the thermal conductivity will be reduced as compared to its bulk value. Figure 11.8 displays a summary of published data of the thickness-dependent thermal conductivity of Si membranes using electrical and optical techniques. It is apparent that a large reduction of k occurs as the thickness of the thin films is reduced. In addition, the data shows that the detailed chemical conformation of the surface becomes important when approaching film thicknesses below 50 nm.[31] In particular, for very thin films of about 10 nm, the data reveals that the presence of the native SiO_2 reduces the thermal conductivity by approximately a factor of two as compared to pure Si-suspended films of equal thickness (see open diamonds in Figure 11.8).[31] All these effects must also be considered in periodic structures such as phononic crystals

since they can play a significant role in the modification of the thermal conductivity and might complicate the differentiation between noncoherent and coherent phonon effects.

11.3.3 Phononic Crystals

Phononic crystals constitute an attractive class of material structures with the potential to control and manipulate the propagation of thermal energy. Comparable to photonic crystals,[54,55] the introduction of a second-order periodicity that fulfills the Bragg condition ($n\lambda \sim 2a$) results in a modification of the phonon dispersion relation and the formation of phononic bandgaps, where the phonon density of states is zero and the propagation of mechanical waves is prohibited in the phononic crystal structure (see Figure 11.9). The reason for the formation of these bandgaps lies within the wave interference of phonons in periodic structures and was comprehensively summarized by Maldovan.[56] An early demonstration of such a phononic bandgap was realized in an artistic sculpture consisting of periodically arranged steel cylinders, which result in a strong attenuation of certain sound frequencies in the low kilohertz range.[57] Since then, the continuing progress in miniaturization and nanofabrication has pushed the accessible frequencies for the formation of phononic bandgaps from a few kilohertz to the edge of the terahertz domain, thus allowing not only the modification of the propagation of sound (kilohertz), ultrasound (megahertz), and hypersound (gigahertz) but also opening the prospect to control the propagation of heat.[58] This development has paved the way towards the realization of novel thermal applications and devices such

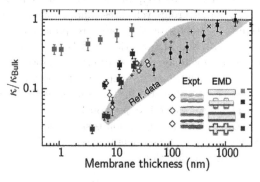

FIGURE 11.8 Normalized computed and measured thermal conductivities (k/k_{Bulk}) of silicon membranes at $T = 300$ K as a function of membrane thickness.[31] The filled squares represent results from equilibrium molecular dynamics simulations computed for smooth crystalline, rough crystalline, oxidized, and rough oxidized silicon membranes. The experimental data (open diamonds) are obtained using 2LRT (Section 11.2.3). (Reprinted with permission from *ACS Nano* 2015, 9, 3820. Copyright (2015) American Chemical Society.)

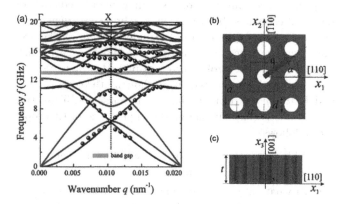

FIGURE 11.9 (a) Phonon dispersion relation of a 2D phononic crystal of air-filled cylinders in a Si-suspended film of 250 nm thickness.[61] Data points represent measured frequency by Brillouin light scattering; solid lines show the calculated phonon dispersion relation for symmetric (light gray) and antisymmetric (dark gray) modes obtained by finite element modeling. The horizontal bar indicates the presence of a phononic band gap in the gigahertz frequency range. Schematic top view (b) and lateral view (c) of the phononic crystals structure and orientation. (Reprinted with permission from *Phys. Rev. B.* 91, 075414. Copyright (2015) American Physical Society.)

as thermal diodes,[6,7] thermal transistors,[59,60] and thermal cloaks.[8-11]

An active research topic in the field of phononic crystals concerns the influence of particle-like incoherent and wave-like coherent phonon heat transport and their individual contributions to the reduction of the thermal conductivity in PnCs.[62-68] The term incoherent phonon scattering refers to all diffusive scattering processes in which the phase information of the phonon is lost. These include phonon–phonon (Umklapp) scattering, phonon–impurity scattering, and diffusive phonon–boundary scattering as discussed in Section 11.3.1. In contrast, coherent phonon effects preserve the phase of the phonons and wave interference occurs. Several recent works have therefore tried to distinguish particle-like effects and wave-like effects in two-dimensional phononic crystals.[62-67,69-72] The majority of these works focus on silicon-based two-dimensional phononic crystals built of air-filled cylinders in a silicon-suspended thin film or membrane where the holes are usually aligned in ordered square, hexagonal, or honey-comb lattices. The main line of arguments to explain the reduction of thermal conductivity in these phononic crystals usually follows the following ideas:

- The introduction of air holes leads to the formation of additional interfaces that increase phonon-boundary scattering and effectively reduce the thermal conductivity.

- For a second-order periodicity below the mean free path of phonons, an additional reduction of the thermal conductivity may occur as the length scale for ballistic propagation of phonons is reduced by the additional interfaces.

- Beyond these purely particle-related effects, coherent wave-like effects may also reduce the thermal conductivity due to the introduction of phononic bandgaps and the reduction of the phonon group velocities.

Although several works have focused on the experimental proof for coherent phonon heat conduction, the absence of a direct measurement technique for thermal phonon coherence fuels controversial discussions up to the present day. So far, most works rely on the measurement of the thermal conductivity in structures with varying geometrical features in order to deduce indirectly the presence of coherent phonon heat conduction.

An elegant approach to study the existence of coherent wave-like phonon effects and the potential impact of phonon coherence on the thermal properties has recently been implemented by the fabrication of 2D PnCs with varying level of disorder in the second-order periodicity. The approach relies on the basic principle that wave interference of phonons will be disturbed by the introduction of disorder, which will nullify a potential reduction of the thermal conductivity by coherent phonon effects. Corresponding TDTR measurements (Section 11.2.2) of PnCs with varying level of disorder were reported by Maire et al. and are displayed in Figure 11.10.[67] A pronounced reduction of the thermal

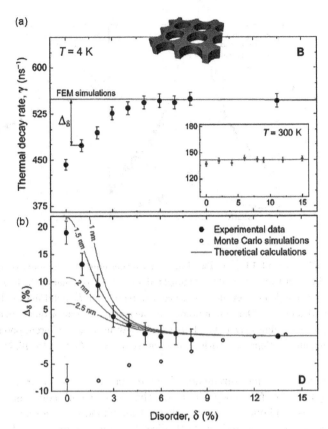

FIGURE 11.10 (a) Thermal decay rate measurements of 2D phononic crystals with varying disorder. At 4 K, the thermal decay rates depend on the level of disorder, whereas at 300 K, heat dissipates through ordered and disordered structures at an equal rate (inset). (b) Theoretically expected disorder dependence alongside the experimentally measured difference between thermal decay rates for different values of the effective surface roughness.[67] (Reprinted with permission of AAAS from *Science Advances* 3, e1700027. © The Authors, some rights reserved; exclusive licensee American Association for the Advancement of Science. Distributed under a Creative Commons Attribution NonCommercial License 4.0 (CC BY-NC).)

decay rate, which corresponds to a reduction in thermal conductivity, is observed in the ordered PnC lattice at low temperatures. With increasing level of disorder, the values approach a constant threshold. In contrast, the thermal decay rate at room temperature (Figure 11.10 inset) is independent of the amount of lattice site disorder in the PnC. Temperature-dependent studies have further shown that the reduced thermal conductivity in the ordered PnCs is only observable for temperatures below about 10 K. At these temperatures, the wavelength of the heat-carrying phonons is of similar size as the periodicity of the phononic crystals lattice so that wave-interference effects in ordered PnCs occur at frequencies of heat-carrying phonons.

These results are in agreement with previously reported measurements of ordered and disordered phononic crystals obtained by 2LRT (Section 11.2.3) that have shown the same thermal conductivity at room temperature (Figure 11.11a,b). Using femtosecond pump-probe

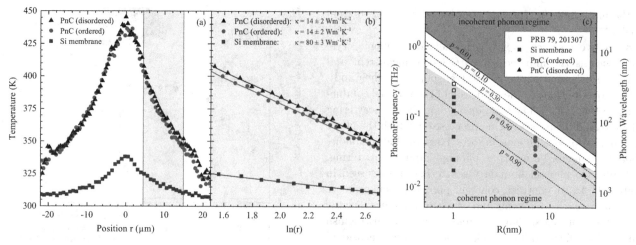

FIGURE 11.11 (a) Two-laser Raman thermometry line scans of an unpatterned Si membrane, an ordered PnC, and a disordered PnC. (b) Logarithmic plot of the highlighted area in (a) to visualize the $\ln(r)$ relation (Section 11.2.3). (c) Phonon frequency/phonon wavelength as a function of disorder and roughness parameter R. Solid lines indicate the dependence for selected specularity parameters p between 0.01 and 0.90. Data points represent measured coherent acoustic phonon frequencies. The light gray-shaded area indicates the coherent phonon regime extrapolated from the highest measured coherent phonon frequencies, the dark-shaded area marks the noncoherent phonon regime.[66] (Reprinted with permission from *Nano Lett.* 2016, 16, 5661. Copyright (2016) American Chemical Society.)

reflectivity measurements, Wagner et al. have investigated the frequency limit for the presence of coherent acoustic phonons.[66] It was shown that phonon interference is strongly affected by both disorder and roughness, and that for a given roughness of 1 nm, coherent phonon effects would play a negligible role for frequencies above about 400 GHz (see Figure 11.11c). Taking into account that most of the heat-carrying phonons at room temperature have frequencies in the low terahertz range, the absence of coherent phonons in this range directly explains why the room temperature thermal conductivity is not affected by disorder in the phononic crystals. Consequently, modifications of the thermal conductivity due to phonon coherence will only occur for very smooth surfaces/interfaces where the limit of the coherent phonon regime reaches the terahertz range or for sufficiently low temperatures where the wavelength of the thermal phonons is significantly larger and thereby greatly exceeds the characteristic roughness of the phononic crystal structure.

A particular case of ordered and disordered 2D phononic crystals is quadratic and rectangular nanomeshes.[71,73,74] One advantage of these nanomeshes is the possibility to adjust the spacing of holes in one direction while leaving it constant in the other direction. In this way, a transition from a regular ordered lattice to an equally spaced nanowire geometry can be realized (see Figure 11.12). Such a design variation can be used to decouple the contributions of wave-related coherence effects and particle-related effects with particular sensitivity to the contributions of phonon backscattering.[74] Analogous to the previously discussed phononic crystals, nanomeshes are typically realized as thin suspended films or membranes with a fabricated periodic mesh of holes.

Figure 11.12 shows the schematic illustrations of periodic and aperiodic nanomeshes with different pitch between the vertically bridging necks and the corresponding measurements of the normalized thermal conductivity and thermal conductance. The modification of the thermal conductivity as a function of pitch in the aperiodic nanomeshes is caused by the change in wall density that can be explained by phonon backscattering. This conclusion is in accordance with several recent works, which mostly agree that coherent phonon effects do not contribute to the reduction of the thermal conductivity in lithographically fabricated 2D phononic crystals at room temperature.

Apart from the investigation of coherent wave-like effects, another promising research development in recent years targets the realization of directional heat propagation by ballistic phonon transport.[75-77] The concept relies on the fact that phonons can travel in straight lines (or undergoing fully specular reflection at surfaces and interfaces) without heat dissipation for hundreds of nanometers as observed in various nanostructures including membranes,[53] holey silicon,[78] and nanowires.[79] However, in order to exploit this property for practical applications, it is crucial to achieve control over the directionality of the phonon propagation. Such a directional control of ballistic phonons was recently demonstrated by Anufriev et al. in patterned silicon nanostructures consisting of aligned and staggered phononic crystals with varying periodicity.[13] Using TDTR measurements (Section 11.2.2), it was shown that the obstruction of ballistic heat conduction in staggered structures results in a significant reduction of the thermal conductivity as compared to the samples with aligned holes. Furthermore, it was demonstrated that heat propagation can be guided by a regular phononic crystal lattice. The alignment of this phononic crystal lattice towards suspended nanowires improves the coupling of heat propagation into the nanowires via ballistic phonons as compared to a misaligned lattice.[13] These effects become even more

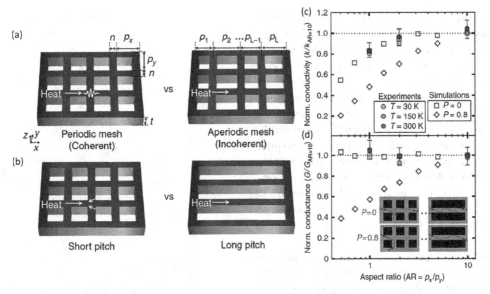

FIGURE 11.12 (a) Periodic and aperiodic nanomesh structures to distinguish between coherence effects and particle backscattering effects, (b) variation of pitch to quantify, the contribution of phonon backscattering at the bridging necks, (c) normalized thermal conductivity and (d) normalized thermal conductance as a function of varying pitch aspect ratio for experimental values at three different temperatures and simulations assuming fully diffusive ($p = 0$) and partly specular ($p = 0.8$) scattering.[74] (Reprinted from *Nature Communications* 8, 14054. Copyright (2017) licensed under CC BY 4.0.)

pronounced at low temperatures due to two different effects: Firstly, the phonon wavelength λ becomes longer at low temperatures, which increases the probability of specular scattering p expressed by

$$p = \exp\left(-\frac{16\pi^2\eta^2\cos^2\alpha}{\lambda^2}\right) \quad (11.11)$$

where η is the surface roughness and α is the normal incidence angle. Hence, the incident angle for specular reflection becomes wider with increasing phonon wavelength, which effectively increases the number of ballistic phonons at low temperatures. Secondly, the phonon mean free path Λ lengthens at lower temperatures[80-82] or, in other words, phonons can travel ballistically over longer distances in the micrometer range.

These recent approaches demonstrate the potential of controlling the directionality of ballistic heat propagation and will further benefit from the realization of smaller nanostructures by advanced nanofabrication techniques and materials with increased phonon mean free path. It therefore becomes conceivable to fabricate phononic structures that can guide, confine, or disperse heat fluxes on length scales of typical nanostructures and microstructures.

11.3.4 Superlattices

The propagation of heat in SLs is determined by a variety of different mechanisms. In particular, the thermal properties are defined by the acoustic impedance mismatch between the different materials of the SL,[83-85] the thermal boundary resistance at the interfaces, alloy scattering,[83,84] the formation of phonon minibands due to the interference of coherently reflected phonons from multiple interfaces,[85,86] phonon tunneling,[87,88] and coherent phonon heat conduction.[89,90]

In contrast to the previously discussed 2D phononic crystals and nanomeshes, SLs constitute a type of 1D phononic crystal, which, in principle, are capable of overcoming the main limitations for coherent phonon heat conduction in lithographically fabricated 2D PnCs. In particular, these are an interface roughness, which is about one order of magnitude smaller than the phonon wavelength and a second-order periodicity of the same order as the wavelength of the heat-carrying phonons (Figure 11.11c). Both conditions are nowadays accessible by epitaxial growth techniques such as molecular beam epitaxy (MBE), metal-organic-chemical vapor deposition (MOCVD) and atomic layer deposition (ALD), which under ideal conditions enable atomically flat SL structures over many SL periods. Consequently, SLs constitute an ideal platform to study the particle and wave nature of phonons via their thermal transport properties.[89-91]

While the experimental studies of coherent phonon effects in 2D phononic crystals are mainly limited to Si caused by the maturity of Si nanofabrication techniques, the epitaxial growth of SLs allows a much wider range of materials that satisfy the requirements on interface roughness and periodicity. These include, e.g., semiconductor-semiconductor SLs (GaAs/AlAs,[89] CaTiO$_3$/SrTiO,[90] SiGe,[83,84] Si/SiOx,[92] AlN/GaN[93]), metal-semiconductor SLs (W/Al$_2$O$_3$,[94] TiN/AlScN,[95,96] ZrN/ScN[97]), and inorganic-organic SLs (ZnO/benzene,[98,99] TiO$_2$/benzene[100]). As the large quantity of recent works renders a comprehensive discussion within the scope of this chapter impossible, we only mention a few selected examples that mainly focus on the possibility of coherent wave-like effects in SLs.

This research field was widely stimulated by the report of coherent phonon heat conduction in epitaxially grown GaAs/AlAs SLs with an individual layer thickness of 12 nm. TDTR measurements (Section 11.2.2) have shown a linear increase of the cross-plane thermal conductivity with increasing number of periods in the temperature range between 30 and 150 K (Figure 11.13), which was attributed to thermal transport by coherent phonons.[89] Above 150 K, the increase of k is less pronounced, which indicates the presence of diffusive interface scattering as the wavelength of the heat-carrying phonons decreases with increasing temperature. However, the persistent increase of k as function of SL periods evidences the existence of partial phonon coherence and their contribution to the heat propagation even at elevated temperatures over 150 K.[89]

In a slightly different approach, Ravichandran et al. have studied the thermal conductivity of perovskite SLs with intrinsically low-phonon Umklapp scattering as a function of periodicity.[90] The authors observed a minimum thermal conductivity for a specific SL periodicity that was attributed to a crossover between coherent wave-like and noncoherent particle-like phonon heat conduction (Figure 11.14a). The minimum forms because of opposing dependences of k as a function of interface density in the coherent and noncoherent transport regime. In the noncoherent regime, a reduction of the period thickness reduces the thermal conductivity as phonons experience diffusive boundary scattering at each interface. In other words, the phonons behave particle-like, and thus, the thermal resistance increases with increasing interface density. A further decrease in the period thickness reverses this trend and results in an increase of k. This behavior is not compatible with only diffusive phonon scattering. Considering that the period thickness is now in the range of the coherence length of thermal phonons, the observed increase in k for small period thicknesses reflects the wave nature of the phonons and was attributed to the presence of coherent phonon heat conduction. The combination of both the effects results in a minimum of k that

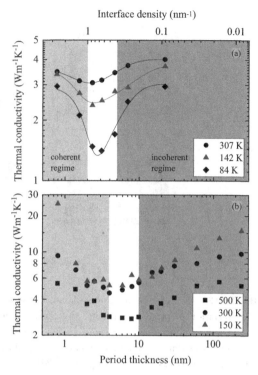

FIGURE 11.14 Measured thermal conductivity values for strontium titanate/calcium titanate $(SrTiO_3)_m/(CaTiO_3)_n$ SLs[90] (a) and TiN/Al$_{0.72}$Sc$_{0.28}$N metal/semiconductor SLs[101] (b) as a function of period thickness (interface density) at different temperatures measured by TDTR. Solid lines (a) are guides to the eye. A minimum in the thermal conductivity is visible in both SL systems that persists over a wide range of temperatures. (Graph adapted from data published in Ravichandran, J. et al. *Nat. Mater.* 13, 168,172 (2014) and Saha, B. et al. *J. Appl. Phys.* 121, 015109 (2017).)

becomes more pronounced at lower temperatures due to the increase of the phonon wavelength and reduction of Umklapp scattering.

Following these initial works on coherent phonon heat conduction, several follow-up works have appeared in recent years that investigate the impact of phonon coherence in various SL structures. These include a variety of theoretical works that investigate phonon coherence in various SLs[102] and random multilayers[103,104] such as Si/Ge,[91] GaAs/AlAs,[105] graphene/h-BN,[106,107] C^{12}/C^{13},[108] and van der Waals SLs.[109] Recently, additional experimental support for coherent phonon heat conduction in SLs was published by Saha et al.[101] Figure 11.14b displays the measured cross-plane thermal conductivity of TiN/Al$_{0.72}$Sc$_{0.28}$N metal/semiconductor SLs as a function of period thickness for temperatures of 150, 300, and 500 K. Similar to the previously discussed work by Ravichandran et al.,[90] the minimum of k becomes more pronounced at lower temperatures and is explained by the crossover between the noncoherent and coherent phonon regimes.

Apart from these works, the lowest in-plane thermal conductivity of Si-based planar SLs was recently reported considering exclusively noncoherent effects.[92] Hybrid

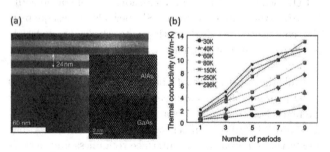

FIGURE 11.13 (a) Cross-sectional transmission electron microscopy (TEM) image of the three-period GaAs/AlAs SL, (inset) high-resolution transmission electron microscopy (HRTEM) image of one of the interfaces. (b) TDTR measurements of thermal conductivity of GaAs/AlAs SLs as a function of number of SL periods for different temperatures between 30 and 296 K.[89] (Reprinted with permission of AAAS from *Science* 338, 936. Copyright (2012) American Association for the Advancement of Science.)

nanomembrane SLs were fabricated by strain-engineered role-up and compression of 20 nm Si and 2 nm amorphous SiO_x layers. Thermal conductivity measurements were conducted using the suspended microchip-suspended platform technique (see Section 11.2.5). The strong reduction of the thermal conductivity (about three times lower than in an individual 20-nm-thick Si layer) was attributed to a dominant phonon scattering in the SiO_2 layers. The results are in agreement with the previously discussed reduction of the thermal conductivity in Si membranes (Section 11.3.2)[31] and phononic crystals[110] by the native oxide and emphasize the crucial role of the surface configuration not only in suspended ultrathin films but also in SL structures.

The here discussed works demonstrate that 1D SLs constitute a versatile class of nanostructures for both, the study of fundamental properties such as the conditions for phonon interference and coherence and for the realization of materials with very low thermal conductivity by the exploitation of coherent and noncoherent phonon effects. Together with their 2D and 3D counterparts, 1D SLs can be used to build phononic metamaterials, which hold the promise of many more interesting discoveries and applications in the field of nanoscale thermal transport in the near future.

11.4   Conclusion

In summary, we have reviewed the state-of-the-art techniques to address thermal properties at the nanoscale level. We have focused on their capabilities, advantages, and drawbacks to address nanoscale phenomena. Each of these techniques is suitable for specific sample geometries mostly depending on the dimensionality of the system under investigation. In particular, we conclude that contactless techniques have received special attention due to their versatility and intrinsic avoidance of thermal interface contacts. However, we note that each material system, dimension, and geometry should be evaluated in detail in order to choose the most suitable experimental methodology for the study of the nanoscale thermal properties.

Furthermore, we have reviewed the main effects, which govern thermal transport in two-dimensional phononic crystals and one-dimensional SLs. We have discussed the relative influence of coherent and noncoherent phonon processes as efficient alternatives to control thermal transport at the nanoscale. Although a large number of studies were already conducted based on noncoherent phonon scattering, the field of coherent heat manipulation is still in a developing stage. We expect that this field will lead to novel approaches in the next years, which could become one of the corner stone of nanoscale heat engineering.

Acknowledgments

We gratefully acknowledge the support of the German Science Foundation within the Collaborative Research Center 787 (CRC 787). The Spanish Ministerio de Economía, Industria y Competitividad is gratefully acknowledged for its support through Grant No. SEV-2015-0496 in the framework of the Spanish Severo Ochoa Centre of Excellence program, and through grant MAT2017-90024-P (TANGENTS).

References

1. Brillouin, L. Diffusion de lumière par un corps transparent homogène. *Ann. Phys.* **17**, 88 (1922).
2. Raman, C. V. A new radiation. *Indian J. Phys.* **2**, 387–398 (1928).
3. Padture, N. P., Gell, M., Jordan, E. H., Sechrist, Z. A. & George, S. M. Thermal barrier coatings for gas-turbine engine applications. *Science* **296**, 280–284 (2002).
4. Siegrist, T., Merkelbach, P. & Wuttig, M. Phase change materials: Challenges on the path to a universal storage device. *Annu. Rev. Condens. Matter Phys.* **3**, 215–237 (2012).
5. Bell, L. E. Cooling, heating, generating power, and recovering waste heat with thermoelectric systems. *Science* **321**, 1457–1461 (2008).
6. Li, B., Wang, L. & Casati, G. Thermal diode: Rectification of heat flux. *Phys. Rev. Lett.* **93**, 184301 (2004).
7. Martínez-Pérez, M. J., Fornieri, A. & Giazotto, F. Rectification of electronic heat current by a hybrid thermal diode. *Nat. Nanotechnol.* **10**, 303–307 (2015).
8. Xu, H., Shi, X., Gao, F., Sun, H. & Zhang, B. Ultrathin three-dimensional thermal cloak. *Phys. Rev. Lett.* **112**, 054301 (2014).
9. Narayana, S. & Sato, Y. Heat flux manipulation with engineered thermal materials. *Phys. Rev. Lett.* **108**, 214303 (2012).
10. Li, Y. et al. Temperature-dependent transformation thermotics: From switchable thermal cloaks to macroscopic thermal diodes. *Phys. Rev. Lett.* **115**, 195503 (2015).
11. Han, T. et al. Experimental demonstration of a bilayer thermal cloak. *Phys. Rev. Lett.* **112**, 054302 (2014).
12. Hatanaka, D., Mahboob, I., Onomitsu, K. & Yamaguchi, H. Phonon waveguides for electromechanical circuits. *Nat. Nanotechnol.* **9**, 520–524 (2014).
13. Anufriev, R., Ramiere, A., Maire, J. & Nomura, M. Heat guiding and focusing using ballistic phonon transport in phononic nanostructures. *Nat. Commun.* **8**, 15505 (2017).
14. Cahill, D. G. Thermal conductivity measurement from 30 to 750 K: The 3ω method. *Rev. Sci. Instrum.* **61**, 802–808 (1990).
15. Cahill, D. G. Analysis of heat flow in layered structures for time-domain thermoreflectance. *Rev. Sci. Instrum.* **75**, 5119–5122 (2004).
16. Schmidt, A. J., Cheaito, R. & Chiesa, M. A frequency-domain thermoreflectance method for

the characterization of thermal properties. *Rev. Sci. Instrum.* **80**, 094901 (2009).

17. Chávez-Ángel, E. et al. Reduction of the thermal conductivity in free-standing silicon nano-membranes investigated by non-invasive Raman thermometry. *APL Mater.* **2**, 012113 (2014).

18. Reparaz, J. S. et al. A novel contactless technique for thermal field mapping and thermal conductivity determination: Two-laser Raman thermometry. *Rev. Sci. Instrum.* **85**, 034901 (2014).

19. Williams, C. C. & Wickramasinghe, H. K. Scanning thermal profiler. *Appl. Phys. Lett.* **49**, 1587–1589 (1986).

20. Rojo, M. M., Calero, O. C., Lopeandia, A. F., Rodriguez-Viejo, J. & Martín-Gonzalez, M. Review on measurement techniques of transport properties of nanowires. *Nanoscale* **5**, 11526–11544 (2013).

21. Zhao, D., Qian, X., Gu, X., Jajja, S. A. & Yang, R. Measurement techniques for thermal conductivity and interfacial thermal conductance of bulk and thin film materials. *J. Electron. Packag.* **138**, 040802 (2016).

22. Regner, K. T., Freedman, J. P. & Malen, J. A. Advances in studying phonon mean free path dependent contributions to thermal conductivity. *Nanoscale Microscale Thermophys. Eng.* **19**, 183–205 (2015).

23. Minnich, A. J. Advances in the measurement and computation of thermal phonon transport properties. *J. Phys. Condens. Matter* **27**, 53202 (2015).

24. Toberer, E. S., Baranowski, L. L. & Dames, C. Advances in thermal conductivity. *Annu. Rev. Mater. Res.* **42**, 179–209 (2012).

25. Cahill, D. G. et al. Nanoscale thermal transport. *J. Appl. Phys.* **93**, 793–818 (2003).

26. Cahill, D. G. et al. Nanoscale thermal transport. II. 2003–2012. *Appl. Phys. Rev.* **1**, 011305 (2014).

27. Cahill, D. G., Goodson, K. E. & Majumdar, A. Thermometry and thermal transport in micro/nanoscale solid-state devices and structures. *J. Heat Transfer* **124**, 223 (2002).

28. Shi, L. Thermal and thermoelectric transport in nanostructures and low-dimensional systems. *Nanoscale Microscale Thermophys. Eng.* **16**, 79–116 (2012).

29. Volz, S. et al. Nanophononics: State of the art and perspectives. *Eur. Phys. J. B* **89**, 15 (2016).

30. Carslaw, H. S. & Jaeger, J. C. *Conduction of Heat in Solids*, 2nd ed. Oxford Science Publications. Oxford University Press, USA (1986).

31. Neogi, S. et al. Tuning thermal transport in ultra-thin silicon membranes by surface nanoscale engineering. *ACS Nano* **9**, 3820–3828 (2015).

32. Kim, K. et al. Quantitative scanning thermal microscopy using double scan technique. *Appl. Phys. Lett.* **93**, 203115 (2008).

33. Gomès, S., Assy, A. & Chapuis, P. O. Scanning thermal microscopy: A review. *Phys. Status Solidi Appl. Mater. Sci.* **212**, 477–494 (2015).

34. Roh, J. W. et al. Observation of anisotropy in thermal conductivity of individual single-crystalline bismuth nanowires. *ACS Nano* **5**, 3954–3960 (2011).

35. Wang, Z. et al. Thermal transport in suspended and supported few-layer graphene. *Nano Lett.* **11**, 113–118 (2011).

36. Xie, G., Ding, D. & Zhang, G. Phonon coherence and its effect on thermal conductivity of nanostructures. *Adv. Phys. X* **3**, 719–754 (2018).

37. Yang, N., Xu, X., Zhang, G. & Li, B. Thermal transport in nanostructures. *AIP Adv.* **2**, 041410 (2012).

38. Luo, T. & Chen, G. Anoscale heat transfer - from computation to experiment. *Phys. Chem. Chem. Phys.* **15**, 3389–3412 (2013).

39. Xu, Z. Heat transport in low-dimensional materials: A review and perspective. *Theor. Appl. Mech. Lett.* **6**, 113–121 (2016).

40. Kim, W. Strategies for engineering phonon transport in thermoelectrics. *J. Mater. Chem. C* **3**, 10336–10347 (2015).

41. Guo, Y. & Wang, M. Phonon hydrodynamics and its applications in nanoscale heat transport. *Phys. Rep.* **595**, 1–44 (2015).

42. Deymier, P. (ed.) *Acoustic Metamaterials and Phononic Crystals*, **173**. Springer: Berlin Heidelberg (2013).

43. Boukai, A. I. et al. Silicon nanowires as efficient thermoelectric materials. *Nature* **451**, 168–171 (2008).

44. Zebarjadi, M., Esfarjani, K., Dresselhaus, M. S., Ren, Z. F. & Chen, G. Perspectives on thermoelectrics: From fundamentals to device applications. *Energy Environ. Sci.* **5**, 5147–5162 (2012).

45. Incropera, F. P., DeWitt, D. P., Bergman, T. L. & Lavine, A. S. *Fundamentals of Heat and Mass Transfer*. Wiley: Hoboken, NJ (2011).

46. Liu, W. & Asheghi, M. Phonon-boundary scattering in ultrathin single-crystal silicon layers. *Appl. Phys. Lett.* **84**, 3819–3821 (2004).

47. Ju, Y. S. & Goodson, K. E. Phonon scattering in silicon films with thickness of order 100 nm. *Appl. Phys. Lett.* **74**, 3005–3007 (1999).

48. Asheghi, M., Leung, Y. K., Wong, S. S. & Goodson, K. E. Phonon-boundary scattering in thin silicon layers. *Appl. Phys. Lett.* **71**, 1798–1800 (1997).

49. Liu, X., Wu, X. & Ren, T. In situ and noncontact measurement of silicon membrane thermal conductivity. *Appl. Phys. Lett.* **98**, 174104 (2011).

50. Graczykowski, B. et al. Thermal conductivity and air-mediated losses in periodic porous silicon membranes at high temperatures. *Nat. Commun.* **8**, 415 (2017).

51. Burzo, M. G., Komarov, P. L. & Raad, P. E. Influence of the metallic absorption layer on the quality

of thermal conductivity measurements by the transient thermo-reflectance method. *Microelectron. J.* **33**, 697–703 (2002).

52. Cuffe, J. et al. Reconstructing phonon mean-free-path contributions to thermal conductivity using nanoscale membranes. *Phys. Rev. B Condens. Matter Mater. Phys.* **91**, 245423 (2015).

53. Johnson, J. A. et al. Direct measurement of room-temperature nondiffusive thermal transport over micron distances in a silicon membrane. *Phys. Rev. Lett.* **110**, 025901 (2013).

54. John, S. Strong localization of photons in certain disordered dielectric superlattices. *Phys. Rev. Lett.* **58**, 2486–2489 (1987).

55. Yablonovitch, E. Inhibited spontaneous emission in solid-state physics and electronics. *Phys. Rev. Lett.* **58**, 2059–2062 (1987).

56. Maldovan, M. Phonon wave interference and thermal bandgap materials. *Nat. Mater.* **14**, 667–674 (2015).

57. Martínez-Sala, R. et al. Sound attenuation by sculpture. *Nature* **378**, 241 (1995).

58. Maldovan, M. Sound and heat revolutions in phononics. *Nature* **503**, 209–217 (2013).

59. Ben-Abdallah, P. & Biehs, S. A. Near-field thermal transistor. *Phys. Rev. Lett.* **112**, 044301 (2014).

60. Joulain, K., Drevillon, J., Ezzahri, Y. & Ordonez-Miranda, J. Quantum thermal transistor. *Phys. Rev. Lett.* **116**, 200601 (2016).

61. Graczykowski, B. et al. Phonon dispersion in hypersonic two-dimensional phononic crystal membranes. *Phys. Rev. B Condens. Matter Mater. Phys.* **91**, 075414 (2015).

62. Alaie, S. et al. Thermal transport in phononic crystals and the observation of coherent phonon scattering at room temperature. *Nat. Commun.* **6**, 7228 (2015).

63. Zen, N., Puurtinen, T. A., Isotalo, T. J., Chaudhuri, S. & Maasilta, I. J. Engineering thermal conductance using a two-dimensional phononic crystal. *Nat. Commun.* **5**, 3435 (2014).

64. Jain, A., Yu, Y. J. & McGaughey, A. J. H. Phonon transport in periodic silicon nanoporous films with feature sizes greater than 100 nm. *Phys. Rev. B Condens. Matter Mater. Phys.* **87**, 195301 (2013).

65. Lim, J. et al. Simultaneous thermoelectric property measurement and incoherent phonon transport in holey silicon. *ACS Nano* **10**, 124–132 (2016).

66. Wagner, M. R. et al. Two-dimensional phononic crystals: Disorder matters. *Nano Lett.* **16**, 5661–5668 (2016).

67. Maire, J. et al. Heat conduction tuning by wave nature of phonons. *Sci. Adv.* **3**, e1700027 (2017).

68. Liao, Y., Shiga, T., Kashiwagi, M. & Shiomi, J. Akhiezer mechanism limits coherent heat conduction in phononic crystals. *Phys. Rev. B* **98**, 134307 (2018).

69. Xie, G. et al. Ultra-low thermal conductivity of two-dimensional phononic crystals in the incoherent regime. *NPJ Comput. Mater.* **4**, 21 (2018).

70. Anufriev, R. & Nomura, M. Reduction of thermal conductance by coherent phonon scattering in two-dimensional phononic crystals of different lattice types. *Phys. Rev. B* **93**, 045410 (2016).

71. Yu, J. K., Mitrovic, S., Tham, D., Varghese, J. & Heath, J. R. Reduction of thermal conductivity in phononic nanomesh structures. *Nat. Nanotechnol.* **5**, 718–721 (2010).

72. Hopkins, P. E. et al. Reduction in the thermal conductivity of single crystalline silicon by phononic crystal patterning. *Nano Lett.* **11**, 107–112 (2011).

73. Ravichandran, N. K. & Minnich, A. J. Coherent and incoherent thermal transport in nanomeshes. *Phys. Rev. B Condens. Matter Mater. Phys.* **89**, 205432 (2014).

74. Lee, J. et al. Investigation of phonon coherence and backscattering using silicon nanomeshes. *Nat. Commun.* **8**, 14054 (2017).

75. Hu, Y., Zeng, L., Minnich, A. J., Dresselhaus, M. S. & Chen, G. Spectral mapping of thermal conductivity through nanoscale ballistic transport. *Nat. Nanotechnol.* **10**, 701–706 (2015).

76. Wilson, R. B. & Cahill, D. G. Anisotropic failure of Fourier theory in time-domain thermoreflectance experiments. *Nat. Commun.* **5**, 5075 (2014).

77. Siemens, M. E. et al. Quasi-ballistic thermal transport from nanoscale interfaces observed using ultrafast coherent soft X-ray beams. *Nat. Mater.* **9**, 26–30 (2010).

78. Lee, J., Lim, J. & Yang, P. Ballistic phonon transport in holey silicon. *Nano Lett.* **15**, 3273–3279 (2015).

79. Hsiao, T. K. et al. Observation of room erature ballistic thermal conduction persisting over 8.3 μm in SiGe nanowires. *Nat. Nanotechnol.* **8**, 534–538 (2013).

80. Minnich, A. J. et al. Thermal conductivity spectroscopy technique to measure phonon mean free paths. *Phys. Rev. Lett.* **107**, 095901 (2011).

81. Regner, K. T. et al. Broadband phonon mean free path contributions to thermal conductivity measured using frequency domain thermoreflectance. *Nat. Commun.* **4**, 1640 (2013).

82. Zeng, L. et al. Measuring phonon mean free path distributions by probing quasiballistic phonon transport in grating nanostructures. *Sci. Rep.* **5**, 17131 (2015).

83. Lee, S. M., Cahill, D. G. & Venkatasubramanian, R. Thermal conductivity of Si-Ge superlattices. *Appl. Phys. Lett.* **70**, 2957–2959 (1997).

84. Huxtable, S. T. et al. Thermal conductivity of Si/SiGe and SiGe/SiGe superlattices. *Appl. Phys. Lett.* **80**, 1737–1739 (2002).

85. Venkatasubramanian, R. Lattice thermal conductivity reduction and phonon localizationlike behavior in

superlattice structures. *Phys. Rev. B Condens. Matter Mater. Phys.* **61**, 3091–3097 (2000).

86. Simkin, M. V. & Mahan, G. D. Minimum thermal conductivity of superlattices. *Phys. Rev. Lett.* **84**, 927–930 (2000).

87. Chen, G. Thermal conductivity and ballistic phonon transport in cross-plane direction of superlattices. *Phys. Rev. B* **57**, 14958–14973 (1998).

88. Ridley, B. K. Optical-phonon tunneling. *Phys. Rev. B* **49**, 17253–17258 (1994).

89. Luckyanova, M. N. et al. Coherent phonon heat conduction in superlattices. *Science* **338**, 936 (2012).

90. Ravichandran, J. et al. Crossover from incoherent to coherent phonon scattering in epitaxial oxide superlattices. *Nat. Mater.* **13**, 168–172 (2014).

91. Tian, Z., Esfarjani, K. & Chen, G. Green's function studies of phonon transport across Si/Ge superlattices. *Phys. Rev. B Condens. Matter Mater. Phys.* **89**, 235307 (2014).

92. Li, G. et al. In-plane thermal conductivity of radial and planar Si/SiO$_x$ hybrid nanomembrane superlattices. *ACS Nano* **11**, 8215–8222 (2017).

93. Koh, Y. K., Cao, Y., Cahill, D. G. & Jena, D. Heat-transport mechanisms in superlattices. *Adv. Funct. Mater.* **19**, 610–615 (2009).

94. Costescu, R. M., Cahill, D. G., Fabreguette, F. H., Sechrist, Z. A. & George, S. M. Ultra-low thermal conductivity in W/Al$_2$O$_3$ nanolaminates. *Science* **303**, 989–990 (2004).

95. Saha, B. et al. Cross-plane thermal conductivity of (Ti,W)N/(Al,Sc)N metal/semiconductor superlattices. *Phys. Rev. B* **93**, 045311 (2016).

96. Saha, B., Shakouri, A. & Sands, T. D. Rocksalt nitride metal/semiconductor superlattices: A new class of artificially structured materials. *Appl. Phys. Rev.* **5**, 021101 (2018).

97. Rawat, V., Koh, Y. K., Cahill, D. G. & Sands, T. D. Thermal conductivity of (Zr,W)N/ScN metal/semiconductor multilayers and superlattices. *J. Appl. Phys.* **105**, 024909 (2009).

98. Giri, A. et al. Heat-transport mechanisms in molecular building blocks of inorganic/organic hybrid superlattices. *Phys. Rev. B* **93**, 115310 (2016).

99. Krahl, F. et al. Thermal conductivity reduction at inorganic-organic interfaces: From regular superlattices to irregular gradient layer sequences. *Adv. Mater. Interfaces* **2018**, 1701692 (2018).

100. Giri, A., Niemelä, J. P., Szwejkowski, C. J., Karppinen, M. & Hopkins, P. E. Reduction in thermal conductivity and tunable heat capacity of inorganic/organic hybrid superlattices. *Phys. Rev. B* **93**, 024201 (2016).

101. Saha, B. et al. Phonon wave effects in the thermal transport of epitaxial TiN/(Al,Sc)N metal/semiconductor superlattices. *J. Appl. Phys.* **121**, 015109 (2017).

102. Latour, B., Volz, S. & Chalopin, Y. Microscopic description of thermal-phonon coherence: From coherent transport to diffuse interface scattering in superlattices. *Phys. Rev. B Condens. Matter Mater. Phys.* **90**, 014307 (2014).

103. Wang, Y., Gu, C. & Ruan, X. Optimization of the random multilayer structure to break the random-alloy limit of thermal conductivity. *Appl. Phys. Lett.* **106**, 073104 (2015).

104. Wang, Y., Huang, H. & Ruan, X. Decomposition of coherent and incoherent phonon conduction in superlattices and random multilayers. *Phys. Rev. B Condens. Matter Mater. Phys.* **90**, 165406 (2014).

105. Yang, B. & Chen, G. Partially coherent phonon heat conduction in superlattices. *Phys. Rev. B Condens. Matter Mater. Phys.* **67**, 195311 (2003).

106. Zhu, T. & Ertekin, E. Phonon transport on two-dimensional graphene/boron nitride superlattices. *Phys. Rev. B Condens. Matter Mater. Phys.* **90**, 195209 (2014).

107. da Silva, C., Saiz, F., Romero, D. A. & Amon, C. H. Coherent phonon transport in short-period two-dimensional superlattices of graphene and boron nitride. *Phys. Rev. B* **93**, 125427 (2016).

108. Mu, X., Zhang, T., Go, D. B. & Luo, T. Coherent and incoherent phonon thermal transport in isotopically modified graphene superlattices. *Carbon N. Y.* **83**, 208–216 (2015).

109. Guo, R., Jho, Y. D. & Minnich, A. J. Coherent control of thermal phonon transport in van der Waals superlattices. *Nanoscale* **10**, 14432–14440 (2018).

110. Verdier, M. et al. Influence of amorphous layers on the thermal conductivity of phononic crystals. *Phys. Rev. B* **97**, 115435 (2018).

Quantum Chaotic Systems and Random Matrix Theory

Akhilesh Pandey, Avanish
Kumar, and Sanjay Puri
Jawaharlal Nehru University

12.1   Introduction

This article provides a pedagogical introduction to the subject of random matrix theory (RMT) and its applications. More advanced readers may refer to several books and reviews which deal extensively with this subject [1–12].

Random matrices were introduced in the context of statistical multivariate analysis by Wishart [13]. They were first used in physics by Wigner to understand energy levels in complex nuclei [2]. In Wigner's prescription, ensembles of random matrices were used to model Hamiltonians of complex systems. The mathematical development of RMT was done by Wigner himself, as well as Dyson [14–18], Mehta [19], Gaudin [20], Porter [1], and several others. The ensemble properties of interest to physicists are energy level spectra, decay widths, and fluctuations of scattering cross sections. In an important discovery [21], it was found that random matrices are applicable to quantum chaotic systems, i.e., quantum systems whose classical counterparts are chaotic. This gave a justification for the utility of RMT to understand complex systems. Interest has also focused on the transmission properties of mesoscopic/nanoscopic systems, quantum field theory, number theory, and communication theory. More recently, there have been many studies of time series invoking the original ideas of Wishart [13].

These have found applications in diverse areas such as econophysics, atmospheric sciences, and biology.

In physics applications, ensembles are classified according to the time-reversal and space-rotation symmetries of the underlying physical systems. The three important classes are Gaussian ensembles of symmetric Hermitian matrices, general complex Hermitian matrices, and quaternion self-dual Hermitian matrices. These are referred to as Gaussian orthogonal ensembles (GOE), Gaussian unitary ensembles (GUE), and Gaussian symplectic ensembles (GSE). (These are also known as *classical ensembles*.) Here, the terms orthogonal, unitary, and symplectic refer to the transformations under which ensembles remain invariant.

Dyson [14–16] introduced analogous ensembles involving unitary matrices. These are referred to as the *circular ensembles*. They have the same three-fold classification as above, giving rise to circular orthogonal ensembles (COE), circular unitary ensembles (CUE), and circular symplectic ensembles (CSE) for symmetric unitary, general unitary, and self-dual unitary matrices, respectively.

This article is organized as follows. In Section 12.2, we present definitions of the ensembles. In Section 12.3, we discuss joint probability distributions (jpd) of the eigenvalues and their fluctuation measures. In Section 12.4, we discuss the statistical properties of eigenvectors.

In Section 12.5, we elucidate the importance of these ensembles in the analysis of nuclear spectra. In Section 12.6, we give a brief history of the connection between quantum chaos and RMT. In this context, we will use quantum kicked rotors (QKR) as a paradigm of quantum chaos. In Section 12.7, we highlight the connection between quantum counterparts of integrable systems and Poisson statistics. In Section 12.8, we introduce the classical kicked rotor and its properties. In Section 12.9, we define the QKR. In Section 12.10, we give results for eigenvalues and eigenvector fluctuations in QKR. In Section 12.11, we discuss transition ensembles, viz., ensembles which are intermediate between the classical ones. In Section 12.12, we turn our attention to mesoscopic systems. We discuss the experimentally important phenomenon of universal conductance fluctuations (UCF) and its understanding via RMT. In Section 12.13, we briefly discuss finite-range Coulomb gas (FRCG) models and their applications to QKR. In Section 12.14, we discuss the Wishart ensemble and its applications. Section 12.15 concludes this chapter.

12.2   Random Matrix Theory

In quantum systems, we encounter two types of matrices: Hermitian matrices which represent Hamiltonians, and unitary matrices which represent scattering matrices and time-evolution operators. Dyson has shown [18] that there are three important classes of matrices, depending on the time-reversal invariance (TRI) and space-rotation invariance (SRI). In Table 12.1, we list these different classes.

In this table, H and U refer to Hermitian and unitary matrices. We also specify the transformation groups under which H and U remain in the same class [1,22]. The parameter β (also known as the Dyson parameter) denotes the number of distinct components in the off-diagonal elements of H or U.

For $\beta = 1$, the matrices are symmetric in both H and U cases. For $\beta = 4$, they are self-dual (see Appendix A). For $\beta = 2$, there are no such constraints on matrices. For other symmetries, the matrices may be of block-diagonal form. For simplicity, we will mostly consider matrices with only a single block.

The jpd of the matrix elements of A, where the N-dimensional matrix A may be of the type H or U, is

$$P(A) = C \exp(-\beta \ tr \ V(A)). \qquad (12.1)$$

Here, $V(A)$ is a positive-definite function of A, and C is the normalization constant. (In this chapter, C and its variants will denote different normalization constants.) The cases studied extensively are (a) the Gaussian ensembles with $V(\xi) = \xi^2/4v^2$ for the Hermitian case (with v setting the scale of matrix elements), and (b) the (uniform) circular ensembles with $V(\xi) = 0$. These are commonly referred to by their acronyms GOE, GUE, and GSE in the Gaussian cases [2,3,9], and COE, CUE, and CSE in the circular cases.

TABLE 12.1   Classification of Random Matrices

TRI	SRI	H or U	β	Transformation group
Yes	Yes	Symmetric	1	Orthogonal (O)
Yes	No but integral spin	Symmetric	1	Orthogonal (O)
Yes	No but half-integral spin	Quaternion self dual	4	Symplectic (S)
No	Irrelevant	General	2	Unitary (U)

The letters O, U, and S specify the transformation invariances in Table 12.1.

The GOE matrices are symmetric, and the distinct matrix elements A_{jk} have independent Gaussian distributions with zero mean and variance $(1 + \delta_{jk})v^2$. For GUE, the matrix elements are complex. The real part has the same distribution as GOE, and the imaginary part is real anti-symmetric with zero mean and the variance $(1 - \delta_{jk})v^2$. For GSE, one needs four matrices. One of these is symmetric as in the GOE, and three are anti-symmetric as described above for GUE. The construction of GSE matrices has been described in Appendix A.

To study the statistical properties of eigenvalues and eigenvectors, we transform the jpd in Eq. (12.1) to a jpd of eigenvalues and eigenvectors. The exponential factor transforms into a corresponding factor for eigenvalues alone as the trace of $V(A)$ is invariant under transformations of the relevant group. There is an additional Jacobian factor which decomposes into a product of eigenvalues and eigenvector-dependent functions. Thus, the eigenvalue and eigenvector distributions are independent of each other.

For $\beta = 4$, the eigenvalues are doubly degenerate. In this case, we have $2N$-dimensional matrices and therefore $2N$ eigenvalues. However, in the discussion of statistical properties, it is standard practice to consider only the N distinct eigenvalues. We will follow this convention in all subsequent sections except in Section 12.11, where the GSE $\rightarrow$ GUE transition requires explicit consideration of the double degeneracy.

12.3   Statistical Properties of Eigenvalues

The transformation from matrix element space to eigenvalue and eigenvector space leads to the jpd of eigenvalues $\{x_i\}$ as

$$p(x_1, \ldots, x_N) = C' \exp(-\beta W), \qquad (12.2)$$

where

$$W = -\sum_{j<k} \log|x_j - x_k| + \sum_j V(x_j). \qquad (12.3)$$

The resultant jpd is

$$p(x_1, \ldots, x_N) = C' \prod_{j<k} |x_j - x_k|^\beta \prod_{j=1}^{N} e^{-\beta V(x_j)}. \qquad (12.4)$$

The logarithmic part of Eq. (12.3) is derived from the Jacobian of the transformation. This contribution is reminiscent of a two-dimensional Coulomb potential, and therefore, the system is termed the Coulomb gas [14]. We will also refer to these ensembles as linear ensembles.

We remark that the eigenvalues for circular ensembles are of the form $e^{i\theta_j}$, and their fluctuations will refer to fluctuations of the eigenangles θ_j. Equations (12.2)–(12.4), with $x_j \to e^{i\theta_j}$, apply to the jpd of eigenangles for circular ensembles. We will use the term level to describe both x and θ.

In our subsequent discussion, we will consider large N, unless otherwise specified. Then, it can be shown that the fluctuation properties of eigenvalues are independent of V for each β. This happens because V provides only the scale of the local spectra, and the fluctuations are studied in terms of local average spacing. Further, H and U give the same fluctuations. Therefore, we will focus mostly on Gaussian linear and uniform circular ensembles here [3,9].

An important quantity is the average density of eigenvalues, which is defined as $\bar{\rho}(x) = \int_{-\infty}^{\infty} \cdots \int_{-\infty}^{\infty} dx_2 \cdots dx_N \ p(x, x_2, \ldots, x_N)$. In the Gaussian case, $\bar{\rho}(x)$ is given by the well-known Wigner's semicircle:

$$\bar{\rho}(x) = \frac{2\sqrt{R^2 - x^2}}{\pi R^2}, \quad R^2 = 4\beta v^2 N. \quad (12.5)$$

Figure 12.1 shows the semicircular density for the GOE. In uniform circular ensembles, the density is

$$\bar{\rho}(\theta) = \frac{1}{2\pi}, \quad 0 \leqslant \theta < 2\pi. \quad (12.6)$$

In this case, the average spacing for eigenvalues is $2\pi/N$ (Figure 12.2). Thus, one needs to study statistical properties of $\xi_j = \theta_j N/(2\pi)$, so that the average spacing is unity everywhere.

In many real systems, the density is non-uniform (as in Eq. (12.5)). In that case, one needs to "unfold" the spectrum so that the average spacing becomes 1 in the entire spectrum. The unfolding function is defined by

$$F(\zeta) = N \int^{\zeta} \bar{\rho}(\zeta') d\zeta', \quad (12.7)$$

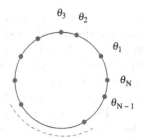

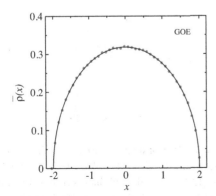

FIGURE 12.1 Eigenvalue density of one GOE matrix for $N = 5000$. The solid line corresponds to Wigner's semicircle in Eq. (12.5).

FIGURE 12.2 A schematic diagram showing a set of N eigenangles $(\theta_1, \ldots, \theta_N)$ in an ordered sequence on a circle. The eigenangles are treated as interacting particles which are restricted to move on the circle.

where ζ' refers to x or θ depending on the system. The unfolding function for the semicircular density of Eq. (12.5) is

$$F(x) = \frac{N}{\pi R^2} \left[x\sqrt{R^2 - x^2} + R^2 \arcsin\left(\frac{x}{R}\right) \right]. \quad (12.8)$$

The unfolded spectrum, ξ_j, is now given by $\xi_j = F(x_j)$ for the Gaussian ensembles (see Figure 12.3). When the Coulomb interaction (i.e., the logarithmic part in Eq. (12.3)) is absent, then the jpd can be written as a product of positive-definite functions $w(x)$:

$$p(x_1, \ldots, x_N) = C' \prod_{j=1}^{N} w(x_j), \quad (12.9)$$

where

$$w(x) = \exp(-\beta V(x)). \quad (12.10)$$

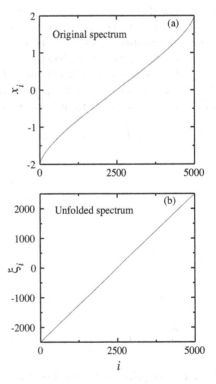

FIGURE 12.3 Plot of (a) original and (b) unfolded spectrum of one GOE matrix with $N = 5000$.

In this case, $\bar{\rho}(x) = w(x)$, and one can use Eq. (12.7) for unfolding the eigenvalue spectra. The jpd of the unfolded eigenvalues ξ_j will be simply 1, with $0 \leqslant \xi_j \leqslant N$. Thus, ξ_j will be distributed independently with uniform probability in $[0, N]$. This is referred to as the *Poisson ensemble*.

The eigenvalue fluctuations are studied in terms of the correlation functions. The most important quantity is the two-level correlation function $R_2(r)$ or, equivalently, the two-level cluster function $Y_2(r) = 1 - R_2(r)$. $R_2(r)dr$ is the conditional probability of finding a level in the interval $[\xi_0 + r, \ \xi_0 + r + dr]$, given that there is an eigenvalue at ξ_0. For the above three classes of ensembles, the (universal) cluster function is given by

$$Y_2(r) = (s(r))^2 + \left(\frac{d}{dr}s(r)\right)\left(\int_r^\infty s(t)dt\right), \quad \beta = 1,$$
$$\tag{12.11}$$

$$Y_2(r) = (s(r))^2, \quad \beta = 2, \tag{12.12}$$

$$Y_2(r) = (s(2r))^2 - \left(\frac{d}{dr}s(2r)\right)\left(\int_0^r s(2t)dt\right), \quad \beta = 4,$$
$$\tag{12.13}$$

where

$$s(r) = \frac{\sin(\pi r)}{\pi r}. \tag{12.14}$$

For $r \gtrsim 1$, $Y_2(r)$ falls off as $1/(\beta\pi r^2)$ after the oscillatory terms have been averaged out. One can also define the n-level cluster function, and the exact results are known for all these ensembles [23,24]. We remark that the correlation functions satisfy the property of stationarity, i.e., independent of ξ_0, and ergodicity, i.e., spectral and ensemble averages are equal [25]. For the Poisson ensemble, i.e., ensemble of independent eigenvalues, $Y_2(r) = 0$ (see Figure 12.4 for plots of $Y_2(r)$ vs. r for all four ensembles).

In actual applications, one usually considers the number variance $\Sigma^2(r)$, viz., the variance of the number of levels in intervals of length r. In terms of Y_2, it is given by

$$\Sigma^2(r) = r - 2\int_0^r (r-s)Y_2(s)ds. \tag{12.15}$$

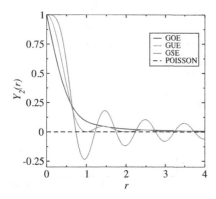

FIGURE 12.4 Two-point cluster function for the classical ensembles corresponding to $\beta = 1, 2, 4$. The dashed line corresponds to the Poisson ensemble.

For the above ensembles,

$$\Sigma^2(r) = \frac{2}{\beta\pi^2}\ln r + C_\beta, \tag{12.16}$$

valid for $r \gtrsim 1$. Here, the constant $C_\beta = 0.4420$, $0.3460, 0.2706$ for $\beta = 1, 2, 4$, respectively. The exact expressions for number variances are [3]

$$\Sigma^2_{\beta=2}(r) = \frac{1}{\pi^2}\left[\ln(2\pi r) + \gamma + 1 - \cos(2\pi r) - \mathrm{Ci}(2\pi r)\right]$$
$$+ r\left[1 - \frac{2\mathrm{Si}(2\pi r)}{\pi}\right], \tag{12.17}$$

$$\Sigma^2_{\beta=1}(r) = 2\Sigma^2_{\beta=2}(r) + \left(\frac{\mathrm{Si}(\pi r)}{\pi}\right)^2 - \frac{\mathrm{Si}(\pi r)}{\pi}, \tag{12.18}$$

$$\Sigma^2_{\beta=4}(r) = \frac{1}{2}\Sigma^2_{\beta=2}(2r) + \left(\frac{\mathrm{Si}(2\pi r)}{2\pi}\right)^2. \tag{12.19}$$

Here, $\gamma \ (= 0.5772\cdots)$ is the Euler constant, and Si and Ci are sine and cosine integrals, respectively. In contrast, for the Poisson ensemble, $\Sigma^2(r) = r$ (see Figure 12.5).

Another widely used fluctuation measure is the nearest-neighbor spacing distribution $p_0(s)$, where s is the spacing between two consecutive (unfolded) eigenvalues, i.e., $s_i = \xi_{i+1} - \xi_i$. For the Poisson ensemble, $p_0(s) = e^{-s}$. For the three random matrix ensembles, the exact results for $p_0(s)$ have been derived by Mehta. These expressions are complicated, but the spacing distributions for two-dimensional matrices give excellent approximations to the exact results. These are (see Figure 12.6)

$$p_0(s) = \frac{\pi}{2}s\exp\left(-\frac{\pi}{4}s^2\right), \qquad \beta = 1, \tag{12.20}$$

$$p_0(s) = \frac{32}{\pi^2}s^2\exp\left(-\frac{4}{\pi}s^2\right), \qquad \beta = 2, \tag{12.21}$$

$$p_0(s) = \frac{2^{18}}{3^6\pi^3}s^4\exp\left(-\frac{64}{9\pi}s^2\right), \qquad \beta = 4. \tag{12.22}$$

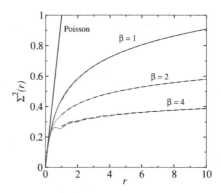

FIGURE 12.5 Plot of number variance, $\Sigma^2(r)$ vs. r. The solid lines, corresponding to $\beta = 1, 2, 4$, have been calculated from Eqs. (12.17)–(12.19). The dashed lines correspond to Eq. (12.16), which is valid for $r \gtrsim 1$. For reference, the Poisson case has also been plotted. For small $r \ (\lesssim 1)$, $\Sigma^2(r)$ for classical ensembles follows linear behavior as in the Poisson case, but for large r, it shows logarithmic behavior.

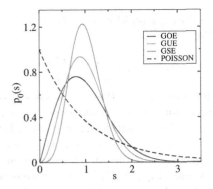

FIGURE 12.6 Nearest-neighbor spacing distributions, $p_0(s)$ vs. s, from Eqs. (12.20)–(12.22) for classical ensembles. The dashed line corresponds to the Poisson ensemble. For small s, $p_0(s) \sim s^{\beta}$, whereas for large s, $p_0(s) \sim \exp(-s^2)$ for $\beta = 1, 2, 4$.

The higher-order spacing distributions $p_k(s)$ have also been studied. These are distributions of spacings between two eigenvalues with k intermediate eigenvalues, i.e., $s_i = \xi_{i+k+1} - \xi_i$. For the nearest-neighbor spacing distribution, $k = 0$. These functions satisfy

$$R_2(s) = \sum_{k=0}^{\infty} p_k(s). \qquad (12.23)$$

Another quantity of interest is the spacing variance $\sigma^2(k)$, which is the variance of $p_k(s)$. This is closely related to $\Sigma^2(k+1)$. These two quantities are equal for the Poisson ensemble. On the other hand, for the three classical ensembles, the difference $\Sigma^2(k+1) - \sigma^2(k)$ is close to $1/6$, e.g., 0.161 for $k = 0$, and approaches $1/6$ rapidly as k increases [3].

To illustrate the difference between random matrix ensembles and the Poisson case, we consider a sequence of, say, 10,000 levels. A segment of this is shown in Figure 12.7. In that case, $\Sigma(r) \simeq 1$ in the random matrix ensembles, whereas $\Sigma(r) \simeq 100$ for the Poisson case. This property of level correlations is referred to as the spectral rigidity and is a consequence of the long-range correlations in the random matrix spectra. In descending order of rigidity, the spectra can be ranked as GSE, GUE, and GOE. Further, we mention that $p_0(s)$ and $R_2(s)$ approach 0 as s^{β} for small s. This is referred to as level repulsion in random matrix ensembles. This should be contrasted with level clustering observed in Poisson ensembles, where $p_0(s) = 1$ for $s \to 0$.

Another frequently used fluctuation measure is the Δ_3 statistic [17]. For an interval $[\xi_0, \xi_0 + r]$, we define the staircase function $N(x)$, which is the number of levels $\leq \xi_0 + x$. The quantity Δ_3 is the least square deviation of $N(x)$ from a best-fit straight line $Ax + B$ (see Figure 12.8):

$$\Delta_3(r) = \frac{1}{r} \operatorname*{Min}_{A,B} \int_{\xi_0}^{\xi_0+r} [N(x) - Ax - B]^2 \, dx. \qquad (12.24)$$

Its average is related to Σ^2, and hence to Y_2, by

$$\overline{\Delta}_3(r) = \frac{2}{r^4} \int_0^r (r^3 - 2r^2 s + s^3) \, \Sigma^2(s) ds. \qquad (12.25)$$

FIGURE 12.7 Comparison of unfolded spectrum for different ensembles. We note that the degree of uniformity increases from the Poisson to GSE ensembles. The arrowheads mark the occurrences of pairs of levels with the spacings smaller than $1/2$ of the average spacing.

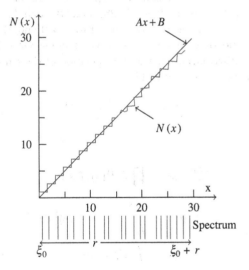

FIGURE 12.8 Schematic diagram for calculation of Δ_3.

Using $\Sigma^2(s) = s$, we find $\overline{\Delta}_3(r) = r/15$ for the Poisson ensemble. Further, for the classical ensembles with $r \geq 1$,

$$\overline{\Delta}_3(r) = \frac{1}{2}\Sigma^2(r) - \frac{9}{4\beta\pi^2}. \qquad (12.26)$$

To compute Δ_3, we consider an ordered sequence of eigenvalues $x_1, \ldots, x_n$ in the interval $[\xi_0, \xi_0 + r]$ containing n eigenvalues. For this sequence, Δ_3 is given by [4]

$$\Delta_3(\xi_0; r) = \frac{n^2}{16} - \frac{1}{r^2} \left(\sum_{j=1}^{n} \tilde{x}_j \right)^2 + \frac{3n}{2r^2} \sum_{j=1}^{n} \tilde{x}_j^2$$

$$- \frac{3}{r^4} \left(\sum_{j=1}^{n} \tilde{x}_j^2 \right)^2 + \frac{1}{r} \left[\sum_{j=1}^{n} (n - 2j + 1) \tilde{x}_j \right], \quad (12.27)$$

where $\tilde{x}_j = x_j - (\xi_0 + \frac{r}{2})$. Note that the number of eigenvalues n has the average value r. Δ_3 is the ensemble average. Fluctuation measures derived from higher-order correlation functions have also been used in numerical analysis [26,27].

The above formulas apply to energy level spectra obtained from a single random matrix ensemble. We can also consider a direct product of l independent random matrices with dimensions $f_j N$ (where $j = 1, \ldots, l$ with $\sum_{j=1}^{l} f_j = 1$). This corresponds to the superposition of l independent spectra. In this case, the fluctuation properties are generalized as

$$Y_2^{(\text{mix})}(r) = \sum_{j=1}^{l} (f_j)^2 Y_2^{(j)}(f_j r), \quad (12.28)$$

and the number variance is given by

$$\Sigma^{2,(\text{mix})}(r) = \sum_{j=1}^{l} \Sigma^{2,(j)}(f_j r). \quad (12.29)$$

Here, $Y_2^{(j)}$ and $\Sigma^{2,(j)}(r)$ are the two-level cluster function and number variance of the j-th component, respectively. There is a similar result for $\bar{\Delta}_3$, which can be derived from Eq. (12.25). In Eqs. (12.28)–(12.29), for l large and $f_j \sim \frac{1}{l}$, one can easily prove that $Y_2(r) \to 0$ and $\Sigma^2(r) \to r$, corresponding to Poisson spectra. The nearest-neighbor spacing distribution for the superposition of the l independent sequences $p_0^{(j)}(s)$ is [9]

$$p_0^{(\text{mix})}(s) = \frac{d^2 E_0^{(\text{mix})}(s)}{ds^2}. \quad (12.30)$$

Here,

$$E_0^{(\text{mix})}(s) = \prod_{j=1}^{l} E^{(j)}(s)(f_j s),$$

$$E^{(j)}(s) = \int_s^{\infty} \left[1 - F^{(j)}(r) \right] dr,$$

$$F^{(j)}(s) = \int_0^s p_0^{(j)}(r) dr. \quad (12.31)$$

For $l = 2$, i.e., superposition of two independent sequences, with $f_1 = f_2 = 1/2$ and $p_0^{(1)}(r) = p_0^{(2)}(r) = p_0(r)$ from Eq. (12.20) for $\beta = 1$, we obtain

$$p_0^{(\text{mix})}(s) = \frac{1}{8} e^{-\pi s^2/8} \left[s\pi \exp\left(\frac{\pi s^2}{16} \right) \text{erfc}\left(\frac{\sqrt{\pi} s}{4} \right) + 4 \right]. \quad (12.32)$$

The number variance for $l = 2$, with $f_1 = f_2 = 1/2$ and $\Sigma^{2,(1)}(r) = \Sigma^{2,(2)}(r) = \Sigma^2_{\beta=1}(r)$ from Eq. (12.18), is obtained from Eq. (12.29) as

$$\Sigma^{2,(\text{mix})}(r) = 2\Sigma^2_{\beta=1}\left(\frac{r}{2} \right). \quad (12.33)$$

Similarly, for the superposition of two GOE, the $Y_2^{(\text{mix})}$ function from Eq. (12.28) is

$$Y_2^{(\text{mix})}(r) = \frac{1}{2} Y_2^{\beta=1}\left(\frac{r}{2} \right). \quad (12.34)$$

Before concluding this section, it is useful to describe numerical procedures for generating matrix ensembles (as in Eq. (12.1)) and their eigenvalue spectra as in Eqs. (12.2)–(12.4). For generating matrices, the Gaussian cases are easiest as only Gaussian random numbers are involved (see the discussion after Eq. (12.1)). The uniform circular ensembles for $\beta = 2$ can be generated by constructing N complex random vectors and orthogonalizing them. For $\beta = 1, 4$ one can construct the matrices from the $\beta = 2$ matrices U by calculating UU^T for $\beta = 1$ and UU^D for $\beta = 4$.

For more general potentials $V(A)$, one can use the Monte Carlo (MC) method to generate matrix ensembles as in Eq. (12.1) and spectra as in Eqs. (12.2)–(12.4). We confine our discussion to the case of eigenvalue ensembles.

In the *linear case* [28], we take a set of N eigenvalues $(x_1, \cdots, x_N)$ ordered sequentially on a real line with fixed boundaries. The boundaries are chosen such that the probability of finding an eigenvalue outside the range is negligible. A stochastic move assigns, to any randomly chosen x_k, the new position x_k' between (x_{k-1}, x_{k+1}) with a uniform probability. The move is accepted with a probability $\exp(-\beta \Delta W)$, where ΔW is the change in the potential after the stochastic move. Time is measured in units of Monte Carlo steps (MCS), with one MCS corresponding to N attempted eigenvalue moves.

In the *circular case* [29], we take a set of N eigenvalues $\left(e^{i\theta_1}, \ldots, e^{i\theta_N} \right)$ ordered sequentially on the unit circle. Again, a stochastic move considers a randomly chosen eigenangle θ_j and assigns it the new position θ_j' between $(\theta_{j-1}, \theta_{j+1})$ with a uniform probability. After each eigenangle movement, we apply periodic boundary conditions; that is, θ_j' is computed modulo 2π. This clearly respects the original order of their positions. Again, the move is accepted with a probability $\exp(-\beta \triangle W)$, where $\triangle W$ is the change in the potential after the stochastic move. After one MCS, the new state with eigenangle positions $(\theta_1', \ldots, \theta_N')$ always has $\theta_i' < \theta_j'$ for $i < j$.

12.4 Statistical Properties of Eigenvectors

The eigenvectors of random matrices are useful in the study of fluctuations of transition widths and expectation values. The eigenvectors are random subject to the conditions of orthonormality. We focus on a typical eigenvector u which consists of N elements u_j. Each element has β components $\{u_j^{(\gamma)} : \gamma = 0, \ldots, \beta - 1\}$ [3]. The jpd of the components and elements of a single eigenvector $\{u_j^{(\gamma)}\}$ is given by

$$Q\left(\{u_j^{(\gamma)}\} \right) = \pi^{-\frac{\beta N}{2}} \Gamma\left(\frac{\beta N}{2} \right) \delta\left(\sum_{j=1}^{N} |u_j|^2 - 1 \right). \quad (12.35)$$

Here, the expression $|u_j|^2$ refers to the sum of absolute squares of $u_j^{(\gamma)}$.

The distribution of a single element of an eigenvector is given by

$$Q(u_j) = \frac{\pi^{-\frac{\beta}{2}} \Gamma\left(\frac{\beta N}{2}\right)}{\Gamma\left(\frac{\beta(N-1)}{2}\right)} \left(1 - |u_j|^2\right)^{\frac{\beta(N-1)}{2} - 1}. \qquad (12.36)$$

For large N, the $u_j^{(\gamma)}$ become independent Gaussian variables with mean 0 and variance $1/(\beta N)$. For a given j, the variable $w = N|u_j|^2$ has a χ_β^2-distribution with mean 1, variance $2/\beta$:

$$f_\beta(w) = \frac{\left(\frac{\beta}{2}\right)^{\frac{\beta}{2}}}{\Gamma(\frac{\beta}{2})} w^{\frac{\beta}{2}-1} \exp\left(-\frac{\beta w}{2}\right). \qquad (12.37)$$

This distribution function is plotted in Figure 12.9 for $\beta = 1, 2, 4$. For $\beta = 1$, it is known as the Porter–Thomas distribution in nuclear physics:

$$f_1(w) = \frac{1}{\sqrt{2\pi}} w^{-\frac{1}{2}} \exp(-w/2). \qquad (12.38)$$

This distribution is realized in physical systems as follows. The transition matrix elements connecting the states in a narrow energy band to a lower energy state may be treated as independent zero-centered Gaussian random variables. Then, Eq. (12.35) describes the distribution of the transition widths ($\propto$ absolute square of the matrix elements) normalized to unity.

12.5   Application to Nuclear Spectra

As mentioned earlier, Wigner introduced RMT to study slow neutron resonances of heavy nuclei. Levels with fixed angular momentum and parity were used (typically, $\frac{1}{2}^+$ levels). The subsequent mathematical developments motivated numerous experiments on nuclear spectra, mostly performed at Columbia University. There were also some experiments on proton resonances in light nuclei. The utility of GOE in understanding nuclear spectra was broadly established by these comparisons. In the early 1980s, the same data was reanalyzed with some new features [26,27,30]:

(a) The nuclear spectra from different experiments were combined to enhance the statistics. The combined data was referred to as "nuclear data ensemble" (NDE).

(b) More sophisticated statistical measures were used to analyze the spectral fluctuations.

(c) The sample errors were carefully calculated and used to quantify the level of agreement.

In this section, we will give a brief review of the NDE analysis. In this analysis, the NDE consisted of 1762 resonance energies corresponding to 36 sequences of 32 different nuclei. Note that, for a single nucleus, the fluctuation measures are calculated as spectral averages. For the NDE, there is further averaging over the ensemble of nuclei. We show the nearest-neighbor spacing histogram of ^{167}Er (obtained from n + ^{166}Er) in Figure 12.10(a) and of the NDE in Figure 12.10(b). One can see that the quality of agreement with GOE improves considerably when the spectra of many nuclei are combined. In Figures 12.11 and 12.12, we show $\Sigma^2(r)$ and $\overline{\Delta}_3(r)$ for the NDE. Again, very good agreement is found with GOE. In Figure 12.13, we show histograms for the distribution of square roots of transition widths for (a) ^{167}Er and (b) the NDE consisting of 1182 widths of 21 sequences. The result is compared with the Gaussian distribution. (Note that the Gaussian distribution arises from the Porter–Thomas distribution in Eq. (12.38) for the variable $x = \sqrt{w}$.)

Finally, we discuss the correlation between the energy levels $\{x_j\}$ and the corresponding normalized widths $\{w_j\}$. RMT predicts the width and energy level fluctuations to be independent. We define the correlation coefficient r (for each sequence) as

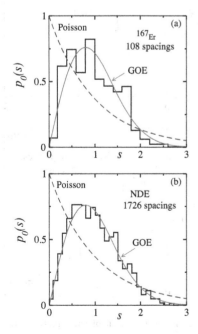

FIGURE 12.10 Nearest-neighbor spacing distribution for (a) ^{167}Er, (b) Nuclear data ensemble (NDE).

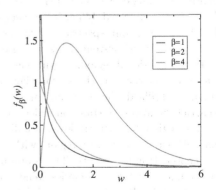

FIGURE 12.9   Plot of $f_\beta(w)$ vs. w for $\beta = 1, 2, 4$.

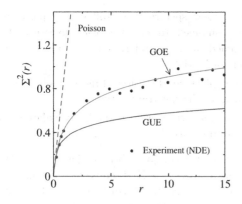

FIGURE 12.11   Number variance $\Sigma^2(r)$ vs. r for the NDE.

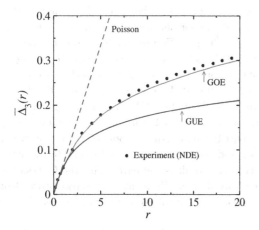

FIGURE 12.12 $\bar{\Delta}_3(r)$ vs. r for the NDE.

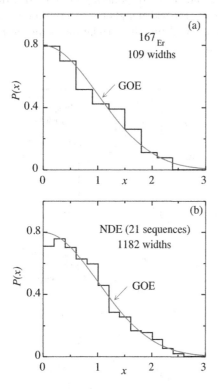

FIGURE 12.13   Distribution of the square roots of transition widths for (a) ^{167}Er, and (b) the NDE. The solid line denotes the Gaussian density: $P(x) = \sqrt{(2/\pi)}\exp(-x^2/2)$.

$$r = \frac{1}{N}\sum_{j=1}^{N}\left(\frac{w_j - \langle w \rangle}{\sigma_w}\right)\left(\frac{x_j - \langle x_j \rangle}{\sigma_x}\right), \qquad (12.39)$$

where N is the number of levels in the sequence. Here, $\langle w \rangle$ and σ_w^2 are the mean and variance of $\{w_j\}$, and

$$\langle x_j \rangle = Aj + B, \quad \sigma_x^2 = \frac{1}{N}\sum_{j=1}^{N}\left(x_j - \langle x_j \rangle\right)^2. \qquad (12.40)$$

The parameters A and B are calculated by minimizing the expression for σ_x^2. For the NDE, one evaluates the average of the correlation coefficient r, weighted according to the size of the sequence. For 1182 widths with 21 sequences, one obtains $r(NDE) = 0.017$, confirming the independence of widths and energy level spectra.

12.6   Quantum Chaos and Random Matrices

It is relevant to ask why the GOE works so well for nuclear spectra. We emphasize that the system's physical symmetries are important in determining the relevant level statistics. Thus, e.g., the GUE is not encountered in TRI systems, e.g., nuclear spectra. Moreover, the quantum numbers should not be mixed as this leads to the superposition of independent spectra leading to Poisson statistics (see Section 12.3). Apart from nuclear spectra, random matrix statistics is also found in complex atomic and molecular spectra. Rosenzweig and Porter [31] have shown that levels having the same quantum numbers show level repulsion, and the spacing distribution follows Wigner's prediction. More generally, it has emerged that quantum chaotic systems, viz., quantum systems whose classical analogs are chaotic, follow random matrix statistics. In this section, we will provide a brief overview of some important developments in quantum chaos.

In early work, Percival [32] introduced the terminology "regular and irregular" spectra, which arise in quantum analogs of integrable and chaotic classical systems, respectively. In 1977, Berry and Tabor showed that the regular spectra follow Poisson statistics and display level clustering [33]. Their proof used semiclassical quantization of integrable systems.

Subsequently, Mcdonald–Kaufman [34] and Berry [35] considered stadium and Sinai billiards, respectively, and showed that their quantum spectra display level repulsion. Classically, a billiard system refers to a free particle moving in a two-dimensional region, obeying classical reflection rules at the boundary of the region. For the corresponding quantum billiard, the eigenfunctions are solutions of the free-particle Schrodinger equation which vanish at the boundary of the region. The energy levels are discrete.

The Sinai billiard is a square region with a circular disk at the center (Figure 12.14). The stadium billiard is a region bounded by two half-circles joined by a rectangle (Figure 12.15). Both these systems exhibit strong classical chaos. Quantum billiards are experimentally realized as

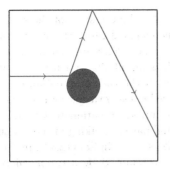

FIGURE 12.14　Sinai billiard.

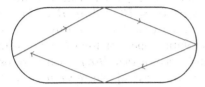

FIGURE 12.15　Stadium billiard.

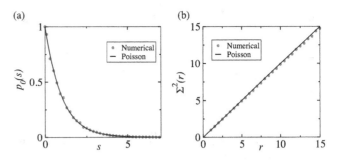

FIGURE 12.16　Spectral analysis of the system corresponding to Eq. (12.41). (a) Nearest-neighbor spacing distribution $p_0(s)$ and (b) number variance $\Sigma^2(r)$.

quantum dots in both mesoscopic and nanoscopic systems. For quantum studies, desymmetrized billiards were used in both cases so that the spectra corresponded to the same quantum numbers.

In 1984, Bohigas et al. [21] considered the same quantum billiards again. However, they discarded the low-lying energy levels to obtain much sharper correspondence with GOE. The corresponding spectra accurately displayed the properties of level repulsion and spectral rigidity found in GOE. The deep connection between quantum chaos and RMT is now known as the Bohigas-Giannoni-Schmit (BGS) conjecture. Seligman et al. [36] showed GOE statistics in a system of coupled quantum nonlinear oscillators. Berry and Robnik [37] found GUE statistics in the Aharonov–Bohm billiard system, where TRI is broken by the introduction of a magnetic field. Another important study in this context is the spectral analysis of the anisotropic Kepler problem [38]. Berry gave a semiclassical theory of spectral rigidity, using Gutzwiller's periodic orbit theory [39].

Finally, we mention the quantum version of periodically kicked rotors, also referred to as QKR. In this system, both GOE and GUE statistics are found [40]. In Sections 12.8 and 12.9, we make a detailed study of the classical and quantum properties of kicked rotor systems. We remark that the QKR has received much attention in the literature because its numerical implementation is much easier than that of the quantum billiard.

12.7　Quantum Integrable Systems and Poisson Statistics

As mentioned above, Poisson statistics is encountered in quantum counterparts of integrable systems. We give here two simple examples to illustrate this. First, we consider the case where the eigenvalue spectrum can be written as

$$x_j = j^2\alpha, \quad (\text{mod } 1), \tag{12.41}$$

where α is an irrational number. In this system, the average density $\bar{\rho}(x)$ is constant. In Figure 12.16, we show that the spacing distribution [$p_0(s)$] and the number variance [$\Sigma^2(r)$] obey Poisson statistics. This spectrum can arise in a two-dimensional system with harmonic binding in one direction and a rigid wall in the other direction.

The system of harmonic oscillators is a major exception to the "integrability implies Poisson" rule. For example, consider the spectrum

$$x_j = j\alpha, \quad (\text{mod } 1), \tag{12.42}$$

which arises in a system of two-dimensional harmonic oscillators. This system does not exhibit Poisson or any other universal statistics [41,42].

Our second example of Poisson statistics is the rectangular billiard [43]. The eigenvalues can be written as

$$E_{n_1,n_2} = \alpha_1 n_1^2 + \alpha_2 n_2^2, \tag{12.43}$$

where $n_1, n_2 = 1, 2, 3, \ldots$ and α_1, α_2 contain information about the lengths of the sides. In this case also, the level density is constant. When α_1, α_2 are not rationally connected, Poisson statistics is obtained.

12.8　Classical Kicked Rotor

Periodically kicked rotors have been widely used in studies of classical and quantum chaos [11,44–47]. In this section, we briefly review the classical kicked rotor.

The Hamiltonian of a periodically kicked rotor can be written as

$$H = \frac{p^2}{2} + V(\phi) \sum_{n=-\infty}^{\infty} \delta(t-n). \tag{12.44}$$

Here, ϕ is the angle of rotation, and p is the angular momentum. Without the loss of generality, we have chosen both the moment of inertia and the time period of kicking to be unity. The classical motion is such that there is free rotation in the time interval $n+1 > t > n$. At each integer time n, the rotor is subjected to an impulse of magnitude $|V'(\phi_n)|$, where ϕ_n is the value of ϕ at $t = n$. Let us also

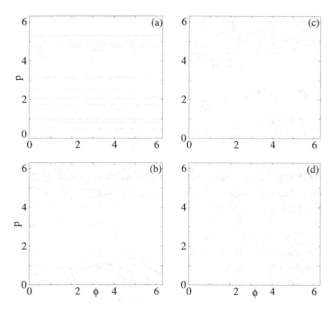

FIGURE 12.17 Phase plot of the Chirikov map for (a) $\alpha = 0$, (b) $\alpha = 0.5$, (c) $\alpha = 1$, and (d) $\alpha = 10$. In each frame, we have taken 100 initial conditions with 250 iterations.

define p_n as the value of the momentum at $t = n - 0$. Note that, at $t = n + 0$, the momentum will be p_{n+1}. Thus, one can write the kicked rotor map as

$$\phi_{n+1} = \phi_n + p_{n+1}, \tag{12.45}$$
$$p_{n+1} = p_n - V'(\phi_n). \tag{12.46}$$

Consider the potential $V(\phi) = \alpha \cos \phi$, where α is the kicking parameter (also called "the chaos parameter"). This gives the Chirikov standard map:

$$\phi_{n+1} = \phi_n + p_{n+1}, \qquad (\text{mod } 2\pi), \tag{12.47}$$
$$p_{n+1} = p_n + \alpha \sin \phi_n, \qquad (\text{mod } 2\pi). \tag{12.48}$$

We have used the operation (mod 2π) for p also since the potential $V(\phi)$ is periodic.

In Figure 12.17, we show the phase plots of the Chirikov map in the (p, ϕ)-plane for several values of α. The smooth curves indicate regular motion, whereas the dotted regions correspond to chaotic motion. The dashed regular curves in all figures correspond to periodic orbits. The different curves for regular motion arise from different initial conditions. On the other hand, even one initial condition can give rise to area-filling trajectories in the chaotic regime. For $\alpha = 0$, we have only regular motion, whereas the trajectories are almost entirely chaotic for large α.

12.9　Quantum Kicked Rotor

We define QKR by using the unit time-evolution operator U to describe the evolution of the wave function. Because of the time-periodicity in the Hamiltonian, U is time independent and is defined by

$$|\psi_{n+1}\rangle = U |\psi_n\rangle. \tag{12.49}$$

We consider $|\psi_n\rangle$ to be the state of the system at time $t = n - 0$, as in the classical case. Eigenvalues of U are of the form $e^{i\theta}$, where θ is called the eigenangle.

In general, U is an infinite-dimensional matrix. However, for the QKR with the cosine potential, one can construct finite-dimensional U using the periodic boundary conditions of p and θ. We introduce parameters γ and ϕ_0 to describe time-reversal and parity breaking in the system by changing $p \to p + \gamma$ and $\phi \to \phi + \phi_0$ in Eq. (12.44). In the classical case, γ and ϕ_0 can be removed by a canonical transformation. However, in the quantum case, both parameters are very important and appear explicitly in the operators.

One can write $U = BG$, where B is the operator for the evolution from time $t = n - 0$ to $t = n + 0$ and G is the free evolution operator from $t = n + 0$ to $t = n + 1 - 0$. It can be shown that $B(\alpha) = \exp[-i\alpha \cos(\phi + \phi_0)]$, and $G = \exp[-i(p + \gamma)^2/2]$, where we have set $\hbar = 1$. For N-dimensional U, the matrix elements of B in the position representation can be written as

$$B_{jk} = \exp\left[-i\alpha \cos\left(\frac{2\pi j}{N} + \phi_0\right)\right] \delta_{jk}. \tag{12.50}$$

Here, $j, k = -N', -N' + 1, \ldots, N'$ with $N' = (N - 1)/2$. On the other hand, G in the momentum representation can be written as

$$G_{mn} = \frac{1}{N} \sum_{l=-N'}^{N'} \exp\left[-i\left(\frac{1}{2}l^2 - \gamma l - \frac{2\pi(m-n)l}{N}\right)\right], \tag{12.51}$$

where $m, n = -N', -N' + 1, \ldots, N'$. Using the transformation properties between position and momentum representations, one can write the matrix elements of U as

$$U_{jk} = \frac{1}{N} \exp\left[-i\alpha \cos\left(\frac{2\pi j}{N} + \phi_0\right)\right]$$
$$\times \sum_{l=-N'}^{N'} \exp\left[-i\left(\frac{l^2}{2} - \gamma l - \frac{2\pi(j-k)l}{N}\right)\right]. \tag{12.52}$$

12.10　Eigenvalue and Eigenvector Fluctuations in QKR

In this section, we will show representative results for eigenvalue and eigenvector statistics in the QKR using the fluctuation measures introduced in Sections 12.3–12.4. We consider $\phi_0 = \pi/(2N)$, corresponding to the case when parity is fully broken. (At some places, we will also consider the case $\phi_0 = 0$, corresponding to parity being preserved.) For γ, we choose two values: $\gamma = 0.0$ and $\gamma = 0.7$, corresponding to TRI and broken TRI, respectively. These two values of γ will give rise to COE (or GOE) and CUE (or GUE) statistics, respectively. We consider $N = 1000$. Since the spectra in the highly chaotic case (α very large) are known to become independent [48] very rapidly with increasing α, we generate spectra for α ranging from 10^4

to 10^6, in steps of 1000. This gives us 1000 independent spectra. As the density is uniform, the unfolded spectrum is obtained by multiplying the original eigenangles by $N/2\pi$.

In Figure 12.18, we plot $p_0(s)$ vs. s for the QKR and compare it with corresponding RMT results. As mentioned earlier, $\gamma = 0.0$ yields the $\beta = 1$ case, and $\gamma = 0.7$ gives the $\beta = 2$ case. We see an excellent agreement between the QKR results and RMT. Figures 12.19 and 12.20 are analogous plots of number variance and the two-point correlation function.

It is also relevant to study the statistics of eigenvectors obtained from the QKR ensembles. In Figure 12.21, we plot $f_\beta(w)$ vs. w for the QKR with $\gamma = 0.0$ ($\beta = 1$) and $\gamma = 0.7$ ($\beta = 2$). Again, the QKR data is in excellent agreement with the RMT results in Eq. (12.37).

Next, let us examine the spectra of mixed ensembles which were introduced in Eq. (12.28). We obtain mixed spectra from QKR by choosing $\gamma = 0.0$, $\phi_0 = 0$. The parameter value $\phi_0 = 0$ corresponds to the sum of 2 independent spectra arising from states of even and odd parity. Clearly, the mixed spectrum is also characterized by TRI. In Figure 12.22, we show results for $p_0(s)$ and $\Sigma^2(r)$ for the mixed spectrum. The QKR data is in excellent agreement with the RMT result in Eqs. (12.32)–(12.33). In Figure 12.23, we show the two-point correlation function.

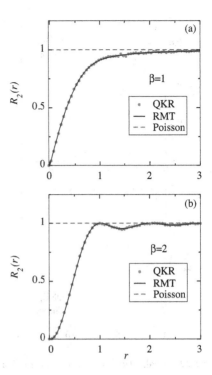

FIGURE 12.20 Two-point correlation function of QKR for (a) $\gamma = 0.0$ ($\beta = 1$) and (b) $\gamma = 0.7$ ($\beta = 2$).

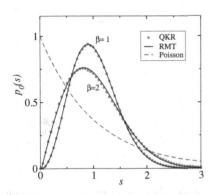

FIGURE 12.18 Spacing distribution of QKR for $\gamma = 0.0$ ($\beta = 1$) and $\gamma = 0.7$ ($\beta = 2$).

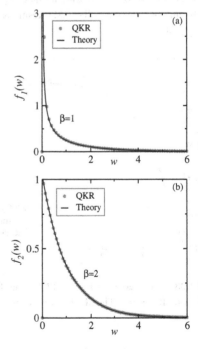

FIGURE 12.21 Distribution of eigenvector components $w = N|u_j|^2$ of the QKR. We show numerical results for (a) $\gamma = 0.0$ ($\beta = 1$) and (b) $\gamma = 0.7$ ($\beta = 2$). The RMT results are plotted from Eq. (12.37).

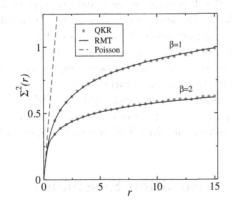

FIGURE 12.19 Analogous to Figure 12.18, but for number variance.

The QKR does not exhibit CSE (or GSE) spectra directly. The system of kicked tops [8] has been investigated for direct realization of all the three random matrix ensembles (COE, CUE, and CSE). We discuss here a method of indirect realization of CSE in QKR.

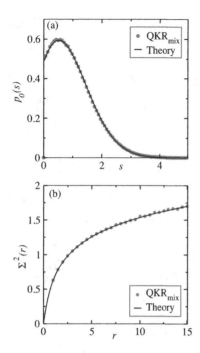

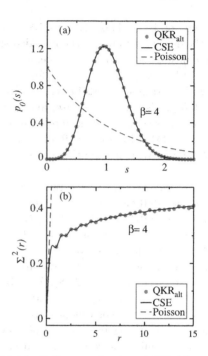

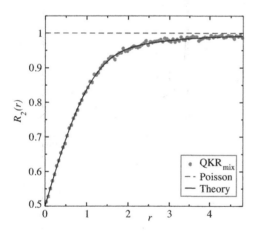

FIGURE 12.22 Realization of mixed spectra from COE in QKR. The parameter values are $\gamma = 0.0$ and $\phi_0 = 0$, which yields a superposition of 2 independent COE spectra (of opposite parity). We show results for (a) Nearest-neighbor spacing distribution and (b) number variance. The RMT result is provided in Eqs. (12.32)–(12.33).

FIGURE 12.24 Indirect realization of CSE ($\beta = 4$) in QKR. Alternate eigenvalues of COE ($\beta = 1$) spectra are analyzed giving rise to CSE statistics. The frames show (a) nearest-neighbor spacing distribution and (b) number variance.

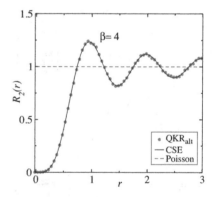

FIGURE 12.25 Two-point correlation function for CSE ($\beta = 4$), obtained indirectly in QKR.

FIGURE 12.23 Analogous to Figure 12.22, but for the two-point correlation function. The RMT result is provided in Eq. (12.34).

There are two remarkable theorems which relate the fluctuations of the three classical ensembles. The first theorem [49] states that the spectra of CSE can be obtained by choosing alternate eigenvalues of COE. The second theorem [16,50] states that the spectra of CUE can be obtained by choosing alternate eigenvalues from a random superposition of two independent COE spectra of the same dimension.

In Figure 12.24, we show the spacing distribution and the number variance of alternate eigenvalues of the QKR ensemble (of size 1000) with matrix dimension $N = 1000$.

The parameter values are $\gamma = 0.0$ and $\phi_0 = \pi/(2N)$. In this case, the dimension of the matrix reduces to 500, and the number of the spectra becomes 2000. The results show excellent agreement with CSE. In Figure 12.25, we show the analogous plot for the two-point correlation function.

In a similar fashion, we consider QKR spectra for $\gamma = 0.0, \phi_0 = 0$. As mentioned earlier, since $\phi_0 = 0$, this leads to a superposition of two independent COE spectra with opposite parities. An analysis of alternate eigenvalues gives excellent agreement with CUE, as seen in Figures 12.26 and 12.27.

12.11 Transition Ensembles

The ensembles introduced in Section 12.2 correspond to exact symmetries; that is, a symmetry is either fully

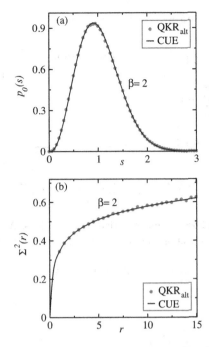

FIGURE 12.26 Indirect realization of CUE ($\beta = 2$) from COE in QKR. In this case, $\gamma = 0.0$ and $\phi_0 = 0$, which preserves both the TRI and parity, giving rise to a superposition of two independent COE spectra. The analysis of alternate eigenvalues of this spectrum gives CUE statistics. We show data for (a) nearest-neighbor spacing distribution and (b) number variance.

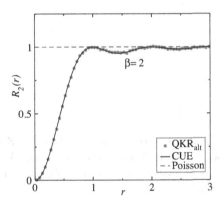

FIGURE 12.27 Analogous to Figure 12.26, but for the two-point correlation function.

preserved or fully broken. In this section, we briefly consider ensembles which have a partially broken symmetry. Let us consider a Gaussian ensemble defined by [51]

$$H_\alpha = H(S) + i\alpha H(A). \qquad (12.53)$$

Here, $H(S)$ are real symmetric matrices, $H(A)$ are real antisymmetric matrices, and α is a real parameter. The distinct matrix elements of $H(S)$ and $H(A)$ are independent Gaussian random variables with zero mean. The variances are v^2 for the off-diagonal elements of $H(S)$ and $H(A)$, and $2v^2$ for the diagonal matrix elements of $H(S)$. The diagonal matrix elements of $H(A)$ are zero. Note that the H_α-ensemble is GOE for $\alpha = 0$ and GUE for $\alpha = 1$. Thus, α is a measure of

TRI breaking in the system and parametrizes the GOE $\rightarrow$ GUE transition.

This problem has been solved approximately in [51] and exactly in [52]. The average density of eigenvalues is again the semicircle of Eq. (12.5) with

$$R^2 = 4Nv^2(1 + \alpha^2). \qquad (12.54)$$

For large N, the exact two-level cluster function is given by [52]

$$Y_2(r, \Lambda) = \left(\frac{\sin \pi r}{\pi r}\right)^2 - \frac{1}{\pi^2} \int_0^\pi dx \, x \, \sin(xr) \exp(2\Lambda x^2)$$
$$\times \int_\pi^\infty dy \, \frac{\sin(yr)}{y} \exp(-2\Lambda y^2), \qquad (12.55)$$

where Λ is the transition parameter:

$$\Lambda = \alpha^2 v^2 / (D(x))^2. \qquad (12.56)$$

Here, $D = 1/[N\bar{\rho}(x)]$ is the average spacing at x. For $\Lambda = 0, \infty$, we obtain the GOE and GUE results of Eqs. (12.11) and (12.12), respectively. Note that, for spectra with finite span, one should choose $v^2 N = 1$ and therefore $\Lambda \propto \alpha^2 N$. Thus, as N increases, α becomes smaller to keep Λ finite. For $N = \infty$, the GOE $\rightarrow$ GUE transition is abrupt at $\alpha = 0$.

The exact spacing distributions are also known for this problem [53]. As in Section 12.3, these are complicated, but the nearest-neighbor spacing distribution for two-dimensional matrices gives an excellent approximation to the $N \rightarrow \infty$ result. The $N = 2$ result is [51]

$$p_0(s) = \frac{s}{4v^2(1 - \alpha'^2)^{1/2}} \exp\left(-\frac{s^2}{8v^2}\right) \operatorname{erf}\left[\left(\frac{1 - \alpha'^2}{8\alpha'^2 v^2}\right) s\right]. \qquad (12.57)$$

Here, instead of Eq. (12.56), one has to use α' as a fitting parameter to obtain the spacing distribution.

In actual applications, Λ should be interpreted as the symmetry-breaking matrix elements of the relevant Hamiltonian. These results also apply to circular ensembles with the Hamiltonian being replaced by the unitary operator [54].

The number variance, $\Sigma^2(r; \Lambda)$, can be obtained by numerical integration of Eq. (12.15) with Y_2 from Eq. (12.55). However, a very good approximation can be written as

$$\Sigma^2(r; \Lambda) = \Sigma^2_{\beta=2}(r) + \frac{1}{2\pi^2} \ln\left[1 + \frac{\pi^2 r^2}{4(\tau + 2\pi^2 \Lambda)^2}\right], \qquad (12.58)$$

with $\tau = 0.615$.

Let us consider some important applications of the GOE $\rightarrow$ GUE transition. This transition has proved to be useful in deriving upper bounds on the TRI breaking part of the nucleon–nucleon interaction [51,55]. Another system where this transition is applicable is the Aharonov–Bohm chaotic billiard, where a single line of magnetic flux passes perpendicular through the plane of the system, breaking TRI [37]. A calculation involving Aharonov–Bohm flux lines in a disordered metallic ring again shows the exact GOE → GUE realization of $\Sigma^2(r)$ and $p_0(s)$ [56].

The parameter Λ also appears in transitions involving breaking of other symmetries. Then, one considers the generalization of Eq. (12.53) as

$$H(\alpha) = A + \alpha B, \tag{12.59}$$

where A and B are independent Gaussian random matrices representing symmetry-preserving and symmetry-breaking parts of the Hamiltonian respectively. In such cases, v^2 in Eq. (12.56) refers to the variance of the symmetry-breaking matrix elements B_{jk} in the A-diagonal representation.

An example is the GSE → GUE transition, which has also been solved exactly [53]. In this case, the exact two-level cluster function is given by

$$Y_2(r, \Lambda) = \left(\frac{\sin \pi r}{\pi r}\right)^2 - \frac{1}{\pi^2} \int_0^\pi dx\, \frac{\sin(xr)}{x} \exp(2\Lambda x^2)$$
$$\times \int_\pi^\infty dy\, y \sin(yr) \exp(-2\Lambda y^2). \tag{12.60}$$

For $\Lambda = 0$, Eq. (12.60) reduces to a modified version of Eq. (12.13), where the double degeneracy of the eigenvalues is explicitly taken into account. Equations (12.55) and (12.60) also apply to transitions in Jacobi ensembles [57].

For studies of parity breaking, one considers the 2GOE → 1GOE transition. More generally, one can consider lGOE → 1GOE transitions for the breaking of symmetries having l quantum numbers. Here, lGOE refers to a direct sum of l independent GOEs. For large l, this becomes the Poisson → GOE transition. An example of this has been found in complex atomic spectra [31], where the Poisson → GOE transition occurs because of the breaking of the LS symmetry.

As mentioned earlier, the Gaussian transition results are obtained for transitions in circular ensembles also (for other transitions, see [58]). Let us illustrate the COE → CUE transition using the QKR. In QKR, γp plays the role of the TRI breaking operator, with p being the momentum operator. For large N, the trace of the matrix γp is $(\gamma^2 N^3)/12$, and the mean square of the matrix elements $\left(\overline{|(\gamma p)_{jk}|^2}\right)$ is $\gamma^2 N/12$. Using $D(x) = 2\pi/N$, Λ in the QKR is given by

$$\Lambda = \frac{\gamma^2 N^3}{48\pi^2}. \tag{12.61}$$

In Figure 12.28, we plot $\Sigma^2(1, \Lambda)$ vs. Λ (see also [59] for a demonstration of this transition). In Figure 12.29, we show the change in the spectrum for the COE → CUE transition with $\Lambda = 0.0, 0.05, 1.0$.

In QKR, one can also find 2COE → 1COE by varying ϕ_0 and keeping $\gamma = 0$ [59]. Here, 2COE corresponds to a direct sum of two independent COEs, as discussed earlier. This kind of transition can be useful in studying parity-breaking in real systems.

Apart from the spectral transitions, we can also study the transitions in the eigenvectors [60]. We will not discuss this subject further here.

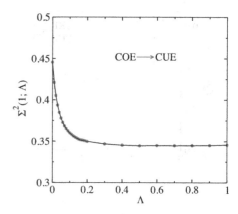

FIGURE 12.28 Number variance $\Sigma^2(r; \Lambda)$ vs. Λ for $r = 1$, showing the transition from COE ($\beta = 1$) → CUE ($\beta = 2$).

FIGURE 12.29 Spectral levels of COE → CUE transition ensemble for $\Lambda = 0.0, 0.05, 1.0$. The arrowheads mark the occurrences of pairs of levels with spacings smaller than $1/2$ of the average spacing.

12.12 Conductance Fluctuations in Mesoscopic Systems

Mesoscopic physics deals with systems which are intermediate in size between the atomic scale and the macroscopic scale. The transport properties of metals and insulators at very small scales have given new insights in understanding mesoscopic physics. The quantities of special interest in this context are the conductance fluctuations and distributions. In this section, we briefly discuss the application of RMT to understand transport properties of mesoscopic systems. There has been extensive study of these systems in the literature, and we refer the interested reader to some important review articles [5–7,57,61].

The most important phenomenon in this context is that of "universal conductance fluctuations" (UCF) in quantum dots. This refers to the independence of conductance fluctuations (measured in units of $e^2/\hbar$) from the sample size, degree of disorder, and other parameters of the system.

The scattering matrix S for the conductance problem, shown schematically in Figure 12.30, has the standard decomposition in terms of reflection matrices r, r' and transmission matrices t, t' as

$$S = \begin{pmatrix} r_{N_1 \times N_1} & t'_{N_1 \times N_2} \\ t_{N_2 \times N_1} & r'_{N_2 \times N_2} \end{pmatrix}. \qquad (12.62)$$

Here, N_1 and N_2 refer to the number of modes in the left and right leads connected to the cavity. The matrices $r_{N_1 \times N_1}$ and $r'_{N_2 \times N_2}$ correspond to the reflection from left to left and from right to right, respectively. Similarly, $t_{N_2 \times N_1}$ and $t'_{N_1 \times N_2}$ represent the transmission from left to right and from right to left, respectively. The scattering matrix S is thus N_s-dimensional, where $N_s = N_1 + N_2$. We also define $N = \min(N_1, N_2)$. As a consequence of unitarity of S, the Hermitian matrices $t^\dagger t$, $t'^\dagger t'$, $1 - r^\dagger r$, and $1 - r'^\dagger r'$ have N common eigenvalues $T_1, \ldots, T_N$ with values between 0 and 1 [62]. We denote the set of these transmission eigenvalues as $\{T_i\}$.

Using the circular ensembles for S, it can be shown that the jpd of transmission eigenvalues $p(\{T_i\})$ has a form analogous to Eqs. (12.2)–(12.4) with

$$p(T_1, \ldots, T_N) = C'' \prod_{j<k} |T_j - T_k|^\beta \prod_{j=1}^{N} T_j^{\frac{\beta}{2}(|N_1 - N_2| + 1 - \frac{2}{\beta})}, \qquad (12.63)$$

and

$$V(T) = -\frac{1}{2}\left(|N_1 - N_2| + 1 - \frac{2}{\beta}\right) \ln T. \qquad (12.64)$$

The dimensionless conductance at zero temperature is related to the eigenvalues by the Landauer formula [63–71]

$$g = \sum_{j=1}^{N} T_j. \qquad (12.65)$$

The average and variance of g can be calculated from the correlation functions of the jpd in Eq. (12.63). For $N_1, N_2 \gg 1$, these are given by

$$\bar{g} = \frac{N_1 N_2}{N_s} \xrightarrow{N_1 = N_2} \frac{N}{2}, \qquad (12.66)$$

and

$$\text{var}(g) = \frac{2N_1^2 N_2^2}{\beta N_s^4} \xrightarrow{N_1 = N_2} \frac{1}{8\beta}. \qquad (12.67)$$

Equation (12.67) for the variance of g describes the UCF.

The dimensionless shot-noise power is given by the Büttiker formula:

$$p = \sum_{j=1}^{N} T_j (1 - T_j). \qquad (12.68)$$

Its average and variance for large $N_1 N_2$ are given by

$$\bar{p} = \frac{N_1^2 N_2^2}{N_s^3} \xrightarrow{N_1 = N_2} \frac{N}{8}, \qquad (12.69)$$

$$\text{var}(p) = \frac{4N_1^4 N_2^4}{\beta N_s^8} \xrightarrow{N_1 = N_2} \frac{1}{64\beta}. \qquad (12.70)$$

As mentioned earlier, the above results apply for quantum dots. There have also been many studies of disordered nanowires, which may be modeled as a sequence of coupled quantum dots (see Figure 12.31). This system was first studied in a different theoretical framework (independent of RMT) using diagrammatic perturbation theory [63].

The corresponding result for conductance fluctuations is

$$\text{var}(g) = \frac{2}{15\beta}, \qquad (12.71)$$

which differs slightly from the RMT result for quantum dots. Subsequently, a Brownian motion model was developed using RMT to re-derive Eq. (12.71). The corresponding diffusion equation is referred to as the Dorokhov–Mello–Pereyra–Kumar (DMPK) equation [5,12].

12.13 Finite Range Coulomb Gas Models

In Eqs. (12.2)–(12.4), all particles have pairwise interactions. In this section, we consider the natural generalization to the case where particles have finite-range interactions. We refer to these ensembles as FRCG models [72]. For linear ensembles, the jpd is described by Eqs. (12.2)–(12.4) with $|j - k| \leq d$, where d is the range of the interaction. A similar definition applies for circular ensembles with $x_j \to e^{i\theta_j}$.

For $N \gg d$, the linear and circular ensembles again give identical fluctuation results. The fluctuation properties are characterized by d. It has been shown that the FRCG models are exactly solvable for each d [72]. We present here some analytic results, supplemented by MC results.

For $d = 0$, there is no interaction between particles, and we find the Poisson results:

$$p_{n-1}(s) = \frac{s^{n-1}}{(n-1)!} e^{-s}, \qquad (12.72)$$

$$\Sigma^2(r) = r. \qquad (12.73)$$

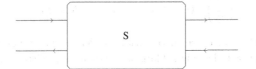

FIGURE 12.30 Schematic of the scattering matrix for a quantum dot.

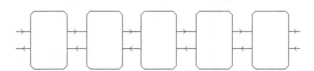

FIGURE 12.31 Schematic of a disordered nanowire. Each segment is analogous to a quantum dot.

For $d = 1$, we obtain

$$p_{n-1}(s) = \frac{(\beta+1)^{n(\beta+1)}}{\Gamma(n(\beta+1))} s^{n(\beta+1)-1} e^{-(\beta+1)s}, \quad (12.74)$$

$$\Sigma^2(r) = \frac{r}{\beta+1} + \frac{\beta(\beta+2)}{6(\beta+1)^2}. \quad (12.75)$$

For $d \geq 2$, analytic results can be derived from an integral equation approach [72]. However, for large k, there is a simple mean-field (MF) approximation, which reduces the arbitrary-d problem to an effective $d = 1$ problem. The MF results are

$$p_{n-1}(s) = [\Gamma(n\xi)]^{-1} \xi^{n\xi} s^{n\xi-1} e^{-\xi s}, \quad (12.76)$$

$$\Sigma^2(r) = \frac{r}{\xi} + \frac{(\xi^2-1)}{6\xi^2}, \quad (12.77)$$

where $\xi = \beta d + 1$.

We have found that the FRCG models describe the spectra of QKR with the identification $d = \alpha^2/N$ [72]. (The importance of the parameter α^2/N was earlier emphasized by Casati and others, who established an empirical relationship between QKR and banded random matrices.) As α is a continuous parameter, this allows d to take non-integer values also. We have formulated FRCG models for fractional d [72], which we do not discuss here.

In Figure 12.32, we plot $p_0(s)$ vs. s for the QKR with $\beta = 1, 2$. The parameter $\alpha = \sqrt{dN}$ with $d = 1$. We also plot the corresponding FRCG result from Eq. (12.74). The agreement between QKR and FRCG results is excellent. In Figure 12.33, we show results for $p_7(s)$ vs. s for $d = 3, \beta = 1$. Here QKR, FRCG, and MF results are compared. Again, the agreement is very good.

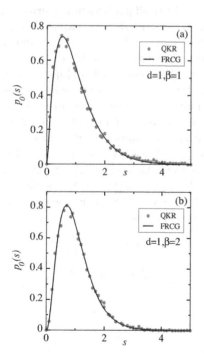

FIGURE 12.32 Nearest-neighbor spacing distribution [$p_0(s)$ vs. s] for $d = 1$, and (a) $\beta = 1$, (b) $\beta = 2$. The QKR data is obtained for $\alpha = \sqrt{dN}$. The FRCG result is given in Eq. (12.74).

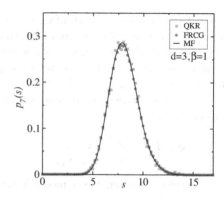

FIGURE 12.33 Plot of $p_7(s)$ vs. s for the QKR and FRCG with $d = 3, \beta = 1$. The FRCG result is obtained via MC. For purposes of comparison, we also show the MF result in Eq. (12.76).

12.14 Wishart Ensembles

As mentioned in Section 12.1, random matrices were first introduced by Wishart in the study of statistical multivariate analysis. Subsequent interest in RMT focused on physics applications. Recently, there has been a resurgence of interest in Wishart ensembles and their generalizations. These have found applications in multivariate analysis of time series which arise in, e.g., chaotic physical systems [73], economics [74,75], disordered solids [76], biological systems [77–80], meteorology [81–83], and communication theory [84–86].

In this section, we briefly review Wishart ensembles and our theoretical understanding of them. We consider matrices of the form

$$H = AA^\dagger, \quad (12.78)$$

where the elements of A can be real ($\beta = 1$), complex ($\beta = 2$), or quaternion ($\beta = 4$). Each of the β components of the matrix elements of A is an independent and identically distributed (i.i.d.) Gaussian random variable with variance $1/2$. For example, in an economics application, each row of A may consist of a time series for a particular stock price. The different rows correspond to different stocks.

In general, A is an $N \times M$ rectangular matrix, where N denotes the number of stocks, and M denotes the number of evenly spaced times. The corresponding H will be an $N \times N$ matrix. The jpd of eigenvalues of H has the form in Eq. (12.4):

$$p(x_1, \ldots, x_N) = C''' \prod_{j<k} |x_j - x_k|^\beta \prod_{j=1}^{N} \left(x_j^\omega e^{x_j} \right), \quad (12.79)$$

where

$$\omega = \left(\frac{\beta}{2} \right)(M - N + 1) - 1, \qquad N \leq M. \quad (12.80)$$

Note that all three distributions which we have mentioned in this article, viz., Gaussian ensembles with $V(x) \sim x^2$ in Eq. (12.4), ensemble of transmission eigenvalues in Eq. (12.63), and the Wishart ensemble in Eq. (12.79), are examples of general Jacobi weight functions.

12.15 Conclusion

In this article, we have attempted to provide a broad overview of RMT and its applications at a pedagogical and accessible level. Let us briefly summarize the topics we have covered. We have defined various types of random matrix ensembles. In physics applications, RMT applies in quantum chaotic systems, i.e., systems whose classical counterparts are chaotic.

Typically, one is interested in the properties of eigenvalues and eigenvectors of these matrix ensembles. We have discussed the various statistical measures used to quantify these properties. In particular, we have shown that the spectra of random matrices display level repulsion and spectral rigidity. Eigenvectors of random matrices also have universal properties, which are experimentally observable.

We have used the system of QKR to illustrate many of the features of random matrix ensembles. Some of the most important applications of RMT have been in the context of quantum chaos in mesoscopic and nanoscopic systems. However, we should stress that present-day applications of RMT are not confined to physics alone. The framework of RMT provides insights on complexity in diverse disciplines.

Appendix A

For $\beta = 4$, one needs to deal with self-dual Hermitian matrices. Consider a $2N \times 2N$ matrix A:

$$A = B_0 e_0 + B_1 e_1 + B_2 e_2 + B_3 e_3. \tag{12.81}$$

Here, the B_i are N-dimensional matrices, and e_0, e_1, e_2, e_3 are a two-dimensional representation of quaternions [1,9]:

$$e_0 = \begin{pmatrix} 1 & 0 \\ 0 & 1 \end{pmatrix}, e_1 = \begin{pmatrix} 0 & -i \\ -i & 0 \end{pmatrix}, e_2 = \begin{pmatrix} 0 & -1 \\ 1 & 0 \end{pmatrix}, e_3 = \begin{pmatrix} -i & 0 \\ 0 & i \end{pmatrix}. \tag{12.82}$$

Thus, in quaternion space, the matrix elements of A can be written as linear combination of quaternions e_j.

The dual of A is defined as

$$A^D = B_0 e_0 - B_1 e_1 - B_2 e_2 - B_3 e_3. \tag{12.83}$$

The transpose of A in the quaternion space is defined as

$$A^T = B_0^T e_0 + B_1^T e_1 + B_2^T e_2 + B_3^T e_3. \tag{12.84}$$

Here, B^T is the transpose of B. A is said to be self-dual if $A^T = A^D$. The matrix A is Hermitian in the $(2N)$-dimensional space if A is real symmetric and B_1, B_2, B_3 are real anti-symmetric. For a self-dual matrix A, the eigenvalues are doubly degenerate.

References

1. C. E. Porter (ed.), *Statistical Theories of Spectra: Fluctuations* (Academic Press, New York, 1965).
2. E. P. Wigner, *SIAM Rev.* **9**, 1 (1967).
3. T. A. Brody, J. Flores, J. B. French, P. A. Mello, A. Pandey and S. S. M. Wong, *Rev. Mod. Phys.* **53**, 385 (1981).
4. O. Bohigas and M. J. Giannoni, *Chaotic Motion and Random Matrix Theories* (World Scientific, Singapore, 1985).
5. C. W. J. Beenakker, Rev. Mod. Phys. **69**, 731 (1997).
6. T. Guhr, A. Müller-Groeling and H. A. Weidenmüller, *Phys. Rep.* **299**, 189 (1998).
7. Y. Alhassid, *Rev. Mod. Phys.* **72**, 895 (2000).
8. F. Haake, *Quantum Signatures of Chaos* (Springer-Verlag, Heidelberg, 2001).
9. M. L. Mehta, *Random Matrices* (Academic Press, New York, 2004).
10. R. Blumel and W. P. Reinhardt, *Chaos in Atomic Physics* (Cambridge University Press, Cambridge, UK, 2005).
11. H. Stockmann, *Quantum Chaos: An Introduction* (Cambridge University Press, Cambridge, UK, 2006).
12. G. Akemann, J. Baik and P. D. Francesco (eds.), *The Oxford Handbook of Random Matrix Theory* (Oxford University Press, New York, 2011).
13. S. S. Wilks, *Mathematical Statistics* (John Wiley & Sons, New York, 1962).
14. F. J. Dyson, *J. Math. Phys.* **3**, 140 (1962).
15. F. J. Dyson, *J. Math. Phys.* **3**, 157 (1962).
16. F. J. Dyson, *J. Math. Phys.* **3**, 166 (1962).
17. F. J. Dyson and M. L. Mehta, *J. Math. Phys.* **4**, 701 (1963).
18. M. L. Mehta and F. J. Dyson, *J. Math. Phys.* **4**, 713 (1963).
19. M. L. Mehta, *Nucl. Phys.* **18**, 395 (1960).
20. M. L. Mehta and M. Gaudin, *Nucl. Phys.* **18**, 420 (1960).
21. O. Bohigas, M. J. Giannoni and C. Schmit, *Phys. Rev. Lett.* **52**, 1 (1984).
22. E. P. Wigner, *Group Theory and its Application to the Quantum Mechanics of Atomic Spectra* (Academic Press, New York, 1959).
23. A. Pandey and S. Ghosh, *Phys. Rev. Lett.* **87**, 024102 (2001).
24. S. Ghosh and A. Pandey, *Phys. Rev. E* **65**, 046221 (2002).
25. A. Pandey, Ann. Phys. (NY) **119**, 170 (1979).
26. O. Bohigas, R. U. Haq and A. Pandey, *Phys. Rev. Lett.* **54**, 1645 (1985).
27. R. U. Haq, A. Pandey and O. Bohigas, *Phys. Rev. Lett.* **48**, 1086 (1982).
28. S. Ghosh, A. Pandey, S. Puri and R. Saha, *Phys. Rev. E* **67**, 025201(R) (2003).
29. S. Kumar and A. Pandey, *Phys. Rev. E* **78**, 026204 (2008).
30. O. Bohigas, R. U. Haq and A. Pandey, fluctuation properties of nuclear energy levels and widths: comparison of theory with experiment, in: K. H. Bockhoff (ed.), *Nuclear Data for Science and Technology* (Reidel, Dordrecht 1983).

31. N. Rosenzweig and C. E. Porter, *Phys. Rev.* **120**, 1698 (1960).

32. I. C. Percival, *J. Phys. B At. Mol. Phys.* **6**, L229 (1973).

33. M. V. Berry and M. Tabor, *Proc. R. Soc. Lond. A* **356**, 375 (1986).

34. S. W. McDonald and A. N. Kaufman, *Phys. Rev. Lett.* **42**, 11891 (1979).

35. M. V. Berry, *Ann. Phys.* (NY) **131**, 163 (1981).

36. T. H. Seligman, J. J. M. Verbaarschot and M. R. Zirnbauer, *Phys. Rev. Lett.* **53**, 215 (1984).

37. M. V. Berry and M. Robnik, *J. Phys. A: Gen.* **17**, 2413 (1984).

38. D. Wintgen and H. Marxer, *Phys. Rev. Lett.* **60**, 971 (1988).

39. M. C. Gutzwiller, *Chaos in Classical, Quantum Mechanics* (Springer, New York, 1990).

40. F. M. Izrailev, *Phys. Rev. Lett.* **56**, 541 (1986).

41. A. Pandey and R. Ramaswamy, *Phys. Rev. A* **43**, 4237 (1991).

42. A. Pandey, O. Bohigas and M. J. Giannoni, *J. Phys. A: Math. Gen.* **22**, 4083 (1989).

43. G. Casati, B. V. Chirikov and I. Guarneri, *Phys. Rev. Lett.* **54**, 1350 (1985).

44. G. Casati and B. Chirikov (eds.), *Quantum Chaos: Between Order and Disorder* (Cambridge University Press, New York, 1995).

45. R. E. Prange and S. Fishman, *Phys. Rev. Lett.* **63**, 704 (1989).

46. F. M. Izrailev, *Phys. Rep.* **196**, 299 (1990).

47. G. Casati, L. Molinari and F. Izrailev, *Phys. Rev. Lett.* **64**, 1851 (1990).

48. Vinayak, S. Kumar and A. Pandey, *Phys. Rev. E* **93**, 032217 (2016).

49. M. L. Mehta and F. J. Dyson, *J. Math. Phys.* **4**, 713 (1962).

50. J. Gunson, *J. Math. Phys.* **3**, 752 (1962).

51. J. B. French, V. K. B. Kota, A. Pandey and S. Tomsovic, *Ann. Phys.* **181**, 198 (1988).

52. A. Pandey and M. L. Mehta, *Comm. Math. Phys.* **87**, 449 (1983).

53. M. L. Mehta and A. Pandey, *J. Phys. A Math. Gen.* **16**, 2655 (1983).

54. A. Pandey and P. Shukla, *J. Phys. A Math. Gen.* **24**, 3907 (1991).

55. J. B. French, V. K. B. Kota, A. Pandey and S. Tomsovic, *Ann. Phys.* **235**, 198 (1988).

56. N. Dupuis and G. Montambaux, *Phys. Rev. B* **43**, 14390 (1991).

57. S. Kumar and A. Pandey, *Ann. Phys.* **326**, 1877 (2011).

58. A. Pandey, *Chaos, Solitons and Fractals* **5**, 1275 (1995).

59. A. Pandey, R. Ramaswamy and P. Shukla, *Pramana* **41**, 75 (1993).

60. P. Shukla and A. Pandey, *Nonlinearity* **10**, 979 (1997).

61. A. D. Mirlin, *Phys. Rep.* **326**, 259 (2000).

62. S. Kumar and A. Pandey, *J. Phys. A: Math. Theor.* **43**, 085001 (2010).

63. B. L. Altshuler and B. I. Shklovskii, *Zh. Eksp. Teor. Fiz.* **91**, 220 (1986).

64. P. A. Lee and A. D. Stone, *Phys. Rev. Lett.* **55**, 1622 (1985).

65. P. A. Lee, A. D. Stone and H. Fukuyama, *Phys. Rev. B* **35**, 1039 (1987).

66. R. Landauer, IBM *J. Res. Dev.* **1**, 223 (1957).

67. R. Landauer, *Philos. Mag.* **21**, 863 (1970).

68. Y. Imry, *Europhys. Lett.* **1**, 249 (1986).

69. Y. Imry and R. Landauer, *Rev. Mod. Phys.* **71**, S306 (1999).

70. K. A. Muttalib, J. L. Pichard and A. D. Stone, *Phys. Rev. Lett.* **59**, 2475 (1987).

71. A. D. Stone, P. A. Mello, K. A. Muttalib and J. L. Pichard. Random matrix theory and maximum entropy models for disordered conductors, in: B. L. Altshuler, P. A. Lee and R. A. Webb (eds.), *Mesoscopic Phenomena in Solids* (Elsevier, Amsterdam, 1991).

72. A. Pandey, A. Kumar and S. Puri, *Phys. Rev. E* **96**, 052211 (2017).

73. Vinayak and A. Pandey, *Phys. Rev. E* **81**, 036202 (2010).

74. V. Plerou, P. Gopikrishnan, B. Rosenow, L. A. N. Amaral and H. E. Stanley, *Phys. Rev. Lett.* **83**, 1471 (1999).

75. L. Laloux, P. Cizeau, J. P. Bouchaud and M. Potters, *Phys. Rev. Lett.* **83**, 1467 (1999).

76. S. K. Sarkar, G. S. Matharoo and A. Pandey, *Phys. Rev. Lett.* **92**, 215503 (2004).

77. F. Luo, J. Zhong, Y. Yang, R. H. Scheuermann and J. Zhou, *Phys. Lett. A* **357**, 420 (2006).

78. F. Luo, Y. Yang, J. Zhong, H. Gao, L. Khan, D. K. Thompson and J. Zhou, *BMC Bioinf.* **8**, 299 (2007).

79. K. Rajan and L. F. Abbott, *Phys. Rev. Lett.* **97**, 188104 (2006).

80. J. Aljadeff, M. Stern and T. Sharpee, *Phys. Rev. Lett.* **114**, 088101 (2015).

81. D. S. Wilks, *Statistical Methods in the Atmospheric Sciences* (Academic Press, San Diego, 2006).

82. D. Wang, X. Zhang, D. Horvatic, B. Podobnik and H. E. Stanley, *Chaos* **27**, 023104 (2017).

83. M. S. Santhanam and P. K. Patra, *Phys. Rev. E* **64**, 016102 (2001).

84. A. M. Tulino and S. Verdú, *Found. Trends Commun. Inf. Theory* **1**, 1 (2004).

85. S. Kumar and A. Pandey, *IEEE Trans. Inf. Theory* **56**, 2360 (2010).

86. R. Couillet and M. Debbah, *Random Matrix Methods for Wireless Communications* (Cambridge University Press, New York, 2011).

Topological Constraint Theory and Rigidity of Glasses

Mathieu Bauchy
University of California

13.1 Introduction

13.1.1 The Glass Age

As illustrated recently by the Materials Genome Initiative, the discovery of new materials with tailored functionalities is essential to economic development and human well-being (Warren, 2012). In fact, 10 of the 14 societal grand challenges identified by the National Academy of Engineering are expected to require the development of novel materials with improved properties (Sciences, 2001). To this end, the discovery of new materials or the optimization of existing ones requires a deep understanding of how the composition and structure of materials control their macroscale properties.

The discovery of new materials has always played a crucial role in human history—to the point that human history time periods are named after materials: Stone Age, Bronze Age, and Iron Age (see Figure 13.1). Among the many materials that have been discovered, glass has been one of the most influential, and its importance keeps increasing (Mauro, 2017). Since the Romans started to use glass to make building windows, glass has defined human progress in many ways (Main, 2018). Spectacles have allowed individuals to

recover their vision. Glass mirrors largely contributed to defining the concept of individual identity. Telescope lenses enabled major discovery in astronomy. Bacteria were discovered thanks to magnifying glasses. Glass was key in the development of light bulbs and television. Today, glasses

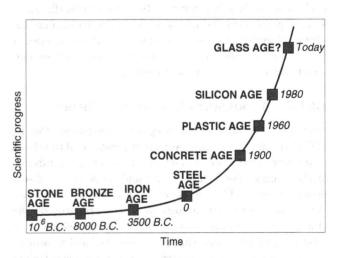

FIGURE 13.1 The discovery of new materials largely defines the progress of our civilization.

are used to immobilize nuclear waste or stimulate bone growth after fracture (Mauro and Zanotto, 2014). Touch-screen display glasses have changed the way humans interact with electronic devices, and virtually, everybody carries a piece of glass in his pocket (Mauro, 2014). Maybe even more importantly, glass optical fibers made it possible for the development of the Internet as we know it today (Mauro et al., 2014). Glass also shows great promises to solve some of tomorrow's grand challenges in clean energy, environment, water treatment, healthcare, information, transportation, etc. (Mauro, 2014). For all these reasons, it has recently been proposed that we are now living in the **Glass Age** (Morse and Evenson, 2016).

13.1.2 Glass Genome and Discovery of New Compositions

Revealing the full potential of glass requires the discovery of new glass formulations showing unique properties. However, this task is especially complicated for non-crystalline glassy materials for several reasons. First, virtually all the elements of the periodic table can be turned into a glass, if cooled fast enough from the liquid state (Zanotto and Coutinho, 2004). Second, unlike crystals, glasses do not have to satisfy any fixed stoichiometry thanks to their out-of-equilibrium nature. As such, the composition of glasses can be continuously changed. For all these reasons, the number of possible glass compositions has been estimated to be around 10^{52}! Yet, only about 200,000 glass compositions have been produced in the last 6000 years of human glass history (Zanotto and Coutinho, 2004). These numbers demonstrate that the range of possible glasses remains largely unexplored—so that there exists an incredible opportunity for the future discovery of new glass formulations with unusual functionalities.

However, although it offers a large room for improvement, the astronomical number of possible glass compositions is also a challenge. Indeed, such a large parametric space renders traditional Edisonian discovery approach based on trial-and-error largely inefficient. To accelerate the discovery of new glass formulations, it is necessary to decode the **Glass Genome**, that is, to decipher how the properties of glasses are controlled by their underlying composition and structural features (the glass "genes").

13.1.3 Topological Constraint Theory

Over the past decades, **topological constraint theory** (TCT) (Phillips, 1979) has been an invaluable tool to help to investigate these problems and gain a better understanding of the linkages between structure and properties in disordered materials. The success of TCT is based on the fact that many macroscopic properties depend primarily on the atomic network topology of materials. In turn, simplifying a complex material into simpler networks of nodes (atoms) that are interconnected to each other via some constraints (chemical bonds) makes it possible to develop models that can be used to analytically predict material properties.

Among others, TCT has been used to predict the properties of glasses (Smedskjaer et al., 2010), elucidate the origin of concrete creep (Pignatelli et al., 2016a), identify durable phase-change semiconducting materials (Micoulaut et al., 2010), understand the effect of irradiation on minerals (Wang et al., 2017a), and examine the origin of protein folding (Phillips, 2004). One of the most popular examples of the power of TCT is offered by Corning® Gorilla® Glass (a scratch- and damage-resistant glass used in more than 5 billion smartphones and tablets), which was designed *in silico* through the optimization of its atomic network topology before anything was actually melted in a laboratory (Ball, 2015; Wray, 2013).

In the following, we provide a general introduction to glass science and TCT, and review how TCT can be used to predict the properties of glasses to pinpoint promising compositions featuring optimal functionalities.

13.2 Introduction to Glass Science

13.2.1 The Glassy State

Despite their critical importance in various applications, the nature of glasses—not truly a solid, neither a liquid—remains poorly understood (Zanotto and Mauro, 2017). The V–T diagram shown in Figure 13.2 is probably the most instructive plot in glass science (Varshneya, 1993). This plot depicts the evolution of volume (or, equivalently, enthalpy) as a function of temperature for glass-forming systems. At high temperature, materials tend to exist in the form of a **liquid** (or eventually gas) state. In this state, due to thermal expansion, liquids typically significantly expand with increasing temperature. At this point, the system is at stable equilibrium. If cooled slowly enough, liquids can undergo crystallization at convert into **crystals**. This transition is sharp (first-order transition), occurs at fixed temperature (melting temperature T_m), and is associated with a discontinuity (decrease) in volume (and enthalpy, see Figure 13.2). Crystals are also at stable equilibrium

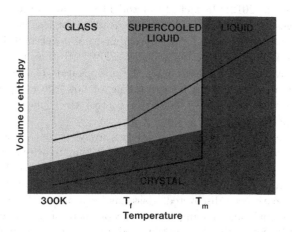

FIGURE 13.2 Schematic showing the evolution of volume (or enthalpy) as a function of temperature in a glass-forming system. T_m and T_f are the melting and fictive temperature, respectively.

and typically represent the most compact and stable state of matter. Crystallization occurs in two stages, namely, nucleation (i.e., the formation of crystal nuclei) and crystal growth. However, since crystallization is not instantaneous (i.e., it requires an energy barrier to be overcome), liquids can be cooled below T_m while not crystallizing. At this point, they are referred to as ***supercooled liquids***. Supercooled liquids are in metastable equilibrium; that is, they occupy a local minimum position in the enthalpy landscape. At this point, there exists a thermodynamic driving force toward crystallization since the free energy of the crystal is lower than that of the supercooled liquid. However, if cooling continues, the viscosity of the melt tends to increase exponentially with decreasing temperature (Mauro et al., 2009b). At some point, the viscosity becomes so high that relaxation becomes kinetically impossible and the melt starts to behave like a solid. A ***glass*** is formed. This transition is continuous and manifests itself by a gradual decrease in the slope of $V(T)$ (i.e., thermal expansion coefficient), which eventually becomes comparable to that of a crystal. The temperature around which the gradual break of slope occurs is referred to as the fictive temperature T_f, as, to the first order, a glass can be considered as a frozen supercooled liquid cooled down to T_f. In the glassy state, the system is out-of-equilibrium; that is, it continually wants to relax toward the metastable supercooled liquid state, but relaxation largely exceeds the observation time (i.e., it can greatly exceed the age of the universe) (Zanotto, 1998). Based on this, glass has been defined as "a nonequilibrium, non-crystalline state of matter that appears solid on a short time scale but continuously relaxes toward the liquid state" (Zanotto and Mauro, 2017). This definition distinguishes glasses (which relax toward the supercooled liquid state upon heating) from amorphous solids (which tend to crystallize upon heating) (Krishnan et al., 2017a).

13.2.2 Network Formers versus Network Modifiers

A fundamental question in glass science is to understand what makes it possible for a substance to avoid crystallization upon cooling below its melting point, that is, to understand the origin of glass-forming ability. For oxide glasses, some useful insights can be gained by considering the chemical composition of the system. The cations A that form the network of oxide glasses can be classified based on the energy of the A–O bonds. Namely, the cations A are classified as network formers and network modifiers if the A–O bond energy is higher than 330 kJ/mol and lower than 250 kJ/mol, respectively (Varshneya, 1993). Other cations are classified as intermediate. Network-forming species comprise Si, B, Ge, Al, P, etc., whereas network-modifying species comprise Ca, Mg, Li, Na, K, etc. This classification is based on the idea that, to avoid crystallization during cooling, supercooled liquids must have a viscosity that is high enough to prevent the atoms from

easily reorganizing. This requires the existence of strong interatomic bonds. As such, network formers form the backbone of glasses, whereas network modifiers tend to depolymerize them.

13.2.3 Zachariasen Rules of Glass Formation

The paper *The Atomic Arrangement in Glass* published by Zachariasen in 1932 is largely considered as the birth of modern glass science (Zachariasen, 1932). In this contribution, Zachariasen introduced a series of rules dictating the ability of a given compound to form a glass. Zachariasen's approach is based on several observations and ideas. First, the properties (stiffness, density) of glasses are fairly similar to those of their isochemical crystals. This suggests that the atomic structure of glasses cannot be too different from that of crystals. Further, although crystals exhibit a lower free energy than glasses, this difference of free energy cannot be too large—otherwise, there would exist a very large driving force promoting crystallization. For all these reasons, Zachariasen proposed that the local atomic structure of glasses must be fairly similar to that of crystals. Second, to avoid crystallization, glasses must be macroscopically rigid. To this end, they must form an extended three-dimensional atomic network; that is, the atomic connectivity must be high enough. Third, glasses do not exhibit any sharp peak upon X-ray diffraction. As such, the atomic network of glasses must be random at distance larger than a few atomic bond distances. This requires an open structure that is flexible enough to yield some long-range randomness. An atomic connectivity that would be too high would not allow such flexibility.

Based on these ideas, Zachariasen proposed a series of four rules to describe the ability of a given oxide compound A_mO_n to form a glass (Zachariasen, 1932):

(1) The oxygen atoms O must be linked to no more than two cations A.

(2) The number of O neighbors around A cations (i.e., their coordination number) must be small, between 3 and 4.

(3) The cation polyhedra must share corners with each other, not edges or faces.

(4) At least three corners of the cation polyhedra must be shared.

Rules (1), (2), and (3) ensure that the atomic network of glasses is flexible enough to enable the formation of a random (i.e., non-periodic) network. In turn, rule (4) ensures that the atomic connectivity is high enough to form a continuous three-dimensional network that is rigid enough to resist crystallization. TCT also relies on this idea that glass must exhibit an atomic connectivity that is neither too low nor too high to be able to avoid crystallization.

13.3 TCT and Glass Rigidity

13.3.1 Stability of Mechanical Trusses

The topological nanoengineering of glasses takes its root in the study of the stability of mechanical trusses, as initially developed by Maxwell and Lagrange (Maxwell, 1864), as described in the following. Let us first consider the three simple trusses presented in Figure 13.3. The left one is clearly flexible as it can be freely deformed with no external energy, whereas the other two trusses are rigid (note that, in this context, rigid means "not flexible" and does not imply that the solid has an infinite stiffness). In general, the degree of flexibility of a given truss can be determined by comparing the number of mechanical constraints N_c (i.e., the number of red sticks in this case) and the initial number of degrees of freedom of the truss nodes. In two dimension, the initial number of degrees of freedom is given by $2N$ (i.e., two translation directions), where N is the number of nodes. The number of remaining internal modes of deformation (or floppy modes) present within the truss F after the application of the constraints is then given by

$$F = 2N - N_c - 3 \tag{13.1}$$

which arises from the fact that, starting from the situation wherein each node can freely move along two directions, each constraint N_c removes one internal degree of freedom. Note that the term "3" corresponds to the three macroscopic degrees of freedom of a rigid structure in two dimension (i.e., two translations and one in-plane rotation). Based on this, one finds $F = 1$, 0, and -1 for the three trusses presented in Figure 13.3 (see Table 13.1). This denotes that the left truss exhibits one internal mode of deformation, whereas the middle truss cannot be deformed. In contrast, the value $F = -1$ for the right truss denotes that, in this structure, there are more constraints than degrees of freedom. In that situation, all the constraints cannot be satisfied at the same time—just like the angles of a triangle cannot be varied to arbitrary values if the dimensions of the three edges are already fixed. In this case, some internal eigenstress will be observed as some of the constraints will be under tension, whereas others will be under compression. Note that such internal eigenstress does not result in any macroscopic stress as the constraints under tension and compression mutually compensate each other—so that the overall structure is at zero macroscopic stress (see Figure 13.4).

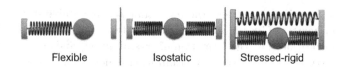

FIGURE 13.4 Illustration of the origin of the internal eigenstress that is present in stressed-rigid structures.

In a similar fashion, the number of internal modes of deformation of a three-dimensional truss is given by

$$F = 3N - N_c - 6 \tag{13.2}$$

where $3N$ is the initial number of degrees of freedom of the nodes (i.e., before the application of the constraints), and the term "6" corresponds to the six macroscopic degrees of freedom of a rigid structure (i.e., three translations and three rotations). In general, mechanical trusses can be classified as (i) flexible if $F > 0$, (ii) stressed-rigid if $F < 0$, and (iii) isostatic (or statically determinate) if $F = 0$, wherein flexible trusses exhibit some internal modes of deformation, stressed-rigid trusses exhibit some internal eigenstress, and isostatic trusses are rigid but free of stress.

13.3.2 Application to Atomic Networks

These concepts of structural stability were used by Phillips to establish TCT in 1979 (Phillips, 1979) and refined by Thorpe in 1983 (Thorpe, 1983). The main idea is that molecular networks can be seen as mechanical trusses, wherein the atoms are the nodes and the chemical bonds are the mechanical constraints (i.e., that prevent the relative motion between the atoms) (Mauro, 2011). In analogy with mechanical trusses, the chemical constraints effectively remove some of the initial degrees of freedom of the atoms (i.e., three per atom in three-dimensional networks). Molecular networks typically comprise two types of constraints: (i) the radial two-body bond-stretching (BS) constraints that keep the interatomic distances fixed around their average values and (ii) the angular three-body bond-bending (BB) constraints that keep the angles fixed around their average values (see Figure 13.5). These bonds can be considered as little springs that prevent the relative motion between the atoms of the network (Bauchy, 2012a). In covalent networks wherein all the BS and BB constraints are intact (e.g., chalcogenide glasses), the number of BS and BB constraints created by each atom depend on its coordination number r. Namely, the number of BS constraints is given by $r/2$ (since each radial constraint is necessarily shared by two atoms), while

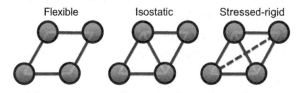

FIGURE 13.3 The three states of rigidity of a mechanical truss. The dashed line denotes a redundant constraint that is here under tension.

TABLE 13.1 Description of the Stability of the Three Trusses Presented in Figure 13.3

Truss Considered	Flexible (Left)	Isostatic (Middle)	Stressed-Rigid (Right)
Initial # of degrees of freedom	8	8	8
# of constraints	4	5	6
# of internal modes of deformation	1	0	0
# of mutually dependent constraints	0	0	1

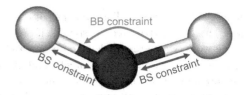

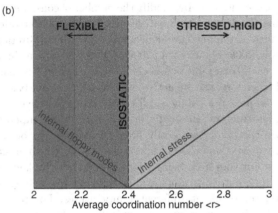

FIGURE 13.5 Schematic illustrating the role of the radial bond-stretching (BS) and angular bond-bending (BB) constraints.

the number of BB constraints is given by $2r - 3$ for $r \geq 2$ (since a triplet of atoms corresponds to one BB constraint and two new BB constraints are then needed to fix to the direction of each additional neighbor).

13.3.3 Mean Field Approximation

Let us consider a network of N atoms, where N_i is the number of atoms having a coordination number $r_i \geq 2$. The total number of constraints N_c is given by

$$N_c = \sum_i \left[N_i \left(\frac{r_i}{2} + 2r_i - 3 \right) \right] = \sum_i \left[N_i \frac{5r_i}{2} \right] - 3N \quad (13.3)$$

Due to the high number of atoms in glass (typically on the order of 10^{23}), it is more convenient to rely on a mean-field approximation and consider the average number of internal modes of deformation per atom $f = F/N$, the average number of constraints per atom $n_c = N_c/N$, and the fraction of each type of atom $x_i = N_i/N$. Following Eq. 13.2, the number of internal modes of deformation and constraints per atoms is then given by

$$f = 3 - n_c - \frac{6}{N} = 3 - n_c \quad (13.4)$$

$$n_c = \sum_i \left[x_i \frac{5r_i}{2} \right] - 3 \quad (13.5)$$

Note that the term "$6/N$" becomes infinitely small for a large number of atoms and, thereby, can be ignored. Following Maxwell's stability criterion, glasses can then be classified as (i) **flexible** if $f > 0$ ($n_c < 3$), (ii) **stressed-rigid** if $f > 0$ ($n_c > 3$), and (iii) **isostatic** if $f = 0$ ($n_c = 3$). In analogy with mechanical trusses, flexible glasses are expected to exhibit some internal floppy modes of deformation (the number of floppy modes per atom being given by $f = 3 - n_c$). In contrast, stressed-rigid glasses are expected to present some internal stress due to the fact that some constraints mutually depend on each other (the number of redundant per atom being given by $n_c - 3$). In turn, isostatic glasses are free of both floppy modes and internal stress.

These relationships can be conveniently expressed in terms of the average coordination number $<r>$:

$$\langle r \rangle = \sum_i x_i r_i \quad (13.6)$$

The number of constraints per atom can then be expressed as

$$n_c = \frac{5 \langle r \rangle}{2} - 3 \quad (13.7)$$

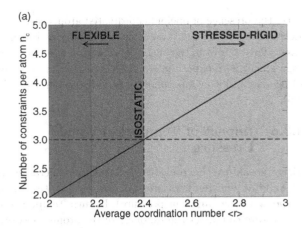

FIGURE 13.6 (a) Schematic illustrating the evolution of the number of constraints per atom and (b) internal floppy modes, and internal stress as a function of the average coordination number in the glass.

The condition of isostaticity $n_c = 3$ is then satisfied for $<r> = 2.4$, that is, the "magic" coordination number featured by isostatic networks (under the assumptions presented above). These results are summarized in Figure 13.6.

13.3.4 Impact of Onefold Coordinated Atoms

Note that Eq. 13.7 is only valid if all BS and BB constraints are active and the coordination number of the atoms is always strictly larger than 1. In the presence of onefold coordinated atoms, the number of constraints per atom can be expressed as (Boolchand et al., 1996)

$$n_c = \sum_{i \neq 1} \left[x_i \left(\frac{r_i}{2} + 2r_i - 3 \right) \right] + \frac{x_1}{2}$$

$$= \sum_i \left[x_i \frac{5r_i}{2} \right] + x_1 - 3 = \frac{5 \langle r \rangle}{2} + x_1 - 3 \quad (13.8)$$

where x_1 is the fraction of onefold coordinated atoms. Consequently, in the presence of onefold coordinated atoms, the isostatic condition $n_c = 3$ is achieved for

$$\langle r \rangle = 2.4 - 0.4 x_1 \quad (13.9)$$

Hence, the presence of onefold coordinated atoms results in a decrease in the average coordination number of isostatic glasses. This shift from the magic coordination number 2.4 is in agreement with experimental observations in amorphous hydrogenated silicon amorphous solids (Boolchand et al., 1996).

13.3.5 Impact of Temperature and Pressure

Strictly speaking, the previous constraint enumeration is only valid at zero temperature (i.e., low temperature) and zero pressure (i.e., ambient pressure). In turn, temperature and pressure can affect the topology of the atomic network of glasses and, thereby, modify the number of constraints per atoms. To describe thermal effects, Mauro et al. introduced the idea of temperature-dependent constraints (Gupta and Mauro, 2009; Mauro et al., 2009a). This is based on the idea that each constraint is associated with a given free energy and, therefore, can be active or thermally broken based on the temperature. Namely, all constraints are active at low temperature and thermally broken at infinite temperature (since the bonds can easily break or reform). In between, there exists an onset temperature T_c at which a given constraint goes from being intact to thermally broken. Note that different constraints exhibit different free energy and, therefore, break at different temperatures. For instance, angular BB constraints are typically weaker than radial BS constraints and, hence, are associated with lower onset temperatures (Bauchy et al., 2015c; Bauchy and Micoulaut, 2011a). In a similar fashion, Bauchy and Micoulaut showed that, by altering the coordination of the atoms, pressure can increase or decrease the number of constraints per atom (Bauchy and Micoulaut, 2015, 2013a; Mantisi et al., 2015).

13.3.6 Intermediate Phase

The previous equations rely on a mean-field approximation (see Section 3.3), which is based on the idea that the glass should be homogeneous—so that the average number of constraints is a representative metric. This yields a single isostatic threshold, that is, a single glass composition for which $n_c = 3$. However, this approach intrinsically cannot capture any local heterogeneity in the atomic topology (Bauchy and Micoulaut, 2013b; Micoulaut and Bauchy, 2017; Wang et al., 2016a). Recently, Boolchand et al. suggested that, in many glass-forming systems, an isostatic state can be achieved for a continuous range of compositions rather than at a fixed threshold (Boolchand et al., 2018, 2001a,b; Rompicharla et al., 2008). Boolchand's results suggest that, rather than a single flexible-to-stressed-rigid transition, glasses can experience two distinct transitions: (i) a flexible-to-rigid transition, that is, when the number of internal floppy modes of deformation becomes zero; and (ii) an unstressed-to-stressed transition, that is, when an onset of eigenstress is observed within the structure. These two transitions define an *intermediate phase*, wherein the atomic network is isostatic, that is, rigid (i.e., free of floppy modes) but unstressed. The existence of the intermediate phase has been attributed to some self-organization within the network, which reorganizes to become rigid while avoiding the onset of internal stress (Bauchy and Micoulaut, 2015). Entropy and weak Van der Waals interactions have also been suggested to play an important role (Yan, 2018). Glasses belonging to the intermediate phase have been shown to exhibit unusual properties, e.g., an optimal space-filling tendency (Rompicharla et al., 2008), a low propensity for relaxation (Bauchy et al., 2017), maximum resistance to creep (Pignatelli et al., 2016a), maximum fracture toughness (Bauchy et al., 2016), and maximum resistance to irradiation (Krishnan et al., 2017c). Note that the existence of the intermediate phase as well as that of potential structural signatures remains debated (Lucas et al., 2009; Wang et al., 2017a; Zeidler et al., 2017).

13.4 Examples of Constraints Enumeration and Application to Glass-Forming Ability Prediction

13.4.1 Topological Description of Glass-Forming Ability

In his pioneering contribution, Phillips predicted that isostatic glasses should exhibit optimal glass-forming ability (Phillips, 1979), i.e., when the number of constraints exactly equals the number of atomic degrees of freedom. This can be explained by the fact that, in the flexible domain ($n_c < 3$), the atoms can easily reorganize toward lower energy states, and hence, glass can easily crystallize. In contrast, in the stressed-rigid domain ($n_c > 3$), the atomic network is locally unstable due to the existence of mutually dependent constraints. In this regime, the redundant constraints exhibit some internal stress; that is, the weaker constraints yield to the stronger constraints by being under tension or compression. The existence of such internal stress acts as a driving force that stimulates relaxation toward lower energy states and, therefore, enhances the thermodynamic propensity for crystallization. This viewpoint is consistent with Zachariasen's description of glass formation, which also relies on the idea that the connectivity of the glass network should be high enough to form a continuous three-dimensional network, but low enough to remain flexible enough to form a random network (see Section 2.3). These ideas can also be expressed in terms of the underlying enthalpy landscape (see Figure 13.7) (Krishnan et al., 2017d; Lacks, 2001; Mauro and Smedskjaer, 2012; Naumis, 2005). Namely, in flexible glasses, the floppy modes of relaxation result in the formation of channels in between local minima within the enthalpy landscape, which facilitates atomic jumps from one state to another (see Figure 13.7a). In contrast, in stressed-rigid glasses, the internal stress results in the existence of some local strain (elastic) energy that can promote the transition from one state to another (Yu

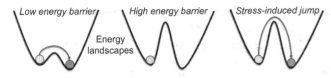

FIGURE 13.7 Schematic illustrating the mechanism of relaxation toward a more stable state in the enthalpy landscape in (a) flexible, (b) isostatic, and (c) stressed-rigid glasses, respectively.

et al., 2018b, 2017a,c, 2015b) (see Figure 13.7c). In turn, isostatic glasses are free of both floppy modes and internal stress and, hence, feature the highest kinetic and thermodynamic resistance to relaxation (and crystallization) (Bauchy et al., 2017).

13.4.2 Chalcogenide Glasses

Historically, TCT was first applied to understand and predict the ability of chalcogenide alloys to form a glass when quenched fast enough (Phillips, 1979). In that sense, TCT offers a natural extension to the Zachariasen rules, which are limited to binary oxide systems (Zachariasen, 1932). Chalcogenide glasses (i.e., alloys of chalcogenide elements, e.g., Ge, Si, As, Sb, S, Se, and Te) constitute an important class of glasses. In contrast to oxide glasses, chalcogenide glasses can form some homopolar bonds (e.g., Se–Se) and, hence, do not have to satisfy a fixed stoichiometry (Bauchy and Micoulaut, 2013c). For instance, in contrast to the Si_xO_{1-x} system (which can only exist for $x = 1/3$), Ge_xSe_{1-x} glasses can be synthesized for continuous values of x (Boolchand et al., 2001a). In many cases (although not always), the coordination number r of the chalcogenide elements present in the glass is given by the $8 - N$ rule (or octet rule) (Hosokawa et al., 2013). For instance, one usually has $r_{Ge} = 4$, $r_{As} = 3$, and $r_{Se} = 2$ (Bauchy et al., 2014a; Micoulaut et al., 2013). In the following, we focus on the glass-forming ability of the Ge–Se and As–Se systems.

Ge Se Glasses

We first focus on Ge–Se glasses. The structure of Ge–Se glasses comprises fourfold coordinated tetrahedral Ge atoms and twofold coordinated Se atoms. Limited chemical order is observed as homopolar Ge–Ge and Se–Se bonds are observed (Salmon and Petri, 2003). Table 13.2 details the constraints enumeration in Ge_xSe_{1-x} glasses. As mentioned in Section 3.2, the number of radial BS and angular BB constraints can be obtained from the coordination number of each atom. The average number of constraints per atom

n_c is given by

$$n_c = 7x + 2(1 - x) = 2 + 5x \qquad (13.10)$$

As expected, the number of constraints per atom increases with the fraction x of highly coordinated Ge atoms. The isostatic composition can be identified by solving $n_c(x_{iso}) = 3$, which yields $x_{iso} = 20\%$. As such, Ge_xSe_{1-x} is flexible for $x < 20\%$ and stressed-rigid for $x > 20\%$. This result can also be obtained by calculating the average coordination number:

$$\langle r \rangle = 4x + 2(1 - x) = 2 + 2x \qquad (13.11)$$

and solving the equation $<r>(x_{iso}) = 2.4$ (see Section 3.3), which also yields $x_{iso} = 20\%$ (i.e., a $GeSe_4$ glass).

This result is a notable success for TCT as it provides an intuitive and elegant explanation to the experimental observation that Ge_xSe_{1-x} exhibits maximum glass-forming ability around $x_{iso} = 20\%$ (see Figure 13.8). For instance, a $GeSe_4$ glass can be formed even by cooling a melt *in situ* in the furnace, whereas air- or water-quenching is required to form Se ($x = 0\%$) and $GeSe_2$ ($x = 33\%$) glasses (Azoulay et al., 1975). These results are also in agreement with the fact that an intermediate phase wherein Ge–Se exhibits minimum non-reversible enthalpy at the glass transition is observed around $x = 20\%$ (Boolchand et al., 2001b) (see Figure 13.9a). This suggests that, around the isostatic threshold, glasses exhibit maximum stability, i.e., minimum relaxation. This has been suggested to arise from the fact that (i) for $n_c < 3$ (flexible domain), relaxation is *facilitated* by the low-energy atomic modes of deformation, (ii) for $n_c > 3$ (stressed-rigid domain), relaxation is *stimulated* by the fact that the glass is unstable due to the presence of internal stress within the network, whereas (iii) for $n_c \approx 3$ (isostatic domain), the glass is free of any internal modes of both deformation and stress (Wang et al., 2017b). It is also interesting to note that

TABLE 13.2 Constraints Enumeration in Ge_xSe_{1-x}

Element	#	r	# BS	# BB	# BS + BB
Ge	x	4	2	5	7
Se	$1 - x$	2	1	1	2

The columns contain the types of atoms, their numbers (#), coordination number (r), number of bond-stretching (BS) constraints, number of bond-bending (BB) constraints, and total number of constraints per atom (BS + BB).

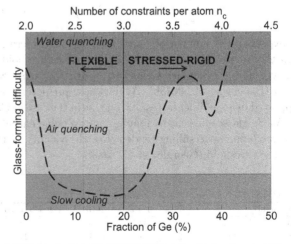

FIGURE 13.8 Glass-forming difficulty of Ge–Se alloys as a function of composition and number of constraints per atom (Azoulay et al., 1975).

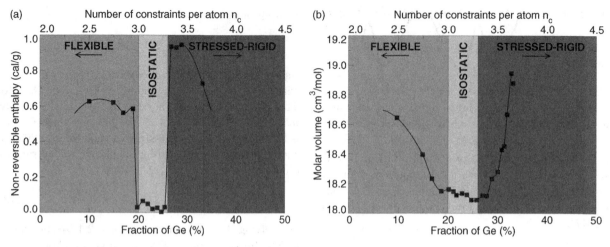

FIGURE 13.9 (a) Non-reversible enthalpy at the glass transition (measured by modulated differential scanning calorimetry (Bechgaard et al., 2018)) and (b) molar volume of Ge–Se glasses as a function of composition and number of constraints per atom (Bhosle et al., 2011).

Ge–Se glasses exhibit minimum molar volume around the isostatic threshold (see Figure 13.9b). Such space-filling tendency has been observed for various glasses, which suggests that it is a generic feature of isostatic glasses (Phillips, 2006; Rompicharla et al., 2008). This can be explained as follows. Flexible Ge–Se glasses present some elongated chains of Se that reduce the local packing efficiency. In turn, stressed-rigid Ge–Se glasses exhibit a locked atomic network that cannot easily reorganize to efficiently fill the space.

As Se Glasses

We now focus on As–Se glasses. The structure of As–Se glasses comprises threefold coordinated pyramidal As atoms and twofold coordinated Se atoms (Bauchy, 2013; Georgiev et al., 2000). Table 13.3 details the constraints enumeration in As_xSe_{1-x} glasses. The average number of constraints per atom n_c is given by

$$n_c = \frac{9}{2}x + 2(1-x) = 2 + \frac{5}{2}x \qquad (13.12)$$

As expected, the number of constraints per atom increases with the fraction x of As atoms. The isostatic threshold is achieved at $n_c = 3$, which yields $x_{iso} = 40\%$ (i.e., a As_2Se_3 glass) (Bauchy et al., 2013b). Again, these predictions are supported by experimental results, and, as in the case of Ge–Se, As–Se glasses present minimum non-reversible enthalpy and maximum space-filling tendency at the vicinity of the isostatic threshold (Georgiev et al., 2000).

13.4.3 Network-Forming Oxides

SiO$_2$ Glass

We now focus on oxide glass-forming systems. We first discuss the case of glassy SiO_2—the archetypal structural basis for all silicate glasses. The structure of SiO_2 is made of fourfold coordinated Si atoms that form some SiO_4 tetrahedra, which are connected to each other via their corners (Grimley et al., 1990; Krishnan et al., 2017b). All the oxygen atoms at the corners of this unit are connected to two Si atoms and are referred to as bridging oxygen (BO) (Wang et al., 2015b). Table 13.4 summarizes the constraints enumeration in SiO_2. As expected, the number of BS constraints created by each atom is half of their coordination number (see Section 3.2). The number of BB constraints created by Si atoms is 5, i.e., the number of independent O–Si–O angles that need to be fixed to define the tetrahedral environment. However, the inter-tetrahedral angle Si–O–Si has been noted to exhibit a broad distribution, which suggests that this angle is poorly constrained (Bauchy et al., 2011; Neuefeind and Liss, 1996; Pettifer et al., 1988; Yu et al., 2016; Yuan and Cormack, 2003). Consequently, no BB constraint is assigned to O atoms (or, in other words, this constraint is considered broken). Altogether, the number of constraints per atom n_c is given by

$$n_c = \frac{1 \times 7 + 2 \times 1}{1 + 2} = 3 \qquad (13.13)$$

which confirms that the excellent glass-forming ability of silica arises from the isostatic nature of its atomic network.

TABLE 13.3   Constraints Enumeration in As_xSe_{1-x}

Element	#	r	# BS	# BB	# BS + BB
As	x	3	3/2	3	9/2
Se	$1-x$	2	1	1	2

The columns contain the types of atoms, their numbers (#), coordination number (r), number of bond-stretching (BS) constraints, number of bond-bending (BB) constraints, and total number of constraints per atom (BS + BB).

TABLE 13.4   Constraints Enumeration in SiO_2

Element	#	r	# BS	# BB	# BS + BB
Si	1	4	2	5	7
O	2	2	1	0	1

The columns contain the types of atoms, their numbers (#), coordination number (r), number of bond-stretching (BS) constraints, number of bond-bending (BB) constraints, and total number of constraints per atom (BS + BB).

Note that the broken nature of the Si–O–Si BB constraint is key in explaining the great glass-forming ability of silica (Zhang and Boolchand, 1994). Such thermal breakage of the BB constraints of O atoms is only observed in pure SiO_2 and was suggested to arise from the high glass transition temperature T_g of silica (1,200°C)—so that this weak constraint is thermally broken at T_g (Mathieu Bauchy et al., 2015c; Gupta and Mauro, 2009). In contrast, this constraint is active in TeO_2, which explains its poor glass-forming ability (Zhang and Boolchand, 1994). Note that, in this case, despite the isostatic nature of SiO_2, the average coordination number is not equal to 2.4 as this relationship assumes that all BB constraints are intact (see Section 13.3.2).

B_2O_3 Glass

We now focus on B_2O_3, that is, the base structural unit of all borate glasses. The structure of B_2O_3 is made of threefold coordinated B atoms that form some trigonal BO_3 units, which are connected to each other via their corners (Varshneya, 1993). Table 13.5 summarizes the constraints enumeration in B_2O_3. In contrast to the case of SiO_2, the BB constraint associated with the B–O–B angle is here considered intact on account of the low glass transition temperature of this glass. The number of constraints per atom n_c is given by

$$n_c = \frac{2 \times 9/2 + 3 \times 2}{2 + 3} = 3 \qquad (13.14)$$

which establishes the isostatic nature of the atomic network of B_2O_3 and explains its excellent glass-forming ability. This is in agreement with the fact that the average coordination number of B_2O_3 is 2.4.

P_2O_5 Glass

Finally, we focus on P_2O_5, which is the base structural unit for all phosphate glasses. The structure of P_2O_5 is made of fourfold tetrahedral PO_4 units (Brow, 2000). However, in contrast to silica, only three of the O corners of the tetrahedra are shared and one O atom is terminating (i.e., non-bridging oxygen, or NBO) (Brow, 2000). As such, there is 1 NBO and 3/2 BO per P atom, respectively. Table 13.6 summarizes the constraints enumeration in P_2O_5. Note that, in this case, since the two types of O atoms (BO and NBO) feature a different topology, they must be treated in a distinct fashion in the enumeration (i.e., as if they were different elements). The number of constraints per atom n_c is given by

TABLE 13.5 Constraints Enumeration in B_2O_3

Element	#	r	# BS	# BB	# BS + BB
B	2	3	3/2	3	9/2
O	3	2	1	1	2

The columns contain the types of atoms, their numbers (#), coordination number (r), number of bond-stretching (BS) constraints, number of bond-bending (BB) constraints, and total number of constraints per atom (BS + BB).

TABLE 13.6 Constraints Enumeration in P_2O_5

Element	#	r	# BS	# BB	# BS + BB
P	2	4	2	5	7
O	5	-	-	-	-
BO	3	2	1	1	2
NBO	2	1	1/2	0	1/2

The columns contain the types of atoms, their numbers (#), coordination number (r), number of bond-stretching (BS) constraints, number of bond-bending (BB) constraints, and total number of constraints per atom (BS + BB).

$$n_c = \frac{2 \times 7 + 3 \times 2 + 2 \times 1/2}{2 + 5} = 3 \qquad (13.15)$$

which, again, demonstrates the isostatic nature of the atomic network of P_2O_5 and explains its good glass-forming ability. Note that, despite the isostatic nature of P_2O_5, the average coordination number is not equal to 2.4 as this relationship assumes that each atom has a coordination number that is equal or larger than two (see Section 3.2).

13.4.4 Impact of Network Modifiers and the Example of Sodium Silicate Glass

In contrast to the network-forming species, network-modifying atoms tend to depolymerize the network by creating some weak bonds (Bauchy et al., 2013a). In the following, we now review how TCT can describe the glass-forming ability of a silicate glass comprising some network modifiers by taking the example of sodium silicate, an archetypical model for all alkali silicate glasses (Bauchy and Micoulaut, 2011b). Such glasses are technologically important as they can be strengthened using ion exchange (e.g., Corning® Gorilla® glass) (Mauro et al., 2013b; Varshneya, 2010; Wang et al., 2017b,c). Starting from the base topology of glassy silica, Na atoms tend to depolymerize the silicate network by creating some NBO atoms (1 NBO per Na atom) (Varshneya, 1993). Table 13.7 summarizes the constraints enumeration in the alloy $(Na_2O)_x(SiO_2)_{1-x}$. As in the case of P_2O_5, it is here necessary to distinctly account for the BO and NBO atoms as they exhibit a different topology. Note that, here, the number of BS constraints created by Na differs from their coordination number (which is found to be around six in sodium silicate glasses) (Bauchy, 2012b; Li et al., 2017; Yu et al., 2017b). Indeed, in this case, atomistic simulations suggested that, despite having a coordination of around six, Na atoms are preferentially bounded

TABLE 13.7 Constraints Enumeration in $(Na_2O)_x(SiO_2)_{1-x}$

Element	#	r	# BS	# BB	# BS + BB
Si	$1-x$	4	2	5	7
O	$2-x$	-	-	-	-
Na	$2x$	1	1/2	0	1/2
BO	$2-3x$	2	1	1	2
NBO	$2x$	1	1	0	1

The columns contain the types of atoms, their numbers (#), coordination number (r), number of bond-stretching (BS) constraints, number of bond-bending (BB) constraints, and total number of constraints per atom (BS + BB).

to the nearest NBO atom and, hence, exhibit only one BS constraint (Bauchy and Micoulaut, 2011a). This illustrates the fact that the number of BS constraints created by an atom is not always given by the geometrical coordination number, which makes it necessary to have an accurate knowledge of the glass structure to meaningfully enumerate the constraints. Also note that even a small fraction x of Na_2O results in a drop of the glass transition temperature of the glass with respect to that of pure silica (Vaills et al., 2005). Consequently, in contrast to silica, the BB constraint associated with BO is not considered as being thermally broken anymore (Bauchy and Micoulaut, 2011a). Finally, note that, due to the ionic and non-directional nature of the Na–O bonds, the Si–NBO–Na angle is very poorly defined (Bauchy, 2012b). Hence, no BB constraint is assigned to this angle. Altogether, the number of constraints per atom n_c is given by

$$n_c = \frac{(1-x) \times 7 + 2x \times 1/2 + (2 - 3x) \times 2 + 2x \times 1}{(1 - x) + (2 - x) + 2x}$$
$$= \frac{11 - 10x}{3} \qquad (13.16)$$

As expected, the rigidity (i.e., n_c) of the glass decreases upon the addition of Na, that is, as the network becomes more and more depolymerized. An isostatic atomic network is then obtained for $n_c = 3$, which is achieved at $x_{iso} = 20\%$. As such, sodium silicate is stressed-rigid ($n_c > 3$) for $x < 20\%$ and flexible ($n_c < 3$) for $x > 20\%$. These predictions are in agreement with experimental results, as sodium silicate exhibits a minimum in enthalpy relaxation around $x_{iso} = 20\%$ (see Figure 13.10) (Micoulaut, 2008; Vaills et al., 2005). Sodium silicate glasses also exhibit a sharp increase in elastic energy (calculated from the variation in density upon annealing) for $x > 20\%$, which denotes an onset of internal flexibility, in agreement with the present predictions (Micoulaut, 2008; Vaills et al., 2005). These results are also

in agreement with the fact sodium silicate glasses tend to show some degree of phase separation (i.e., low glass-forming ability) for $x < 20\%$ (Kreidl, 1991), which may arise from the presence of internal stress within the structure. Altogether, this analysis illustrates how TCT can be used to describe the rigidity of modified silicate glasses.

13.4.5 Interest of Molecular Dynamics Simulations

Although the number of constraints per atom can be calculated analytically for select glasses, it requires an accurate knowledge of the glass structure that is not always available. In addition, several issues can render the constraints enumeration challenging.

1. The coordination number of each species is not always known. For instance, in the case of borosilicate glasses, B atoms can be threefold or fourfold coordinated depending on composition (Smedskjaer et al., 2011; Wang et al., 2018), and Ca atoms typically exhibit a coordination that is between four and eight, thereby not following the Octet rule (Bauchy, 2014). Moreover, it has been shown that the effective number of BS constraints created by each atom is not always equal to their coordination number (i.e., as calculated by enumerating the number of neighbors inside the first coordination shell). In particular, Na atoms in sodium silicate glass have a coordination number around six but only show one active BS constraint (Bauchy and Micoulaut, 2011a).

2. Isolated atoms or molecules (e.g., water molecules (Qomi et al., 2014a; Mathieu Bauchy et al., 2014b; Krishnan et al., 2016)) are not part of the atomic network and, hence, do not contribute to its rigidity. Therefore, they should not be taken into account in the constraints enumeration.

3. Each constraint is associated with a given energy and can consequently be intact or broken depending on temperature, i.e., the amount of available thermal energy (Gupta and Mauro, 2009). Hence, weaker angular constraints (like the Si–O–Na bond in sodium silicate) are broken even at 300 K (Bauchy and Micoulaut, 2011a; Wang et al., 2016b). For these reasons, one cannot just rely on unproven guesses to enumerate the number of constraints.

Molecular dynamics simulations, which offer a full access to the structure and dynamics of the atoms, provide a valuable tool to tackle these difficulties. To this end, a general enumeration method has been developed, which allows one to compute the number of constraints per atom in network glasses. This method has been widely applied to chalcogenide and oxide glasses (Bauchy, 2012a; Bauchy et al., 2013b, 2011; Bauchy and Micoulaut, 2013a, 2011a;

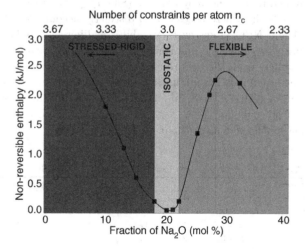

FIGURE 13.10 Non-reversible heat flow at the glass transition (measured by modulated differential scanning calorimetry) in $(Na_2O)_x(SiO_2)_{1-x}$ glasses as a function of composition and number of constraints per atom (Vaills et al., 2005).

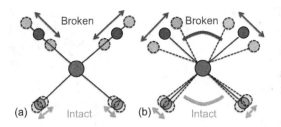

FIGURE 13.11 Illustration of the usage of molecular dynamics to compute the number of (a) radial BS and (b) angular BB constraints per atom (Bauchy et al., 2014b; Bauchy and Micoulaut, 2011a).

Micoulaut and Bauchy, 2013) and, more recently, to atomic-scale models of cement hydrates (Bauchy et al., 2016, 2014b). This enumeration method is based on the analysis of atomic trajectories obtained through molecular dynamics simulations. Since the nature of the constraints imposed on the atomic motion is not known *a priori*, the opposite approach is adopted, that is, by looking at the motion of each atom and deducing the underlying constraints that cause this motion. In other words, an active constraint would maintain bond lengths or angles fixed around their average values, whereas a large atomic motion implies the absence of any underlying constraint (see Figure 13.11). Specifically, to assess the number of BS constraints that apply to a central atom, the radial excursion of each neighbor is computed. A small radial excursion implies the existence of an underlying constraint that maintains the bond length fixed around its average value. On the contrary, a large radial excursion implies a broken constraint. The number of BB constraints can be accessed in the same fashion, that is, by analyzing the angular excursion of each neighbor. The detailed implementation of this method is reported in (Bauchy et al., 2014b; Bauchy and Micoulaut, 2011a; Micoulaut et al., 2015) and unambiguously discriminates intact from broken constraints, including those created by Na and Ca atoms. As such, molecular dynamics can be used to inform TCT so that the constraints enumeration is based on an accurate structural base rather than simple guesses.

13.5 Applications of TCT for the Prediction of Glass Properties

13.5.1 General Principles

Besides glass-forming ability, TCT has been extensively applied toward the prediction of the compositional dependence of glass properties. These topological models are based on the following approaches:

1. By capturing the important connectivity of glasses while filtering out less relevant structural details, TCT can be used to capture the first-order contribution of the structure of the atomic network on macroscopic properties. This makes it possible to simplify complex disordered network into more simple mechanical trusses, thereby allowing the development of analytical predictive models. These models rely on (i) the knowledge of the glass structure as a function of composition, so that one can analytically calculate the number of constraints per atom $n_c(x)$ as a function of the composition x of the glass, and (ii) a model relating n_c to a given macroscopic property $P(n_c)$. The combination of (i) and (ii) yields an analytical model $P(x)$. Hence, the number of constraints per atom acts as a reduced-order parameter to facilitate the understanding of the linkages between structure and macroscopic properties. For instance, this type of models is used for predicting glass hardness as a function of composition (see Section 5.2).

2. For select properties, no direct analytical model relating n_c and P is yet available. However, P is sometimes known to be minimized or maximized for optimally constrained isostatic glasses, e.g., $P_{\max} = P(n_c = 3)$. As such, this type of model cannot predict the compositional dependence of a property but can be used to pinpoint promising compositions exhibiting optimal properties. These models rely on (i) the knowledge of the glass structure as a function of composition so that one can analytically calculate the number of constraints per atom $n_c(x)$ as a function of the composition x of the glass and (ii) the ability to solve $n_c(x_{\mathrm{iso}}) = 3$. To predict optimal composition(s) $\{x_{\mathrm{iso}}\}$. For instance, this type of models is used for predicting glass compositions exhibiting maximum fracture toughness (see Section 5.3).

In the following, we briefly review a selection of the topological models that are available in the literature.

13.5.2 Hardness

Hardness characterizes the resistance of materials to permanent deformations (Oliver and Pharr, 1992; Smedskjaer et al., 2015). Smedskjaer and Mauro proposed that the hardness of glasses can be predicted from the knowledge of the number of constraints per atom (Smedskjaer et al., 2010). They introduced the following formula:

$$H = \left(\frac{\partial H}{\partial n_c} \right) [n_c - n_{\mathrm{crit}}] \qquad (13.17)$$

This formula is based on the idea that (i) a glass needs a critical minimum number of constraints n_{crit} to be cohesive and that (ii) each additional constraints should contribute to increase hardness. The value $n_{\mathrm{crit}} = 2.5$ (i.e., the minimum number of constraints that are needed to achieve rigidity in two dimensions) was found to be appropriate (Smedskjaer

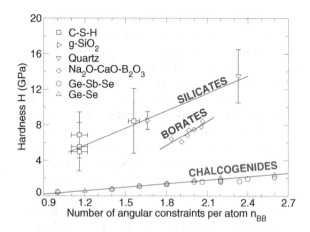

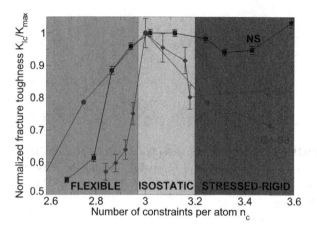

FIGURE 13.12 Hardness of various oxide and chalcogenide phases as a function of their number of angular BB constraints per atom (Qomi et al., 2017, 2015a, 2014b, 2013, 2015b; Bauchy, 2017; Bauchy et al., 2015c, 2014b–d; Smedskjaer et al., 2010, 1991; Swiler et al., 1990; Varshneya and Mauro, 2007).

FIGURE 13.13 Fracture toughness (normalized by the fracture toughness at $n_c = 3$) of densified sodium silicate glasses (NS), calcium–silicate–hydrates (C–S–H), and Ge–Se glasses as a function of their number of constraints per atom (Bauchy et al., 2016, 2015a; Guin et al., 2002; Yu et al., 2015a).

et al., 2010), whereas the $\partial H/\partial n_c$ term is a fitting parameter that depends on the indenter geometry, the indentation load, and the glass family being considered (but not the specific glass composition). This model has been found to yield accurate hardness predictions for various oxide glasses (Smedskjaer, 2014; Smedskjaer et al., 2010). Recently, it was suggested that BS and BB constraints may not contribute in the same fashion to hardness. Specially, the hardness of various materials has been found to preferentially scale with the number of angular BB constraints per atom rather than the total number of constraints per atom (see Figure 13.12) (Bauchy et al., 2015c). This has been attributed to the fact that, upon indentation loading, the atomic network deforms by following the lowest energy path, that is, by breaking the weaker angular BB constraints rather than the stronger radial BS constraints. It has also recently been suggested that hardness should depend on the number of constraints per unit of volume rather than the number of constraints per atom (Zheng et al., 2017).

13.5.3 Fracture Toughness

Fracture toughness characterizes the resistance of materials to fracture (Bauchy et al., 2015a; Frederiksen et al., 2018; Januchta et al., 2017a,b; Rouxel and Yoshida, 2017; Wang et al., 2015a; Yu et al., 2015a). Although no topological model predicting the compositional dependence of the fracture of glasses has been proposed thus far, various glasses have been noted to exhibit maximum fracture toughness at the isostatic threshold (see Figure 13.13) (Bauchy, 2017; Bauchy et al., 2016). This has been suggested to arise from the fact that (i) flexible glasses ($n_c < 3$) exhibit low cohesion (low surface energy) due to their low connectivity, (ii) stressed-rigid glasses ($n_c > 3$) break in a brittle fashion as their high connectivity prevents any ductile atomic reorganization, whereas (iii) isostatic glasses ($n_c = 3$) exhibit the best balance between cohesion and ability

to plastically deform (Bauchy et al., 2016). Specifically, isostatic glasses have been noted to exhibit a maximum propensity for crack blunting, which contributes to postponing fracture (Bauchy et al., 2016).

13.5.4 Viscosity, Fragility, and Glass Transition Temperature

Predicting the viscosity of glass-forming melts is critical for various applications (Bauchy et al., 2013a; Mauro et al., 2009b). However, such predictions are complex as (i) the viscosity varies by many orders of magnitude with temperature, and (ii) it is sensitive to small changes in composition or structure (Bauchy et al., 2013a). Most mathematical models used to describe the viscosity of supercooled liquids are empirical and based on the Vogel–Fulcher–Tamman (VFT) equation:

$$\eta\left(T\right) = A\exp\left[\frac{B}{R\left(T - T_0\right)}\right] \qquad (13.18)$$

where R is the perfect gas constant, T is the temperature, and A and T_0 are some empirical parameters, which depend on the composition of the glass. Although the VFT equation has met notable success, it gives an incorrect description of the asymptotic Arrhenius behavior of viscosity at low temperature. To overcome this limitation, the Mauro–Yue–Ellison–Gupta–Allan (MYEGA) equation (Mauro et al., 2009b) was recently proposed to describe the temperature dependence of liquid viscosity:

$$\eta\left(T\right) = \log\eta_\infty + \left(12 - \log\eta_\infty\right)\frac{T_g}{T}\exp$$
$$\times \left[\left(\frac{m}{12 - \log\eta_\infty} - 1\right)\left(\frac{T_g}{T} - 1\right)\right] \qquad (13.19)$$

where T_g is the glass transition temperature, defined by $\log\eta\left(T_g\right) = 10^{12}$ Pa s, η_∞ is the extrapolated infinite-temperature limit of liquid viscosity (universally found to be equal to $10^{-2.9}$ Pa s), and m is the liquid fragility index (Angell, 1991; Bechgaard et al., 2017):

$$m = \frac{d \log \eta}{d(T_g/T)} \bigg| T = T_g \qquad (13.20)$$

As opposed to the VFT equation, this model has a clear physical foundation based on the temperature dependence of the configurational entropy (given by the Adam–Gibbs equation) and offers more accurate predictions at low temperature. In addition, it only relies on two physical parameters, namely, the fragility index and the glass transition temperature. As such, the composition dependence of the viscosity is entirely captured by that of T_g and m. Analytical models have been developed to predict the composition dependence of the viscosity, T_g, and m through the use of temperature-dependent TCT and the Adam–Gibbs equation (Gupta and Mauro, 2009; Mauro et al., 2016, 2013a, 2009). Therefore, the prediction of the compositional dependence of these dynamical properties only relies on the mere knowledge of the number of topological constraints per atom n_c in the network with respect to composition and temperature.

13.5.5 Dissolution Kinetics

Predicting glass corrosion is important for various applications, including nuclear waste immobilization glasses, bioactive glasses, or supplementary cementitious materials (Gin, 2014; Jones, 2013; Tandré Oey et al., 2017b). It was recently proposed that the dissolution rate K of glasses in dilute conditions (i.e., forward rate, far from saturation) is controlled by the topology of the atomic network (Pignatelli et al., 2016b) as follows:

$$K = K_0 \exp \left(-\frac{n_c E_0}{RT} \right) \qquad (13.21)$$

where K_0 is a rate constant that depends on the solution phase chemistry (i.e., the barrier-less dissolution rate of a completely depolymerized material for which $n_c = 0$), and $E_0 = 20$–25 kJ/mol is an energy barrier that needs to be overcome to break a unit atomic constraint. Based on this equation, the dissolution process is characterized by the effective activation energy:

$$E_A^{\text{eff}} = n_c E_0 \qquad (13.22)$$

The following atomistic picture was suggested to explain this model: starting from $n_c = 0$ (i.e., which would correspond to a fully depolymerized material), each new constraint per atom effectively reduces the dissolution kinetics by increasing the associated activation energy needed for bond rupture (Pignatelli et al., 2016b). In detail, it was proposed that n_c serves as an indicator of steric effects acting in the atomic network, which prevent the reorganization and internal motion of the constituent species. Indeed, whether it occurs by hydrolysis or ion exchange, corrosion results in the formation of some local stress within the network.

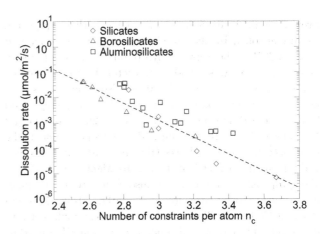

FIGURE 13.14 Dissolution rate of various silicate phases as a function of the number of constraints per atom (Oey et al., 2017a,b; Mascaraque et al., 2018, 2017; Pignatelli et al., 2016b).

Namely, hydrolysis requires the formation of larger intermediate over-coordinated species (fivefold coordinated Si or threefold coordinated O), whereas ion exchange requires some local opening of the network to enable the jump of mobile cations from one pocket to another (Bunker, 1994). In any case, the activation energy associated with these processes is controlled by the ability of the atomic network to locally reorganize to accommodate these local strains. In details, the resulting activation energy takes the form of the strain elastic energy that is applied by the rest of the network to resist the creation of this local defect (Guo et al., 2016). This strain elastic energy is controlled by the local number of constraints per atom n_c, since each constraint acts as a little spring connecting the atoms. Therefore, n_c characterizes the local stiffness of the atomic network, which tends to prevent the accommodation of local defects. This picture is in line with results from density functional theory, which have shown that the activation energy associated with the hydrolysis of inter-tetrahedra bridging oxygen atoms increases with the network connectivity and, therefore, rigidity (Pelmenschikov et al., 2000). As shown in Figure 13.14, this model has since then been extensively validated over a broad range of silicate phases and is able to predict the dissolution kinetics of silicate phases over four orders of magnitude (Pignatelli et al., 2016a,c; Oey et al., 2017a–c, 2015; Hsiao et al., 2017; Mascaraque et al., 2018, 2017).

13.5.6 Other Properties

Several other properties have been shown to be correlated with the topology of the atomic network (Micoulaut and Yue, 2017). For instance, stiffness was found to scale with the number of constraints per atom in select systems (Thorpe, 1983; Thorpe and Duxbury, 1999). Isostatic glasses were found to exhibit a stress-free character (Boolchand and Goodman, 2017; Wang et al., 2005). Isostatic glasses have also been found to exhibit optimal strengthening upon ion exchange (Wang et al., 2017b,c; Wang and Bauchy, 2015;

Svenson et al., 2016). Sub-critical crack growth was reported to be controlled by the atomic topology (Smedskjaer and Bauchy, 2015). The degree of hydrophilicity/hydrophobicity of silica surfaces was found to be dictated by the topology of the glass surface (Yu et al., 2018a). Isostatic glasses were noted to exhibit maximum elastic volume recovery upon loading–unloading cycles (Bauchy et al., 2017; Mauro and Varshneya, 2007). Relaxation and aging were found to be minimized in isostatic glasses (Bauchy et al., 2017; Bauchy and Micoulaut, 2015; Chen et al., 2010; Mantisi et al., 2015). The propensity for creep was also found to be minimal in isostatic phases (Bauchy et al., 2017, 2015b; Masoero et al., 2015; Pignatelli et al., 2016a). The thermal, mechanical, electrical, and optical properties of a-SiC:H thin films were demonstrated to be influenced by the topology of their atomic network (King et al., 2013). Susceptibility was found to be maximal in isostatic granular systems (Moukarzel, 2002, 1998). The resistance to irradiation of silicate hydrates was found to be optimal in isostatic systems (Krishnan et al., 2017c; Wang et al., 2017a). Fast-ion conduction was observed to offer a signature of the intermediate phase (Micoulaut et al., 2009; Novita et al., 2009, 2007). The performances of phase-change materials were noted to be controlled by their atomic topology (Micoulaut et al., 2010). Finally, protein folding was found to be controlled by the topology of their molecular architecture (Phillips, 2009, 2004; Rader et al., 2002). The wide variety of systems and properties for which TCT has been successfully applied denotes its generic nature and suggests that the topological nanoengineering of materials is likely to yield exciting new developments in material science and engineering.

13.6 Conclusion

The various examples covered in the chapter highlight how TCT can be used as an effective tool to predict the properties of disordered materials or pinpoint optimized material compositions with tailored functionalities. It is especially well suited for disordered systems with no fixed stoichiometry, for which traditional trial-and-error approaches are rendered inefficient due to the virtually infinite number of possible compositions. By capturing relevant structural features while filtering out the less relevant ones, topological approaches can largely simplify complex disordered structures and, thereby, facilitate the development of analytical predictive models—by capturing the essential structural features of atomic networks and using them as a reduced-order parameter to inform predictive models linking composition to properties. Topological models only rely on the accurate knowledge of the atomic structure of materials, which can be obtained by high-resolution experiments or atomistic simulations. As such, TCT exemplifies how a synergetic combination of experiments, simulations, and theoretical models can be used to decode the genome of glass, that is, by deciphering how glass' basic structural units control its macroscopic properties.

Acknowledgments

Matthieu Micoulaut, John C. Mauro, Morten M. Smedskjaer, Punit Boolchand, and Gaurav N. Sant are gratefully acknowledged for numerous discussions that form the basis of this chapter. This work was supported by the National Science Foundation under Grants No. 1562066, 1762292, 1826420, and 1928538.

References

Angell, C.A., 1991. Relaxation in liquids, polymers and plastic crystals - strong fragile patterns and problems. *J. Non-Cryst. Solids* 131, 13–31. doi: 10.1016/0022-3093(91)90266-9.

Azoulay, R., Thibierge, H., Brenac, A., 1975. Devitrification characteristics of Ge_xSe_{1-x} glasses. *J. Non-Cryst. Solids* 18, 33–53. doi: 10.1016/0022-3093(75)90006-X.

Ball, P., 2015. Material witness: Concrete mixing for gorillas. *Nat. Mater.* 14, 472–472. doi: 10.1038/nmat4279.

Bauchy, M., 2017. Nanoengineering of concrete via topological constraint theory. *MRS Bull.* 42, 50–54. doi: 10.1557/mrs.2016.295.

Bauchy, M., 2014. Structural, vibrational, and elastic properties of a calcium aluminosilicate glass from molecular dynamics simulations: The role of the potential. *J. Chem. Phys.* 141, 024507. doi: 10.1063/1.4886421.

Bauchy, M., 2013. Structure and dynamics of liquid $AsSe_4$ from ab initio molecular dynamics simulation. *J. Non-Cryst. Solids* 377, 39–42. doi: 10.1016/j.jnoncrysol.2012.12.018.

Bauchy, M., 2012a. Topological constraints and rigidity of network glasses from molecular dynamics simulations. *Am. Ceram. Soc. Bull.* 91, 34–38A.

Bauchy, M., 2012b. Structural, vibrational, and thermal properties of densified silicates: Insights from molecular dynamics. *J. Chem. Phys.* 137, 044510. doi: 10.1063/1.4738501.

Bauchy, M., Guillot, B., Micoulaut, M., Sator, N., 2013a. Viscosity and viscosity anomalies of model silicates and magmas: A numerical investigation. *Chem. Geol.* 346, 47–56. doi: 10.1016/j.chemgeo.2012.08.035.

Bauchy, M., Kachmar, A., Micoulaut, M., 2014a. Structural, dynamic, electronic, and vibrational properties of flexible, intermediate, and stressed rigid As-Se glasses and liquids from first principles molecular dynamics. *J. Chem. Phys.* 141, 194506. doi: 10.1063/1.4901515.

Bauchy, M., Laubie, H., Qomi, M.J.A., Hoover, C.G., Ulm, F.-J., Pellenq, R.J.-M., 2015a. Fracture toughness of calcium–silicate–hydrate from molecular dynamics simulations. *J. Non-Cryst. Solids* 419, 58–64. doi: 10.1016/j.jnoncrysol.2015.03.031.

Bauchy, M., Masoero, E., Ulm, F.-J., Pellenq, R., 2015b. Creep of bulk C-S-H: Insights from molecular dynamics simulations. *Proceedings of the 10th*

International Conference on Mechanics and Physics of Creep, Shrinkage, and Durability of Concrete and Concrete Structure. American Society of Civil Engineers, Vienna, Austria, pp. 511–516.

Bauchy, M., Micoulaut, M., 2015. Densified network glasses and liquids with thermodynamically reversible and structurally adaptive behaviour. *Nat. Commun.* 6, 6398. doi: 10.1038/ncomms7398.

Bauchy, M., Micoulaut, M., 2013a. Transport anomalies and adaptive pressure-dependent topological constraints in tetrahedral liquids: Evidence for a reversibility window analogue. *Phys. Rev. Lett.* 110, 095501. doi: 10.1103/PhysRevLett.110.095501.

Bauchy, M., Micoulaut, M., 2013b. Percolative heterogeneous topological constraints and fragility in glass-forming liquids. *EPL Europhys. Lett.* 104, 56002. doi: 10.1209/0295-5075/104/56002.

Bauchy, M., Micoulaut, M., 2013c. Structure of As_2Se_3 and AsSe network glasses: Evidence for coordination defects and homopolar bonding. *J. Non-Cryst. Solids* 377, 34–38. doi: 10.1016/j.jnoncrysol.2013.01.019.

Bauchy, M., Micoulaut, M., 2011a. Atomic scale foundation of temperature-dependent bonding constraints in network glasses and liquids. *J. Non-Cryst. Solids* 357, 2530–2537. doi: 10.1016/j.jnoncrysol.2011.03.017.

Bauchy, M., Micoulaut, M., 2011b. From pockets to channels: Density-controlled diffusion in sodium silicates. *Phys. Rev. B* 83, 184118. doi: 10.1103/PhysRevB.83.184118.

Bauchy, M., Micoulaut, M., Boero, M., Massobrio, C., 2013b. Compositional thresholds and anomalies in connection with stiffness transitions in network glasses. *Phys. Rev. Lett.* 110, 165501. doi: 10.1103/PhysRevLett.110.165501.

Bauchy, M., Micoulaut, M., Celino, M., Le Roux, S., Boero, M., Massobrio, C., 2011. Angular rigidity in tetrahedral network glasses with changing composition. *Phys. Rev. B* 84, 054201. doi: 10.1103/PhysRevB.84.054201.

Bauchy, M., Qomi, M.J.A., Bichara, C., Ulm, F.-J., Pellenq, R.J.-M., 2014b. Nanoscale structure of cement: Viewpoint of rigidity theory. *J. Phys. Chem. C* 118, 12485–12493. doi: 10.1021/jp502550z.

Bauchy, M., Qomi, M.J.A., Pellenq, R.J.M., Ulm, F.J., 2014c. Is cement a glassy material? *Comput. Model. Concr. Struct.* 169, 169–176.

Bauchy, M., Qomi, M.J.A., Ulm, F.-J., Pellenq, R.J.-M., 2014d. Order and disorder in calcium–silicate–hydrate. *J. Chem. Phys.* 140, 214503. doi: 10.1063/1.4878656.

Bauchy, M., Qomi, M.J.A., Bichara, C., Ulm, F.-J., Pellenq, R.J.-M., 2015c. Rigidity transition in materials: Hardness is driven by weak atomic constraints. *Phys. Rev. Lett.* 114, 125502. doi: 10.1103/PhysRevLett.114.125502.

Bauchy, M., Wang, B., Wang, M., Yu, Y., Qomi, M.J.A., Smedskjaer, M.M., Bichara, C., Ulm, F.-J., Pellenq, R., 2016. Fracture toughness anomalies: Viewpoint of topological constraint theory. *Acta Mater.* 121, 234–239. doi: 10.1016/j.actamat.2016.09.004.

Bauchy, M., Wang, M., Yu, Y., Wang, B., Krishnan, N.M.A., Masoero, E., Ulm, F.-J., Pellenq, R., 2017. Topological control on the structural relaxation of atomic networks under stress. *Phys. Rev. Lett.* 119, 035502. doi: 10.1103/PhysRevLett.119.035502.

Bechgaard, T.K., Gulbiten, O., Mauro, J.C., Hu, Y., Bauchy, M., Smedskjaer, M.M., 2018. Temperature-modulated differential scanning calorimetry analysis of high-temperature silicate glasses. ArXiv180501433 Cond-Mat.

Bechgaard, T.K., Mauro, J.C., Bauchy, M., Yue, Y., Lamberson, L.A., Jensen, L.R., Smedskjaer, M.M., 2017. Fragility and configurational heat capacity of calcium aluminosilicate glass-forming liquids. *J. Non-Cryst. Solids* 461, 24–34. doi: 10.1016/j.jnoncrysol.2017.01.033.

Bhosle, S., Gunasekera, K., Chen, P., Boolchand, P., Micoulaut, M., Massobrio, C., 2011. Meeting experimental challenges to physics of network glasses: Assessing the role of sample homogeneity. *Solid State Commun.* 151, 1851–1855. doi: 10.1016/j.ssc.2011.10.016.

Boolchand, P., Bauchy, M., Micoulaut, M., Yildirim, C., 2018. Topological phases of chalcogenide glasses encoded in the melt dynamics. *Phys. Status Solidi B.* doi: 10.1002/pssb.201800027.

Boolchand, P., Feng, X., Bresser, W.J., 2001a. Rigidity transitions in binary Ge-Se glasses and the intermediate phase. *J. Non-Cryst. Solids* 293–295, 348–356. doi: 10.1016/S0022-3093(01)00867-5.

Boolchand, P., Georgiev, D.G., Goodman, B., 2001b. Discovery of the intermediate phase in chalcogenide glasses. *J. Optoelectron. Adv. Mater.* 3, 703–720.

Boolchand, P., Goodman, B., 2017. Glassy materials with enhanced thermal stability. *MRS Bull.* 42, 23–28. doi: 10.1557/mrs.2016.300.

Boolchand, P., Zhang, M., Goodman, B., 1996. Influence of one-fold-coordinated atoms on mechanical properties of covalent networks. *Phys. Rev. B* 53, 11488–11494. doi: 10.1103/PhysRevB.53.11488.

Brow, R.K., 2000. Review: The structure of simple phosphate glasses. *J. Non-Cryst. Solids* 263, 1–28. doi: 10.1016/S0022-3093(99)00620-1.

Bunker, B.C., 1994. Molecular mechanisms for corrosion of silica and silicate glasses. *J. Non-Cryst. Solids, Proceedings of the First PAC RIM Meeting on Glass and Optical Materials* 179, 300–308. doi: 10.1016/0022-3093(94)90708-0.

Chen, P., Boolchand, P., Georgiev, D.G., 2010. Long term aging of selenide glasses: Evidence of sub-Tg endotherms and pre-Tg exotherms. *J. Phys. Condens. Matter* 22. doi: 10.1088/0953-8984/22/6/065104.

Frederiksen, K.F., Januchta, K., Mascaraque, N., Youngman, R.E., Bauchy, M., Rzoska, S.J., Bockowski, M., Smedskjaer, M.M., 2018. Structural compromise between high hardness and crack resistance in aluminoborate glasses. *J. Phys. Chem. B.* doi: 10.1021/acs.jpcb.8b02905.

Georgiev, D.G., Boolchand, P., Micoulaut, M., 2000. Rigidity transitions and molecular structure of As_xSe_{1-x} glasses. *Phys. Rev. B* 62, R9228–R9231. doi: 10.1103/PhysRevB.62.R9228.

Gin, S., 2014. Open scientific questions about nuclear glass corrosion. *Procedia Mater. Sci., 2nd International Summer School on Nuclear Glass Wasteform: Structure, Properties and Long-Term Behavior, SumGLASS 2013* 7, 163–171. doi: 10.1016/j.mspro.2014.10.022.

Grimley, D.I., Wright, A.C., Sinclair, R.N., 1990. Neutron scattering from vitreous silica IV. Time-of-flight diffraction. *J. Non-Cryst. Solids* 119, 49–64. doi: 10.1016/0022-3093(90)90240-M.

Guin, J.-P., Rouxel, T., Sanglebuf, J.-C., Melscoët, I., Lucas, J., 2002. Hardness, toughness, and scratchability of germanium–selenium chalcogenide glasses. *J. Am. Ceram. Soc.* 85, 1545–1552. doi: 10.1111/j.1151-2916.2002.tb00310.x.

Guo, P., Wang, B., Bauchy, M., Sant, G., 2016. Misfit stresses caused by atomic size mismatch: The origin of doping-induced destabilization of dicalcium silicate. *Cryst. Growth Des.* 16, 3124–3132. doi: 10.1021/acs.cgd.5b01740.

Gupta, P.K., Mauro, J.C., 2009. Composition dependence of glass transition temperature and fragility. I. A topological model incorporating temperature-dependent constraints. *J. Chem. Phys.* 130, 094503. doi: 10.1063/1.3077168.

Hosokawa, S., Koura, A., Bérar, J.-F., Pilgrim, W.-C., Kohara, S., Shimojo, F., 2013. Does the 8-N bonding rule break down in As_2Se_3 glass? *EPL Europhys. Lett.* 102, 66008. doi: 10.1209/0295-5075/102/66008.

Hsiao, Y.-H., La Plante, E.C., Krishnan, N.M.A., Le Pape, Y., Neithalath, N., Bauchy, M., Sant, G., 2017. Effects of irradiation on Albite's chemical durability. *J. Phys. Chem. A* 121, 7835–7845. doi: 10.1021/acs.jpca.7b05098.

Januchta, K., Youngman, R.E., Goel, A., Bauchy, M., Logunov, S.L., Rzoska, S.J., Bockowski, M., Jensen, L.R., Smedskjaer, M.M., 2017a. Discovery of ultra-crack-resistant oxide glasses with adaptive networks. *Chem. Mater.* 29, 5865–5876. doi: 10.1021/acs.chemmater.7b00921.

Januchta, K., Youngman, R.E., Goel, A., Bauchy, M., Rzoska, S.J., Bockowski, M., Smedskjaer, M.M., 2017b. Structural origin of high crack resistance in sodium aluminoborate glasses. *J. Non-Cryst. Solids* 460, 54–65. doi: 10.1016/j.jnoncrysol.2017.01.019.

Jones, J.R., 2013. Review of bioactive glass: From hench to hybrids. *Acta Biomater.* 9, 4457–4486. doi: 10.1016/j.actbio.2012.08.023.

King, S.W., Bielefeld, J., Xu, G., Lanford, W.A., Matsuda, Y., Dauskardt, R.H., Kim, N., Hondongwa, D., Olasov, L., Daly, B., Stan, G., Liu, M., Dutta, D., Gidley, D., 2013. Influence of network bond percolation on the thermal, mechanical, electrical and optical properties of high and low-k a-SiC:H thin films. *J. Non-Cryst. Solids* 379, 67–79. doi: 10.1016/j.jnoncrysol.2013.07.028.

Kreidl, N., 1991. Phase separation in glasses. *J. Non-Cryst. Solids* 129, 1–11. doi: 10.1016/0022-3093(91)90074-G.

Krishnan, N.M.A., Wang, B., Falzone, G., Le Pape, Y., Neithalath, N., Pilon, L., Bauchy, M., Sant, G., 2016. Confined water in layered silicates: The origin of anomalous thermal expansion behavior in calcium-silicate-hydrates. *ACS Appl. Mater. Interfaces* 8, 35621–35627. doi: 10.1021/acsami.6b11587.

Krishnan, N.M.A., Wang, B., Le Pape, Y., Sant, G., Bauchy, M., 2017a. Irradiation-driven amorphous-to-glassy transition in quartz: The crucial role of the medium-range order in crystallization. *Phys. Rev. Mater.* 1, 053405. doi: 10.1103/PhysRevMaterials.1.053405.

Krishnan, N.M.A., Wang, B., Le Pape, Y., Sant, G., Bauchy, M., 2017b. Irradiation- vs. vitrification-induced disordering: The case of alpha-quartz and glassy silica. *J. Chem. Phys.* 146, 204502. doi: 10.1063/1.4982944.

Krishnan, N.M.A., Wang, B., Sant, G., Phillips, J.C., Bauchy, M., 2017c. Revealing the effect of irradiation on cement hydrates: Evidence of a topological self-organization. *ACS Appl. Mater. Interfaces* 9, 32377–32385. doi: 10.1021/acsami.7b09405.

Krishnan, N.M.A., Wang, B., Yu, Y., Le Pape, Y., Sant, G., Bauchy, M., 2017d. Enthalpy landscape dictates the irradiation-induced disordering of quartz. *Phys. Rev. X* 7, 031019. doi: 10.1103/PhysRevX.7.031019.

Lacks, D.J., 2001. Energy landscapes and the non-newtonian viscosity of liquids and glasses. *Phys. Rev. Lett.* 87, 225502. doi: 10.1103/PhysRevLett.87.225502.

Li, X., Song, W., Yang, K., Krishnan, N.M.A., Wang, B., Smedskjaer, M.M., Mauro, J.C., Sant, G., Balonis, M., Bauchy, M., 2017. Cooling rate effects in sodium silicate glasses: Bridging the gap between molecular dynamics simulations and experiments. *J. Chem. Phys.* 147, 074501. doi: 10.1063/1.4998611.

Lucas, P., King, E.A., Gulbiten, O., Yarger, J.L., Soignard, E., Bureau, B., 2009. Bimodal phase percolation model for the structure of Ge-Se glasses and the existence of the intermediate phase. *Phys. Rev. B* 80, 214114. doi: 10.1103/PhysRevB.80.214114.

Main, D., 2018. Humankind's most important material. *The Atlantic*, April 7, 2018.

Mantisi, B., Bauchy, M., Micoulaut, M., 2015. Cycling through the glass transition: Evidence for reversibility windows and dynamic anomalies. *Phys. Rev. B* 92, 134201. doi: 10.1103/PhysRevB.92.134201.

Mascaraque, N., Bauchy, M., Fierro, J.L.G., Rzoska, S.J., Bockowski, M., Smedskjaer, M.M., 2017. Dissolution kinetics of hot compressed oxide glasses. *J. Phys. Chem. B* 121, 9063–9072. doi: 10.1021/acs.jpcb.7b04535.

Mascaraque, N., Januchta, K., Frederiksen, K.F., Youngman, R.E., Bauchy, M., Smedskjaer, M.M.,

2018. Structural dependence of chemical durability in modified aluminoborate glasses. ArXiv180505191 Cond-Mat.

Masoero, E., Bauchy, M., Gado, E.D., Manzano, H., Pellenq, R.M., Ulm, F.-J., Yip, S., 2015. Kinetic simulations of cement creep: Mechanisms from shear deformations of glasses. *Proceedings of the 10th International Conference on Mechanics and Physics of Creep, Shrinkage, and Durability of Concrete and Concrete Structures*. American Society of Civil Engineers, Vienna, Austria, pp. 555–564.

Mauro, J.C., 2017. Decoding the glass genome. *Curr. Opin. Solid State Mater. Sci.* doi: 10.1016/j.cossms.2017.09.001.

Mauro, J.C., 2014. Grand challenges in glass science. *Fontiers Mater.* 1, 20. doi: 10.3389/fmats.2014.00020.

Mauro, J.C., 2011. Topological constraint theory of glass. *Am. Ceram. Soc. Bull.* 90, 31–37.

Mauro, J.C., Ellison, A.J., Allan, D.C., Smedskjaer, M.M., 2013a. Topological model for the viscosity of multicomponent glassforming liquids. *Int. J. Appl. Glass Sci.* 4, 408–413. doi: 10.1111/ijag.12009.

Mauro, J.C., Ellison, A.J., Pye, L.D., 2013b. Glass: The nanotechnology connection. *Int. J. Appl. Glass Sci.* 4, 64–75. doi: 10.1111/ijag.12030.

Mauro, J.C., Gupta, P.K., Loucks, R.J., 2009a. Composition dependence of glass transition temperature and fragility. II. A topological model of alkali borate liquids. *J. Chem. Phys.* 130, 234503. doi: doi:10.1063/1.3152432.

Mauro, J.C., Philip, C.S., Vaughn, D.J., Pambianchi, M.S., 2014. Glass science in the United States: Current status and future directions. *Int. J. Appl. Glass Sci.* 5, 2–15. doi: 10.1111/ijag.12058.

Mauro, J.C., Smedskjaer, M.M., 2012. Minimalist landscape model of glass relaxation. *Phys. Stat. Mech. Appl.* 391, 3446–3459. doi: 10.1016/j.physa.2012.01.047.

Mauro, J.C., Tandia, A., Vargheese, K.D., Mauro, Y.Z., Smedskjaer, M.M., 2016. Accelerating the design of functional glasses through modeling. *Chem. Mater.* doi: 10.1021/acs.chemmater.6b01054.

Mauro, J.C., Varshneya, A.K., 2007. Modeling of rigidity percolation and incipient plasticity in germanium-selenium glasses. *J. Am. Ceram. Soc.* 90, 192–198. doi: 10.1111/j.1551-2916.2006.01374.x.

Mauro, J.C., Yue, Y., Ellison, A.J., Gupta, P.K., Allan, D.C., 2009b. Viscosity of glass-forming liquids. *Proc. Natl. Acad. Sci.* 106, 19780–19784. doi: 10.1073/pnas.0911705106.

Mauro, J.C., Zanotto, E.D., 2014. Two centuries of glass research: Historical trends, current status, and grand challenges for the future. *Int. J. Appl. Glass Sci.* 5, 313–327. doi: 10.1111/ijag.12087.

Maxwell, J.C., 1864. L. On the calculation of the equilibrium and stiffness of frames. *Philos. Mag. Ser.* 27(4), 294–299. doi: 10.1080/14786446408643668.

Micoulaut, M., 2008. Constrained interactions, rigidity, adaptative networks, and their role for the description of silicates. *Am. Mineral.* 93, 1732–1748. doi: 10.2138/am.2008.2903.

Micoulaut, M., Bauchy, M., 2017. Evidence for anomalous dynamic heterogeneities in isostatic supercooled liquids. *Phys. Rev. Lett.* 118, 145502. doi: 10.1103/PhysRevLett. 118.145502.

Micoulaut, M., Bauchy, M., 2013. Anomalies of the first sharp diffraction peak in network glasses: Evidence for correlations with dynamic and rigidity properties. *Phys. Status Solidi B* 250, 976–982. doi: 10.1002/pssb.201248512.

Micoulaut, M., Bauchy, M., Flores-Ruiz, H., 2015. Topological constraints, rigidity transitions, and anomalies in molecular networks. *Mol. Dyn. Simul. Disord. Mater.* 215, 275–311, Springer Series in Materials Science.

Micoulaut, M., Kachmar, A., Bauchy, M., Le Roux, S., Massobrio, C., Boero, M., 2013. Structure, topology, rings, and vibrational and electronic properties of Ge_xSe_{1-x} glasses across the rigidity transition: A numerical study. *Phys. Rev. B* 88, 054203. doi: 10.1103/PhysRevB.88.054203.

Micoulaut, M., Malki, M., Novita, D.I., Boolchand, P., 2009. Fast-ion conduction and flexibility and rigidity of solid electrolyte glasses. *Phys. Rev. B* 80, 184205. doi: 10.1103/PhysRevB.80.184205.

Micoulaut, M., Raty, J.-Y., Otjacques, C., Bichara, C., 2010. Understanding amorphous phase-change materials from the viewpoint of Maxwell rigidity. *Phys. Rev. B* 81, 174206. doi: 10.1103/PhysRevB.81.174206.

Micoulaut, M., Yue, Y., 2017. Material functionalities from molecular rigidity: Maxwell's modern legacy. *MRS Bull.* 42, 18–22. doi: 10.1557/mrs.2016.298.

Morse, D.L., Evenson, J.W., 2016. Welcome to the glass age. *Int. J. Appl. Glass Sci.* 7, 409–412. doi: 10.1111/ijag.12242.

Moukarzel, C.F., 2002. Granular matter instability: A structural rigidity point of view, in: Thorpe, M.F., Duxbury, P.M. (Eds.), *Rigidity Theory and Applications, Fundamental Materials Research*. Springer International Publishing, Boston, MA, pp. 125–142. ISBN: 10.1007/0-306-47089-6.8.

Moukarzel, C.F., 1998. Isostatic phase transition and instability in stiff granular materials. *Phys. Rev. Lett.* 81, 1634–1637. doi: 10.1103/PhysRevLett.81.1634.

Naumis, G.G., 2005. Energy landscape and rigidity. *Phys. Rev. E* 71, 026114. doi: 10.1103/PhysRevE.71.026114.

Neuefeind, J., Liss, K.D., 1996. Bond angle distribution in amorphous germania and silica. *Berichte Bunsen-Ges. Phys. Chem.* 100, 1341–1349.

Novita, D.I., Boolchand, P., Malki, M., Micoulaut, M., 2009. Elastic flexibility, fast-ion conduction, boson and floppy modes in AgPO 3 –AgI glasses. *J. Phys. Condens. Matter* 21, 205106. doi: 10.1088/0953-8984/21/20/205106.

Novita, D.I., Boolchand, P., Malki, M., Micoulaut, M., 2007. Fast-ion conduction and flexibility of glassy networks. *Phys. Rev. Lett.* 98, 195501. doi: 10.1103/PhysRevLett. 98.195501.

Oey, T., Hsiao, Y.-H., Callagon, E., Wang, B., Pignatelli, I., Bauchy, M., Sant, G.N., 2017a. Rate controls on silicate dissolution in cementitious environments. *RILEM Tech. Lett.* 2, 67–73. doi: 10.21809/rilemtechlett.2017.35.

Oey, T., Huang, C., Worley, R., Ho, S., Timmons, J., Cheung, K.L., Kumar, A., Bauchy, M., Sant, G., 2015. Linking fly ash composition to performance in cementitious systems. *World of Coal Ash (WOCA) Conference*, Nashville, TN.

Oey, T., Kumar, A., Pignatelli, I., Yu, Y., Neithalath, N., Bullard, J.W., Bauchy, M., Sant, G., 2017b. Topological controls on the dissolution kinetics of glassy aluminosilicates. *J. Am. Ceram. Soc.* 100, 5521–5527. doi: 10.1111/jace.15122.

Oey, T., Timmons, J., Stutzman, P., Bullard, J.W., Balonis, M., Bauchy, M., Sant, G., 2017c. An improved basis for characterizing the suitability of fly ash as a cement replacement agent. *J. Am. Ceram. Soc.* 100, 4785–4800. doi: 10.1111/jace.14974.

Oliver, W.C., Pharr, G.M., 1992. Improved technique for determining hardness and elastic modulus using load and displacement sensing indentation experiments. *J. Mater. Res.* 7, 1564–1583.

Pelmenschikov, A., Strandh, H., Pettersson, L.G.M., Leszczynski, J., 2000. Lattice resistance to hydrolysis of Si−O−Si bonds of silicate minerals: Ab initio calculations of a single water attack onto the (001) and (111) β-cristobalite surfaces. *J. Phys. Chem. B* 104, 5779–5783. doi: 10.1021/jp994097r.

Pettifer, R.F., Dupree, R., Farnan, I., Sternberg, U., 1988. NMR determinations of Si-O-Si bond angle distributions in silica. *J. Non-Cryst. Solids* 106, 408–412. doi: 10.1016/0022-3093(88)90299-2.

Phillips, J.C., 2009. Scaling and self-organized criticality in proteins: Lysozyme C. *Phys. Rev. E* 80. doi: 10.1103/PhysRevE.80.051916.

Phillips, J.C., 2006. Microscopic reversibility, space-filling, and internal stress in strong glasses. ArXivcond-Mat0606418.

Phillips, J.C., 2004. Constraint theory and hierarchical protein dynamics. *J. Phys. Condens. Matter* 16, S5065–S5072. doi: 10.1088/0953-8984/16/44/004.

Phillips, J.C., 1979. Topology of covalent non-crystalline solids 1: Short-range order in chalcogenide alloys. *J. Non Cryst. Solids* 34, 153–181. doi: 10.1016/0022-3093(79)90033-4.

Pignatelli, I., Kumar, A., Alizadeh, R., Pape, Y.L., Bauchy, M., Sant, G., 2016a. A dissolution-precipitation mechanism is at the origin of concrete creep in moist environments. *J. Chem. Phys.* 145, 054701. doi: 10.1063/1.4955429.

Pignatelli, I., Kumar, A., Bauchy, M., Sant, G., 2016b. Topological control on silicates' dissolution kinetics. *Langmuir* 32, 4434–4439. doi: 10.1021/acs.langmuir.6b00359.

Pignatelli, I., Kumar, A., Field, K.G., Wang, B., Yu, Y., Le Pape, Y., Bauchy, M., Sant, G., 2016c. Direct experimental evidence for differing reactivity alterations of minerals following irradiation: The case of calcite and quartz. *Sci. Rep.* 6, 20155. doi: 10.1038/srep 20155.

Qomi, M.J.A., Bauchy, M., Pellenq, R.J.-M., Ulm, F.-J., 2013. Applying tools from glass science to study calcium-silicate-hydrates. *Mechanics and Physics of Creep, Shrinkage, and Durability of Concrete: A Tribute to Zdenek P. Bazant: Proceedings of the Ninth International Conference on Creep, Shrinkage, and Durability Mechanics (CONCREEP-9)*, September 22–25, 2013, Cambridge, MA. ASCE Publications, pp. 78–85. doi: 10.1061/9780784413111.008.

Qomi, M.J.A., Bauchy, M., Ulm, F.-J., Pellenq, R., 2015a. Polymorphism and its implications on structure-property correlation in calcium-silicate-hydrates, in: Sobolev, K., Shah, S.P. (Eds.), *Nanotechnology in Construction*. Springer International Publishing, Cham, Switzerland, pp. 99–108.

Qomi, M.J.A., Bauchy, M., Ulm, F.-J., Pellenq, R.J.-M., 2014a. Anomalous composition-dependent dynamics of nanoconfined water in the interlayer of disordered calcium-silicates. *J. Chem. Phys.* 140, 054515. doi: 10.1063/1.4864118.

Qomi M.J.A., Ebrahimi D., Bauchy, M., Pellenq, R., Ulm, F.-J., 2017. Methodology for estimation of nanoscale hardness via atomistic simulations. *J. Nanomech. Micromech.* 7, 04017011. doi: 10.1061/(ASCE)NM.2153-5477.0000127.

Qomi, M.J.A., Krakowiak, K.J., Bauchy, M., Stewart, K.L., Shahsavari, R., Jagannathan, D., Brommer, D.B., Baronnet, A., Buehler, M.J., Yip, S., Ulm, F.-J., Van Vliet, K.J., Pellenq, R.J.-M., 2014b. Combinatorial molecular optimization of cement hydrates. *Nat. Commun.* 5, 4960. doi: 10.1038/ncomms5960.

Qomi, M.J.A., Masoero, E., Bauchy, M., Ulm, F.-J., Gado, E.D., Pellenq, R.J.-M., 2015b. C-S-H across length scales: From nano to micron. *Proceedings of the 10th International Conference on Mechanics and Physics of Creep, Shrinkage, and Durability of Concrete and Concrete Structure*. American Society of Civil Engineers, Vienna, Austria, pp. 39–48.

Rader, A.J., Hespenheide, B.M., Kuhn, L.A., Thorpe, M.F., 2002. Protein unfolding: Rigidity lost. *Proc. Natl. Acad. Sci.* 99, 3540–3545. doi: 10.1073/pnas.0624 92699.

Rompicharla, K., Novita, D.I., Chen, P., Boolchand, P., Micoulaut, M., Huff, W., 2008. Abrupt boundaries of intermediate phases and space filling in oxide glasses. *J. Phys. Condens. Matter Inst. Phys. J.* 20. doi: 10.1088/0953-8984/20/20/202101.

Rouxel, T., Yoshida, S., 2017. The fracture toughness of inorganic glasses. *J. Am. Ceram. Soc.* doi: 10.1111/jace.15108.

Salmon, P.S., Petri, I., 2003. Structure of glassy and liquid GeSe$_2$. *J. Phys. Condens. Matter* 15, S1509–S1528. doi: 10.1088/0953-8984/15/16/301.

National Research Council, 2001. *Grand Challenges in Environmental Sciences.* National Academy Press, Washington, D.C.

Smedskjaer, M.M., 2014. Topological model for boroaluminosilicate glass hardness. *Front. Mater.* 1, 23. doi: 10.3389/fmats.2014.00023.

Smedskjaer, M.M., Bauchy, M., 2015. Sub-critical crack growth in silicate glasses: Role of network topology. *Appl. Phys. Lett.* 107, 141901. doi: 10.1063/1.4932377.

Smedskjaer, M.M., Bauchy, M., Mauro, J.C., Rzoska, S.J., Bockowski, M., 2015. Unique effects of thermal and pressure histories on glass hardness: Structural and topological origin. *J. Chem. Phys.* 143, 164505. doi: 10.1063/1.4934540.

Smedskjaer, M.M., Mauro, J.C., Youngman, R.E., Hogue, C.L., Potuzak, M., Yue, Y., 2011. Topological principles of borosilicate glass chemistry. *J. Phys. Chem. B* 115, 12930–12946. doi: 10.1021/jp208796b.

Smedskjaer, M.M., Mauro, J.C., Yue, Y., 2010. Prediction of glass hardness using temperature-dependent constraint theory. *Phys. Rev. Lett.* 105, 115503. doi: 10.1103/PhysRevLett.105.115503.

Sreeram, A.N., Varshneya, A.K., Swiler, D.R., 1991. Microhardness and indentation toughness versus average coordination number in isostructural chalcogenide glass systems. *J. Non-Cryst. Solids* 130, 225–235. doi: 10.1016/0022-3093(91)90358-D.

Svenson, M.N., Thirion, L.M., Youngman, R.E., Mauro, J.C., Bauchy, M., Rzoska, S.J., Bockowski, M., Smedskjaer, M.M., 2016. Effects of thermal and pressure histories on the chemical strengthening of sodium aluminosilicate glass. *Glass Sci.* 14. doi: 10.3389/fmats.2016.00014.

Swiler, D., Varshneya, A.K., Callahan, R., 1990. Microhardness, surface toughness and average coordination number in chalcogenide glasses. *J. Non-Cryst. Solids* 125, 250–257.

Thorpe, M.F., 1983. Continuous deformations in random networks. *J. Non-Cryst. Solids* 57, 355–370. doi: 10.1016/0022-3093(83)90424-6.

Thorpe, M.F., Duxbury, P.M. (Eds.), 1999. *Rigidity Theory and Applications*, 1st ed. Springer, New York.

Vaills, Y., Qu, T., Micoulaut, M., Chaimbault, F., Boolchand, P., 2005. Direct evidence of rigidity loss and self-organization in silicate glasses. *J. Phys. Condens. Matter* 17, 4889–4896. doi: 10.1088/0953-8984/17/32/003.

Varshneya, A.K., 2010. Chemical strengthening of glass: Lessons learned and yet to be learned. *Int. J. Appl. Glass Sci.* 1, 131–142. doi: 10.1111/j.2041-1294.2010.00010.x.

Varshneya, A.K., 1993. *Fundamentals of Inorganic Glasses.* Academic Press Inc., New York.

Varshneya, A.K., Mauro, D.J., 2007. Microhardness, indentation toughness, elasticity, plasticity, and brittleness of Ge–Sb–Se chalcogenide glasses. *J. Non-Cryst. Solids* 353, 1291–1297. doi: 10.1016/j.jnoncrysol.2006.10.072.

Wang, B., Krishnan, N.M.A., Yu, Y., Wang, M., Le Pape, Y., Sant, G., Bauchy, M., 2017a. Irradiation-induced topological transition in SiO_2: Structural signature of networks' rigidity. *J. Non-Cryst. Solids* 463, 25–30. doi: 10.1016/j.jnoncrysol.2017.02.017.

Wang, B., Yu, Y., Lee, Y.J., Bauchy, M., 2015a. Intrinsic nano-ductility of glasses: The critical role of composition. *Front. Mater.* 2, 11. doi: 10.3389/fmats.2015.00011.

Wang, B., Yu, Y., Pignatelli, I., Sant, G., Bauchy, M., 2015b. Nature of radiation-induced defects in quartz. *J. Chem. Phys.* 143, 024505. doi: 10.1063/1.4926527.

Wang, B., Yu, Y., Wang, M., Mauro, J.C., Bauchy, M., 2016a. Nanoductility in silicate glasses is driven by topological heterogeneity. *Phys. Rev. B* 93, 064202. doi: 10.1103/PhysRevB.93.064202.

Wang, F., Mamedov, S., Boolchand, P., Goodman, B., Chandrasekhar, M., 2005. Pressure Raman effects and internal stress in network glasses. *Phys. Rev. B* 71, 174201. doi: 10.1103/PhysRevB.71.174201.

Wang, M., Krishnan, N.M.A., Wang, B., Smedskjaer, M.M., Mauro, J.C., Bauchy, M., 2018. A new transferable interatomic potential for molecular dynamics simulations of borosilicate glasses. *J. Non-Cryst. Solids.* doi: 10.1016/j.jnoncrysol.2018.04.063.

Wang, M., Bauchy, M., 2015. Ion-exchange strengthening of glasses: Atomic topology matters. ArXiv150507880 Cond-Mat.

Wang, M., Smedskjaer, M.M., Mauro, J.C., Sant, G., Bauchy, M., 2017b. Topological origin of the network dilation anomaly in ion-exchanged glasses. *Phys. Rev. Appl.* 8, 054040. doi: 10.1103/PhysRevApplied.8.054040.

Wang, M., Wang, B., Bechgaard, T.K., Mauro, J.C., Rzoska, S.J., Bockowski, M., Smedskjaer, M.M., Bauchy, M., 2016b. Crucial effect of angular flexibility on the fracture toughness and nano-ductility of aluminosilicate glasses. *J. Non-Cryst. Solids* 454, 46–51. doi: 10.1016/j.jnoncrysol.2016.10.020.

Wang, M., Wang, B., Krishnan, N.M.A., Yu, Y., Smedskjaer, M.M., Mauro, J.C., Sant, G., Bauchy, M., 2017c. Ion exchange strengthening and thermal expansion of glasses: Common origin and critical role of network connectivity. *J. Non-Cryst. Solids* 455, 70–74. doi: 10.1016/j.jnoncrysol.2016.10.027.

Warren, J., 2012. Materials genome initiative. *AIP Conference Proceedings.* American Institute of Physics, Ste. 1 NO 1 Melville NY 11747-4502 United States.

Wray, P., 2013. Gorilla glass 3 explained (and it is a modeling first for corning!). *Ceram. Tech Today.*

Yan, L., 2018. Entropy favors heterogeneous structures of networks near the rigidity threshold. *Nat. Commun.* 9, 1359. doi: 10.1038/s41467-018-03859-9.

Yu, Y., Krishnan, N.M.A., Smedskjaer, M.M., Sant, G., Bauchy, M., 2018a. The hydrophilic-to-hydrophobic transition in glassy silica is driven by the atomic topology of its surface. *J. Chem. Phys.* 148, 074503. doi: 10.1063/1.5010934.

Yu, Y., Mauro, J.C., Bauchy, M., 2017a. Stretched exponential relaxation of glasses: Origin of the mixed-alkali effect. *Am. Ceram. Soc. Bull.* 96, 34–36.

Yu, Y., Wang, B., Lee, Y.J., Bauchy, M., 2015a. Fracture toughness of silicate glasses: Insights from molecular dynamics simulations. *Symposium UU – Structure-Property Relations in Amorphous Solids*, MRS Online Proceedings Library. doi: 10.1557/opl.2015.50.

Yu, Y., Wang, B., Wang, M., Sant, G., Bauchy, M., 2017b. Reactive molecular dynamics simulations of sodium silicate glasses—toward an improved understanding of the structure. *Int. J. Appl. Glass Sci.* 8, 276–284. doi: 10.1111/ijag.12248.

Yu, Y., Wang, B., Wang, M., Sant, G., Bauchy, M., 2016. Revisiting silica with ReaxFF: Towards improved predictions of glass structure and properties via reactive molecular dynamics. *J. Non-Cryst. Solids* 443, 148–154. doi: 10.1016/j.jnoncrysol.2016.03.026.

Yu, Y., Wang, M., Krishnan, N.M.A., Smedskjaer, M.M., Deenamma Vargheese, K., Mauro, J.C., Balonis, M., Bauchy, M., 2018b. Hardness of silicate glasses: Atomic-scale origin of the mixed modifier effect. *J. Non-Cryst. Solids* 489, 16–21. doi: 10.1016/j.jnoncrysol.2018.03.015.

Yu, Y., Wang, M., Smedskjaer, M.M., Mauro, J.C., Sant, G., Bauchy, M., 2017c. Thermometer effect: Origin of the mixed Alkali effect in glass relaxation. *Phys. Rev. Lett.* 119, 095501. doi: 10.1103/PhysRevLett.119.095501.

Yu, Y., Wang, M., Zhang, D., Wang, B., Sant, G., Bauchy, M., 2015b. Stretched exponential relaxation of glasses at low temperature. *Phys. Rev. Lett.* 115, 165901. doi: 10.1103/PhysRevLett.115.165901.

Yuan, X.L., Cormack, A., 2003. Si-O-Si bond angle and torsion angle distribution in vitreous silica and sodium silicate glasses. *J. Non-Cryst. Solids* 319, 31–43. doi: 10.1016/S0022-3093(02)01960-9.

Zachariasen, W.H., 1932. The atomic arrangement in glass. *J. Am. Chem. Soc.* 54, 3841–3851. doi: 10.1021/ja01349a006.

Zanotto, E.D., 1998. Do cathedral glasses flow? *Am. J. Phys.* 66, 392–395. doi: 10.1119/1.19026.

Zanotto, E.D., Coutinho, F.A.B., 2004. How many non-crystalline solids can be made from all the elements of the periodic table? *J. Non-Cryst. Solids* 347, 285–288. doi: 10.1016/j.jnoncrysol.2004.07.081.

Zanotto, E.D., Mauro, J.C., 2017. The glassy state of matter: Its definition and ultimate fate. *J. Non-Cryst. Solids* 471, 490–495. doi: 10.1016/j.jnoncrysol.2017.05.019.

Zeidler, A., Salmon, P.S., Whittaker, D.A.J., Pizzey, K.J., Hannon, A.C., 2017. Topological ordering and viscosity in the glass-forming Ge–Se system: The search for a structural or dynamical signature of the intermediate phase. *Front. Mater.* 4. doi: 10.3389/fmats.2017.00032.

Zhang, M., Boolchand, P., 1994. The central role of broken bond-bending constraints in promoting glass formation in the oxides. *Science* 266, 1355–1357. doi: 10.1126/science.266.5189.1355.

Zheng, Q., Yue, Y., Mauro, J.C., 2017. Density of topological constraints as a metric for predicting glass hardness. *Appl. Phys. Lett.* 111, 011907. doi: 10.1063/1.4991971.

Topological Descriptors of Carbon Nanostructures

Sakander Hayat
GIK Institute of Engineering Sciences and Technology

14.1 Introduction

A broad spectrum of physico-chemical characteristics, specifically the characterization and modelling of molecular structures, has found graph theory especially degree- and distance-based numerical graph invariants, significantly useful. The fact that the underlying activities and properties of molecules are closely related to their connectivities in its chemical graph/network, originates the graph-theoretic applications in chemistry and drug research. The study of topological connectivity and characterization of a chemical structure find a deep concern with graph theory. The exploration of relationships between graph-theoretic topological indices and physico-chemical properties of underlying chemical structures has been a focus of research over the years [14].

As certain physico-chemical characteristics can be derived from their chemical structures, the relation with the quantitative structure-activity relationship (QSAR) and quantitative structure-property relationship (QSPR) models is developed. The integration of concepts from chemistry, mathematics and information science forms an emerging field nowadays called cheminformatics. The quantification of the chemical structure is the key step in QSAR/QSPR study so as building a close correlation model between the physico-chemical and biological properties and the corresponding chemical structures of a wide range of chemical compounds [14,44].

A topological descriptor/index is a numerical quantity which is calculated based on the underlying chemical graph of a chemical compound. These molecular descriptors are an important part of chemical graph theory. Therefore,

a topological index simultaneously provides quantitative characterization of a chemical graph which is topologically invariant to labeling and differentiate properties of isomers. There are a number of such graph-related numerical invariants, which are of a key importance in reticular chemistry and nanotechnology, and therefore, computation of these topological indices is one of the recent areas of research [4]. We refer to the book by Karelson [44] for more details on this topic.

The most common terminology which is being used for these numerical invariants in the literature is "topological indices". Thus, we will pursue with this terminology throughout the chapter.

14.2 Topological Indices

A graph G is a mathematical object written as $G = (V, E)$, where V is the set of points called vertices and E is the set of lines between those points. A chemical graph is a representation of the structural formula of a chemical compound in terms of graph theory. In other words, a chemical graph is a graph whose vertices correspond to the atoms of the compound and edges correspond to chemical bonds. A graph is called bipartite if it contains no cycle of odd length. In a graph G, the number of edges connected to a fixed vertex u is called its degree and denoted by d_u. In a chemical graph, the degree of a vertex can be at most four. The distance $d(u,v)$ between two vertices u and v is the length of a shortest path between them. The eccentricity of a vertex u, denoted by e_u, is the maximum distance between u and any other vertex of G.

A topological index is a map from the set of finite connected graphs to the set of real numbers with a property that it has significant applications in chemistry. A topological index is called distance-based (resp. degree-distance-based), if it is defined based on the degrees (resp. degrees and distances) of/between vertices in a graph. Eccentricity-based topological indices are a subclass of distance-based topological indices. They are defined based on the eccentricity of vertices. We refer to the mathematical chemistry books [15,32] for related concepts.

In Table 14.1, we present some well-known degree-based topological indices of graphs. The corresponding references in the table comprise some basic mathematical properties and chemical significance of these indices. By conducting a comparative testing, Hayat et al. [40] recently showed that among the existing degree-based indices, only R_{-1}, ABC, AZI, GA and SCI_3 should be preferred for the studies of computational and application perspectives.

The transmission $Tr(u)$ of a vertex u is defined as $Tr(u) = \sum_{v \in V} d(u, v)$. For an edge $uv \in E(G)$, we define the quantities n_u, n_v and n_0 as follows:

$$n_u = | \{x \in V(G) \mid d(x, u) < d(x, v)\} |$$

$$n_v = | \{x \in V(G) \mid d(x, v) < d(x, u)\} |$$

$$n_0 = | \{x \in V(G) \mid d(x, v) = d(x, u)\} | .$$

In Table 14.2, we define some distance-based topological indices. Table 14.3 presents some eccentricity-based topological indices. The corresponding references in the table comprises some basic mathematical properties and chemical significance of these indices.

Now we define two matrices based on an n-vertex connected graph G. The adjacency matrix A_G of G is an $n \times n$ symmetric matrix defined as

$$(A_G)_{u,v} = \begin{cases} 1, & uv \in E(G); \\ 0, & \text{Otherwise.} \end{cases}$$

TABLE 14.1 Some Degree-Based Topological Indices

Topological Index	Definition
Randić index [53]	$R(G) = \sum_{uv \in E(G)} \frac{1}{\sqrt{d_u d_v}}$
General Randić index [3,12]	$R_\alpha(G) = \sum_{uv \in E(G)} (d_u d_v)^\alpha$, where $\alpha \in \mathbb{R}$
Atom-bond connectivity (ABC) index [18]	$ABC(G) = \sum_{uv \in E(G)} \sqrt{\frac{d_u + d_v - 2}{d_u d_v}}$
Augmented Zagreb index (AZI) [24]	$AZI(G) = \sum_{uv \in E(G)} \left(\frac{d_u d_v}{d_u + d_v - 2}\right)^3$
Geometric-arithmetic (GA) index [60]	$GA(G) = \sum_{uv \in E(G)} \frac{2\sqrt{d_u d_v}}{d_u + d_v}$
Sum-connectivity index (SCI) [65]	$SCI(G) = \sum_{uv \in E(G)} \frac{1}{\sqrt{d_u + d_v}}$
General SCI index [66]	$SCI_\alpha(G) = \sum_{uv \in E(G)} (d_u + d_v)^\alpha$, where $\alpha \in \mathbb{R}$

TABLE 14.2 Some Distance-Based Topological Indices

Topological Index	Definition
Wiener index [61]	$W(G) = \sum_{\{u,v\} \subset E(G)} d(u, v)$
Szeged index [29]	$Sz(G) = \sum_{uv \in E(G)} n_u n_v$
Padmakar–Ivan (PI) index [45]	$PI(G) = \sum_{uv \in E(G)} [n_u + n_v]$
Revised Szeged index [54]	$RSz(G) = \sum_{uv \in E(G)} \left[n_u + \frac{n_0}{2}\right]\left[n_v + \frac{n_0}{2}\right]$
Second ABC index [26]	$ABC_2(G) = \sum_{uv \in E(G)} \sqrt{\frac{n_u + n_v - 2}{n_u n_v}}$
Second GA index [21]	$GA_2(G) = \sum_{uv \in E(G)} \frac{2\sqrt{n_u n_v}}{n_u + n_v}$

TABLE 14.3 Some Eccentricity-Based Topological Indices

Topological Index	Definition
Eccentric-connectivity index [56]	$EC(G) = \sum_{u \in V(G)} d_u e_u$
Total eccentricity index [57]	$TE(G) = \sum_{u \in V(G)} e_u$
Fourth GA index [25]	$GA_4(G) = \sum_{uv \in E(G)} \sqrt{\frac{e_u + e_v - 2}{e_u e_v}}$
Fifth ABC index [19]	$ABC_5(G) = \sum_{uv \in E(G)} \frac{2\sqrt{e_u e_v}}{e_u + e_v}$

The distance matrix D_G of G is also an $n \times n$ symmetric matrix which is defined as follows:

$$(D_G)_{u,v} = \begin{cases} k, & d(u, v) = k; \\ 0, & \text{Otherwise.} \end{cases}$$

In Figure 14.1, we give an example of a graph and its adjacency and distance matrices.

Let $\mathbf{j}$ be the $n \times 1$ all-ones vector and $\mathbf{k} = A_G \mathbf{j}$. Note that $\mathbf{k}_u = d_u$ for any $u \in V(G)$. In Table 14.4, certain degree-distance-based topological indices are defined. The references in the table contain chemical importance of these indices in QSPR/QSAR studies.

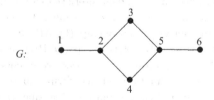

G:

$$A_G = \begin{pmatrix} 0 & 1 & 0 & 0 & 0 & 0 \\ 1 & 0 & 1 & 1 & 0 & 0 \\ 0 & 1 & 0 & 0 & 1 & 0 \\ 0 & 1 & 0 & 0 & 1 & 0 \\ 0 & 0 & 1 & 1 & 0 & 1 \\ 0 & 0 & 0 & 0 & 1 & 0 \end{pmatrix} \quad D_G = \begin{pmatrix} 0 & 1 & 2 & 2 & 3 & 4 \\ 1 & 0 & 1 & 1 & 2 & 3 \\ 2 & 1 & 0 & 2 & 1 & 2 \\ 2 & 1 & 2 & 0 & 1 & 2 \\ 3 & 2 & 1 & 1 & 0 & 1 \\ 4 & 3 & 2 & 2 & 1 & 0 \end{pmatrix}$$

FIGURE 14.1 A graph G with its adjacency matrix A_G and distance matrix D_G.

TABLE 14.4 Some Degree-Distance-Based Topological Indices

Topological Index	Definition
Degree distance index [16]	$DD(G) = \sum_{u \neq v} (d_u + d_v) d(u,v)$
Gutman index [30]	$Gut(G) = \sum_{u \neq v} d_u d_v d(u,v)$
Molecular topological index [55]	$MTI(G) = \sum_{i=1}^{n} [\mathbf{k}(A_G + D_G)]$
Additively weighted Harary index [2]	$H_A(G) = \sum_{u \neq v} \frac{d_u + d_v}{d(u,v)}$
Multiplicatively weighted Harary index [41]	$H_M(G) = \sum_{u \neq v} \frac{d_u d_v}{d(u,v)}$

14.3 Computational Techniques

QSAR and QSPR of molecular structures require expressions for the topological properties of these structures. Structure-based topological indices of chemical networks provide these expressions which in result enable prediction of certain physico-chemical properties and the bioactivities of these compounds through QSAR/QSPR methods. These important facts gave birth to the computational theory of topological indices. There are hundreds of papers which have focused on computing certain topological indices of chemical graphs and networks. Bača et al. [10,11] computed certain degree-based topological indices of some infinite families of fullerenes and carbon nanotubes. Hayat et al. [34–40] and Imran et al. [43] also studied certain degree-based topological indices of certain graphs and networks. We also refer [6,48] for computation of topological indices of certain inorganic networks. For the computational study of degree-based and distance-based topological indices, we refer to [1,5,7,9,20,27,62–64].

All these aforementioned papers have focused on computing certain degree-based indices of some special families and classes of graphs. In view of this, it is natural to ask for developing a general technique to compute some topological indices for general graphs. Ashrafi and coauthors [8,50,51] were the first researchers who responded to this question and developed a computational technique for the Szeged, revised Szeged and PI indices of chemical graphs. Their technique works for any infinite family of chemical graphs to compute these three indices. In the following subsection, we explain their method.

14.3.1 The Technique by Ashrafi and Coauthors

For a given chemical graph L, this method can be described by the following three steps:

Step 1: Draw the chemical graph L on Hyperchem[1] [42]. It outputs a .hin file for the drawing of L.

Step 2: Input the .hin file of L to Topocluj[2] [59]. Compute the adjacency and distance matrices of L. Topocluj produces .m files of these matrices.

Step 3: Input the .m files of both adjacency and distance matrices of L in a GAP[3] [28] program developed by Ashrafi and coauthors [8,50,51]. GAP will output the Szeged, revised Szeged and PI indices of L.

In Figure 14.2, we present a flowchart of the technique by Ashrafi and coauthors.

14.3.2 Our Modified Technique

Certain inorganic chemical compounds cannot be modeled into a chemical graph as the valency of an atom could be larger than four. Thus, it is natural to develop a computational technique for general graphs rather than just chemical graphs. Note that the set of chemical graphs is a proper subset of the set of general graphs. We proposed a technique for general graphs which can compute various distance-based, eccentricity-based and degree-distance-based topological indices. For a general graph G, our technique can be illustrated in the following two steps:

Step 1: Draw L on newGraph[4] [58]. Compute the distance matrix D of L.

Step 2: Input the matrix D in our MATLAB [49] program and compute various degree-based (Table 14.1), distance-based (Table 14.2), eccentricity-based (Table 14.3) and degree-distance-based (Table 14.4) topological indices.

In Figure 14.3, a flowchart of our proposed technique is shown.

A public repository has been created on GitHub where our MATLAB code, a ReadMe file with instructions and a PDF file to explain the working pattern of our technique with a minimal working example (MWE) have been uploaded. Go to the webpage https://github.com/Sakander/Degree-and-distance-related-topological-indices.git to get access to the page on GitHub.

By means of our technique, it is very convenient to compute topological indices from Tables 14.1–14.5 for a finite graph. However, computing these indices for an infinite family of graphs is a bit tedious and requires some extra work. In the following subsection, we explain our computational technique to compute the aforementioned topological indices for a somewhat simple infinite family of graphs.

---

[1]Hyperchem is a computational and quantum chemistry software to perform various theoretical analyses of chemical compounds.

[2]Topocluj is a software of molecular topology and mathematical chemistry. It derives about 12 important graph-theoretic and chemical invariants for a given .hin file of a chemical graph.

[3]Groups, Algorithms and Programming (GAP) is a mathematical software to perform certain analyses for various mathematical objects such as groups, sets and graphs.

[4]newGRAPH is a fully integrated environment used for improving a research process in graph theory. It is compatible both with conveniently drawing a general graph and then computing various graph-theoretic parameters including the adjacency and distance matrices.

The technique by Ashrafi and coauthors

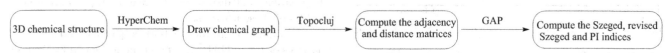

FIGURE 14.2 Flowchart of the technique by Ashrafi and coauthors.

Our modified technique

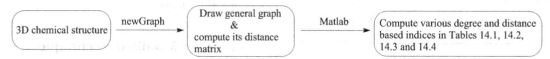

FIGURE 14.3 Flowchart of our modified technique.

It is important to note that Hayat [33] also extended the technique of Ashrafi and coauthors to some other distance-based topological indices.

Fibonacenes

Benzenoid systems are divided [31] into catacondensed and pericondensed systems. In a pericondensed system, there exist three hexagons share common vertex. In catacondensed systems, no three hexagons sharing a common vertex. Catacondensed benzenoids are further classified into non-branched (in which no hexagon has more than two neighbors) and branched (in which at least one hexagon has three neighbors). Thus, Fibonacenes is an infinite family of non-branched catacondensed benzenoids. It is constructed simply by attaching hexagons following a zigzag pattern, see Figure 14.4. We denote an n-dimensional Fibonacenes by $\mathbb{F}_n$, where $n \geq 1$.

The following lemma counts the number of vertices and edges in an n-dimensional Fibonacene $\mathbb{F}_n$.

LEMMA 14.1 Let n (resp. m) be the number of vertices (resp. edges) in an n-dimensional Fibonacene $\mathbb{F}_n$. Then, $n = 4n + 2$ and $m = 5n + 1$.

The following proposition presents explicit expressions of certain degree-based topological indices for Fibonacenes $\mathbb{F}_n$.

PROPOSITION 14.1 For an n-dimensional Fibonacene $\mathbb{F}_n$ ($n \geq 2$), the following hold:

$$ABC(\mathbb{F}_n) = \left(\frac{8 + 9\sqrt{2}}{6}\right)n + 2\sqrt{2} - 2;$$

$$GA(\mathbb{F}_n) = \left(\frac{15 + 4\sqrt{6}}{5}\right)n + 1;$$

$$AZI(\mathbb{F}_n) = \frac{1497}{32}n + \frac{4235}{64}; R_{-1}(\mathbb{F}_n) = \frac{29}{36}n + \frac{2}{3};$$

$$SCI_{-3}(\mathbb{F}_n) = \left(\frac{43}{1728} + \frac{2}{125}\right)n + \frac{7}{144}.$$

The following proposition presents explicit formulas of certain degree- and distance-based topological indices for Fibonacenes $\mathbb{F}_n$.

PROPOSITION 14.2

(a) For an n-dimensional Fibonacene $\mathbb{F}_n$ ($n \geq 2$), the following hold:

$$Sz(\mathbb{F}_n) = RSz(\mathbb{F}_n) = \frac{40}{3}n^3 + 20n^2 + \frac{107}{3}n - 15;$$

$$PI(\mathbb{F}_n) = 20n^2 + 14n + 2;$$

$$W(\mathbb{F}_n) = \frac{16}{3}n^3 + 8n^2 + \frac{62}{3}n - 7;$$

$$DD(\mathbb{F}_n) = \frac{80}{3}n^3 + 28n^2 + \frac{274}{3}n - 38;$$

$$Gut(\mathbb{F}_n) = \frac{100}{3}n^3 + 20n^2 + \frac{311}{3}n - 49;$$

$$MTI(\mathbb{F}_n) = \frac{80}{3}n^3 + 28n^2 + \frac{352}{3}n - 40.$$

(b) For an n-dimensional Fibonacene $\mathbb{F}_n$ ($n \geq 4$), the following hold:

$$EC(\mathbb{F}_n) = 15n^2 + 9n + 4;$$

$$TE(\mathbb{F}_n) = 6n^2 + 6n + 2.$$

By Proposition 14.2, we see that $Sz(\mathbb{F}_n) = RSz(\mathbb{F}_n)$. That is because the graph $\mathbb{F}_n$ is bipartite for any $n \geq 1$. Note that if a graph G is bipartite, then $Sz(G) = RSz(G)$ as $n_0 = 0$ for any edge of G. Based on the defining structures of indices from Section 14.2, we define certain vertex and edge partitions which will be helpful in proving Propositions 14.1 and 14.2. Table 14.5 shows these partitions, their types and examples of indices for which the corresponding partition is necessary. For any given graph, these partitions can be derived by using our technique from Subsection 14.3.2 with an assistance of cftoolbox by MATLAB.

Note that if a graph G is regular, then the ε-partition and δ-partition can be obtained directly from the ω-partition and τ-partition, respectively, and therefore, the ν-partition is not needed.

Next, we compute these partitions for Fibonacenes $\mathbb{F}_n$. Table 14.6 presents the ν-partition of Fibonacenes $\mathbb{F}_n$.

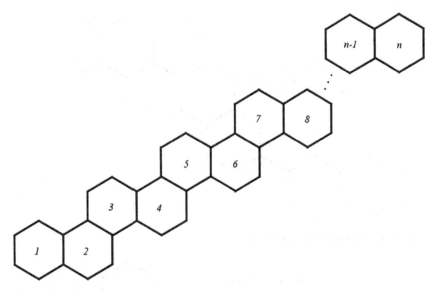

FIGURE 14.4   Fibonacene of dimension n.

TABLE 14.5   Some Edge and Vertex Partition of a Graph G

Partition	Type	Definition	Examples
ν-partition	edge	(d_u, d_v) for any $uv \in E(G)$	ABC, GA, AZI
ψ-partition	edge	(n_u, n_v, n_0) for any $uv \in E(G)$	Szeged, PI
ω-partition	vertex	$Tr(u)$ for any $u \in V(G)$	Wiener
δ-partition	vertex	$d_u Tr(u)$ for any $u \in V(G)$	Degree distance
ε-partition	vertex	$d_u e_u$ for any $u \in V(G)$	Eccentric connectivity
τ-partition	vertex	e_u for any $u \in V(G)$	Total eccentricity

TABLE 14.6   The ν-Partition of $\mathbb{F}_n$

Types	(d_u, d_v) for any $uv \in E(\mathbb{F}_n)$	Cardinality
I	(2,2)	$n+4$
II	(2,3)	$2n$
III	(3,3)	$2n-3$

TABLE 14.7   The ψ-Partition of $\mathbb{F}_n$

Types	(n_u, n_v) for any $uv \in E(\mathbb{F}_n)$	Cardinality
I	$(4n-1, 3)$	8
II	$(4n-2i+1, 2i+1)$, $2 \leq i \leq n$, $i = 1, 2, 3, \ldots$	$\beta = \begin{cases} \beta_1, & 2 \leq i \equiv 0 (\bmod\ 2); \\ \beta_2, & 3 \leq i \equiv 1 (\bmod\ 2), \end{cases}$ where $\beta_1 = \begin{cases} 3, & n = i; \\ 6, & n \geq i+1, \end{cases}$ and $\beta_2 = \begin{cases} 2, & n = i; \\ 4, & n \geq i+1. \end{cases}$

Note that it is convenient to prove Proposition 14.1 by using the definitions of those indices from Table 14.1 and the ν-partition from Table 14.6.

The three types of edges in this partition are illustrated in Figure 14.5.

We derive its ψ-partition based on quantities n_u, n_v only as $\mathbb{F}_n$ is a bipartite graph for any $n \geq 1$. Table 14.7 presents this partition for $\mathbb{F}_n$.

The two types of edges in the ψ-partition of $\mathbb{F}_n$ are shown in Figure 14.6.

Now we prove first two equations of Proposition 14.2. The corresponding ψ-partition in Table 14.7 is needed. Let $m, n \in \mathbb{Z}$ and $c \in \mathbb{R}$. Then, the following equations hold:

$$\sum_{i=m}^{n} c = c(n+1-m),$$

$$\sum_{i=m}^{n} i = \frac{(n+1-m)(n+m)}{2},$$

$$\sum_{\substack{i=m \\ i \equiv 0 (\bmod\ 2)}}^{n} i = \left(\frac{m+n}{2}\right)\left(\frac{m+n-2}{2}\right),$$

$$\sum_{\substack{i=m \\ i \equiv 1 (\bmod\ 2)}}^{n} i = \left(\frac{m+n}{2}\right)\left(\frac{m+n-2}{2}\right). \tag{14.1}$$

We also need some special cases of above series.

$$\sum_{i=0}^{n} c = c(n+1),$$

$$\sum_{i=0}^{n} i = \frac{n(n+1)}{2},$$

$$\sum_{i=0}^{n} i^2 = \frac{n(n+1)(2n+1)}{6}. \tag{14.2}$$

In what follows, we use the ψ-partition in Table 14.7 and formulas in (14.1) and (14.2) to prove the first two equations of Proposition 14.2. Note that the Szeged, revised Szeged and PI indices have a similar defining structure based on quantities n_u, n_v and n_0; therefore, we only prove the Szeged index expression. By definition of the Szeged index, we have

$$Sz(G) = \sum_{uv \in E(G)} n_u n_v.$$

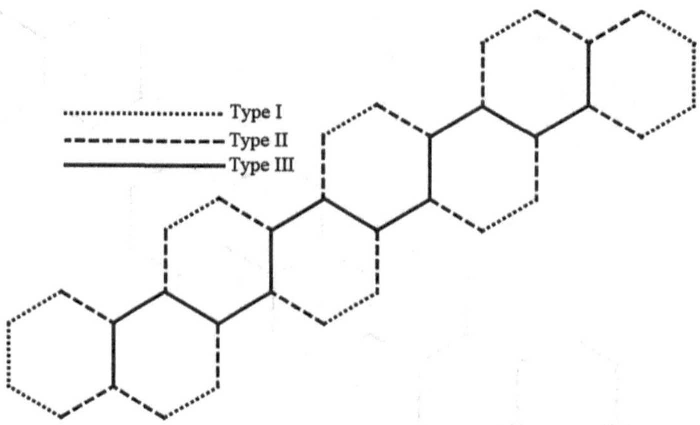

FIGURE 14.5   Three types of edges in the ν-partition of $\mathbb{F}_n$.

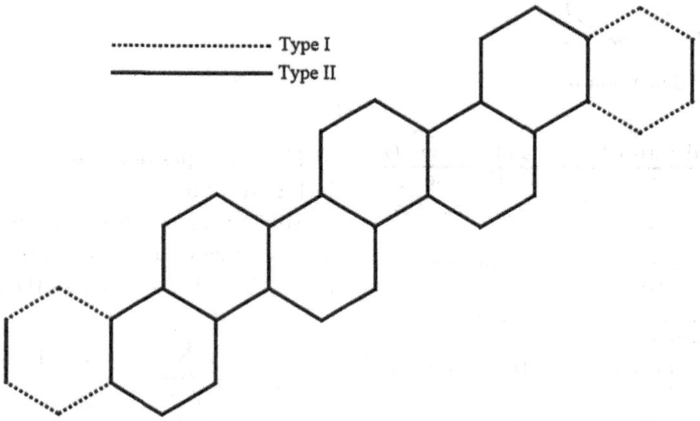

FIGURE 14.6   Two types of edges in the ψ-partition of $\mathbb{F}_n$.

By using the ψ-partition in Table 14.7, we obtain

$$Sz(\mathbb{F}_n) = 24(4n - 1) + 3(2n + 1)^2$$
$$+ 6 \sum_{\substack{i=2 \\ i \equiv 0(\text{mod } 2)}}^{n-1} (4n - 2i + 1)(2i + 1),$$

and

$$Sz(\mathbb{F}_n) = 24(4n - 1) + 2(2n + 1)^2$$
$$+ 4 \sum_{\substack{i=3 \\ i \equiv 1(\text{mod } 2)}}^{n-1} (4n - 2i + 1)(2i + 1).$$

By using formulas in (14.1) and (14.2) and doing some routine calculations, we obtain the following same solution for both of the above equations.

$$Sz(\mathbb{F}_n) = \frac{40}{3}n^3 + 20n^2 + \frac{107}{3}n - 15 = RSz(\mathbb{F}_n).$$

This shows the first equation of Proposition 14.2. The PI index i.e. the second equation of Proposition 14.2, can be proved similarly. Note that, in order to compute the indices in the fourth column from Table 14.5, we only need the corresponding edge/vertex partition in the first column of

the table. Therefore, we only present the remaining four partition of $\mathbb{F}_n$ and skip proving other equations of Proposition 14.2. Table 14.8 presents the τ-partition of $\mathbb{F}_n$. The last expression in Proposition 14.2 can be proved by using this partition.

Table 14.9 shows the ε-partition of $\mathbb{F}_n$. The second last formula in Proposition 14.2 can be proved by using this partition.

In a similar manner, Tables 14.10 and 14.11 show the ω-partition and δ-partition respectively. These partitions are necessary to compute Wiener, Degree distance, Gutman and MTI indices of Fibonacenes $\mathbb{F}_n$.

Consequently, in order to compute certain degree and distance related topological indices from Tables 14.2–14.4

TABLE 14.8   The τ-Partition of $\mathbb{F}_n$

Types	e_u for any $u \in V(\mathbb{F}_n)$	Cardinality
I	$2n + 1$	2
II	$2n - i$, $0 \leq i \leq n - 1$	4

TABLE 14.9   The ε-Partition of $\mathbb{F}_n$

Types	e_u for any $u \in V(\mathbb{F}_n)$	d_u for any $u \in V(\mathbb{F}_n)$	Cardinality
I	$2n + 1$	2	2
II	$2n$	2	4
III	$2n - i$, $1 \leq i \leq n - 1$	2	2
IV	$2n - i$, $1 \leq i \leq n - 1$	3	2

TABLE 14.10 The ω-Partition of $\mathbb{F}_n$

Types	$Tr(u)$ for any $u \in V(\mathbb{F}_n)$	Cardinality
I	$4n^2 + 4n + 1$	2
II	$4n^2 + 9$	2
III	$4n^2 - 4n + 13$	2
IV	$2n^2 + 2n + 17$	2
V	$2n^2 + 2n + 4i + 1,\ i = 1,2,3,4$	2
VI	$4n^2 - (4i - 16)n + 2i^2 - 14i + 29,$ $4 \le i \le n$ and $i = 1,2,3,\ldots$	2
VII	$4n^2 - (4i - 12)n + 2i^2 - 10i + 25,$ $5 \le i \le n$ and $i = 1,2,3,\ldots$	2

TABLE 14.11 The δ-Partition of $\mathbb{F}_n$

Types	$Tr(u)$ for any $u \in V(\mathbb{F}_n)$	d_u for any $u \in V(\mathbb{F}_n)$	Cardinality
I	$4n^2 + 4n + 1$	2	2
II	$4n^2 + 9$	2	2
III	$4n^2 - 4n + 13$	2	2
IV	$2n^2 + 2n + 17$	2	2
V	$2n^2 + 2n + 13$	2	2
VI	$2n^2 + 2n + 4i + 1,\ i = 1,2,4$	3	2
VII	$4n^2 - (4i - 16)n + 2i^2 - 14i + 29,$ $4 \le i \le n$ and $i = 1,2,3,\ldots$	3	2
VIII	$4n^2 - (4i - 12)n + 2i^2 - 10i + 25,$ $5 \le i \le n$ and $i = 1,2,3,\ldots$	2	2

for a general graph, we only need the edge/vertex partitions in Table 14.5. Our computational method in Subsection 14.3.2, with some assistance of cftoolbox by MATLAB, compute those partitions for an infinite family of general graphs.

In the next three sections, we compute certain degree and distance related topological indices of certain infinite families of fullerenes, carbon nanotubes and carbon nanocones. We skip all the details of how to obtain the results by using our proposed method since these details are similar to what we have done for Fibonacenes $\mathbb{F}_n$ in this subsubsection.

14.4 Fullerenes

In 1985, Kroto and Smalley [46,47] discovered a new form of existence of carbon apart from diamond, graphite and amorphous carbon. Initially, they discovered the fullerene C_{60} also called the Buckminsterfullerene (Figure 14.7), and studied its remarkable stability and other physico-chemical properties. Later on, researchers discovered other types of fullerenes and thus, a new field of research called "nanotechnology" started. Besides studying its chemical properties, a number of mathematicians studied its theoretical properties by considering it a cubic graph, see, for example, [13,22,52].

A fullerene graph is a cubic planar graph with only pentagonal and hexagonal faces. In a fullerene graph, we denote the number of pentagons, hexagons, carbon atoms and bonds between them by $\mathfrak{p}$, $\mathfrak{h}$, ν and $\mathfrak{e}$, respectively. Since every carbon atom is shared by three faces and on the other hand each edge is shared by exactly two faces, the number of faces are $\mathfrak{f} = \mathfrak{p} + \mathfrak{h}$, the number of vertices are $\nu = \frac{5\mathfrak{p} + 6\mathfrak{h}}{3}$ and the number of edges are $\mathfrak{e} = \frac{3}{2}\nu = \frac{5\mathfrak{p} + 6\mathfrak{h}}{2}$. Note that any planar graph satisfies the Euler's identity $\nu - \mathfrak{e} + \mathfrak{f} = 2$, thus we obtain $\frac{5\mathfrak{p} + 6\mathfrak{h}}{3} + \frac{5\mathfrak{p} + 6\mathfrak{h}}{2} + \mathfrak{p} + \mathfrak{h} = 2$, and this implies that $\mathfrak{p} = 12$, $\nu = 2\mathfrak{h} + 20$, and $\mathfrak{e} = 3\mathfrak{h} + 30$. Consequently, for a natural number $\nu \ge 22$, a fullerene is entirely made up of

ν carbon atoms, 12 pentagonal faces and $\frac{\nu}{2} + 10$ hexagonal faces [23].

It is believed that number of fullerene graphs on ν vertices is of order $\Theta(\nu^9)$, see Fowler & Manolopoulos [23] for more details. In this section, we consider an infinite family of fullerene on $\nu = 10n$, $n \ge 2$ vertices such that for any fixed n, the fullerene graph is isomorphic to a unique graph of family $F[n]$, see Figure 14.8. Thus for a fixed $n \ge 2$, the family $F[n]$ comprises one graph among $\Theta(n^9)$ graphs.

In what follows, we apply our method to compute certain degree and distance-based topological indices for the infinite fullerenes family $F[n]$, $n \ge 2$. We calculated the partitions in Table 14.5 and used those partitions to prove general expressions of topological indices for $F[n]$.

The following lemma counts the number of vertices and edges in a n-dimensional fullerene $F[n]$.

LEMMA 14.2 Let n (resp. m) be the number of vertices (resp. edges) in a n-dimensional fullerene $F[n]$. Then, $n = 10n$ and $m = 15n$.

In the following proposition, we compute certain degree-based topological indices for the family of fullerenes $F[n]$.

PROPOSITION 14.3 For an n-dimensional fullerene graph $F[n]$, $n \ge 2$, the following equations hold:

$$ABC(F[n]) = 10n, \quad GA(F[n]) = 15n,$$
$$AZI(F[n]) = \frac{10935}{64}n, \quad R_{-1}(F[n]) = \frac{5}{3}n,$$
$$SCI_{-3}(F[n]) = \frac{5}{72}n.$$

The following theorem presents explicit expressions for certain degree- and distance-based topological indices of the family of fullerenes $F[n]$.

THEOREM 14.1

(a) *The Szeged and revised Szeged indices of fullerene graph $F[n]$, $n \ge 9$, are computed as follows:*

$$Sz(F[n]) = 250n^3 + 3075n - 13800;$$
$$RSz(F[n]) = 250n^3 + 250n^2 + 4275n - 15650.$$

(b) *The eccentric connectivity and total connectivity indices of fullerene graph $F[n]$, $n \ge 8$, are computed as follows:*

$$EC(F[n]) = 45n^2 - 15n;$$
$$TE(F[n]) = \frac{1}{3}EC(F[n]) = 15n^2 - 5n.$$

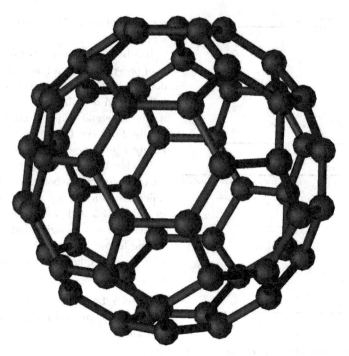

FIGURE 14.7 A 3D representation of the Buckminsterfullerene.

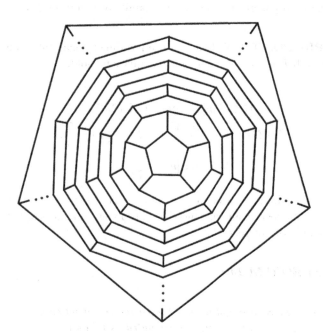

FIGURE 14.8 2D Schlegel representation of fullerene graph $F[n]$.

(c) *For an n-dimensional fullerene graph* $F[n]$, $n \geq 5$,
the following equations hold:

$$PI(F[n]) = 150n^2 - 100n;$$
$$W(F[n]) = \frac{100}{3}n^3 + \frac{1175}{3}n - 670;$$
$$MTI(F[n]) = 200n^3 + 2440n - 4020;$$
$$DD(F[n]) = 6(W(F[n])) = 200n^3 + 2350n - 4020;$$
$$Gut(F[n]) = 18(W(F[n])) = 300n^3 + 3525n - 6030.$$

14.5 Carbon Nanotubes

Nanotechnology works with structures of size in the range 1–100 nm. Nanotechnology creates many new materials and devices with a variety of applications in medicine, electronics and computer. From the applications point of view, carbon nanotubes have been the most significant nanostructures so far [17].

Carbon nanotubes are types of nanostructure which are allotropes of carbon and having a cylindrical shape. They are a type of fullerene and have potential applications in fields such as nanotechnology, electronics, optics, materials science and architecture. Carbon nanotubes provide a certain potential for metal-free catalysis of inorganic and organic reactions. They have applications in fields of electronic and electrochemical and mechanical reinforcements in high-performance composites. Nanotube-based field emitters have applications as nanoprobes in metrology and biological and chemical investigations and as templates for the creation of other nanostructures.

A C_4C_8 nanotube is a trivalent decoration made by alternating squares C_4 and octagons C_8. There are two types of nanotubes which possess this trivalent decoration of squares and octagons. The first nanotube is $TUC_4C_8(S)[p,q]$ in which C_4 shapes a square on the cylindrical wall of the nanotube, and the other nanotube is $TUC_4C_8(R)[p,q]$ in which C_4 forms a rhombus. The defining parameters p and q in these nanotubes denote the number octagons in a row and column, respectively. Figure 14.9 presents two different 3D views of the family of $TUC_4C_8(R)[p,q]$ nanotubes, whereas Figure 14.10 depicts a 2D graph of this nanotube.

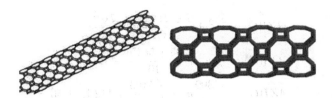

FIGURE 14.9 Two different 3D views of $TUC_4C_8(S)[p,q]$ nanotubes.

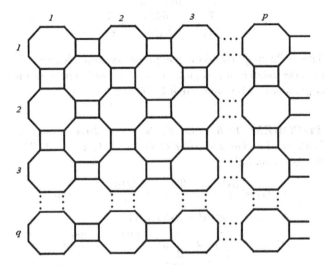

FIGURE 14.10 2D graph of the (p,q)-dimensional $TUC_4C_8(S)[p,q]$ nanotube.

This section focuses on computing the degree- and distance-based topological indices from Tables 14.2–14.4 for the family of $TUC_4C_8(S)[p,q]$ nanotubes.

The following lemma counts the number of vertices and edges in a $TUC_4C_8(S)[p,q]$ nanotube.

LEMMA 14.3 Let n (resp. m) be the number of vertices (resp. edges) in a (p,q)-dimensional $TUC_4C_8(S)[p,q]$ nanotube. Then, $n = 8pq$ and $m = 12pq - 2p$.

The following theorem provides explicit formulas of certain degree-based topological indices of $TUC_4C_8(S)[p,q]$ nanotubes.

THEOREM 14.2 Let G be the (p,q)-dimensional $TUC_4C_8(S)[p,q]$ carbon nanotube, where $p,q \geq 2$. Then, the following hold:

$$ABC(G) = 8pq + \left(3\sqrt{2} - \frac{16}{3}\right)p;$$

$$GA(G) = 12pq + \left(\frac{8\sqrt{6}}{5} - 6\right)p;$$

$$AZI(G) = \frac{2187}{16}pq - \frac{345}{8}p;$$

$$R_{-1}(G) = \frac{4}{3}pq + \frac{5}{16}p;$$

$$SCI_{-3}(G) = \frac{1}{16}pq + \left(\frac{253}{4000} - \frac{1}{27}\right)p.$$

In the following theorem, we compute certain degree- and distance-related topological indices for the (p,q)-dimensional $TUC_4C_8(S)[p,q]$ carbon nanotube. Since the graph of $TUC_4C_8(S)[p,q]$ nanotube is bipartite, we find that the Szeged and revised Szeged indices are same.

THEOREM 14.3 Let G be the (p,q)-dimensional $TUC_4C_8(S)[p,q]$ carbon nanotube, where $p \geq 2q - 1$. Then, the following hold:

$$Sz(G) = \left(\frac{512}{3}q^3 - \frac{32}{3}q\right)p^3 - \left(\frac{64}{3}q^5 - \frac{16}{3}q^3\right)p;$$

$$PI(G) = (96q^2 - 16q)p^2;$$

$$W(G) = 32q^2p^3 + \left(\frac{64}{3}q^3 - \frac{16}{3}q\right)p^2 + \left(\frac{32}{3}q^4 - \frac{8}{3}q^2\right)p;$$

$$MTI(G) = (192q^2 - 32q)p^3 + (128q^3 - 32q^2 - 16q)p^2$$
$$+ \left(\frac{413}{6}q^4 - \frac{207}{2}q^3 + \frac{1460}{3}q^2 - 1218q + 1140\right)p;$$

$$DD(G) = (192q^2 - 32q)p^3 + (128q^3 - 32q^2 - 16q)p^2$$
$$+ \left(64q^4 - \frac{64}{3}q^3 - 16q^2 + \frac{16}{3}q\right)p;$$

$$Gut(G) = (288q^2 - 96q + 8)p^3 + (192q^3 - 96q^2$$
$$+ 8q - 4)p^2 + \left(96q^4 - 64q^3 - 16q^2 + 16q - \eta\right)p,$$

$$\text{where } \eta = \begin{cases} 4, & p \equiv 1 \pmod 2; \\ 0, & p \equiv 0 \pmod 2. \end{cases}$$

THEOREM 14.4 Let G be the (p,q)-dimensional $TUC_4C_8(S)[p,q]$ carbon nanotube, where $p \geq 2q$. Then, the following hold:

$$EC(G) = (48q - 8)p^2 + (36q^3 - 20q^2 + 4)p;$$

$$TE(G) = 16qp^2 + (12q^2 - 4q)p.$$

14.6 Carbon Nanocones

The occurrence of hollow carbon structures is a fascinating phenomenon. Except for fullerenes and nanotubes, carbon nanocones have been observed as caps on the ends of the nanotubes or as free-standing structures on a flat graphite surface. A cone can be modeled by extracting a fan from a graphene sheet, rolling the remained sector around its apex and joining the two open sides. Especially, when the fan is above 60°, a heptagon is presented at the tip apex of the nanocone (see Figures 14.11 and 14.12). Continuing this process would produce more pentagons and simultaneously reduce the opening angle of the cone. A nanocone with six pentagons has an opening angle of zero, which is generally viewed as a nanotube with one end open.

An n-dimensional one-heptagonal nanocone is usually denoted by $CNC_7[n]$, where n is the number of hexagon

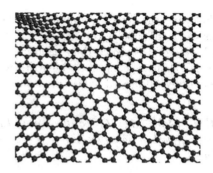

FIGURE 14.11   2D sheet of one-heptagonal nanocone.

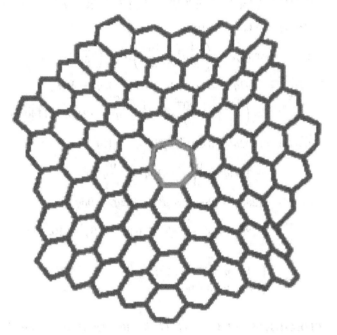

FIGURE 14.12   A 4D one-heptagonal carbon nanocone, i.e., $CNC_7[4]$.

layers encompassing the conical surface of the nanocone and 7 denotes that there is a heptagon on the tip called its core. In this section, we compute degree- and distance-based topological indices from Tables 14.1–14.4 for the family of one-heptagonal carbon nanocones.

The number of vertices and edges in $CNC_7[n]$ nanocones is given by the following lemma.

LEMMA 14.4   Let n (resp. m) be the number of vertices and edges in an n-dimensional one-heptagonal carbon nanocone. Then, $n = 7(n+1)^2$ and $m = \frac{21}{2}n^2 + \frac{35}{2}n + 7$.

The following theorem presents explicit formulas for certain degree-based topological indices of one-heptagonal carbon nanocones.

THEOREM 14.5   Let G be the n-dimensional one-heptagonal carbon nanocone $CNC_7[n]$, where $n \geq 1$. Then, the following hold:

$$ABC(G) = 7n^2 + \frac{7 + 21\sqrt{2}}{3}n + \frac{7\sqrt{2}}{2};$$

$$GA(G) = \frac{21}{2}n^2 + \frac{35 + 56\sqrt{6}}{10} + 7;$$

$$AZI(G) = \frac{15309}{128}n^2 + \left(\frac{5103}{128} + 112\right)n + 56;$$

$$R_{-1}(G) = \frac{7}{6}n^2 + \frac{49}{18}n + \frac{7}{4};$$

$$SCI_{-3} = \frac{7}{144}n^2 + \frac{6923}{54000}n + \frac{7}{64}.$$

The following theorem computes certain degree- and distance-related topological indices of n-dimensional one-heptagonal carbon nanocone $CNC_7[n]$.

THEOREM 14.6   Let G be the n-dimensional one-heptagonal carbon nanocone $CNC_7[n]$, where $n \geq 2$. Then, the following hold:

$$Sz(G) = \frac{315}{2}n^6 + \frac{903}{2}n^5 + \frac{4319}{4}n^4 + 1379n^3 + 994n^2 + 385n + 63;$$

$$PI(G) = \frac{147}{2}n^4 + \frac{525}{2}n^3 + \frac{693}{2}n^2 + \frac{399}{2}n + 42;$$

$$RSz(G) = \frac{315}{4}n^6 + \frac{1897}{4}n^5 + \frac{2387}{2}n^4 + \frac{3213}{2}n^3 + \frac{2443}{2}n^2 + \frac{1995}{4}n + \frac{343}{4};$$

$$EC(G) = 70n^3 + \frac{343}{2}n^2 + \frac{287}{2}n + 42;$$

$$TE(G) = \frac{70}{3}n^3 + \frac{133}{2}n^2 + \frac{385}{6}n + 21;$$

$$W(G) = \frac{238}{5}n^5 + 238n^4 + \frac{2821}{6}n^3 + \frac{917}{2}n^2 + \frac{3311}{15}n + 42;$$

$$MTI(G) = \frac{1428}{5}n^5 + 1309n^4 + \frac{4739}{2}n^3 + 2184n^2 + \frac{10339}{10}n + 196;$$

$$DD(G) = \frac{1428}{5}n^5 + 1309n^4 + \frac{4739}{2}n^3 + 2121n^2 + \frac{9429}{10}n + 108;$$

$$Gut(G) = \frac{2142}{5}n^5 + 1785n^4 + \frac{8834}{3}n^3 + \frac{4837}{2}n^2 + \frac{6989}{7}n + 168.$$

14.7   Concluding Remarks

This chapter surveys the known computational techniques for structure-based topological indices of graphs. A new computational technique is presented for the classes of degree-based, distance-based and degree-distance-based topological indices. By applying the proposed technique, various degree-based, distance-based and degree-distance-based indices are computed for certain infinite families of fullerenes, carbon nanotubes and carbon nanocones.

Extending these techniques to other classes of topological indices such as spectrum-based topological indices and counting related topological indices is open for future study.

Acknowledgments

Sakander Hayat is supported by the Startup Research Grant Program of Higher Education Commission (HEC) Pakistan under Project# 2285 and grant No. 21-2285/SRGP/R&D/HEC/2018.

Bibliography

1. M.A. Alipour and A.R. Ashrafi. A numerical method for computing the wiener index of one-heptagonal carbon nanocone. *Journal of Computational and Theoretical Nanoscience*, 6(5):1204–1207, 2009.

2. Y. Alizadeh, A. Iranmanesh, and T. Došlić. Additively weighted Harary index of some composite graphs. *Discrete Mathematics*, 313(1):26–34, 2013.

3. D. Amić, D. Bešlo, B. Lucić, S. Nikolić, and N. Trinajstić. The vertex-connectivity index revisited. *Journal of Chemical Information and Computer Sciences*, 38(5):819–822, 1998.

4. M. Arockiaraj, J. Clement, and K. Balasubramanian. Analytical expressions for topological properties of polycyclic benzenoid networks. *Journal of Chemometrics*, 30(11):682–697, 2016.

5. M. Arockiaraj, S. R. J. Kavitha, and K. Balasubramanian. Vertex cut method for degree and distance-based topological indices and its applications to silicate networks. *Journal of Mathematical Chemistry*, 54(8):1728–1747, 2016.

6. M. Arockiaraj, S. R. J. Kavitha, K. Balasubramanian, and J.-B. Liu. On certain topological indices of octahedral and icosahedral networks. *IET Control Theory & Applications*, 12(2):215–220, 2018.

7. A.R. Ashrafi, T. Došlic, and M. Saheli. The eccentric connectivity index of $TUC_4C_8(R)$ nanotubes. *MATCH Communications in Mathematical and in Computer Chemistry*, 65(1):221–230, 2011.

8. A.R. Ashrafi, M. Ghorbani, and M. Jalali. The PI and edge Szeged polynomials of an infinite family of fullerenes. *Fullerenes, Nanotubes and Carbon Nanostructures*, 18(2):107–116, 2010.

9. A.R. Ashrafi, M.A. Iranmanesh, and Z. Yarahmadi. Study of fullerenes by some new topological index. In *Topological Modelling of Nanostructures and Extended Systems*, pages 473–485. Springer, 2013.

10. M. Bača, J. Horváthová, M. Mokrišová, A. Semaničová-Feňovčíková, and A. Suhányiová.

11. M. Bača, J. Horváthová, M. Mokrišová, and A. Suhányiová. On topological indices of fullerenes. *Applied Mathematics and Computation*, 251:154–161, 2015.

12. B. Bollobás and P. Erdos. Graphs of extremal weights. *Ars Combinatoria*, 50:225–233, 1998.

13. S. Daugherty, W. Myrvold, and P.W. Fowler. Backtracking to compute the closed-shell independence number of a fullerene. *MATCH Communications in Mathematical and in Computer Chemistry*, 58(2):385–401, 2007.

14. M.V. Diudea. *Nanomolecules and nanostructures: polynomials and indices*. University of Kragujevac. Faculty of Science, 2010.

15. M.V. Diudea, I. Gutman, and L. Jantschi. *Molecular Topology*. Nova Science Publishers, Huntington, NY, 2001.

16. A.A. Dobrynin and A.A. Kochetova. Degree distance of a graph: A degree analog of the Wiener index. *Journal of Chemical Information and Computer Sciences*, 34(5):1082–1086, 1994.

17. M.S. Dresselhaus, G. Dresselhaus, and P.C. Eklund. *Science of Fullerenes and Carbon Nanotubes: Their Properties and Applications*. Elsevier, Burlington, 1996.

18. E. Estrada, L. Torres, L. Rodriguez, and I. Gutman. An atom-bond connectivity index: Modelling the enthalpy of formation of alkanes. *Indian Journal of Chemistry*, 37A:849–855, 1998.

19. M.R. Farahani. Eccentricity version of atom-bond connectivity index of benzenoid family $ABC_5(H_k)$. *World Applied Sciences Journal*, 21(9):1260–1265, 2013.

20. M.R. Farahani. Fifth geometric-arithmetic index of $TURC_4C_8(S)$ nanotubes. *Journal of Chemica Acta*, 2(2):62–64, 2013.

21. G. Fath-Tabar, B. Furtula, and I. Gutman. A new geometric–arithmetic index. *Journal of Mathematical Chemistry*, 47(1):477, 2010.

22. P.W. Fowler, D. Horspool, and W. Myrvold. Vertex spirals in fullerenes and their implications for nomenclature of fullerene derivatives. *Chemistry-A European Journal*, 13(8):2208–2217, 2007.

23. P.W. Fowler and D.E. Manolopoulos. *An Atlas of Fullerenes*. Courier Corporation, North Chelmsford, MA, 2006.

24. B. Furtula, A. Graovac, and D. Vukičević. Augmented Zagreb index. *Journal of Mathematical Chemistry*, 48:370–380, 2010.

25. M. Ghorbani and A. Khaki. A note on the fourth version of geometric-arithmetic index. *Optoelectronics and Advanced Materials-Rapid Communication*, 4(12):2212–2215, 2010.

26. A. Graovac and M. Ghorbani. A new version of atom-bond connectivity index. *Acta Chimica Slovenica*, 57(3):609, 2010.

27. A. Graovac, O. Ori, M. Faghani, and A.R. Ashrafi. Distance property of fullerenes. *Iranian Journal of Mathematical Chemistry*, 2:99–107, 2011.

28. GAP Group et al. Gap–groups, algorithms, and programming, version 4.7. 5; 2014, 2012.

29. I. Gutman. A formula for the Wiener number of trees and its extension to graphs containing cycles. *Graph Theory Notes NY*, 27(1):9–15, 1994.

30. I. Gutman. Selected properties of the Schultz molecular topological index. *Journal of Chemical Information and Computer Sciences*, 34(5):1087–1089, 1994.

31. I. Gutman and S.J. Cyvin. *Advances in the Theory of Benzenoid Hydrocarbons*. Springer Publishing Company, Incorporated, Heidelberg, 2013.

32. I. Gutman and O.E. Polansky. *Mathematical Concepts in Organic Chemistry*. Springer Science & Business Media, Heidelberg, 2012.

33. S. Hayat. Computing distance-based topological descriptors of complex chemical networks: New theoretical techniques. *Chemical Physics Letters*, 688:51–58, 2017.

34. S. Hayat and M. Imran. Computation of topological indices of certain networks. *Applied Mathematics and Computation*, 240:213–228, 2014.

35. S. Hayat and M. Imran. On topological properties of nanocones $CNC_k[n]$. *Studia UBB Chemia*, 59:113–128, 2014.

36. S. Hayat and M. Imran. Computation of certain topological indices of nanotubes. *Journal of Computational and Theoretical Nanoscience*, 12(1):70–76, 2015.

37. S. Hayat and M. Imran. Computation of certain topological indices of nanotubes covered by C_5 and C_7. *Journal of Computational and Theoretical Nanoscience*, 12(4):533–541, 2015.

38. S. Hayat and M. Imran. On degree based topological indices of certain nanotubes. *Journal of Computational and Theoretical Nanoscience*, 12(8):1599–1605, 2015.

39. S. Hayat, M.A. Malik, and M. Imran. Computing topological indices of honeycomb derived networks. *Romanian Journal of Information Science and Technology*, 18:144–165, 2015.

40. S. Hayat, S. Wang, and J.-B. Liu. Valency-based topological descriptors of chemical networks and their applications. *Applied Mathematical Modelling*, 2018.

41. H. Hua and S. Zhang. On the reciprocal degree distance of graphs. *Discrete Applied Mathematics*, 160(7-8):1152–1163, 2012.

42. Package release HyperChem. 7.5 for windows hypercube inc. *Florida, USA*, 2002.

43. M. Imran, S. Hayat, and M.Y.H. Mailk. On topological indices of certain interconnection networks. *Applied Mathematics and Computation*, 244:936–951, 2014.

44. M. Karelson. *Molecular Descriptors in QSAR/QSPR*. Wiley-Interscience, New York, s2000.

45. P.V. Khadikar, S. Karmarkar, and V.K. Agrawal. A novel PI index and its applications to QSPR/QSAR studies. *Journal of Chemical Information and Computer Sciences*, 41(4):934–949, 2001.

46. H.W. Kroto, J.E. Fichier, and D.E. Cox. *The Fullerene*. Pergamon Press, New York, 1993.

47. H.W. Kroto, J.R. Heath, S.C. ÓBrien, R.F. Curl, and Smalley R.E. C_{60}: Buckminsterfullerene. *Journal of Chemical Information and Computer Sciences*, 318:162–163, 1985.

48. J.-B. Liu, S. Wang, C. Wang, and S. Hayat. Further results on computation of topological indices of certain networks. *IET Control Theory & Applications*, 11(13):2065–2071, 2017.

49. Matlab-Version Matlab. 8.0.0.783 (R2012b). *Natick, Massachusetts: The MathWorks Inc*, 82, 2012.

50. Z. Mehranian, A. Mottaghi, and A.R. Ashrafi. The topological study of IPR fullerenes by Szeged and revised Szeged indices. *Journal of Theoretical and Computational Chemistry*, 11(03):547–559, 2012.

51. A. Mottaghi and A.R. Ashrafi. Topological edge properties of C_{12n+60} fullerenes. *Beilstein Journal of Nanotechnology*, 4:400–405, 2013.

52. W. Myrvold, B. Bultena, S. Daugherty, B. Debroni, S. Girn, M. Minchenko, J. Woodcock, and P.W. Fowler. Fuigui: A graphical user interface for investigating conjectures about fullerenes. *MATCH Communications in Mathematical and in Computer Chemistry*, 58(2):403–422, 2007.

53. M. Randic. Characterization of molecular branching. *Journal of the American Chemical Society*, 97(23):6609–6615, 1975.

54. M. Randić. On generalization of Wiener index for cyclic structures. *Acta Chimica Slovenica*, 49:483–496, 2002.

55. H.P. Schultz. Topological organic chemistry. 1. graph theory and topological indices of alkanes. *Journal of Chemical Information and Computer Sciences*, 29(3):227–228, 1989.

56. V. Sharma, R. Goswami, and A.K. Madan. Eccentric connectivity index: a novel highly discriminating topological descriptor for structure- property and structure-activity studies. *Journal of Chemical Information and Computer Sciences*, 37(2):273–282, 1997.

57. V.A. Skorobogatov and A.A. Dobrynin. Metrical analysis of graphs. *MATCH Communications in Mathematical and in Computer Chemistry*, 23:105–155, 1988.

58. D. Stevanović, V. Brankov, D. Cvetković, and S. Simić newGraph. A fully integrated environment used for research process in graph theory,

http://www.mi.sanu.ac.rs/newgraph/(accessed July 2019), 2003.

59. O. Ursu, M.V. Diudea, and Cs.L. Nagy. *Topocluj Software Program*. Babes-Bolyai University, Cluj, 2005.

60. D. Vukievié and B. Furtula. Topological index based on the ratios of geometrical and arithmetical means of end-vertex degrees of edges. *Journal of Mathematical Chemistry*, 46(4):1369–1376, 2009.

61. H. Wiener. Structural determination of paraffin boiling points. *Journal of the American Chemical Society*, 69(1):17–20, 1947.

62. R. Xing and B. Zhou. On the revised Szeged index. *Discrete Applied Mathematics*, 159(1):69–78, 2011.

63. J. Yazdani and A. Bahrami. Topological descriptors of H-naphtalenic nanotubes. *Digest Journal of Nanomaterials & Biostructures*, 4(1), 2009.

64. B. Zhou and Z. Du. On eccentric connectivity index. *MATCH Communications in Mathematical and in Computer Chemistry*, 63:181–198, 2010.

65. B. Zhou and N. Trinajstić. On a novel connectivity index. *Journal of Mathematical Chemistry*, 46(4):1252–1270, 2009.

66. B. Zhou and N. Trinajstić. On general sum-connectivity index. *Journal of Mathematical Chemistry*, 47(1):210–218, 2010.

15

Numerical Methods for Large-Scale Electronic State Calculation on Supercomputer

Takeo Hoshi
Tottori University

Yusaku Yamamoto
The University of Electro-Communications

Tomohiro Sogabe
Nagoya University

Kohei Shimamura
Kobe University

Fuyuki Shimojo
Kumamoto University

Aiichiro Nakano, Rajiv Kalia, and Priya Vashishta
University of Southern California

15.1 Introduction

Computer was invented in the 20th century, and a vast number of numerical algorithms have been developed so far. For example, Francis Sullivan and Jack Dongarra listed up 'The Top 10 Algorithms' in the 20th century [55]. Following is the list in chronological order: The Metropolis Algorithm, Simplex Method, Krylov Subspace Method, The Decompositional Approach to Matrix Computations, The Fortran Optimizing Compiler, QR Algorithm, Quicksort, Fast Fourier Transform, Integer Relation Detection, and Fast Multipole Method. Many algorithms in the list are well used in computational physics areas, including nanoscience.

In the 21th century, however, the emergence of massively parallel supercomputers drove a paradigm shift of computational science, and novel numerical algorithms have been proposed for parallelism. The collaboration of researchers among computational science, applied mathematics, and computer science is of great importance and is called Application–Algorithm–Architecture co-design. Figure 15.1 is an overview of important algorithms in the four computational science areas of fluid dynamics, structural analysis, electronic state calculation, and lattice quantum chromo dynamics (QCD). One can find that many aspects are common among the four areas. In addition to efficient algorithms, high-performance computing techniques such

as parallelization and auto-tuning are important to exploit the computing power of modern many-core and massively parallel machines. Standard textbooks on these aspects include [18,32,53].

The present chapter is devoted to a tutorial of novel numerical methods for large-scale electronic state calculations on current and next-generation supercomputers. The present chapter is written for nanoscience researchers who are not familiar to numerical algorithms and supercomputers.

This chapter is organized as follows: Section 15.2 explains the numerical foundations of large-scale electronic state calculations and related physical theories. Section 15.3 is an overview of linear algebraic solvers and their numerical libraries. Two application studies in nanoscience appear in Section 15.4.

15.2 Numerical Foundations of Large-Scale Electronic State Calculations

This section explains the numerical foundations of large-scale electronic state calculations, that is, generalized eigenvalue problems (GEPs) and the order-N concept.

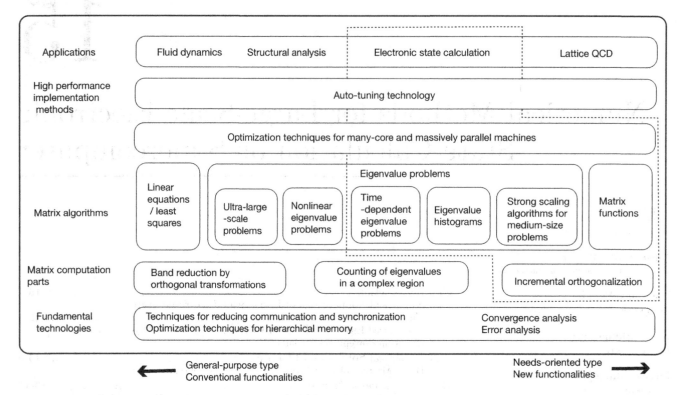

FIGURE 15.1 An overview of important algorithms in the four computational science areas of fluid dynamics, structural analysis, electronic state calculation, lattice QCD. The elements strongly related to electronic state calculation are surrounded with a dashed square.

15.2.1 Generalized Eigenvalue Problem

This section introduces GEP as a numerical foundation of large-scale electronic state calculations. The fundamental Schrödinger-type equation, a partial differential equation in real space $\boldsymbol{r}$, is written for an electronic wavefunction $\phi(\boldsymbol{r})$ as

$$\hat{H}\phi(\boldsymbol{r}) = \lambda\phi(\boldsymbol{r}) \qquad (15.1)$$

with the Hamilton operator of

$$\hat{H} \equiv -\frac{\hbar^2}{2m}\Delta + V_{\text{eff}}(\boldsymbol{r}). \qquad (15.2)$$

Here, Δ is Laplacian, m is the mass of electron, and $\hbar$ is the Planck constant, a physical constant ($\hbar \approx 1.05^{-34}$ Js). $V_{\text{eff}}(\boldsymbol{r})$ is the effective potential, a scalar function. The normalization condition of

$$\int |\phi(\boldsymbol{r})|^2 = 1 \qquad (15.3)$$

is imposed and stems from the fact that the sum of the weight distribution of one electron should be the unity.

An eigenvalue of λ means the energy of an electron in the material and is called eigenenergy. The k-th eigenpair of $(\lambda_k, \phi_k(\boldsymbol{r}))$ is defined for $k = 1, 2, .., M$ in the order of $\lambda_1 \leq \lambda_2 \leq \cdots \leq \lambda_M$. Each material has a specific integer of k_{HO} called the highest occupied eigenenergy, and the eigenpairs for $k = 1, 2, ...k_{\text{HO}}$ are occupied by the electrons. A para-spin material with N_{elec} electrons, for example, gives the value

of $k_{\text{HO}} = N_{\text{elec}}/2$, if N_{elec} is even. Semiconductor material has a finite energy gap between the k_{HO}-th and $(k_{\text{HO}}+1)$-th eigenenergies ($\lambda_{k_{\text{HO}}+1} - \lambda_{k_{\text{HO}}} > 0$).

Now we consider, as a typical case, that $\phi(\boldsymbol{r})$ is expressed as a linear combination of given basic functions

$$\phi(\boldsymbol{r}) = \sum_j^M c_j \chi_j(\boldsymbol{r}). \qquad (15.4)$$

The basis functions $\{\chi_j(\boldsymbol{r})\}$ are normalized to be

$$\int \chi_j^*(\boldsymbol{r})\chi_j(\boldsymbol{r})d\boldsymbol{r} = 1. \qquad (15.5)$$

A typical function is called atomic orbital and is localized near the position of an atomic nucleus. Since each basis function belongs to one atom, the basis index i is equivalent to the composite indices of an atom index I and an orbital index α ($i \equiv (I, \alpha)$). The orbital index α distinguishes the basis functions that belong to the same atom but different in their shape.

A generalized eigenvalue equation appears, when Eq. (15.4) is substituted for Eq. (15.1);

$$A\boldsymbol{y}_k = \lambda_k B\boldsymbol{y}_k \qquad (15.6)$$

with the $M \times M$ matrices of

$$A_{ij} \equiv \int \chi_i^*(\boldsymbol{r})\hat{H}\chi_j(\boldsymbol{r})d\boldsymbol{r} \qquad (15.7)$$

$$B_{ij} \equiv \int \chi_i^*(\boldsymbol{r})\chi_j(\boldsymbol{r})d\boldsymbol{r}. \qquad (15.8)$$

The matrices A and B are Hermitian. The matrix B is positive definite and satisfies $B_{jj} = 1$ and $|B_{ij}| < 1 (i \neq j)$. Hereafter, we consider, as among many researches, that the basis functions are real and the matrices A and B are real symmetric. The normalization condition of Eq. (15.3) is reduced to

$$\boldsymbol{y}_k^{\mathrm{T}} B \boldsymbol{y}_k = 1, \qquad (15.9)$$

which is called B-normalization. Sparsity of the matrices of A_{ij} and B_{ij} is explained briefly. As explained in the previous subsection, the indices i and j are the composite indices of the atom indices I and J and the orbital indices α and β, respectively $(i \equiv i(I, \alpha), j \equiv j(J, \beta)))$. Therefore, an element of the matrices A and B is expressed by the four indices as $A_{I\alpha;J\beta}$ and $B_{I\alpha;J\beta}$, respectively. Since a matrix element value decreases quickly and monotonically as the function of the inter-atomic distance between the I-th and J-th atoms (r_{IJ}), a cutoff distance r_{cut} can be introduced. A matrix element, $A_{I\alpha;J\beta}$ or $B_{I\alpha;J\beta}$, is ignored, if $r_{IJ} > r_{\mathrm{cut}}$, which makes the matrices to be sparse.

15.2.2 Order-N Approach

A crucial approach for large-scale electronic state calculations is the concept of order-N method. Order-N method is the general name of the methods, in which the computational time T is order-N or linearly proportional to the number of atoms in the system N $(T = O(N))$. Since the conventional methods consume an $O(N^3)$ computational time or an heavier one $(O(N^q), q > 3)$, the order-N method will be much faster than the conventional ones in a large system $(N \rightarrow \infty)$. Many methods and codes were developed as order-N electronic state calculations and are found, for example, in Refs. [16,23,33,35,46,47,52,62]. Figure 15.2 demonstrates the benchmark of an order-N calculation, in which the elapsed time is proportional to the number of atoms between $N = 10^5 - 10^7$ atoms [23]. The benchmark was carried out by extra-large-scale electronic

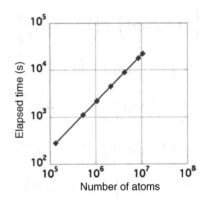

FIGURE 15.2 A benchmark of order-N calculations between $N = 10^5 - 10^7$ atoms. Calculation of the amorphous-like conjugated polymers of poly-(9,9 dioctyl-fluorene) with up to 10^7 atoms by ELSES [23].

structure calculation (ELSES), an order-N calculation code. See Section 15.4.1 for an application.

A fundamental concept of the order-N method was proposed by Walter Kohn, a Nobel Prize winner on Chemistry at 1998. The concept is called 'nearsightedness principle' [26], which is closely related to the Hohenberg–Kohn theorem [20]. Now we can recall that the Hohenberg–Kohn theorem gives the ground state energy as a functional of the charge density $n(\boldsymbol{r})$ and that the charge density is the *diagonal* elements of the density matrix $(n(\boldsymbol{r}) = \rho(\boldsymbol{r}, \boldsymbol{r}))$. Instead of such a general theory, we discuss the free electron system, as an example, since the system is well familiar to nanoscientists.

In the free electron system, the Hamiltonian is simply the kinetic energy part

$$H_0 \equiv -\frac{1}{2}\Delta, \qquad (15.10)$$

and the ground state is characterized by the Fermi wavenumber k_{F}. The total energy per volume is calculated in the reciprocal space by

$$\frac{E}{V} = \int_{k<k_{\mathrm{F}}} \frac{d\boldsymbol{k}}{(2\pi)^3} \frac{1}{2} k^2 = \frac{k_{\mathrm{F}}^5}{20\pi^2}. \qquad (15.11)$$

The corresponding one-body density matrix is defined as

$$\rho(\boldsymbol{r}_1, \boldsymbol{r}_2) \equiv \int \frac{d\boldsymbol{k}}{(2\pi)^3} \frac{e^{i\boldsymbol{k}\cdot\boldsymbol{r}_1}}{\sqrt{V}} \frac{e^{-i\boldsymbol{k}\cdot\boldsymbol{r}_2}}{\sqrt{V}}$$
$$= \int \frac{d\boldsymbol{k}}{(2\pi)^3} \frac{e^{i\boldsymbol{k}\cdot(\boldsymbol{r}_1 - \boldsymbol{r}_2)}}{V}. \qquad (15.12)$$

Due to the uniform property, the density matrix is reduced to that of the function of the distance $r \equiv |\boldsymbol{r}_1 - \boldsymbol{r}_2|$. Without the volume factor $1/V$, we redefine the density matrix and calculate

$$\rho(r) \equiv \int_{k<k_{\mathrm{F}}} \frac{d\boldsymbol{k}}{(2\pi)^3} e^{i\boldsymbol{k}\cdot\boldsymbol{r}}$$
$$= \frac{2}{(2\pi)^2} \left\{ -\frac{k_{\mathrm{F}}}{r^2} \cos k_{\mathrm{F}}r + \frac{1}{r^3} \sin k_{\mathrm{F}}r \right\}. \qquad (15.13)$$

The resultant density matrix shows a long-range oscillation with the Fermi wavenumber k_{F}

$$\rho(r) \propto \frac{\cos k_{\mathrm{F}}r}{r^2} \quad (r \rightarrow \infty), \qquad (15.14)$$

which is known as the Friedel oscillation. The short-range behavior, on the other hand, is given by the Taylor expansion as

$$\rho(r) = \frac{2}{(2\pi)^2} \left(C_0 - \frac{C_2}{2} r^2 + O(r^4) \right), \qquad (15.15)$$

using the zero-th and second Taylor coefficients

$$C_0 \equiv \frac{k_{\mathrm{F}}^3}{6}, \quad C_2 \equiv \frac{k_{\mathrm{F}}^5}{15}. \qquad (15.16)$$

The total energy per volume can be also calculated by the density matrix as

$$\frac{E}{V} \equiv \frac{1}{V}\text{Tr}[\rho H_0] = \frac{1}{V}\int dr_1 \left.\frac{-\Delta_{r_1}}{2}\rho(r_1, r_2)\right|_{r_1=r_2}$$
$$= \lim_{\varepsilon\to 0}\frac{E_{\text{sphere}}(\varepsilon)}{(4\pi\varepsilon^3/3)} \quad (15.17)$$

Here, $E_{\text{sphere}}(\varepsilon)$ is the energy of a tiny (real-space) sphere with the radius of ε. Using Eq. (15.15) and the calculation

$$E_{\text{sphere}}(\varepsilon) \equiv \int_{r<\varepsilon} dr \frac{-\Delta_r}{2}\rho(r) = \frac{C_2}{\pi}\varepsilon^3 + O(\varepsilon^5), \quad (15.18)$$

the energy per volume is given by

$$\frac{E}{V} = \lim_{\varepsilon\to 0}\frac{E_{\text{sphere}}(\varepsilon)}{(4\pi\varepsilon^3/3)} = \frac{3C_2}{4\pi^2}. \quad (15.19)$$

With the definition $C_2 \equiv k_{\text{F}}^5/15$, the above result reproduces that of Eq. (15.11). This shows that the total energy is determined only by the second-order Taylor coefficient C_2. This statement is understandable, because the present Hamiltonian, the Laplacian operator, is the *second* order derivative. In other words, the total energy is determined explicitly by the *short-range* behavior of the density matrix. Here, it should be emphasized that the above density matrix $\rho(r)$ has the off-diagonal long-range components, as in Eq. (15.14), and the system is metallic. In short, the total energy is governed by the *short-range* behavior of the density matrix, while the transport property is governed by the off-diagonal *long-range* behavior.

15.2.3 Algorithm for Quantum Wavepacket Dynamics

This subsection picks out the methods of quantum wavepacket dynamics of (excited) electron or hole, as a topic that is interesting from the numerical viewpoint. First of all, the wavepacket dynamics is much faster that of the atomic motion. A typical time scale is given by the atomic time unit of

$$\tau_{\text{au}} \equiv \frac{\hbar^3}{m}\left(\frac{e^2}{4\pi\varepsilon_0}\right)^{-2} \approx 0.024189\,\text{fs}, \quad (15.20)$$

with the electron mass m, the elementary charge e, Planck's constant $\hbar$, and the dielectric constant of vacuum ε_0. Therefore, the time interval δt in the wavepacket dynamics simulation is usually set to be $\delta t \approx 0.1$ fs or less and is much smaller than that in the molecular dynamics simulation ($\delta t_{\text{MD}} \approx 1$fs).

In general, the mathematical foundation of wavepacket dynamics is the time-dependent Schrödinger equation and a prototypical numerical problem appears, when the wavefunction is reduced to a vector of $\boldsymbol{u}(t) \equiv (u_1(t), u_2(t), ..., u_M(t))$ and the Hamiltonian is reduced to an $M \times M$ Hermitian matrix H;

$$i\frac{d\boldsymbol{u}}{dt} = H\boldsymbol{u}. \quad (15.21)$$

When H depends neither on t nor $\boldsymbol{u}$, the formal solution is expressed as

$$\boldsymbol{u}(t + \Delta t) = e^{-i\Delta t H}\boldsymbol{u}(t) \quad (15.22)$$

with the finite time interval of Δt. When a first-order approximation of Taylor's expansion is used, for example, one obtains a simple numerical method of

$$\boldsymbol{u}(t + \Delta t) := (I - i\Delta t H)\boldsymbol{u}(t), \quad (15.23)$$

where I is the identity matrix of order M. The local (per step) error of this method is $O((\Delta t)^2)$, and its global error (for time evolution over a fixed period, say, $[0, T]$) is $O(\Delta t)$. This method is classified as an *explicit method* since one only needs to multiply the matrix $I - i\Delta t H$ to advance the solution by one time step. However, this method is rarely used in practice, because its time evolution operator $I - i\Delta t H$ is not unitary and it does not preserve the norm $\|\boldsymbol{u}(t)\|^2 = \sum_{i=1}^{M}|u_i(t)|^2$. In fact, by denoting the Hermitian conjugate of a matrix A by $A^\dagger$, we have

$$(I - i\Delta t H)^\dagger(I - i\Delta t H) = (I + i\Delta t H)(I - i\Delta t H)$$
$$= I + (\Delta t)^2 H^2, \quad (15.24)$$

which shows that the deviation from unitarity, and therefore the possible increase in the norm, is $O((\Delta t)^2)$ per time step.

The simplest method that preserves the norm is the *Crank–Nicolson method* [19]. It is based on the $(1, 1)$ Padé approximation of the exponential function, $e^x \simeq (1 - \frac{x}{2})^{-1}(1 + \frac{x}{2})$, and computes the solution at the next time step as follows:

$$\boldsymbol{u}(t + \Delta t) := \left(I + \frac{i}{2}\Delta t H\right)^{-1}\left(I - \frac{i}{2}\Delta t H\right)\boldsymbol{u}(t). \quad (15.25)$$

The local and global errors of this method are $O((\Delta t)^3)$ and $O((\Delta t)^2)$, respectively. In Eq. (15.25), the application of $\left(I + \frac{i}{2}\Delta t H\right)^{-1}$ to $\boldsymbol{y} \equiv \left(I - \frac{i}{2}\Delta t H\right)\boldsymbol{u}(t)$ can be realized by solving the linear simultaneous equation $\left(I + \frac{i}{2}\Delta t H\right)\boldsymbol{x} = \boldsymbol{y}$. The Crank–Nicolson method is classified as an *implicit method*. It is easy to see that the time evolution operator in this method is unitary:

$$\left\{\left(I + \frac{i}{2}\Delta t H\right)^{-1}\left(I - \frac{i}{2}\Delta t H\right)\right\}^\dagger$$
$$\times \left(I + \frac{i}{2}\Delta t H\right)^{-1}\left(I - \frac{i}{2}\Delta t H\right)$$
$$= \left(I + \frac{i}{2}\Delta t H\right)\left(I - \frac{i}{2}\Delta t H\right)^{-1}$$
$$\times \left(I + \frac{i}{2}\Delta t H\right)^{-1}\left(I - \frac{i}{2}\Delta t H\right)$$
$$= \left(I - \frac{i}{2}\Delta t H\right)^{-1}\left(I + \frac{i}{2}\Delta t H\right)$$
$$\times \left(I + \frac{i}{2}\Delta t H\right)^{-1}\left(I - \frac{i}{2}\Delta t H\right)$$
$$= I, \quad (15.26)$$

where we used the fact that $\left(I - \frac{i}{2}\Delta t H\right)^{-1}$ and $I + \frac{i}{2}\Delta t H$ commute in the second inequality. Since the method is norm-preserving, it is frequently used for wavepacket dynamics [57]. When H depends on t and $\boldsymbol{u}(t)$, one can replace H in Eq. (15.25) with $\frac{1}{2}\left\{H(t, \boldsymbol{u}(t)) + H(t + \Delta t, \boldsymbol{u}(t + \Delta t))\right\}$ to obtain a method that is norm-preserving and has the global error of $O((\Delta t)^2)$.[1] However, it requires solving a nonlinear equation in $\boldsymbol{u}(t + \Delta t)$.[2] If one uses $H(t, \boldsymbol{u}(t))$ instead of H, the resulting equation becomes linear in $\boldsymbol{u}(t + \Delta t)$ and the method is still norm-preserving. However, the global error increases to $O(\Delta t)$.

15.3 Numerical Solvers and Libraries

This section is devoted to an overview of the parallel numerical solvers and libraries useful to nanoscience simulations. The solvers of linear equation and eigenvalue problems are focused here.

Table 15.1 shows a classification of the solvers into the three categories: direct solver, sparse-direct solver, and Krylov solver. The solvers are classified from the two viewpoints; The first view point classifies the solvers into those for dense and sparse matrices. A sparse matrix has a small number of nonzero elements N_{NZ} $(N_{\mathrm{NZ}} \ll M^2)$, while in a dense matrix, most of the elements are nonzero $(N_{\mathrm{NZ}} \simeq M^2)$. The second viewpoint classifies the solvers into 'exact' and 'approximate' solvers. Here, the word of 'exact' means that the result has no error, except for the rounding errors arising from floating-point arithmetic.

15.3.1 Dense Matrix Solvers

Linear Equation Solvers

We first consider the linear simultaneous equations $A\boldsymbol{x} = \boldsymbol{b}$, where $A \in \mathbb{R}^{M \times M}$ is a dense matrix and $\boldsymbol{x}, \boldsymbol{b} \in \mathbb{R}^M$. In the standard procedure for solving these equations, A is first decomposed into the product of a lower triangular matrix L and an upper triangular matrix U as $A = LU$ using *Gaussian elimination*, and then the equations $L\boldsymbol{y} = \boldsymbol{b}$ and $U\boldsymbol{x} = \boldsymbol{y}$ are solved successively [17]. The first step is

called the *LU decomposition*, and the second and third steps are called the *forward/backward substitution*. The first step requires $\frac{2}{3}M^3$ floating-point operations, while the second and the third steps require M^2 operations each. In some applications, one needs to solve multiple linear systems with the same coefficient matrix A and different right-hand-sides $\boldsymbol{b}_1, \ldots, \boldsymbol{b}_K$. In that case, the LU decomposition needs to be computed only once, since it does not depend on the right-hand-side vectors, and only the forward/backward substitution steps are performed K times. This technique reduces the required computational work from $K\left(\frac{2}{3}M^3 + 2M^2\right)$ to $\frac{2}{3}M^3 + 2KM^2$, which is a great saving. It can be used, for example, to compute the time evolution of the Schrödinger equation $i\frac{d\boldsymbol{u}}{dt} = H\boldsymbol{u}$, where the Hamiltonian H does not depend on time, by the Crank–Nicolson method. When A is symmetric positive definite (SPD), the *Cholesky decomposition* $A = R^{\top}R$, where R is an upper triangular matrix, can be used in place of the LU decomposition. The Cholesky decomposition is more economical since its computational work and memory requirement is half that of the LU decomposition.

In high-performance implementations of the LU decomposition, the *blocked algorithm* is used [17]. In the blocked algorithm, the matrix A is partitioned into square blocks of appropriate size, say $b \times b$ (Figure 15.3a), and the LU decomposition is performed by regarding each block as a matrix element. Thus, the multiplication of two elements in the original algorithm translates into a matrix–matrix multiplication. If b is chosen so that three blocks can be stored in the cache memory, this matrix multiplication can be performed entirely within the cache memory and therefore access to the slower main memory is reduced. Blocked algorithms are used extensively in *LAPACK* [2], which is the de facto standard matrix library for sequential or shared-memory parallel machines. Note that LAPACK has various routines for linear equation solution, depending on whether the matrix A is nonsymmetric or SPD, and whether it is fully dense or banded (i.e., $a_{ij} = 0$ unless $-\mathrm{NLD} \leq j - i \leq \mathrm{NUD}$ for some nonnegative integers NLD and NUD). LAPACK routines can also deal with the equations with multiple right-hand-sides, written as $AX = B$.

Blocked algorithms are also used for distributed-memory parallel machines. In this case, each block is allocated to a process running on a computational node using the

TABLE 15.1 Classification of Numerical Linear Algebraic Solvers

	Dense or Sparse	Exact or Approximate
Direct solver	Dense	Exact
Sparse-direct solver	Sparse	Exact
Krylov solver	Sparse	Approximate

---

[1]Note that this is different from the Crank–Nicolson method. The latter method can be written as $\boldsymbol{u}(t + \Delta t) := \left\{I + \frac{i}{2}\Delta t H(t + \Delta t, \boldsymbol{u}(t + \Delta t))\right\}^{-1}\left\{I - \frac{i}{2}\Delta t H(t, \boldsymbol{u}(t))\right\}\boldsymbol{u}(t)$ in this case. Although it has the global error of $O(\Delta t^2)$, it does not preserve the norm of $\boldsymbol{u}(t)$.

[2]When H depends only on t, this nonlinear equation reduces to a linear equation.

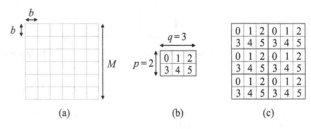

FIGURE 15.3 (a) The coefficient matrix partitioned into blocks, (b) process grid for six processes, and (c) two-dimensional block cyclic distribution.

two-dimensional block cyclic distribution. More specifically, assume that the processes are arranged as a *process grid* of size $p \times q$, where $P = pq$ is the total number of processes. Then, the (I, J) block is allocated to the process at the position $(\mathrm{mod}(I-1, p)+1, \mathrm{mod}(J-1, q)+1)$ in the process grid (Figure 15.3b and c). For distributed-memory machines, *ScaLAPACK* [6] is the de facto matrix library. Although it was developed in the 1990s and some of its routines are no longer suited for modern supercomputers, it can still be used efficiently for solving linear simultaneous equations with a fully dense coefficient matrix. In ScaLAPACK, the distribution of A, as well as that of b, is specified by an argument called the *descriptor*. On the other hand, the process grid is specified by a separate initialization routine. The actual matrix/vector distribution is determined by the combination of them. It is user's responsibility to distribute the matrix/vector across the nodes according to this distribution scheme prior to calling ScaLAPACK solvers. ScaLAPACK offers various auxiliary routines to help the user accomplish this task.

Standard Eigenvalue Solvers

Next, consider the standard eigenvalue problem (SEP) $Ax = \lambda x$. Here, we limit ourselves to the symmetric eigenvalue problem, where A is an $M \times M$ real symmetric or complex Hermitian matrix. In this case, the standard algorithm consists of the three steps: (i) reduction of A to a tridiagonal matrix by orthogonal transformations, $Q^{\top}AQ = T$; (ii) solution of the tridiagonal eigenvalue problem, $Ty_i = \lambda_i y_i$ ($i = 1, 2, \ldots, K$); and (iii) back-transformation of the eigenvectors, $x_i = Qy_i$ ($i = 1, 2, \ldots, K$) [17]. Here, K is the number of wanted eigenvectors. Since Q is chosen to be an orthogonal (or unitary when A is Hermitian) matrix, T has the same eigenvalues as A and $\{\lambda_i\}_{i=1}^{K}$ can be used as the eigenvalue of A. The first step is carried out using the so called *Householder method* and requires $\frac{4}{3}M^3$ work. The third step requires $2M^2K$ work. As for the second step, there are several algorithms such as the *QR algorithm*, *bisection and inverse iteration*, the *divide-and-conquer (DC) algorithm*, and the *MRRR algorithm* [8]. Among these, the DC seems to be the algorithm of choice considering the balance between speed and stability. The computational work of the DC algorithm varies between $O(M^2)$ and $O(M^3)$

depending on the eigenvalue distribution and other factors. ScaLAPCK offers both QR-based (PDSYEV) and DC-based (PDSYEVD) routines. The distribution of the matrix A, as well as that of the eigenvector matrix $X = \{x_1, \ldots, x_K\}$, is the same as that in the linear equation solver.

However, unlike in the case of linear equation solvers, the standard algorithm implemented in ScaLAPACK is not very efficient on modern distributed-memory supercomputers. The main reason is that the tridiagonalization step (i) cannot be cast into a block algorithm appropriately and thus, the cache memory cannot be used efficiently. Also, the non-blocked algorithm results in many interprocessor communications, which causes a severe performance bottleneck. Several new algorithms have been proposed to overcome this difficulty and are adopted by newer eigenvalue solver libraries such as *ELPA* [10] and *EigenExa* [9]. ELPA uses a *two-stage tridiagonalization* approach, which first transforms A into a banded matrix B and then reduces B to a tridiagonal matrix T. In this approach, the first stage can be cast into a fully blocked algorithm. The second stage cannot be cast into a block algorithm, but its computational work is $O(M^2b)$, where b is the bandwidth of B, and accounts for only a small fraction of the total computational work. Once the matrix has been reduced to the tridiagonal form, any of the solvers for the tridiagonal eigenvalue problem can be applied. EigenExa also transforms the input matrix A into a banded matrix B, but it computes the eigenvalues and eigenvectors of B directly using a specially designed DC algorithm without further reducing B to a tridiagonal matrix. The current version of EigenExa uses a penta-diagonal matrix as an intermediate band matrix B. The computational procedures of ScaLAPACK, ELPA and EigenExa are depicted in Figure 15.4. Thanks to the new algorithms, ELPA and EigenExa outperform ScaLAPACK on a modern massively parallel supercomputer like the K computer [13,58]. ELPA uses the same two-dimensional block cyclic distribution as adopted by ScaLAPACK, while EigenExa uses the two-dimensional cyclic distribution.

There are other emerging algorithms for the symmetric eigenvalue problem such as the *block Jacobi method* [27,34] and the *spectral DC* [31]. While not yet publically available as library routines, they have an attractive feature of being *communication-avoiding*, which means that they require much smaller number or amount of interprocessor

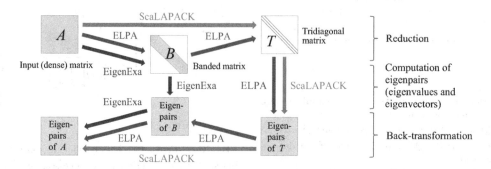

FIGURE 15.4 Computational procedures of ScaLAPACK, ELPA and EigenExa.

communications. Since the cost of communication relative to computation is increasing, they are promising candidates for next-generation supercomputers. In addition, the block Jacobi method has a potential to compute small eigenvalues of a certain kind of matrices to high relative accuracy. Investigation into such a super-accurate solver is an active area of research in numerical linear algebra.

Solvers for the GEP

The symmetric definite GEP $A\boldsymbol{x} = \lambda B\boldsymbol{x}$, where both A and B are real symmetric or complex Hermitian and B is positive definite, also arises frequently in electronic structure calculations. Solution of this problem consists of four steps, namely, (a) Cholesky decomposition of B; $B = R^\top R$, where R is an upper triangular matrix; (b) reduction to the SEP $A'\boldsymbol{z}_i = \lambda \boldsymbol{z}_i$, where $A' = R^{-\top} A R^{-1}$; (c) solution of the SEP; and (d) transformation of the eigenvectors, $\boldsymbol{x}_i = R^{-1}\boldsymbol{z}_i$. We call steps (a), (b), and (d) the *reducer* from GEP to SEP and step (c) the *SEP solver*. ScaLAPACK and ELPA provide the reducer, while all of ScaLAPACK, ELPA, and EigenExa provide the SEP solver.

It is in principle possible to choose the reducer from one library and the SEP solver from another library. Such a hybrid selection can offer a performance benefit in some cases. An obstacle to this is the difference of matrix distribution schemes between different libraries. To resolve this problem, a middleware for the GEP named *EigenKernel* has been developed [58]. EigenKernel provides data conversion routines and allows the user to choose the reducer and the SEP solver freely from different libraries. A performance prediction function, which helps the user to choose the best routines and an appropriate number of computational nodes depending on the target architecture and the problem specification, will be available in the future version of EigenKernel.

15.3.2 Sparse-Direct Solvers

When $A \in \mathbb{R}^{M \times M}$ is a sparse matrix, it is known that its LU factors L and U or the Cholesky factor R (when A is SPD) often inherit the sparsity to some extent. This property can be exploited to save the computational work and memory requirement of the LU or Cholesky decomposition. The resulting linear equation solver is referred to as the *sparse-direct solver* [7,14]. In the following, we focus on the case where A is SPD and discuss sparse-direct solvers based on the Cholesky decomposition.

Structure of the Sparse-Direct Solver

The recurrence formulas to compute the elements of the Cholesky factor R of A can be written as follows:

$$r_{ii} = \sqrt{a_{ii} - \sum_{k=1}^{i-1} r_{ki}^2}, \quad r_{ij} = \left(a_{ij} - \sum_{k=1}^{i-1} r_{ki}r_{kj} \right)/r_{ii} \ (i < j).$$
$$(15.27)$$

These formulas can be derived from the (i,i) and (i,j) entries of $A = R^\top R$, respectively. It is clear from these equations that $r_{ij} \neq 0$ whenever $a_{ij} \neq 0$ unless accidental cancellation occurs. Moreover, it can occur that $r_{ij} \neq 0$ even if $a_{ij} = 0$. Such a nonzero element of R is called *fill-in*. Due to the fill-ins, R usually has much more nonzero elements than the upper triangular part of A. Since fill-ins lead to the increase of the computational cost and memory requirement, it is important in the sparse solver to minimize the number of fill-ins. To this end, *ordering* is applied to A prior to the Cholesky decomposition. In the ordering step, simultaneous permutation of the rows and columns of A is performed with the aim of minimizing the fill-ins. This corresponds to replacing A with $A' = P^\top A P$, where P is some permutation matrix, and does not change the SPD property of A. When A is a matrix arising from discretization of a partial differential equation on a grid or a mesh, ordering is equivalent to renumbering the grid/mesh points. Representative ordering algorithms include *minimum degree ordering* and *nested dissection ordering*.

The step after ordering is *symbolic factorization*. Here, the Cholesky decomposition is performed symbolically without doing floating-point computations and the positions of the fill-ins are calculated. Based on this, the data structure to store the sparse Cholesky factor R is constructed. The computational work and memory requirement for the (sparse) Cholesky decomposition are also estimated at this step. Next comes the *numeric factorization* step, where the Cholesky decomposition is performed numerically using the sparse data structure prepared in the symbolic factorization step. During the decomposition, computations involving zero elements are skipped, and thus the sparsity of the matrix is fully exploited. Finally, *forward/backward substitution* steps are performed. Of course, the sparsity of the Cholesky factor is fully exploited also in these steps. In summary, a typical sparse-direct solver consists of five steps: ordering, symbolic factorization, numeric factorization, and forward/backward substitution.

Parallelization of the Sparse-Direct Solver

To parallelize the sparse-direct solver, we can use the nested dissection ordering. To be concrete, let us consider the Poisson equation discretized on a two-dimensional grid using 5-point finite difference formula. Assume that the grid is partitioned into two regions **A** and **B** and a border **S** (called *separator*) in such a way that the grid points in **A** and **B** are not directly connected. Further, assume that the grid points in region **A** are numbered first, those in **B** are numbered next, and those in **S** are numbered last (Figure 15.5a). Then, it is easy to see that the resulting coefficient matrix has the bordered block diagonal form as shown in Figure 15.5b. Now, apply Gaussian elimination (which is equivalent to the Cholesky decomposition) to the first block row of A, which corresponds to the grid points in region **A**. Then, it is observed that the second block row, corresponding to the grid points in **B**, is not modified due to this elimination

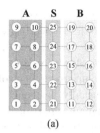
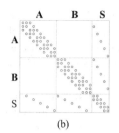
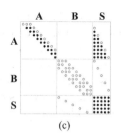

(a)   (b)   (c)

FIGURE 15.5 (a) Ordering of the grid points, (b) the coefficient matrix, where circles represent nonzero elements, and (c) the matrix after the elimination of the first block row, where black circles represent modified elements or fill-ins.

(Figure 15.5c). This means that we do not need to wait until the first block row has been eliminated to start the elimination of the second block row. In other words, the elimination of the first block row and the second block row can be done *in parallel*. This is the source of parallelism of the nested dissection ordering. Note that the last block row corresponding to the separator **S** can be eliminated only after both the first and the second block rows have been eliminated. Thus, this part is sequential in nature.

This parallelization strategy can be applied recursively by subdividing each of **A** and **B** into two subregions and a separator. Then, the matrix will have a recursive bordered block diagonal form, and the parallelism of the Cholesky decomposition can be represented by a binary tree, in which the leaves correspond to the smallest subregions that can be eliminated in parallel and the root corresponds to the coarsest border **S**. Since the separators cannot be eliminated in parallel, it is optimal for parallel efficiency to partition the grid so that the number of grid points belonging to the separators is minimized and each subregion has roughly the same number of grid points. This problem is studied well under the name of *graph partitioning*, and there are efficient libraries such as *METIS* [30], *ParMETIS* [37], and *SCOTCH* [43].

Representative parallel sparse-direct solver libraries include *SuperLU* [56], *UMFPACK*, which is part of the SuiteSparse package [54], and *PARDISO* [36]. In addition to SPD linear systems, they can solve linear systems with symmetric indefinite or nonsymmetric coefficient matrices. Among these libraries, a version of PARDISO implemented in *Intel Math Kernel Library* (MKL) is the easiest to use for beginners because the subroutine interface is simple and all matrix types—SPD, symmetric indefinite and nonsymmetric—can be dealt with via only one subroutine.

In sparse-direct solvers, most of the storage and computational effort are consumed to store and update the *frontal matrix*, which is a part of the entire matrix currently being updated. As a technique to reduce this storage requirement and computational work, use of the *block low-rank approximation* of the frontal matrix is advocated recently [1]. While this technique is expected to greatly increase the size of the matrices solvable by sparse-direct methods, the resulting *LU* or Cholesky decompositions will be only approximate ones. Hence, they will be combined with the Krylov subspace solvers and will be used as a (very efficient) preconditioner for the latter.

15.3.3 Krylov Solvers

Krylov solvers are iterative methods that generate a sequence of improving approximate solutions, and find approximate solutions over the following Krylov subspaces[3] (see Figure 15.6):

$$K_n(A, \boldsymbol{v}) = \text{span}\{\boldsymbol{v}, A\boldsymbol{v}, A^2\boldsymbol{v}, \ldots, A^{n-1}\boldsymbol{v}\}, \quad (15.28)$$

where $A \in \mathbb{R}^{M \times M}$ and $\boldsymbol{v} \in \mathbb{R}^M$. Krylov solvers have recently received attention, possibly due to the following features:

1. Approximate solution with variable accuracy

   It is possible to control the accuracy of the computed solution by monitoring the accuracy at each iteration step. When one desires low accuracy (e.g., several digits), the computation time tends to be much faster than that of sparse-direct solvers by adopting a small number of iterations.

2. No need for explicitly storing matrix

   When considering extremely large matrix so that it is impossible to store the matrix in the memory in a computer, direct methods and sparse-direct methods become unavailable. Only choice would be Krylov solvers because the solvers only require the result of matrix-vector multiplication $A\boldsymbol{v}$, which indicates no requirement of storing A in the memory.

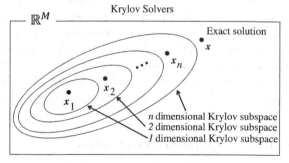

$$x_n \begin{cases} n\text{th approximate solution of } A\boldsymbol{x} = \boldsymbol{b} \\ n\text{th approximate eigenvector of } A\boldsymbol{x} = \lambda\boldsymbol{x} \end{cases}$$

FIGURE 15.6 The idea of Krylov solvers.

---

[3]Krylov subspace can also be written as $K_n(\boldsymbol{v}, \boldsymbol{v}) = \{c_1\boldsymbol{v} + c_2 A\boldsymbol{v} + \cdots + c_{n-1}A^{n-1}\boldsymbol{v} : c_1, c_2, \ldots, c_{n-1} \in \mathbb{R}\}$.

This subsection provides the brief theory of Krylov solvers for linear systems and eigenvalue problems. In this subsection, $\|\boldsymbol{v}\|$ is a vector 2-norm defined by $\|\boldsymbol{v}\| = \sqrt{\boldsymbol{v}^{\mathrm{T}}\boldsymbol{v}}$.

Krylov Solvers for Linear Systems $A\boldsymbol{x} = \boldsymbol{b}$

Let $\boldsymbol{x}_n$ be an approximate solution at nth iteration step. Then, the residual vector is defined as $\boldsymbol{r}_n = \boldsymbol{b} - A\boldsymbol{x}_n$. If $\boldsymbol{x}_n$ is the exact solution of the linear systems $A\boldsymbol{x} = \boldsymbol{b}$, then $\boldsymbol{r}_n = \boldsymbol{0}$.

Krylov solvers, usually known as Krylov subspace methods, for solving linear systems find nth approximate solution $\boldsymbol{x}_n$ over Krylov subspace:

$$\boldsymbol{x}_n = \boldsymbol{x}_0 + \boldsymbol{z}_n, \quad \boldsymbol{z}_n \in K_n(A, \boldsymbol{r}_0),$$

where $\boldsymbol{x}_0$ is an initial guess, often chosen as either the zero vector or a random vector.

Determining $\boldsymbol{z}_n \in K_n(A, \boldsymbol{r}_0)$ yields various Krylov solvers. When matrix A is symmetric (positive definite[4]), the best known Krylov solver is *the conjugate gradient (CG) method*. The CG method determines $\boldsymbol{z}_n$ such that $\boldsymbol{r}_n \perp K_n(A, \boldsymbol{r}_0)$. (This is referred to as *Ritz–Galerkin* approach.) The convergence rate of the CG method depends on the condition number $\kappa(A)$[5]:

$$\|\boldsymbol{x} - \boldsymbol{x}_n\|_A \le 2\left(\frac{\sqrt{\kappa(A)} - 1}{\sqrt{\kappa(A)} + 1}\right)^n \|\boldsymbol{x} - \boldsymbol{x}_0\|_A, \quad (15.29)$$

where $\|\boldsymbol{x} - \boldsymbol{x}_n\|_A$ is a kind of a distance between the exact solution $\boldsymbol{x}$ and the nth approximate solution $\boldsymbol{x}_n$.[6] This indicates that if $\kappa(A) = \lambda_{\max}/\lambda_{\min} = 1$, the CG method yields the exact solution at only one iteration.

As may be expected from Eq. (15.29), the convergence rate of Krylov solvers depends strongly on spectral property (distribution of eigenvalues) of the coefficient matrix. It is therefore natural to try to transform the original system into one having the same solution but more favorable spectral properties. If K is a nonsingular matrix, the transformed linear system

$$K^{-1}A\boldsymbol{x} = K^{-1}\boldsymbol{b}$$

has the same solution as the original one. Hence, we can obtain the solution by applying Krylov solvers to the above system. The matrix K is referred to as *preconditioner*, and a good preconditioner satisfies the following properties:

1. K is close to A;
2. $K^{-1}\boldsymbol{z}$ is readily obtained for any vector $\boldsymbol{z}$.

Of a large number of preconditioners, they can be roughly classified into three groups: *incomplete matrix factorization*

preconditioner, polynomial preconditioners, approximate inverse preconditioners. For details of the preconditioners, see, for example, [5].

If matrix A is real nonsymmetric (or complex nonsymmetric), the best known Krylov solvers are *Bi-Conjugate Gradient Stabilized (BiCGStab) method* [59] and *the Generalized Minimal Residual (GMRES) method* [41]. If matrix A is complex symmetric ($A = A^{\mathrm{T}} \ne \bar{A}^{\mathrm{T}}$), well-known Krylov solvers are *the Conjugate Orthogonal Conjugate Gradient (COCG) method* [60] and *the Conjugate Orthogonal Conjugate Residual (COCR) method* [51]. For further details of Krylov solvers for linear systems, see [4,40].

In nanomaterials science and particle physics (lattice quantum chromodynamics [12]), the solution of shifted linear systems

$$(A + \sigma_k I)\boldsymbol{x}^{(k)} = \boldsymbol{b} \quad (k = 1, 2, \ldots, m) \quad (15.30)$$

has received much attention, where σ_k is scaler, I is the identity matrix, and m is the number of linear systems. In this case, Krylov solvers are significantly useful because of the so-called shift invariance property of Krylov subspaces: $K_n(A + \sigma I, \boldsymbol{v}) = K_n(A, \boldsymbol{v})$. The property implies that in order to obtain all the solutions $\boldsymbol{x}^{(1)}, \boldsymbol{x}^{(2)}, \ldots, \boldsymbol{x}^{(m)}$, Krylov solvers, under an ideal condition, require the computational cost of solving only one linear system, whereas direct solvers require the computational cost of solving not one but all the linear systems. The trick is that once we generate a Krylov subspace $K_n(A + \sigma_1 I, \boldsymbol{v})$ for solving $(A + \sigma_1 I)\boldsymbol{x}^{(1)} = \boldsymbol{b}$, we can reuse $K_n(A + \sigma_1 I, \boldsymbol{v})$ to solve the rest of linear systems $(A + \sigma_k I)\boldsymbol{x}^{(k)} = \boldsymbol{b}$ $(k = 2, 3, \ldots, m)$. For the theory and a generalization, see [50] and references therein.

Below is the list of software libraries of Krylov solvers for linear systems.

- PETSc: Portable, Extensible Toolkit for Scientific Computation [38]
- Lis: Library of Iterative Solvers for Linear Systems [29].

Krylov Solvers for Eigenvalue Problems $A\boldsymbol{x} = \lambda B\boldsymbol{x}$

In this subsection, we consider eigenvalue problems $A\boldsymbol{x} = \lambda B\boldsymbol{x}$. The problem is referred to as *generalized eigenvalue problems* (GEP) and the problem with $B = I$ (the identity matrix) is referred to as *standard eigenvalue problems* (SEP). Solution of large eigenvalue problems plays a fundamental role in a rich variety of scientific fields such as computational science, data science, artificial intelligence, and many others. Many scientific applications have various needs to compute eigenvalues (and the corresponding eigenvectors). Below is the list of the needs and the solutions.

P1: all the eigenvalues

If one desires all the eigenvalues, direct solvers such as the QR method for $B = I$ and the QZ method are the best. A large number of studies have been developed in accelerating manner and parallelization manner.

---

[4] If matrix A is symmetric (Hermitian) and all the eigenvalues are positive, then it is referred to as symmetric positive definite matrix.

[5] The condition number of symmetric positive definite matrix A is given by $\lambda_{\max}/\lambda_{\min}$, where $\lambda_{\max}$ and $\lambda_{\min}$ are maximum/minimum eigenvalues of A.

[6] $\|\boldsymbol{v}\|_A$ is defined by $\sqrt{\boldsymbol{v}^{\mathrm{T}}A\boldsymbol{v}}$.

P2: all the eigenvalues in a given region of the complex plane

For this problem, Sakurai–Sugiura method [42] and FEAST [39] are well known. These methods are based on contour integral of the resolvent of $(A - \sigma B)$.

If $B = I$, shifted linear systems (15.30) emerge as a subproblem, and the shifted linear systems are efficiently solved by Krylov solvers for shifted linear systems. If $B \neq I$, the corresponding linear systems are referred to as *generalized shifted linear systems*. See [50] for efficient solution of generalized shifted linear systems.

P3: largest/smallest eigenvalues

For problem P3, the Lanczos method for symmetric (Hermitian) matrix A and the Arnoldi method for nonsymmetric (non-Hermitian) matrix are the best known Krylov solvers. The Lanczos/Arnoldi methods find approximate eigenvectors over Krylov subspace $K_n(A, \boldsymbol{v})$.

The key idea is to generate matrix $V_n = [\boldsymbol{v}_1, \ldots, \boldsymbol{v}_n]$ such that $\boldsymbol{v}_1, \ldots, \boldsymbol{v}_n$ are orthonormal basis vectors of $K_n(A, \boldsymbol{v})$. Then, the Lanczos/Arnoldi methods find approximate eigenvectors from a linear combination of the basis vectors, that is, $\boldsymbol{x}_n = \sum_{k=1}^{n} c_k \boldsymbol{v}_k$. In order to determine the coefficients $c_1, \ldots, c_n$, *Rits-Galerkin* approach is adopted, that is, determining $c_1, \ldots, c_n$ such that the residual vector $\boldsymbol{r}_n (= A\boldsymbol{x}_n - \theta \boldsymbol{x}_n)$ satisfies $\boldsymbol{r}_n \perp K_n(A, \boldsymbol{v})$, resulting in much smaller eigenvalue problems of the form $V^{\mathrm{T}} A V \boldsymbol{c} = \theta \boldsymbol{c}$, where the vector $\boldsymbol{c} = [c_1, \ldots, c_n]^{\mathrm{T}}$ corresponds to the coefficients $c_1, \ldots, c_n$ of the linear combination. Largest/smallest eigenvalues of $V^{\mathrm{T}} A V \boldsymbol{c} = \theta \boldsymbol{c}$ well approximate the largest/smallest eigenvalues of $A\boldsymbol{x} = \lambda \boldsymbol{x}$. It is known that the Lanczos/Arnoldi methods also well approximate the second, third, and several largest/smallest eigenvalues. The iteration number n is increased until the residual norm $\|\boldsymbol{r}_n\|$ is sufficiently small. In practice, $V^{\mathrm{T}} A V$ is not computed. For details of Lanczos/Arnoldi methods, see, for example, [3]. If A is symmetric and B is symmetric positive definite, another well-known algorithm is *Locally Optimal Block Preconditioned Conjugate Gradient (LOBPCG) method* [25].

P4: some eigenvalues around a desired target point

For problem P4, when one desires eigenvalues (and eigenvectors) around the target point σ in the complex plane, useful approach is to consider applying the Lanczos/Arnoldi methods to the transformed eigenvalue problem $(A - \sigma I)^{-1} \boldsymbol{x} = (\lambda - \sigma)^{-1} \boldsymbol{x} (\Leftrightarrow A\boldsymbol{x} = \lambda \boldsymbol{x})$.

Let $\theta = (\lambda - \sigma)^{-1}$ be the largest eigenvalue θ of the transformed eigenvalue problem, then λ closest to σ is given by $\lambda = \theta^{-1} + \sigma$. In this case,

the Lanczos/Arnoldi methods require the solution of linear systems of the form $(A - \sigma I)\boldsymbol{x} = \boldsymbol{b}$ at each iteration step, which is usually solved by Krylov solvers with some preconditioners. The resulting algorithm is referred to as *the shift-and-invert Lanczos method* and *the shift-and-invert Arnoldi method*. Another well-known algorithm is the *Jacobi–Davidson method* [48].

P5: the k-th eigenvalue with a given number k

For problem P5, the explanation is given in detail. Let $\lambda_1 \leq \lambda_2 \leq \cdots \leq \lambda_k \leq \cdots \leq \lambda_M$ be the eigenvalues of M-by-M symmetric matrix. Then, problem P5 is to compute λ_k and the corresponding eigenvector. Naive approach is to compute all eigenvalues by direct solvers. However, when matrix is large and sparse, direct solvers do not work due to memory limitation.

In order to solve the problem, when matrix A is symmetric and B is symmetric positive definite (e.g., $B = I$), the combination of *Sylvester's law of inertia*,[7] sparse-direct solvers, and the shift-and-invert Lanczos method for problem P4 is promising. The key idea is to compute the negative inertia of $A - \sigma I$ (i.e., $\nu(A - \sigma I)$) with an initial guess σ in order to know that the target eigenvalue λ_k is less than or greater than σ. If $k < \nu(A - \sigma I)$, then σ_{new} is chosen as $\sigma_{\mathrm{new}} < \sigma$ and then compute $\nu(A - \sigma_{\mathrm{new}} I)$. If $k > \nu(A - \sigma I)$, then σ_{new} is chosen as $\sigma_{\mathrm{new}} > \sigma$ and then compute $\nu(A - \sigma_{\mathrm{new}} I)$. As a result, it is possible to shrink the interval including λ_k. When the interval shrinks enough, the shift-and-invert Lanczos method is used to obtain all eigenvalues in the small interval. Sparse-direct solvers play an important role in computing the inertia because the sparse-direct solvers yield $A - \sigma I = LDL^{\mathrm{T}}$, where D is the diagonal matrix whose diagonal elements have full information of the inertia; that is, the number of negative (or positive) diagonal elements in D corresponds to $\nu(A - \sigma I)$ (or $\pi(A - \sigma I)$). For the solution of the GEPs, see [28].

For details of algorithms and theory of large and sparse eigenvalue problems, see [3], and below is the list of software libraries of Krylov solvers for eigenvalue problems.

- SLEPc—Scalable Library for Eigenvalue Problem Computations [49]
- FEAST eigensolver [11]
- z-Pares: Parallel Eigenvalue Solver [63].

---

[7]Let $\nu(A), \zeta(A), \pi(A)$ be the number of negative, zero, positive eigenvalues of A, and let X be nonsingular matrix. Then, Sylvester's law of inertia guarantees that $\nu(A) = \nu(X^{\mathrm{T}} A X)$, $\zeta(A) = \zeta(X^{\mathrm{T}} A X)$, and $\pi(A) = \pi(X^{\mathrm{T}} A X)$.

15.4 Application in Nanoscience

This section reviews two application studies of large-scale electronic state calculations on supercomputer.

15.4.1 Organic Device Material

The first example is organic device material calculated by a large-scale electronic state calculation code ELSES [21–24], an order-N electronic state calculation code with ab-initio-based modeled (tight-binding) theory. The order-N benchmark is shown in Figure 15.2 of Section 15.2.2. In short, the code is not based on the GEP of Eq. (15.6) but on the generalized shifted linear equations (GSLE) of

$$(zB - A)\boldsymbol{x} = \boldsymbol{b}, \tag{15.31}$$

a theoretical extension of Eq. (15.30). Here, z is a (complex) energy value. The vector $\boldsymbol{b}$ is an input, and the vector $\boldsymbol{x}$ is the solution vector. The GSLE are solved by an iterative Krylov subspace solver [23]. The use of GSLE results in the Green's (propagation) function formalism, since the solution of Eq. (15.31) is written formally as $\boldsymbol{x} = G\boldsymbol{b}$ with the Green's function $G \equiv (zB - A)^{-1}$. The Green's function holds the relationship of

$$G(z) = \sum_{k}^{M} \frac{\boldsymbol{y}_k \boldsymbol{y}_k^{\mathrm{T}}}{z - \lambda_k} \tag{15.32}$$

and gives the physical quantities.

Figure 15.7 shows the strong scaling benchmark on Oakforest-PACS, a supercomputer in Joint Center for Advanced High Performance Computing (JCAHPC; http://jcahpc.jp/). The Oakforest-PACS system has 8,208 compute nodes with Intel Xeon Phi (Knights Landing) processor. The benchmark was carried out with up to 2,048 nodes or a quarter of Oakforest-PACS. The simulated system is a crystalline diamond sample with N=106,168,320

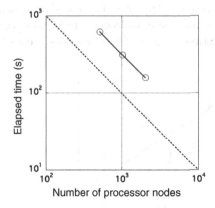

FIGURE 15.7 The strong scaling benchmark of ELSES with P = 512, 1024, 2048 nodes on OFP. The simulated system is a crystalline diamond sample with $N = 106, 168, 320$ atoms in the simulation cell. The solid line is the benchmark data and the ideal scaling line is drawn as dotted line.

atoms in the simulation cell. The elapsed times at $P = 512$ and $P = 2048$ give the parallel efficiency of the strong scaling α as

$$\alpha = \frac{T(512)}{T(2048)} \bigg/ \frac{2048}{512} = 0.974. \tag{15.33}$$

The strong scaling property is found also in the benchmark of the K computer with up to the whole system [21].

Application studies were carried out for organic device material. Organic device material is semiconductor and forms the foundation of flexible devices such as flexible displays [15], flexible solar cells [61], and human-friendly wearable electronics [44]. Since complicated disordered structure is crucial for device properties such as device performance or device lifetime, large-scale simulations in 10–100 nm scales are required. Figure 15.8a shows disordered condensed polymers of poly-(phenylene-ethynylene), a typical organic material. The data analysis reveals that local networks with a couple of polymers play a crucial role in the propagation of electronic wave through a disordered structure. In addition, Figure 15.8b shows a simulation result quantum wavepacket dynamics of hole, in which the fundamental equation is an effective time-dependent equation in the form of $\partial_t \Psi = H_{\mathrm{eff}} \Psi$ for the hole wavepacket of $\Psi \equiv \Psi(\boldsymbol{r}, t)$. More recently, a data scientific research was carried out for the classification of disordered polymers [22]. The methods will realize the simulation of whole organic devices in near future with next-generation supercomputers.

15.4.2 Hydrogen Production from Water with Metal Nanoparticles

The second application study is the hydrogen production from water with metal nanoparticles [45]. The used code is LDC-DFT (Lean Divide-and-Conquer Density Functional Theory), a quantum molecular dynamics simulator based on lean DC DFT algorithm [33,46]. Briefly, the algorithm represents the three-dimensional space Ω as a union of overlapping spatial domains, $\Omega = \cup_\alpha \Omega_\alpha$. Global charge density ρ is calculated by a real-space multigrid method as linear combinations of domain local densities ρ_α. On the other hand, a plane-wave basis is used to represent local charge density ρ_α within each domain Ω_α, which takes advantage of a highly efficient numerical implementation based on fast Fourier transform. Each domain Ω_α is further decomposed into core domain $\Omega_{0\alpha}$ and the surrounding buffer layer Γ_α. $\Omega_{0\alpha}$ is a non-overlapping with other core domains (i.e., $\Omega_{0\alpha} \cap \Omega_{0\beta} = 0$ $(\alpha \neq \beta)$). By introducing a buffer layer of an appropriate size, it is possible not only to obtain accurate ρ_α but also to reduce the calculation cost.

Figure 15.9 shows a scalability test on Oakforest-PACS (OFP). The model is explained here. The total system contains liquid water consisting of 648 atoms in the cubic cell of side length 1.9 nm. We first took time to choose the appropriate size of buffer layer that enable the LDC-DFT potential energy to converge within 10^{-3} a.u. per atom. The total system Ω was divided into $4 \times 4 \times 4 = 64\Omega_{0\alpha}$

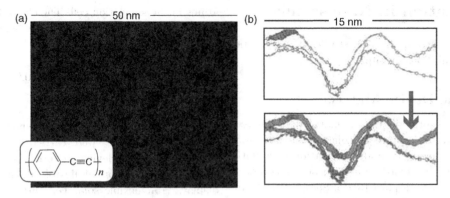

FIGURE 15.8   (a) Disordered condensed polymers of poly-(phenylene-ethynylene) (PPE) [21]. The atomic structure of the polymer is shown in the inset. (b) Example of quantum wavepacket dynamics of hole through three polymers of PPE, or a part of (a) [21]. The initial wavepacket is localized on one polymer at $t = 0$ (upper panel) and propagates into three other two polymers at $t = 5$ ps (lower panel).

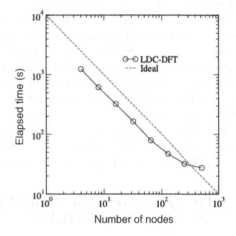

FIGURE 15.9   Strong scaling benchmark of LDC-DFT on OFP. The ideal scaling line is drawn for eye guide.

core domains. The size of the domain containing the buffer layer (i.e., Ω_α) was a cube with a side of 1.3 nm. Each Ω_α domain is solved in parallelism. Fixing the domain size, we investigated the strong scaling property. The elapsed time was measured for the electronic state calculation for a given atomic structure. The numbers of the used processor

nodes are $P = 4, 8, 16, 32, 64, 128, 256,$ and 512. The calculations were carried out with different numbers of OMP threads. The number of MPI processes is set to be $64 \times$ (# nodes) / (# OMP threads). It was found that our code shows the fastest result, when the number of OMP threads is two.

Figure 15.10 shows the application study of the hydrogen production from water with metal nanoparticles [45]. Hydrogen production from water using Al particles could provide a renewable energy cycle. However, its practical application is hampered by the low reaction rate and poor yield. The simulation result is shown in Figure 15.10a, in which hydrogen production is observed from water using Li-Al particles. The total number of atoms is 16,611 atoms. The produced hydrogens are drawn as red balls. As a systematic investigation, simulations in different temperatures were carried out and the hydrogen production rate is plotted as a function of temperature, which is shown in Figure 15.10b. The result indicates that orders-of-magnitude faster reactions with higher yields can be achieved by alloying Al particles with Li. A key nanostructural design is identified as the abundance of neighboring Lewis acid–base pairs, where water-dissociation and hydrogen production require very small activation energies. These reactions are facilitated by

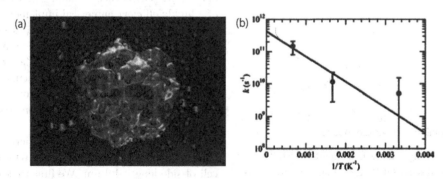

FIGURE 15.10   (a) Hydrogen production from water using Li-Al particles. The total number of atoms is 16,611 atoms. The produced hydrogens are drawn as balls. (b) Hydrogen production rate as a function of temperature (circles with error bars), where the solid line is the best fit to the Arrhenius equation.

charge pathways across Al atoms that collectively act as a "superanion" and a surprising autocatalytic behavior of bridging Li-O-Al products. Furthermore, the dissolution of Li atoms into water produces a corrosive basic solution that inhibits the formation of a reaction-stopping oxide layer on the particle surface, thereby increasing the yield. These atomistic mechanisms not only explain recent experimental findings but also predict the scalability of this hydrogen-on-demand technology at industrial scales.

15.5 Conclusion and Future Aspect

The present chapter gives a brief overview of the mathematical foundation of large-scale electronic state calculations and their nanomaterial application studies on supercomputers. The application studies demonstrate the importance of large-scale electronic state calculation for the industrial material design. In future or in the coming post-Moore era, the application-algorithm-architecture co-design and the knowledge of numerical methods will be more important among nanoscience researchers for fruitful simulations.

References

1. P. Amestoy, C. Ashcraft, O. Boiteau, A. Buttari, J.-Y. L'Excellent, and C. Weisbecker. Improving multifrontal methods by means of block low-rank representations. *SIAM J. Sci. Comput.*, 37(3)/A1451–A1474, 2015.

2. E. Anderson, Z. Bai, C. Bischof, S. Blackford, J. Demmel, J. Dongarra, J. Du Croz, A. Greenbaum, S. Hammarling, A. McKenney, and D. Sorensen. *LAPACK Users' Guide*. SIAM, Philadelphia, PA 1999.

3. Z. Bai, J. Demmel, J. Dongarra, A. Ruhe, and H. A. Van der Vorst (Eds.) *Templates for the Solution of Algebraic Eigenvalue Problems: A Practical Guide*. SIAM, Philadelphia, PA, 2000.

4. R. Barrett, M. Berry, T. F. Chan, J. Demmel, J. Donato, J. Dongarra, V. Eijkhout, R. Pozo, C. Romine, and H. A. Van der Vorst. *Templates for the Solution of Linear Systems: Building Blocks for Iterative Methods, 2nd Edition*. SIAM, Philadelphia, PA, 1994.

5. M. Benzi. Preconditioning techniques for large linear systems: A survey. *J. Comput. Phys.*, 182(2)/418–477, 2002.

6. L. S. Blackford, J. Choi, A. Cleary, E. D'Azevedo, J. Demmel, I. Dhillon, J. Dongarra, S. Hammarling, G. Henry, A. Petitet, K. Stanley, D. Walker, and R. C. Whaley. *ScaLAPACK Users' Guide*. SIAM, Piladelphia, PA 1997.

7. T. A. Davis. *Direct Methods for Sparse Linear Systems*. SIAM, Piladelphia, PA 2006.

8. I. S. Dhillon, B. N. Parlett, and C. Vömel. The design and implementation of the MRRR algorithm. *ACM Trans. Math. Softw*, 32(4)/533–560, 2006.

9. EigenExa. www.r-ccs.riken.jp/labs/lpnctrt/en/projects/eigenexa/.

10. ELPA. https://elpa.mpcdf.mpg.de/.

11. FEAST. http://www.feast-solver.org/.

12. A. Frommer, T. Lippert, B. Medeke, and K. Schilling (Eds.) *Numerical Challenges in Lattice Quantum Chromodynamics*. Springer-Verlag, Berlin, Heidelberg, 2000.

13. T. Fukaya and T. Imamura. Performance evaluation of the EigenExa eigensolver on Oakleaf-FX: Tridiagonalization versus pentadiagonalization. In *IPDPS Workshops*, pp. 960–969. IEEE Computer Society, 2015.

14. K. Gallivan, C. Romine, R. Plemmons, A. Sameh, James M. Ortega, Robert G. Voigt, M. Heath, and E. Ng. *Parallel Algorithms for Matrix Computations*. SIAM, Piladelphia, PA, 1990.

15. G. H. Gelinck, H. E. Huitema, E. van Veenendaal, E. Cantatore, L. Schrijnemakers, J. B. van der Putten, T. C. Geuns, M. Beenhakkers, J. B. Giesbers, B. H. Huisman, E. J. Meijer, E. M. Benito, F. J. Touwslager, A. W. Marsman, B. J. van Rens, and D. M. de Leeuw. Flexible active-matrix displays and shift registers based on solution-processed organic transistors. *Nat. Mater.*, 36/106–110, 2004.

16. M. J. Gillan, D. R. Bowler, A. S. Torralba, and T. Miyazaki. Order-N first-principles calculations with the conquest code. *Comp. Phys. Commun.*, 177/14, 2007.

17. G. H. Golub and C. F. Van Loan. *Matrix Computations*, 3rd edition. Johns Hopkins University Press, Baltimore, MD 2012.

18. G. Hager and G. Wellein. *Introduction to High Performance Computing for Scientists and Engineers*. CRC Press, Boca Raton, FL, 2010.

19. E. Hairer, S. P. Nørsett, and G. Wanner. *Solving Ordinary Differential Equations I*. Springer, Heidelberg, 1993.

20. P. Hohenberg and W. Kohn. Inhomogeneous electron gas. *Phys. Rev.*, 136:B864–B871, Nov 1964.

21. T. Hoshi, H. Imachi, K. Kumahata, M. Terai, K. Miyamoto, K. Minami, and F. Shoji. Extremely scalable algorithm for 10^8-atom quantum material simulation on the full system of the K computer. *Proceedings Of the ScalA16 in SC16*, pp. 33–40, 2016.

22. T. Hoshi, H. Imachi, A. Kuwata, K. Kakuda, T. Fujita, and H. Matsui. Numerical aspect of large-scale electronic state calculation for flexible device material. *Preprint: https://arxiv.org/abs/1808.02027*.

23. T. Hoshi, S. Yamamoto, T. Fujiwara, T. Sogabe, and S.-L. Zhang. An order-N electronic structure theory

with generalized eigenvalue equations and its application to a ten-million-atom system. *J. Phys. Condens. Matter*, 24/165502, 1–5, 2012.

24. H. Imachi, S. Yokoyama, T. Kaji, Y. Abe, T. Tada, and T. Hoshi. One-hundred-nm-scale electronic structure and transport calculations of organic polymers on the k computer. *AIP Conference Proceedings*, pp. 1790/020010, 1–4, 2016.

25. A. Knyazev. Toward the optimal preconditioned eigensolver: Locally optimal block preconditioned conjugate gradient method. *SIAM J. Sci. Comput.*, 23(2)/517–541, 2001.

26. W. Kohn. Density functional and density matrix method scaling linearly with the number of atoms. *Phys. Rev. Lett.*, 76/3168–3171, 1996.

27. S. Kudo, Y. Yamamoto, M. Bečka, and M. Vajteršic. Performance analysis and optimization of the parallel one-sided block Jacobi SVD algorithm with dynamic ordering and variable blocking. *Concurr. Comput. Pract. Exp.*, 29(9)/24pp, 2017.

28. D. Lee, T. Hoshi, T. Sogabe, Y. Miyatake, and S.-L. Zhang. Solution of the k-th eigenvalue problem in large-scale electronic structure calculations. *J. Comput. Phys.*, 371(15)/618–632, 2018.

29. Lis. https://www.ssisc.org/lis/.

30. METIS. http://glaros.dtc.umn.edu/gkhome/metis/metis/overview.

31. Y. Nakatsukasa and N. J. Higham. Stable and efficient spectral divide and conquer algorithms for the symmetric eigenvalue decomposition and the SVD. *SIAM J. Sci. Comput*, 35(3)/A1325–A1349, 2013.

32. K. Naono, K. Teranishi, J. Cavazos, and R. Suda (eds.) *Software Automatic Tuning: From Concepts to State-of-the-Art Results*. Springer, New York, 2010.

33. K. Nomura, R. K. Kalia, A. Nakano, P. Vashishta, Shimamura K., F. Shimojo, Kunaseth M., P. C. Messina, and N. A. Romero. Metascalable quantum molecular dynamics simulations of hydrogen-on-demand. *Proceedings Of the SC14*, pp. 661–673, 2014.

34. G. Okša, Y. Yamamoto, M. Bečka, and M. Vajteršic. Asymptotic quadratic convergence of the parallel block-Jacobi EVD algorithm with dynamic ordering for Hermitian matrices. *BIT Numerical Mathematics*, 58/10991123, 2018.

35. T. Ozaki. O(N) Krylov-subspace method for large-scale ab initio electronic structure calculations. *Phys. Rev. B*, 74/245101, 2006.

36. PARDISO. https://www.pardiso-project.org/.

37. ParMETIS. http://glaros.dtc.umn.edu/gkhome/metis/parmetis/overview.

38. PETSc. http://www.mcs.anl.gov/petsc/.

39. E. Polizzi. Density-matrix-based algorithm for solving eigenvalue problems. *J. Comput. Appl. Math.*, 79/115112/6pp, 2009.

40. Y. Saad. *Iterative Methods for Sparse Linear Systems*. SIAM, Philadelphia, PA, 2003.

41. Y. Saad and M. H. Schultz. GMRES: A generalized minimal residual algorithm for solving nonsymmetric linear systems. *SIAM J. Sci. and Stat. Comput.*, 7(3)/856–869, 1986.

42. T. Sakurai and H. Sugiura. A projection method for generalized eigenvalue problems using numerical integration. *J. Comput. Appl. Math.*, 159(1)/119–128, 2003.

43. SCOTCH. https://www.labri.fr/perso/pelegrin/scotch/.

44. T. Sekitani and T. Someya. Human-friendly organic integrated circuits. *Mater. Today*, 14/398–407, 2011.

45. K. Shimamura, F. Shimojo, R. K. Kalia, A. Nakano, K. Nomura, and P. Vashishta. Hydrogen-on-demand using metallic alloy nanoparticles in water. *Nano Lett.*, 14/4090–4096, 2014.

46. F. Shimojo, Hattori S., R. K. Kalia, Kunaseth M., Mou W., A. Nakano, K. Nomura, Ohmura S., Rajak P., Shimamura K., and P. Vashishta. A divide-conquer-recombine algorithmic paradigm for large spatiotemporal quantum molecular dynamics simulations. *J. Chem. Phys.*, 140/18A529, 2014.

47. C.-K. Skylaris, P. D. Haynes, A. A. Mostofi, and M. C. Payne. Introducing onetep: Linear-scaling density functional simulations on parallel computers. *J. Chem. Phys.*, 122/084119, 2005.

48. G. Sleijpen. A Jacobi–Davidson iteration method for linear eigenvalue problems. *SIAM J. Matrix Anal. & Appl.*, 17(2)/401–425, 1996.

49. SLEPc. http://slepc.upv.es/.

50. T. Sogabe, T. Hoshi, S.-L. Zhang, and T. Fujiwara. Solution of generalized shifted linear systems with complex symmetric matrices. *J. Comput. Phys.*, 231(17)/5669–5684, 2012.

51. T. Sogabe and S.-L. Zhang. A COCR method for solving complex symmetric linear systems. *J. Comput. Appl. Math.*, 199(2)/297–303, 2007.

52. J. M. Soler, E. Artacho, J. D. Gale, A. Garcia, J. Junquera, P. Ordejon, and D. Sanchez-Portal. The SIESTA method for ab initio order-N materials simulation. *J. Phys. Condens. Matter*, 14/2745, 2002.

53. T. Sterling, M. Anderson, and M. Brodowicz. *High Performance Computing: Modern Systems and Practices*. Morgan Kaufmann, Burlington, MA, 2017.

54. SuiteSparse. http://faculty.cse.tamu.edu/davis/research.html.

55. F. Sullivan and J. Dongarra. Guest editors' introduction: The top 10 algorithms. *Comput. Sci. Eng.*, 2/22–23, 2000.

56. SuperLU. http://crd-legacy.lbl.gov/xiaoye/SuperLU/.

57. T. Tada, T. Yamamoto, and S. Watanabe. Molecular orbital concept on spin-flip transport in molecular junctions: Wave-packet scattering approach and Green's function method. *Theor. Chem. Acc.*, 130(4–6)/775–788, 2011.

58. K. Tanaka, H. Imachi, T. Fukumoto, T. Fukaya, Y. Yamamoto, and T. Hoshi. EigenKernel - A middleware for parallel generalized eigenvalue solvers to attain high scalability and usability. *ArXiv e-prints*, 2018.

59. H. A. van der Vorst. BI-CGSTAB: a fast and smoothly converging variant of BI-CG for the solution of nonsymmetric linear systems. *SIAM J. Sci. Stat. Comput.*, 13(2)/631–644, 1992.

60. H. A. van der Vorst and J. B. M. Melissen. A Petrov-Galerkin type method for solving $Ax = b$, where A is symmetric complex. *IEEE Trans. Magn.*, 26(2)/706–708, 1990.

61. X. Xu, K. Fukuda, A. Karki, S. Park, H. Kimura, H. Jinno, N. Watanabe, S. Yamamoto, S. Shimomura, D. Kitazawa, T. Yokota, S. Umezu, T-Q. Nguyen, and T. Someya. Thermally stable, highly efficient, ultraflexible organic photovoltaics. *PNAS*, pp. 4589–4594, 2018.

62. V. W.-Z. Yu, F. Corsetti, A. Garcia, W. P. Huhn, M. Jacquelin, W. Jia, B. Lange, L. Lin, J. Lu, W. Mi, A. Seifitokaldani, Á. Vazquez-Mayagoitia, C. Yang, H. Yang, and V. Blum. *Comput. Phys. Commun.*, pp. 222/267–285, 2018.

63. z-Pares. http://zpares.cs.tsukuba.ac.jp/.

Atomistic Simulation of Disordered Nano-electronics

Youqi Ke
Shanghaitech University

16.1 Introduction

The progress in modern semiconductor technology has been driven by the miniaturization of electronic devices. The most remarkable example is the continued reduction in the size of transistor by the application of nano-fabrication and nano-materials, pushing the field into the nano-electronics. Today, a modern computer processor can integrate billion-plus nano-transistors, and each nano-transistor has barely tens of atoms long. Nowadays, the term "nano-electronics" has been developed to cover vast application areas including information processing and storage, optoelectronic devices, displays, solar energy transformation, and so on. However, since device features are already pushed down to the nanometer scale, the traditional semiconductor industry faces important technological and fundamental challenges for continuing the miniaturization, such as the increasingly difficult heat dissipation, random dopant fluctuation, and gate leakage. As devices approach nanoscale, many important effects emerge to influence or even determine the functionality of the devices. For example, (i) at the small scale, the quantum confinement effect can become prominent to significantly modulate the material's electronic and optical properties. As an example, energy spectrum at nanoscale becomes discrete, measured as quanta, while spectra of bulk material are continuous. As a result, the quantized conductance can be observed if the typical length is small enough; (ii) the quantum tunneling effect becomes important at nanoscale, giving rise to the serious gate-leakage problem that affects the reliability of nano-transistor;

(iii) at nanoscale, coupling of the device properties to the structural and chemical details becomes important. As a result, randomly distributed defects/dopants can significantly influence the device property, presenting a large variance of device characteristics. The large device-to-device variability presents a great challenge for large-scale device integration. However, due to the lack of effective theoretical method, the effects of disorders on electron transport remain largely unexplored or poorly understood; (iv) as the device becomes smaller and smaller, interfaces and surfaces become more and more important, and can even dominate the device properties. Therefore, for the emergent behaviors of nanoelectronics, the understanding of electron transport on the atomic scale is desirable and important for the science and engineering.

The transport model of a device is usually determined by the behavior of carriers, namely classical particle or quantum wave. The particle–wave duality is the most basic property of the electron. However, when the device size L is long enough, the electron suffers from the inelastic scattering like electron–phonon scattering during transport through the device, and the phase coherence of the wave can be completely destroyed (L is much larger than the coherence length of the carrier). As a consequence, the nature of electron wave is eliminated, and the electron, thus, behaves like effective classical particle with specified coordinates and momentum. For such cases, the classical Drift–Diffusion model and semiclassical Bloch–Boltzmann transport equation can be developed to describe the electron transport in large devices. However, in mesoscopic and nanoscale systems, the system size is so small that the phase coherence of electron wave is preserved and thus the quantum interference effect becomes important in the electron transport so that the semiclassical device models must be abandoned. Due to the fact that the wave interference is sensitive to the scattering process that causes phase shifts, the random scattering of disorders and scattering of boundaries can have important influences on the electron transport through nanoscale devices. As a result, to describe the electron transport through nano-electronics, a fully quantum mechanical method is required.

However, the development of quantum transport method for simulating realistic nano-electronics faces with several challenges: (i) since electron transport in current flow is an intrinsically nonequilibrium process, the nonequilibrium quantum statistics must be correctly treated for simulating the operating devices; (ii) the strong coupling of transport properties to the structural, chemical, and materials details at nanoscale requires accurate atomic-level description without using any empirical parameter; (iii) due to the quantum uncertainty of electron, the electron transport at nanoscale is a stochastic process, and thus, studying nanoelectronic device requires transport statistics which is usually approximated by a set of cumulants, such as, from low to high orders, current, shot noise, and skewness. For example, the shot noise contains important information of the transport that cannot be obtained by the current flow. (iv) The inevitable disorders, such as intentional dopants, interfacial defects, and impurities, break translational invariance and render many well-established state-of-the-art computational methods useless. Moreover, the presence of disorders lifts the limitation of symmetry that is conserved for transport in perfect device and thus allows the electron to transport through new channels formed by the interchannel scattering, presenting profound effects of disorders in nanoelectronics. Therefore, accurate treatment of disorder effects is required to describe the quantum transport statistics of realistic nano-electronics. (v) At nanoscale, devices with different disorder configurations can behave very differently, the theoretical transport properties, such as current and shot noise, must be averaged over a large ensemble of disorder configurations to be physically meaningful, and the computation of statistical fluctuation is important to measure the device-to-device variability, presenting important challenges for nano-electronics simulation. Above five difficulties involve different areas' physics. A quantum transport method, which combines first-principles approach, nonequilibrium statistics and effective treatment of disorder average, holds the promise for simulating the electron transport in realistic nano-electronics.

In this chapter, we will introduce a coherent medium theory for simulating quantum transport through disorders and its combination with first-principles method for realizing atomistic simulation of disordered nanoelectrnics. This method combines Kohn–Sham density functional theory (DFT) with the Keldysh's nonequilibrium Green's function (NEGF) formalism and uses the generalized nonequilibrium

vertex correction (NVC) in coherent potential approximation (CPA) to account for nonequilibrium multiple disorder scattering. The generalized CPA–NVC in combination with NEGF–DFT method provides a unified algorithm for calculating the disorder-averaged nonequilibrium density matrix, current flow, shot noise, and current fluctuations of disordered nano-electronics from first principles. Due to the fact that quantum transport involves vast areas of physics, we limit our discussion on the coherent transport in noninteracting system in steady state in this chapter. However, in principle, including the physics of dephasing mechanism, strong correlations and transient transport in present quantum transport method are possible but requires further development. This chapter is organized as follows: In Section 16.2, we introduce the NEGF formalism for treating general nonequilibrium problem and NEGF-based quantum transport method; Section 16.3 introduces the generalized CPA–NVC method for treating disorder average of various single-particle Green's functions, and any two-Green's-function correlator, and discusses how to realize a first-principles self-consistent simulation of electron transport through disordered device; Section 16.4 introduces possible ways to account for the short-range cluster effects and nonlocal correlations, beyond the present CPA–NVC method.

16.2 Nonequilibrium Green's Function-Based Quantum Transport Theory

To deal with a nonequilibrium quantum many-body problem, the most general and rigorous theoretical framework is provided by the NEGF theory, which has been discussed in many reviews and books. [1–7] The NEGF theory provides a unified algorithm for treating quantum-transport problem with complex scattering mechanisms including electron–phonon, random impurity scattering, etc. In this section, we will first introduce the key ideas of the NEGF theory including the closed-time contour and the contour-ordered GF, introduce the Keldysh's NEGF formalism and the 2×2-matrix representation, and then derive the electron current flow and shot noise formulas for a biased device within the NEGF theory.

16.2.1 Closed-Time Contour for Nonequilibrium Problem

In general, the nonequilibrium problem can be described by the time-dependent Hamiltonian:

$$\mathcal{H}(t) = H_0 + H_i + H'(t) \qquad (16.1)$$

where H_0 represents the solvable noninteracting part, H_i the intrinsic interaction between particles (such as electron-electron, electron–phonon, and electron–photon, which are difficult to treat even in equilibrium physics) and $H'(t)$ is the time-dependent external driving field, which is switched

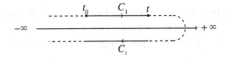

FIGURE 16.1 Keldysh's closed time contour.

on at some initial time t_0 and drive the system away from equilibrium. We assume the system at the distant past $t = -\infty$ is in the equilibrium state of the noninteracting H_0 with a density matrix $\rho(-\infty)$, the interaction H_i can be adiabatically switched on to reach the full strength at t_0. Same as the equilibrium, the task of nonequilibrium quantum theory is to calculate the expectation value of a given observable O of the system at $t \geq t_0$:

$$\langle O \rangle(t) = \frac{Tr[\rho(-\infty)O_{\mathcal{H}}(t)]}{Tr[\rho(-\infty)]}$$
$$= \frac{Tr[\rho(-\infty)U(-\infty,t)OU(t,-\infty)]}{Tr[\rho(-\infty)]}, \quad (16.2)$$

where the operator $O_{\mathcal{H}}(t)$ is in the Heisenberg representation, the evolution operator $U(t,t_0) = T[e^{-\frac{i}{\hbar}\int_{t_0}^{t} d\bar{t}\mathcal{H}(\bar{t})}]$. Here, the idea of closed-time contour appears naturally. Eq. 16.2 (reading from right to left) describes time-evolution process along a closed contour C, namely $-\infty \to t \to -\infty$, so-called the Keldysh's contour as depicted in Figure 16.1. Evolution along the closed-time contour is the central idea for treating nonequilibrium quantum problem. It should be emphasized that the initial density matrix $\rho(-\infty)$ and also the time-evolution operator U, required for evaluation of Eq. 16.2, have a strong dependence on the details of system Hamiltonians.

16.2.2 Keldysh's Nonequilibrium Green's Function Formalism

In the following, based on the Keldysh's closed-time contour, we introduce the contour-ordered Green's function method for nonequilibrium many-body problem. By using the contour-ordered operator T_c, we can define the quantity (note that we will use the atomic unit in the rest of the chapter)

$$G_c(\vec{r},t;\vec{r}\,',t') = -i\langle\Phi|T_c[\psi_{\mathcal{H}}(\vec{r},t)\psi_{\mathcal{H}}^{\dagger}(\vec{r}\,',t')]|\Phi\rangle \quad (16.3)$$

where $\psi_{\mathcal{H}}$ and $\psi_{\mathcal{H}}^{\dagger}$ are the respective elimination and creation field operators defined in the Heisenberg picture. $|\Phi\rangle$ refers to the normalized ground state of the initial system with the Hamiltonian $H_0 + H_i$. T_c arranges the time-dependent operators according to their order on the closed-time contour. $G_c(\vec{r},t;\vec{r}\,',t')$ is the so-called contour-ordered GF, in which the evolution starts from remote past, passes through t and t', and finally returns to the remote past again. By changing from Heisenberg to interaction representation [7], one obtains:

$$G(\vec{r},t;\vec{r}\,',t') = -i\frac{\langle\Phi_0|T_c[S_c(-\infty,-\infty)\widehat{\psi}(\vec{r},t)\widehat{\psi}^{\dagger}(\vec{r}\,',t')]\Phi_0\rangle}{\langle\Phi_0|S_c(-\infty,-\infty)|\Phi_0\rangle}$$
$$(16.4)$$

where $S_c(-\infty,-\infty)$ is the time-evolution operator on the closed-time contour. One can find the contour-ordered GF in Eq. 16.4 has the same mathematical structure as the time-ordered equilibrium GF defined on the real-time axis $-\infty \to +\infty$.[2] Therefore, the perturbation expansion technique, namely the Dyson equation, can be directly applied to the contour-ordered Green's function

$$G(t,t') = G_0(t,t') + \int_c dt_1 \int_c dt_2 G_0(t,t_1)\Sigma(t_1,t_2)G(t_2,t')$$
$$(16.5)$$

where G is the Green function of the nonequilibrium sytem with interactions, G_0 is the bared Green's function determined by the noninteracting H_0, and $\Sigma(t_1,t_2)$, the self-energy containing the contribution of the external driving fields $H'(t)$ and complex interaction H_i. Eq. 16.5 provides a general and powerful many-body perturbation method for calculating the complex nonequilibrium quantum system. Because the times t and t' can lie on the two branches C_1 and C_2 of the Keldysh's contour as shown in Figure 16.1, the contour-ordered Green's function comprises four different real-time GF components, namely

$$G(t,t') = \begin{cases} G^t(t,t') & t \in C_1, t' \in C_1, \\ G^<(t,t') & t \in C_1, t' \in C_2, \\ G^>(t,t') & t \in C_2, t' \in C_1, \\ G^{\bar{t}}(t,t') & t \in C_2, t' \in C_2, \end{cases} \quad (16.6)$$

which are called time-ordered, lesser, greater, and anti-time-ordered GFs, respectively. These four real-time GFs are interconnected by the Eq. 16.5 and are not linearly independent since they always satisfy

$$G^t + G^{\bar{t}} = G^< + G^>. \quad (16.7)$$

Consequently, G can be equivalently represented by three linearly independent GFs as

$$\begin{cases} G^R = G^t - G^< = G^> - G^{\bar{t}}, \\ G^A = G^t - G^> = G^< - G^{\bar{t}}, \\ G^K = G^t + G^{\bar{t}} = G^> + G^<, \end{cases} \quad (16.8)$$

which are called the respective Retarded, Advanced, and Keldysh Green's function with the symmetry relations $G^R = [G^A]^{\dagger}$ and $G^K = -[G^K]^{\dagger}$. It should be emphasized that we are forced to resort to the contour-ordered Green's function, not because it is directly related to physical observables but because it can be expanded in a perturbation series, whereas no such expansion exists for the real-time GFs, such as $G^<, G^>, G^K$. To make contact with physical quantities at nonequilibrium, these real-time GFs must be extracted from the contour-ordered Green's function. In practical applications, one way to realize this extraction is based on Langreth's theorem, so-called analytical continuation method.[1,3] The other realization is by constructing a

2×2 real-time matrix representation of the contour-ordered quantity. As proposed by Craig [8], the matrix

$$G_c = \begin{pmatrix} G^t & -G^< \\ G^> & -G^{\bar{t}} \end{pmatrix} \qquad (16.9)$$

contains the same amount of information as the contour-ordered Green's function in Eq. 16.6. One can check, for contour convolution product $c = ab$, the results of c by analytical continuation are the same as the product of the 2×2 matrices of a and b. [9] To eliminate the redundancy in Craig's matrix in Eq. 16.9, an equivalent Keldysh's matrix representation can be introduced by a unitary transformation

$$\bar{G} = R^{-1}GR = \begin{pmatrix} G^A & 0 \\ G^K & G^R \end{pmatrix}, R = \frac{1}{\sqrt{2}}\begin{pmatrix} 1 & 1 \\ -1 & 1 \end{pmatrix}. \quad (16.10)$$

Note that the Keldysh matrix in Eq. (16.10) features a lower triangular matrix, simplifying matrix operations. The various matrix operations with the Keldysh's matrices preserve the mathematical structure. Therefore, it is usually convenient to work with the Keldysh's matrix representation and then transform to the Craig's matrix to obtain all the real-time GFs in Eq. 16.6.[9] As an important consequence, with the matrix representation of contour-ordered quantity, the perturbation expansion, namely Dyson equation in Eq. 16.5, can be written in a compact form

$$G = G_0 + G_0 \Sigma G. \qquad (16.11)$$

with the Keldysh's representation, we can obtain

$$G^R = G_0^R(I - \Sigma^R G_0^R)^{-1}, \qquad (16.12a)$$

$$G^A = G_0^A(I - \Sigma^A G_0^A)^{-1}, \qquad (16.12b)$$

$$G^K = G^R \Sigma^K G^A + (I + G^R \Sigma^R)G_0^K(I + \Sigma^A G^A), \quad (16.12c)$$

16.2.3 Current Formula

We now apply the Keldysh's NEGF formalism to derive the current formula for a nanoelectronic device under external bias. As illustrated in Figure 16.2, we here consider a two-probe device containing a central scattering region sandwiched by two leads. Each of the leads is at its own chemical potential μ. To apply the bias, we can assume $\mu_L > \mu_R$ to drive the system out of equilibrium, creating the electron current flow through the device. Electron transport is an extremely complicated problem because it can couple

FIGURE 16.2 Two-probe device model with a central device region sandwiched by left and right leads.

with various difficult interactions and transient physics. We thus only consider a noninteracting system at nonequilibrium steady state in this chapter. To use the Keldysh's NEGF to study the device, we assume, at time t_0, the central region and two leads are brought into contact to establish the coupling between them, to allow electrons to tunnel from the leads to the central region and vice versa. The Hamiltonian for such a system, consisting of the left/right leads and the central device region, is written as:

$$H = H_L + H_R + H_D + H_T. \qquad (16.13)$$

where H_L, H_R, and H_D are the respective Hamiltonians for the left and right leads and central device, and H_T describes the coupling between device and leads. Here, we adopt

$$H_\alpha = \sum_k \epsilon_{k\alpha} c_{k\alpha}^\dagger c_{k\alpha} \qquad \alpha = L, R, \qquad (16.14)$$

$$H_T = \sum_{kn} \sum_\alpha \left(V_{k\alpha,n} c_{k\alpha}^\dagger d_n + h.c. \right) \qquad \alpha = L, R, \quad (16.15)$$

$$H_D = \sum_n \epsilon_n d_n^\dagger d_n, \qquad (16.16)$$

where the spin index is omitted for simplicity, and k and n are the indices for the eigenstates of the Hamiltonians for lead and central device, respectively. The matrix element $V_{k\alpha,n}$ describes the hopping of an electron from the state $|k\rangle$ of the lead α into the state $|n\rangle$ of the central region. Without the coupling H_T, the leads and central device are independent and are in equilibrium states. The presence of the coupling drives the system out of equilibrium. The current flow through the device can be defined by the changing rate of the number of electrons in the left lead

$$I_L(t) \equiv -\langle \dot{N}_L \rangle. \qquad (16.17)$$

where, the number operator $N_L = \sum_k c_{kL}^\dagger c_{kL}$. Since $i\dot{N}_L = [N_L, H]$, and N_L commutes with H_L, H_R, and H_D,

$$
\begin{aligned}
I_L(t) &= i\langle[N_L, H]\rangle = [N_L, H_T]\rangle \\
&= i \sum_{kn} \left(V_{kL,n} \left\langle c_{kL}^\dagger(t)d_n(t) \right\rangle - h.c. \right) \\
&= \sum_{kn} \left(V_{kL,n} G_{n,kL}^<(t;t) - V_{kL,n}^* G_{kL,n}^<(t;t) \right) \\
&= 2Re\left[\sum_{kn} V_{kL,n} G_{n,kL}^<(t;t) \right] \qquad (16.18)
\end{aligned}
$$

where we have used the definitions $G_{n,kL}^<(t;t') = i\left\langle c_{kL}^\dagger(t')d_n(t) \right\rangle$ and $G_{kL,n}^<(t;t') = i\left\langle d_n^\dagger(t')c_{kL}(t) \right\rangle$, and the relation $[G_{n,kL}^<(t;t)]^* = -G_{kL,n}^<(t;t)$. To find the lesser GF $G_{n,kL}^<(t;t)$ to obtain the current, we utilize the contour-ordered Green's function defined as

$$G_{n,kL}(t;t') = -i\langle T_c[d_n(t)c_{kL}^\dagger(t')]\rangle, \qquad (16.19)$$

and with the Dyson equation in Eq. 16.5, we obtain

$$G_{n,kL}(t;t') = \sum_m \int_c dt_1 G_{nm}(t;t_1) V^*_{kL,m} G^0_{kL,kL}(t_1;t').$$

$$(16.20)$$

where $G^0(t_1;t')$ correspond to the system without H_T (thus $G^0_{nk} = 0$), and the contour-ordered GFs $G^0_{kL}(t_1;t') = -i\langle T_c c_{kL} c^\dagger_{kL}(t')\rangle_0$ and $G_{nm}(t;t_1) = -i\langle T_c d_n(t) d^\dagger_m(t')\rangle$. Now we apply the Langreth rule and obtain

$$G^<_{n,kL}(t;t') = \sum_m \int_{-\infty}^{\infty} dt_1 \left[G^R_{nm}(t;t_1) G^{0<}_{kL}(t_1;t') \right.$$
$$\left. + G^<_{nm}(t;t_1) G^{0A}_{kL}(t_1;t') \right] V^*_{kL,m}. \quad (16.21)$$

As a result, we can obtain the current formula for steady state by Fourier transformation to frequency domain,

$$I_L = \int_{-\infty}^{\infty} \frac{dE}{\pi} Re \sum_{nmk} V_{kL,n} V^*_{kL,m} \left[G^R_{nm}(E) G^{0<}_{kL}(E) \right.$$
$$\left. + G^<_{nm}(E) G^{0A}_{kL}(E) \right]. \quad (16.22)$$

By applying the relation for the left lead at equilibrium

$$G^{0<}_{kL}(E) = 2\pi i f_L(E) \delta(E - \epsilon_{kL})$$
$$G^{0A}_{kL}(E) = \frac{1}{E - \epsilon_{kL} - i0^+}, \quad (16.23)$$

where $f_L(E)$ is the Fermi–Dirac distribution function in the left lead, the current can be rewritten as

$$I_L = i \int_{-\infty}^{\infty} \frac{dE}{2\pi} Tr \left\{ \Gamma^L(E)[f_L(E)(G^R(E) - G^A(E)) \right.$$
$$\left. + G^<(E)] \right\} \quad (16.24)$$

where $\Gamma^L_{mn}(E) = 2\pi \sum_k V^*_{kL,m} V_{kL,n} \delta(E - \epsilon_{kL})$, called the line-width function. In Eq. 16.24, matrix notations are adopted for the quantities Γ^L, G^R, G^A and $G^<$. Because the system is in the steady state, $I = I_L = -I_R$, where I_R is the current from right lead. Thus, the current can also be written as $I = (I_L - I_R)/2$, namely the Meir–Wingreen formula for steady state,

$$I = \frac{i}{4\pi} \int_{-\infty}^{\infty} dE Tr\{[f_L(E)\Gamma^L(E) - f_R(E)\Gamma^R(E)]$$
$$\times [G^R(E) - G^A(E)] + [\Gamma^L(E) - \Gamma^R(E)]G^<(E)\}. \quad (16.25)$$

This formula expresses the current in terms of the Green's functions of the central scattering region. In general, the calculation of these functions is nontrivial. For the noninteracting central scattering region, with the NEGF method, one can obtain

$$G^R - G^A = -iG^R(\Gamma^L + \Gamma^R)G^A,$$
$$G^< = iG^R(f_L\Gamma^L + f_R\Gamma^R)G^A \quad (16.26)$$

Putting these equations into the equation for the current, we obtain the compact formula

$$I = \frac{1}{2\pi} \int_{-\infty}^{+\infty} dE[f_L(E) - f_R(E)]T(E) \quad (16.27)$$

which is known as the Landauer formula and the so-called transmission function

$$T(E) = Tr\left[\hat{T}\right]. \quad (16.28)$$

with the transmission matrix
$$\hat{T} = G^R(E)\Gamma^R(E)G^A(E)\Gamma^L(E) = -iG^<\Gamma^L$$

16.2.4 Quantum Shot Noise

In realistic nanoelectronics, the electrical current continually fluctuates in time so that it carries noise. The sources of noise in a electronic device includes the external including $1/f$ and telegraph noises and the internal including the thermal and shot noises. "The noise is the signal" was a saying of Rolf Landauder. Due to the importance of studying noise, we here focus on the shot noise in nano-electronics, which measures the fluctuation of current in time due to the quantum nature of electron. The shot noise is not only important for the noise–signal ratio in applications but also provides important information about the nature of transport channels, unit of transferred charge, and other diagnostic information. (For reviews, see Refs. 10 and 11) Shot noise, as the second cumulant of quantum transport statistics, carries important information for understanding quantum transport mechanisms. In the following, we will focus our discussion on shot noise in the coherent transport regime. To derive the shot noise formula, we first consider the electron transport through a single channel with scattering by a potential barrier. Then it has a transmission probability T and a reflection probability $R = 1-T$. Assume that the average occupation number for the incoming flux is $\langle n_i \rangle = f$, and then the occupation $\langle n_T \rangle = fT$ and $\langle n_R \rangle = fR$ for the transmitted and reflected states, respectively (note $n = n(E,t)$ is time dependent with energy E.). Because the electron is either transmitted or reflected, the electron transport satisfies the binomial distribution. As a result, we can obtain that

$$\langle (\delta n_T)^2 \rangle = \langle n_T \rangle - \langle n_T \rangle^2 = fT(1 - fT) \quad (16.29)$$
$$\langle (\delta n_R)^2 \rangle = \langle n_R \rangle - \langle n_R \rangle^2 = fR(1 - fR), \quad (16.30)$$

where $\delta n_T = n_T - \langle n_T \rangle$ and $\delta n_R = n_R - \langle n_R \rangle$. If we assume the incident state is populated with probability 1 ($f = 1$ at zero temperature), then the above expressions reduce to

$$\langle (\delta n_T)^2 \rangle = \langle (\delta n_R)^2 \rangle = TR = T(1 - T) \quad (16.31)$$

As known, the current across the device relates to the occupation number by $\hat{I}(t) = \int n_T(E,t)dE/2\pi$. The shot noise, as the second cumulant of fluctuating current, can be defined as the correlation $S(t - t') = \langle \delta\hat{I}(t)\delta\hat{I}(t') + \delta\hat{I}(t')\delta\hat{I}(t)\rangle/2$. In the zero-frequency limit, the shot noise is related to the fluctuation of n and thus the transmission T[12]

$$S = \frac{1}{\pi} \int dE \langle \delta n_T \delta n_T \rangle = \frac{1}{\pi} \int dE T(E)(1 - T(E)) \quad (16.32)$$

The shot noise power formula for single channels can be easily extended to the case of multiple channels. The total shot noise power is the sum of the contribution from each single transport channel. Thus, we have the total shot noise for a multichannel conductor

$$S = \frac{1}{\pi} \int dE \sum_n T_n (1 - T_n) = \frac{1}{\pi} \int dE Trace[\hat{T}(1 - \hat{T})],$$
(16.33)

where T_n is the eigenvalue of the transmission matrix $\hat{T}$ which can be defined as $\hat{T} = \Gamma^L G^R \Gamma^R G^A$ with the NEGF method.

Usually, without correlations, the transport can be described by the Poissonian statistics, the shot noise can be found as

$$S_{poisson} = 2e\langle I \rangle. \quad (16.34)$$

It will be very instructive to define a dimensionless quantity, called the Fano factor $F = S/2eI$, that quantifies the deviation from the Poissonion noise. Importantly, the coherency and correlations of electrons, which are not accessible by the time-averaged current, can be reflected by the deviation of Fano factor from $F = 1.0$.

16.3 Coherent Medium Theory for Quantum Transport Simulation of Disordered Nanoelectronics

The NEGF-based quantum transport formalism introduced so far is only applicable to a specific system. However, for the device containing atomic disorders, its transport properties vary from sample to sample, the meaningful theoretical results have to be obtained by disorder average over a large number of configurations. One way to do disorder average is to carry out simulation for a number of samples one by one, and then obtain the averaged result at final (so-called supercell approach). The computational cost of this type of brute-force average is prohibitively large for calculating a disordered nanoelectronics. In this section, a coherent medium theory is introduced to simulate the disordered nanoelectronics at nonequilibrium. We will first introduce CPA to obtain the disorder averaged one-particle GFs by self-consistently constructing an ordered effective medium.[15,16] Then, we introduce the generalized NVC to handle the disorder average of any two-GF correlators required for quantum transport calculation.[9,17] At last, the combination with DFT is introduced to realize the first-principles simulation of disordered nanoelectronics.[18]

16.3.1 Coherent Potential Approximation

One important step towards the simulation of nonequilibrium-disordered nanoelectronics is to calculate the disorder averaged single-particle Keldysh's NEGF, which we introduce in the following. We consider the system with a disordered central device region sandwiched by two ordered leads as shown in Figure 16.3. Because of the random distribution of disorders, the potential of

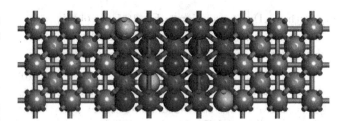

FIGURE 16.3 Atomic structure of a Fe/MgO(5ML)/Fe(001) magnetic tunnel junction. The spheres are Fe, O, Mg, and randomly distributed defects. Leads extend to infinity.

the system V is a random quantity, and it contains the contribution from each cell, namely $V = \sum_n v_n$, where v_n is the random potential in the cell centered on atomic site R. For the Green's function of the central device, we can introduce an equivalent effective Hamiltonian,

$$H = H_0 + \Sigma_{ld} + V, \quad (16.35)$$

where H_0 is an ordered Hamiltonian of the central region, and Σ_{ld} is the self-energy describing the effects of coupling with the leads. The central idea of CPA is to self-consistently construct a coherent effective medium with the GF $\bar{G}$, which is the same as the disorder averaged GF $\langle G \rangle$, namely

$$\bar{G} = \langle G \rangle. \quad (16.36)$$

This effective medium can be described by the Hamiltonian $\bar{H} = H_0 + \Sigma_{ld} + \Sigma_{im}$, where $\Sigma_{im} = \sum_n \Sigma_{im,n}$ is due to the effects of disorders. Then, we can rewrite the Hamiltonian of a disordered device in Eq. 16.35 as follows

$$H = (H_0 + \Sigma) + (V - \Sigma_{im}), \quad (16.37)$$

where $\Sigma = \Sigma_{ld} + \Sigma_{im}$, and the second bracket contains the deviation of random potential from Σ_{im}, which can be rewritten as $V - \Sigma_{im} = \sum_n (v_n - \Sigma_{n,im})$. With the perturbation technique of GF in Eq. 16.11, one can obtain the following two Dyson equations

$$G = \bar{G} + \bar{G}(V - \Sigma_{im})G = \bar{G} + G(V - \Sigma_{im})\bar{G}, \quad (16.38)$$

$$\bar{G} = G_0 + G_0 \Sigma \bar{G} = G_0 + \bar{G}\Sigma G_0, \quad (16.39)$$

where G, $\bar{G}$ and G_0 are the GFs corresponding to the Hamiltonians H, $H_0 + \Sigma$ and H_0, respectively. Here, one can check that Eq. (16.38) can be rewritten in another form

$$G = \bar{G} + \bar{G}T\bar{G}, \quad (16.40)$$

where T is called T-matrix given as

$$T \equiv (V - \Sigma_{im}) + (V - \Sigma_{im})\bar{G}(V - \Sigma_{im}) + \cdots$$
$$= (V - \Sigma_{im})(I + \bar{G}T) = (I + T\bar{G})(V - \Sigma_{im}). \quad (16.41)$$

We can see T-matrix contains all the randomness and complexity of a disordered device. By taking average on both sides of Eq. (16.40), the condition of Eq. (16.36) requires

$$\langle T \rangle = 0, \quad (16.42)$$

which provides an important condition to solve for Σ_{im}. Equations 16.42 and (16.39) forms a closed set of self-consistent equations to solve for Σ_{im} and $\bar{G}$ of the effective medium. However, to evaluate $\langle T \rangle$ in Eq. (16.42), one needs to enumerate all possible configurations of the disorders, which is computationally prohibitive. Therefore, for practical applications, further approximation to the average of T-matrix is required to establish CPA self-consistent calculation.

16.3.2 Single-Site Approximation

To make the CPA condition, namely Eq. (16.42), practical, single-site approximation (SSA)[13] was introduced to decouple all the disorder scattering events contained in the T, transforming to a single-site disorder average problem. To introduce SSA, one can rewrite Eq. (16.41) as

$$T = \sum_n Q_n, \tag{16.43}$$

where $Q_n \equiv (v_n - \Sigma_{n,im})(I + \bar{G}T)$. Q_n can be written in the form

$$Q_n = t_n(I + \bar{G} \sum_{m \neq n} Q_m), \tag{16.44}$$

$$t_n \equiv [I - (v_n - \Sigma_{n,im})\bar{G}]^{-1}(v_n - \Sigma_{n,im}). \tag{16.45}$$

Here, t_n describes the scattering event on the single site n ($t_n = 0$ at the site without random occupations). recursively substituting Eq. (16.44) into Eq. (16.43) obtains the multiple-scattering equation

$$T = \sum_n t_n + \sum_n \sum_{m \neq n} t_n \bar{G} t_m + \cdots, \tag{16.46}$$

in which the total disorder scattering effect can be expressed as successive multiple scattering processes from one site to another. By here, we have not introduce extra approximation. The SSA is introduced in the disorder average on Eq. (16.44) by neglecting the correlation between t_n and Q_m, namely

$$\langle Q_n \rangle = \langle t_n \rangle (I + \bar{G} \sum_{m \neq n} \langle Q_m \rangle), \tag{16.47}$$

where we neglect the fluctuation $\langle t_n \bar{G} \sum_{m \neq n} (Q_m - \langle Q_m \rangle) \rangle$. In the SSA, the successive scattering events are uncorrelated. As a result, we can write

$$\langle T \rangle = \sum_n \langle Q_n \rangle = \sum_n \langle t_n \rangle + \sum_{n \neq m} \langle t_n \rangle \bar{G} \langle t_m \rangle + \cdots. \tag{16.48}$$

As an immediate result, the CPA self-consistent condition $\langle T \rangle = 0$ is reduced to $\langle Q \rangle_n = 0$ and thus

$$\langle t_n \rangle \equiv \sum_Q c_n^Q t_n^Q = 0, \tag{16.49}$$

where c_n^Q is the concentration of Q element on the site n. Combining the above single-site equation and Eq. (16.39),

the on-site self-energy $\Sigma_{n,im}$ can be self-consistently solved for each site of the system. In such a way, the effective medium described by $\Sigma_{im} = \sum_n \Sigma_{n,im}$ can be efficiently constructed, and thus $\bar{G}$ is obtained. Since the probability is small for scattering off multiple impurities at the same time, the SSA is a good approximation and becomes accurate at low impurity concentration.

16.3.3 CPA in Keldysh's Representation

The formalism of CPA and SSA introduced above is generally applicable to the GFs satisfying the mathematical structure of Eq. 16.11, namely Dyson equation. Here, we can apply the Keldysh's representation of the contour-ordered GFs to treat a nonequilibrium disordered system, aiming to the disorder average of all the real-time GFs introduced in Section 16.2.2. To do this, we need to write the quantities G, Σ, V, T and their single-site counterparts (such as Σ_n, v_n and t_n) in the form of the Keldysh's 2×2 matrix. For example, $\Sigma = \begin{pmatrix} \Sigma^A & 0 \\ \Sigma^K & \Sigma^R \end{pmatrix}$, $T = \begin{pmatrix} T^A & 0 \\ T^K & T^R \end{pmatrix}$, and $V = \begin{pmatrix} V^A & 0 \\ V^K & V^R \end{pmatrix} = \begin{pmatrix} V & 0 \\ 0 & V \end{pmatrix}$ since the potential is Hermitian. Replacing the quantities in Eq. (16.39) with Keldysh's matrices leads to the following equations,

$$\bar{G}^R = G_0^R (I - \Sigma^R G_0^R)^{-1}, \tag{16.50a}$$

$$\bar{G}^A = G_0^A (I - \Sigma^A G_0^A)^{-1}, \tag{16.50b}$$

$$\bar{G}^K = \bar{G}^R \Sigma^K \bar{G}^A + (I + \bar{G}^R \Sigma^R) G_0^K (I + \Sigma^A \bar{G}^A), \tag{16.50c}$$

where $\Sigma = \Sigma_{ld} + \Sigma_{im}$. Equations (16.50a) and (16.50b) for retarded and advanced GFs are the same as the conventional CPA. Equation (16.50c) is usually called the Keldysh's formula for G^K which relates $\bar{G}^K$ to $\bar{G}^{R/A}$ and different components of Σ. From the above equation, we can see the three components of G, namely $G^{R/A/K}$, are not independent of each other. Given the self-energy components $\Sigma^{R/A/K}$, G^R provides the sufficient knowledge to compute all other GFs. The second term in Eq. 16.50c accounts for the initial correlations, which are important for a transient problem. However, for the long-time limit, or steady state, the initial correlations can be neglected,[1,14] then the Keldysh's formula becomes

$$\bar{G}^K = \bar{G}^R \Sigma^K \bar{G}^A, \tag{16.51}$$

Actually, by inserting the Keldysh's matrices in the definition of T, we can obtain the three components

$$T^R = [I - (V - \Sigma_{im}^R)\bar{G}^R]^{-1}(V - \Sigma_{im}^R), \tag{16.52a}$$

$$T^A = [I - (V - \Sigma_{im}^A)\bar{G}^A]^{-1}(V - \Sigma_{im}^A), \tag{16.52b}$$

$$T^K = T^R \bar{G}^K T^A - (I + T^R \bar{G}^R)\Sigma_{im}^K (I + \bar{G}^A T^A). \tag{16.52c}$$

By applying the general CPA condition $\langle T \rangle$, one can find the Σ^K can be given as

$$\Sigma^K = \langle T^R \bar{G}^R \Sigma_{ld} \bar{G}^A T^A \rangle \tag{16.53}$$

Similarly, for the single-site t_n in Eq. (16.45), we can obtain

$$t_n^R = [I - (v_n - \Sigma_{n,im}^R)\bar{G}^R]^{-1}(v_n - \Sigma_{n,im}^R), \qquad (16.54a)$$

$$t_n^A = [I - (v_n - \Sigma_{n,im}^A)\bar{G}^A]^{-1}(v_n - \Sigma_{n,im}^A), \qquad (16.54b)$$

$$t_n^K = t_n^R \bar{G}^K t_n^A - (I + t_n^R \bar{G}^R)\Sigma_{n,im}^K(I + \bar{G}^A t_n^A). \qquad (16.54c)$$

Here the Eqs.(16.52c) and (16.54c) can be called the Keldysh's formula for T and t_n. Similar to G, we find that the retarded, advanced and Keldysh components of T or t_n are also not independent. Based on the single-site CPA condition, namely $\langle t_n^K \rangle = 0$, we can find

$$\Sigma_{n,im}^K = \langle t_n^R \bar{G}^K t_n^A \rangle - \langle t_n^R \bar{G}^R \Sigma_{n,im}^K \bar{G}^A t_n^A \rangle \qquad (16.55)$$

Above equations in combination with CPA conditions enable the self-consistent computation of the self-energy $\Sigma_{im}^{R/A/K}$ and $\bar{G}^{R/A/K}$. As an important result, for a disordered non-equilibrium system, the average of any single-particle real-time GFs can be easily obtained with the linear combination of the averaged $\bar{G}^{R/A/K}$.

16.3.4 Generalized Non-equilibrium Vertex Correction

The CPA method introduced above only provides an effective way to calculate disorder averaged single-particle GFs. However, many physical quantities contain the product of two GFs, such as all the quantum transport properties including the transmission operator, shot noise, current fluctuation. Because the two GFs describing the same disordered system are internally correlated by the multiple disorder scattering, the two-GF correlator $\langle GCG \rangle$ is not simply equal to $\langle G \rangle \mathcal{C} \langle G \rangle$ where $\mathcal{C}$ is an arbitrary disorder independent constant. In this section, the generalized NVC is formulated to compute $\langle GCG \rangle$, so that the disorder average of any two-GF correlator can be obtained, such as $\langle G^< \mathcal{C} G^< \rangle$.

Here, we consider a two-GF correlator

$$K = \langle G(z_1)\mathcal{C}G(z_2) \rangle, \qquad (16.56)$$

where C is an arbitrary constant. In Eq. (16.56), the GFs can be at two different energies. For simplicity, these energy indices will be suppressed in the rest of the derivation. To evaluate K, we insert Eq. (16.40) into Eq. (16.56) and apply the CPA condition $\langle T \rangle = 0$, and then obtain

$$\langle GCG \rangle = \bar{G}(\mathcal{C} + \Omega)\bar{G}, \qquad (16.57)$$

where

$$\Omega \equiv \langle T\bar{G}\mathcal{C}\bar{G}T \rangle \qquad (16.58)$$

is the generalized NVC, containing all the effects of disorders on the two-GF correlator.

In order to compute Ω, one can substitute the T with Eq. (16.43) and then obtain

$$\Omega = \sum_n \sum_m \langle Q_n \bar{G}\mathcal{C}\bar{G}\tilde{Q}_m \rangle. \qquad (16.59)$$

For terms with $n \neq m$, SSA leads to $\langle Q_n \bar{G}\mathcal{C}\bar{G}\tilde{Q}_m \rangle = 0$. Consequently, Eq. (16.59) is simplified to

$$\Omega = \sum_n \Omega_n, \qquad (16.60)$$

where the single-site quantity $\Omega_n \equiv \langle Q_n \bar{G}\mathcal{C}\bar{G}\tilde{Q}_n \rangle$. To proceed, we apply the relation in Eq. (16.44), $Q_n = t_n(I + \bar{G}\sum_{p\neq n} Q_p)$, and the counterpart $\tilde{Q}_n = (I + \sum_{q\neq n} \tilde{Q}_q \bar{G})t_n$, and obtain

$$\Omega_n = \langle t_n(I + \bar{G}\sum_{p\neq n} Q_p)\bar{G}\mathcal{C}\bar{G}(I + \sum_{q\neq n} \tilde{Q}_q \bar{G})t_n \rangle. \qquad (16.61)$$

By applying the SSA assumptions, Eq. (16.61) finally becomes

$$\Omega_n = \langle t_n \bar{G}\mathcal{C}\bar{G}t_n \rangle + \sum_{p\neq n}\langle t_n \bar{G}\Omega_p \bar{G}t_n \rangle, \qquad (16.62)$$

which forms a closed set of linear equations for the unknown Ω_n. In Eq. (16.62), the average is over pairs of scattering events on the same site. In other words the scattering from different sites is regarded as statistically uncorrelated and the motion of two particles, represented by the two GFs, in the medium is correlated only if they both scatter from the same site. To apply the Keldysh's representation for contour-ordered quantities, we consider the following four cases for the constant $\mathcal{C}$

$$\mathcal{C}^{(1)} = \begin{pmatrix} C & 0 \\ 0 & 0 \end{pmatrix}, \quad \mathcal{C}^{(2)} = \begin{pmatrix} 0 & C \\ 0 & 0 \end{pmatrix},$$

$$\mathcal{C}^{(3)} = \begin{pmatrix} 0 & 0 \\ C & 0 \end{pmatrix}, \quad \mathcal{C}^{(4)} = \begin{pmatrix} 0 & 0 \\ 0 & C \end{pmatrix}.$$

By applying these four $\mathcal{C}^{(i)}$s to Eq. (16.58), we obtain four different $\Omega^{(i)}$s as follows,

$$\Omega^{(1)} = \begin{pmatrix} \Omega^{AA} & 0 \\ \Omega^{KA} & 0 \end{pmatrix}, \quad \Omega^{(2)} = \begin{pmatrix} \Omega^{AK} & \Omega^{AR} \\ \Omega^{KK} & \Omega^{KR} \end{pmatrix},$$

$$\Omega^{(3)} = \begin{pmatrix} 0 & 0 \\ \Omega^{RA} & 0 \end{pmatrix}, \quad \Omega^{(4)} = \begin{pmatrix} 0 & 0 \\ \Omega^{RK} & \Omega^{RR} \end{pmatrix}.$$

Here, we obtain 9 generalized NVCs Ω^{XY} ($X, Y = A, R, K$). From Eq. (16.62), nine equations can be found for the 9 Ω^{XY}

$$\Omega_n^{RR} = \langle t_n^R \bar{G}^R C \bar{G}^R t_n^R \rangle + \sum_{p\neq n}\langle t_n^R \bar{G}^R \Omega_p^{RR} \bar{G}^R t_n^R \rangle,$$

$$\Omega_n^{RA} = \langle t_n^R \bar{G}^R C \bar{G}^A t_n^A \rangle + \sum_{p\neq n}\langle t_n^R \bar{G}^R \Omega_p^{RA} \bar{G}^A t_n^A \rangle,$$

$$\Omega_n^{AR} = \langle t_n^A \bar{G}^A C \bar{G}^R t_n^R \rangle + \sum_{p\neq n}\langle t_n^A \bar{G}^A \Omega_p^{AR} \bar{G}^R t_n^R \rangle,$$

$$\Omega_n^{AA} = \langle t_n^A \bar{G}^A C \bar{G}^A t_n^A \rangle + \sum_{p\neq n}\langle t_n^A \bar{G}^A \Omega_p^{AA} \bar{G}^A t_n^A \rangle,$$

$$\Omega_n^{RK} = \langle t_n^R \bar{G}^R C \bar{G}^R t_n^K \rangle + \langle t_n^R \bar{G}^R C \bar{G}^K t_n^A \rangle$$
$$+ \sum_{p\neq n}\left[\langle t_n^R \bar{G}^R \Omega_p^{RK} \bar{G}^A t_n^A \rangle + \langle t_n^R \bar{G}^R \Omega_p^{RR} \bar{G}^R t_n^K \rangle\right]$$

$$+ \langle t_n^R \bar{G}^R \Omega_p^{RR} \bar{G}^K t_n^A \rangle \Big],$$

$$\Omega_n^{AK} = \langle t_n^A \bar{G}^A C \bar{G}^R t_n^K \rangle + \langle t_n^A \bar{G}^A C \bar{G}^K t_n^A \rangle$$
$$+ \sum_{p \neq n} \Big[\langle t_n^A \bar{G}^A \Omega_p^{AK} \bar{G}^A t_n^A \rangle + \langle t_n^A \bar{G}^A \Omega_p^{AR} \bar{G}^R t_n^K \rangle$$
$$+ \langle t_n^A \bar{G}^A \Omega_p^{AR} \bar{G}^K t_n^A \rangle \Big],$$

$$\Omega_n^{KR} = \langle t_n^K \bar{G}^A C \bar{G}^R t_n^R \rangle + \langle t_n^R \bar{G}^K C \bar{G}^R t_n^R \rangle$$
$$+ \sum_{p \neq n} \Big[\langle t_n^R \bar{G}^R \Omega_p^{KR} \bar{G}^R t_n^R \rangle + \langle t_n^K \bar{G}^A \Omega_p^{AR} \bar{G}^R t_n^R \rangle$$
$$+ \langle t_n^R \bar{G}^K \Omega_p^{AR} \bar{G}^R t_n^R \rangle \Big],$$

$$\Omega_n^{KA} = \langle t_n^K \bar{G}^A C \bar{G}^A t_n^A \rangle + \langle t_n^R \bar{G}^K C \bar{G}^A t_n^A \rangle$$
$$+ \sum_{p \neq n} \Big[\langle t_n^R \bar{G}^R \Omega_p^{KA} \bar{G}^A t_n^A \rangle + \langle t_n^K \bar{G}^A \Omega_p^{AA} \bar{G}^A t_n^A \rangle$$
$$+ \langle t_n^R \bar{G}^K \Omega_p^{AA} \bar{G}^A t_n^A \rangle \Big],$$

$$\Omega_n^{KK} = \langle t_n^K \bar{G}^A C \bar{G}^R t_n^K \rangle + \langle t_n^K \bar{G}^A C \bar{G}^K t_n^A \rangle$$
$$+ \langle t_n^R \bar{G}^K C \bar{G}^R t_n^K \rangle + \langle t_n^R \bar{G}^K C \bar{G}^K t_n^A \rangle$$
$$+ \sum_{p \neq n} \Big[\langle t_n^R \bar{G}^R \Omega_p^{KK} \bar{G}^A t_n^A \rangle + \langle t_n^K \bar{G}^A \Omega_p^{AK} \bar{G}^A t_n^A \rangle$$
$$+ \langle t_n^R \bar{G}^K \Omega_p^{AK} \bar{G}^A t_n^A \rangle + \langle t_n^R \bar{G}^R \Omega_p^{KR} \bar{G}^R t_n^K \rangle$$
$$+ \langle t_n^K \bar{G}^A \Omega_p^{AR} \bar{G}^R t_n^K \rangle + \langle t_n^R \bar{G}^K \Omega_p^{AR} \bar{G}^R t_n^K \rangle$$
$$+ \langle t_n^R \bar{G}^R \Omega_p^{KR} \bar{G}^K t_n^A \rangle + \langle t_n^K \bar{G}^A \Omega_p^{AR} \bar{G}^K t_n^A \rangle$$
$$+ \langle t_n^R \bar{G}^K \Omega_p^{AR} \bar{G}^K t_n^A \rangle \Big].$$

By solving the nine equations from top to down, we can obtain the 9 generalized NVCs to account for the multiple impurity scattering under non-equilibrium condition. As a result, the nine pairwise combinations of $G^{R,A,K}$ can be computed with the generalized NVCs

$$\langle G^R C G^R \rangle = \bar{G}^R (C + \Omega^{RR}) \bar{G}^R, \tag{16.63}$$

$$\langle G^R C G^A \rangle = \bar{G}^R (C + \Omega^{RA}) \bar{G}^A, \tag{16.64}$$

$$\langle G^A C G^R \rangle = \bar{G}^A (C + \Omega^{AR}) \bar{G}^R, \tag{16.65}$$

$$\langle G^A C G^A \rangle = \bar{G}^A (C + \Omega^{AA}) \bar{G}^A, \tag{16.66}$$

$$\langle G^R C G^K \rangle = \bar{G}^R \Omega^{RK} \bar{G}^A + \bar{G}^R (C + \Omega^{RR}) \bar{G}^K, \tag{16.67}$$

$$\langle G^A C G^K \rangle = \bar{G}^A \Omega^{AK} \bar{G}^A + \bar{G}^A (C + \Omega^{AR}) \bar{G}^K, \tag{16.68}$$

$$\langle G^K C G^R \rangle = \bar{G}^R \Omega^{KR} \bar{G}^R + \bar{G}^K (C + \Omega^{AR}) \bar{G}^R, \tag{16.69}$$

$$\langle G^K C G^A \rangle = \bar{G}^R \Omega^{KA} \bar{G}^A + \bar{G}^K (C + \Omega^{AA}) \bar{G}^A, \tag{16.70}$$

$$\langle G^K C G^K \rangle = \bar{G}^R \Omega^{KK} \bar{G}^A + \bar{G}^K \Omega^{AK} \bar{G}^A$$
$$+ \bar{G}^R \Omega^{KR} \bar{G}^K + \bar{G}^K (C + \Omega^{AR}) \bar{G}^K. \tag{16.71}$$

As a result, the average of any two-GF correlator can be obtained as the linear combination of the average of these nine two-GF correlators, For example

$$\langle G^< C G^< \rangle = \frac{1}{4} [\langle G^R C G^R \rangle - \langle G^R C G^A \rangle - \langle G^R C G^K \rangle$$
$$- \langle G^A C G^R \rangle + \langle G^A C G^A \rangle + \langle G^A C G^K \rangle$$
$$- \langle G^K C G^R \rangle + \langle G^K C G^A \rangle + \langle G^K C G^K \rangle]. \tag{16.72}$$

Consequently, with the generalized NVC, the averaged physical properties which contain two-GF correlators, such as averaged non-equilibrium electron density, averaged current, current fluctuation and averaged shot noise can all be computed in a unified way.

16.3.5 Conditionally Averaged Green's Function

In previous sections, we have introduced the generalized CPA-NVC algorithm to calculate various quantum transport properties through disorders, based on the given random potential of the system. In this section, we will discuss how to combine the generalized CPA-NVC with DFT to realize a self-consistent calculation of disordered nano-electronics on the atomic scale. For a disordered system, the central quantity for realizing DFT self-consistent calculation is the conditionally averaged lesser Green's function $\bar{G}^{<,Q}$, which provides the ρ_n^Q required for updating v_n^Q for the Q element on site n in each self-consistent iteration. In general, the conditionally averaged GF $\bar{G}^Q$ is defined with the system in which the n-th site is occupied by the fixed Q element, and the disorder average is carried out for the rest of the disordered sites. Thus, $\bar{G}^Q$ corresponds to the effective medium with Q element occupying the site n, as shown in Figure 16.4(b). In order to calculate $\bar{G}^Q$, we expand it with reference to $\bar{G}$ shown in Figure 16.4(a) by using Eq. (16.40), and obtain

$$\bar{G}^Q = \bar{G} + \bar{G} t_n^Q \bar{G}, \tag{16.73}$$

where

$$t_n^Q = [I - (v_n^Q - \Sigma_{n,im}) \bar{G}]^{-1} (v_n^Q - \Sigma_{n,im}). \tag{16.74}$$

Note that we have used $T = t_n^Q$ since there is only one scattering center in the effective system. One can check that

$$\sum_Q c^Q \bar{G}^Q = \bar{G} \tag{16.75}$$

by applying the single-site CPA condition $\langle t_n \rangle = 0$ in

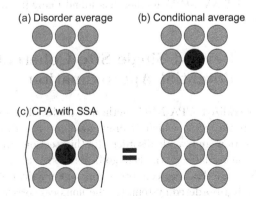

(a) Disorder average   (b) Conditional average

(c) CPA with SSA

FIGURE 16.4 (a) A fully disorder averaged system. (b) A conditionally averaged system. (c) Schematic illustration of CPA with SSA.

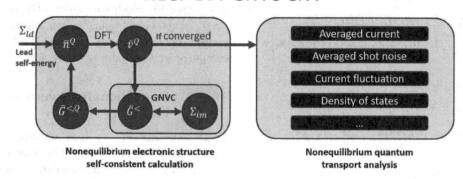

FIGURE 16.5 Schematic description of first-principles simulation of disordered nanoelectronics.

Eq. (16.73). Figure 16.4(c) provides a schematic illustration of the above equation. By substituting with Keldysh's matrices in Eq. (16.73), we obtain

$$\bar{G}^{R,Q} = \bar{G}^R + \bar{G}^R t_n^{R,Q} \bar{G}^R, \tag{16.76a}$$

$$\bar{G}^{A,Q} = \bar{G}^A + \bar{G}^A t_n^{A,Q} \bar{G}^A, \tag{16.76b}$$

$$\bar{G}^{K,Q} = \bar{G}^K + \bar{G}^R t_n^{K,Q} \bar{G}^A$$
$$+ \bar{G}^K t_n^{A,Q} \bar{G}^A + \bar{G}^R t_n^{R,Q} \bar{G}^K, \tag{16.76c}$$

where the matrices $t_n^{R/A/K}$ are defined in Eq. (16.54). With above three conditionally averaged GFs, $\bar{G}^{<,Q}$ can be calculated by the relation

$$\bar{G}^{<,Q} = \frac{1}{2}(-\bar{G}^{R,Q} + \bar{G}^{A,Q} + \bar{G}^{K,Q}). \tag{16.77}$$

The conditionally averaged $\bar{G}^{<,Q}$ provides the nonequilibrium density matrix $\bar{\rho}^Q$ for each disordered element in the system. In combination with DFT, the potential v_n^Q can be computed from the electron density in a self-consistent way. As a result, the effects of disorders on the quantum transport properties can be simulated on the atomic scale from first principles. Figure 16.5 provides the schematic description of the first-principles simulation of disordered nanoelectronics. Presently, we have realized a the Tight-Binding Muffin-Tin Orbital method based first-principles quantum transport simulation package for disordered nanoelectronics. [17,18] Presently, the NEGF-DFT-NVC-CPA method has found many important applications.

16.4 Beyond Single-Site Coherent Potential Approximation

The generalized CPA-NVC method introduced above is based on the SSA, which is only good for low disorder concentration and weak disorder. In this section, we will discuss the available methods to go beyond SSA. SSA-based CPA–NVC method neglects two important physical effects in strongly disordered system: (i) the nonlocal correlations between multiple scattering from different site; (ii) the effect of short-range order including the formation of special clusters and local distortion induced by defects. As the disorder

scattering becomes strong, it becomes crucial to account for the nonlocal correlations and short-range order to accurately simulate the material properties.

A straightforward approach to include effect of short-range order is the so-called molecular-CPA (MCPA) method,[19,20] in which the CPA condition is taken on a cluster of sites instead of a single site. MCPA embeds a real-space cluster into the effective medium. Short-range correlations can be gradually incorporated by increasing the size of the clusters. However, the major problem of MCPA lies in the fact that the self-energy describing the effective medium is only related to the sites inside the cluster, neglecting the coupling or correlation between clusters, and thus breaking the translational invariance of underlying lattice.

To preserve the symmetry of the underlying lattice, Jarrell and Krishnamurthy [21] introduced modified dynamical cluster approximation (DCA) method, also known as the nonlocal CPA (NLCPA). [22] Translational invariance of the effective medium is preserved as a result of consistent treatment of the cluster in real space and coarse-grained Brillouin zone in reciprocal space. The Brillouin zone is sampled over a number of nonoverlapping tiles, labeled by 'cluster momenta', same as the number of sites in the cluster. Then, the self-consistent impurity problem is solved by iterating back and forth via repeated lattice Fourier transforms. The NLCPA method fulfills all the requirements of a successful cluster method [23]: (i) it yields analytic, physically meaningful results; (ii) it preserves the full point-group symmetry of the underlying lattice; (iii) it provides a rigorous, self-consistent treatment of short-range order effects; (iv) it becomes exact in the limit of large cluster sizes and recover the SSA–CPA when the cluster size becomes one; (v) it becomes exact in all physical limits, e.g. dilute and concentrated; (vi) it is computationally feasible. This method has been combined with DFT in the framework of Korringa-Kohn-Rostoker scattering theory[24,25] to study the electronic properties of alloys.

Despite the success of NLCPA, it is still unable to capture the divergent behavior at the Anderson localization transition,[26] which is an important effect in disordered system. This is due to the fact that these effective medium theories adopt the arithmetically averaged density of states (DOS), which does not become critical at the Anderson transition and hence cannot s erve as an order parameter. [27,28] In order to develop an order parameter formalism for Anderson localization, Dobrosavljević et al formulated an effective mean-field theory, called typical-medium theory (TMT), which uses the geometrically averaged (typical) DOS in its self-consistency approach. The typical DOS vanishes at the localization transition and thus can act as an order parameter for Anderson localization. Recently, the typical-medium DCA (TMDCA) [29] extends the single-site TMT to cluster and thereby allows for a systematic inclusion of nonlocal correlations. It is reported that the TMDCA is able to qualitatively and quantitatively describe Anderson localization. [30–33]

Although the above methods are developed to study the equilibrium electronic properties, they all have the potential to be applied to study the nonequilibrium transport properties. Alternatively, diagrammatic technique has been employed to study the nonequilibrium quantum statistics in disordered system, efforts beyond the single-site CPA can be made in this route. For example, including more diagrammatic corrections, which describe short-range correlations, such as the maximally crossed diagram[34] which is responsible for the localization effect. Also, nonlocal correlation has been one of the challenges in the study of strongly correlated electronic system,[35] the diagrammatic-expansion methods developed in that field can be introduced to deal with the disorder problem.

Acknowledgment

This work is supported by ShanghaiTech University start-up fund (Y. Ke).

References

1. H. Haug, A.-P. Jauho, and M. Cardona, *Quantum Kinetics in Transport and Optics of Semiconductors*, Vol. 2. (Springer, Berlin, 2008)
2. J. Rammer and H. Smith. *Rev. Mod. Phys.* **58**, 2 (1986).
3. R. A. Jishi, *Feynman Diagram Techniques in Condensed Matter Physics*. (Cambridge University Press, Cambridge, 2013).
4. A. Kamenev, *Field, Theory of Non-equilibrium Systems*. (Cambridge University Press, Cambridge, 2011).
5. G. Stefanucci, R. van Leeuwen, *Nonequilibrium Many-Body Theory of Quantum Systems*. (Cambridge University Press, Cambridge, 2013).
6. M. Bonitz, *Quantum Kinetic Theory*. (Springer, Berlin, 1998).
7. Van-Nam Do. *Adv. Nat. Sci. Nanosci. Nanotechnol.* **5** 033001 (2014).
8. R. A. Craig, *J. Math. Phys.* **9**, 605 (1968).
9. J. Yan and Y. Ke, *Phys. Rev. B* **94**, 045424 (2016).
10. Ya.M. Blanter, M. Bugttiker, *Phys. Rep.* **336** (2000).
11. C. Beenakker, and C. Schönenberger, *Phys. Today* **56**, 5, 37 (2003).
12. M. Büttiker, *Phys. Rev. B* **46**, 12485 (1992).
13. B. Velicky, S. Kirkpatrick, and H. Ehrenreich, *Phys. Rev.* **175**, 747 (1968).
14. F. S. Khan, J. H. Davies, and J. W. Wilkins, *Phys. Rev. B* **36**, 2578 (1987).
15. P. Soven, *Phys. Rev.* **156**, 809 (1967).
16. D. W. Taylor, *Phys. Rev.* **156**, 1017 (1967).
17. Y. Ke, K. Xia, and H. Guo, *Phys. Rev. Lett.* **100**, 166805 (2008).
18. J. Yan, S. Wang, K. Xia, and Y. Ke, *Phys. Rev. B* **95**, 125428 (2017).
19. M. Tsukada, *J. Phys. Soc. Jpn.* **26**, 684–696 (1969).
20. M. Tsukada, *J. Phys. Soc. Jpn.* **32**, 1475–1485 (1972).
21. M. Jarrell and H. R. Krishnamurthy, *Phys. Rev. B* **63**, 125102 (2001).
22. D. A. Rowlands, *Rep. Prog. Phys.* **72**, 086501 (2009).
23. A. Gonis, *Green Functions for Ordered and Disordered Systems*. Studies in Mathematical Physics, vol. 4. (North-Holland, Amsterdam, 1992).
24. D. A. Rowlands, et al., *Phys. Rev. B* **72**,045101 (2005).
25. D. A. Rowlands, et al., *Phys. Rev. B* **73**, 165122 (2006).
26. P. W. Anderson, *Phys. Rev.* **109**, 1492 (1958).
27. V. Dobrosavljević, A. A. Pastor and B. K. Nikolić, *Europhys. Lett.* **62**, 76 (2003).
28. V. Dobrosavljević, *Int. J. Mod. Phys. B* **24**, 1680 (2010).
29. C. E. Ekuma, et al., *Phys. Rev. B* **89**, 081107(R) (2014).
30. H. Terletska, et al., *Phys. Rev. B* **90**, 094208 (2014).
31. Y. Zhang, et al., *Phys. Rev. B* **92**, 205111 (2015).
32. C. E. Ekuma, et al., *Phys. Rev. B* **92**, 201114(R) (2015).
33. H. Terletska, et al., *Phys. Rev. B* **95**, 134204 (2017).
34. P. A. Lee and T. V. Ramakrishnan, *Rev. Mod. Phys.* **57**, 287 (1985).
35. G. Rohringer, et al., arXiv:1705.00024v2.

Ab Initio Simulations of Carboxylated Nanomaterials

Vivian Machado de Menezes
Universidade Federal da Fronteira Sul (UFFS)

Ivi Valentini Lara
Pontiffcia Universidade Católica do Rio Grande do Sul (PUCRS)

17.1 Introduction

Currently, graphene, carbon nanotubes (CNTs) and phosphorene have gained prominence as potential candidates for nanoscale use, especially in electronic devices (Kauffman and Star 2010, Yáñez-Sedeño et al., 2010, Liu et al., 2015) due to their favorable electronic, chemical and transport characteristics, in addition to its high chemical stability (Saito et al. 1998, Tans et al. 1997, Sun et al. 2011, Carvalho et al. 2016, Liu et al. 2014). Because of these interesting properties, these nanomaterials arouse the interest of the scientific community for a variety of applications in several areas, but these applications often require control or some changes in the properties of pristine nanostructures.

An interesting tool to control the properties of nanomaterials is the functionalization process (de Menezes 2016), which can be performed through the walls or ends by encapsulation (in the case of CNTs), adsorption of atoms or molecules directly, substitutional doping (de Souza Filho and Fagan 2007), deformation (Fagan et al. 2003) and adsorption of chemical functional groups (Veloso et al. 2006).

Nanotechnology represents a promise to improve people's quality of life, whose expectation is to produce more efficient and cost-effective products, with less aggression to the environment. However, for the "nano" products to reach the shelves, much research is necessary to ensure the safety and efficiency of these products (Vieira Segundo and Vilar 2016). In this context, modeling and computational simulation play an essential role, from which many of the currently available data are obtained, and from which it is predicted, with high accuracy for many cases, results that could not be well described experimentally (Menezes and Fagan 2012).

In this chapter, we will discuss the role of molecular modeling to predict changes in electronic and structural properties of CNTs, graphene and phosphorene caused by functionalization with carboxyl functional groups.

17.2 Nanomaterials

The new electronic, mechanical, magnetic and vibrational properties of nanostructures are both the effect of decreasing their size and its association with the different allotropic forms of atoms (we should highlight nanomaterials based on carbon (Choudhary, Hwang and Choi 2014), phosphorus (Lee, Kim and Won 2016), silicon (Peng et al. 2014), boron (Li 2009) and the transition metal dichalcogenides (TMDs) (Liu et al. 2015)). These properties are especially related to the hybridization between the atoms and their electronic confinement. The scheme in Figure 17.1 represents the dimensionality of such structures as a function of its continuous reduction, from macroscopic dimensions to dimensions smaller than 100 nm. Table 17.1 describes the electronic quantum confinement in one, two and three dimensions:

It is evident that this reduction in the materials' dimensionality implies directly in its electronic behavior and, consequently, mechanical, thermal and magnetic ones. Therefore, new properties are observed according to the type of confinement. Here, we present three of the most studied nanomaterials nowadays: graphene, CNT and phosphorene.

17.2.1 Graphene

Graphene is a very versatile carbon nanomaterial. This important feature is mainly due to its large surface area (Qian, Ismail and Stein 2014), high intrinsic mobility

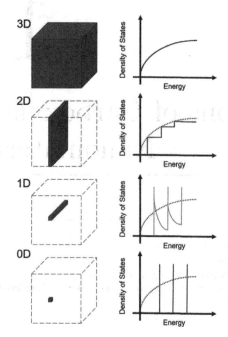

FIGURE 17.1 Reduction on materials' dimensionality and respective density of states quantum confinement.

TABLE 17.1 Definition of 3D, 2D, 1D and 0D Materials in Terms of its Electronic Confinement

3D structures	Composed of a large number of atoms; electrons in each atom are subject to interactions with neighboring atoms, and their energy levels add up to form energy bands. Example: diamond
2D structures	The dimension in one direction is reduced to the nanoscale, in this case, the electrons move freely in states of energy allowed in the x and y directions (for instance), but their density of electronic states becomes quantized in the z direction. Example: graphene
1D structures	The dimension is reduced in two directions, the electrons move freely in only one direction (in the allowed energy states) and in the y and z directions (for instance), their density of states is quantized. Example: CNT
0D structures	Confinement is present in the three space directions and the electrons present their energies fully quantized. Example: fullerene

(Bolotin et al. 2008), high Young's modulus (Memarian, Fereidoon and Ganji 2015), high thermal conductivity (Pop, Varshney and Roy 2012), optical transmittance (Zhu, Yuan and Janessen 2014) and good electrical conductivity (Marinho et al. 2012). All these properties permit many potential applications, and for this reason, it has been extensively studied both experimentally and theoretically, since it was first obtained as a carbon monolayer by Novoselov in 2004. Researchers have scrutinized not only its pristine form but also the use of involving chemically modified graphene to make new materials, for example, the graphene oxide (GO) (Chen, Feng and Li 2012) and graphene nanoribbons (GNRs) (Zhihong et al. 2007).

A large number of applications refer to graphene, both on its pristine form or as GO and GNR. Besides its well-known huge electronic potential applications in the semiconductors

and transistors industry, this nanostructure is predicted to be used as membranes for water filtration and desalination (Homaeigohar and Elbahri 2017), gas separation (Yoo et al. 2017) and as composites and coatings for the development of lighter, stronger and safer materials (Singh et al. 2011).

From the energy point of view, some researchers show graphene-based materials as promising candidates for supercapacitors (Ke and Wang 2016) and solar cells (Lin et al. 2016). In the biomedical context, graphene has a potential for targeted drug delivery, as well as biochemical sensors, i.e., working as detectors of individual events on a molecular level (Tomita, Ishihara and Kurita 2017).

In order to understand the structure of the graphene, it is convenient that its lattice representation be made in the reciprocal space (space of $\vec{k}$ wave vectors). Its periodicity allows its electronic properties (as for the other nanostructures that will be presented in this chapter) to be described in a much simpler and more concise manner in this space. Figure 17.2a illustrates the vectors in real space and Figure 17.2a in the reciprocal space. Thus, in the transformation of these vectors from direct space to reciprocal, we have that

$$\vec{a_1} = \left(\frac{\sqrt{3}a}{2}, \frac{a}{2} \right) \text{ and } \vec{a_2} = \left(\frac{\sqrt{3}a}{2}, -\frac{a}{2} \right) \quad (17.1)$$

become

$$\vec{b_1} = \left(\frac{2\pi}{\sqrt{3}a}, \frac{2\pi}{a} \right) \text{ and } \vec{b_2} = \left(\frac{2\pi}{\sqrt{3}a}, -\frac{2\pi}{a} \right), \quad (17.2)$$

once, a is the interatomic distance and its lattice constant is equal to 2.46 Å.

The description of the unique electronic structure of graphene is done in terms of the energy dispersion relations in its two-dimensional lattice. The relationships in the first Brillouin zone (the Wigner–Seitz cell of the reciprocal space) are made along the high symmetry axes for the bonding and antibonding band.

Considering temperature of 0 K, all electrons occupy the valence band, and the conduction band remains empty. Since the states' density at the Fermi level is zero, graphene is assumed as a null gap semiconductor (Dresselhaus, Dresselhaus and Avouris 2000).

The calculation of these relationships is done using the tight-binding energy model, and its approximation to the electronic structure of the graphene lattice is

$$E_{2D}(k_x, k_y)$$
$$= \pm t \sqrt{1 + 4\cos\left(\frac{\sqrt{3}k_x a}{2}\right)\cos\left(\frac{k_y a}{2}\right) + 4\cos^2\left(\frac{k_y a}{2}\right)}, \quad (17.3)$$

where, k_x and k_y are the allowed wave vectors in the first Brillouin zone, t is the transfer integral and a is the graphene lattice constant.

When graphene is a narrow ribbon, a GNR, the 1D confinement takes place and the electronic confinement modules its band (similar to CNTs, which will be presented in the next section), and the energy gap depends on the GNR width and type (Son, Cohen and Louie 2007). The GNR can present zigzag, armchair or chiral edges, and different electronic (Yu et al. 2008), structural (Faccio et al. 2009) and magnetic (Son, Cohen and Louie 2006) properties can be explored.

17.2.2 Carbon Nanotubes

In 1991, multiwall CNTs (MWCNTs) were observed by Iijima, and two years later, single-walled CNTs (SWCNTs) were studied for the first time by Iijima and Ichihashi (1993) and Bethune et al. (1993). The SWCNT finding is of remarkable importance because a single-layer nanotube represents the basis of theoretical studies and predictions that precede experimental observations.

MWCNTs present an elevated elastic modulus ($\sim$1 TPa) and tensile strength (Khandoker et al. 2011), which is higher than the fiber used in industry actually. SWCNTs can present elevated thermal conductivity (Han and Fina 2011) higher than the diamonds' conductivity. These features permit the usage of CNTs in composite materials (Senokos et al. 2018), in thin films and in special coats (which give more lightness, resistance and flexibility

to the material (Iijima, Brabec and Maiti 1996)), in microelectronics through the supercapacitors (Pan, Li and Feng 2010) and transistors development (Kim et al. 2016), as well as in biotechnology, participating in the process of drug delivery (Kushwaha et al. 2013) and sensing biomolecules (Satishkumar et al. 2007).

CNTs are theoretically conceived as a graphene sheet properly wrapped in a cylinder of nanometer diameter, so that the electronic properties of nanotubes can be determined in consideration of those from graphene. These properties are of special interest, since their prediction can show metallic or semiconducting CNTs depending only on their diameter and chirality (Dresselhaus, Dresselhaus and Avouris 2000).

Conveniently, the SWCNT structure is described in terms of its 1D unit cell, defined as the rectangle formed by the combination of the $\vec{T}$ translational vector and the $\vec{C}_h$ chiral vector (Figure 17.3).

$$\vec{C}_h = n\vec{a_1} + m\vec{a_2}, \qquad (17.4)$$

(with n and m integers and $n \geq m \geq 0$), which connects two crystallographically equivalent points, being a combination of the $\vec{a_1}$ and $\vec{a_2}$ graphene base vectors,

$$|\vec{a_1}| = |\vec{a_2}| = 3^{\frac{1}{2}}(a) = 3^{\frac{1}{2}}(1.42) = 2.46\text{Å}. \qquad (17.5)$$

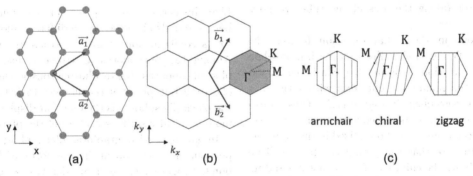

FIGURE 17.2 Graphene vectors in real space (a) and in the reciprocal space (b). Cut lines regarding each CNT's electronic character (c).

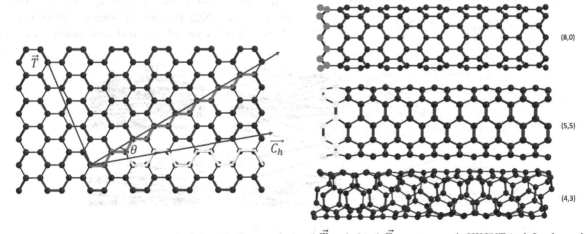

FIGURE 17.3 Graphene structure on the left with the translational $\vec{T}$ and chiral $\vec{C}_h$ vectors, each SWCNT is defined, on the right, as the rolled rectangle formed from the combination of these vectors.

The CNT structure depends on the n and m indexes (n,m), which specify the chiral vector. On the same way, the $\vec{T}$ vector is

$$\vec{T} = t_1 \vec{a_1} + t_2 \vec{a_2}, \tag{17.6}$$

with t_1 and t_2 as integers. Since $\vec{C_h}$ and $\vec{T}$ are perpendicular to each other $\left(\vec{C_h} \cdot \vec{T} = 0 \right)$,

$$t_1 = \frac{2m + n}{x} \text{ and } t_2 = -\frac{2n + m}{x} \tag{17.7}$$

where x is the maximum common divisor of $2n + m$ and $2m + n$. Therefore, if we know $\vec{C_h}$ and $\vec{T}$, the N number of hexagons contained in the rectangle can be determined

$$N = \frac{\left| \vec{C_h} \times \vec{T} \right|}{|\vec{a_1} \times \vec{a_2}|} = \frac{2 \left(n^2 + m^2 + nm \right)}{x}. \tag{17.8}$$

Three distinct types of structures for CNTs can be generated through the wrapping of the graphene sheet. The zigzag $(n,0)$ and armchair (n,n) types, respectively, correspond to the chiral angles $\theta = 0°$ and $\theta = 30°$, whereas the chiral nanotubes (n,m) correspond to $0° < \theta < 30°$, as it can be seen in Figure 17.3, on the right.

For the CNT, considered to be infinitely long, periodic contour conditions in the circumferential direction (due to the surface curvature effects) denoted by the chiral vector are used, the wave vector associated with this direction becomes quantized, whereas the vector associated to the translational vector (along the axis of the tube) remains continuous.

The confinement in the radial direction limited the number of states allowed in this direction much more than that in the axial direction. This number can be evaluated as a function of the number of cut lines within the first Brillouin zone of the graphene in Figure 17.2c. These states are parallel lines in k space, being continuous in the direction of growth of the tube (z) and quantized in the xy plane. It is important to note that the permitted states will be determined mainly by the chirality of the nanotube and its diameter.

Thus, with periodic boundary conditions, the allowed values for k_x and k_y in the circumferential direction are

$$k_x^q = \frac{q2\pi}{n\sqrt{3}a} \text{ and } k_y^q = \frac{q2\pi}{na} \tag{17.9}$$

for the armchair and zigzag nanotube, respectively. So, the dispersion relations are

$$E_q^{\text{armchair}} = \pm t \sqrt{1 + 4\cos\left(\frac{q\pi}{n}\right)\cos\left(\frac{ka}{2}\right) + 4\cos^2\left(\frac{ka}{2}\right)} \tag{17.10}$$

and

$$E_q^{\text{zigzag}} = \pm t \sqrt{1 + 4\cos\left(\frac{\sqrt{3}ka}{2}\right)\cos\left(\frac{q\pi}{n}\right) + 4\cos^2\left(\frac{q\pi}{n}\right)} \tag{17.11}$$

with $-\pi < ka < \pi$ and $q = 1, \ldots, 2n$.

If for a given nanotube (n,m) the cut lines pass over the k-point, the energy bands will have zero gap and the nanotube will have a metallic character. Otherwise, if the cut lines do not pass over the k-point, the nanotube will have a semiconductor character, with a finite energy gap between the valence and conduction bands. In particular, armchair nanotubes will always present a metallic character, whereas zigzag nanotubes are usually semiconductors, and they will only have semimetallic character when n is a multiple of three. Figure 17.4 illustrates the dispersion relations of nanotubes with semiconducting, metallic and semiconducting character with a certain gap.

17.2.3 Phosphorene

Like the carbon atom, phosphorus is found in all living beings and also is presented in different allotropic forms due to its configuration of five electrons in the valence layer. Among the phosphorus allotropic forms: red, black and blue phosphorus, white phosphorous is the most common allotrope but presents the highest toxicity. Among these, the black phosphorus is the one that displays the most thermodynamically stable structure of the phosphorus atom.

In analogy with graphite, black phosphorus and blue phosphorus also appear in the form of layers of atoms bonded together through van der Waals bonds. Each layer of phosphorus is called black phosphorene and blue phosphorene. The black phosphorus has an orthorhombic structure and was obtained in 1916 by heating white phosphorus under high pressure (Bridgman 1914), but only in 1960s, when the monolayers had been isolated, the inherent

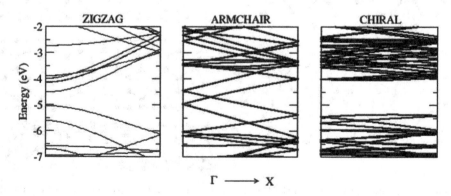

FIGURE 17.4 SWNTs' band structure for an (8,0) armchair, (5,5) zigzag and a (4,3) chiral CNT.

interesting properties of the black phosphorus were actually observed (Warschauer, 1962). Predicted only theoretically, recently in 2017, blue phosphorene was first synthesized through a (111) gold substrate adopting the process of molecular beam epitaxy (Han, Gao and Zhao 2017).

Black phosphorene has been investigated for its potential application in electronic devices and sensors (Cui et al. 2015). It is known that phosphorene-based devices can exhibit high charge carrier mobility (Liu et al. 2014), indicating that this nanostructure can be used in field-effect transistors (FETs). The structure of phosphorene may show high photo-sensitivity, and for this reason, high-performance spectral photodetectors can be developed (Li et al. 2018). In addition, the sensing property of this nanomaterial has been extensively studied, and it seems to be one of the most sensitive materials in the gas analysis (Kou, Frauenheim and Chen 2015). Finally, the phosphorene nanoribbons are predicted to have even better electrochemical performance than the monolayer.

Due to phosphorus sp^3 hybridization in phosphorene, each of the p orbitals retains a solitary pair of electrons. Each layer of the honeycomb network contains two atomic layers with phosphorus distance in the plane equals to 2.22 Å and between top and bottom atoms equals to 2.24 Å. These distances reflect the connections between the 3p orbitals (Cho, Yang and Lu 2017). The black phosphorene has also an orthorhombic structure, and its lattice vectors $\vec{a_1}$ and $\vec{a_2}$, with $|\vec{a_1}| \neq |\vec{a_2}|$, are presented in Figure 17.5a, whereas the blue phosphorene evidenced in Figure 17.5b has a hexagonal structure with $|\vec{a_1}| = |\vec{a_2}|$, both perpendicular to each other.

Both blue and black phosphorene have a semiconducting character, and the estimated energy band gap varies between 0.3 and 2 eV (Ferreira and Ribeiro 2017).

In addition to the 3D and 2D allotropic phosphorous structures, nanoribbons (Zhang et al. 2014) and phosphorene nanotubes (Hongyan et al. 2014) are already known, and theoretically, a new phase of phosphorus, green phosphorus, is predicted, i.e., a new phosphorus allotrope with a direct band gap and high mobility (Woo et al. 2017). These structures open new perspectives in the development of new materials with properties still unexplored.

17.3 The First Principles Method

In an attempt to predict more accurately the physical and chemical properties of solids and molecules, quantum mechanics methods have been used for the study of systems containing up to a few hundred atoms. These systems require a precise solution of the Schrödinger equation, which itself is easily constructed for a multibody system. However, due to methodological limitations, it is extremely complicated or even impossible to directly solve the Schrödinger equation beyond the simpler systems without the use of approximations (Menezes and Fagan 2012).

In this section, we will describe the method used to perform the *ab initio* molecular modeling of the studied systems (pristine and functionalized nanostructures). To describe the structural and electronic properties of the systems as accurately as possible, we will use a first principles approach, which corresponds to the state of the art in computational physics, where we must solve the Schrödinger equation for many electron systems.

The nonrelativistic and time-independent Schrödinger equation for a system composed of electrons and nuclei can be written generally as

$$\hat{H}\left(\vec{r}, \vec{R}\right) \Psi\left(\vec{r}, \vec{R}\right) = E\Psi\left(\vec{r}, \vec{R}\right), \qquad (17.12)$$

where

$$\hat{H}\left(\vec{r}, \vec{R}\right) = \hat{T}_i\left(\vec{R}\right) + \hat{V}_i\left(\vec{R}\right) + \hat{T}_e\left(\vec{r}\right) + \hat{V}_e\left(\vec{r}\right)$$
$$+ \hat{V}_{ie}\left(\vec{r}, \vec{R}\right), \qquad (17.13)$$

being $\hat{H}(\vec{r}, \vec{R})$ the Hamiltonian operator of the system, $\hat{T}_i(\vec{R}) = \sum_k \frac{P_k^2}{2M_k}$ the ionic kinetic energy operator,

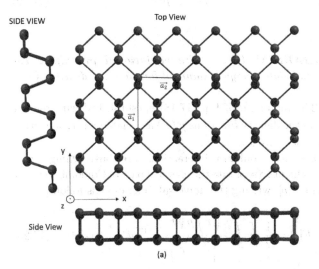

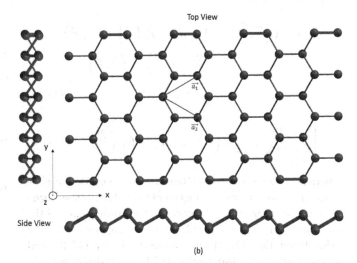

FIGURE 17.5 Atomic structures for (a) black and (b) blue phosphorene.

$\hat{V}_i(\vec{R}) = \frac{1}{2}\sum_{kk'}\frac{Z_k Z_{k'}}{\left|\vec{R}_k - \vec{R}_k'\right|}$ the ion–ion interaction potential,

$\hat{T}_e(\vec{r}) = \sum_i \frac{P_i^2}{2}$ the electronic kinetic energy operator,
$\hat{V}_e(\vec{r}) = \frac{1}{2}\sum_{ij}\frac{1}{\left|\vec{r}_i - \vec{r}_j\right|}$ the electron–electron interaction

energy, $\hat{V}_{ie}(\vec{r}, \vec{R}) = \sum_{ki}\frac{Z_k}{\left|\vec{R}_k - \vec{r}_i\right|}$ the ion–electron inter-

action potential and E the numerical value of the energy described by the state $\Psi(\vec{r}, \vec{R})$.

In solving the Schrödinger equation, each solution is given by a wave function Ψ, which corresponds to a quantized level of energy. This solution consists of a mathematical tool used by quantum mechanics to determine the physical and chemical properties of the electron at that level, the state and the expectation values of the physical quantities of any physical system (Eisberg and Resnick 1979).

Considering the fact that matter is composed of positively charged nuclei and electrons, and that nuclei are thousands of times heavier than electrons, one might think that electrons are much faster than nuclei, "realizing" nuclei as immobile. As a good approximation, one can consider that the electrons in a molecule move in a fixed field formed by the nuclei. This enables us to approach electronic and nuclear motions separately. This is the basic hypothesis of the Born–Oppenheimer approximation (Born and Oppenheimer 1927), and it is almost invariably the first simplification in any application of quantum mechanics to molecules and crystals. By decoupling the wave function from many bodies as a product of wave functions, we have

$$\Psi\left(\vec{r}, \vec{R}\right) = \varphi\left(\vec{R}\right)\chi\left(\vec{r}, \vec{R}\right), \qquad (17.14)$$

where $\varphi(\vec{R})$ is the wave function dependent on nuclear positions (ionic wave function) and $\chi(\vec{r}, \vec{R})$ is the function of the electrons distribution for a fixed nuclear arrangement (electronic wave function).

Substituting (17.14) into equation (17.12) and separating variables, we have

$$\hat{H}_{e+i}\chi\left(\vec{r}, \vec{R}\right) = \left[\hat{V}_i\left(\vec{R}\right) + \hat{T}_e\left(\vec{r}\right) + \hat{V}_e\left(\vec{r}\right)\right.$$
$$\left. + \hat{V}_{ie}\left(\vec{r}, \vec{R}\right)\right]\chi\left(\vec{r}, \vec{R}\right)$$
$$= E_{e+i}\chi\left(\vec{r}, \vec{R}\right), \qquad (17.15)$$

$$\left[\sum_k \frac{P_k^2}{2M_k} + E_{e+i}\left(\vec{R}\right)\right]\varphi\left(\vec{R}\right) = E\varphi\left(\vec{R}\right), \qquad (17.16)$$

where $\hat{H}_{e+i}$ is the electronic Hamiltonian operator (for electrons in a fixed nuclear arrangement) and E_{e+i} is the Born–Oppenheimer energy of the system. The consequence of the imposition of expression (17.14) is the separation of the Schrödinger Eq. (17.12) into an electronic Eq. (17.15) and an equation for nuclear motion (17.16) (classically solved), being the description of the nuclear and electronic motions

made separately. Equation (17.15) describes the electrons dynamics, but it cannot be solved without the use of approximations either. The Born–Oppenheimer approximation is extremely useful and used in most theoretical studies (Soler et al. 2002, Vianna et al. 2004).

In a system with a large number of constituents, to address the electronic problem, the wavefunction Ψ plays a fundamental role. However, to solve this problem, the N electrons in Schrödinger equation with wave function of $3N$ variable can be written as an electronic density equation with only three variables. This solution of the Schrödinger equation for many electron systems, whose fundamental variable is the electronic density in detriment of the wavefunction, constitutes the density functional theory (DFT).

Hohenberg and Kohn (1964) established the connection between electronic density and the Schrödinger equation for many bodies. They have shown that the ground-state energy can be obtained by calculating a universal form for the density functional of a ground-state electron system interacting with an external potential. The DFT is based on Hohenberg and Kohn theorems.

THEOREM 17.1   *The density as basic variable:* *The external potential $v(\vec{r})$ is a unique functional of electronic density, apart from an additive constant.*

The charge density is represented by

$$\rho\left(\vec{r}\right) = \sum_{i=1}^{N}\left|\psi_i\left(\vec{r}\right)\right|^2, \qquad (17.17)$$

and the energy as density functional for a given external potential $v(\vec{r})$ becomes

$$E[\rho] = \int v\left(\vec{r}\right)\rho\left(\vec{r}\right)d^3\vec{r} + F[\rho], \qquad (17.18)$$

where $F[\rho]$ is a universal functional, valid for any electronic system and any external potential and is given by

$$F[\rho] = T[\rho] + U[\rho]. \qquad (17.19)$$

THEOREM 17.2   *The variational principle:* *The ground-state energy is minimal for the exact density $\rho(\vec{r})$.*

The minimum of Eq. (17.18) is established with respect to all the density functions $\rho'(\vec{r})$ associated with some other external potential $v'(\vec{r})$.

Because Coulomb interactions are far-reaching, it is convenient to separate the classical part of Coulomb's energy from $F[\rho]$, writing the universal functional as follows:

$$F[\rho] = \frac{1}{2}\int\int\frac{\rho\left(\vec{r}\right)\rho\left(\vec{r}'\right)}{\left|\vec{r} - \vec{r}'\right|}d^3\vec{r}\,d^3\vec{r}' + G[\rho], \qquad (17.20)$$

where $G[\rho]$ is a universal functional like $F[\rho]$.

$E[\rho]$ then becomes

$$E[\rho] = \int v\left(\vec{r}\right) \rho\left(\vec{r}\right) d^3\vec{r}$$
$$+ \frac{1}{2} \int \int \frac{\rho\left(\vec{r}\right)\rho\left(\vec{r}'\right)}{\left|\vec{r}-\vec{r}'\right|} d^3\vec{r}\, d^3\vec{r}' + G[\rho]. \quad (17.21)$$

In 1965, Kohn and Sham (1965) proposed that the functional $G[\rho]$ could be written in the form

$$G[\rho] = T[\rho] + E_{xc}[\rho], \quad (17.22)$$

where $T[\rho]$ is the kinetic energy of a noninteracting electronic system with density $\rho(\vec{r})$, and $E_{xc}[\rho]$ is, by the definition of Kohn and Sham, the exchange and correlation energy of an interacting system with density $\rho(\vec{r})$. The exchange term comes from Pauli Exclusion Principle, which says that two electrons cannot simultaneously occupy the same quantum state, being the wave function antisymmetric, and the correlation comes from the fact that as a consequence of the Coulombian interaction, the electrons tend to repel each other, causing the motion of one electron to be correlated with the others.

The minimum condition for the energy functional must be restricted by a condition, causing the number of electrons of the system to be given correctly. According to the variational principle, the condition of the electronic charge can be expressed by

$$\int \rho\left(\vec{r}\right) d^3\vec{r} = N \Leftrightarrow \int \rho\left(\vec{r}\right) d^3\vec{r} - N = 0. \quad (17.23)$$

Minimizing the functional Eq. (17.21) (including the condition of constant number of particles) in relation to the electronic density,

$$\delta\left\{ E[\rho] - \mu\left[\int \rho\left(\vec{r}\right) d^3\vec{r} - N \right] \right\} = 0, \quad (17.24)$$

we obtain

$$\int \delta\rho\left(\vec{r}\right) \left\{ \frac{\delta T}{\delta\rho} + v\left(\vec{r}\right) + \int \frac{\rho\left(\vec{r}'\right)}{\left|\vec{r}-\vec{r}'\right|} d^3\vec{r}' \right.$$
$$\left. + v_{xc}[\rho] - \mu \right\} d^3\vec{r} = 0, \quad (17.25)$$

where μ is a Lagrange multiplier and represents the chemical potential of the system and v_{xc} is the exchange and correlation potential, given by

$$v_{xc}(\rho) = \frac{\delta E_{xc}}{\delta\rho}. \quad (17.26)$$

The solution of the Eq. (17.25) can be obtained by solving the Schrödinger equation of a particle, which satisfies Eq. (17.17)

$$\hat{H}_{KS}\psi_i\left(\vec{r}\right) = \left(-\frac{1}{2}\nabla^2 + v_{eff}[\rho] \right)\psi_i\left(\vec{r}\right) = \varepsilon_i \psi_i\left(\vec{r}\right), \quad (17.27)$$

where

$$v_{eff} = v\left(\vec{r}\right) + \int \frac{\rho\left(\vec{r}'\right)}{\left|\vec{r}-\vec{r}'\right|} d^3\vec{r}' + v_{xc}(\rho). \quad (17.28)$$

Equations (17.17), (17.27) and (17.28) are known as Kohn–Sham equations, and their solutions (the Kohn–Sham orbitals ψ_i) are obtained by self-consistent calculations. Through the Kohn and Sham formulation (Kohn and Sham 1965), the possibility that the multielectronic problem is represented by an equivalent set of self-consistent equations of an electron that moves under the action of an effective potential produced by the others electrons is confirmed.

Although the Hohenberg and Kohn theorems show that the total energy can be written as a unique functional of the electronic density of the ground state, to solve this functional, we still have the problem of exchange and correlation energy, which does not have its exact expression known. The terms of exchange and correlation are the terms with the most difficult physical interpretation, and to overcome the difficulty of their imprecision, several different schemes have been developed in order to obtain their functional approximation, being the most used: local density approximation – LDA (Ceperly and Alder 1980), which assumes that the electronic *density* can be treated *locally* and can also be applied to spin-polarized systems; generalized gradient approximation – GGA (Perdew et al. 1996), which considers second-order aspects of electronic density, where the energy will depend on the electronic density at the point as well as the gradient of the electronic density at the point in question; meta-GGA functional (Tao et al. 2003), which includes the Laplacian of the density or the kinetic energy density in addition to the density and magnitude of the gradient of the density and hybrid functionals, which are a mixture of a DFT and a Hartree–Fock calculations, B3LYP being (Stephens et al. 1994) a very widely used functional.

In the calculations of this chapter, we used LDA for the exchange and correlation term, which is one of the most used approaches in DFT and also the simplest one. It is considered here that electrons behave as a homogeneous gas with few variations in density. By approximation, each volume element would contribute to the exchange and correlation term as if it were a system with the same density.

In the functional form, for this approximation, the exchange and correlation energy is given by

$$E_{xc}[\rho] = \int \rho(\vec{r})\varepsilon_{xc}(\rho(\vec{r}))d\vec{r}, \quad (17.29)$$

where $\varepsilon_{xc}(\rho)$ is the exchange energy plus the correlation energy per electron of a homogeneous electron gas with density ρ.

The LDA is based on the uniform electron gas model, assuming that the functional exchange and correlation energy is purely local, and it is exact for systems with uniform electronic density. Ceperly and Alder (1980) were able to accurately estimate the correlation effects for a free electron gas system using Quantum Monte Carlo methods, whose results are parameterized by Perdew and Zunger (1981). This parameterization has been shown to be very efficient, being the scheme adopted in this work in the LDA approach.

To simplify the electronic structure calculations, the pseudopotential method is introduced, where the substitution of the core electrons (electrons of the innermost layers of the atom) and the strong ionic potential by a pseudopotential that acts on valence wave pseudofunctions is performed. In other words, the strong Coulombian potential is replaced by a smoother potential, the pseudopotential, and the valence wave functions are replaced by a smooth and nodeless pseudofunction, equal to wavefunction of the valence electrons (electrons of the atomic outer layer), besides a cutoff radii. As a consequence, less computational effort is required (Menezes and Fagan 2012).

Troullier and Martins (1991) made an implementation to generate a refined pseudopotential that leads to a rapid convergence in the calculated total energy of the system, and consequently, a fast convergence of the properties related to the basis functions, being considered a soft pseudopotential. This pseudopotential was the type used in performing the calculations in this chapter.

To perform the *ab initio* simulations, we used the SIESTA code (Spanish Initiative for Electronic Simulation with Thousands of Atoms) (Soler et al. 2002), which is a computational code based on DFT for the realization of first principles calculations of electronic structure and molecular dynamics simulations of solids and molecules. Written in FORTRAN90, it is an open source program that allows solutions with up to thousands of atoms. It solves the Kohn–Sham equations in a self-consistent way, using *norm-conserving* pseudopotentials. It uses periodic boundary conditions (to make systems periodically infinite, which are the case for most solids and crystalline structures) and strictly localized basis constructed as a linear combination of atomic orbitals (LCAOs).

In the calculations of this chapter, double-ζ plus polarization (DZP) basis sets were used, whose confinement of these basis functions was given in terms of an energy shift of 0.05 eV. For the structural analysis, the studied systems had their geometries optimized through the Conjugate Gradient (CG) method, an iterative method that tends to find the minimum energy. This technique provides an efficient method to locate the minimum of the energy functional of Kohn–Sham formulation, allowing the calculation of the atomic forces and the total energy of the systems (Soler et al. 2002, Martin 2004).

17.4   Study of Functionalized Nanomaterials' Properties

Once CNTs have apolar character, high stability and chemical inertia, its application in systems that need their dispersion in polar solvents is hampered, making its biological application impossible, what makes necessary to functionalize it, for instance. At the same time, even without hydrophobic character, the interaction of phosphorene with other molecules is facilitated when its structure is modified through the functionalization process. The COOH group is considered as a standard group for the functionalization of different nanostructures, and its interaction with graphene, CNTs and phosphorene will be analyzed sequentially in this work.

In order to characterize the interaction between the systems, the binding energy (E_{B}) between the COOH groups and each nanostructure studied was quantified by the following equation:

$$E_{\mathrm{B}} = -[E_{\mathrm{Total}} - E_{\mathrm{Nanostructure}} - E_{\mathrm{COOH}}], \qquad (17.30)$$

where E_{Total} is the total energy of the functionalized nanosystem, $E_{\mathrm{Nanostructure}}$ is the energy of pristine nanostructure and E_{COOH} is the energy of an isolated COOH group. For all configurations, the charge transfer and the presence or absence of spin polarization were analyzed.

17.4.1   Carboxylated Graphene

One of the methods for obtaining graphene on a large scale is by reducing the GO. This method allows its production in pristine form or, previously, in its form containing groups with oxygen, carbon and hydrogen (Chen, Feng and Li 2012). In addition, its interaction with biological molecules and their potential in application in polar media is facilitated when its structure is carboxylated. Theoretical studies show that the position in which the COOH groups bind to carbon nanostructures influences the electronic properties for both the graphene, CNTs and phosphorene (shown in the sequence) (Saidi 2013, Zanella et al. 2008, Garcia-Lastra et al. 2008, Santos, Sanchez-Portal and Ayuela 2011, Tonel et al. 2016).

The dimensions of the supercell used are 25.94 × 20 × 14.98 Å^3 in the graphene calculations and 40 × 40 × 14.98 Å^3 for the x, y and z axes for the nanoribbon, which is a hydrogenated zigzag nanoribbon (ZGNR). For the integration of the Brillouin zone, 3 × 1 × 3 k-points were generated through the Monkhorst–Pack scheme (Monkhorst and Pack 1976) for both graphene and GNR.

Figure 17.6 shows the optimized structure of (a) pristine and (b) carboxylated graphene and in (c) pristine and (d) carboxylated ZGNR. The binding distances between the carbon atoms change so that the carboxylic group deforms the nanostructures locally by pulling the carbon up the plane, increasing the C–C binding length from 1.44 to 1.52 Å at the carbon atoms in graphene lattice where the functional group is connected.

Table 17.2 shows values for distance $d_{\mathrm{C-C}}$ (between COOH carbon atom and the nanostructures carbon atom) and binding angle θ_{CCC} between the carbon atoms of the carboxyl group and graphene, binding energy E_{B}, charge transfer and the resulting spin polarization for each

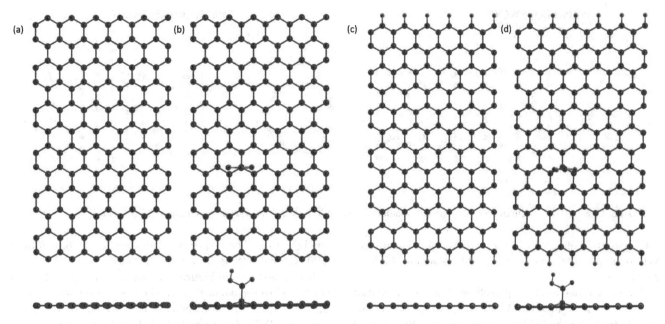

FIGURE 17.6 Atomic structures obtained for (a) pure and (b) carboxylated graphene and (c) pure and (d) carboxylated ZGNR.

TABLE 17.2 Bond Distance, Binding Angle, Binding Energies for the COOH Group, Electronic Charge Transfer (the "+" Signal Indicates that the Charge Is Transferred from the Graphene/ZGNR to the COOH Group) and the Spin Polarization for Each Functionalized Relaxed Configuration

Configuration	d_{C-C} (Å)	θ_{CCC} (°)	E_B (Å)	Charge Transfer (e)	Spin Polarization (μ_B)
Carboxylated graphene	1.58	107.54	1.23	+0.12	-
Carboxylated graphene nanoribbon	1.58	107.16	1.16	+0.13	4.27

configuration. For both graphene and nanoribbon, the values for binding energy are, respectively, 1.23 and 1.16 eV, what characterizes, thus, a chemical interaction through covalent bonds. These energies in addition to the values for the distances and angle of connection between the COOH group and the graphene/ZGNR indicate systems with hybridization of the sp³ type.

Concerning the band structure presented by both graphene and nanoribbon, it can be observed that, for both carboxylated nanostructures, there is a formation of a defect level with little dispersion (Figure 17.7), breaking the symmetry of the graphene and inducing a more localized level above the Fermi level for the ZGNR, which also characterizes a sp³ hybridization.

Table 17.2 also shows the spin polarization presented by GNR. In its pristine form, the ZGNR has a spin polarization equals to 3.33 μ_B due to edge states, which are hydrogenated and, when carboxylated, the nanoribbon presents a polarization equals to 4.27 μ_B due to the presence of the COOH radical in the structure, differently from the null spin polarization presented by graphene, even when functionalized with the COOH group.

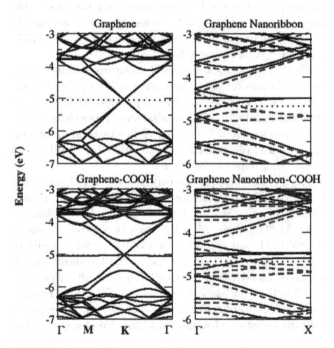

FIGURE 17.7 Electronic band structure for pristine and carboxylated graphene and nanoribbon. The band levels represent the up and down spins. Fermi's energy is displayed in dotted line.

17.4.2 Carboxylated Carbon Nanotubes

The CNTs functionalization is presented as an alternative in making these nanostructures disperse in water, and the most used strategy to do so is the use of oxidative treatments that generate COOH groups on the surface of the nanotubes (Zhao et al. 2004, Chiu et al. 2002). The carboxyl groups bind covalently to the wall of the nanotube allowing it to carry out different reactions and interactions

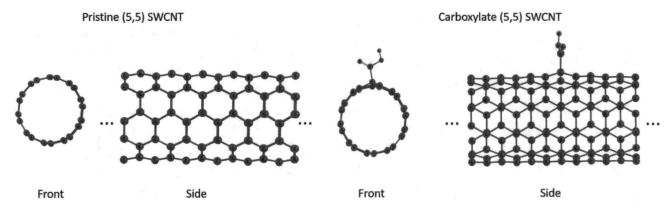

FIGURE 17.8 Atomic structure obtained in the simulations for (5,5) pristine and carboxylated zigzag (5,5) CNT.

with various molecules (de Souza Filho and Fagan 2007, Veloso et al. 2006, Saidi 2013, Milowska and Majewski 2013).

The dimensions of the supercell used in the calculations are equal to $40 \times 40 \times 8.6$ Å^3 for the (8,0) nanotube, $40 \times 40 \times 14.96$ Å^3 for the (5,5) nanotube and $40 \times 40 \times 26.23$ Å^3 for the (4,3) nanotube, regarding to the x, y and z axes. For the integration of the Brillouin zone along the axis of the nanotube, $1 \times 1 \times 9$, $1 \times 1 \times 30$ and $1 \times 1 \times 30$ k-points were generated and used in the calculations for (8,0), (5,5) and (4,3) CNTs, respectively.

Figure 17.8 shows the relaxed structures obtained from the simulations for the (5,5) pristine and the (5,5) carboxylated CNT. The same procedure was performed for the (8,0) armchair and (4,3) chiral pristine and carboxylated CNT. For all analyzed systems, as expected, structural distortions in the wall of the nanotube in the radial direction induced by COOH groups were observed.

The distance between the carbon atom of the nanotubes to the carboxyl carbon atom d_{C-C} is 1.55 Å, on average, which is close enough to the C–C distance in the diamond (1.54 Å). The angle θ_{CCC} (between two carbon atoms of the SWCNT and one atom of COOH) found is equal to 109.27°, very close to the value of the tetrahedral angle in the diamond (109.5°) as well. These results show that the COOH groups are linked to the SWCNT through covalent bonds with sp^3 type hybridization and are in agreement with the literature (Tournus et al. 2010, Doudou et al. 2011, 2012).

Table 17.3 shows the structural values, binding energy, charge transfer and the resulting spin polarization for each

TABLE 17.3 Bond Distance, Binding Angle, Binding Energies for the COOH Group, Electronic Charge Transfer (the "+" Signal Indicates that the Charge is Transferred from the Nanotube to the COOH Group) and the Spin Polarization for Each Functionalized Relaxed Configuration

Configuration	d_{C-C}(Å)	θ_{CCC}(°)	E_B(Å)	Charge Transfer (e)	Spin Polarization (μ_B)
(8,0)-COOH CNT	1.55	110.28	1.88	+0.13	0.84
(5,5)-COOH CNT	1.55	108.97	2.09	+0.15	-
(4,3)-COOH CNT	1.54	108.56	2.09	+0.13	0.96

configuration. The respective diameters are 6.38, 6.85 and 4.90 Å, for the (8.0), (5.5) and (4.3) CNTs.

It can be drawn in Figure 17.9 that for a functional group connected to the nanotube, like the case of graphene, a half-filled level with little dispersion is observed for the three types of CNTs. Only the (5,5) SWCNT presents metallic character in its pristine form and does not present spin polarization, even when carboxylated (Lara, Zanella and Fagan 2014).

In this analysis, the dual behavior of these systems is clear, i.e., a pure semiconductor nanotube connected to a carboxyl presents a half-filled level in the gap, which is induced by the sp^3 defect, as well as a metallic nanotube maintains its electronic character and presents the half-filled level in its band structure (Lara et al. 2014).

17.4.3 Carboxylated Phosphorene

The functionalization with the different functional groups, such as NH and O (Dai and Zeng 2014), generate modification in the arrangement of the P atoms of the black phosphorene structure and can alter its stability. Furthermore, degradation effects in the phosphorus structure caused by oxidation can be overcome using encapsulation and surface passivation techniques (Kuriakose et al. 2018). In this sense, the interaction with other molecules and structures may be affected by the carboxylated phosphorene structure, and this specific interaction must be well known.

The dimensions of the supercell used in the calculations for the black phosphorene monolayer are equal to $20 \times 13.62 \times 40$ Å^3 and for the nanoribbon, which is an hydrogenated armchair nanoribbon (APNR), are equal to $40 \times 13.62 \times 40$ Å^3, respectively, for the x, y and z axes. For the integration of the Brillouin zone, $11 \times 11 \times 1$ k-points were generated and used in the calculations for both the black phosphorene and the nanoribbon.

Figure 17.10 shows the structures obtained from the energy minimization calculations for pristine and carboxylated black phosphorene and APNR. It is noted that the binding distances between the carbon and phosphorus atoms are modified, so that the phosphorus atoms (neighboring the phosphorus atom which is bound to the carboxyl) move

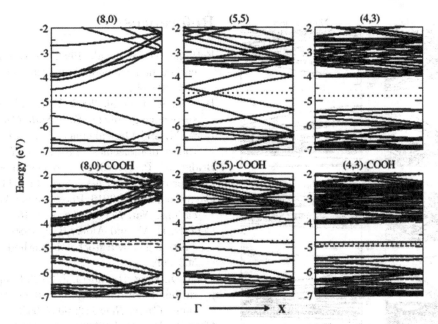

FIGURE 17.9 Figure 17.9 Electronic bands structure for (8,0), (5,5) and (4,3) pristine and COOH functionalized SWCNTs. The band levels represent the up and down spins. Fermi's energy is displayed in dotted line.

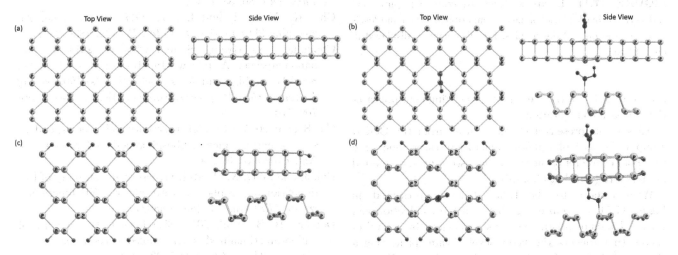

FIGURE 17.10 Atomic structure for (a) pristine and (b) carboxylated black phosphorene and for (c) pristine and (d) carboxylated nanoribbon obtained from the simulations.

away and increase the P–P bond distance from 2.27 to 2.65 Å.

For black phosphorene and APNR, the values for binding energy are, respectively, equal to 1.73 and 1.57 eV, characterizing in both cases a chemical interaction through covalent bonds and, these energies together with the values for the interatomic distances, indicating systems with sp^3-type hybridization. However, the angle of connection between the atoms of phosphorus and carbon is lower than that for carbon nanostructures. It is also observed that the structural deformation caused by this functionalization in the phosphorus nanostructures is different from the deformation in a carbon lattice. For phosphorene (monolayer and nanoribbon), the COOH group pushes phosphorus atoms that are located in the opposite "layer" to that in which the group is bound. This buckling effect does not appear on

TABLE 17.4 Bond Distance, Binding Angle, Binding Energies for the COOH Group, Electronic Charge Transfer (the "−" Signal Indicates that the Charge Is Transferred to the Phosphorene from the COOH Group) and the Spin Polarization for Each Functionalized Relaxed Configuration

Configuration	d_{C-P}(Å)	θ_{CPP}(°)	E_B(Å)	Charge Transfer (e)	Spin Polarization (μ_B)
Black Phosphorene-COOH	1.93	92.22	1.73	−0.13	0.99
Nanoribbon-COOH	1.93	89.80	1.57	−0.04	–

graphene or CNT and has been treated in literature (Dai and Zeng 2014).

Table 17.4 shows the values for the distance d_{C-P} and bonding angle between the carbon atoms θ_{CPP} of the carboxyl group and phosphorous of phosphorene, bonding

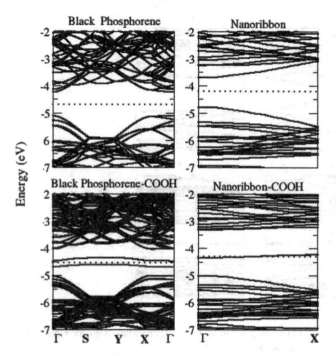

FIGURE 17.11 Electronic band structure for pure and carboxylated black phosphorene and nanoribbon. The band levels represent the up and down spins. Fermi's energy is displayed in dotted line.

energy E_B, charge transfer and the resulting spin polarization for each configuration.

All systems present a charge transfer from the COOH group to the black phosphorus nanostructures, as opposed to the donor character of the carbon nanostructures, as referred in literature (Tonel et al. 2016).

With respect to the band structure presented in Figure 17.11, the gap energy between the conduction and valence bands of pristine systems is replaced by a level of energy that touches the Fermi level. Again, presenting a characteristic level of the hybridization of the COOH group with the nanostructures, the carboxylated black phosphorene has a spin polarization equal to 0.99 μ_B.

This chapter has presented the effect of COOH functionalization on different nanostructures. For the 2D systems, namely, graphene and phosphorene, and 1D systems, SWNTs, phosphorus and carbon nanorribons, structural, magnetic and electronic properties were analyzed. The carboxylation process deforms locally the atomic structure and has different effects on the carbon structure compared to the phosphorus structure. Spin polarization can be displayed in some carboxylated structures depending on its chirality, quantum confinement and the atom type nanostructure. Finally, for all the systems, the COOH binding is responsible for a defect level introduction over the Fermi level, as shown in the energy bands. This electronic behavior is typical of a sp³-type hybridization, and it does not depend on the atom nanostructure modulating the electronic character of each nanostructure.

References

Bethune, D. S. et al. 1993. Cobalt-catalysed growth of carbon nanotubes with singleatomic-layer walls. *Nature* 363: 605–607.

Bolotin, K. I. et al. 2008. Ultrahigh electron mobility in suspended graphene. *Solid State Commun.* 146: 351–355.

Born, M. and Oppenheimer, J. R. 1927. On the quantum theory of molecules. *Ann. Phys. (Leipzig)* 84: 457–484.

Bridgman, P. W. 1914. Two new modifications of phosphorus. *J. Am. Chem. Soc.* 36: 1344–1363.

Carvalho, A. et al. 2016. Phosphorene: From theory to applications. *Nat. Rev. Mater.* 1: 1–16.

Ceperly, D. M. and Alder, B. J. 1980. Ground state of the electron gas by a stochastic method. *Phys. Rev. Lett.* 45: 566–569.

Chen, D., Feng, H. and Li, J. 2012. Graphene oxide: Preparation, functionalization, and electrochemical applications. *Chem. Rev.* 112: 6027–6053.

Chiu, P. W. et al. 2002. Interconnection of single-walled carbon nanotubes by chemical functionalization. *Appl. Phys. Lett.* 80: 3811–3813.

Cho, K., Yang, J. and Lu, Y. 2017. Phosphorene: An emerging 2D material. *J. Mater. Res.* 32: 2839–2847.

Choudhary, N., Hwang, S. and Choi W. 2014. Carbon nanomaterials: A review. In: Bhushan B., Luo D., Schricker S., Sigmund W., Zauscher S. (eds) *Handbook of Nanomaterials Properties*. Springer: Berlin, Heidelberg 709–770.

Cui, S. et al. 2015. Ultrahigh sensitivity and layer-dependent sensing performance of phosphorene-based gas sensors. *Nat. Commun.* 6: 1–9.

Dai, J. and Zeng, X. C. 2014. Structure and stability of two dimensional phosphorene with =O or =NH functionalization. *RSC Adv.* 4: 48017–48021.

Doudou, B. B. et al. 2011. Size-dependent properties of amino-functionalized single walled carbon nanotubes. *Comput. Theor. Chem.* 967: 231–234.

Doudou, B. B. et al. 2012. *Ab initio* study of the size dependent effect on the covalent functionalization of single walled carbon nanotubes with hydroxyl, amine and carboxyl groups. *J. Nanosci. Nanotechnol.* 12: 8635–8639.

Dresselhaus, M. S., Dresselhaus, G. and Avouris, P. 2000. *Carbon Nanotubes.* Springer: New York.

Eisberg, R. and Resnick, R. 1979. *Física Quântica – Átomos, Moléculas, Sólidos, Núcleos e Partículas.* Campus: Rio de Janeiro.

Fagan, S. B., da Silva, L. B. and Mota, R. 2003. *Ab initio* study of radial deformation plus vacancy on carbon nanotubes: Energetics and electronic properties. *Nano Lett.* 3: 289–291.

Faccio, R. et al. 2009. Mechanical properties of graphene nanoribbons. *J. Phys. Condens. Matter* 21: 285304–285312.

Ferreira, F. and Ribeiro, R. M. 2017. Improvements in the GW and BSE calculations on phosphorene. *Centro* de Física and Departamento de Física and QuantaLab, Universidade do Minho, Portugal.

Garcia-Lastra, J. M. et al. 2008. Conductance of sidewall-functionalized carbon nanotubes: Universal dependence on adsorption sites. *Phys. Rev. Lett.* 101: 236806–236804.

Han, N., Gao, N. and Zhao, J. 2017. Initial growth mechanism of blue phosphorene on Au(111) surface *J. Phys. Chem. C.* 121: 17893–17899.

Han, Z. and Fina, A. 2011. Thermal conductivity of carbon nanotubes and their polymer nanocomposites: A review. *Prog. Polym. Sci.* 36: 914–944.

Hohenberg, P. and Kohn, W. 1964. Inhomogeneous electron gas. *Phys. Rev. B* 136: 864–871.

Homaeigohar, S. and Elbahri, M. 2017. Graphene membranes for water desalination. *NPG Asia Mater.* 9: 427–443.

Hongyan, G. et al. 2014. Phosphorene nanoribbons, phosphorus nanotubes, and van der Waals multilayers. *J. Phys. Chem. C.* 118: 14051–14059.

Iijima, S. 1991. Helical microtubules of graphitic carbon. *Nature* 354: 56–58.

Iijima, S. and Ichihashi, T. 1993. Single-shell carbon nanotubes pf 1-nm diameter. *Nature* 363: 603–605.

Iijima, S., Brabec, C. and Maiti, A. 1996. Structural flexibility of carbon nanotubes. *J. Chem. Phys.* 104: 2089–2092.

Kauffman, D. R. and Star, A. 2010. Graphene *versus* carbon nanotubes for chemical sensor and fuel cell applications. *Analyst* 135: 2790–2797.

Ke, Q. and Wang, J. 2016. Graphene-based materials for supercapacitor electrodes – a review. *J. Materiomics* 2: 37–54.

Khandoker, N. et al. 2011. Tensile strength of spinnable multiwall carbon nanotubes. *Procedia Eng.* 10: 2572–2578.

Kim, U. J. et al. 2016. Electric characteristics of the carbon nanotube network transistor with directly grown ZnO nanoparticles. *J. Nanosci. Nanotechnol.* 16: 2887–2890.

Kohn, W. and Sham, L. J. 1965. Self-consistent equations including exchange and correlation effects. *Phys. Rev. A* 140: 1133–1138.

Kou, L., Frauenheim, T. and Chen, C. 2015. Phosphorene as a superior gas sensor: Selective adsorption and distinct I–V response. *J. Phys. Chem. Lett.*, 5: 2675–2681.

Kuriakose, S. et al. 2018. Black phosphorus: Ambient degradation and strategies for protection. *2D Mater.* 5: 032001.

Kushwaha, S. K. S. et al. 2013. Carbon nanotubes as a novel drug delivery system for anticancer therapy: A review. *Braz. J. Pharm. Sci.* 49: 629–643.

Lara, I. V. et al. 2014. Influence of concentration and position of carboxyl groups on the electronic properties of single-walled carbon nanotubes. *Phys. Chem. Chem. Phys.* **16**: 21602–21608.

Lara, I. V., Zanella, I. and Fagan, S. B. 2014. Functionalization of carbon nanotube by carboxyl group under radial deformation. *Chem. Phys.* 428: 117–120.

Lee, T. H., Kim, S. Y. and Won, H. 2016. Black phosphorus: Critical review and potential for water splitting photocatalyst. *Nanomaterials* 6: 194–210.

Li, G. Q. 2009. The transport properties of boron nanostructures. *Appl. Phys. Lett.* 94: 193116–193119.

Li, S. et al. 2018. Self-powered photogalvanic phosphorene photodetectors with high polarization sensitivity and suppressed dark current. *Nanoscale* 26: 7694–7701.

Lin, X. F. et al. 2016. Graphene-based materials for polymer solar cells. *Chin. Chem. Lett.* 27: 1–12.

Liu, G. B. et al. 2015. Electronic structures and theoretical modelling of two-dimensional group-VIB transition metal dichalcogenides. *Soc. Rev.* 44: 2643–2663.

Liu, H. et al. 2014. Phosphorene: An unexplored 2D semiconductor with a high hole mobility. *ACS Nano* 8: 4033–4041.

Liu, H. et al. 2015. Semiconducting black phosphorus: Synthesis, transport properties and electronic applications. *Chem. Soc. Rev.* 44: 2732–2743.

Marinho, B. et al. 2012. Electrical conductivity of compacts of graphene, multi-wall carbon nanotubes, carbon black, and grafite poder. *Powder Technol.* 221: 351–358.

Martin, R. M. 2004. *Electronic Structure: Basic Theory and Practical Methods.* Cambridge University Press: Urbana-Champaign, IL.

Memarian, F., Fereidoon. A. and Ganji, M. D. 2015. Graphene Young's modulus: Molecular mechanics and DFT treatments. *Superlattices and Microstruct.* 85: 348–356.

Menezes, V. M. and Fagan, S. B. 2012. Modelagem molecular de nanossistemas de interesse biológico para aplicações em carreamento de fármacos. In: Sausen, P. and Sausen, A. (eds) *Pesquisas Aplicadas em Modelagem Matemática*, 323–350. Editora Unijuí: Ijuí.

de Menezes, V. M. 2016. Functionalized carbon nanotubes. In: Sattler, K. D. (ed.) *Carbon Nanomaterials Sourcebook: Graphene, Fullerene, Nanotubes, and Nanodiamonds,* 415–428. Taylor & Francis: Boca Raton, FL.

Milowska, K. Z., Malewski, J. A. 2013. Functionalization of carbon nanotubes with ch(n), -nh(n) fragments, -cooh and -oh groups. *J. Chem. Phys.* 138: 194704–194714.

Monkhorst, H. J. and Pack, J. D. 1976. On special points for brillouin zone integrations. *Phys. Rev. B* 13: 5188–5192.

Novoselov, K. S. et al. 2004. Electric field effect in atomically thin carbon films. *Science* 306: 666–669.

Pan, H., Li, J. and Feng, Y. P. 2010. Carbon nanotubes for supercapacitor. *Nanoscale Res. Lett.* 5: 654–668.

Peng, F. et al. 2014. Silicon nanomaterials platform for bioimaging, biosensing, and cancer therapy. *Acc. Chem. Res.* 18: 612–623.

Perdew, J. P. and Zunger, A. 1981. Self-interaction correction to density-functional approximations for many-electron systems. *Phys. Rev.* 23: 5048–5079.

Perdew, J. P., Burke, K. and Ernzerhof, M. 1996. Generalized gradient approximation made simple. *Phys. Rev. Lett.* 77: 3865–3868.

Pop, E., Varshney, V. and Roy, A. K. 2012. Thermal properties of graphene: Fundamentals and applications. *MRS Bull.* 37: 1273–1281.

Qian, Y., Ismail, I. M. and Stein, A. 2014. Ultralight, high-surface-area, multifunctional graphene-based aerogels from self-assembly of graphene oxide and resol. *Carbon* 68: 221–231.

Saidi, W. 2013. A. Functionalization of single-wall zigzag carbon nanotubes by carboxyl groups: Clustering effect. *J. Phys. Chem. C* 117: 9864–9871.

Saito, R., Dresselhaus, G. and Dresselhaus, M. S. 1998. *Physical Properties of Carbon Nanotubes.* Imperial College Press: London.

Santos, E. J. G., Sanchez-Portal, D., Ayuela, A. 2011. Magnetism of covalently functionalized carbon nanotubes. *Appl. Phys. Lett.* 99: 062503–062503.

Satishkumar, B. C. et al. 2007. Reversible fluorescence quenching in carbon nanotubes for biomolecular sensing. *Nat. Nanotechnol.* 2: 560–564.

Senokos, E. et al. 2018. Energy storage in structural composites by introducing CNT fiber/polymer electrolyte interleaves. *Sci. Rep.* 8: 3407–3417.

Singh, V. et al. 2011. Graphene based materials: Past, present and future. *Prog. Mater. Sci.* 56: 1178–1271.

Stephens, P. J. et al. 1994. *Ab initio* calculation of vibrational absorption and circular dichroism spectra using density functional force fields. *J. Phys. Chem.* 98: 11623–11627.

Soler, J. M. et al. 2002. The SIESTA method for *ab initio* order-N materials simulation. *J. Phys. Condens. Matter.* 14: 2745–2779.

Son, Y. W., Cohen, M. L. and Louie, S.G. 2006. Half-metallic graphene nanoribbons. *Nature* 16: 347–349.

Son, Y. W., Cohen, M. L. and Louie, S. G. 2007. Energy gaps in graphene nanoribbons. *Phys. Rev. Lett.* 98: 089901–089905.

de Souza Filho, A. G. and Fagan, S. B. 2007. Funcionalização de nanotubos de carbono. *Quim. Nova* 30: 1695–1703.

Sun, Y., Wu, Q. and Shi, G. 2011. Graphene based new energy materials. *Energy Environ. Sci.* 4: 1113–1132.

Tans, S. T. et al. 1997. Individual single-wall carbon nanotubes as quantum wires. *Nature* 386: 474–477.

Tao, J. et al. 2003. Climbing the density functional ladder: Nonempirical meta–generalized gradient approximation designed for molecules and solids. *Phys. Rev. Lett.* 91: 146401.

Tomita, S., Ishihara, S. and Kurita, R. 2017. A multi-fluorescent DNA/graphene oxide conjugate sensor for signature-based protein discrimination. *Sensors* 17: 2194–2260.

Tonel, M. Z. et al. 2016. An *ab initio* study of carboxylated graphene and black - and blue – phosphorene. *Disciplinarum Scientia* 17: 349–356.

Tournus, F. et al. 2010. π-stacking interaction between carbon nanotubes and organic molecules. *Carbon* 48: 1961–1965.

Troullier, N. and Martins, L. 1991. Efficient pseudopotentials for plane-wave calculations. *Phys. Rev. B* 43: 1993–2006.

Veloso, M. V. et al. 2006. *Ab initio* study of covalently functionalized carbon nanotubes. *Chem. Phys. Lett.* 430: 71–74.

Vianna, J. D. M., Fazzio, A. and Canuto, S. 2004. *Teoria Quântica de Moléculas e Sólidos: Simulação Computacional.* Livraria da Física: São Paulo.

Vieira Segundo, J. E. D. and Vilar E. O. 2016. Grafeno: Uma revisão sobre propriedades, mecanismos de produção e potenciais aplicações em sistemas energéticos. *Revista Eletrônica de Materiais e Processos* 11: 54–57.

Warschauer, D. 1962. Electrical and optical properties of crystalline black phosphorus. *J. Appl. Phys.* 34: 1853–1860.

Woo, H. H. et al. 2017. Prediction of green phosphorus with tunable direct band gap and high mobility. *J. Phys. Chem. Lett.* 8: 4627–4632.

Yáñez-Sedeño, P. et al. 2010. Electrochemical sensing based on carbon nanotubes. *Trends Anal. Chem.* 29: 939–953.

Yoo, B. et al. 2017. Graphene and graphene oxide membranes for gas separation applications. *Curr. Opin. Chem. Eng.* 16: 39–47.

Yu, S. S. et al. 2008. Electronic properties of graphene nanoribbons with armchair-shaped edges. *J. Mol. Simul.* 34: 1085–1090.

Zanella, I. et al. 2008. Chemical doping-induced gap opening and spin polarization in graphene. *Phys. Rev. B* 77: 073404.

Zhang, J. et al. 2014. Phosphorene nanoribbon as a promising candidate for thermoelectric applications. *Sci. Rep.* 4: 1–7.

Zhao, J. et al. 2004. Electronic properties of carbon nanotubes with covalent sidewall functionalization. *J. Phys. Chem. B* 108: 4227–4230.

Zhihong, C. et al. 2007. Graphene nano-ribbon electronics. *Physica E* 40: 228–232.

Zhu, S. E., Yuan, S. and Janessen, G. C. A. M. 2014. Optical transmittance of multilayer graphene. *EPL* 108: 17007–17011.

Phase Behavior of Atomic and Molecular Nanosystems

R. Stephen Berry
The University of Chicago

18.1 Introduction

The physical and chemical properties of aggregates of matter, typically at a scale we can see and handle but extending to micron scales, even sometimes to nanometer scales, are the substance of thermodynamics and classical mechanics. We recognize that classical Newtonian mechanics must be replaced by quantum mechanics when we want to describe very small systems and the phenomena they exhibit. We can often explain how the passage occurs between the scale that requires a quantum description and the larger scale where classical Newtonian mechanics is completely adequate. We have learned to recognize many ways that very small systems can exhibit different chemical behavior from their macroscopic counterparts, largely through quantum-based behavior. However there has been a general (and tacit) assumption that classical thermodynamics would be valid for all systems, whatever their size. As we shall see, this is not strictly correct, although the explanation for the apparent violations becomes clear when we turn to statistical mechanics, specifically statistical thermodynamics, for the needed insight.

In this chapter, we focus primarily on two main ways that very small systems, aggregates of several or even hundreds of atoms or molecules, differ from their bulk counterparts, sometimes in ways that seem outwardly to violate deep, widely accepted principles. One is the behavior of very small systems regarding the phases they exhibit, the changes they undergo and the kinds of coexistence they can show. The other aspect of this apparent discrepancy concerns the differences between the physical structures of very small systems and the macroscopic size objects we encounter in everyday life. A third main difference is the chemical behavior some very small systems exhibit, particularly their behavior as catalysts. This is described in another chapter.

We begin by reviewing one of the very widely used generalizations of traditional thermodynamics and probably the simplest basic equation in science, the Gibbs Phase Rule. This is the relation connecting the number of degrees of freedom f, the number of phases p present in equilibrium, and the number c of components or substances comprising the system we are studying. That relation is simply

$$f = c - p + 2.$$

The number of things we can vary, f, is most limited if our system consists of a single substance, meaning $c = 1$. With two substances, we can change their relative amounts. Each phase, solid, liquid or gas, must satisfy its own equation of state for the conditions we have chosen, so each phase p brings a constraint. Thus, for a single phase of a single substance, $f = 1 - 1 + 2 = 2$, so we can vary temperature and pressure, two degrees of freedom, and keep the substance in equilibrium. This is the situation with liquid water, for example, or for ice. However if we want to have liquid water and ice together in equilibrium, we have two phases, thereby setting $p = 2$ so $f = 1 - 2 + 2 = 1$. We have only one degree of freedom; if we want to change the temperature while the ice and liquid water remain in equilibrium, we must adjust the pressure to the specific value at which solid–liquid equilibrium holds at the new temperature. This means that if we want to show this relation graphically, we can choose temperature and pressure as the variables defining the horizontal and vertical axes, and have entire areas corresponding to solid, to liquid and to gas or vapor. These areas are separated by simple curves which show the specific temperature-pressure conditions along which two phases can coexist in equilibrium, whether solid and liquid, solid and vapor or liquid and vapor. Carrying this argument one step further, se see immediately that there can be only a single point, with no degrees of freedom, at which solid, liquid and vapor can coexist together in equilibrium, the "triple point". Figure 18.1 is a schematic representation of this graph for water, called the "phase diagram".

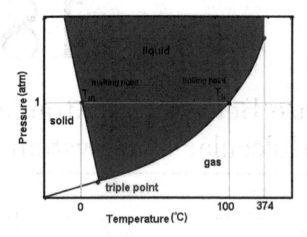

FIGURE 18.1 A phase diagram for water. The diagram illustrates one of the unusual properties of water, the consequence of liquid water being denser than ice; this property causes the solid–liquid boundary to slope downward with decreasing pressure and increasing temperature, since increasing pressure converts the solid to the liquid.

All substances conform to similar phase diagrams, with simple curves separating regions of pure phases and points of intersection of the curves at the triple points where three phases can coexist in equilibrium. The phase diagram for the inert gas Argon, Figure 18.2, is very similar, qualitatively, to that of water, except that the slope of the line separating solid and liquid has positive slope, that is, it goes up to the right, indicating higher pressures make for more solid. The corresponding line for water has opposite slope and goes up to the left, because water has the unusual property that it has a higher density as a liquid than as a solid, so increasing the pressure on an ice-water mixture squeezes some ice to become liquid. We mention Argon because a model system resembling this substance has become a widely used device to simulate simple atomic clusters, one that has become a guide and basis for many of our ideas about clusters and a stimulus for many experiments as well. By and large, those experiments have turned out to be quite consistent with the results of the simulations; in other words, the model used to simulate Argon turns out to be remarkably reliable.

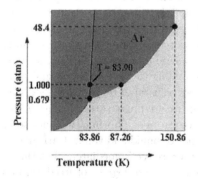

FIGURE 18.2 The phase diagram for Argon. Similar in many ways to that of water, the solid–liquid boundary line has positive slope for this substance.

18.2 Structural Differences

Now we can turn to two related phenomena that reveal ways that small systems may behave in ways different from their bulk counterparts. The first is the most stable form of the solid, and here we can use Argon as our main illustration, because of its simplicity. Bulk solid argon has a lattice structure, a close-packed form called "face-centered cubic" abbreviated fcc. Eight atoms form the corners of the basic cubic building block, and each of the six faces of that cube holds an atom, which comprise an interpenetrating network of cubes with the first set. This lattice is what is called "close-packed" because it is as dense as a cubic lattice can be. But small clusters do not have structures anything like that lattice. Argon and many other substances take polyhedral structures, especially based on the 12-vertex, 20-faced icosahedron. A cluster of 13 Argon atoms has, as its most stable structure, an icosahedron with one atom at the center and 12 atoms occupying the vertices of the icosahedron. It was in 1971 and 1972 when Hoare and Pal established polyhedral structures as the primary forms of such clusters (Hoare and Pal, 1972), through simulations using the interparticle force field with an attraction between two particles varying with their distance r as r^{-6} and their repulsion, as r^{-12}. This is known as the Lennard-Jones potential, named for the British theorist Sir John Lennard-Jones. This elegant structure is exceptionally stable (and hence relatively inert) and is called a "magic number structure". A cluster of 12 atoms has almost the same structure but with one vacancy in the shell of 12 vertices. A cluster of 14 atoms has the 14th resting on one of the 20 faces of the "closed shell" of 12 atoms. A cluster of 55 atoms is the next larger that has a complete filled outer shell; it has the 13 in the inner core, 12 outer-shell vertices, and another 30 atoms that reside on the 30 edges of the icosahedron. Hence the cluster of 55 atoms is also unusually stable so 55 is the second "magic number". The next is 147; we leave it to the reader to determine where those 92 outer-shell atoms reside to close that third shell.

An open, unsolved challenge for which some headway has been made is determining the smallest sizes of systems for which the most stable structure is that of the bulk lattice. There is even some suggestive work that would indicate that there might be an intermediate size range, probably several thousand atoms, for which the most stable form of solid is a lattice but different from that of the bulk.

So we see that small systems can differ from their bulk counterparts in structure. However that need not be the situation for all substances. Sodium chloride, composed of positive sodium ions and negative chloride ions, has a cubic structure with alternating positive and negative ions along all its edges. This is the structure of the bulk and of a cluster of only four NaCl molecules, the smallest complete cube this system can make. Clusters of all sizes of this substance and of all the other alkali halides such as KBr and LiI exhibit this behavior: their stable structures are lattices, even for very small systems. The comparison of simple salts such as these with Argon provides a warning to be careful about generalizing any findings in our comparisons of bulk and small systems.

18.3 Solid Liquid Coexistence

Now we turn to the subject of solid–liquid equilibrium of small clusters and its comparison with its counterpart for macroscopic systems. Experiments showed that small clusters can exhibit both solid (Farges et al., 1975, 1977; Bartell, 1986) and liquid (Bartell et al., 1988, 1989) forms. Argon clusters containing a molecule of dimethyl ether were used by Stace to show that if the clusters were prepared cold, from a high-pressure jet, they would be solid, but if they came warm from a low-pressure jet, they would be liquid (Stace, 1983). How does the solid–liquid equilibrium behavior of bulk Argon compare with that of small clusters of Argon atoms? We know that in the bulk system, solid and liquid can coexist in equilibrium only at the temperature-pressure combinations that lie on the steep line separating the stable regions of those two phases; one temperature for any pressure on that line, i.e. above 0.679 atmospheres, where the melting point is actually the triple point of Argon, 83.86 K or −189°C. Below that temperature, Argon is either solid or gas, never liquid.

As early as 1963, Terrill Hill showed (Hill, 1963) that the solid–liquid phase transition is actually smooth and not discontinuous, but for bulk matter, occurs over an extremely narrow region of temperature and pressure. However we see such transitions as abrupt, sharp changes at a single sharp temperature at any given pressure, at least for the macroscopic systems we normally encounter. But simulations by molecular dynamics (Alder and Wainwright, 1957) and reinterpreted Monte Carlo calculations (Wood and Jacobson, 1957) indicated that there could be a range of temperatures in which collections of 32 and of 108 hard spheres could exhibit both solid and liquid-like forms. This became clearer and more persuasive with extensive simulations by Briant and Burton (1973, 1975), Etters and Kaelberer (1975, 1977) and Etters et al. (1977). These theoretical results indicated unambiguously that small clusters of argon atoms would exhibit a solid polyhedral structure at low temperatures, a nonrigid liquid form at higher temperatures, and, between these, a finite region of temperature in which solid and liquid forms would coexist in dynamic equilibrium.

That clusters exhibit bands of coexisting solid and liquid is a phenomenon not limited to the rare gases. Bachels showed (Bachels et al., 2000) that clusters of 430 atoms of tin have a range of temperatures in which solid and liquid forms coexist in equilibrium. But tin clusters, particularly in the size range of 10–30 atoms, show another "anomalous" kind of behavior. As early as 1909, Pawlow had shown that small aggregates of a material should melt at a lower temperature than the corresponding bulk substance, primarily because of the high surface and curvature of the small systems (Pawlow, 1909a,b). But almost a century later, Shvartsburg

and Jarrold demonstrated a violation of that "rule"; they showed that clusters of 10–30 tin atoms melt at temperatures *significantly higher* than the bulk (Shvartsburg and Jarrold, 2000). The difference between geometric structures of the clusters and the bulk, and perhaps associated differences between electronic structures are somehow presumably responsible for this remarkable behavior. In fact, tin is not unique; small gallium clusters also melt at temperatures above the bulk melting point (Breaux et al., 2003). This apparently anomalous behavior is quite a puzzle, both as to its origin and the extent to which it may occur in other systems.

Before we return to explaining the behavior of clusters that violates the Gibbs Phase Rule, we examine one other related phenomenon which clusters sometimes exhibit, that doesn't fit at all with our conventional expectations. This is the behavior of the heat capacity of many clusters. This behavior, we must make clear, appears under conditions of *constant energy*, not constant temperature. The heat capacity is the change in a system's temperature when a specific amount of heat energy is added to that system. We of course expect heat capacities to be positive; adding heat makes the temperature go up. And this is what we expect and find for all systems whose temperature is a primary specification of their condition. We speak of hypothetical ensembles of macroscopically identical systems all at the same temperature, large imagined collections of copies of a system we are studying, as a *canonical ensemble* and infer the properties of our system from the behavior expected of the majority of the hypothetical copies in the ensemble—when we are keeping our systems at the fixed temperature set by the amount of heat we have put into them. But there is an important, widely-applied alternative to the canonical ensemble, based on keeping the energy fixed at the value determined by the energy we have put into the system. The hypothetical collection of copies all with the same energy is called a *microcanonical ensemble*. Remarkably, systems represented by microcanonical ensembles, i.e. systems whose energy, rather than temperature, specifies their condition, can exhibit *negative heat capacities*. That is, under certain circumstances, adding heat can make the temperature of these systems drop. This was demonstrated theoretically by Bixon and Jortner (1989) and Carignano and Gladich (2010), and was first reported from experiments by Schmidt et al. (2001) using clusters of 147 sodium atoms.

The seemingly bizarre idea that temperature could drop when heat is added to a system is, when examined carefully, understandable and not so surprising. It is something that can happen when there is a phase change, notably from solid to liquid, and the two phases are present in equilibrium. The energy of the tightly-bound solid is of course lower than that of the more loosely bound liquid. (It is, of course, the *total* energy of the system that is held constant under these conditions.) With the energy of a system (and its subsystems) fixed, this means that the effective temperatures of the solid and liquid must be different; the low-energy solid must have a higher temperature than the higher-energy liquid. Hence adding heat energy to the system converts some of the solid to liquid; that is, the added heat transforms some low-energy, warm solid to higher-energy, cooler liquid—so the overall mean temperature of the system drops! And this corresponds precisely to a negative heat capacity. This can only happen with systems small enough to exhibit detectable ranges of temperature, at a given pressure, in which solid and liquid can coexist in equilibrium. No such behavior would be observable for a macroscopic system with its sharp melting point of temperature. And of course the phenomenon of a negative heat capacity appears in systems held at constant energy, but not at systems held at constant temperature (at each infinitesimal step in the addition of heat).

Now we turn to examination of that apparent anomaly of coexisting phases over ranges of temperature and pressure, as exhibited by small clusters of atoms and molecules. Let us begin by seeing why, despite Hill's recognition of the smooth nature of solid–liquid phase changes, these seem to be sharp and discontinuous. We can do this by examining the behavior of the ratio of the concentrations of the two species when they are present in equilibrium. This ratio, [liquid]/[solid], is precisely the *equilibrium constant*, K_{eq}, which tells us the equilibrium ratio of any two species in dynamic equilibrium. This quantity is a sensitive function of the difference ΔF between the free energies of the two forms; this difference can be expressed in terms of the number N of particles comprising the total system, multiplied by $\Delta\mu$, the free energy difference per particle. (Recall that the free energy relevant here is the system's actual energy E, minus the temperature T times the entropy S, so $F = E - TS$, and $\Delta F = \Delta E - T\Delta S$. Thus $\Delta F/N = \Delta\mu$). Now we need the precise relation between K_{eq}, N and $\Delta\mu$. This is an exponential: $K_{eq} = e^{(N\Delta\mu)/kT}$. The exponent must, of course, be a dimensionless number; this is attained by measuring the energy units of the free energy difference in the energy units of temperature T, multiplied by the Boltzmann constant k, whose value is approximately 1.38×10^{-23} J/K.

Now imagine a very small but macroscopic system, say 10^{20} atoms, roughly the size of a grain of sand. If the two phases have the same free energy per particle so the exponent is zero, the amounts of the two phases are equal; their ratio is 1. And let us ask what the ratio of the two phases will be if $\Delta\mu/kT$ is nonzero but very, very small, say $\pm10^{-10}$. The exponent of K_{eq} becomes $\pm10^{10}$ which makes K_{eq} either a huge number or an extremely small one. This tells us that the ratio of the two species, under conditions unmeasurably close to that of true, exact equilibrium, is either so large or so small that we would only be able to observe the favored, stable species. In short, the transition is so abrupt for our small macroscopic system that it looks discontinuous.

In contrast, let us now look at the behavior of a system of only a few particles, say a cluster of 10 atoms, so $N = 10$, and K_{eq} is $e^{(10\Delta\mu)/kT}$. Then if $\Delta\mu/kT$ is ±0.1, i.e. the free energy difference per particle $\Delta\mu/k$ is equal to a tenth of the temperature, the ratio of the two forms, K_{eq}, is $e^{(\pm10 \times 0.1)}$

or either e or $1/e$, approximately 2.7 or 1/2.7, clearly a readily-observable ratio. In other words, if a system is small enough, the free energy difference between the favored and unfavored forms is sufficiently small that both favored and unfavored forms can be present in observable but unequal amounts, in equilibrium. So long as the conditions allow for *local* stability of the unfavored form as well as the thermodynamically favored form, we will be able to see both forms present.

Thus we see that the Gibbs Phase Rule is valid for macroscopic systems because of the behavior of the equilibrium of two species when the number of elements comprising them is macroscopically large, and that the Rule simply loses its applicability when we deal with very small numbers of elements. As we said earlier, an open question at this time is "At approximately what size is the smallest system whose most stable solid is that of the bulk material?" It may even be that there is an intermediate size range for some substances whose most stable structure differs from that of the bulk and from that of small cluster.

This reasoning carries further. It goes on to imply that small systems need not be restricted to just two species or two phases or two forms in equilibrium. We can observe three phase-like forms in equilibrium if a system allows for such a third species. And for some atomic clusters, there is indeed the equivalent of a third phase. Clusters of 55 Argon atoms, for example, have a closed-shell icosahedral solid form with a central atom, an icosahedral shell of 12 atoms around that central one, and a second icosahedral shell of 42 atoms. Simulations reveal not only the closed-shell solid and a 55-atom liquid but also an intermediate with a solid, 13-atom icosahedral core and a liquid outer shell! This was recognized, almost in passing, by Hoare and Pal in 1971 (Hoare and Pal, 1971) and explored in detail considerably later (Nauchitel and Pertsin, 1980; Cheng and Berry, 1992). It remains to be seen how general the surface-melting form may be; in the case of the 55-atom Lennard-Jones representation of Ar_{55}, the surface-melted form acts as an independent phase. Moreover the small-system behavior that allows solid and liquid to coexist over a range of temperatures and pressures also allows the coexistence of more than just two phases, so that at least this example has a range of temperatures and pressures in which solid, liquid and surface-melted phases all coexist. This was demonstrated through the time dependence of the internal energy at various temperatures; at 35 K, the distribution displays three distinct levels, corresponding to the solid, surface-melted and liquid phases, but at 33 K, only a single surface atom is promoted to move freely, and at 40 K, the cluster is entirely liquid (Kunz and Berry, 1994). The band of temperature is quite narrow, in which solid, surface-melted and liquid clusters of 55 atoms represented by Lennard-Jones interaction potentials can coexist, but it is nonetheless there, and it is likely that a similar narrow band exists for clusters of 55 Argon atoms. Figure 18.3 is a kind of phase diagram that illustrates the stable forms this cluster can assume.

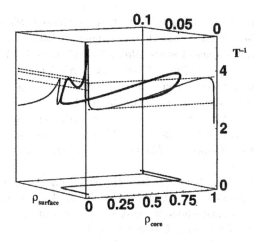

FIGURE 18.3 A three-dimensional phase diagram for the Ar_{55} cluster. The vertical axis, labeled b, is the inverse of the temperature; the axis labeled ρ_c measures the density of defects or vacancies in the core; the axis labeled ρ_s measures the corresponding density on the surface. Thus, near the top, at low temperatures, only the defect-free form exists; as the system warms, some surface defects appear, corresponding to a pure surface-melted state coexisting with the solid. The heavy curve then moves upward, corresponding to a negative heat capacity, and then swings downward, still with no core melting; following that, the heavy curve moves upward again, this time to steadily larger densities of defects in the core—i.e. core-melting. And finally the curve swings downward again as the cluster has melted completely.

18.4 Perspective

What can we learn about science generally, from our experience with these examples of perhaps-unexpected behavior? They do illustrate an inherent but often overlooked aspect to science and scientific understanding. These examples illustrate and teach us that we must never suppose that the scientific picture and concepts familiar to us are "final words," that they are ultimate and universal truths. We must recognize that concepts and ideas we often call "laws" may well be correct for describing some phenomena, but that the range of validity of those concepts may well be limited to some, perhaps many situations but not all that we may encounter.

Moreover as our technological and experimental skills improve and expand, enabling us to study nature under more and more conditions, we encounter more and more circumstances in which new or refined scientific concepts, new theories, new intellectual models are required to understand what we observe. When the laws and rules of thermodynamics were being developed, we had no way to observe or study such things as atomic clusters. At that time, we had no way to observe matter at that scale. It was probably Einstein's 1903 interpretation of Brownian motion, the fluctuating. Random wiggles of very small colloidal particles, visible under a microscope, and the subsequent experimental verification of that interpretation by Jean Perrin in 1908 that unambiguously established the atomic nature of matter and the size scale of atoms. However it was almost a century later that we could first observe and manipulate individual atoms and small aggregates of them.

This experience with the structures and solid–liquid equilibrium of small atomic clusters tells us that we must never suppose that the science we know and use is *the ultimate, universal* truth. What we know may well be entirely valid for the circumstances and systems from which we learned and derived that science, but, with time, we have become ever more skilled at observing and studying nature and natural phenomena at new scales, under new conditions and in new ways. And sometimes, as we have seen here, these new scales of observation require deeper, more careful ways of using earlier concepts and generalizations, and sometimes even of introducing altogether new concepts, as happened with the introduction of quantum theory, to resolve an obvious flaw in the theory of radiation from warm and hot objects. We learn here that science is constantly evolving and growing, and always will do that. It will never be finished and complete.

References

Alder, B. J.; Wainwright, T. E., 1957. Phase transition for a hard sphere system. *J. Chem. Phys.* 27: 1208–1209.

Bachels, T.; Schäfer, R.; Güntherodt, H. -J., 2000. Dependence of formation energies of Tin nanoclusters on their size and shape. *Phys. Rev. Lett.* 84: 4890-4893.

Bartell, L. S., 1986. Diffraction studies of clusters generated in supersonic flow. *Chem. Rev.* 86: 492–505.

Bartell, L. S.; Sharkey, L. R.; Shi, X., 1988. Electron diffraction and Monte Carlo studies of liquids. 3. Supercooled benzene. *J. Am. Chem. Soc.* 110: 7006–7013.

Bartell, L. S.; Harsanyi, L.; Valente, E. J., 1989. Phases and phase changes of molecular clusters generated in supersonic flow. *J. Phys. Chem.* 93: 6201–6205.

Bixon, M.; Jortner, J., 1989. Energetic and thermodynamic size effects in molecular clusters. *J. Chem. Phys.* 91: 1631–1642.

Breaux, G. A.; Benirschke, R. C.; Sugai, T.; Kinnear, B. S.; Jarrold, M. F., 2003. Hot and solid gallium clusters: Too small to melt. *Phys. Rev. Lett.* 91: 215508.

Briant, C. L.; Burton, J. J., 1973. Thermodynamics— melting of small clusters of atoms. *Nat. Phys. Sci.* 243: 100–102.

Briant, C. L; Burton, J. J., 1975. Molecular dynamics study of the structure and thermodynamic properties of argon microclusters. *J. Chem. Phys.* 63: 2045–2058.

Carignano, M. A.; Gladich, I., 2010. Negative heat capacity of small systems in the microcanonical ensemble. *Eur. Phys. Lett.* 90: 5.

Cheng, H.-P.; Berry, R. S., 1992. Surface melting of clusters and implications for bulk matter. *Phys. Rev. A* 45: 7969–7980.

Etters, R. D.; Kaelberer, J. B., 1975. Thermodynamic properties of small aggregates of rare-gas atoms. *Phys. Rev. A* 11: 1068–1079.

Etters, R. D.; Kaelberer, J. B., 1977. On the character of the melting transition in small atomic aggregates. *J. Chem. Phys.* 66: 5112–5116.

Etters, R. D.; Danilowicz, R.; Kaelberer, J. B., 1977. Metastab le states of small rare gas crystallites. *J. Chem. Phys.* 67: 4145–4148.

Farges, J.; DeFeraudy, M. F.; Roualt, B.; Torchet, G., 1975. Structure compacte désordonnée et agrégats moléculaires models polytétraédriques. *J. Phys. Paris* 36: 13–17.

Farges, J., DeFeraudy, M. F., Roualt, B., Torchet, G., 1977. Transition dans l'ordre local des agrégats de quelques dizaines d'atomes. *J. Phys. Paris* 38: 47–51.

Hill, T. L., 1963. *The Thermodynamics of Small Systems, Part 1.* New York: W. A. Benjamin.

Hoare, M. R.; Pal, P., 1971. Physical cluster mechanics: Statics and energy surfaces for monatomic systems. *Adv. Phys.* 20: 161–196.

Hoare, M. R.; Pal, P., 1972. Geometry and stabiliy of "spherical" f.c.c. microcrystallites. *Nat. Phys. Sci.* 238: 35–37.

Kunz, R. E.; Berry, R. S., 1994. Multiple emery 12,18apr18 phase coexistence in finite systems. *Phys. Rev. E* 49: 1895–1908.

Nauchitel, V. V.; Pertsin, A. J., 1980. A Monte Carlo study of the structure and thermodynamic behavior of small Lennard-Jones clusters. *Mol. Phys.* 40: 1341–1355.

Pawlow, P., 1909a. Uber die Abhängigkeit des Schmelz;unkte von der Oberflächenenergie eines festen Körpers. *Z. Phys. Chem.* 65: 1–35.

Pawlow, P., 1909b. Uber die Abhängigkeit des Schmelz;unkte von der Oberflächenenergie eines festen Körpers (Zusatz). *Z. Phys. Chem.* 65: 545–548.

Schmidt, M.; Kusche, R.; Hippler, T.; Donges, J.; Krobnmüller, W.; von Issendorff, B.; Haberland, H., 2001. Negative heat capacity for a cluster of 147 sodium atoms. *Phys. Rev. Lett.* 86: 1191–1194.

Shvartsburg, A. A.; Jarrold, M. F., 2000. Solid clusters above the bulk melting point. *Phys. Rev. Lett.* 85: 2530–2532.

Stace, A. J., 1983. Experimental evidence of a phase transition in a microcluster. *Chem. Phys. Lett.* 99: 470–474.

Wood, W. W.; Jacobson, J. D., 1957. Preliminary results from a recalculation of the Monte Carlo equation of state of hard spheres. *J. Chem. Phys.* 27: 1207–1208.

Exact solutions in the Density Functional Theory (DFT) and Time-Dependent DFT of Mesoscopic Systems

Vladimir U. Nazarov
Research Center for Applied Sciences,
Academia Sinica, Taiwan

Density functional theory (DFT) and its time-dependent counterpart TDDFT are formally rigorous approaches to theoretically study the ground-states and the electronic excitations, respectively, of atomic, molecular, and condensed matter systems, including the mesoscopic ones. Practically, the usefulness of DFT and TDDFT is determined by the accuracy and the computational efficiency of the available approximations to the corresponding *exchange-correlation* (xc) functionals. In the three-dimensional, as well as in the purely low-dimensional, cases, the local-density approximation (LDA) to the xc functional of the corresponding dimensionality is traditionally and predominantly used. However, LDA, as well as its extensions of generalized gradient approximation (GGA) and meta-GGA, encounter severe difficulties when applied to the systems of intermediate dimensionality, i.e., mesoscopic systems. In this chapter, we review the recent progress in the use of the static and dynamic exact exchange (EXX) functional, which we show to be well fit to capture the characteristic features of mesoscopic systems. On this way, we find exact analytical solutions to the exchange-only DFT and TDDFT problems for specific systems of reduced-dimensionality: those of the quasi-2(1)D electron gas with one filled subband. These solutions provide us with important insights into the role of the interparticle interactions, allowing, in particular, to

identify and separate the collective and the single-electron regimes in the quantum dynamics of the systems under study.

19.1 Introduction

In this chapter, we will be concerned with the ground-state and dynamic properties of *mesoscopic* systems, the latter systems defined as those having, at the same time, microscopic and macroscopic dimensionalities. The typical examples of such systems are schematically shown in Figure 19.1.

The general formulation of (TD)DFT does not, of course, differ in the mesoscopic case from that of the corresponding problems for any other many-body system. In this chapter, we will highlight special and instructive features which arise in the *applications* of DFT and TDDFT to mesoscopic systems. We will see that those features can lead to both complications in some cases and simplifications in the others, as compared to atomic and extended 3D systems. The general source of the complications is that the LDA [adiabatic LDA (ALDA), in the time-dependent case] fails at the strong confinement of the Q2(1)DEG along the direction of the low dimensionality [1,9,10,21], calling for more elaborate, dimensionality-independent exchange-correlation functionals to be used. On the other hand,

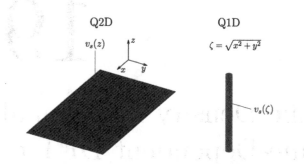

FIGURE 19.1 Schematic illustration of quasi-2D and quasi-1D systems.

examples of drastic simplifications, when analytical solutions within rather sophisticated approximations become possible, will be devoted special attention.

In Section 19.2, we give a brief reminder of the general DFT and TDDFT. For self-containedness and to prepare a general reader to easily understand our main material on Q2(1)DEG, in Section 19.3, we present the theory of the EXX, which is given with all the particulars necessary for the seamless application to the mesoscopic case. The reader fluent in (TD)DFT and (TD)EXX but interested in their application to Q2(1)DEG can skip everything prior to Section 19.4. Considering the quasi-low-dimensional electron gas, in Sections 19.4 and 19.5 we find the analytical EXX solutions to the DFT problem in the static and the time-dependent cases, respectively, discuss the properties of the obtained results, as well as their significance and implications. Our conclusions are collected in Section 19.6. We use atomic units unless otherwise indicated.

19.2 Basic Facts from DFT and TDDFT

Within the framework of the non-relativistic quantum mechanics, the stationary N-body problem of electrons moving in the field of the external potential $v_{ext}(\mathbf{r})$ and interacting with each other via the pairwise Coulomb potential $1/|\mathbf{r}_i - \mathbf{r}_j|$ is defined by the many-body Hamiltonian [12]

$$\hat{H} = \sum_i \left[-\frac{1}{2}\Delta_i + v_{ext}(\mathbf{r}_i) \right] + \frac{1}{2}\sum_{i \neq j} \frac{1}{|\mathbf{r}_i - \mathbf{r}_j|}. \quad (19.1)$$

The ground-state energy of this system is determined as the lowest eigenvalue E of the Schrödinger equation

$$\hat{H}\Psi(\mathbf{r}_1, \ldots, \mathbf{r}_N) = E\Psi(\mathbf{r}_1, \ldots, \mathbf{r}_N), \quad (19.2)$$

where the eigenfunction $\Psi(\mathbf{r}_1, \ldots, \mathbf{r}_N)$ is the many-body wave-function, and the electron density is found as

$$n(\mathbf{r}) = N \int |\Psi(\mathbf{r}, \mathbf{r}_2, \ldots, \mathbf{r}_N)|^2 d\mathbf{r}_2, \ldots, d\mathbf{r}_N. \quad (19.3)$$

In the case of $N > 1$, the original many-body formulation of Eqs. (19.1)–(19.2) becomes prohibitively complicated for practical calculations. An attractive alternative approach is

the reformulation of the problem, exact or approximate, in terms of the independent-particles motion. Within the Kohn–Sham (KS) DFT, we find the single-particle orbitals $\phi_\alpha(\mathbf{r})$ by solving the system of the KS equations [11]

$$\left[-\frac{1}{2}\Delta + v_s(\mathbf{r}) \right] \phi_\alpha(\mathbf{r}) = \epsilon_\alpha \phi_\alpha(\mathbf{r}), \quad (19.4)$$

where

$$v_s(\mathbf{r}) = v_{ext}(\mathbf{r}) + v_H(\mathbf{r}) + v_{xc}(\mathbf{r}) \quad (19.5)$$

is the KS potential, $v_{ext}(\mathbf{r})$, $v_H(\mathbf{r})$, and $v_{xc}(\mathbf{r})$ are the externally applied, the Hartree

$$v_H(\mathbf{r}) = \int \frac{n(\mathbf{r}')}{|\mathbf{r} - \mathbf{r}'|} d\mathbf{r}', \quad (19.6)$$

and the *exchange-correlation* potentials, respectively, and ϵ_α are KS eigenenergies. The particle-density $n(\mathbf{r})$ is determined self-consistently as a sum over the occupied orbitals

$$n(\mathbf{r}) = \sum_{\alpha \in occ} |\phi_\alpha(\mathbf{r})|^2. \quad (19.7)$$

The exchange-correlation potential $v_{xc}(\mathbf{r})$ is defined as the functional derivative of the exchange-correlation energy E_{xc} with respect to the density, meaning that

$$\delta E_{xc} = \int v_{xc}(\mathbf{r})\delta n(\mathbf{r})d\mathbf{r}, \quad (19.8)$$

which is written using the notation

$$v_{xc}(\mathbf{r}) = \frac{\delta E_{xc}}{\delta n(\mathbf{r})}. \quad (19.9)$$

The density (19.7) is equal (exact DFT) or must be close (approximation to DFT) to the electron density (19.3) found by the solution of the original many-body problem.

In the time-dependent case, KS equations take the form

$$i\frac{\partial \phi_\alpha(\mathbf{r}, t)}{\partial t} = \left[-\frac{1}{2}\Delta + v_{ext}(\mathbf{r}, t) + v_H(\mathbf{r}, t) + v_{xc}(\mathbf{r}, t) \right] \phi_\alpha(\mathbf{r}, t), \quad (19.10)$$

while all the quantities acquire an additional time argument. The proper choice of an approximation to $v_{xc}(\mathbf{r}, t)$, making the density of the reference KS system as close as possible to that of the original many-body one, while not too complicated and resources consuming, is the main question to be answered in every particular physical situation.

An important specific case of a many-body system excitation is the *linear response regime*, when one considers weak deviations of the density $\delta n(\mathbf{r}, t)$ from its ground-state distribution under the action of a weak external perturbation $\delta v_{ext}(\mathbf{r}, t)$. In this case, an exhaustive characteristic of the system's reaction is the *interacting density-response function*, which is defined, in the frequency domain, as

$$\delta n(\mathbf{r}, \omega) = \int \chi(\mathbf{r}, \mathbf{r}', \omega)\delta v_{ext}(\mathbf{r}', \omega)d\mathbf{r}'. \quad (19.11)$$

According to the definition of the functional derivative, Eq. (19.11) can be written as

$$\chi(\mathbf{r}, \mathbf{r}', \omega) = \frac{\delta n(\mathbf{r}, \omega)}{\delta v_{ext}(\mathbf{r}', \omega)}. \qquad (19.12)$$

The dynamic *exchange-correlation kernel* $f_{xc}(\mathbf{r}, \mathbf{r}', \omega)$ is introduced as

$$f_{xc}(\mathbf{r}, \mathbf{r}', \omega) = \frac{\delta v_{xc}(\mathbf{r}, \omega)}{\delta n(\mathbf{r}', \omega)}. \qquad (19.13)$$

When an approximation to f_{xc} is found, the interacting density-response function χ is determined from [7]

$$\chi^{-1} = \chi_s^{-1} - f_H - f_{xc}, \qquad (19.14)$$

where

$$f_H(\mathbf{r}, \mathbf{r}') = \frac{1}{|\mathbf{r} - \mathbf{r}'|} \qquad (19.15)$$

is the Hartree kernel, and χ_s is the independent-particle density-response function

$$\chi_s(\mathbf{r}, \mathbf{r}', \omega) = \frac{\delta n(\mathbf{r}, \omega)}{\delta v_s(\mathbf{r}', \omega)}, \qquad (19.16)$$

which has an explicit representation in terms of the single-particle (KS) orbitals and orbital energies [14]

$$\chi_s(\mathbf{r}, \mathbf{r}', \omega) =$$
$$\sum_{\alpha\beta} \frac{f_\alpha - f_\beta}{\omega - \epsilon_\beta + \epsilon_\alpha + 0_+} \phi_\alpha^*(\mathbf{r}) \phi_\beta(\mathbf{r}) \phi_\beta^*(\mathbf{r}') \phi_\alpha(\mathbf{r}'), \qquad (19.17)$$

where f_α are the occupation numbers, and 0_+ stands for an infinitesimal positive. We point out that (19.14) is an operator equation, with each term being a kernel of an integral transformation, while χ^{-1} and χ_s^{-1} denote the inverse of the corresponding operators.

19.3 Exact Exchange–Optimized Effective Potential

In DFT, the notion of the *EXX* has proven very useful conceptually. The corresponding potential can be introduced in three, seemingly independent, ways as (I) the functional derivative of the Hartree–Fock (HF) exchange energy [25] with respect to density; (II) as the optimized effective potential (OEP) [24,26]; and (III) as the first-order term in the series in the powers of the interaction constant at fixed density, the latter approach known as the adiabatic connection perturbation theory [5,6].

Since we will need both methods (I) and (III) in the application to the Q2(1)DEG, below we introduce these methods and establish the general equivalence of their results. For the sake of completeness and logical consistency, the method (II) is presented as well.

19.3.1 Exact Exchange

The exchange energy (as obtained from the HF theory) is defined as [25]

$$E_x = -\frac{1}{2} \int \frac{|\rho(\mathbf{r}, \mathbf{r}')|^2}{|\mathbf{r} - \mathbf{r}'|} d\mathbf{r} d\mathbf{r}', \qquad (19.18)$$

where

$$\rho(\mathbf{r}, \mathbf{r}') = \sum_{\alpha \in occ} \phi_\alpha(\mathbf{r}) \phi_\alpha^*(\mathbf{r}') \qquad (19.19)$$

is the single-particle density-matrix [12], the summation in Eq. (19.19) running over the occupied states only. Using the definition (19.9) of the exchange-correlation potential, at the level of the exchange only, we can write by virtue of Eqs. (19.18) and (19.19)

$$v_x(\mathbf{r}) = \frac{\delta E_x}{\delta n(\mathbf{r})}$$
$$= -\sum_{\alpha, \beta \in occ} \int \frac{\phi_\alpha^*(\mathbf{r}'') \phi_\beta(\mathbf{r}'') \phi_\beta^*(\mathbf{r}')}{|\mathbf{r}' - \mathbf{r}''|} \frac{\delta \phi_\alpha(\mathbf{r}')}{\delta n(\mathbf{r})} d\mathbf{r}' d\mathbf{r}'' + c.c. \qquad (19.20)$$

Using the chain rule

$$\frac{\delta \phi_\alpha(\mathbf{r}')}{\delta n(\mathbf{r})} = \int \frac{\delta \phi_\alpha(\mathbf{r}')}{\delta v_s(\mathbf{r}'')} \frac{\delta v_s(\mathbf{r}'')}{\delta n(\mathbf{r})} d\mathbf{r}'' = \int \frac{\delta \phi_\alpha(\mathbf{r}')}{\delta v_s(\mathbf{r}'')} \chi_s^{-1}(\mathbf{r}'', \mathbf{r}) d\mathbf{r}'', \qquad (19.21)$$

where χ_s is the Lindhard density-response function of Eq. (19.17), taken at $\omega = 0$, we can rewrite Eq. (19.20) as

$$\int \chi_s(\mathbf{r}, \mathbf{r}') v_x(\mathbf{r}') d\mathbf{r}'$$
$$= -\sum_{\alpha, \beta \in occ} \int \frac{\phi_\alpha^*(\mathbf{r}'') \phi_\beta(\mathbf{r}'') \phi_\beta^*(\mathbf{r}')}{|\mathbf{r}' - \mathbf{r}''|} \frac{\delta \phi_\alpha(\mathbf{r}')}{\delta v_s(\mathbf{r})} d\mathbf{r}' d\mathbf{r}'' + c.c. \qquad (19.22)$$

Furthermore, by the first-order perturbation theory [12]

$$\frac{\delta \phi_\alpha(\mathbf{r}')}{\delta v_s(\mathbf{r})} = \sum_{\gamma \neq \alpha} \frac{\phi_\gamma(\mathbf{r}') \phi_\alpha(\mathbf{r}) \phi_\gamma^*(\mathbf{r})}{\epsilon_\alpha - \epsilon_\gamma}, \qquad (19.23)$$

which allows us to write Eq. (19.22) as[1]

$$\int \chi_s(\mathbf{r}, \mathbf{r}') v_x(\mathbf{r}') d\mathbf{r}'$$
$$= -\sum_{\substack{\alpha, \beta \in occ \\ \gamma \in unocc}} \int \frac{\phi_\alpha^*(\mathbf{r}'') \phi_\beta(\mathbf{r}'') \phi_\beta^*(\mathbf{r}') \phi_\gamma(\mathbf{r}') \phi_\alpha(\mathbf{r}) \phi_\gamma^*(\mathbf{r})}{|\mathbf{r}' - \mathbf{r}''|(\epsilon_\alpha - \epsilon_\gamma)} d\mathbf{r}' d\mathbf{r}'' + c.c. \qquad (19.24)$$

The integral equation (19.24) is to be solved to find the exchange potential $v_x(\mathbf{r})$.

For further references, we note that, due to Eq. (19.17), $\chi_s(\mathbf{r}, \mathbf{r}')$ can be conveniently written as

$$\chi_s(\mathbf{r}, \mathbf{r}') = \sum_{\substack{\alpha \in occ \\ \gamma \in unocc}} \frac{\phi_\alpha^*(\mathbf{r}) \phi_\gamma(\mathbf{r}) \phi_\gamma^*(\mathbf{r}') \phi_\alpha(\mathbf{r}')}{\epsilon_\alpha - \epsilon_\gamma} + c.c. \qquad (19.25)$$

---

[1]In Eq. (19.24), the summation over γ has been restricted to the unoccupied states, since the corresponding sum over $\gamma \in occ$ is zero identically, as can be shown by the simultaneous interchanging the indices α and γ and the integration variables $\mathbf{r}'$ and $\mathbf{r}''$.

19.3.2 Optimized Effective Potential

The OEP [24,26] is defined as such a single-particle potential $v_{OEP}(\mathbf{r})$ that the expectation value $\langle \Phi_s | \hat{H} | \Phi_s \rangle$ of the many-body Hamiltonian of Eq. (19.1) in the Slater-determinant [12] state Φ_s, built with the orbitals $\phi_\alpha(\mathbf{r})$ of $v_{OEP}(\mathbf{r})$, is minimal. Since

$$\langle \Phi_s | \hat{H} | \Phi_s \rangle = \sum_{\alpha \in occ} \int \phi_\alpha^*(\mathbf{r}) \left[-\frac{1}{2}\Delta + v_{ext}(\mathbf{r}) \right] \phi_\alpha(\mathbf{r}) d\mathbf{r}$$
$$+ \frac{1}{2} \int \frac{n(\mathbf{r})n(\mathbf{r}')}{|\mathbf{r} - \mathbf{r}'|} d\mathbf{r}d\mathbf{r}'$$
$$- \frac{1}{2} \sum_{\alpha,\beta \in occ} \int \frac{\phi_\alpha^*(\mathbf{r})\phi_\beta^*(\mathbf{r}')\phi_\alpha(\mathbf{r}')\phi_\beta(\mathbf{r})}{|\mathbf{r} - \mathbf{r}'|} d\mathbf{r}d\mathbf{r}'$$

$$(19.26)$$

and since the orbitals satisfy the equations

$$\left[-\frac{1}{2}\Delta + v_{OEP}(\mathbf{r}) \right] \phi_\alpha(\mathbf{r}) = \epsilon_\alpha \phi_\alpha(\mathbf{r}), \qquad (19.27)$$

equation (19.26) can be rewritten as

$$\langle \Phi_s | \hat{H} | \Phi_s \rangle = \sum_{\alpha \in occ} \epsilon_\alpha - \int n(\mathbf{r}) \left[v_{OEP}(\mathbf{r}) - v_{ext}(\mathbf{r}) \right] d\mathbf{r}$$
$$+ \frac{1}{2} \int \frac{n(\mathbf{r})n(\mathbf{r}')}{|\mathbf{r} - \mathbf{r}'|} d\mathbf{r}d\mathbf{r}'$$
$$- \frac{1}{2} \sum_{\alpha,\beta \in occ} \int \frac{\phi_\alpha^*(\mathbf{r})\phi_\beta^*(\mathbf{r}')\phi_\alpha(\mathbf{r}')\phi_\beta(\mathbf{r})}{|\mathbf{r} - \mathbf{r}'|} d\mathbf{r}d\mathbf{r}'.$$

$$(19.28)$$

Taking the functional derivative of Eq. (19.28) with respect to $v_{OEP}(\mathbf{r})$ and equating it to zero, we find

$$0 = \sum_{\alpha \in occ} \frac{\delta\epsilon_\alpha}{\delta v_{OEP}(\mathbf{r})} - n(\mathbf{r})$$
$$- \int \chi_s(\mathbf{r},\mathbf{r}') \left[v_{OEP}(\mathbf{r}') - v_{ext}(\mathbf{r}') \right] d\mathbf{r}'$$
$$+ \int \chi_s(\mathbf{r},\mathbf{r}') \frac{n(\mathbf{r}'')}{|\mathbf{r}'-\mathbf{r}''|} d\mathbf{r}' d\mathbf{r}''$$
$$- \left[\sum_{\alpha,\beta \in occ} \int \frac{\phi_\alpha^*(\mathbf{r}'')\phi_\beta^*(\mathbf{r}')\phi_\beta(\mathbf{r}'')}{|\mathbf{r}'-\mathbf{r}''|} \frac{\delta\phi_\alpha(\mathbf{r}')}{\delta v_{OEP}(\mathbf{r})} d\mathbf{r}'d\mathbf{r}'' + c.c. \right]$$

$$(19.29)$$

By the perturbation theory [12]

$$\frac{\delta\epsilon_\alpha}{\delta v_{OEP}(\mathbf{r})} = |\phi_\alpha(\mathbf{r})|^2, \qquad (19.30)$$

and, with account of Eq. (19.7), the first two terms in Eq. (19.29) cancel each other. Therefore,

$$\int \chi_s(\mathbf{r},\mathbf{r}') \left[v_{OEP}(\mathbf{r}') - v_{ext}(\mathbf{r}') - v_H(\mathbf{r}') \right] d\mathbf{r}'$$
$$= - \sum_{\alpha,\beta \in occ} \int \frac{\phi_\alpha^*(\mathbf{r}'')\phi_\beta^*(\mathbf{r}')\phi_\beta(\mathbf{r}'')}{|\mathbf{r}'-\mathbf{r}''|} \frac{\delta\phi_\alpha(\mathbf{r}')}{\delta v_{OEP}(\mathbf{r})} d\mathbf{r}'d\mathbf{r}'' + c.c.$$

$$(19.31)$$

Noting that, by the chain rule,

$$\frac{\delta\phi_\alpha(\mathbf{r}')}{\delta v_{OEP}(\mathbf{r})} = \int \chi_s(\mathbf{r},\mathbf{r}'') \frac{\delta\phi_\alpha(\mathbf{r}')}{\delta n(\mathbf{r}'')} d\mathbf{r}'' \qquad (19.32)$$

and canceling $\chi_s(\mathbf{r},\mathbf{r}'')$, as an operator, from both sides of Eq. (19.31), we arrive at

$$v_{OEP}(\mathbf{r}) - v_{ext}(\mathbf{r}) - v_H(\mathbf{r})$$
$$= -\sum_{\alpha,\beta \in occ} \int \frac{\phi_\alpha^*(\mathbf{r}'')\phi_\beta^*(\mathbf{r}')\phi_\beta(\mathbf{r}'')}{|\mathbf{r}'-\mathbf{r}''|} \frac{\delta\phi_\alpha(\mathbf{r}')}{\delta n(\mathbf{r})} d\mathbf{r}'d\mathbf{r}'' + c.c.$$

$$(19.33)$$

Comparing Eqs. (19.20) and (19.33), we conclude that

$$v_{OEP}(\mathbf{r}) = v_{ext}(\mathbf{r}) + v_H(\mathbf{r}) + v_x(\mathbf{r})., \qquad (19.34)$$

proving that v_{OEP} coincides with the KS potential within the EXX theory of the previous section.

19.3.3 Adiabatic Connection Perturbation Series

The adiabatic connection perturbation method provides a systematic way to treat exchange and correlation within DFT, in the first and higher orders in the inter-electron interaction, respectively, while keeping the density fixed [6]. The many-body Hamiltonian (19.1) is rewritten with the scaled interaction term as

$$\hat{H}_\lambda = \sum_i \left[-\frac{1}{2}\Delta_i + v_\lambda(\mathbf{r}_i) \right] + \frac{1}{2} \sum_{i \neq j} \frac{\lambda}{|\mathbf{r}_i - \mathbf{r}_j|}, \quad (19.35)$$

and one tunes $v_\lambda(\mathbf{r})$ so that the electron density corresponding to the Hamiltonian (19.35) remains the same for all $\lambda \in [0,1]$. Obviously,

$$v_0(\mathbf{r}) = v_s(\mathbf{r}), \qquad (19.36)$$
$$v_1(\mathbf{r}) = v_{ext}(\mathbf{r}), \qquad (19.37)$$

where v_s is the KS potential. It will be convenient to work in the density-matrix representation [12]. The many-body density-matrix $\hat{\rho}_\lambda$ commutes with the Hamiltonian (19.35)

$$\left[\hat{H}_\lambda, \hat{\rho}_\lambda \right] = 0. \qquad (19.38)$$

We expand to the first order in λ

$$\hat{H}_\lambda = \hat{H}_0 + \lambda\hat{H}^{(1)}, \qquad (19.39)$$
$$\hat{\rho}_\lambda = \hat{\rho}_0 + \lambda\hat{\rho}^{(1)}, \qquad (19.40)$$
$$v_\lambda(\mathbf{r}) = v_0(\mathbf{r}) + \lambda v^{(1)}(\mathbf{r}), \qquad (19.41)$$

where

$$\hat{H}_0 = \sum_i \left[-\frac{1}{2}\Delta_i + v_0(\mathbf{r}_i) \right], \qquad (19.42)$$

$$\hat{H}^{(1)} = \sum_i v^{(1)}(\mathbf{r}_i) + \frac{1}{2} \sum_{i \neq j} \frac{1}{|\mathbf{r}_i - \mathbf{r}_j|}. \qquad (19.43)$$

Substituting $\lambda = 1$ in Eq. (19.41), and using Eqs. (19.36), (19.37), and (19.5), we conclude that

$$v^{(1)}(\mathbf{r}) = -v_H(\mathbf{r}) - v_x(\mathbf{r}). \qquad (19.44)$$

To the first-order in λ, Eq. (19.38) gives

$$\left[\hat{H}_0, \hat{\rho}^{(1)}\right] + \left[\hat{H}^{(1)}, \hat{\rho}_0\right] = 0. \qquad (19.45)$$

Introducing the full orthonormal set of the Slater-determinant eigenfunctions $|n\rangle$ of the Hamiltonian $\hat{H}_0$ and their corresponding eigenenergies E_n, and taking the matrix elements of Eq. (19.45), we have

$$(E_n - E_m)\langle n|\hat{\rho}^{(1)}|m\rangle + (\delta_{m0} - \delta_{n0})\langle n|\hat{H}^{(1)}|m\rangle = 0, \qquad (19.46)$$

where we have taken into account that

$$\langle n|\hat{\rho}_0|m\rangle = \delta_{n0}\delta_{m0}, \qquad (19.47)$$

the latter being the consequence of our KS system residing in its ground-state. If $E_n \neq E_m$, Eq. (19.46) leads to

$$\langle n|\hat{\rho}^{(1)}|m\rangle = \frac{\delta_{n0} - \delta_{m0}}{E_n - E_m}\langle n|\hat{H}^{(1)}|m\rangle. \qquad (19.48)$$

Assuming that our KS system has a non-degenerate ground-state, we have

$$\langle n \neq 0|\hat{\rho}^{(1)}|0\rangle = \langle 0|\hat{\rho}^{(1)}|n\rangle^* = \frac{\langle n|\hat{H}^{(1)}|0\rangle}{E_0 - E_n}. \qquad (19.49)$$

In order to prove that all other matrix elements of $\hat{\rho}^{(1)}$ vanish, we note that the density matrix is idempotent in the pure state we are considering, i.e.,

$$\hat{\rho}^2 = \hat{\rho}. \qquad (19.50)$$

$$\sum_{\substack{\alpha \in occ \\ \gamma \in unocc}} \int \frac{\phi_\alpha^*(\mathbf{r})\phi_\gamma(\mathbf{r})\phi_\gamma^*(\mathbf{r}')\phi_\alpha(\mathbf{r}')v^{(1)}(\mathbf{r}')}{\epsilon_\alpha - \epsilon_\gamma}d\mathbf{r}' + c.c. =$$

$$\sum_{\substack{\alpha,\beta \in occ \\ \gamma \in unocc}} \int \frac{\phi_\alpha^*(\mathbf{r})\phi_\gamma(\mathbf{r})\phi_\gamma^*(\mathbf{r}')\phi_\beta^*(\mathbf{r}'')[\phi_\alpha(\mathbf{r}'')\phi_\beta(\mathbf{r}') - \phi_\alpha(\mathbf{r}')\phi_\beta(\mathbf{r}'')]}{(\epsilon_\alpha - \epsilon_\gamma)|\mathbf{r}' - \mathbf{r}''|}d\mathbf{r}'d\mathbf{r}'' + c.c., \qquad (19.60)$$

Substituting Eq. (19.40) into Eq. (19.50) and collecting the terms at λ, we have

$$\hat{\rho}^{(1)} = \hat{\rho}_0\hat{\rho}^{(1)} + \hat{\rho}^{(1)}\hat{\rho}_0. \qquad (19.51)$$

Further, taking the matrix elements of Eq. (19.51), we can write with the use of Eq. (19.47)

$$(\delta_{n0} + \delta_{m0} - 1)\langle n|\hat{\rho}^{(1)}|m\rangle = 0, \qquad (19.52)$$

from which it follows that

$$\langle n|\hat{\rho}^{(1)}|m\rangle = 0, \text{ if } n = m = 0 \text{ or } (n \neq 0 \text{ and } m \neq 0). \qquad (19.53)$$

Equations (19.49) and (19.53) can be written together as

$$\langle n|\hat{\rho}^{(1)}|m\rangle = \frac{\delta_{n0} - \delta_{m0}}{E_n - E_m + i0_+}\langle n|\hat{H}^{(1)}|m\rangle. \qquad (19.54)$$

We require the KS density to be equal to the many-body one. Then,

$$n^{(1)}(\mathbf{r}) = \text{Sp}\left[\hat{\rho}^{(1)}\hat{n}(\mathbf{r})\right] = 0, \qquad (19.55)$$

where

$$\hat{n}(\mathbf{r}) = \sum_i \delta(\mathbf{r} - \mathbf{r}_i) \qquad (19.56)$$

is the density operator. In terms of the matrix elements, Eq. (19.55) reads

$$\sum_{nm}\langle n|\hat{\rho}^{(1)}|m\rangle\langle m|\hat{n}(\mathbf{r})|n\rangle = 0, \qquad (19.57)$$

which, with the use of Eqs. (19.49) and (19.53), can be rewritten as

$$\sum_{n\neq 0}\frac{\langle 0|\hat{n}(\mathbf{r})|n\rangle\langle n|\hat{H}^{(1)}|0\rangle}{E_0 - E_n} + c.c. = 0, \qquad (19.58)$$

or, using Eq. (19.43) and with account of the identity of the particles,

$$\left[\sum_{n\neq 0}\frac{\langle 0|\delta(\mathbf{r} - \mathbf{r}_1)|n\rangle\langle n|v^{(1)}(\mathbf{r}_1)|0\rangle}{E_0 - E_n}\right.$$
$$\left. + (N-1)\sum_{n\neq 0}\frac{\langle 0|\delta(\mathbf{r} - \mathbf{r}_1)|n\rangle\langle n|\frac{1}{|\mathbf{r}_1 - \mathbf{r}_2|}|0\rangle}{E_0 - E_n}\right] + c.c. = 0. \qquad (19.59)$$

Substituting the Slater-determinant wave-functions $|0\rangle$ and $|n\rangle$ into Eq. (19.59), we can rewrite it as

which, with the use of Eq. (19.7), can be also written as

$$\sum_{\substack{\alpha \in occ \\ \gamma \in unocc}} \int \frac{\phi_\alpha^*(\mathbf{r})\phi_\gamma(\mathbf{r})\phi_\gamma^*(\mathbf{r}')\phi_\alpha(\mathbf{r}')[v^{(1)}(\mathbf{r}') + v_H(\mathbf{r}')]}{\epsilon_\alpha - \epsilon_\gamma}d\mathbf{r}' + c.c. =$$

$$\sum_{\substack{\alpha,\beta \in occ \\ \gamma \in unocc}} \int \frac{\phi_\alpha^*(\mathbf{r})\phi_\gamma(\mathbf{r})\phi_\gamma^*(\mathbf{r}')\phi_\beta^*(\mathbf{r}'')\phi_\alpha(\mathbf{r}'')\phi_\beta(\mathbf{r}')}{(\epsilon_\alpha - \epsilon_\gamma)|\mathbf{r}' - \mathbf{r}''|}d\mathbf{r}'d\mathbf{r}'' + c.c. \qquad (19.61)$$

Finally, using Eqs. (19.25) and (19.44), from Eq. (19.61) we retrieve the EXX expression (19.24) for the potential $v_x(\mathbf{r})$, which concludes our proof that EXX, OEP, and adiabatic connection perturbation theory all produce the same exchange potential.

19.4 Exact Exchange DFT in the Quasi-Low-Dimensional Setup

We are now ready to apply the formalism of the EXX theory, as described above, to the case of the quasi-low-dimensional electron gas. However, we immediately face a difficulty: the EXX energy functional (19.18) and the EXX potential (19.24) are, generally speaking, *orbital-dependent*, meaning that, while written in terms of KS orbitals and eigenenergies, they cannot be expressed explicitly via the density only (of course, implicitly, through the orbitals' dependence on the density, they are uniquely defined by the latter,[2] as the basic principles of DFT stipulate [8,11]).

Working with orbital-dependent EXX DFT functional is possible only numerically, is cumbersome, computationally expensive, and usually does not provide much physical insight into the corresponding problem. For these reasons, we take a different pathway: We will focus attention on specific systems and regimes, when explicit analytical solutions in terms of the density are possible. We will see that a great deal of insight into the ground-state and dynamic properties of quasi-low-dimensional systems occurs possible on this way. In order to quickly introduce these ideas, we start with the trivial case of the singlet of the two-electron atom.

19.4.1 Singlet State of a Two-Electron Atom as a Prototype of the Quasi-Low-Dimensional Electron Gas with One Filled Subband

Let us consider the ground-state of a two-electron system with anti-parallel spin orientations. This, e.g., may be the helium atom. By Eqs. (19.18) and (19.19), we can write in this case

$$E_x = -\int \frac{|\phi(\mathbf{r})|^2 |\phi(\mathbf{r}')|^2}{|\mathbf{r} - \mathbf{r}'|} d\mathbf{r} d\mathbf{r}', \qquad (19.62)$$

where $\phi(\mathbf{r})$ is the coordinate part of the orbital, which is the same for both electrons. On the other hand, by Eq. (19.7)

$$n(\mathbf{r}) = 2|\phi(\mathbf{r})|^2, \qquad (19.63)$$

and, therefore, we can write

$$E_x = -\frac{1}{4} \int \frac{n(\mathbf{r})n(\mathbf{r}')}{|\mathbf{r} - \mathbf{r}'|} d\mathbf{r} d\mathbf{r}'. \qquad (19.64)$$

For the exchange potential, we then readily find

$$v_x(\mathbf{r}) = \frac{\delta E_x}{\delta n(\mathbf{r})} = -\frac{1}{2} \int \frac{n(\mathbf{r}')}{|\mathbf{r} - \mathbf{r}'|} d\mathbf{r}', \qquad (19.65)$$

which, by Eq. (19.6), shows that, in the ground-state of a singlet of two electrons, the exchange potential is *minus one half of the Hartree potential*.

The above example illustrates drastic simplifications arising in the case when orbitals can be expressed through the density, as in Eq. (19.63). We will see below that, in the less trivial cases of Q2(1)DEG with one subband filled, similar simplifications occur, though Eq. (19.65) does not hold any more.

One might wonder: While we have arrived at Eq. (19.65) directly, does the general, orbital-dependent formula (19.24) still hold? The answer is, of course, yes. Indeed, Eqs. (19.24) and (19.25) can be written in this case as

$$\int \chi_s(\mathbf{r}, \mathbf{r}') v_x(\mathbf{r}') d\mathbf{r}'$$
$$= -2 \sum_{\gamma \neq 0} \int \frac{|\phi_0(\mathbf{r}'')|^2 \phi_0^*(\mathbf{r}') \phi_\gamma(\mathbf{r}') \phi_0(\mathbf{r}) \phi_\gamma^*(\mathbf{r})}{|\mathbf{r}' - \mathbf{r}''|(\epsilon_0 - \epsilon_\gamma)} d\mathbf{r}' d\mathbf{r}'' + c.c.$$
$$(19.66)$$

$$\chi_s(\mathbf{r}, \mathbf{r}') = 2 \sum_{\gamma \neq 0} \frac{\phi_0^*(\mathbf{r}) \phi_\gamma(\mathbf{r}) \phi_\gamma^*(\mathbf{r}') \phi_0(\mathbf{r}')}{\epsilon_0 - \epsilon_\gamma} + c.c. \qquad (19.67)$$

Rewriting Eq. (19.66)

$$\int \chi_s(\mathbf{r}, \mathbf{r}') v_x(\mathbf{r}') d\mathbf{r}'$$
$$= -\sum_{\gamma \neq 0} \int \frac{\phi_0^*(\mathbf{r}') \phi_\gamma(\mathbf{r}') \phi_0(\mathbf{r}) \phi_\gamma^*(\mathbf{r})}{\epsilon_0 - \epsilon_\gamma} v_H(\mathbf{r}') d\mathbf{r}' + c.c.,$$
$$(19.68)$$

we have with the use of Eq. (19.67)

$$\int \chi_s(\mathbf{r}, \mathbf{r}') v_x(\mathbf{r}') d\mathbf{r}' = -\frac{1}{2} \int \chi_s(\mathbf{r}, \mathbf{r}') v_H(\mathbf{r}') d\mathbf{r}'. \qquad (19.69)$$

The latter equation can only hold if $v_x(\mathbf{r})$ differs from $-\frac{1}{2}v_H(\mathbf{r})$ by a constant, which constant does not matter. QED.

19.4.2 Quasi-Two-Dimensional Electron Gas

In this section, we consider the same systems and arrive at the same results as in Ref. [16], although here, in a more pedagogical way, we do this using a different method of applying the results of the general EXX theory of Section 19.3. Let us consider quasi-2D electron gas confined within its z-dimension by an external potential and homogeneous in the xy plane, as illustrated in the left panel of Figure 19.1. The in-plane potential being flat, variables z and x, y separate, and, assuming additionally that only one orbital $\mu_0^\sigma(z)$ for each spin orientation σ is occupied in the z-direction, we can write the KS wave-functions as

$$\phi_{\mathbf{k}_\parallel}^\sigma(\mathbf{r}) = \frac{1}{\sqrt{\Omega}} e^{i\mathbf{k}_\parallel \cdot \mathbf{r}_\parallel} \mu_0^\sigma(z), \qquad (19.70)$$

where $\mathbf{r}_\parallel = (x, y)$, $\mathbf{k}_\parallel$ is the conserved parallel component of the wave-vector, and Ω is the normalization area in the xy plane. The KS eigenenergies are, accordingly,

$$\epsilon_{n\mathbf{k}_\parallel}^\sigma = \lambda_n^\sigma + \frac{k_\parallel^2}{2}, \qquad (19.71)$$

---

[2]To within an additive constant, for the potential.

where λ_n^σ are the eigenenergies of the perpendicular motion, described by the one-dimensional KS equations

$$\hat{h}_s^\sigma \mu_n^\sigma(z) \equiv \left[-\frac{1}{2}\frac{d^2}{dz^2} + v_{ext}(z) + v_H(z) + v_x^\sigma(z) \right] \mu_n^\sigma(z)$$
$$= \lambda_n^\sigma \mu_n^\sigma(z). \qquad (19.72)$$

Substituting Eq. (19.70) into Eqs. (19.24) and (19.25), and taking into account that v_x is a function of z only, we have

$$\int \chi_s^\sigma(\mathbf{r},\mathbf{r}')v_x^\sigma(z')d\mathbf{r}'_\parallel dz' = -\frac{1}{\Omega^3} \sum_{\substack{k_{\alpha\parallel},k_{\beta\parallel}\leq k_F^\sigma \\ \gamma \in unocc}} \int d\mathbf{r}'d\mathbf{r}''$$

$$\times \frac{e^{i(\mathbf{k}_{\beta\parallel}-\mathbf{k}_{\alpha\parallel})\cdot\mathbf{r}''_\parallel}|\mu_0^\sigma(z'')|^2 e^{i(\mathbf{k}_{\gamma\parallel}-\mathbf{k}_{\beta\parallel})\cdot\mathbf{r}'_\parallel}\mu_0^\sigma(z')\mu_{n_\gamma}^\sigma(z')e^{e(\mathbf{k}_{\alpha\parallel}-\mathbf{k}_{\gamma\parallel})\cdot\mathbf{r}_\parallel}\mu_0^\sigma(z)\mu_{n_\gamma}^\sigma(z)}{(\lambda_0 + \frac{k_{\alpha\parallel}^2}{2} - \lambda_{n_\gamma} - \frac{k_{\gamma\parallel}^2}{2})|\mathbf{r}'-\mathbf{r}''|} + c.c., \qquad (19.73)$$

$$\chi_s(\mathbf{r},\mathbf{r}') = \frac{1}{\Omega^2} \sum_{\substack{k_{\alpha\parallel}\leq k_F^\sigma \\ \gamma \in unocc}} \frac{e^{i(\mathbf{k}_\alpha-\mathbf{k}_\gamma)\cdot(\mathbf{r}'_\parallel-\mathbf{r}_\parallel)}\mu_0^\sigma(z)\mu_{n_\gamma}^\sigma(z)\mu_0^\sigma(z')\mu_{n_\gamma}^\sigma(z')}{\lambda_0 + \frac{k_{\alpha\parallel}^2}{2} - \lambda_{n_\gamma} - \frac{k_{\gamma\parallel}^2}{2}} + c.c. \qquad (19.74)$$

With the variable substitution $\mathbf{r}''_\parallel = \mathbf{r}'_\parallel + \mathbf{r}'''_\parallel$ and integrating Eq. (19.73) over $\mathbf{r}'_\parallel$, we obtain

$$\int \chi_s^\sigma(\mathbf{r},\mathbf{r}')v_x^\sigma(z')d\mathbf{r}'_\parallel dz' = -\frac{1}{\Omega^2} \sum_{\substack{k_{\alpha\parallel},k_{\beta\parallel}\leq k_F^\sigma \\ n_\gamma \geq 1}} \int dz'dz''d\mathbf{r}'''_\parallel$$

$$\times \frac{e^{i(\mathbf{k}_{\beta\parallel}-\mathbf{k}_{\alpha\parallel})\cdot\mathbf{r}'''_\parallel}|\mu_0^\sigma(z'')|^2\mu_0^\sigma(z')\mu_{n_\gamma}^\sigma(z')\mu_0^\sigma(z)\mu_{n_\gamma}^\sigma(z)}{(\lambda_0 - \lambda_{n_\gamma})\sqrt{(z'-z'')^2 + (r'''_\parallel)^2}} + c.c., \qquad (19.75)$$

where only $\mathbf{k}_{\gamma\parallel} = \mathbf{k}_{\alpha\parallel}$ has given a nonzero contribution. On the other hand, by Eq. (19.74),

$$\int \chi_s^\sigma(\mathbf{r},\mathbf{r}')d\mathbf{r}'_\parallel = \frac{2N_\sigma}{\Omega}\chi_s^\sigma(z,z'), \qquad (19.76)$$

where N_σ is the number of electrons with the spin direction σ, and we have introduced the perpendicular KS density-response function

$$\chi_s(z,z') = \sum_{n\geq 1} \frac{\mu_0^\sigma(z)\mu_n^\sigma(z)\mu_0^\sigma(z')\mu_n^\sigma(z')}{\lambda_0 - \lambda_n}. \qquad (19.77)$$

With the use of Eq. (19.77), Eq. (19.75) can be rewritten as

$$\int \chi_s^\sigma(z,z')v_x^\sigma(z')dz' = -\frac{1}{N_\sigma\Omega}\int \chi_s^\sigma(z,z')$$

$$\times \left| \sum_{k_\parallel \leq k_F^\sigma} e^{i\mathbf{k}_\parallel\cdot\mathbf{r}_\parallel} \right|^2 \frac{|\mu_0^\sigma(z'')|^2}{\sqrt{(z'-z'')^2 + r_\parallel^2}}dz''d\mathbf{r}_\parallel dz'. \qquad (19.78)$$

Canceling[3] the operator $\chi_s^\sigma(z,z')$ from the both sides of Eq. (19.78) and evaluating the sum

$$\Omega\rho^\sigma(\mathbf{r}_\parallel) = \sum_{k_\parallel \leq k_F^\sigma} e^{i\mathbf{k}_\parallel\cdot\mathbf{r}_\parallel} = \frac{\Omega k_F^\sigma J_1(k_F^\sigma r_\parallel)}{2\pi r_\parallel}, \qquad (19.79)$$

where $J_1(u)$ is the Bessel function of the first order, we can write Eq. (19.78) as

$$v_x^\sigma(z) = -\frac{\Omega(k_F^\sigma)^2}{2\pi N_\sigma}\int \frac{J_1^2(k_F^\sigma r_\parallel)}{r_\parallel}\frac{|\mu_0^\sigma(z')|^2}{\sqrt{(z-z')^2+r_\parallel^2}}dz'dr_\parallel, \qquad (19.80)$$

and after the integration over $r_\parallel$, we arrive at [16]

$$v_x^\sigma(z) = -\int \frac{F_2(k_F^\sigma|z-z'|)|\mu_0^\sigma(z')|^2}{|z-z'|}dz', \qquad (19.81)$$

where

$$F_2(u) = 1 + \frac{L_1(2u) - I_1(2u)}{u}, \qquad (19.82)$$

$L_1(u)$ and $I_1(u)$ are the first-order modified Struve and Bessel functions [22,28], respectively, and we have taken into account that

$$\frac{N_\sigma}{\Omega} = n_s^\sigma = \frac{(k_F^\sigma)^2}{4\pi}. \qquad (19.83)$$

Since the spin-density is

$$n^\sigma(z) = n_s^\sigma|\mu_0^\sigma(z)|^2, \qquad (19.84)$$

Eq. (19.81) can be alternatively written as [16]

$$v_x^\sigma(z) = -\frac{1}{n_s^\sigma}\int \frac{F_2(k_F^\sigma|z-z'|)n^\sigma(z')}{|z-z'|}dz', \qquad (19.85)$$

which shows that the exchange potential is, in this case, explicitly expressed via the spin-density.[4]

The function $F_2(u)$ is shown in Figure 19.2 in solid line. Its near-zero and asymptotic behaviors are

$$F_2(u) \underset{u\to 0}{\longrightarrow} \frac{8u}{3\pi} - \frac{u^2}{2} + \ldots, \qquad (19.86)$$

$$F_2(u) \underset{u\to\infty}{\longrightarrow} 1 - \frac{2}{\pi u} + \ldots. \qquad (19.87)$$

---

[3]It can be proven that the equation $\int \chi_s^\sigma(z,z')y(z')dz' = 0$ has the only $y = const$ solution, and we omit an arbitrary constant in the potential as giving rise to no physical consequence.

[4]The exchange potential of Q2DEG with one filled subband was earlier obtained in Ref. [23], however, in a less useful form than Eq. (19.85), since it involved more unfulfilled quadratures.

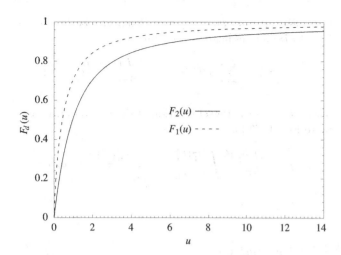

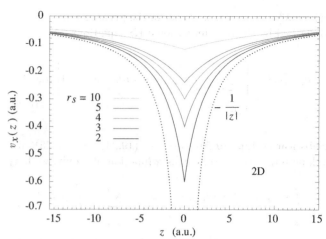

FIGURE 19.2 Functions $F_2(u)$ and $F_1(u)$ of Eqs. (19.82) and (19.92), which determine the quasi-2D and quasi-1D exchange potentials, respectively, [16].

FIGURE 19.3 EXX potential of the spin-unpolarized 2DEG, obtained with Eqs. (19.88) and (19.82), versus the distance from the EG plane z, with the density parameter $r_s = (\pi n_s)^{-1/2}$ changing from 10 a.u. (top) to 2 a.u. (bottom). The dotted lines show the asymptotic $-1/|z|$ [16].

Since $\lim_{u \to \infty} F_2(u) = 1$ and $\int n^\sigma(z)dz = n_s^\sigma$, we conclude from Eq. (19.85) that our $v_x^\sigma(z)$ respects the $-1/z$ asymptotic behavior, which is a requirement to the EXX potential of an arbitrary slab [3].

It is very instructive to consider the limit of the strictly $2D$ (zero-thickness) electron gas. Then, $|\mu_0^\sigma(z)|^2 = \delta(z)$, and Eq. (19.81) reduces to

$$v_x^\sigma(z) = -\frac{F_2(k_F^\sigma|z|)}{|z|}. \qquad (19.88)$$

We see that, although in this limit the density is zero everywhere, except for $z = 0$ $[n^\sigma(z) = n_s^\sigma \delta(z)]$, the exchange potential of Eq. (19.88) is finite everywhere. With the use of Eqs. (19.86), (19.87), and (19.88), we obtain that

$$v_x^\sigma(z = 0) = -\frac{8k_F^\sigma}{3\pi}, \qquad (19.89)$$

$$v_x^\sigma(z) \xrightarrow[z \to \infty]{} -\frac{1}{z} + \frac{2}{\pi k_F^\sigma z^2} + ... \qquad (19.90)$$

In Figure (19.3), we plot the exchange potential of Eq. (19.88) of the spin-neutral strictly 2D EG for several values of the density parameter r_s between 10 and 2. Although the asymptotic of the potential $-1/|z|$ is valid for an arbitrary r_s, the distance z, from where the asymptotic behavior starts, depends on the EG density, being very different for $r_s = 10$ and $r_s = 2$. From Eq. (19.90), we see that the condition of the asymptotic regime is $z \gg (k_F^\sigma)^{-1}$.

Although the limit of the full planar confinement presents a theoretical interest, the realistic quasi-2D EG has a finite z-extent. In Figure 19.4, we plot the exchange potential calculated self-consistently with the density using Eq. (19.85) and compare it with the case of the full 2D confinement. For the former, the confining potential is chosen as that of the strictly 2D uniform distribution of the positive charge at the $z = 0$ plane, with the same value of r_s, ensuring the charge neutrality of the system.

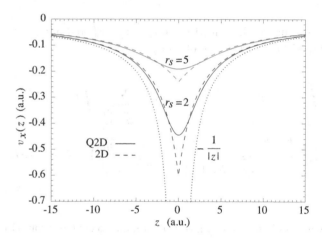

FIGURE 19.4 The EXX potential of spin-neutral Q2D EG with one subband filled (solid lines) compared to the analytical solution for the zero-thickness limit (dashed lines) for two values of the density parameter $r_s = 2$ and 5. The dotted lines show the asymptotic $-1/|z|$ [16].

19.4.3 Quasi-One-Dimensional Electron Gas

Since all the principle arguments of the previous subsection apply to the case of quasi-1D EG with one filled subband, here we only list the corresponding results [16]. The exchange potential is given by

$$v_x^\sigma(\zeta) = -\frac{1}{n_l^\sigma} \int \frac{F_1(k_F^\sigma|\zeta - \zeta'|)n^\sigma(\zeta')}{|\zeta - \zeta'|} d\zeta', \qquad (19.91)$$

where $\zeta = (x, y)$, n_l^σ is the linear spin-density,

$$F_1(u) = \frac{1}{2\pi} G_{2,4}^{2,2} \left[u^2 \left| \begin{matrix} \frac{1}{2}, 1 \\ \frac{1}{2}, \frac{1}{2}, -\frac{1}{2}, 0 \end{matrix} \right. \right], \qquad (19.92)$$

and $G_{p,q}^{m,n}\left[u\ \middle|\ \begin{matrix} a_1,\ldots,a_p \\ b_1,\ldots,b_q \end{matrix}\right]$ is the Meijer G-function [22,28]. The function $F_1(u)$ is plotted in Figure 19.2 in dashed line. The strict 1D limit is obtained from Eq. (19.91), and it is

$$v_x^\sigma(\zeta) = -\frac{F_1(k_F^\sigma \zeta)}{\zeta}. \tag{19.93}$$

The following limiting behavior holds in the 1D case

$$v_x^\sigma(\zeta) \xrightarrow[\zeta \to 0]{} \frac{k_F^\sigma}{\pi}[2\log(k_F^\sigma \zeta) + 2\gamma - 3], \tag{19.94}$$

$$v_x^\sigma(\zeta) \xrightarrow[\zeta \to \infty]{} -\frac{1}{\zeta} + \frac{1}{\pi k_F^\sigma \zeta^2} + ..., \tag{19.95}$$

where $\gamma \approx 0.5772$ is the Euler's constant. The strict 1D exchange potential of Eq. (19.91) and its comparison with the results of the self-consistent calculations with the use of Eq. (19.91) are shown in Figures 19.5 and 19.6, respectively.

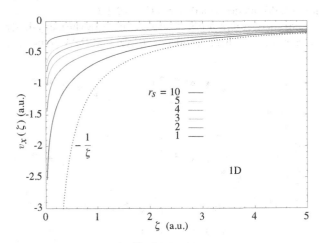

FIGURE 19.5 EXX potential of the spin-unpolarized 1DEG, obtained with Eqs. (19.88) and (19.92), versus the distance from the EG line ζ, with the density parameter $r_s = (2n)^{-1}$ changing from 10 (top) to 1 a.u. (bottom). The dotted line shows the asymptotic $-1/\zeta$ [16].

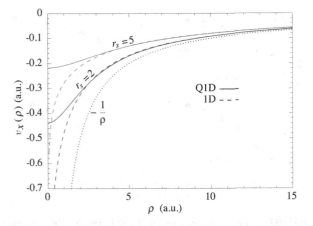

FIGURE 19.6 The same as Figure 19.5 but for 1D case [16].

19.5 Quasi-Low-Dimensional Electron Gas with One Filled Subband: Time-Dependent Case

Having developed the ground-state EXX theory of Q2(1)DEG with one filled subband, we would, naturally, like to extend it to study the excitations in the same system. This occurs, indeed, possible, as we will see in this section. Here, we essentially follow the method of Ref. [17].

19.5.1 EXX TDDFT of Q2DEG

We start from the ground-state of a Q2DEG with one filled subband, as described in Section 19.4, with the orbitals given by Eq. (19.70). To this system, we apply a time-dependent potential, which *depends on the z-coordinate only*, as is schematically shown in Figure 19.7. This particular setup ensures that no in-plane component of the wave-vector is transferred to the system, and it is chosen because it admits the analytical solution of the problem, while a wealth of insight to physical phenomena is still preserved [17].

We will prove that the time-dependent EXX potential in this system and under the said conditions is

$$v_x^\sigma(z,t) = -\frac{1}{n_s^\sigma} \int \frac{F_2(k_F^\sigma|z-z'|)}{|z-z'|} n^\sigma(z',t)dz', \tag{19.96}$$

where $F_2(u)$ is given by Eq. (19.82). While Eq. (19.96) may look like a trivial generalization of Eq. (19.85) to the time-dependent case, it is far from being a such. Indeed, there is no way to directly extend Eq. (19.85) to the time-dependent case without introducing an additional *adiabatic* approximation. What we claim here is a stronger statement that Eq. (19.96) holds without any additional approximation and, therefore, it is the truly *dynamic* EXX potential for the considered system under the specified excitation geometry. A formal proof of Eq. (19.96) is given in Appendix 19.1.

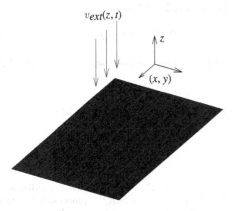

FIGURE 19.7 Q2DEG under the action of a time-dependent external potential.

In the *linear-response regime*, Eq. (19.96) immediately yields for the *exchange kernel* [4,7]

$$f_x^{\sigma\sigma'}(z,z',\omega) = \frac{\delta v_x^\sigma(z,\omega)}{\delta n^{\sigma'}(z',\omega)} = -\frac{1}{n_{2D}^\sigma}\frac{F_2(k_F^\sigma|z-z'|)}{|z-z'|}\delta_{\sigma\sigma'}. \tag{19.97}$$

In the following, we use the kernel of Eq. (19.97) with the basic linear-response TDDFT equality [4,7]

$$\left(\chi^{-1}\right)^{\sigma\sigma'}(z,z',\omega) = \left(\chi_s^{-1}\right)^{\sigma\sigma'}(z,z',\omega)$$
$$- f_H(z,z') - f_x^{\sigma\sigma'}(z,z',\omega), \tag{19.98}$$

where χ and χ_s are the interacting-electrons and KS spin-density-response functions, respectively. The latter is given, in the case of one filled subband and $\mathbf{q}_\parallel = 0$, by

$$\chi_s^{\sigma\sigma'}(z,z',\omega) = n_s^\sigma \mu_0^\sigma(z)\mu_0^\sigma(z')\sum_{n=1}^\infty\left(\frac{1}{\omega + \lambda_0^\sigma - \lambda_n^\sigma + i0_+}\right.$$
$$\left. - \frac{1}{\omega - \lambda_0^\sigma + \lambda_n^\sigma + i0_+}\right)\mu_n^\sigma(z)\mu_n^\sigma(z')\delta_{\sigma\sigma'}, \tag{19.99}$$

where λ_n^σ and $\mu_n^\sigma(z)$ are the eigenenergies and the eigenfunctions of the perpendicular motion, respectively, and the summation over the states includes the integration over the continuous spectrum. A remarkable property of the KS response function (19.99) is that it can be analytically inverted, resulting in (see Appendix 19.2)

$$(\chi_s^{-1})^{\sigma\sigma'}(z,z',\omega) = \frac{\omega^2 X_1^\sigma(z,z') - X_2^\sigma(z,z')}{2n_s^\sigma\mu_0^\sigma(z)\mu_0^\sigma(z')}\delta_{\sigma\sigma'}, \tag{19.100}$$

with

$$X_1^\sigma(z,z') = \sum_{n=1}^\infty\frac{\mu_n^\sigma(z)\mu_n^\sigma(z')}{\lambda_n^\sigma - \lambda_0^\sigma} = \left(\hat{h}_s^\sigma - \lambda_0^\sigma\right)^{-1}$$
$$\times\left[\delta(z-z') - \mu_0^\sigma(z)\mu_0^\sigma(z')\right], \tag{19.101}$$

$$X_2^\sigma(z,z') = \sum_{n=1}^\infty(\lambda_n^\sigma - \lambda_0^\sigma)\mu_n^\sigma(z)\mu_n^\sigma(z') = \left(\hat{h}_s^\sigma - \lambda_0^\sigma\right)\delta(z-z'), \tag{19.102}$$

where $\hat{h}_s^\sigma$ is the static KS Hamiltonian of Eq. (19.72).[5] The Hartree part of the kernel is

$$f_H^{\sigma\sigma'}(z,z') = -2\pi|z-z'|. \tag{19.103}$$

The many-body excitation energies ω can be found from the equation

$$\sum_{\sigma'}\int\left(\chi^{-1}\right)^{\sigma\sigma'}(z,z',\omega)\delta n^{\sigma'}(z',\omega)dz' = 0, \tag{19.104}$$

where $\delta n^\sigma(z,\omega)$ is the self-oscillation of the spin-density. With the use of Eqs. (19.98) and (19.100)–(19.102), Eq. (19.104) turns into the following eigenvalue problem

$$\left(\hat{h}_s^\sigma - \lambda_0^\sigma\right)\left[\left(\hat{h}_s^\sigma - \lambda_0^\sigma\right)y^\sigma(z) + 2n_s^\sigma\right.$$
$$\times\int\mu_0^\sigma(z)\sum_{\sigma'}f_H^{\sigma\sigma'}(z,z')\mu_0^{\sigma'}(z')y^{\sigma'}(z')dz' + 2n_s^\sigma$$
$$\left.\times\int\mu_0^\sigma(z)f_x^{\sigma\sigma'}(z,z')\mu_0^\sigma(z')y^\sigma(z')dz'\right] = \omega^2 y^\sigma(z), \tag{19.105}$$

where we have performed the substitution $y^\sigma(z) = \delta n^\sigma(z)/\mu_0^\sigma(z)$.

The excitation energies obtained from the numerical solution of Eq. (19.105) on a z-axis grid for various densities of the EG are presented in Figures 19.8 and 19.9, for the spin-neutral and fully spin-polarized cases, respectively. TDEXX is compared to the random-phase approximation (RPA) [f_x set to zero in Eq. (19.105)] and to the KS transitions [both f_x and f_H set to zero in Eq. (19.105)]. The static KS problem was solved self-consistently with the use of the EXX potential (19.85), as described in Section 19.4.2.

The first excited state is affected strongly by the many-body interactions, resulting in both TDEXX and RPA being very

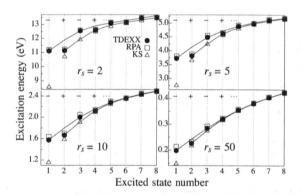

FIGURE 19.8 Excitation energies of a spin-neutral Q2DEG with one subband filled. Circles, squares, and triangles are TDEXX, RPA, and KS excitation energies, respectively. Plus and minus signs mark even and odd excitations, respectively. Lines connect eigenvalues of the even and odd self-oscillations, separately [17].

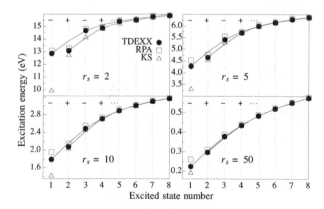

FIGURE 19.9 The same as Figure 19.8, but for the fully spin-polarized Q2DEG [17].

---

[5]The operator $\hat{h}_s^\sigma - \lambda_0^\sigma$ is invertible on the subspace of functions orthogonal to $\mu_0^\sigma(z)$, to which the function in the brackets on the right-hand side of Eq. (19.101) belongs. The orthonormality of the set $\mu_n^\sigma(z)$ has been used in Eqs. (19.101) and (19.102) to write the second equalities on the right-hand sides.

different from the single-particle KS transition. This effect, however, weakens for higher excited states. Second, the difference between the TDEXX and RPA increases with the growth of r_s (decrease of the density), the former moving closer to the KS values, which is more pronounced for the spin-polarized rather than for the spin-neutral EG. This effect is shown important below, and it is explained as follows.

Using Eq. (19.86), we can expand Eq. (19.97) at

$$k_F^\sigma |z - z'| \ll 1 \qquad (19.106)$$

as

$$f_x^{\sigma\sigma'}(z, z', \omega) \approx \left(-\frac{32}{3k_F^\sigma} + 2\pi |z - z'| \right) \delta_{\sigma\sigma'}. \qquad (19.107)$$

The first term in Eq. (19.107) is a constant, and consequently, it does not play a role in f_x.[6] Comparing the second term in Eq. (19.107) with Eq. (19.103), we conclude that, in the regime of Eq. (19.106), the exchange part of the kernel *cancels the Hartree part, by a half and completely, for the spin-neutral and fully spin-polarized EG, respectively*. In the fully spin-polarized case, this brings the many-body excitation energies back to the KS values.

The regime of Eq. (19.106) can realize for two particular reasons: (i) Q2DEG is dilute (k_F^σ is small), and (ii) k_F^σ is arbitrary, but the Q2DEG is strongly confined in the z-direction (then only small z and z', and, hence, small $|z - z'|$ matter). The first case is shown in Figures 19.8 and 19.9, lower right panels, and it has been already discussed. Following Ref. [18], we now consider the second case, i.e., the confinement-dependence of the excitation spectra of Q2DEG.

In Figure 19.10, the electron-density distribution confined in the z-direction with the purely 2D positive uniform charge is plotted for several values of the density of the latter, which is characterized by the density-parameter r_s^+.[7] The corresponding excitation spectra are shown in Figure 19.11. A tendency of the many-body excitation energy getting closer to the single-particle KS values with the decreasing r_s^+ (hence, with the increasing confinement of the EG) can be clearly seen in this figure. We note that this is not the case for the charge-neutral system, as seen in the upper left panel in Figure 19.11, due to the insufficient confinement at $r_s^+ = r_s$ (see the broadest curve in Figure 19.10).

19.5.2 Quasi-1D Case

As in the static case, the dynamic problem of quasi-1D electron gas admits the same treatment as the Q2DEG one. Below, we only summarize the results.

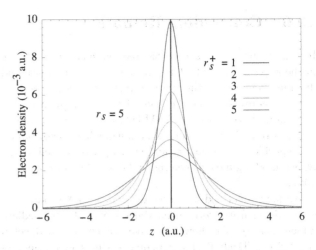

FIGURE 19.10 Ground-state EXX electron density distribution of the fully spin-polarized quasi-2DEG with the density parameter $r_s = 5$. The confining potential is that of the strictly 2D positive charge in the $z = 0$ plane with the density corresponding to the parameter r_s^+ [18].

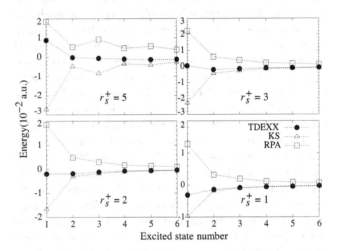

FIGURE 19.11 TDEXX, KS, and RPA excitation energies of the fully spin-polarized quasi-2DEG of the density parameter $r_s = 5$, confined with the strictly 2D uniform positive charge of the density parameter r_s^+. For better visualization, each point is presented relative to the arithmetic mean of the TDEXX, KS, and RPA values [18].

The TD exchange potential is given by

$$v_x^\sigma(\zeta, t) = -\frac{1}{n_l^\sigma} \int \frac{F_1(k_F^\sigma |\zeta - \zeta'|)}{|\zeta - \zeta'|} n^\sigma(\zeta', t) d\zeta', \qquad (19.108)$$

and the corresponding exchange kernel is

$$f_x^{\sigma\sigma'}(\zeta, \zeta', \omega) = -\frac{1}{n_l^\sigma} \frac{F_1(k_F^\sigma |\zeta - \zeta'|)}{|\zeta - \zeta'|} \delta_{\sigma\sigma'}, \qquad (19.109)$$

where $F_1(u)$ is given by Eq. (19.92) (see also Figure 19.2). The Hartree kernel in the Q1D case is

$$f_H^{\sigma\sigma'}(\zeta, \zeta') = -2 \log \left(k_F^\sigma |\zeta - \zeta'| \right). \qquad (19.110)$$

---

[6]Indeed, let $f(\mathbf{r}, \mathbf{r}', \omega) = \tilde{f}(\mathbf{r}, \mathbf{r}', \omega) + const.$ Then $\delta v_{xc}(\mathbf{r}, \omega) = \int f(\mathbf{r}, \mathbf{r}', \omega) \delta n(\mathbf{r}', \omega) d\mathbf{r}' = \int \tilde{f}(\mathbf{r}, \mathbf{r}', \omega) \delta n(\mathbf{r}', \omega) d\mathbf{r}' + const \int \delta n(\mathbf{r}', \omega) d\mathbf{r}'$, where the second term disappears due to the conservation of the number of particles.

[7]The system is necessarily charged at $r_s^+ \neq r_s$.

19.6   Concluding Remarks

In this chapter, we have considered the challenges arising in the application of the DFT, in both its ground-state and dynamic versions, to the study of mesoscopic systems. The problem addressed was the general failure of (semi-) local approximations (LDA, GGA, ALDA, *etc.*), which has motivated us to look around for a functional, which is not bound to a specific dimensionality and which performs well at the transitional regime between the higher and lower dimensionalities.

We have identified the (time-dependent) EXX theory as a promising framework to handle systems of intermediate dimensionality. We, however, have not chosen to develop the EXX (TD)DFT of mesoscopic systems in its general form, which would be a formidable task. Rather, we focused on a specific, but conceptually important case, when exact solutions within EXX occur possible. This is the case of a low-dimensional electron gas with only one subband occupied in the direction perpendicular to the EG extent. By this, on one hand, we were able to obtain clear-cut analytical solutions, which admit profound insights into the many-body processes in mesoscopic systems. On the other, the sacrifice of the generality was not too big, since the regime of the one occupied subband realizes in the wide density ranges of the low-dimensional EG, $r_s > 1.46$ and $r_s > 0.72$, for the Q2D and Q1D cases, respectively, [16], which spans almost the entire range of interesting cases of not too dense EG.

The most striking conclusion derived from the analytical solutions was that, for Q2DEG with one filled subband, the exchange and Hartree kernels cancel each other, by half and completely, in the spin-neutral and fully spin-polarized cases, respectively, if (I) the EG is either dilute or strongly confined in the z-direction and (II) the excitation is carried out without the transfer of the xy-component of the wave-vector. In particular, in this regime, our findings clearly discard the, conventionally accepted, interpretation of the corresponding excitations as the inter-subband plasmons [27], since plasmons need the returning Hartree force [20], which is shown to be canceled or half canceled by the force of the exchange potential.

Because of our aim to focus on the very essence of the subject, while at the same time to remain self-contained, this chapter is by no means exhaustive in its coverage. In particular, such interesting and important issues as the interrelation of the EXX and the localized HF [2,15,19] potentials in the case of Q2(1)EG, as well their relevance to the direct energy functionals of DFT [13] are left out by this review. For these aspects of the theory, the reader is referred to the original literature [16,17].

Obviously, one of the most important directions to proceed in the future will be to go beyond the exchange only theory, i.e., to include correlations. We expect that a progress on this way is achievable by means of the adiabatic connection perturbation method we have been using above, including higher order contributions in λ. We finally express conviction that the approaches described here and the results obtained will prove useful in the mesoscopic physics in general, since they provide us with the guidelines and form our expectations in situations also lying outside the particular framework of this study.

Appendices

Proof of Eq. (19.96) for Time-Dependent Exchange Potential

We use the time-dependent adiabatic connection perturbation method [5], however, doing this beyond the linear-response regime. The adiabatic connection Hamiltonian is

$$\hat{H}^\lambda(t) = \sum_i \left[-\frac{1}{2}\Delta_i + v_{ext}(\mathbf{r}_i, t) + \tilde{v}^\lambda(\mathbf{r}_i, t) \right] + \frac{1}{2}\sum_{i \neq j} \frac{\lambda}{|\mathbf{r}_i - \mathbf{r}_j|}, \quad (19.111)$$

where, for brevity, we are omitting the spin index. The N-body density-matrix $\hat{\rho}^\lambda(t)$ satisfies the Liouville's equation [12]

$$i\frac{\partial \hat{\rho}^\lambda(t)}{\partial t} = [\hat{H}^\lambda(t), \hat{\rho}^\lambda(t)]. \quad (19.112)$$

Expanding up to the first order in λ

$$\hat{H}^\lambda(t) = \hat{H}_0(t) + \lambda\hat{H}_1(t), \quad (19.113)$$

$$\hat{\rho}^\lambda(t) = \hat{\rho}_0(t) + \lambda\hat{\rho}_1(t), \quad (19.114)$$

$$\tilde{v}^\lambda(\mathbf{r}, t) = \tilde{v}_0(\mathbf{r}, t) + \lambda\tilde{v}_1(\mathbf{r}, t), \quad (19.115)$$

we can write due to Eqs. (19.111) and (19.112)

$$i\frac{\partial \hat{\rho}_0(t)}{\partial t} = [\hat{H}_0(t), \hat{\rho}_0(t)], \quad (19.116)$$

$$i\frac{\partial \hat{\rho}_1(t)}{\partial t} = [\hat{H}_0(t), \hat{\rho}_1(t)] + [\hat{H}_1(t), \hat{\rho}_0(t)], \quad (19.117)$$

$$\hat{H}_0(t) = \sum_i \left[-\frac{1}{2}\Delta_i + v_{ext}(\mathbf{r}_i, t) + \tilde{v}_0(\mathbf{r}_i, t) \right], \quad (19.118)$$

$$\hat{H}_1(t) = \sum_i \tilde{v}_1(\mathbf{r}_i, t) + \frac{1}{2}\sum_{i \neq j} \frac{1}{|\mathbf{r}_i - \mathbf{r}_j|}. \quad (19.119)$$

We assume that at $t \leq 0$ the system was in its ground-state (for both the many-body and the KS systems), when the external potential did not depend on time. Let $|n\rangle$ be the complete orthonormal set of the Slater-determinant eigenfunctions of the Hamiltonian $\hat{H}_0(t = 0)$. At $t = 0$, the time-dependence of the external potential is switched on, and the independent-particle wave-functions propagate as

$$i\frac{\partial |n(t)\rangle}{\partial t} = \hat{H}_0(t)|n(t)\rangle, \quad (19.120)$$

while

$$|n(0)\rangle = |n\rangle. \quad (19.121)$$

Due to Eqs. (19.120) and (19.121) with the Hermitian $\hat{H}_0(t)$, the set $|n(t)\rangle$ is complete and orthonormal as well at an arbitrary t.

We proceed by taking matrix elements of Eq. (19.117) and obtain with the use of Eq. (19.120)

$$i\frac{\partial}{\partial t}\langle n(t)|\hat{\rho}_1(t)|m(t)\rangle = (\delta_{m0} - \delta_{n0})\langle n(t)|\hat{H}_1(t)|m(t)\rangle.$$
(19.122)

On the other hand, at $t = 0$, by Eq. (19.54)

$$\langle n|\hat{\rho}_1|m\rangle = \frac{\delta_{m0} - \delta_{n0}}{E_m - E_n + i0_+}\langle n|\hat{H}_1|m\rangle,$$
(19.123)

where E_n are the eigenvalues of $\hat{H}_0$. Together, Eqs. (19.122) and (19.123) give

$$\langle n(t)|\hat{\rho}_1(t)|m(t)\rangle = (\delta_{m0} - \delta_{n0})$$
$$\times \left[\frac{\langle n|\hat{H}_1|m\rangle}{E_m - E_n + i0_+} - i\int_0^t \langle n(t')|\hat{H}_1(t')|m(t')\rangle dt'\right].$$
(19.124)

The first-order correction in λ to the density is

$$n_1(\mathbf{r},t) = \mathrm{Sp}\left\{\hat{\rho}_1(t)\sum_i \delta(\mathbf{r} - \mathbf{r}_i)\right\}$$
(19.125)

or

$$n_1(\mathbf{r},t) = N\sum_{nm}\langle n(t)|\hat{\rho}_1(t)|m(t)\rangle\langle m(t)|\delta(\mathbf{r} - \mathbf{r}_1)|n(t)\rangle,$$
(19.126)

where we have taken account of the identity of electrons. Further, we can write the last equation as

$$n_1(z,t) = N\sum_{n\neq 0}\langle n(t)|\hat{\rho}_1(t)|0(t)\rangle\langle 0(t)|\delta(z - z_1)|n(t)\rangle + c.c.,$$
(19.127)

where the structure of Eq. (19.124) as well as the symmetry of our system and that of the external perturbation was taken into account. During the time-evolution, the orbitals remain of the form

$$\phi_i(\mathbf{r},t) = \mu_{n_i}(z,t)\frac{e^{i\mathbf{k}_{i\parallel}\cdot\mathbf{r}_{i\parallel}}}{\Omega^{1/2}},$$
(19.128)

from which we can conclude that only the matrix elements $\langle n_{ns}(t)|\hat{\rho}_1(t)|0(t)\rangle$ and their Hermitian conjugates contribute to Eq. (19.127), where we have introduced the notation

$$|n_{ns}(t)\rangle = \frac{1}{(N!\Omega^N)^{1/2}}$$
$$\times \begin{vmatrix} \mu_0(z_1,t)e^{i\mathbf{k}_1\cdot\mathbf{r}_{1\parallel}} & \cdots & \mu_0(z_N,t)e^{i\mathbf{k}_1\cdot\mathbf{r}_{N\parallel}} \\ \vdots & \cdots & \vdots \\ \mu_0(z_1,t)e^{i\mathbf{k}_{s-1}\cdot\mathbf{r}_{1\parallel}} & \cdots & \mu_0(z_N,t)e^{i\mathbf{k}_{s-1}\cdot\mathbf{r}_{N\parallel}} \\ \mu_n(z_1,t)e^{i\mathbf{k}_s\cdot\mathbf{r}_{1\parallel}} & \cdots & \mu_n(z_N,t)e^{i\mathbf{k}_s\cdot\mathbf{r}_{N\parallel}} \\ \mu_0(z_1,t)e^{i\mathbf{k}_{s+1}\cdot\mathbf{r}_{1\parallel}} & \cdots & \mu_0(z_N,t)e^{i\mathbf{k}_{s+1}\cdot\mathbf{r}_{N\parallel}} \\ \vdots & \cdots & \vdots \\ \mu_0(z_1,t)e^{i\mathbf{k}_N\cdot\mathbf{r}_{1\parallel}} & \cdots & \mu_0(z_N,t)e^{i\mathbf{k}_N\cdot\mathbf{r}_{N\parallel}} \end{vmatrix},$$
(19.129)

where $n = 1, 2, \ldots$ and $s = 1\ldots N$. In other words, in $|n_{ns}(t)\rangle$ of Eq. (19.129) the only $\mu_{n\neq 0}$ stands together with the plane-wave of the $\mathbf{k}_s$ vector. The ground-state determinant is

$$|0(t)\rangle = \frac{1}{(N!\Omega^N)^{1/2}}$$
$$\times \begin{vmatrix} \mu_0(z_1,t)e^{i\mathbf{k}_1\cdot\mathbf{r}_{1\parallel}} & \cdots & \mu_0(z_N,t)e^{i\mathbf{k}_1\cdot\mathbf{r}_{N\parallel}} \\ \vdots & \cdots & \vdots \\ \mu_0(z_1,t)e^{i\mathbf{k}_N\cdot\mathbf{r}_{1\parallel}} & \cdots & \mu_0(z_N,t)e^{i\mathbf{k}_N\cdot\mathbf{r}_{N\parallel}} \end{vmatrix}.$$
(19.130)

We can write by Eq. (19.119)

$$\langle n_{ns}(t)|\hat{H}_1(t)|0(t)\rangle = \langle n_{ns}(t)|N\tilde{v}_1(\mathbf{r}_1,t) + \frac{N(N-1)}{2|\mathbf{r}_1 - \mathbf{r}_2|}|0(t)\rangle,$$
(19.131)

and we evaluate

$$\langle n_{ns}(t)|\tilde{v}_1(z_1,t)|0(t)\rangle = \frac{1}{N}\int \mu_n^*(z_1,t)\tilde{v}_1(z_1,t)\mu_0(z_1,t)dz_1,$$
(19.132)

$$\langle n_{ns}(t)|\frac{1}{|\mathbf{r}_1 - \mathbf{r}_2|}|0(t)\rangle = \frac{2}{\Omega^2 N(N-1)}$$
$$\times \int \frac{\mu_n^*(z_1,t)\mu_0(z_1,t)|\mu_0(z_2,t)|^2}{|\mathbf{r}_1 - \mathbf{r}_2|}$$
$$\times \left[N - \Omega e^{i\mathbf{k}_s\cdot(\mathbf{r}_{2\parallel} - \mathbf{r}_{1\parallel})}\rho^*(\mathbf{r}_{2\parallel} - \mathbf{r}_{1\parallel})\right]d\mathbf{r}_1 d\mathbf{r}_2,$$
(19.133)

$$\langle n_{ns}(t)|\delta(z - z_1)|0(t)\rangle = \frac{1}{N}\mu_n^*(z,t)\mu_0(z,t),$$
(19.134)

where $\rho(\mathbf{r}_\parallel)$ is given by Eq. (19.79). Then, by Eqs. (19.131)–(19.133), and with an integration variable substitution,

$$\langle n_{ns}(t)|\hat{H}_1(t)|0(t)\rangle = \int \mu_{nn}^*(z,t)\mu_0(z,t)\left\{\tilde{v}_1(z,t)\right.$$
$$\left. + \frac{1}{\Omega}\int \frac{|\mu_0(z',t)|^2}{\sqrt{(z-z')^2 + r_\parallel^2}}\left[N - \Omega e^{i\mathbf{k}_s\cdot\mathbf{r}_\parallel}\rho^*(\mathbf{r}_\parallel)\right]d\mathbf{r}_\parallel dz'\right\}dz.$$
(19.135)

By virtue of Eqs. (19.124) and (19.135), we arrive at

$$\langle n_{ns}(t)|\hat{\rho}_1(t)|0(t)\rangle = -i\int_0^t dt'\int \mu_n^*(z,t')\mu_0(z,t')$$
$$\times \left\{\tilde{v}_1(z,t') + \frac{1}{\Omega}\int \frac{|\mu_0(z',t')|^2}{\sqrt{(z-z')^2 + r_\parallel^2}}\left[N - \Omega e^{i\mathbf{k}_s\cdot\mathbf{r}_\parallel}\rho^*(\mathbf{r}_\parallel)\right]d\mathbf{r}_\parallel dz'\right\}dz$$
$$+ \frac{1}{\lambda_0 - \lambda_n}\int \mu_n^*(z)\mu_0(z)$$
$$\times \left\{\tilde{v}_1(z) + \frac{1}{\Omega}\int \frac{|\mu_0(z')|^2}{\sqrt{(z-z')^2 + r_\parallel^2}}\left[N - \Omega e^{i\mathbf{k}_s\cdot\mathbf{r}_\parallel}\rho^*(\mathbf{r}_\parallel)\right]d\mathbf{r}_\parallel dz'\right\}dz.$$
(19.136)

With the use of Eqs. (19.127) and (19.134), we can write

$$n_1(z,t) = \sum_{n=1}^{\infty} \sum_{s=1}^{N} \langle n_{ns}(t)|\hat{\rho}_1(t)|0(t)\rangle \, \mu_n(z,t)\mu_0^*(z,t) + c.c.$$

(19.137)

By the adiabatic connection perturbation method, with the change of λ within $[0,1]$, the density should remain unchanged. Since, by Eqs. (19.79) and (19.136),

$$\sum_{s=1}^{N} \langle n_{ns}(t)|\hat{\rho}_1(t)|0(t)\rangle = -i\int_0^t dt' \int \mu_n^*(z,t')\mu_0(z,t')$$

$$\times \left\{ N\tilde{v}_1(z,t') + \frac{1}{\Omega}\int \frac{|\mu_0(z',t')|^2}{\sqrt{(z-z')^2+r_\parallel^2}}\left[N^2-\Omega^2|\rho(\mathbf{r}_\parallel)|^2\right] d\mathbf{r}_\parallel dz' \right\} dz$$

$$+ \frac{1}{\lambda_0-\lambda_n}\int \mu_n^*(z)\mu_0(z)$$

$$\left\{ N\tilde{v}_1(z) + \frac{1}{\Omega}\int \frac{|\mu_0(z')|^2}{\sqrt{(z-z')^2+r_\parallel^2}}\left[N^2-\Omega^2|\rho(\mathbf{r}_\parallel)|^2\right] d\mathbf{r}_\parallel dz' \right\} dz,$$

(19.138)

we see from Eqs. (19.137) and (19.138) that $n_1(z,t) = 0$ if

$$\tilde{v}_1(z,t) = -\frac{1}{\Omega N}\int \frac{|\mu_0(z',t)|^2}{\sqrt{(z-z')^2+r_\parallel^2}}$$

$$\times \left[N^2-\Omega^2|\rho(\mathbf{r}_\parallel)|^2\right] d\mathbf{r}_\parallel dz'.$$

(19.139)

According to Eq. (19.111), $\tilde{v}^0(\mathbf{r},t) = v_H(\mathbf{r},t) + v_{xc}(\mathbf{r},t)$ and $\tilde{v}^1(\mathbf{r},t) = 0$, where $v_H(\mathbf{r},t)$ and $v_{xc}(\mathbf{r},t)$ are Hartree and the exchange-correlations potentials, respectively. Then, by Eq. (19.115), we can write

$$\tilde{v}_1(\mathbf{r},t) = -v_H(\mathbf{r},t) - v_x(\mathbf{r},t),$$

(19.140)

where in the notation of v_x, we have taken into account that to the first order in λ we have, by definition, exchange only [5,6]. On the other hand, for the Hartree potential, we have

$$v_H(\mathbf{r},t) = \frac{N}{\Omega}\int \frac{|\mu_0(z',t)|^2}{\sqrt{(z-z')^2+r_\parallel^2}} d\mathbf{r}_\parallel dz',$$

(19.141)

which leads us to

$$v_x(z,t) = -\frac{1}{n_s}\int \frac{|\mu_0(z',t)|^2}{\sqrt{(z-z')^2+r_\parallel^2}}|\rho(\mathbf{r}_\parallel)|^2 d\mathbf{r}_\parallel dz',$$

(19.142)

We conclude the proof of Eq. (19.96) by integrating explicitly over $\mathbf{r}_\parallel$ in Eq. (19.142) with the account of Eq. (19.79) and noting that

$$n(z,t) = n_s|\mu_0(z,t)|^2.$$

(19.143)

KS Spin-Density-Response Function and Its Inverse

In this Appendix, we prove Eqs. (19.100)–(19.102) and (19.105). We construct the operator

$$(\chi_s^\sigma)^{-1}(z,z',\omega) = \frac{1}{n_s^\sigma}$$

$$\times \sum_{n=1}^{\infty} \left(\frac{1}{\omega+\lambda_0^\sigma-\lambda_n^\sigma+i0_+} - \frac{1}{\omega-\lambda_0^\sigma+\lambda_n^\sigma+i0_+} \right)^{-1}$$

$$\times \frac{\mu_n^\sigma(z)\mu_n^\sigma(z')}{\mu_0^\sigma(z)\mu_0^\sigma(z')}\delta_{\sigma\sigma'}$$

(19.144)

and directly check that for an arbitrary function $g(z)$ such that

$$\int g(z)dz = 0,$$

(19.145)

the equality holds

$$\int \chi_s^\sigma(z,z'',\omega)(\chi_s^\sigma)^{-1}(z'',z',\omega)g(z')dz''dz' = g(z),$$

(19.146)

where χ_s is given by Eq. (19.99). In arriving at Eq. (19.146), we have used the completeness relation

$$\sum_{n=0}^{\infty} \mu_n^\sigma(z)\mu_n^\sigma(z') = \delta(z-z').$$

(19.147)

On the other hand, the operator $(\chi_s^\sigma)^{-1}\chi_s^\sigma$ is defined on any function $h(z)$ of the Hilbert space, and

$$\int (\chi_s^\sigma)^{-1}(z,z'',\omega)\chi_s^\sigma(z'',z',\omega)h(z')dz''dz'$$

$$= h(z) - \int [\mu_0^\sigma(z')]^2 h(z')dz',$$

(19.148)

where the second term on the right-hand side is a constant. Equations (19.146) and (19.148) prove that χ_s of Eq. (19.99) and χ_s^{-1} of Eq. (19.144) are inverse to each other in the sense as the density-response function and its inverse should be.

From Eq. (19.144), we arrive at Eqs. (19.100)–(19.102) by simple algebraic manipulations with the expression in the parentheses on the right-hand side and by using again the completeness relation (19.147).

With the use of Eqs. (19.98) and (19.100)–(19.102), we write Eqs. (19.104) as

$$\frac{1}{2n_s^\sigma\mu_0^\sigma(z)}\int \left\{ \omega^2\left(\hat{h}_s^\sigma-\lambda_0^\sigma\right)^{-1}[\delta(z-z')-\mu_0^\sigma(z)\mu_0^\sigma(z')] \right.$$

$$\left. - \left(\hat{h}_s^\sigma-\lambda_0^\sigma\right)\delta(z-z') \right\}\frac{\delta n^\sigma(z')}{\mu_0^\sigma(z')}dz'$$

$$- \int \sum_{\sigma'} f_H^{\sigma\sigma'}(z,z')\delta n^{\sigma'}(z')dz'$$

$$- \int f_x^{\sigma\sigma}(z,z')\delta n^\sigma(z')dz' = 0.$$

(19.149)

By integrating in the first line of Eq. (19.149) over z', taking account of $\int \delta n^\sigma(z)dz = 0$, multiplying both sides by $2n_s^\sigma\mu_0^\sigma(z)$, and finally applying $\left(\hat{h}_s^\sigma-\lambda_0^\sigma\right)$, we arrive at Eq. (19.105).

Acknowledgments

Our work was partially supported by Ministry of Science and Technology, Taiwan (Grant No. 106-2923-M-001 -002-MY3 and 107-2112-M-001 -033).

Bibliography

1. L. A. Constantin, J. P. Perdew, and J. M. Pitarke. Collapse of the electron gas to two dimensions in density functional theory. *Phys. Rev. Lett.*, 101:016406, 2008.

2. F. Della Sala and A. Görling. Efficient localized Hartree-Fock methods as effective exact-exchange Kohn-Sham methods for molecules. *J. Chem. Phys.*, 115(13):5718–5732, 2001.

3. E. Engel. Exact exchange plane-wave-pseudopotential calculations for slabs. *J. Chem. Phys.*, 140(18):18A505, 2014.

4. G. F. Giuliani and G. Vignale. *Quantum Theory of the Electron Liquid*. Cambridge University Press, Cambridge, 2005.

5. A. Görling. Time-dependent Kohn-Sham formalism. *Phys. Rev. A*, 55:2630–2639, 1997.

6. A. Görling and M. Levy. Exact Kohn-Sham scheme based on perturbation theory. *Phys. Rev. A*, 50:196–204, 1994.

7. E. K. U. Gross and W. Kohn. Local density-functional theory of frequency-dependent linear response. *Phys. Rev. Lett.*, 55(26):2850–2852, 1985.

8. P. Hohenberg and W. Kohn. Inhomogeneous electron gas. *Phys. Rev.*, 136:B864–B871, 1964.

9. S. Karimi and C. A. Ullrich. Three- to two-dimensional crossover in time-dependent density-functional theory. *Phys. Rev. B*, 90:245304, 2014.

10. Y.-H. Kim, I.-H. Lee, S. Nagaraja, J.-P. Leburton, R. Q. Hood, and R. M. Martin. Two-dimensional limit of exchange-correlation energy functional approximations. *Phys. Rev. B*, 61:5202–5211, 2000.

11. W. Kohn and L. J. Sham. Self-consistent equations including exchange and correlation effects. *Phys. Rev.*, 140:A1133–A1138, 1965.

12. L. D. Landau and E. M. Lifshitz. *Quantum Mechanics: The Non-Relativistic Theory*. Butterworth-Heinemann, London, 1981.

13. M. Levy and F. Zahariev. Ground-state energy as a simple sum of orbital energies in Kohn-Sham theory: A shift in perspective through a shift in potential. *Phys. Rev. Lett.*, 113:113002, 2014.

14. J. Lindhard. On the properties of a gas of charged particles. *K. Dan. Vidensk. Selsk. Mat. Fys. Medd.*, 28(8):1, 1954.

15. V. U. Nazarov. Time-dependent effective potential and exchange kernel of homogeneous electron gas. *Phys. Rev. B*, 87:165125, 2013.

16. V. U. Nazarov. Exact exact-exchange potential of two- and one-dimensional electron gases beyond the asymptotic limit. *Phys. Rev. B*, 93:195432, 2016.

17. V. U. Nazarov. Quasi-low-dimensional electron gas with one populated band as a testing ground for time-dependent density-functional theory of mesoscopic systems. *Phys. Rev. Lett.*, 118:236802, 2017.

18. V. U. Nazarov. Crossover between collective and independent-particle excitations in quasi-2D electron gas with one filled subband. *Eur. Phys. J. B*, 91(6):95, 2018.

19. V. U. Nazarov and G. Vignale. Derivative discontinuity with localized Hartree-Fock potential. *J. Chem. Phys.*, 143(6):064111, 2015.

20. D. Pines. *Elementary Excitations in Solids*. W. A. Benjamin, Inc., New York, 1963.

21. L. Pollack and J. P. Perdew. Evaluating density functional performance for the quasi-two-dimensional electron gas. *J. Phys. Condens. Matter*, 12:1239, 2000.

22. A. P. Prudnikov, O. I. Marichev, and Yu. A. Brychkov. *Integrals and Series*, volume 3: *More Special Functions*. Gordon and Breach, Newark, NJ, 1990.

23. F. A. Reboredo and C. R. Proetto. Exact-exchange density functional theory for quasi-two-dimensional electron gases. *Phys. Rev. B*, 67:115325, 2003.

24. R. T. Sharp and G. K. Horton. A variational approach to the unipotential many-electron problem. *Phys. Rev.*, 90:317–317, 1953.

25. J. C. Slater. A simplification of the Hartree-Fock Method. *Phys. Rev.*, 81(3):385–390, 1951.

26. J. D. Talman and W. F. Shadwick. Optimized effective atomic central potential. *Phys. Rev. A*, 14:36–40, 1976.

27. C. A. Ullrich. *Time-Dependent Density-Functional Theory: Concepts and Applications*. Oxford University Press, Oxford, 2012.

28. Wolfram Research. *Mathematica*. Champaign, Illinios, version 11.0 edition, 2016.

20

Molecular Simulation of Porous Graphene

Ziqi Tian, Haoran Guo, and
Liang Chen
Ningbo Institute of Materials Technology &
Engineering, Chinese Academy of Sciences

Chad Priest
Sandia National Laboratories

20.1 Introduction

In 1996, Nobel laureate Richard E. Smalley predicted, "Carbon has this genius of making a chemically stable two-dimensional, one-atom-thick membrane in a three-dimensional world. And that, I believe, is going to be very important in the future of chemistry and technology in general". As foreshadowed, the isolation and characterization of freestanding graphene sheet from graphite was accomplished by Novoselov et al. (2004) immediately opened our 3D world to the practicality of the two-dimensional material by unveiling its fundamental physics, materials science, and nanodevices (Ji et al. 2016, Solis-Fernandez et al. 2017, Wang et al. 2017, Yu et al. 2017). It has been discovered that graphene's pragmatism is contributed to its unique structure, along with its many uncommon properties, which include strong mechanical strength, high thermal electric conductivity, and nonlinear diamagnetism. In the past decade, fabrication techniques have been developed and refined to synthesize graphene nanosheet from different methods of exfoliation, chemical vapor deposition (CVD), and reduction of graphene oxide sheet (Allen et al. 2010, Sun et al. 2018).

To meet quotidian (daily) practical utilization, the tailoring of intrinsic properties is required, usually by structure modification of pristine graphene. A common method is by introducing nanopores on the basal plane or introducing vacancies to design new types of graphene-based materials (Liang et al. 2017, Yang et al. 2017, Yao and Zhao 2017, Ito et al. 2018). By creating carbon atom vacancies or nanoscale pores, the electronic structure of porous graphene can be greatly modified, leading to wide applications on nanoelectronics and energy storage. Simultaneously, this nanoporous two-dimensional structure can provide an outstanding starting point to investigate

advanced separation technology. They can perform as molecular sieve to distinguish various components via size effect and thus serve as ultimate membrane for mixture separation, e.g., gas purification and desalination (Wang et al. 2017).

There are many experimental efforts on the production of porous graphene, which can be generally divided into two classes: the top-down method and bottom-up approach. The top-down method is to punch nanopores on the freestanding basal plane. Currently, large-scale synthesis of graphene sheet has become a relatively mature technology. Starting from the graphene sheet, one can etch pores with electron beam, heavy atom/molecule, controlled oxidization, and other techniques (Fischbein and Drndic 2008, Koenig et al. 2012, Russo and Golovchenko 2012). The apertures of created pores range from few angstroms to several nanometers. Another way, the bottom-up approach constructs two-dimensional polymeric structure from isolated building blocks. For example, Bieri et al. produced poly-phenylene on silver through surface-promoted aryl–aryl coupling, in which the hexaiodo-substituted cyclohexa-m-phenylene was used as the building unit (Bieri et al. 2009). In comparison, the top-down method is more suitable for large-scale preparation, whereas the bottom-up approach facilitates precise control of the aperture in nanoscale. For more details on the utility and synthesis of porous graphene, we refer to other chapter of this book.

Despite the great development of nanotechnology, it is still difficult to introduce numerous uniform nanopores on large area of the single-layered graphene. Some pioneering investigations to guide experiments are in high demand. Advanced computational tools can effectively help researchers understand the properties of prepared porous graphene, provide insight into their impressive performance, and forecast the undiscovered materials (Jiang and Chen 2013). First, we will overview the computational methodologies for the simulation of porous graphene system, including density functional theory (DFT) and classical molecular dynamics (CMD). Subsequently, the applications of these methods on specific fields will be summarized, which show that theoretical studies can not only elucidate the experimental phenomena, but also propose new concept of novel material and guide the design in practice. Finally, we offer a perspective for future computational studies on porous graphene.

20.2 Theoretical Methods

In regard to the simulated properties and modeling scale, we mainly introduce two important theoretical methodologies: (i) DFT method for studying geometric and electronic structure of porous graphene model; and (ii) CMD simulation for reproducing dynamic process of complicate system containing single or few layers of porous graphene.

20.2.1 DFT to Describe the Electronic Structure

First-principle calculations for periodic systems, usually based on DFT, are able to describe the electronic structure without higher order parameters, e.g. material properties. Once the geometry of porous graphene is determined, its properties depending on electronic structure, such as conductivity and interaction to small molecules, can be evaluated. DFT is a quantum mechanical (QM) calculation method used to study the electronic structure of atoms, molecules, and periodic condensed phases (Parr and Yang 1989). This theory illustrates that the properties of a many-electron system are a function of electron density, which is another function of spatial coordinates. That is where the name density functional comes from. Although the first attempt of DFT method, Thomas–Fermi model, was proposed in 1927, which is a qualitative method than a quantitative method, it is possible to develop contemporary DFT methods until Walter Kohn and Pierre Hohenberg demonstrated two Hohenberg–Kohn (HK) theorems in 1964 (Hohenberg and Kohn 1964). The first HK theorem represents that the ground state energy of a many-electron system is a unique functional of electron density function that only depends on atomic coordinates. The second HK theorem proves that the energy functional for the system should obtain the minimum energy, if and only if the electron density is the correct ground state density. Based on HK theorems, Walter Kohn and Lu-Jeu Sham proposed well-known Kohn–Sham (KS) equations (Kohn and Sham 1965), in which electron energy is expressed as the sum of kinetic energy, E^T, potential energy from the attraction of nuclei, E^V, electron–electron Coulomb repulsion, E^J, and exchange-correlation potential, E^{XC}:

$$E = E^T + E^V + E^J + E^{XC}$$

All four of these terms are the functionals of electron density. The first three terms can be expressed by electron density explicitly and relate to most of the total energy. The fourth term, E^{XC}, includes all the rest many-particle interactions. Although this term doesn't contribute much to the total energy, it determines the accuracy of KS equation. Its exact expression is unknown and has to be approximated. There have been several approximations which allow acceptable accuracy of certain calculated quantities, for instance, local density approximation (LDA), local spin-density approximation (LSDA), and generalized gradient approximations (GGA). From 1980s, tens of functionals have been developed to describe electronic structure of various systems. At the same time, there are lots of available software packages to perform DFT calculation on periodic system, such as Vienna Ab initio simulation package (VASP) (Kresse and Furthmuller 1996), Quantum ESPRESSO (Giannozzi et al. 2009), and CP2K (Hutter et al. 2014). Nowadays, DFT method has been a significant tool for the calculation of solid-state physics and quantum chemistry.

An exemplary application of DFT is its ability to pry into the applicability of two-dimensional nanodevice by calculating its band structure. It has been determined pristine graphene layer lacks a band gap, limiting its usage in nanoelectronics. As shown in Figure 20.1c, the computed band structure indicates graphene as a semimetallic material, which is not suitable for utilization in semiconductor industry. To open or tune its band gap, researchers have explored various ways, including but not limited to doping (Balog et al. 2010), chemical functionalization (Cervantes-Sodi et al. 2008), cutting into nanoribbon (Yang et al. 2007), and creating pore on the basal plane. De La Pierre et al. simulated the band structures of two porous graphenes, named as hydrogenated porous graphene (HPG) and biphenylene carbon (BPC) (De La Pierre et al. 2013). They showed that the modification of structure leads to apparent band gap, as plotted in Figure 20.1. For HPG and BPC, the calculated band gaps are 3.95 and 1.08 eV, respectively. The broad band gap of HPG benefits from its peculiar aromatic structure. These C–C single bonds interconnecting benzene rings isolate conjugated units and prohibit electron delocalization between neighboring groups. As a result, π conjugation system only exists in the benzene unit. While enhancing the connection between benzene-like units can effectively narrow the band gap, as in the case of BPC. In light of theoretical results, experimental studies have been conducted to utilize porous graphene in nanoelectronics. For more examples, please see Section 3.1.

Despite the great progress in DFT methods, there are some issues to accurately compute the band gap and ferromagnetism in semiconductors. Most pure functionals account electron correlation too much, resulting in significantly underestimated band gap and magnetism. Hybrid functionals, such as Heyd–Scuseria–Emzerhof (HSE) functional (Heyd et al. 2003), address this problem via

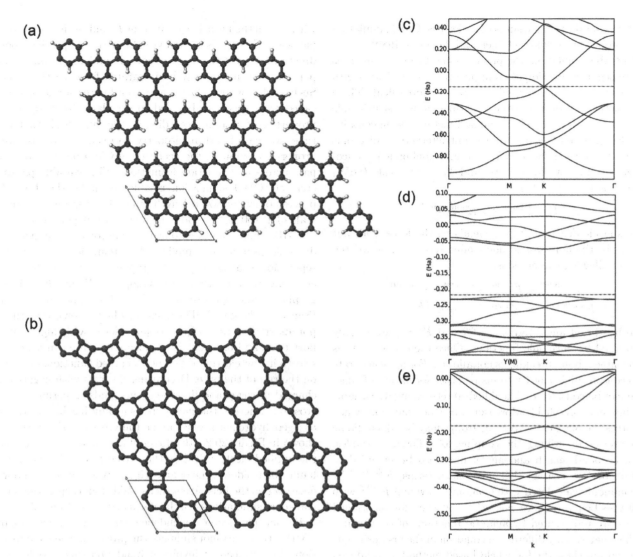

FIGURE 20.1 Top view of (a) HPG and (b) BPC. Gray and white balls represent C and H atoms, respectively. The simulated band structures of (c) pristine graphene, (d) HPG, and (e) BPC. (Reproduced with permission from De La Pierre et al. (2013). Copyright 2013, American Chemical Society.)

introducing Hartree–Fock (HF) exchange. But for the calculation of solid-state properties, the consideration of HF term in real space is always expensive. GW approximation is another choice, which estimates system energy as the lowest term of the expansion in terms of single-electron Green's function, G, and screened Coulomb interaction, W (Aryasetiawan and Gunnarsson 1998). It could be more accurate to picture the electronic structure, but also more time-consuming than hybrid functional. Recently, Ferreira et al. revisited DFT technique of half-occupation, named as DFT-1/2 method (Ferreira et al. 2011). With similar computational consumption, calculated band gap using GGA-1/2 or LDA-1/2 is far better than that computed by standard functional. It may be a choice in the high-throughput screening of potential materials.

One critical drawback of standard DFT methods is its lack of considering long-range weak interactions (dispersion interaction), especially when dispersion dominates most interaction between graphene-based material and other small molecules. To overcome this issue, many dispersion corrections have been developed for years (Tsuneda and Hirao 2014). Due to the $1/r^6$ scaling of dispersion interaction (r is the distance between two particles), total energy can be regulated by adding an empirical term to yield reasonable correction. DFT-D (Grimme et al. 2010) and vdW-TS (Tkatchenko and Scheffler 2009) methods are two examples with empirical correction. Besides, some DFT approaches, such as vdW-DF (Dion et al. 2004) and VV10 (Vydrov and Van Voorhis 2010), can consider nonlocal correlation terms into correlation functional; herein, empirical parameters are not required.

20.2.2 CMD to Simulate the Dynamics Process

Limited by hardware and software, contemporary first-principle methods can handle systems containing hundreds of atoms. To mimic the dynamic process of a real-life system containing thousands of molecules, CMD simulation is appropriate workhorse (Allen and Tildesley 1989).

In CMD simulation, the potential energy of system, E, is expressed as a function of atom (or group) coordinates with parameter sets, which is also known as force field. These parameters are derived from experiments or first-principle calculations, or both. The common functional form contains bonded (covalent) energy to account interactions of atoms linked by covalent bonds (E^{bond}), and nonbonded (noncovalent) terms for electrostatic and Van der Waals (vdW) interaction ($E^{\text{nonbonded}}$):

$$E = E^{\text{bonded}} + E^{\text{nonbonded}}$$

The specific decomposition depends on the force field, but generally, the bonded and nonbonded terms can be written as the following summations:

$$E^{\text{bonded}} = E^{\text{bond}} + E^{\text{angle}} + E^{\text{dihedral}}$$
$$E^{\text{nonbonded}} = E^{\text{electrostatic}} + E^{\text{vdW}}$$

The bond and angle terms (E^{bond} and E^{angle}) are usually computed by harmonic functions. Once the connectivity is determined, bond breaking is not allowed. For more accurate description, higher order terms or other formation of function can be considered. The dihedral term is highly variable in different force field. Moreover, improper function which enforces the planarity of certain groups may be added. Some cross-terms describing the coupling of different variables, such as bond length and angle, may also be included to obtain better accuracy. For nonbonded terms, $E^{\text{electrostatic}}$ is commonly calculated with Coulomb's law and E^{vdW} with pairwise Lennard-Jones potential. All the aforementioned terms are represented by analytic functions of coordinates rather than solving differential equation in the first-principle calculation; thus, the force field-based method, also labeled as molecular mechanics (MM), are much faster than DFT.

Once the potential energy of system is represented by "classical function", the force on each particle, F_i, and its acceleration, a_i, can be obtained from Newton's laws of motion:

$$F_i = -\nabla_i E$$
$$a_i = F_i / m_i$$

where F_i is the negative gradient of E, and m_i is the mass of particle i. Since E is an analytical function of particle coordinates, the velocity of each particle and evolution of the particle coordinates may be integrated for each time step. Such evolution of velocity and coordinates for long time, called trajectory, can be used to calculate thermodynamic properties of system, according to the hypothesis that the time average of certain property corresponds to ensemble property. Obviously, the accuracy of CMD depends on the parameters and function formation. With suitable parameter set, CMD simulation is able to model the dynamic process of system on a length scale of 1–100 nm in a time scale of 1–100 ns in good agreement with experiment.

CMD simulation is employed ubiquitously to mimic the dynamic process of complicated system, dealing with the separation process of porous graphene as the media. Sint et al. designed functionalized nanopores (Figure 20.2a,b) in graphene monolayers to mimic the ionic channels in solution (Sint et al. 2008). CMD simulations have shown that these porous structures perform as ions sieves with high transport rates and excellent ionic selectivity. One can tune the selectivity by electing differently charged functional groups on the rim of the pore. Decorating the rims with negatively charged nitrogen and fluorine atoms tends to permit cation crossing, whereas the nanopores surrounding by positively charged hydrogen atoms prefer the crossing of anions. As shown in Figure 20.2c, the fluxes of hydrated ions grow with the applied electric field as the driven force. On the other hand, the resident time of anion in the positive charged pore decreases as the field strengthens. Additional separation and transport simulations will be discussed in Section 3.3.

Beyond parameter dependence, one critical drawback of CMD is that it cannot simulate the process of chemical reaction. The trajectory involving bond breaking and formation can be reproduced if the system energy is yielded from first-principle calculation which doesn't rely on connectivity. This approach is called *ab initio* molecular dynamics (AIMD), which is much more time-consuming than CMD (Tuckerman 2002). Moreover, one can treat the reactive part with quantum mechanics and other part of system with MM, known as QM/MM method (Lin and Truhlar 2007). Some specific force fields have been developed to model the

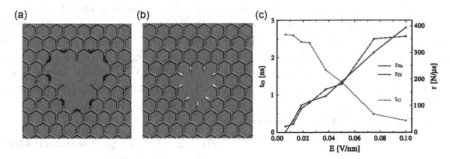

FIGURE 20.2 Two porous graphene models with (a) F-N-terminated nanopore and (b) H-terminated nanopore. (c) The field dependences of residence time of Cl⁻ inside the H-pore (t_{Cl}, left axis), and flow rates of the hydrated Na⁺ and Cl⁻ through the F-N- and H-pores, respectively (r_{Na} and r_{Cl}, right axis). (Reproduced with permission from Sint et al. (2008). Copyright 2008, American Chemical Society.)

reaction process, such as ReaxFF (van Duin et al. 2001). These computational techniques have been used to reproduce the experimental study on the formation of single-layered reduced graphene oxide and to construct the initial structure for following simulation on gas separation or water desalination (Lin and Grossman 2015).

20.3 Simulation on Typical Applications of Porous Graphene

With the established theoretical tools, researchers are able to give insight into properties of porous graphene system and design unknown material for specific purpose. In this section, we will review simulations on three types of applications of porous graphene, involving nano-electronics/nanosensors, energy storage unit and novel separation.

20.3.1 Nanoelectronics and Nanosensors

As described in Section 2.1, the zero-band gap of perfect graphene sheet is a formidable issue that hinders the practical usage of graphene-based nanoelectronics as well as nanosensors. As previously mentioned, creating impurities onto the 2D graphene material, such as nanopores at different shape and sizes, thickness, doping, and even conjugating the 2D material has shown to be an effective approach to open the band gap and expanding its practicality. By extenuating these defects, it provides a mean for theorists to lead experimentalists in designing novel nanoelectronics and nanosensors.

Band gap plays an essential role on the application of graphene-based nanodevices, including nanoelectronics and nanosensors. Based on computational methods, like a tight-binding (TB) description, Pederson et al. showed that graphene sheets with regularly spaced pores ("antidots") have a non-zero band gap (Pedersen et al. 2008). Vanevic extended the understanding of antidots by examining the electronic states of various antidot lattices, and discovering zero-energy flat bands and quasiflat bands at low energies (Vanević et al. 2009). Du et al. studied HPG by performing DFT and predicted that high porosity opens a direct band gap (Du et al. 2010). With HSE06 hybrid functional, calculated band gap was 3.2 eV, comparable to those of TiO_2 and C_3N_4. Continually, doped 2D graphene was studied by Ding et al. from the comparison of the band gaps of boron- and nitrogen-doped structures with the carbon-based membrane (Ding et al. 2011). The dopants cause tunable band gaps and direct-to-indirect band-gap transitions. Lu et al. examined various nitrogen-substitution sites, which could tailor the band gap either (Lu et al. 2013). Brunetto et al. designed the dehydrogenated porous graphene, named BPC, which is comprised of 4-membered, 6-membered, and 8-membered carbon rings in the sheet (Brunetto et al. 2012). This two-dimensional material was observed to have

a narrow band gap of 0.8 eV as well as well-delocalized frontier orbitals.

Another class of nanoporous graphene are porous graphene derivatives fabricated from conjugated microporous polymer (CMP) that is prepared via bottom-up method. Yang et al. employed demonstrated with DFT and nonequilibrium Green's function (NEGF) method to how electronic structures of benzo-CMP, aza-CMP, and BN-doped CMP (Yang et al. 2014) effect the electronic properties of the graphene material. The DFT results showed all three structures increased the band gap rendering them semiconductors, with band gaps of 0.92, 1.07, and 0.47 eV, respectively. It was later revealed that pore size influences the electronic properties as well (Li et al. 2015). Their work illustrated an increased in pore size resulted in a decreased band gap.

Nanopore can not only modify the energy band of graphene sheet but also localize small molecules via physical or chemical interaction which further affect the band structure of the graphene substrate. Therefore, porous graphene is also a candidate as nanosensor. Using DFT, Zhang et al. showed strong adsorption of CO_2, NO, and NO_2 on the defective graphene (Zhang et al. 2009). Small gas molecules tend to be attracted around the nanopore edges. The electronic properties of the porous graphene-gas adducts are strongly dependent on the adsorption configuration. Besides gases, DNA nucleobases interact with the nanopore intensely. As depicted in Figure 20.3a,b, Saha et al. theoretically demonstrated the four nucleobases inserted into the nanopore led to a unique change in the conductance of the porous nanoribbon (Saha et al. 2012). Inserted nucleobases in the nanopore change the charge density in the surrounding area. As a result, the edge conduction current could be modified on the order of milliampere at bias voltage 0.1 V. Simulation supported that the differential conductance of two-dimensional molecular electronics spectroscopy could recognize various nucleobases for DNA sequencing, including these related to cancerous cell growth (Rajan et al. 2014). Girdhar et al. presented a computational model for porous graphene ribbon as a gate electrode (Girdhar et al. 2014). Both the carrier concentration and the conductance have electrical response when RNA or DNA cross the nanopore. Combined NEGF, DFT, and CMD, Chang investigated the pore size effect on the current within the graphene plane (Chang et al. 2014). Based on *ab initio* calculations, McFarland et al. designed three nanoporous graphene-based devices, which could distinguish four nucleobases according to transverse currents using voltages of 0.5, 1.3, and 1.6 V (McFarland et al. 2015).

Beyond identifying nucleobases from surface current, the dynamic process of transport of single-strained DNA is also critical to DNA sequencing (Figure 20.3a). Wells et al. performed CMD and Brownian dynamics, which showed that the translocation of single DNA strand across nanopores took place in single nucleotide steps (Wells et al. 2012). The fluctuation of nucleotides could be strictly constrained in a well-defined nanopore. As shown in

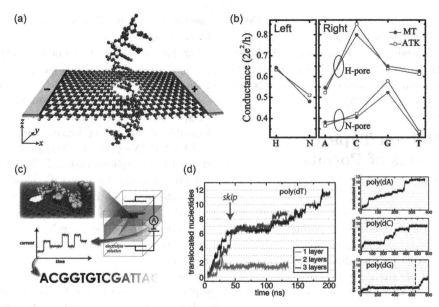

FIGURE 20.3 (a) Schematic illustration of the porous graphene-based device where DNA molecule crosses the pore to affect edge current. (b) Left: The conductance of the device with nanopore whose edge is passivated by either hydrogen (H−) or nitrogen (N−). Right: The calculated conductance when one of the four nucleobases (A, adenine; C, cytosine; G, guanine; and T, thymine) is located in the center of the nanopore. MT and ATK indicate that the computational codes are MT-NEGF-DFT and ATK, respectively. (Reproduced with permission from Saha et al. (2012). Copyright 2012, American Chemical Society.) (c) Schematic view of device for sequencing DNA. (d) Number of translocated nucleotides versus simulation time for systems containing one-, two-, and three-layered porous membranes and poly(dT)$_{20}$, poly(dA)$_{20}$, poly(dC)$_{20}$, and poly(dG)$_{20}$ under a 500 mV bias. The dashed line indicates a bias increase from 500 to 800 mV. (Reproduced with permission from Wells et al. (2012). Copyright 2012, American Chemical Society.)

Figure 20.3d, the translocation rate is related to the thickness of porous membrane. In three-layered porous graphene model, the translocation times of four nucleobases are quite different. Herein, one can sequence DNA via measuring the nucleobase- and conformation-dependent ionic current blockades. From CMD simulation, Avdoshenko et al. determined the relation between the DNA translocation time crossing nanopore and the intensity of electric field which is used as driven force (Avdoshenko et al. 2013). DNA transport could be precisely controlled if four additional graphene sheets are placed above the porous graphene substrate. Similarly, Shi et al. also suggested that the translocation time could be employed to determine DNA sequence (Shi et al. 2015). Zhang et al. illustrated the circular pore is the optimal geometry to differentiate various bases (Zhang et al. 2014b). Shankla and Aksimentiev further addressed the essential role of surface charge on the motion of DNA (Shankla and Aksimentiev 2014). Their group systematically showed how the blockade ionic current is affected by DNA sequence, group conformation, and pore geometry (Comer and Aksimentiev 2016). Differently, the stretch-induced stepwise translocation is less dependent of the pore (Qiu et al. 2015). The vertical conformational fluctuation is reduced by adhesion, and at the same time, the translocation speed is slowed down, leading to greatly improved signal-to-noise ratio. BN and other porous graphene derivatives with functionalized nanopores for DNA sequencing have also been theoretically studied (Gu et al. 2016, 2017, Yu et al. 2018). These works indicate porous graphene as a superior

sensor for next-generation rapid DNA sequencing biotechnology. Moreover, the practice has been extended to larger protein systems. Utilizing CMD simulation, Aksimentiev's group revealed the possibility of porous graphene as protein sequencing tool (Wilson et al. 2016).

20.3.2 Energy Storage Units

Benefits from the unique electrical and structural properties, porous graphene offers an ideal platform to build up advanced energy storage units, such as lithium battery, supercapacitor, and sorbent for hydrogen storage.

Lithium Battery

Lithium-ion battery is one of the most important systems among electrochemical energy storage technologies. Conductivity of carbonaceous material makes it possible be used as electrode in battery. Porous structure provides active sites for Li-ion loading and facilitates mass transfer. Correspondingly, theoretical simulations are carried out to elucidate the ion capacity as well as ion diffusion rate in electrode.

Xiao et al. reported porous graphene as a promising lithium-air battery electrode with high capacity of 15,000 mA h/g (Xiao et al. 2011). Their DFT calculation indicated that Li_2O_2 unit tends to nucleate near defective sites with functional groups (Figure 20.4b). Free-energy changes as a function of the size of Li_2O_2 cluster, in Figure 20.4c, show that compared with aggregation of relatively large Li_2O_2

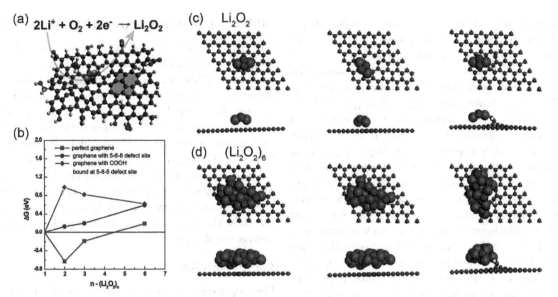

FIGURE 20.4 (a) Schematic structure of a functionalized graphene with typical defects. (b) Top and side views of optimized structures of Li_2O_2 on three porous graphene substrates. From left to right, the substrates of three columns are pristine graphene, defect graphene, and defect graphene with COOH group, respectively. (c) Simulated free-energy change as a function of $(Li_2O_2)_n$ cluster size. (d) The optimized geometries of $(Li_2O_2)_6$ cluster on various substrates. (Reproduced with permission from Xiao et al. (2011). Copyright 2011, American Chemical Society.)

clusters, the formation of isolated Li_2O_2 nano-clusters is energetically favorable on porous substrates, in agreement with experiment. Limited cluster size prevents O_2 blocking the channel for gas transport during the discharge process, like that occurs in Figure 20.4d, thus achieving improved reversibility. DFT simulation performed by Mukherjee et al. elucidated that high lithium capacity was achieved in a porous graphene network as the defect density was increased (Mukherjee et al. 2014).

Continually, Li-S battery is an attractive system with high theoretical capacity, in which conductive porous graphene can be used as anode material. Combing experimental and theoretical studies, Gu et al. showed that phosphorous dopant could enhance the interaction between lithium polysulfides and porous graphene, improving the capacity and reversibility (Gu et al. 2015). Based on DFT, Liang et al. illustrated that the adsorption of Li_nS clusters ($n = 1$–4) is coupled with charge transfer to graphene layer (Liang et al. 2016). The double vacancies allow Li ions crossing the graphene plane but imped the transport of large Li_nS clusters. Liu et al. prepared Co-embedded N-doped porous graphene as anode (Liu et al. 2018). Their DFT results suggested that Co substrate and N dopant promote the reservation of Li-S cluster cooperatively and accelerate the redox reaction on sulfur cathode. In addition, derivatives of porous graphene, such as g-C_3N_4 and SiC_2, have been proposed as the anode material for Li-S battery theoretically (He et al. 2018, Zhao et al. 2018).

Beyond explaining the experimental phenomena, DFT calculations are carried out to predict unknown electrode materials. Wang et al. designed a porous graphene model as Li-ion battery anode (Wang et al. 2016). There exists strong interaction between Li and porous graphene. Holes with

specific size result in dramatically improved mobility with moderate barriers of 0.37–0.39 eV and capacity of 2857.7 mA h/g. Interactions between Li and anode could be further enhanced by applying external strain. Liu et al. explored the possibility of using topological semimetal porous carbon, bco-C16, as lithium battery anode (Liu et al. 2017). In comparison of graphite used in practice, its specific capacity is increased by 50% (Li-C4 vs. Li-C6). The porous structure forms one-dimensional channels for Li-ion migration, which are robust against strains during operation. Interestingly, the diffusion energy barrier decreases as Li-ion concentration increases. At the same time, the volume change is comparable to that of graphite during charging/discharging cycle. It can be concluded that this kind of topological carbon materials would inspire experimental finding on advanced battery materials.

Supercapacitor

Supercapacitor is another promising electrochemical energy storage technology with numerous advantages, including fast charging rates, high power density, and almost unlimited charging/discharging cycle (Zhang et al. 2014a). The total capacitance of an electrode (C) can be simulated by the following equation:

$$\frac{1}{C} = \frac{1}{C_{EDL}} + \frac{1}{C_Q}$$

in which C_{EDL} and C_Q denote the electric double-layer capacitance and quantum capacitance, respectively. C_{EDL} originates from the dynamic rearrangement of electrolyte close to the electrode surface as a response to charging/discharging process. It is determined by the morphology of the surface and component of electrolyte, and can be modeled by CMD or other classical method. C_Q is from the intrinsic quantum effects of electrode that is able to be calculated using first-principle approach.

Paek et al. combined CMD and DFT calculations to reproduce capacitance between pristine graphene and [BMIM][PF$_6$] ionic liquid, involving both C_{EDL} and C_Q (Paek et al. 2012). They highlighted the importance of C_Q in the overall capacitance, and pointed out the possibility to achieve high capacitance by improving C_Q of graphene-based electrode. Later, Pak et al. showed topological defects with special geometry could efficiently enhance C_Q compared to pristine graphene, because of the defect-induced quasi-localized states near the Fermi level (Pak et al. 2014). Various heteroatom dopants can lead to vacancy nearby. DFT calculations illustrated that the combination of doping and vacancies is a promising strategy to increase the capacitance of graphene-based materials. Especially, the nitrogen-doping effect was investigated in detail that one could effective optimize the capacitance by controlling N-doping (Zhan et al. 2016b). Chen et al. reported strong experimental evidence of defect effect, also supported by their collaborated theoretical calculations (Chen et al. 2016). Furthermore, Zhan et al. predicted that capacitance of porous two-dimensional boron nanosheets could be three times higher than that of graphene (Zhan et al. 2016a). Thus, the boron-based nanosheet might be better material for the supercapacitor. For more comprehensive understanding on the theoretical studies of supercapacitor, we refer reader to a specific review (Zhan et al. 2017).

Hydrogen Storage Material

Apart from electrochemical energy storage, porous structure is promising for storing fuel gas, especially hydrogen. Coupled with high-accuracy *ab initio* calculation, Cabria et al. studied the hydrogen storage capacities in two parallel flat graphene layers, as a model of porous carbon (Cabria et al. 2008). However, the weak non-bonded interaction between hydrogen and carbon results in low gas uptake, even with the optimal structure. To yield higher hydrogen loading at ambient condition, interaction between porous structure and gas molecule has to be strengthened. Transition metal (TM)-dispersed materials are considered as the candidates for large-capacity hydrogen storage, due to the relatively strong metal-hydrogen binding energy. The binding strength between dispersed metal atom and dispersant material must be increased to improve the stability in TM dispersion and prevent TM aggregation. From DFT calculation, Kim et al. illustrated that the porous graphene with pyridinic nitrogen-dopant bonded dispersed TM intensely, thus could serve as hydrogen storage mediums (Kim et al. 2008). Furthermore, a series of alkaline, alkaline-earth, and transition metals were screened as the adatom on various porous graphene substrates. Ca-decorated materials showed the highest capacity for hydrogen storage (Reunchan and Jhi 2011, Lu et al. 2012, Fair et al. 2013, Seenithurai et al. 2014).

20.3.3 Ultimate Separation Membrane

The dense delocalized π-electron cloud above the aromatic ring leads to the pristine graphene sheet impermeable to any molecules. While on the other hand, this two-dimensional structure offers an ideal platform to control transport of specific molecule or ion by introducing well-defined nanopore. Thanks to its one-atom thickness, the separation efficiency can be extremely high. In this part, simulation works on pore-size-induced separation are summarized, which include gas separation and water desalination. DFT calculation is carried out to estimate the permeation energy barrier, as well as CMD simulation is performed to study the dynamic process of particle passing through porous membrane.

Gas Separation

Gas separation technology plays a critical role on both chemical industry and environment remediation. In 2009, Jiang et al. first showed the potential capability of porous graphene for gas separation by DFT calculations (Jiang et al. 2009). They designed two subnanometer-sized pores on graphene sheet, which are plotted in Figure 20.5. A naphthalene-like ring is removed to create nanopores, generating eight unsaturated carbon atoms. Then the porous structure is passivated by two methods: first, all the eight sp^2 hybridized carbon atoms are saturated with hydrogen atoms, leading to an approximately rectangular pore with dimension of 2.5 Å × 3.8 Å determined by electron density isosurface; second, the four carbon atoms on the zigzag edges are replaced by nitrogen atoms, and the other four carbons are saturated with hydrogen atoms, resulting in a little larger pore with dimension of 3.0 Å ×3.8 Å. Calculation based on vdW-DF was employed to map potential energy surfaces of hydrogen and methane crossing these two nanopores. Derived energy barriers of H$_2$ and CH$_4$ are 0.22 and 1.60 eV through the first pore, and 0.04 and 0.51 eV for the second pore, respectively. According to Arrhenius Equation, the ideal selectivities of H$_2$/CH$_4$ at room temperature (300 K) across the two pores are estimated on the order of 108 and 1,023, respectively, which are much higher than that of commercial materials. In 2012, Bunch et al. first realized size-controlled nanopores in double-layered graphene by the top-down approach via ultraviolet-induced oxidative (UV/ozone) etching (Koenig et al. 2012). Their prepared ultrathin membrane could separate H$_2$ and CO$_2$ from other larger molecules efficiently, supporting the theoretical prediction. Inspired by the experimental progress, Jiang's group further performed CMD to estimate the permeances of a series of gas molecules (Liu et al. 2013a,b). Coupled with umbrella sampling technique to evaluate free-energy barrier across the nanopore, they found the trend of flux as: H$_2$ > CO$_2$ ≫ N$_2$ > Ar > CH$_4$, generally following the sequence of kinetic diameters. Later, many porous graphene models have been theoretically explored. Liu et al. demonstrated that membrane strength was related to pore size, shape, and porosity that guide the design of porous graphene in

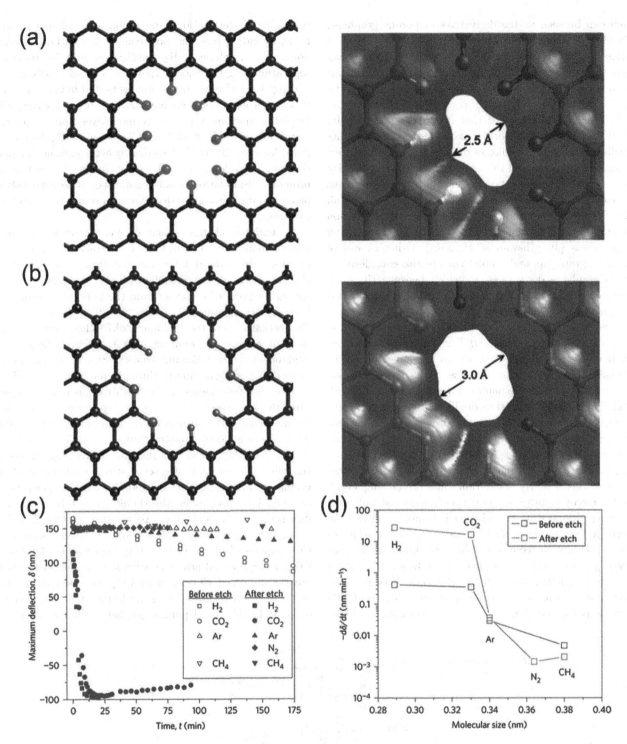

FIGURE 20.5 Two models of proposed porous graphene. (a and b) H-saturated pore and its electron density isosurface (isovalue = 0.003 a.u.). (c and d) N-functionalized pore and its electron density isosurface (isovalue = 0.003 a.u.). (Reproduced with permission from Jiang et al. (2009). Copyright 2009, American Chemical Society.)

practical (Liu and Chen 2014). Continually, Du et al. found that compared with H_2, relatively strong surface adsorption of N_2 leads to higher permeance when nanopores are slightly bigger (Du et al. 2011). Furthermore, Hauser et al. showed that suitable porous structures could also be used for separation of helium isotopic mixture (Hauser and Schwerdtfeger

2012a,b). Many other functionalized nanoporous graphene models with affinities to specific gas molecules have been reported for high-performance gas separation (Qin et al. 2013, Ambrosetti and Silvestrelli 2014, Lei et al. 2014).

Beyond the top-down method, two-dimensional polymer models are designed on the basis of bottom-up approach, which can be seen as the derivatives of porous graphene. Poly-phenylene (aforementioned HPG structure) is a fabricated porous two-dimensional material. Employing dispersion-corrected DFT, Blankenburg et al. studied the crossing barriers for several gas molecules, including H_2, He, Ne, O_2, CO_2, NH_3, CO, N_2 and Ar (Blankenburg et al. 2010). They suggested that it could be used for H_2 and He separation from other gases. Independent studies utilizing different methodologies by Li and Shrier's also drew the similar conclusion (Li et al. 2010b, Schrier 2010). By doping with nitrogen or boron, selectivity to gases might be modified (Schrier 2011, Hankel et al. 2012, Lu et al. 2014). Moreover, Jungthawan et al. showed that permeation rate and selectivity of gases could be controlled by applying tensile stress (Jungthawan et al. 2013). Other classes of graphene, graphyne, and graphdiyne are also considered as porous graphene derivatives, in which benzene rings are connected with one and two acetenyl chains, as depicted in Figure 20.6a. Graphdiyne film has been successfully prepared in square-centimeter scale by bottom-up method (Li et al. 2010a). Based on DFT calculation, Jiao et al. first demonstrated graphdiyne as an ideal membrane for hydrogen purification from fuel gas, which was confirmed by reactive molecular dynamics investigation (Jiao et al. 2011). Their computational model illustrated that nitrogen-doping could enhance efficiency of hydrogen separation (Jiao and Xu 2015). Additionally, Zhang et al. examined the hydrogen crossing barriers through graphyne, graphdiyne, and rhombic-graphyne, showing that the rhombic-graphyne possesses the best H_2 separation capacity (Zhang et al. 2012). Moreover, other organic two-dimensional polymers have been designed as the advanced membrane either. As shown in Figure 20.6b, Shrier et al. proposed a series of planar polymer by connecting benzene units with other planar groups, such as phenyl and ethenyl (Brockway and Schrier 2012, Schrier 2012). The pore sizes in these planar polymers are controllable by structure design. CMD simulations showed that this kind of membrane could be

promising materials for various purposes of gas separations, including pre- and post-combustion carbon capture, noble gas enrichment, H_2 purification and hydrocarbon separation (Figure 20.6c). Apart from purely carbon- and hydrogen-based membranes, nitrogen- and oxygen-enriched two-dimensional polymers have been designed theoretically, involving graphite C_3N_3, C_2N monolayer and porphyrin-based polymer (Ma et al. 2014, Chen et al. 2015, Tian et al. 2015, Xu et al. 2015). By introducing hetero-atoms, the pore size can be tailored as well as the dipolar interaction be promoted. Simulation results supported these materials to be candidates for high-performance H_2 separation and CO_2 capture.

To understand mechanism of gas separation through porous monolayer, Drahushuk et al. constructed a model in which gas crossed ultra-thin membrane via five steps: absorption to surface, association to a pore, translocation through a pore, dissociation from the pore, and desorption from surface (Drahushuk and Strano 2012). The schematic illustration is depicted in Figure 20.7. They derived equations of various rate-limiting steps to offer prediction and guidance for the following studies. Similarly, Sun et al. proposed that gas fluxes through membrane could be divided into two classes: in the first flux, gas molecules pass through pore directly from one side to the other side; and in the second flux, adsorption on membrane surface takes place before molecule crossing (Sun et al. 2014). From CMD simulations, they illustrated that direct flux plays an essential role in permeation of gases that don't adsorb much on graphene surface, such as He and H_2, whereas permeation of CO_2, CH_4 and other absorbable gases rely on surface flux rather than direct flux. To enhance surface flux, Lee et al. developed water-coated porous graphene model for CO_2 capture. Due to the relatively high water-solubility of CO_2, water-covered graphene with large nanopores showed greatly improved CO_2/O_2 and CO_2/N_2 selectivities (Lee and Aluru 2013). Jiang et al. performed CMD simulation on ionic liquid-coated porous graphene (Tian et al. 2017).

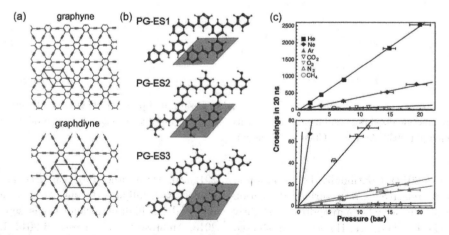

FIGURE 20.6 (a) Structures of graphyne and graphdiyne. (b) Three derivatives of poly-phenylene: PG-ES1, PG-ES2, and PG-ES3. (c) Permeation processes of various gases through PG-ES1 based on CMD simulations and the data with an emphasis on large species. (Reproduced with permission from Brockway and Schrier (2012). Copyright 2013, American Chemical Society.)

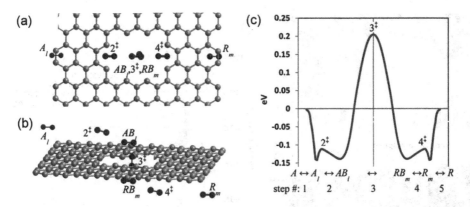

FIGURE 20.7 Five steps of gas permeation through a nanopore in which nitrogen is taken as an example. (a) Top view, (b) side view, and (c) calculated vdW energy curve. (Reproduced with permission from Drahushuk and Strano (2012). Copyright 2012, American Chemical Society.)

The ionic liquid layer could accumulate much more CO_2 on surface than N_2 and CH_4. Furthermore, ions located in the relatively large pore to modulate the pore size, hindering the larger CH_4 molecule passing through.

Water Purification

Similar to gas separation, theoretical investigations have been performed on transport process of water and hydrated ions through porous graphene, which is determined by size effect as well. The vdW diameter of water is about 2.8 Å, whereas other hydrated ions are much larger. As we mentioned as an example in Section 2.2, Sint et al. carried out CMD to show the possibility of porous graphene as membrane for desalination (Sint, Wang et al. 2008). Later, Suk et al. compared water transport through a porous graphene and that through thin carbon nanotube (CNT) (Suk and Aluru 2010). They illustrated that smaller pores accommodate a single water molecule and thus exhibit single-file movement of water. In this case CNT resulted in a higher water flux. For larger pores where the bulk-like water can go through (diameter >8 Å), porous graphene offers higher flux. Konatham et al. indicated that pore with diameter under 7.5 Å effectively rejects hydrated ions (Konatham et al. 2013). Carboxyl groups promote ion exclusion, especially for low ion concentration and small pore diameter. Considering external electrostatic field as driven force, Suk et al. drew the similar conclusion that the mobility of ion would sharply drop when the pore radius was smaller than 9 Å (Suk and Aluru 2014). For pores with diameters larger than 15 Å, Zhu et al. showed a linear dependence of water flux on the pore area (Zhu et al. 2014). Beyond pore size, Cohen-Tanugi et al. presented that pore chemistry is essential to desalination performance (Cohen-Tanugi and Grossman 2012). They examined hydrogenated and hydroxylated pores with similar pore size below 5.5 Å (Figure 20.8a). Although the hydrophilic hydroxylated pores increase the water flux, functional groups substitute for water in the hydration shell of the ions, leading to less salt rejection. On the contrast, hydrophobic H-termination on

the pore rim imposes additional conformational order of water flow, in which the constraint hydrogen bonding allows for stronger salt rejection. As compared in Figure 20.8c, by applying external pressure, the water permeability of porous graphene would be several orders of magnitude higher than current reverse osmosis membranes (Cohen-Tanugi and Grossman 2014a). Furthermore, they illustrated that porous graphene maintains the mechanical strength under high pressure if the substrate is appropriate (Cohen-Tanugi and Grossman 2014b). The substrate with gaps smaller than 1 μm would support porous graphene under pressures exceeding 57 MPa. Greater porosity might improve the stability under even higher pressures. Besides, ultra-high permeability for water could be kept even at low pressures. In 2015, Surwade et al. realized single-layer porous graphene as a desalination membrane in experiment (Surwade et al. 2015). Oxygen plasma was used to etch controllable pores in a graphene monolayer, preparing membranes with a nearly 100% salt rejection rate and high water flux (10^6 g/m^2 s) under regular osmotic pressure (70 g/m^2 s atm) and room temperature. Additionally, the fluorine-functionalized porous graphene was modeled for water purification (Azamat et al. 2015).

Inspired by the biological ion channel, He et al. theoretically designed functionalized porous graphene models (Figure 20.9a), which are used to identify hydrated K^+ and Na^+ cations (He et al. 2013). Using an electrostatic potential as driven force, CMD results in Figure 20.9b show that four carbonyl groups or negatively charged carboxylate groups on the relatively large pore mimic the K^+ channel in biological system, which favorably conducts K^+ over Na^+. While for a smaller pore with three carboxylate groups, the selectivity is tuned by changing the magnitude of the applied voltage. As shown in Figure 20.9d, lower voltage bias relates to a single-file manner of Na^+ permeation. Transporting Na^+ locates in the pore that blocks K^+ permeation. Whereas the nanopore under higher voltage is K^+-selective, since the affinity to Na^+ is not strong enough to prohibit K^+ passage, slows the Na^+ permeation instead. Kang et al. designed oxygen-doped porous graphene models

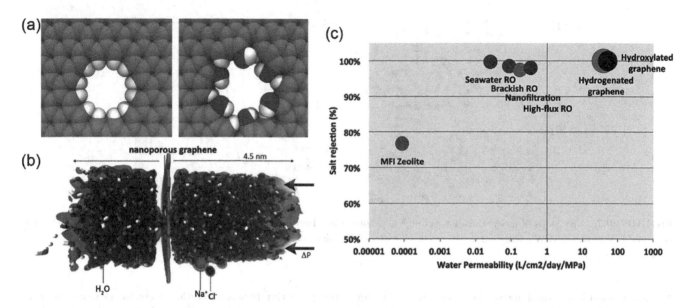

FIGURE 20.8 (a) Hydrogenated and hydroxylated graphene nanopores. (b) The side view of the simulated desalination system. (c) Comparison of desalination performance of porous graphene and other existing technologies. (Reproduced with permission from Cohen-Tanugi and Grossman (2012). Copyright 2012, American Chemical Society.)

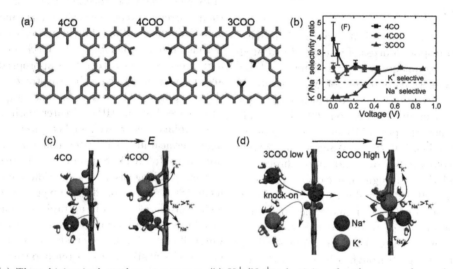

FIGURE 20.9 (a) Three bioinspired graphene nanopores. (b) K^+/Na^+ selectivity of each pore under various bias voltages. The mechanisms of ion selectivity through (c) relatively large pores (4CO and 4COO) and (d) small pore under different voltages. (Reproduced with permission from He et al. (2013). Copyright 2013, American Chemical Society.)

with proper pore sizes to obtain high K^+ selectivity over Na^+ (Kang et al. 2014). Zhao et al. demonstrated that introducing negative partial charges at the edge of pore could remarkably hinder the transport of anion but promote the crossing of cation (Zhao et al. 2013). Chen et al. showed the ion separation might in turn result in modification of the hydrated layer and asymmetrical charged graphene (Chen and Ruckenstein 2014). Sahu et al. established a dehydration-only mechanism for considering ion selectivity, which depends on the pore geometry and dehydration (Sahu et al. 2017).

Studies on separation on water purification with derivatives of porous graphene have been examined. According to CMD simulations, Lin and Kou et al. independently

proposed graphynes as water filters, and predicted graphtriyne with an optimal performance on removing salt, heavy metal cations and organic contaminant from water (Lin and Buehler 2013, Kou et al. 2014). From first-principle calculations on permeation barriers, Bartolomei et al. suggested that water permeation should be blocked through graphyne but feasible across graphtriyne (Bartolomei et al. 2014). For graphdiyne with medium pore size, water exhibits a lower barrier. Such barrier would disappear if additional water forms active hydrogen bonds on the opposite side of the membrane. Thus, graphdiyne is a promising membrane for water purification. To consider the correct dynamic process of water permeation through graphdiyne, common force fields have to be adjusted based on QM calculation. BN monolayer is another candidate for desalination. Garnier et al. showed that surface tension decreased on both graphene and BN layers, due to long-range wetting of water (Garnier et al. 2016). Compared with graphene monolayer, the low surface tension on BN monolayer leads to increased water permeation through BN.

Moreover, few-layered graphene oxide can serve for water filtration which distinguishes ions via interlayer spacing between sheets rather than pore size. Chen et al. synthesized cation-controlled graphene oxide membranes (Chen et al. 2017). The interlayer spacing is tailored by hydrated cations within angstrom precision. Interlayered spacing tuned by one cation can exclude other cations with larger hydrated volumes. Interestingly, the K^+-controlled graphene oxide rejected almost all cations in experiment, including K^+. DFT calculations revealed that oxide-containing groups and aromatic rings adsorb cation together. Commonly, the cation's hydration energy is stronger than the interaction between cation and membrane. But for K^+ only, these two interactions are comparable, implying that the hydrated K^+ in the interlayer has a distorted hydration structure and then narrows the interlayer spacing. Here, simulation offered insight into the mechanism of cationic control of interlayer spacing.

20.4 Summary and Perspective

In this chapter, we summarize the recent progress on simulations of porous graphene systems. First, two widely used theoretical methods, i.e., DFT and CMD, are briefly described. The parameter-free DFT calculation provides the picture of the band structure and evaluates the interaction between surface and adsorbed molecules. Simple "ab initio" functional significantly underestimates band gap and poorly describes long-range interaction; herein, several improved approaches have been developed, such as hybrid functional, GW, and DFT-1/2. Current DFT calculation can handle the systems containing hundreds of atoms in the scale of several nanometers. In comparison, CMD is more appropriate to model the dynamic process of much larger system, in which potential energy is expressed as an analytical function of coordinates. By solving classical Newton's equation, the trajectory as well as thermodynamic properties can be derived. The two major drawbacks of CMD are dependence of parameter and inability to consider chemical reaction. There have been several reactive MD methods to mimic the process of bond breaking and formation.

Using theoretical tools, the working principle of porous graphene-based devices can be elucidated, and novel materials are further predicted. By introducing nanopore in the graphene sheet, the band gap is opened, facilitating graphene using in semiconductor industry. Moreover, other molecules in the pore affect the surface current and electron density; thus, porous graphene is a potential platform for the construction of high-resolution sensor. A series of investigations have shown its ability for next-generation DNA sequencing. Porous structure also allows it to accommodate various particles, such as lithium cation, electrolyte, and fuel gases. As a result, simulations have been performed to demonstrate the possibility of porous graphene as energy storage units, including lithium battery, supercapacitor, and hydrogen sorbent. Furthermore, the well-defined nanopore is able to distinguish molecules or ions by size effect. Therefore, porous graphene and its derivatives are theoretically designed as the high-performance membrane for gas separation and desalination. Some of the predictions have been realized in experiments.

Although great progress has been made in both theoretical understanding and designing of materials based on porous graphene system, it is still challenging to prepare large area of single or few-layered graphene sheet with uniform nanopores. Novel synthesis and characterization techniques are highly demanded to combine with simulations. Besides, there is ample room for the improvement of computational approaches. How to quickly and accurately calculate the band structure is always worth exploring. Recently, graphics processing unit (GPU)-accelerated simulation offers an attractive opportunity on large-scale calculations. The GPU can easily handle multiple tasks in parallel; thus, its employment accelerates computing effectively, making it possible to model large system with high-level method. Several popular codes have supported GPU-accelerated computing, including but not limited to Amber, Lammps, and Gaussian.

Acknowledgment

This work was supported by the National Natural Science Foundation of China (No. 51672287). Z. T is supported by the startup fund from Ningbo Institute of Materials Technology & Engineering, Chinese Academy of Sciences.

References

Allen, M. J., Tung, V. C. and Kaner, R. B. 2010. Honeycomb carbon: A review of graphene. *Chemical Reviews* 110: 132–45.

Allen, M. P. and Tildesley, D. J. 1989. *Computer Simulation of Liquids*, Oxford University Press, Oxford.

Ambrosetti, A. and Silvestrelli, P. L. 2014. Gas separation in nanoporous graphene from first principle calculations. *The Journal of Physical Chemistry C* 118: 19172–9.

Aryasetiawan, F. and Gunnarsson, O. 1998. The GW method. *Reports on Progress in Physics* 61: 237–312.

Avdoshenko, S. M., Nozaki, D., Gomes da Rocha, C., et al. 2013. Dynamic and electronic transport properties of DNA translocation through graphene nanopores. *Nano Letters* 13: 1969–76.

Azamat, J., Khataee, A. and Joo, S. W. 2015. Molecular dynamics simulation of trihalomethanes separation from water by functionalized nanoporous graphene under induced pressure. *Chemical Engineering Science* 127: 285–92.

Balog, R., Jorgensen, B., Nilsson, L., et al. 2010. Bandgap opening in graphene induced by patterned hydrogen adsorption. *Nature Materials* 9: 315–9.

Bartolomei, M., Carmona-Novillo, E., Hernandez, M. I., et al. 2014. Penetration barrier of water through Graphynes' pores: First-principles predictions and force field optimization. *The Journal of Physical Chemistry Letters* 5: 751–5.

Bieri, M., Treier, M., Cai, J. M., et al. 2009. Porous graphenes: Two-dimensional polymer synthesis with atomic precision. *Chemical Communications* (45): 6919–21.

Blankenburg, S., Bieri, M., Fasel, R., et al. 2010. Porous graphene as an atmospheric nanofilter. *Small* 6: 2266–71.

Brockway, A. M. and Schrier, J. 2012. Noble gas separation using PG-ESX (X = 1, 2, 3) nanoporous two-dimensional polymers. *The Journal of Physical Chemistry C* 117: 393–402.

Brunetto, G., Autreto, P. A. S., Machado, L. D., et al. 2012. Nonzero gap two-dimensional carbon allotrope from porous graphene. *The Journal of Physical Chemistry C* 116: 12810–3.

Cabria, I., Lpez, M. J. and Alonso, J. A. 2008. Hydrogen storage capacities of nanoporous carbon calculated by density functional and Mller-Plesset methods. *Physical Review B* 78: 075415.

Cervantes-Sodi, F., Csanyi, G., Piscanec, S. and Ferrari, A. C. 2008. Edge-functionalized and substitutionally doped graphene nanoribbons: Electronic and spin properties. *Physical Review B* 77: 165427.

Chang, P.-H., Liu, H. and Nikoli, B. K. 2014. First-principles versus semi-empirical modeling of global and local electronic transport properties of graphene nanopore-based sensors for DNA sequencing. *Journal of Computational Electronics* 13: 847–56.

Chen, H. and Ruckenstein, E. 2014. Nanomembrane containing a nanopore in an electrolyte solution: A molecular dynamics approach. *The Journal of Physical Chemistry Letters* 5: 2979–82.

Chen, J., Han, Y., Kong, X., et al. 2016. The origin of improved electrical double-layer capacitance by inclusion of topological defects and dopants in graphene for supercapacitors. *Angewandte Chemie International Edition* 55: 13822–7.

Chen, L., Shi, G., Shen, J., et al. 2017. Ion sieving in graphene oxide membranes via cationic control of interlayer spacing. *Nature* 550: 380–3.

Chen, Z., Li, P. and Wu, C. 2015. A uniformly porous 2D CN (1 : 1) network predicted by first-principles calculations. *RSC Advances* 5: 11791–6.

Cohen-Tanugi, D. and Grossman, J. C. 2012. Water desalination across nanoporous graphene. *Nano Letters* 12: 3602–8.

Cohen-Tanugi, D. and Grossman, J. C. 2014a. Mechanical strength of nanoporous graphene as a desalination membrane. *Nano Letters* 14: 6171–8.

Cohen-Tanugi, D. and Grossman, J. C. 2014b. Water permeability of nanoporous graphene at realistic pressures for reverse osmosis desalination. *Journal of Chemical Physics* 141: 074704.

Comer, J. and Aksimentiev, A. 2016. DNA sequence-dependent ionic currents in ultra-small solid-state nanopores. *Nanoscale* 8: 9600–13.

De La Pierre, M., Karamanis, P., Baima, J., et al. 2013. *Ab initio* periodic simulation of the spectroscopic and optical properties of novel porous graphene phases. *The Journal of Physical Chemistry C* 117: 2222–9.

Ding, Y., Wang, Y., Shi, S. and Tang, W. 2011. Electronic structures of porous graphene, BN, and BC2N sheets with one- and two-hydrogen passivations from first principles. *The Journal of Physical Chemistry C* 115: 5334–43.

Dion, M., Rydberg, H., Schroder, E., Langreth, D. C. and Lundqvist, B. I. 2004. Van der Waals density functional for general geometries. *Physical Review Letters* 92: 246401.

Drahushuk, L. W. and Strano, M. S. 2012. Mechanisms of gas permeation through single layer graphene membranes. *Langmuir* 28: 16671–8.

Du, A. J., Zhu, Z. H. and Smith, S. C. 2010. Multifunctional porous graphene for nanoelectronics and hydrogen storage: New properties revealed by first principle calculations. *Journal of the American Chemical Society* 132: 2876–7.

Du, H., Li, J., Zhang, J., et al. 2011. Separation of hydrogen and nitrogen gases with porous graphene membrane. *The Journal of Physical Chemistry C* 115: 23261–6.

Fair, K. M., Cui, X. Y., Li, L., et al. 2013. Hydrogen adsorption capacity of adatoms on double carbon vacancies of graphene: A trend study from first principles. *Physical Review B* 87: 014102.

Ferreira, L. G., Marques, M. and Teles, L. K. 2011. Slater half-occupation technique revisited: The LDA-1/2 and GGA-1/2 approaches for atomic ionization energies and band gaps in semiconductors. *Aip Advances* 1: 032119.

Fischbein, M. D. and Drndic, M. 2008. Electron beam nanosculpting of suspended graphene sheets. *Applied Physics Letters* 93: 113107.

Garnier, L., Szymczyk, A., Malfreyt, P. and Ghoufi, A. 2016. Physics behind water transport through nanoporous boron nitride and graphene. *The Journal of Physical Chemistry Letters* 7: 3371–6.

Giannozzi, P., Baroni, S., Bonini, N., et al. 2009. QUANTUM ESPRESSO: A modular and open-source software project for quantum simulations of materials. *Journal of Physics-Condensed Matter* 21: 395502.

Girdhar, A., Sathe, C., Schulten, K. and Leburton, J. P. 2014. Gate-modulated graphene quantum point contact device for DNA sensing. *Journal of Computational Electronics* 13: 839–46.

Grimme, S., Antony, J., Ehrlich, S. and Krieg, H. 2010. A consistent and accurate *ab initio* parametrization of density functional dispersion correction (DFT-D) for the 94 elements H-Pu. *Journal of Chemical Physics* 132: 154104.

Gu, X., Tong, C.-J., Lai, C., et al. 2015. A porous nitrogen and phosphorous dual doped graphene blocking layer for high performance Li–S batteries. *Journal of Materials Chemistry A* 3: 16670–8.

Gu, Z., Zhang, Y., Luan, B. and Zhou, R. 2016. DNA translocation through single-layer boron nitride nanopores. *Soft Matter* 12: 817–23.

Gu, Z., Zhao, L., Liu, S., et al. 2017. Orientational binding of DNA guided by the C2N template. *ACS Nano* 11: 3198–206.

Hankel, M., Jiao, Y., Du, A., Gray, S. K. and Smith, S. C. 2012. Asymmetrically decorated, doped porous graphene as an effective membrane for hydrogen isotope separation. *The Journal of Physical Chemistry C* 116: 6672–6.

Hauser, A. W. and Schwerdtfeger, P. 2012a. Methane-selective nanoporous graphene membranes for gas purification. *Physical Chemistry Chemical Physics* 14: 13292–8.

Hauser, A. W. and Schwerdtfeger, P. 2012b. Nanoporous graphene membranes for efficient 3He/4He separation. *The Journal of Physical Chemistry Letters* 3: 209–13.

He, F., Li, K., Yin, C., et al. 2018. A combined theoretical and experimental study on the oxygenated graphitic carbon nitride as a promising sulfur host for lithium–sulfur batteries. *Journal of Power Sources* 373: 31–9.

He, Z., Zhou, J., Lu, X. and Corry, B. 2013. Bioinspired graphene nanopores with voltage-tunable ion selectivity for Na+ and K+. *ACS Nano* 7: 10148–57.

Heyd, J., Scuseria, G. E. and Ernzerhof, M. 2003. Hybrid functionals based on a screened Coulomb potential. *Journal of Chemical Physics* 118: 8207–15.

Hohenberg, P. and Kohn, W. 1964. Inhomogeneous electron gas. *Physical Review B* 136: B864.

Hutter, J., Iannuzzi, M., Schiffmann, F. and Vande-Vondele, J. 2014. CP2K: Atomistic simulations of condensed matter systems. *Wiley Interdisciplinary Reviews-Computational Molecular Science* 4: 15–25.

Ito, Y., Tanabe, Y., Sugawara, K., et al. 2018. Three-dimensional porous graphene networks expand graphene-based electronic device applications. *Physical Chemistry Chemical Physics* 20: 6024–33.

Ji, L. W., Meduri, P., Agubra, V., Xiao, X. C. and Alcoutlabi, M. 2016. Graphene-based nanocomposites for energy storage. *Advanced Energy Materials* 6: 1502159.

Jiang, D. E. and Chen, Z. 2013. *Graphene Chemistry: Theoretical Perspectives*, John Wiley & Sons, Ltd, West Sussex.

Jiang, D. E., Cooper, V. R. and Dai, S. 2009. Porous graphene as the ultimate membrane for gas separation. *Nano Letters* 9: 4019–24.

Jiao, S. and Xu, Z. 2015. Selective gas diffusion in graphene oxides membranes: A molecular dynamics simulations study. *ACS Applied Materials and Interfaces* 7: 9052–9.

Jiao, Y., Du, A., Hankel, M., et al. 2011. Graphdiyne: A versatile nanomaterial for electronics and hydrogen purification. *Chemical Communications* 47: 11843–5.

Jungthawan, S., Reunchan, P. and Limpijumnong, S. 2013. Theoretical study of strained porous graphene structures and their gas separation properties. *Carbon* 54: 359–64.

Kang, Y., Zhang, Z., Shi, H., et al. 2014. Na(+) and K(+) ion selectivity by size-controlled biomimetic graphene nanopores. *Nanoscale* 6: 10666–72.

Kim, G., Jhi, S.-H. and Park, N. 2008. Effective metal dispersion in pyridinelike nitrogen doped graphenes for hydrogen storage. *Applied Physics Letters* 92: 013106.

Koenig, S. P., Wang, L. D., Pellegrino, J. and Bunch, J. S. 2012. Selective molecular sieving through porous graphene. *Nature Nanotechnology* 7: 728–32.

Kohn, W. and Sham, L. J. 1965. Self-consistent equations including exchange and correlation effects. *Physical Review* 140: 1133–8.

Konatham, D., Yu, J., Ho, T. A. and Striolo, A. 2013. Simulation insights for graphene-based water desalination membranes. *Langmuir* 29: 11884–97.

Kou, J., Zhou, X., Lu, H., Wu, F. and Fan, J. 2014. Graphyne as the membrane for water desalination. *Nanoscale* 6: 1865–70.

Kresse, G. and Furthmuller, J. 1996. Efficient iterative schemes for *ab initio* total-energy calculations using a plane-wave basis set. *Physical Review B* 54: 11169–86.

Lee, J. and Aluru, N. R. 2013. Water-solubility-driven separation of gases using graphene membrane. *Journal of Membrane Science* 428: 546–53.

Lei, G., Liu, C., Xie, H. and Song, F. 2014. Separation of the hydrogen sulfide and methane mixture by the porous graphene membrane: Effect of the charges. *Chemical Physics Letters* 599: 127–32.

Li, G. X., Li, Y. L., Liu, H. B., et al. 2010a. Architecture of graphdiyne nanoscale films. *Chemical Communications* 46: 3256–8.

Li, S., Yang, Z.-D., Zhang, G. and Zeng, X. C. 2015. Electronic and transport properties of porous graphene sheets and nanoribbons: Benzo-CMPs and BN codoped derivatives. *Journal of Materials Chemistry C* 3: 9637–49.

Li, Y., Zhou, Z., Shen, P. and Chen, Z. 2010b. Two-dimensional polyphenylene: Experimentally available porous graphene as a hydrogen purification membrane. *Chemical Communications* 46: 3672–4.

Liang, L., Shen, J. W., Zhang, Z. and Wang, Q. 2017. DNA sequencing by two-dimensional materials: As theoretical modeling meets experiments. *Biosensors and Bioelectronics* 89: 280–92.

Liang, Z., Fan, X., Singh, D. J. and Zheng, W. T. 2016. Adsorption and diffusion of Li with S on pristine and defected graphene. *Physical Chemistry Chemical Physics* 18: 31268–76.

Lin, H. and Truhlar, D. G. 2007. QM/MM: What have we learned, where are we, and where do we go from here? *Theoretical Chemistry Accounts* 117: 185–99.

Lin, L. C. and Grossman, J. C. 2015. Atomistic understandings of reduced graphene oxide as an ultrathin-film nanoporous membrane for separations. *Nature Communications* 6: 8335.

Lin, S. and Buehler, M. J. 2013. Mechanics and molecular filtration performance of graphyne nanoweb membranes for selective water purification. *Nanoscale* 5: 11801–7.

Liu, H., Dai, S. and Jiang, D. E. 2013a. Permeance of H2 through porous graphene from molecular dynamics. *Solid State Communications* 175–176: 101–5.

Liu, H., Dai, S. and Jiang, D. E. 2013b. Insights into CO_2/N_2 separation through nanoporous graphene from molecular dynamics. *Nanoscale* 5: 9984–7.

Liu, J., Wang, S. and Sun, Q. 2017. All-carbon-based porous topological semimetal for Li-ion battery anode material. *Proceedings of the National Academy of Sciences of the United States of America* 114: 651–6.

Liu, S., Li, J., Yan, X., et al. 2018. Superhierarchical cobalt-embedded nitrogen-doped porous carbon nanosheets as two-in-one hosts for high-performance lithium-sulfur batteries. *Advanced Materials* 30: e1706895.

Liu, Y. L. and Chen, X. 2014. Mechanical properties of nanoporous graphene membrane. *Journal of Applied Physics* 115: 034303.

Lu, R., Meng, Z., Kan, E., et al. 2013. Tunable band gap and hydrogen adsorption property of a two-dimensional porous polymer by nitrogen substitution. *Physical Chemistry Chemical Physics* 15: 666–70.

Lu, R., Meng, Z., Rao, D., et al. 2014. A promising monolayer membrane for oxygen separation from harmful gases: Nitrogen-substituted polyphenylene. *Nanoscale* 6: 9960–4.

Lu, R., Rao, D., Lu, Z., et al. 2012. Prominently Improved hydrogen purification and dispersive metal binding for hydrogen storage by substitutional doping in porous graphene. *The Journal of Physical Chemistry C* 116: 21291–6.

Ma, Z., Zhao, X., Tang, Q. and Zhou, Z. 2014. Computational prediction of experimentally possible g-C_3N_3 monolayer as hydrogen purification membrane. *International Journal of Hydrogen Energy* 39: 5037–42.

McFarland, H. L., Ahmed, T., Zhu, J. X., Balatsky, A. V. and Haraldsen, J. T. 2015. First-principles investigation of nanopore sequencing using variable voltage bias on graphene-based nanoribbons. *The Journal of Physical Chemistry Letters* 6: 2616–21.

Mukherjee, R., Thomas, A. V., Datta, D., et al. 2014. Defect-induced plating of lithium metal within porous graphene networks. *Nature Communications* 5: 3710.

Novoselov, K. S., Geim, A. K., Morozov, S. V., et al. 2004. Electric field effect in atomically thin carbon films. *Science* 306: 666–9.

Paek, E., Pak, A. J. and Hwang, G. S. 2012. A computational study of the interfacial structure and capacitance of graphene in [BMIM][PF6] ionic liquid. *Journal of the Electrochemical Society* 160: A1–10.

Pak, A. J., Paek, E. and Hwang, G. S. 2014. Tailoring the performance of graphene-based supercapacitors using topological defects: A theoretical assessment. *Carbon* 68: 734–41.

Parr, R. G. and Yang, W. 1989. *Density-Functional Theory of Atoms and Molecules*, Oxford University Press, New York.

Pedersen, T. G., Flindt, C., Pedersen, J., et al. 2008. Graphene antidot lattices: Designed defects and spin qubits. *Physical Review Letters* 100: 136804.

Qin, X., Meng, Q., Feng, Y. and Gao, Y. 2013. Graphene with line defect as a membrane for gas separation: Design via a first-principles modeling. *Surface Science* 607: 153–8.

Qiu, H., Sarathy, A., Leburton, J. P. and Schulten, K. 2015. Intrinsic stepwise translocation of stretched ssDNA in graphene nanopores. *Nano Letters* 15: 8322–30.

Rajan, A. C., Rezapour, M. R., Yun, J., et al. 2014. Two dimensional molecular electronics spectroscopy for molecular fingerprinting, DNA sequencing, and cancerous DNA recognition. *ACS Nano* 8: 1827–33.

Reunchan, P. and Jhi, S.-H. 2011. Metal-dispersed porous graphene for hydrogen storage. *Applied Physics Letters* 98: 093103.

Russo, C. J. and Golovchenko, J. A. 2012. Atom-by-atom nucleation and growth of graphene nanopores. *Proceedings of the National Academy of Sciences of the United States of America* 109: 5953–7.

Saha, K. K., Drndic, M. and Nikolic, B. K. 2012. DNA base-specific modulation of microampere transverse edge currents through a metallic graphene nanoribbon with a nanopore. *Nano Letters* 12: 50–5.

Sahu, S., Di Ventra, M. and Zwolak, M. 2017. Dehydration as a universal mechanism for ion selectivity in graphene and other atomically thin pores. *Nano Letters* 17: 4719–24.

Schrier, J. 2010. Helium separation using porous graphene membranes. *The Journal of Physical Chemistry Letters* 1: 2284–7.

Schrier, J. 2011. Fluorinated and nanoporous graphene materials as sorbents for gas separations. *ACS Applied Materials and Interfaces* 3: 4451–8.

Schrier, J. 2012. Carbon dioxide separation with a two-dimensional polymer membrane. *ACS Applied Materials and Interfaces* 4: 3745–52.

Seenithurai, S., Pandyan, R. K., Kumar, S. V., Saranya, C. and Mahendran, M. 2014. Li-decorated double vacancy graphene for hydrogen storage application: A first principles study. *International Journal of Hydrogen Energy* 39: 11016–26.

Shankla, M. and Aksimentiev, A. 2014. Conformational transitions and stop-and-go nanopore transport of single-stranded DNA on charged graphene. *Nature Communications* 5: 5171.

Shi, C., Kong, Z., Sun, T., et al. 2015. Molecular dynamics simulations indicate that DNA bases using graphene nanopores can be identified by their translocation times. *RSC Advances* 5: 9389–95.

Sint, K., Wang, B. and Krl, P. 2008. Selective ion passage through functionalized graphene nanopores. *Journal of the American Chemical Society* 130: 16448–9.

Solis-Fernandez, P., Bissett, M. and Ago, H. 2017. Synthesis, structure and applications of graphene-based 2D heterostructures. *Chemical Society Reviews* 46: 4572–613.

Suk, M. E. and Aluru, N. R. 2010. Water transport through ultrathin graphene. *The Journal of Physical Chemistry Letters* 1: 1590–4.

Suk, M. E. and Aluru, N. R. 2014. Ion transport in sub-5-nm graphene nanopores. *Journal of Chemical Physics* 140: 084707.

Sun, C., Boutilier, M. S., Au, H., et al. 2014. Mechanisms of molecular permeation through nanoporous graphene membranes. *Langmuir* 30: 675–82.

Sun, Q., Zhang, R. Y., Qiu, J., Liu, R. and Xu, W. 2018. On-surface synthesis of carbon nanostructures. *Advanced Materials* 30: 1705630.

Surwade, S. P., Smirnov, S. N., Vlassiouk, I. V., et al. 2015. Water desalination using nanoporous single-layer graphene. *Nature Nanotechnology* 10: 459–64.

Tian, Z., Dai, S. and Jiang, D. E. 2015. Expanded porphyrins as two-dimensional porous membranes for CO_2 separation. *ACS Applied Materials and Interfaces* 7: 13073–9.

Tian, Z. Q., Mahurin, S. M., Dai, S. and Jiang, D. E. 2017. Ion-gated gas separation through porous graphene. *Nano Letters* 17: 1802–7.

Tkatchenko, A. and Scheffler, M. 2009. Accurate molecular van der Waals interactions from ground-state electron density and free-atom reference data. *Physical Review Letters* 102: 073005.

Tsuneda, T. and Hirao, K. 2014. Long-range correction for density functional theory. *Wiley Interdisciplinary Reviews-Computational Molecular Science* 4: 375–90.

Tuckerman, M. E. 2002. *Ab initio* molecular dynamics: Basic concepts, current trends and novel applications. *Journal of Physics-Condensed Matter* 14: R1297–355.

van Duin, A. C. T., Dasgupta, S., Lorant, F. and Goddard, W. A. 2001. ReaxFF: A reactive force field for hydrocarbons. *Journal of Physical Chemistry A* 105: 9396–409.

Vanevi, M., Stojanovi, V. M. and Kindermann, M. 2009. Character of electronic states in graphene antidot lattices: Flat bands and spatial localization. *Physical Review B* 80: 045410.

Vydrov, O. A. and Van Voorhis, T. 2010. Nonlocal van der Waals density functional: The simpler the better. *Journal of Chemical Physics* 133: 244103.

Wang, L., Boutilier, M. S. H., Kidambi, P. R., et al. 2017. Fundamental transport mechanisms, fabrication and potential applications of nanoporous atomically thin membranes. *Nature Nanotechnology* 12: 509–22.

Wang, Y., Zhang, Q., Jia, M., et al. 2016. Porous graphene for high capacity lithium ion battery anode material. *Applied Surface Science* 363: 318–22.

Wells, D. B., Belkin, M., Comer, J. and Aksimentiev, A. 2012. Assessing graphene nanopores for sequencing DNA. *Nano Letters* 12: 4117–23.

Wilson, J., Sloman, L., He, Z. and Aksimentiev, A. 2016. Graphene nanopores for protein sequencing. *Advanced Functional Materials* 26: 4830–8.

Xiao, J., Mei, D., Li, X., et al. 2011. Hierarchically porous graphene as a lithium-air battery electrode. *Nano Letters* 11: 5071–8.

Xu, B., Xiang, H., Wei, Q., et al. 2015. Two-dimensional graphene-like C2N: An experimentally available porous membrane for hydrogen purification. *Physical Chemistry Chemical Physics* 17: 15115–8.

Yang, L., Park, C. H., Son, Y. W., Cohen, M. L. and Louie, S. G. 2007. Quasiparticle energies and band gaps in graphene nanoribbons. *Physical Review Letters* 99: 186801.

Yang, T., Lin, H., Zheng, X., Loh, K. P. and Jia, B. 2017. Tailoring pores in graphene-based materials: From generation to applications. *Journal of Materials Chemistry A* 5: 16537–58.

Yang, Z.-D., Wu, W. and Zeng, X. C. 2014. Electronic and transport properties of porous graphenes: Two-dimensional benzo- and aza-fused π-conjugated-microporous-polymer sheets and boron–nitrogen co-doped derivatives. *Journal of Materials Chemistry C* 2: 2902–7.

Yao, X. and Zhao, Y. 2017. Three-dimensional porous graphene networks and hybrids for lithium-ion batteries and supercapacitors. *Chem* 2: 171–200.

Yu, X. W., Cheng, H. H., Zhang, M., et al. 2017. Graphene-based smart materials. *Nature Reviews Materials* 2: 17046.

Yu, Y. S., Lu, X., Ding, H. M. and Ma, Y. Q. 2018. Computational investigation on DNA sequencing using functionalized graphene nanopores. *Physical Chemistry Chemical Physics* 20: 9063–9.

Zhan, C., Lian, C., Zhang, Y., et al. 2017. Computational insights into materials and interfaces for capacitive energy storage. *Advanced Science* 4: 1700059.

Zhan, C., Zhang, P. F., Dai, S. and Jiang, D. E. 2016a. Boron supercapacitors. *ACS Energy Letters* 1: 1241–6.

Zhan, C., Zhang, Y., Cummings, P. T. and Jiang, D. E. 2016b. Enhancing graphene capacitance by nitrogen: Effects of doping configuration and concentration. *Physical Chemistry Chemical Physics* 18: 4668–74.

Zhang, H., He, X., Zhao, M., et al. 2012. Tunable hydrogen separation in sp–sp^2 hybridized carbon membranes: A first-principles prediction. *The Journal of Physical Chemistry C* 116: 16634–8.

Zhang, X., Zhang, H., Li, C., et al. 2014a. Recent advances in porous graphene materials for supercapacitor applications. *RSC Advances* 4: 45862–84.

Zhang, Y. H., Chen, Y. B., Zhou, K. G., et al. 2009. Improving gas sensing properties of graphene by introducing dopants and defects: A first-principles study. *Nanotechnology* 20: 185504.

Zhang, Z., Shen, J., Wang, H., et al. 2014b. Effects of graphene nanopore geometry on DNA sequencing. *The Journal of Physical Chemistry Letters* 5: 1602–7.

Zhao, S., Xue, J. and Kang, W. 2013. Ion selection of charge-modified large nanopores in a graphene sheet. *Journal of Chemical Physics* 139: 114702.

Zhao, Y., Zhao, J. and Cai, Q. 2018. SiC 2 siligraphene as a promising anchoring material for lithium-sulfur batteries: A computational study. *Applied Surface Science* 440: 889–96.

Zhu, C., Li, H. and Meng, S. 2014. Transport behavior of water molecules through two-dimensional nanopores. *Journal of Chemical Physics* 141: 18C528.

Metallic Nanoglasses Investigated by Molecular Dynamics Simulations

D. Şopu
Technische Universität Darmstadt & Austrian Academy of Sciences

O. Adjaoud
Technische Universität Darmstadt

21.1 Metallic Glasses

The history of metallic glasses starts in 1960, at California Institute of Technology, when Duwez et al. [1] discovered the first binary metallic amorphous alloy $Au_{80}Si_{20}$. They developed rapid quenching techniques for cooling metallic liquids at very high rates of about 10^6 K/s. At this cooling rate, atoms do not have enough time or energy to rearrange for crystal nucleation and the material is frozen in an amorphous state. The atoms retain an amorphous distribution, i.e., random packing with no long-range order. The very high cooling rate required to produce metallic glasses restricts the specimen geometry to thin ribbons, foils and powders with sizes of the order of microns [2] and hinder their application range [3]. In order to slow down the crystallization kinetics when quenching a melt and form metallic glasses with high glass-forming ability, multiple components alloys are required. Following, in the early 1980s, glassy ingots were casted with thickness of 1 cm and obtained the first bulk metallic glass (BMG) [4]. Perhaps the most popular commercial BMG is a pentary alloy based on Zr-Ti-Cu-Ni-Be with a critical casting thickness of up to 10 cm [5] and became known as Vitreloy 1, developed by Johnson and Peker in a project funded by NASA to develop new aerospace materials. Besides the technique of rapid quenching from the liquid state, other production routes such as vapor deposition (inert-gas condensation) [6], solid-state amorphization reactions, e.g., ball milling [7], or amorphization by high energy radiation [8] have been extensively developed and elaborated for the purpose of producing a wide variety of metallic glasses.

Metallic glasses possess interesting properties in comparison with the crystalline state. Due to the amorphous structure (absence of long-range order), the displacement of atoms, e.g., to accommodate a dislocation is obstructed. With no crystal defects, metallic glasses show special properties such as high strength (twice that of steel, but lighter), increased hardness, high toughness and elasticity [9,10]. Also, some metallic glasses have high corrosion resistance [11] and interesting magnetic properties [12]. Holding these unique characteristics, metallic glasses are interesting candidates for structural and functional applications. However, the current limits of these materials are tied to the poor plasticity at room temperature. The lack of plastic strain is a consequence of shear softening originating from shear-induced dilatation that causes plastic strain to be highly localized in shear bands [13]. The detrimental nature of these defects will promote premature fracture of BMGs. In order to avoid the formation of critical shear bands and consequently to stabilize the glass against catastrophic failure, soft heterogeneities, ranging from intrinsic structural fluctuations [14,15] to secondary phases [16,17], must be generated into the structure. In general, the role of heterogeneities is to introduce strain concentrations which act as initiation sites for shear bands and effectively distribute the applied strain. Consequently, the generation of a high density of shear bands which in turn interact in a very complex way with each other or other heterogeneities would hinder the early failure of BMGs.

21.2 Nanoglasses: A New Kind of Noncrystalline Material

Different strategies to improve the ductility of BMGs have been proposed in the last decades such as synthesizing composite materials reinforced with a crystalline phase [16–21], or artificial creation of heterogeneous microstructures with controlled morphology by using a variety of mechanical pre-treatments, such as cold rolling [22], high-pressure torsion [23] or imprinting [24], and chemical treatments to form porous metallic glasses [25–27]. However, all these techniques to fabricate ductile BMGs do not allow controlled variation of structural features. It is the idea of nanoglasses to generate new kinds of noncrystalline materials that permit the controlled modification of their defect microstructures and chemical microstructures by methods that are comparable to the methods used today for crystalline materials [28]. For crystalline materials, it is a well-established concept to tune materials properties by varying the grain size. Cold compaction of nanocrystallites as shown in Figure 21.1 (a) leads to microstructures where a substantial fraction of the atoms are located in grain boundaries [29], which strongly influences the mechanical properties of a material. In general, grain boundaries are defined as two-dimensional planar defects in crystalline materials separating domains of different crystallographic orientation but similar crystal structure [30]. The mismatch between adjacent grains is accommodated in the interface, and therefore, grain boundaries are typically less ordered regions. For the case of glasses, in contrast, there is no established concept of grain boundaries, due to the lack of translational symmetry and long-range order. If one considers, however, that a high degree of short- and medium-range order is found in glassy materials [31,32], as well, the definition of a grain boundary in a glass as an internal interface enclosing a domain of atoms becomes conceivable. Moreover, in BMGs, it is a well proven fact that planar defects do exist, namely in form of shear bands induced by plastic deformation (see Figure 21.1 (c)). These shear bands exhibit an enhanced free volume and a modified local order [13,33]. The question is whether planar defects in metallic glasses can be introduced by other means than deformation. Experimentally, indications for the existence of glass–glass interfacial areas with identical or different chemical compositions have been found by consolidating nanometer-sized metallic glass clusters [34–36], which were proposed to form a so-called nanoglass [37,38] (Figure 21.1 (b)). The presence of interfaces in nanoglasses can possibly provide an important benefit by improving the mechanical properties, if one assumes that they act similar to shear bands in pre-deformed metallic glasses [22].

So far, nanoglasses have been produced by inert-gas condensation [28,34,35,40] and magnetron sputtering [41–43]. The inert-gas condensation process involves the following two steps. During the first step, nanometer-sized glassy clusters are generated by evaporating (or sputtering)

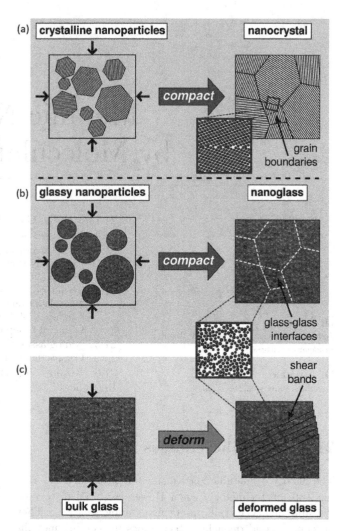

FIGURE 21.1 Different strategies to control the structure of metallic glasses: nanoglasses containing grain-boundary-like interfaces can be obtained by powder consolidation (b), similar to the synthesis of nanocrystalline materials (a). Another approach to modify the glass structure is to introduce shear bands by preplastic deformation (c). Both shear bands and interfaces are planar defects with modified structure. Figures taken with permission from Ref. [39]

the material in an inert-gas atmosphere. The resulting clusters are subsequently consolidated at pressures of up to 5 GPa into a pellet-shaped nanoglass. Scanning tunneling microscopy image of the polished surface of the $Fe_{90}Sc_{10}$ nanoglass is presented in Figure 21.2(a) [44]. It can be readily seen that glass–glass interfaces were formed between the glassy nanoparticles.

The magnetron sputtering is another way to prepare nanoglasses. Figure 21.2(b) displays scanning electron microscope (SEM) image of the surface morphology of 1 μm thick of a $Ni_{50}Ti_{45}Cu_5$ nanoglass produced by this method [43]. Aligned chains of humps are observed on the surface. In contrast to the nanoglasses prepared by inert-gas condensation, where the glassy grains form three-dimensional distribution connected with glass–glass interfaces [45], the

$Ni_{50}Ti_{45}Cu_5$ nanoglass consists of columnar glassy grains with a diameter of about 8 nm also separated with glass–glass interfaces. Although the microstructure of nanoglasses prepared by the two methods is slightly different, both types of nanoglasses show higher hardness and Young's modulus under nanoindentation as compared to metallic glasses with the same chemical composition [41,46].

21.3 Structure of Glass Glass Interfaces

Experimental evidences for the existence of glass–glass interfaces in nanoglass were provided by means of Möss-bauer spectroscopy, wide and small angle X-ray diffraction [34,40,47,48] scanning tunneling microscopy [45], transmission electron microscopy, positron annihilation spectroscopy, elemental mapping, scanning tunneling electron microscopy as well as atomic force microscopy [28] (for a review of these studies, see reference [49]). These interfaces can be characterized by an excess volume (low density) and different structure in comparison with the atomic structure in the center of the glassy particles. Besides the high number of methods available today to investigate the microstructure of metallic nanoglasses, a number of details are still puzzling. Still it is not clear if the difference in the Möss-bauer spectra and X-ray diffraction pattern of a nanoglass compared to those of melt-spun glass with the same chemical composition is due to the presence of interfaces or other effects. Recent small-angle X-ray scattering measurements of $Sc_{75}Fe_{25}$ nanoglasses, with an average grain size of about 7 nm, revealed that the relative electron density difference between the interfaces and glassy regions is about 17% which cannot solely be explained by the presence of an excess volume in the interface area [45]. Moreover, significant local variations in the elastic response of nanoglass surface have been reported, which also can hardly be explained by local topological variations, but could be due to

compositional variations [46], and indeed, element mapping data on $Fe_{90}Sc_{10}$ nanoglass revealed differences in the chemical composition of glass–glass interfaces and glassy regions (Figure 21.2) [45,50].

In order to gain a better understanding of the structure and composition of glass–glass interfaces and the impact of these planar defects on mechanical properties of nanoglasses, molecular dynamics (MD) computer simulations have been performed, a powerful tool to study the atomic-level structure and properties of metallic glasses and nanoglasses, respectively. This chapter reports the results of MD simulations which revisit the structural model of nanoglasses and provide an atomistic understanding of the relationship between the structure of glass–glass interfaces and mechanical properties of nanoglasses.

21.3.1 Chemically Homogeneous Glass Glass Interfaces

Even though there is seldom direct experimental evidence for the existence and the structure of interfaces in a nanoglass, in metallic glasses it is a well-proven fact that planar defects exist in form of shear bands introduced by plastic deformation. These shear bands exhibit an excess free volume and a modified local structure [13,33]. Next, the structure of a glass–glass interface is analyzed and compared to the one of an individual shear band.

A planar glass–glass interface is prepared by joining two planar relaxed glass surfaces of $Cu_{64}Zr_{36}$ (the prototype material used in the present study). The sample geometry is similar to a bicrystal-setup with three-dimensional periodic boundary conditions containing one interface parallel to the yz-plane. The Voronoi tessellation method was involved [51] to divide amorphous samples in Voronoi polyhedra around each atom, yielding local volumes and information about the nearest-neighbor environments. The occupation of certain types of Voronoi polyhedra was defined as a fingerprint of

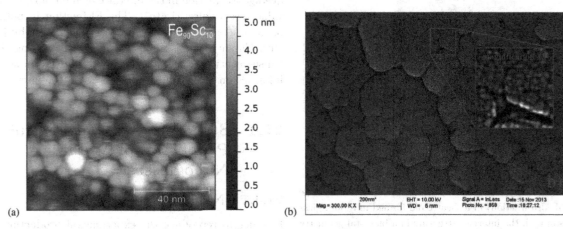

FIGURE 21.2 (a) Constant-current scanning tunneling electron micrograph (STEM) of the polished surface of a $Fe_{90}Sc_{10}$ nanoglass specimen. The STEM reveals the granular structure of the $Fe_{90}Sc_{10}$ nanoglass produced by consolidating $Fe_{90}Sc_{10}$ glassy clusters with a pressure of 4.5 GPa. (b) SEM images of the surface morphology of 1 μm thick $Ni_{50}Ti_{45}Cu_5$ sputtered film. The smallest building blocks are shown in the magnified view in the inset of panel b. Figures taken with permission from Ref. [42,43].

the short-range order of the amorphous alloy. In the case of $Cu_{64}Zr_{36}$ metallic glass, the Cu-centered full icosahedron (FI) [0,0,12,0] is the most prominent Voronoi polyhedra with a population of about 22% (with respect to the number of Cu atoms in the system), which corresponds to a volume fraction of approximately 15%. This Voronoi polyhedron is known to be a key structural motif in amorphous $Cu_{64}Zr_{36}$ alloys, characterized by high packing density [52] and high shear resistance [53].

Figure 21.3 shows the distribution of Voronoi volumes and the fraction of Cu-centered FI in a $Cu_{64}Zr_{36}$ glass sample with one planar interface (located at $x = 0$) in comparison with a shear band in a bulk glass. In a region of about 1 nm width (middle area in Figure 21.3), the Voronoi volumes are increased by about 1–2% with respect to the average bulk Voronoi volume Ω_0. Given the free volume expression by Turnbull and Cohen [55], who defined the free volume as the difference between the specific volume (analogous to Voronoi volume) and the molecule volume (here: atomic volume) and assuming a constant atomic volume, the increase in Voronoi volume can be directly translated into an increase in free volume. Hence, the interface between two glassy grains is characterized by an excess free volume of 1–2% in our model system. These interfaces are stable if the local strain induced by the enhanced free volume within the interfaces does not exceed the flow strain of the bulk glass [38]. The localized free volume in the interface is related to the atomic structure. By evaluating the topology of the occurring Voronoi polyhedra, their frequency and spatial distribution in the planar interface and the grain interior structural differences are obvious. In the case of the grain interior or bulk glass, respectively, the volume fraction of Cu-centered FI is about 22%, while in the interface, this structural backbone [56] is defective, since the fraction of the FI structural motif is as low as 10% (see Figure 21.3, light grey data points in the middle area) which yields an average fraction in the interface

of only 45%. Although the width of the planar interface is significantly smaller than that of the shear band (large grey area in Figure 21.3), both show the same topological features in their inner parts, namely, a deficiency in the icosahedra fraction and a slightly increased atomic volume.

21.3.2 Glass Glass Interfaces with Modified Chemical Composition

Apart from an excess free volume and a lack in short-range order, glass–glass interfaces can have different chemical composition as compared to the bulk value. In the case of amorphous CuZr-alloys, Cu-atoms were found to segregate to the surface at an elevated temperature [57–59]. When consolidating a glassy nanopowder, which was subjected to a pre-annealing process such as inert-gas condensation, nanoglasses characterized by interfaces with an increased Cu concentration would form.

In order to elucidate the structure and properties of Cu-enriched interfaces, a planar glass–glass interface was prepared by contacting two annealed surfaces. Initially, the glass was prepared following the procedure presented before for an interface with a homogeneous element distribution. Afterwards, the glass was annealed for 10 ns at 800 K (below transition temperature, $T_g \approx 950$ K), leading to a thin layer with enhanced Cu concentration on the surface. Contacting two such surfaces at low temperature (50 K), an interface characterized by an increased Cu concentration forms. Figure 21.4(a) shows the Cu-atoms density relative to the bulk density. A line pattern characterized by an increased Cu concentration in the center of the sample can be seen. Moreover, in the vicinity of this interface, a low fraction of Cu-atoms can be observed, since Cu-atoms have migrated to the surface during the segregation process. Besides an increased population of Cu-atoms, these interfaces have a lower fraction of Cu-centered FI relative to the number of Cu atoms. The average FI-fraction in the interface is about 12% of the bulk value which yields an average FI-fraction in the interface of about 45% of the bulk value (see Figure 21.4(b)). The FI-fraction in both kinds of interfaces (with homogeneous element distribution and with Cu-enriched) is almost the same, indicating that the Cu-segregation to the interface does not have a significant influence on atomic short-range order in the interface.

21.4 Structure and Mechanical Properties of Metallic Nanoglasses

21.4.1 Nanoglass Compaction

In order to reproduce the experimental production routes of nanoglasses, two types of geometry for the glassy particles have been used: columnar grains with a hexagonal cross section and spherical grains cut from a cubic $Cu_{64}Zr_{36}$ glass quenched from the melt. The nanoglasses with an

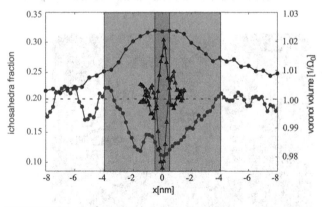

FIGURE 21.3 Local atomic Voronoi volume and short-range order (icosahedra fraction) in a shear band formed in a bulk glass in comparison with the interface structure in a bicrystal-geometry of $Cu_{64}Zr_{36}$. Although the width of the planar interface is significantly smaller than that of the shear band, both show the same topological features in their inner parts. Figure taken with permission from Ref. [54].

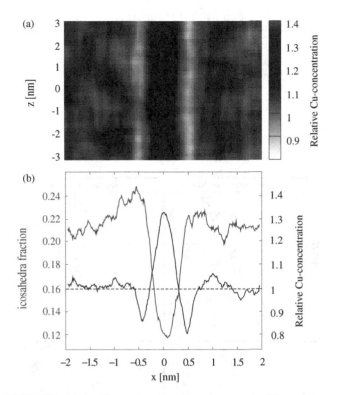

(a)

(b)

FIGURE 21.4 Glass–glass interface characterized by a lower fraction of Cu-centered full icosahedra and a higher fraction of Cu-atoms. In a 2d-plot of the Cu-density distribution (a) and linear scans of the FI-density and Cu-density distribution (b), the interface of about 1 nm width is clearly detectable.

idealized nanostructure consisting of columnar grains with a hexagonal cross section resemble the geometry of those obtained in magnetron sputtering, while the nanoglasses with spherical grains are prepared to investigate the structure and properties of nanoglasses prepared by inert-gas condensation. Additionally, the impact of chemical composition is investigated by considering chemically homogeneous glassy powder and glassy particles with Cu atoms segregated to the surfaces which are generated by pre-annealing the particles before compaction. All metallic glass powders, with and without surface segregation, are compacted by applying an external hydrostatic pressure of 3 GPa (for droplets with a honeycomb geometry) and a higher value of 5 GPa (for glassy grains with spherical geometry) at 50 K to obtain a nanoglass free of pores.

21.4.2 Deformation Behavior of As-Prepared Nanoglasses

The reported observations about the new atomic structure of the glass–glass interfaces predict already a different mechanical response for nanoglasses under deformation when compared to the properties of the metallic glasses available today. By introducing shear band precursors like interfaces, shear localization can be controlled to achieve a more homogeneous plasticity. For studying the deformation mechanism of nanoglasses in comparison with

a homogeneous BMG, inhomogeneous (with Cu-enriched interfaces) and homogeneous (with homogeneous elements distribution) nanoglasses consisting of columnar grains have been deformed under uniaxial tension parallel to the z-direction. All structures were deformed at 50 K with a constant strain rate of 4×10^7 1/s in z-direction. Periodic boundary conditions were applied in all three dimensions, and the pressure in x- and y-directions was adjusted to 0 kbar allowing for lateral contraction.

Figure 21.5(a) shows the stress–strain curves for both nanoglasses and the BMG under tensile deformation. Up to a strain of about 2%, all three curves show a similar slope and, thus, the interfaces have no strong influence on the elastic deformation. The stress at which the curves deviate from a linear behavior is defined to be the yield stress of the samples, which is significantly lower for both nanoglasses at about 2.4 GPa compared to the bulk glass with a yield stress of approximately 3.3 GPa. The same trend is found for the maximum stress, which is 3.9 GPa for the

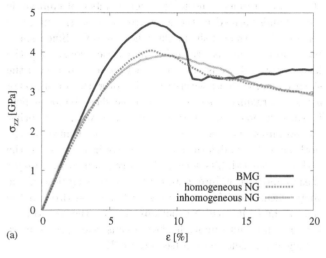

(a)

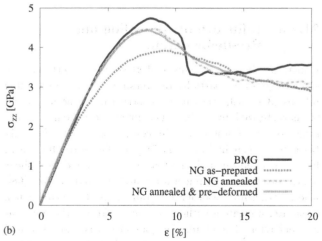

(b)

FIGURE 21.5 Tensile stress–strain curves for (a) BMG, and as-prepared homogeneous and inhomogeneous nanoglasses; (b) BMG, As-prepared, annealed, annealed and pre-deformed inhomogeneous nanoglass, at a constant strain rate of 4×10^7 1/s Figures taken with permission from Ref. [60].

inhomogeneous nanoglass and 4.0 GPa for the homogeneous nanoglass compared to 4.7 GPa in the case of the BMG. This can be explained by the lower activation barrier for shear transformation zones (STZs) in the nanoglass interfaces due to a lower population of FI compared to the bulk glass. Cu-centered FIs are Voronoi polyhedra with a high packing density [52] and high shear resistance [61], and therefore, those regions in a glass with a lower content of FIs are easier to deform plastically. These results are supported by an analysis of the atomic scale deformation mechanism of nanoglasses investigated and visualized using local atomic shear strain [62] calculated with the OVITO analysis and visualization software [63]. Up to a strain of 8%, where the maximum stress is reached, STZs, the elementary units of plasticity in metallic glasses consisting of clusters of atoms that cooperatively rearrange under the action of an applied stress [64], are only activated in the soft interface regions (see Figure 21.6). With increasing strain, in both nanoglasses, multiple embryonic shear bands are formed along the interfaces and eventually start to propagate through the grain interiors. Due to the high number of embryonic shear bands formed in the interfaces, the elastic energy is released homogeneously in the whole sample. Since shear band propagation is driven by the elastic energy [65], the local energy release is not sufficient to accelerate one of the shear bands, so that goes critical. Moreover, the formation of a highly organized pattern of multiple shear bands which, in turn, interact in a very complex way with each other compromises the shear band propagation. Consequently, both nanoglasses deform homogeneously in contrast to the BMG, which exhibits localized deformation in one major shear band (see Figure 21.6, top). The propagation of one single shear band leads to a significant stress drop in the case of the BMG, which is not observed for the NGs due to the branching into multiple shear bands leading to a more homogeneous deformation (see Figure 21.5).

21.4.3 Influence of Annealing and Pre-deformation

Under the thermal treatment of glass–glass interfaces, the short-range order partially recovers and the excess volume relaxes [39]. Still, the interface structure after annealing is characterized by a difference in the icosahedral short-range order of about 2% in comparison with the bulk (as can be seen in Figure 21.7) [66]. These results are in line with experimental observations on pre-induced shear bands produced by indentation [67] or cold rolling [68], where structural disorder is retained in the shear bands even after annealing without strongly affecting the mechanical behavior. Therefore, the questions arise how thermal annealing prior to deformation does affect the plastic deformation mechanisms operating in a nanoglass. In order to answer this question, both types of nanoglasses have been subjected to an annealing process at 700 K (0.75 T_g) for 2 ns. After annealing, the nanoglasses have been cooled down to 50 K and subsequently deformed in uniaxial tension

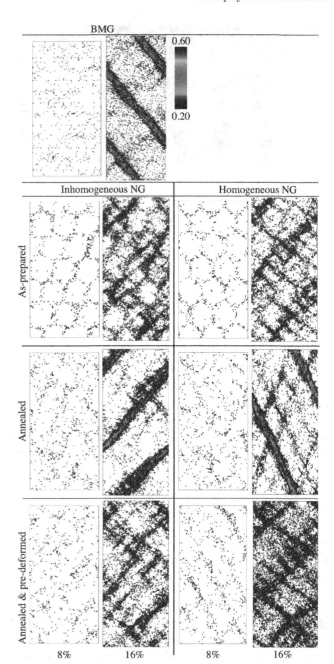

FIGURE 21.6 Local atomic shear strain for an inhomogeneous (with Cu-enriched interfaces) and homogeneous (with homogeneous elements distribution) nanoglasses As-prepared, annealed, annealed and pre-deformed in comparison with a BMG, under tensile deformation. Figures taken with permission from Ref. [60]

following the same procedure as before. At a strain of 8%, shear transformations occur in the interfaces similar to the previous case of non-annealed NGs. When the strain is increased, however, the deformation mechanism changes. In agreement with experimental observation [68], the plastic deformation no longer takes place in a network of multiple shear bands. Instead, the deformation is more localized and at a strain of about 16%, one dominant shear band is formed. Nevertheless, the plastic deformation mechanism of the annealed nanoglasses deviates from what has been

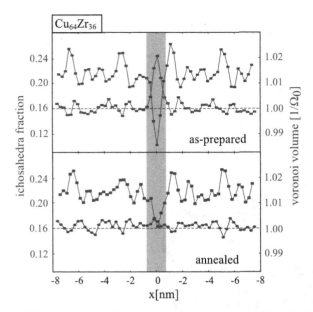

FIGURE 21.7 The distribution of Cu-centered full icosahedra and Voronoi volume in the $Cu_{64}Zr_{36}$ metallic glasses with a planar glass-glass interface before and after annealing.

21.4.4 Influence of Grain Size

For studying the effect of grain size on the mechanical properties, the deformation of Cu-rich glasses with grain sizes of $\langle d \rangle = 4$, 10 and 16 nm is simulated. All nanoglasses and a BMG are deformed in tension following the procedure described in the previous section. To study how the deformation mechanism of nanoglasses changes by varying the glassy grain size, the local atomic shear strain is calculated for each of these structures (Figure 21.8). Up to a strain of 8% in all three nanoglasses, the STZs are mostly activated in the soft interface regions. Although the shear band nucleation process is similar in all three nanoglasses, the shear band propagation in case of the nanoglass with the largest grain size strongly differs from the other two nanoglasses. It can be seen in Figure 21.8 that the plastic deformation is more localized at a strain of 16% in case of the nanoglass with a grain size of 16 nm. This can be explained by the lower fraction of soft interfaces with respect to the nanoglass volume. Up to a strain of 8%, the elastic energy is released locally in embryonic shear bands formed by precipitation of the STZs along the interfaces (see Figure 21.8).

After a strain of 8%, the local energy released is sufficient to accelerate one of these embryonic shear bands, so that it goes critical. However, together with a dominant shear band, many secondary shear bands mediate the plastic deformation of the nanoglass (see Figure 21.8). This observation is supported by the calculated strain localization parameter ψ [56]. A larger ψ value indicates larger fluctuations in the atomic strain and a more localized deformation mode. The ψ values for all three nanoglasses and BMG at a strain of 16% are plotted in Figure 21.9. It can be seen that the ψ value of the nanoglass with a grain size of 16 nm is higher than for the other two nanoglasses but still lower than for the BMG. For the other two nanoglasses with smaller glassy grains, the ψ values are much lower indicating a more homogeneous deformation mode.

21.4.5 Chemical Composition Effects

Next question addressed here is whether the variation in the chemical composition affects the mechanical properties of nanoglasses. Thus, the composition of the glassy grains is varied, and Zr-rich composition is studied, $Cu_{36}Zr_{64}$. The short-to-medium-range structural order varies with alloy composition [69] and is assumed to play a major role in controlling the macroscopic properties of metallic glasses, particularly the plastic deformation [53,70]. Therefore, one can expect that $Cu_{36}Zr_{64}$ nanoglass shows a different deformation behavior when compared to the Cu-rich nanoglass.

Hence, the chemical composition effect on the plastic behavior of nanoglasses is investigated by using two alloy compositions: a Zr-rich metallic glass ($Cu_{36}Zr_{64}$) and the Cu-rich metallic glass ($Cu_{64}Zr_{36}$). Both types of nanoglasses have an average grain size of 10 nm. The operating deformation mechanisms in these two nanoglasses and the corresponding BMGs is analyzed under tensile deformation. In

observed for the BMG. The shear band in the annealed NGs is not as well localized as in the BMG. At a strain of 16%, even indications of shear band branching are observed (see Figure 21.6), a mechanism which has been reported for BMGs with pre-induced SBs, as well [22]. Moreover, comparing the stress–strain behavior for the BMG and the annealed nanoglasses (see Figure 21.5(b)), the differences are still significant. Even though a shear band is formed in the annealed nanoglasses, the stress–strain behavior hardly deviates from the BMG case and no sudden stress drop occurs. Therefore, the interfaces in the annealed nanoglasses still can have an impact on plastic deformation.

The next question addressed is whether a homogeneous deformation mode can be recovered after annealing. Therefore, the annealed nanoglasses have been preloaded to a strain level of 8%, just where the plastic deformation already starts to initiate in the interfaces, but no localized shear bands occur (see Figure 21.6). After unloading to zero strain, the nanoglasses are deformed for a second time. Interestingly, the plastic deformation behavior is almost identical to that before annealing. The nanoglasses undergo plastic deformation mediated by multiple shear bands, uniformly distributed over the sample volume. This is attributed to STZs which affect the short-range order and, at a strain of 8%, occur mainly in the soft interfaces. Consequently, the short-range order in the interfaces, which had been recovered during annealing, is damaged during preloading, and again, the interfaces act as shear band nucleation sites preventing the plastic deformation from localizing in a single shear band as in the case of BMGs or annealed nanoglasses. Interestingly, the annealed and then preloaded nanoglass shows an almost identical stress–strain behavior to the annealed nanoglass, despite the observed differences in deformation mode (Figure 21.5(b)).

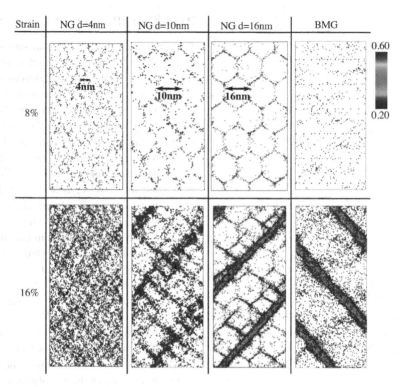

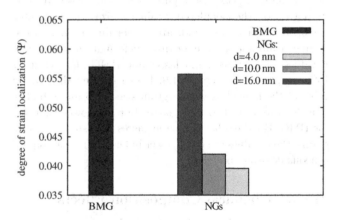

FIGURE 21.8   Local atomic shear strain for $Cu_{64}Zr_{36}$ BMG and nanoglasses with grain sizes of 4 nm, 10 nm and 16 nm, respectively. Figures taken with permission from Ref. [66].

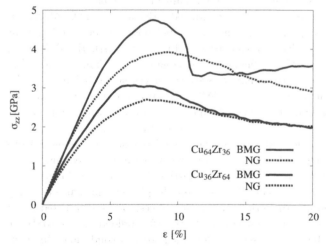

FIGURE 21.9   The ψ values for a BMG in comparison with three NGs with grain sizes of 4, 10 and 16 nm, respectively, after plastic deformation to an overall strain of 16%. Figures taken with permission from Ref. [66]

FIGURE 21.10   Tensile stress–strain curves for $Cu_{36}Zr_{64}$ and $Cu_{64}Zr_{36}$ BMGs and nanoglasses with the same chemical composition and a grain size of 10 nm, at a constant strain rate of 4×10^7 1/s.

Figure 21.10, it can be observed that the Zr-rich BMG shows a lower yield stress compared to the yield stress of the Cu-rich BMG. The population of Cu-centered FI in the Cu-rich alloy of about 22% compared to only 5% in the Zr-rich alloy (see Figure 21.12) [66]. FI-clusters have a high shear resistance [52], and thus, this explains the lower yielding stress of the Zr-rich BMG. Moreover, it can be seen in Figure 21.11 that also the plastic behavior of the $Cu_{36}Zr_{64}$ BMG differs completely from the one of the Cu-rich BMG. The Cu-rich BMG exhibits localized deformation in one dominant shear band, while the Zr-rich BMG exhibits a homogeneous deformation.

The difference in the short-to-medium structural order for these two alloy compositions can serve as an explanation. In the $Cu_{36}Zr_{64}$ alloy, the low fraction of FI-clusters leaves space for the formation of a high fraction of Voronoi polyhedra characterized by a low packing density and low shear resistance. Hence, under loading in Zr-rich BMGs, more STZs should activate and more strain can be carried. On the other hand, the Cu-rich BMG contains a high fraction of FI-clusters, and therefore, a high degree of medium

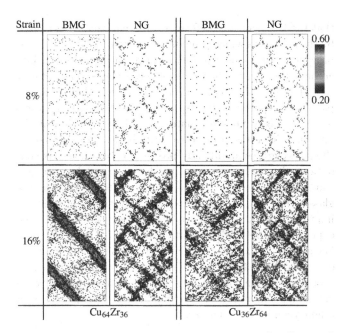

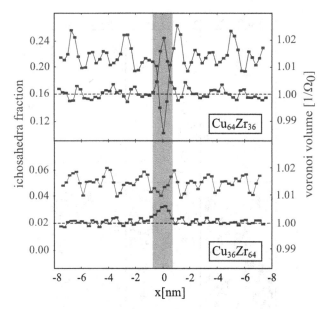

FIGURE 21.11 Local atomic shear strain for $Cu_{36}Zr_{64}$ and $Cu_{64}Zr_{36}$ BMGs and nanoglasses with a grain size of 10 nm.

FIGURE 21.12 The fraction of Cu-centered full icosahedra and Voronoi volume in the $Cu_{64}Zr_{36}$ and $Cu_{36}Zr_{64}$ metallic glass with a planar glass–glass interface.

range order formed from the interconnection of FI-units. Due to the smaller flow region within this alloy, also the number of activated STZ is low, and hence, $Cu_{64}Zr_{36}$ BMG is unable to mediate large plastic strains. Consequently, the applied elastic energy cannot be further released, and the already activated STZs start percolate along one dominant shear band (see Fig 21.11, lower panels).

Next, the impact of the glass–glass interface structure on the plastic behavior of Cu-rich and Zr-rich nanoglasses is investigated in comparison. Previously, it has been shown that glass–glass interfaces in $Cu_{64}Zr_{36}$ nanoglass are characterized by a low fraction of FI-clusters and enhanced free volume. On the other hand, in case of the $Cu_{36}Zr_{64}$ alloy, the FI density is constant in the whole sample, and no deviation at the position of the interface is found (see Figure 21.12). The population of Cu-centered FIs in case of the Zr-rich glass is only 5%. This value is similar to the one found in other studies [53,71]. Surprisingly, in Figure 21.12, it can be seen that the density in the interface is decreased although the FI density is not affected. However, the fraction of about 0.7% excess free volume detected for the interface in Zr-rich glass is much lower compared to the case of Cu-rich metallic glass of ≈2%. The increase in the free volume in the interface is not related to the decrease of densely packed FI-cluster as found for the case of Cu-rich metallic glass. Here, the volume expansion inside the interface is related to the destruction (when the planar surfaces are created) and formation (when joining the resulting glassy surfaces) of new FI-clusters which show a lower packing density [71]. The creation of these new less densely packed FI-clusters fully compensates the destruction of the original icosahedra short range order (SRO). Although the excess free volume of the planar interface in the Zr-rich glass is below 1%, it can be seen in Figure 21.11 that also the plastic behavior of

$Cu_{36}Zr_{64}$ nanoglass deviates from the homogeneous BMG. Also, the stress–strain curves show a lower maximum stress for both nanoglasses when compared to the BMG, and for $Cu_{64}Zr_{36}$ nanoglass, the maximum stress is lower (about 4.0 GPa) compared to the BMG (about 4.7 GPa), while the maximum stress of $Cu_{36}Zr_{64}$ nanoglass is about 2.6 GPa compared to 3.0 GPa found for the BMG. This can be explained by the lower activation barrier for STZs in the soft interfaces of nanoglasses [60], and it is supported when calculating the local atomic shear strain. Up to a strain of 8%, the STZs are only activated in the soft interface regions of both nanoglasses (see Figure 21.11). Increasing the strain to 16%, embryonic shear bands are formed along the interfaces and propagate through the grain interiors. In all cases, the embryonic shear bands are blocked, and no dominant shear band is formed.

21.4.6 Dual-Phase Metallic Nanoglasses

In a next step, the impact of chemical composition on the deformation behavior of metallic nanoglasses is investigated. Hence, glassy grains of different chemical composition ($Cu_{64}Zr_{36}$ and $Cu_{36}Zr_{64}$ grains) are compacted, and a designed dual-phase metallic nanoglass is created (Figure 21.13 middle column panels). In this way, heterogeneities in terms of regions of harder and softer chemical compositions can be generated. The deformation behavior of the nanoglass composite is investigated in comparison with the nanoglasses of the Cu-rich and Zr-rich composition. For a layered nanoglass composite, the mechanical properties can be seen as average of the constituent glassy phases. Snapshots of the distribution of atomic shear strains are given in Figure 21.13 for 8% and 16% total strain. In all cases, STZs are activated in the interface areas as can be

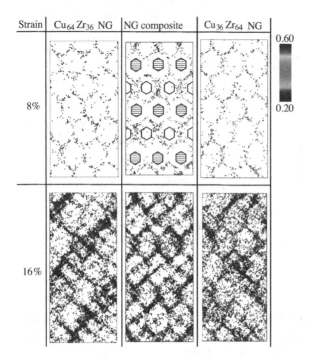

FIGURE 21.13 Left: Atomic shear strain in $Cu_{64}Zr_{36}$ nanoglass. Right: Atomic shear strain in $Cu_{36}Zr_{64}$ nanoglass. Center: Atomic shear strain in composite nanoglass. All nanoglasses have a grain diameter of 10 nm. Symbols show the position of Cu- and Zr-rich grains, respectively. Figures taken with permission from Ref. [54]

seen from configurations at 8% strain, which is close to the maximum stress for these glasses. At larger strains (16%), the Cu-rich glass $Cu_{64}Zr_{36}$ has the strongest tendency for shear localization, whereas the Zr-rich glass and the nanoglass composite are more homogeneously strained [54]. The strain distribution in the nanoglass composite also reflects the compositional gradient between adjacent grains. The hard and soft regions created in the nanoglass composite are characterized by different elastic properties. This is an important point of interest because the elastic mismatch arising between these regions during deformation may induce stress concentration at the hard–soft interfaces that, in turn, may influence shear band formation and evolution. Thus, nanoglass structures consolidated from glassy grains of different size and composition are another means to tune mechanical properties of glassy materials.

21.5 Nanoglasses Prepared by Inert-Gas Condensation

In the atomistic model presented before, to systematically analyze the structure–property relationship in metallic glass, nanoglasses with an idealized quasi-two-dimensional columnar nanostructure consisting of hexagonal grains have been considered. This model could show that glass–glass interfaces are characterized by a defective short-range order and an excess free volume. Moreover, it could also explain

that the homogeneous deformation in nanoglasses is caused by the interfaces which promote the formation of STZs, thus preventing strain localization. However, it does not take into account the thermal history of glassy spheres. Therefore, it cannot explain the compositional variation within the glassy particles prepared in inert-gas condensation [50]. Furthermore, it might underestimate the volume fraction of the interfaces in terms of defective SRO because in this model, the glassy grains are already pre-shaped before the consolidation, although the glassy powder prepared by inert-gas condensation, most common technique to produce nanoglasses, has a spherical shape [45]. Therefore, in order to improve this model and to be more consistent with the experimental setup, the particle-derived nanoglass model is elaborated [72]. The particle-derived nanoglass model is described in the following, as well as the microstructure and properties of nanoglasses obtained by this model are investigated and compared to those found for the nanoglasses with a honeycomb-like geometry.

21.5.1 Compaction and microstructure

To investigate the atomic structure of the glassy nanoparticles prepared by inert-gas condensation, $Cu_{64}Zr_{36}$ glassy powders were simulated in the gas phase by MD simulations [72,73]. To this end, a glassy sphere with a diameter of about 7 nm was cut from a metallic glass and heated above T_g and then cooled down to low temperature. The glassy sphere was kept long enough above T_g, so that the atoms could diffuse within the glassy sphere. Structural analysis shows clear evidence for compositional variations within the glassy sphere due to surface segregation effects. The glassy sphere is separated into a core region and a shell region with different compositions (Figure 21.14(c)), indicating that the glassy clusters prepared by inert-gas condensation have a heterogeneous composition. This is in agreement with experimental results showing that surface segregation of Fe occurs in primary glassy powder of a $Fe_{90}Sc_{10}$ nanoglass [50].

Recently, element mapping data on $Sc_{75}Fe_{25}$ nanoglasses by means of energy-dispersive X-ray spectroscopy (EDX) showed that the interfaces are Sc-rich. Moreover, atom probe tomography revealed that Cu-rich interfaces in $Cu_{50}Zr_{50}$ nangolasses [74]. To elucidate the origin of this difference of composition between the interfaces and glassy regions, MD simulation was performed to simulate the procedure of preparing a nanoglass. Several $Cu_{64}Zr_{36}$ glassy spheres with diameter ranging from 6 nm to 8 nm were consolidated, and the microstructure of the resulting nanoglass is shown in Figure 21.14(b). MD simulation results reveal that the core and shell regions of the glassy spheres are not mixed after the consolidation and shell regions are now forming the glass–glass interfaces of the nanoglass. The fact that the glassy spheres undergo surface segregations during the inert-gas condensation stage, the interfaces have a different composition as compared to the glassy regions. In the case of the $Cu_{64}Zr_{36}$ nanoglass, the glassy regions and the interfaces

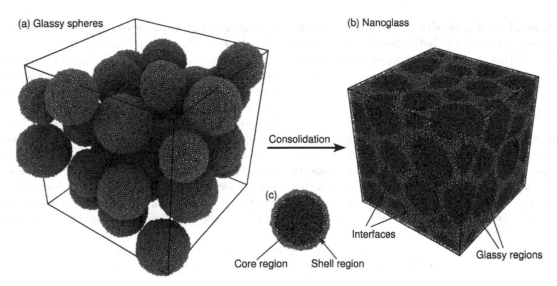

FIGURE 21.14 Procedure of preparing a $Cu_{64}Zr_{36}$ nanoglass. (a) Simulation box containing polydisperse glassy spheres with diameter ranging from 6 nm to 8 nm. The glassy spheres are consolidated to form (b) a nanoglass. (c) Slice showing core and shell regions of a glassy sphere. After the consolidation, the core and shell regions of the glassy spheres form the glassy regions and interfaces of the nanoglass, respectively. Figure taken with permission from [72].

have compositions of $Cu_{61}Zr_{39}$ and $Cu_{72}Zr_{28}$, respectively (Figure 21.14(b)). Based on the MD simulation results, the interface composition of about $Sc_{85}Fe_{15}$ observed in the $Sc_{75}Fe_{25}$ nanoglass comes from the compositional variation in the glassy nanoparticles which occurs during the inert-gas condensation.

As can be seen in Figure 21.15 (a), the interfaces in terms of composition have a width of about 1 nm. This is in agreement with the experimental value 0.8 nm reported for $Sc_{75}Fe_{25}$ nanoglasses [45]. By analyzing the Voronoi volumes of these interfacial and bulk atoms [51], the as-produced $Cu_{64}Zr_{36}$ nanoglass consists of about 37 vol.% interfaces and 63 vol.% glassy regions, which is consistent with 35 vol.% interfaces and 65 vol.% glassy regions observed in the

$Sc_{75}Fe_{25}$ nanoglass [45]. Figure 21.15 (b) shows the same snapshot as in Figure 21.15, (a) but the atoms are colored according to their atomic shear strain. It can be seen that the region with massively sheared atoms, atoms with shear strain higher than 0.2, is significantly larger than the interfaces which are defined in terms of composition. Moreover, MD simulation results have shown that this region is characterized by a defective SRO [72]. Therefore, the interfaces in terms of defective SRO have a width of at least 2 nm (21.15 (b)).

To see whether the pores between the glassy spheres can be closed after the consolidation, the density as a function of consolidation pressure for the $Cu_{64}Zr_{36}$ glassy spheres was calculated and plotted relative to the bulk glass

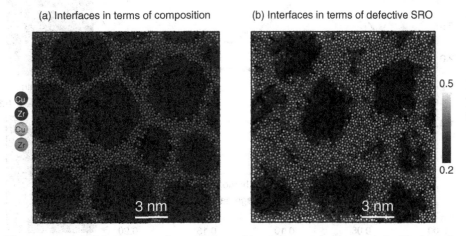

FIGURE 21.15 Microstructure of $Cu_{64}Zr_{36}$ nanoglass, which consists of glassy regions connected with glass–glass interfaces. (a) Thin slice of about 2 nm thickness showing the microstructure of nanoglasses in terms of composition: dark colors are atoms in the glassy regions; bright colors are atoms in the interfaces with a width of about 1 nm. (b) Thin slice of about 2 nm thickness showing shear localization after the consolidation of glassy spheres at 50 K. It can be seen that the width of the region, with the atomic strain higher than 0.2, is larger than 1 nm. The average glassy grain size is about 7 nm. Figure reproduced with permission from [72].

value in Figure 21.16. The density increases with pressure and approaches to that of the bulk glass at a pressure of about 5 GPa. At this pressure, the entire porosity is closed (Figure 21.16).

21.5.2 Deformation behavior

The results presented above show that the volume fraction of the interfaces in the particle-derived nanoglass is significantly larger than that in the honeycomb-like geometry model with the same average grain size. To see whether this large volume fraction of the interfaces affect the deformation behavior, the tensile deformation of particle-derived

nanoglass as well as the deformation of a homogeneous bulk glass for comparison were simulated.

The stress–strain curves are plotted in Figure 21.17. It can be seen that the metallic glass shows a stress strop after the maximum stress of about 3.1 GPa and the flow stress drops to 2 GPa. This behavior is known to be accompanied by the formation of a critical shear band [60,66,71]. On the other hand, the nanoglass does not exhibit such a stress drop; it shows a ductile behavior with the steady-state flow stress of about 1.75 GPa. This can be explained by the lower activation barrier for STZs in the nanoglass due to the defective SRO at the interfaces. Figure 21.17 also shows snapshots of the atomic configurations according to the local atomic

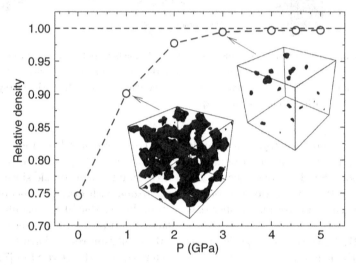

FIGURE 21.16   Variation of density as a function pressure during the consolidation of $Cu_{64}Zr_{36}$ glassy spheres to produce a nanoglass. Density is given relative to the metallic glass value. Inset: remaining porosity at 1 GPa and 3 GPa. Figures taken with permission from Ref. [72].

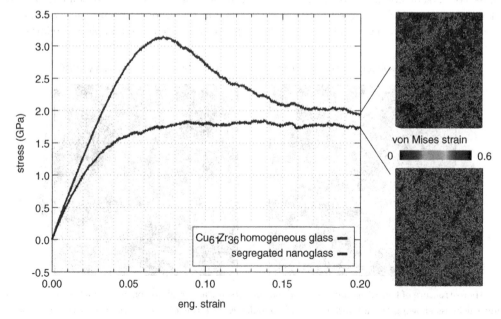

FIGURE 21.17   Tensile tests at 50 K and constant engineering strain rate of a homogeneous MG of composition $Cu_{61}Zr_{39}$ and a segregated NG, where the glassy grains have the same composition as the homogeneous glass. The snapshots of the MG and NG show the local atomic von Mises strain. Figures taken with permission from Ref. [75].

von Mises strain at 20% macroscopic strain. A shear band is visible in the homogeneous glass, whereas the strain localization in the nanoglass is more homogeneous.

21.6 Acknowledgments

The authors acknowledge the financial support through the Deutsche Forschungsgemeinschaft (DFG) (Grants: SO 1518/1-1, AL 578/6 and AL 578/15).

References

1. W. Klement, R. H. Willens, P. Duwez, Non-crystalline structure in solidified gold-silicon alloys, *Nature* 187 (4740) (1960) 869–870.

2. J. Basu, S. Ranganathan, Bulk metallic glasses: A new class of engineering materials, *Sadhana* 28 (2003) 783–98.

3. W. H. Wang, C. Dong, C. H. Shek, Bulk metallic glasses, *Mater. Sci. Eng. R-reports* 44 (2-3) (2004) 45–89.

4. H. W. Kui, A. L. Greer, D. Turnbull, Formation of bulk metallic-glass by fluxing, *Appl. Phys. Lett.* 45 (6) (1984) 615–616.

5. A. Peker, W. L. Johnson, A Highly Processable Metallic Glass - Zr41.2Ti13.8Cu12.5Ni10.0Be22.5, *Appl. Phys. Lett.* 63 (17) (1993) 2342–2344.

6. A. L. Greer, Metallic glasses, *Science* 267 (5206) (1995) 1947–1953.

7. J. Eckert, L. Schultz, E. Hellstern, K. Urban, Glass-forming range in mechanically alloyed Ni-Zr and the influence of the milling intensity, *J. Appl. Phys.* 64 (6) (1988) 3224–3228.

8. R. W Cahn, A. L. Greer, Metastable states of alloys. In: Cahn, R.W, Hassen, P., eds. *Physical Metallurgy.* Elsevier Science, Amsterdam (1996) 1724.

9. A. Inoue, Stabilization of metallic supercooled liquid and bulk amorphous alloys, *Acta Mater.* 48 (1) (2000) 279–306. doi:10.1016/S1359-6454(99)00300-6. www.sciencedirect.com/science/article/pii/S1359645499003006

10. M. Ashby, A. Greer, Metallic glasses as structural materials, *Scr. Mater.* 54 (3) (2006) 321–326. www.sciencedirect.com/science/article/pii/S135964 6205006111

11. A. Gebert, K. Buchholz, A. Leonhard, K. Mummert, J. Eckert, L. Schultz, Investigations on the electrochemical behaviour of Zr-based bulk metallic glasses, *Mater. Sci. Eng., A* 267 (2) (1999) 294–300.

12. J. F. Loffler, Bulk metallic glasses, *Intermetallics* 11 (6) (2003) 529–540.

13. A. J. Cao, Y. Q. Cheng, E. Ma, Structural processes that initiate shear localization in metallic glass, *Acta Mater.* 57 (17) (2009) 5146–5155.

14. D. Magagnosc, G. Kumar, J. Schroers, P. Felfer, J. Cairney, D. Gianola, Effect of ion irradiation on tensile ductility, strength and fictive temperature in metallic glass nanowires, *Acta Mater.* 74 (2014) 165–182. doi:10.1016/j.actamat.2014.04.002. www.sciencedirect.com/science/article/pii/S1359645414002419

15. S. V. Ketov, Y. H. Sun, S. Nachum, Z. Lu, A. Checchi, A. R. Beraldin, H. Y. Bai, W. H. Wang, D. V. Louzguine-Luzgin, M. A. Carpenter, A. L. Greer, Rejuvenation of metallic glasses by non-affine thermal strain, *Nature* 524 (7564) (2015) 200–203. http://dx.doi.org/10.1038/nature14674

16. C. Fan, A. Inoue, Improvement of mechanical properties by precipitation of nanoscale compound particles in Zr-Cu-Pd-Al amorphous alloys, *Mater. Trans. JIM* 38 (12) (1997) 1040–1046.

17. J. Eckert, J. Das, S. Pauly, C. Duhamel, Mechanical properties of bulk metallic glasses and composites, *J. Mater. Res.* 22 (2007) 285–301. doi:10.1557/jmr.2007.0050. http://journals.cambridge.org/article'S0884291400092554

18. G. He, J. Eckert, W. Löser, L. Schultz, Novel ti-base nanostructure-dendrite composite with enhanced plasticity, *Nat. Mater.* 2 (1) (2003) 33–7.

19. C. C. Hays, C. P. Kim, W. L. Johnson, Microstructure controlled shear band pattern formation and enhanced plasticity of bulk metallic glasses containing in situ formed ductile phase dendrite dispersions, *Phys. Rev. Lett.* 84 (13) (2000) 2901–2904.

20. C. C. Hays, C. P. Kim, W. L. Johnson, Improved mechanical behavior of bulk metallic glasses containing in situ formed ductile phase dendrite dispersions, *Mater. Sci. Eng. A* 304 (2001) 650–655.

21. S. Pauly, S. Gorantla, G. Wang, U. Kuhn, J. Eckert, Transformation-mediated ductility in CuZr-based bulk metallic glasses, *Nat. Mater.* 9 (6) (2010) 473–477.

22. M. H. Lee, K. S. Lee, J. Das, J. Thomas, U. Kühn, J. Eckert, Improved plasticity of bulk metallic glasses upon cold rolling, *Scr. Mater.* 62 (9) (2010) 678–681.

23. S.-H. Joo, D.-H. Pi, A. D. H. Setyawan, H. Kato, M. Janecek, Y. C. Kim, S. Lee, H. S. Kim, Work-hardening induced tensile ductility of bulk metallic glasses via high-pressure torsion, *Sci. Rep.* 5 (2015) 9660. http://www.ncbi.nlm.nih.gov/pmc/articles/PMC5386117

24. S. Scudino, B. Jerliu, S. Pauly, K. Surreddi, U. Kühn, J. Eckert, Ductile bulk metallic glasses produced through designed heterogeneities, *Scr. Mater.* 65 (9) (2011) 815–818. doi:10.1016/j.scriptamat.2011.07.039. www.sciencedirect.com/science/article/pii/S1359646211004386

25. A. Brothers, D. Dunand, Amorphous metal foams, *Scr. Mater.* 54 (4) (2006) 513–520, viewpoint set no. 38 on: Frontiers on fabrication and properties of porous and cellular metallic materials. doi:10.1016/j.scriptamat.2005.10.048. www.sciencedirect.com/science/article/pii/S1359646205006871

26. T. Wada, A. Inoue, A. L. Greer, Enhancement of room-temperature plasticity in a bulk metallic glass by finely dispersed porosity, *Appl. Phys. Lett.* 86 (25) (2005) 251907. doi:10.1063/1.1953884. http://scitation.aip.org/content/aip/journal/apl/86/25/10.1063/1.1953884

27. A. Inoue, T. Wada, D. V. Louzguine-Luzgin, Improved mechanical properties of bulk glassy alloys containing spherical pores, *Mater. Sci. Eng. A* 471 (12) (2007) 144–150. doi:10.1016/j.msea.2006.10.172. www.sciencedirect.com/science/article/pii/S0921509307004571

28. H. Gleiter, Nanoglasses: A new kind of noncrystalline material and the way to an age of new technologies?, *Small* 12 (16) (2013) 2225–2233. doi:10.1002/smll.201500899.

29. H. S. Kim, Y. Estrin, M. B. Bush, Plastic deformation behaviour of fine-grained materials, *Acta Mater.* 48 (2) (2000) 493–504. www.sciencedirect.com/science/article/pii/S1359645499003535

30. R. Sutton, A.P. Balluffi, *Interfaces in Crystalline Materials*, Oxford Science Publications, Oxford (1996).

31. D. B. Miracle, The efficient cluster packing model - an atomic structural model for metallic glasses, *Acta Mater.* 54 (16) (2006) 4317–4336.

32. M. Li, C. Z. Wang, S. G. Hao, M. J. Kramer, K. M. Ho, Structural heterogeneity and medium-range order in ZrxCu100-x metallic glasses, *Phys. Rev. B* 80 (18) (2009) 184201.

33. Q. K. Li, M. Li, Atomic scale characterization of shear bands in an amorphous metal, *Appl. Phys. Lett.* 88 (24) (2006) 241903.

34. J. Jing, A. Kramer, R. Birringer, H. Gleiter, U. Gonser, Modified atomic structure in a Pd-Fe-Si nanoglass : A Mössbauer study, *J. Non-Cryst. Solids* 113 (2-3) (1989) 167–170. www.sciencedirect.com/science/article/B6TXM-48CXMM7-V3/2/0f05e9baad09654aa30abff391ba440a

35. H. Gleiter, Nanocrystalline solids, *J. Appl. Crystallogr.* 24 (2) (1991) 79–90. doi:10.1107/S0021889890011013.

36. H. Kato, Y. Kawamura, A. Inoue, T. Masumoto, Bulk glassy Zr-based alloys prepared by consolidation of glassy alloy powders in supercooled liquid region, *Mater. Sci. Eng. A* 226 (1997) 458–462.

37. H. Gleiter, Our thoughts are ours, their ends none of our own: Are there ways to synthesize materials beyond the limitations of today?, *Acta Mater.* 56 (19) (2008) 5875–5893.

38. D. Şopu, K. Albe, Y. Ritter, H. Gleiter, From nanoglasses to bulk massive glasses, *Appl. Phys. Lett.* 94 (19) (2009) 191911. doi:10.1063/1.3130209.

39. Y. Ritter, D. opu, H. Gleiter, K. Albe, Structure, stability and mechanical properties of internal interfaces in Cu64Zr36 nanoglasses studied by –MD simulations, *Acta Mater.* 59 (17) (2011) 6588–6593. doi:10.1016/j.actamat.2011.07.013. www.sciencedirect.com/science/article/pii/S1359645411004782

40. J. Weissmueller, R. Birringer, H. Gleiter, Nanostructured crystalline and amorphous solids, *Key Eng. Mater.* 77-78 (77-78) (1993) 161.

41. N. Chen, R. Frank, N. Asao, D. Louzguine-Luzgin, P. Sharma, J. Wang, G. Xie, Y. Ishikawa, N. Hatakeyama, Y. Lin, M. Esashi, Y. Yamamoto, A. Inoue, Formation and properties of Au-based nanograined metallic glasses, *Acta Mater.* 59 (16) (2011) 6433–6440. www.sciencedirect.com/science/article/pii/S1359645411004721

42. H. Gleiter, Nanoglasses: A new kind of noncrystalline materials, *Beilstein J. Nanotechnol.* 4 (2013) 517–533. www.ncbi.nlm.nih.gov/pmc/articles/PMC3778333/

43. Z. Sniadecki, D. Wang, Y. Ivanisenko, V. Chakravadhanula, C. Kbel, H. Hahn, H. Gleiter, Nanoscale morphology of Ni50Ti45Cu5 nanoglass, *Mater. Charact.* 113 (2016) 26–33. doi:10.1016/j.matchar.2015.12.025. www.sciencedirect.com/science/article/pii/S1044580315300942

44. R. Witte, T. Feng, J. X. Fang, A. Fischer, M. Ghafari, R. Kruk, R. A. Brand, D. Wang, H. Hahn, H. Gleiter, Evidence for enhanced ferromagnetism in an iron-based nanoglass, *Appl. Phys. Lett.* 103 (7) (2013) 073106. doi:10.1063/1.4818493.

45. J. X. Fang, U. Vainio, W. Puff, R. Würschum, X. L. Wang, D. Wang, M. Ghafari, F. Jiang, J. Sun, H. Hahn, H. Gleiter, Atomic structure and structural stability of Sc75Fe25 nanoglasses, *Nano Lett.* 12 (1) (2012) 458–463. doi:10.1021/nl2038216.

46. O. Franke, D. Leisen, H. Gleiter, H. Hahn, Thermal and plastic behavior of nanoglasses, *J. Mater. Res.* 29 (10) (2014) 12101216. doi:10.1557/jmr.2014.101.

47. J. Weissmueller, R. Birringer, G. H, Microcomposites and nanophase materials. In: *Proceedings of the TMS Annual Meeting New Orleans* (1991) 291.

48. J. Weissmueller, P. Schubert, H. Franz, R. Birringer, H. Gleiter, Nanostructured amorphous solids. In: *Proceedings Of the VII National Conference on the Physics of Non-crystalline* Solids Cambridge, England August 4-9.

49. H. Gleiter, T. Schimmel, H. Hahn, Nanostructured solids - from nano-glasses to quantum transistors, *Nano Today* 9 (1) (2014) 17–68. http://www.sciencedirect.com/science/article/pii/S1748013214000243

50. C. Wang, D. Wang, X. Mu, S. Goel, T. Feng, Y. Ivanisenko, H. Hahn, H. Gleiter, Surface segregation of primary glassy nanoparticles of Fe90Sc10 nanoglass, *Mater. Lett.* 181 (2016) 248–252. doi:10.1016/j.matlet.2016.05.189. www.sciencedirect.com/science/article/pii/S0167577X16309557

51. G. Z. Voronoi, Nouvelles applications des parametres continus a la theorie des formes quadratiques, *Journal für die reine und angewandte Mathematik* 134 (1908) 199–287.

52. J. C. Lee, K. W. Park, K. H. Kim, E. Fleury, B. J. Lee, M. Wakeda, Y. Shibutani, Origin of the plasticity in bulk amorphous alloys, *J. Mater. Res.* 22 (11) (2007) 3087–3097.

53. Y. Q. Cheng, H. W. Sheng, E. Ma, Relationship between structure, dynamics, and mechanical properties in metallic glass-forming alloys, *Phys. Rev. B* 78 (1) (2008) 014207.

54. K. Albe, Y. Ritter, D. Şopu, Enhancing the plasticity of metallic glasses: Shear band formation, nanocomposites and nanoglasses investigated by molecular dynamics simulations, *Mech. Mater.* 67 (0) (2013) 94–103. doi:10.1016/j.mechmat.2013.06.004. www.sciencedirect.com/science/article/pii/S0167663613001117

55. D. Turnbull, M. H. Cohen, Free-volume model of amorphous phase - glass transition, *J. Chem. Phys.* 34 (1) (1961) 120–125.

56. Y. Q. Cheng, A. J. Cao, E. Ma, Correlation between the elastic modulus and the intrinsic plastic behavior of metallic glasses: The roles of atomic configuration and alloy composition, *Acta Mater.* 57 (11) (2009) 3253–3267.

57. M. Kilo, M. Hund, G. Sauer, A. Baiker, A. Wokaun, Reaction induced surface segregation in amorphous CuZr, NiZr and PdZr alloys - An XPS and SIMS depth profiling study, *J. Alloys Compd.* 236 (1-2) (1996) 137–150.

58. J. Bukowska, A. Kudelski, M. JanikCzachor, SERS on modified amorphous Cu-Zr alloys, *Chem. Phys. Lett.* 268 (5-6) (1997) 481–484.

59. R. Novakovic, M. L. Muolo, A. Passerone, Bulk and surface properties of liquid X-Zr (X = Ag, Cu) compound forming alloys, *Surf. Sci.* 549 (3) (2004) 281–293.

60. D. Şopu, Y. Ritter, H. Gleiter, K. Albe, Deformation behavior of bulk and nanostructured metallic glasses studied via molecular dynamics simulations, Phys. Rev. B 83 (10) (2011) 100202. doi:10.1103/PhysRevB.83.100202.

61. Y. Q. Cheng, A. J. Cao, H. W. Sheng, E. Ma, Local order influences initiation of plastic flow in metallic glass: Effects of alloy composition and sample cooling history, *Acta Mater.* 56 (18) (2008) 5263–5275.

62. F. Shimizu, S. Ogata, J. Li, Theory of shear banding in metallic glasses and molecular dynamics calculations, *Mater. Trans.* 48 (11) (2007) 2923–2927.

63. A. Stukowski, Visualization and analysis of atomistic simulation data with ovito-the open visualization tool, Modell. *Simul. Mater. Sci. Eng.* 18 (1) (2010) 015012.

64. A. S. Argon, Plastic-deformation in metallic glasses, *Acta. Metall.* 27 (1) (1979) 47–58.

65. C. Q. Chen, Y. T. Pei, J. T. M. De Hosson, Effects of size on the mechanical response of metallic glasses investigated through in situ tem bending and compression experiments, *Acta Mater.* 58 (1) (2010) 189–200.

66. D. Şopu, K. Albe, Influence of grain size and composition, topology and excess free volume on the deformation behavior of CuZr nanoglasses, *Beilstein J. Nanotech.* 6 (2015) 537–545. doi:10.3762/bjnano.6.56.

67. S. Xie, E. P. George, Size-dependent plasticity and fracture of a metallic glass in compression, *Intermetallics* 16 (3) (2008) 485–489.

68. W. H. Jiang, F. E. Pinkerton, M. Atzmon, Mechanical behavior of shear bands and the effect of their relaxation in a rolled amorphous Al-based alloy, *Acta Mater.* 53 (12) (2005) 3469–3477.

69. H. W. Sheng, W. K. Luo, F. M. Alamgir, J. M. Bai, E. Ma, Atomic packing and short-to-medium-range order in metallic glasses, *Nature* 439 (7075) (2006) 419–425.

70. M. Lee, C. M. Lee, K. R. Lee, E. Ma, J. C. Lee, Networked interpenetrating connections of icosahedra effects on shear transformations in metallic glass, *Acta Mater.* 59 (1) (2011) 159–170.

71. Y. Ritter, K. Albe, Chemical and topological order in shear bands of Cu64Zr36 and Cu36Zr64 glasses, *J. Appl. Phys.* 111 (10) (2012) 103527. doi:10.1063/1.4717748.

72. O. Adjaoud, K. Albe, Microstructure formation of metallic nanoglasses: Insights from molecular dynamics simulations, *Acta Mater.* 145 (2018) 322–330. doi:10.1016/j.actamat.2017.12.014. http://www.sciencedirect.com/science/article/pii/S1359645417310194

73. O. Adjaoud, K. Albe, Interfaces and interphases in nanoglasses: Surface segregation effects and their implications on structural properties, *Acta Mater.* 113 (2016) 284–292. doi:10.1016/j.actamat.2016.05.002. http://www.sciencedirect.com/science/article/pii/S1359645416303299

74. S. H. Nandam, Y. Ivanisenko, R. Schwaiger, Z. Sniadecki, X. Mu, D. Wang, R. Chellali, T. Boll, A. Kilmametov, T. Bergfeldt, H. Gleiter, H. Hahn, Cu-Zr nanoglasses: Atomic structure,

thermal stability and indentation properties, *Acta Mater.* 136 (2017) 181–189. doi:10.1016/j.actamat. 2017.07.001. www.sciencedirect.com/science/ article/pii/S1359645417305505

75. C. Kalcher, O. Adjaoud, J. Rohrer, A. Stukowski, K. Albe, Reinforcement of nanoglasses by interface strengthening, Scripta Materialia 141 (2017) 115–119. doi:10.1016/j.scriptamat.2017. 08.004. www.sciencedirect.com/science/article/ pii/S135964621730458X

Index

9781032337319